Trigonometric Identities

Symmetry	$\sin(-\theta) = -\sin(\theta)$
	$\cos(-\theta) = \cos(\theta)$
Angle addition	$\sin(\theta + \phi) = \sin(\theta)\cos(\phi) + \cos(\theta)\sin(\phi)$
	$\cos(\theta + \phi) = \cos(\theta)\cos(\phi) - \sin(\theta)\sin(\phi)$
Double angle	$\sin(2\theta) = 2\sin(\theta)\cos(\theta)$
	$\cos(2\theta) = \cos^2(\theta) - \sin^2(\theta)$
Basic Identity	$\sin^2(\theta) + \cos^2(\theta) = 1$

Basic Discrete-Time Dynamical Systems

Description	Updating function	Solution or behavior
Constant per capita reproduction r	$b_{t+1} = rb_t$	Exponential growth $b_t = r^t b_0$
Constant growth by amount c	$h_{t+1} = h_t + c$	Linear growth $h_t = h_0 + ct$
Lung model with fraction q exchanged and ambient concentration γ	$c_{t+1} = (1-q)c_t + q\gamma$	Approaches equilibrium at $c^* = \gamma$
Selection model with mutant reproduction s wild-type reproduction r	$p_{t+1} = \dfrac{sp_t}{sp_t + r(1-p_t)}$	Approaches $p^* = 1$ if $s > r$ and $p^* = 0$ if $s < r$
Logistic model with per capita reproduction r and carrying capacity 1	$x_{t+1} = rx_t(1 - x_t)$	Stable equilibrium if $r \leq 3$

Basic Autonomous Differential Equations

Description	Equation	Solution or behavior
Constant per capita reproduction rate λ	$\dfrac{db}{dt} = \lambda$	Exponential growth $b(t) = b(0)e^{\lambda t}$
Constant immigration at rate c	$\dfrac{dV}{dt} = c$	Linear growth $V(t) = V(0) + ct$
Newton's Law of Cooling with cooling rate α and ambient temperature A	$\dfrac{dH}{dt} = \alpha(A - H)$	Approaches equilibrium at $H^* = A$
Selection model with mutant reproduction rate μ wild-type reproduction rate λ	$\dfrac{dp}{dt} = (\mu - \lambda)p(1 - p)$	Approaches $p^* = 1$ if $\mu > \lambda$ $p^* = 0$ if $\mu < \lambda$

Basic Theorems of Calculus

Intermediate Value Theorem: A continuous function f defined for $a \leq x \leq b$ takes on each value between $f(a)$ and $f(b)$ at least once.

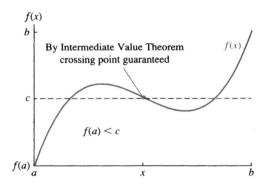

Extreme Value Theorem: A continuous function f defined for $a \leq x \leq b$ has a maximum.

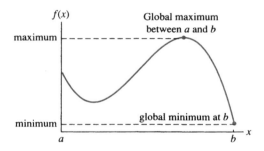

Mean Value Theorem: A differentiable function f defined for $a \leq x \leq b$ has a point c where $f'(c)$ is equal to the slope of the secant connecting $(a, f(a))$ to $(b, f(b))$.

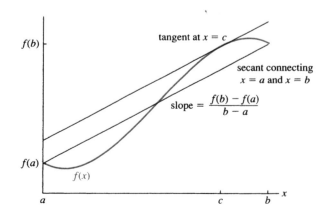

Fundamental Theorem of Calculus: If $F(x) = \int f(x)\,dx$ then $\int_a^b f(x)\,dx = F(b) - F(a) = F(x)|_a^b$.

Modeling the Dynamics of Life
CALCULUS AND PROBABILITY FOR LIFE SCIENTISTS

Frederick R. Adler
University of Utah

Brooks/Cole Publishing Company

 An International Thomson Publishing Company

Pacific Grove • Albany • Belmont • Bonn • Boston • Cincinnati • Detroit • Johannesburg • London
Madrid • Melbourne • Mexico City • New York • Paris • Singapore • Tokyo • Toronto • Washington

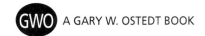 A GARY W. OSTEDT BOOK

Publisher: *Gary W. Ostedt*
Marketing Team: *Caroline Croley, Christine Davis*
Editorial Associate: *Carol Ann Benedict*
Production Editor: *Marlene Thom*
Production and Typesetting: *Integre Technical Publishing Company, Inc.*

Interior Design: *Spectrum Publisher Services, Inc.*
Interior Illustration: *Scientific Illustrators*
Cover Design: *Roy R. Neuhaus*
Cover Printing: *Phoenix Color Corporation, Inc.*
Printing and Binding: *The Courier Company, Inc., Kendallville*

COPYRIGHT © 1998 by Brooks/Cole Publishing Company
A division of International Thomson Publishing Inc.
I(T)P The ITP logo is a registered trademark under license.

For more information, contact:

BROOKS/COLE PUBLISHING COMPANY
511 Forest Lodge Road
Pacific Grove, CA 93950
USA

International Thomson Publishing Europe
Berkshire House 168–173
High Holborn
London WC1V 7AA
England

Thomas Nelson Australia
102 Dodds Street
South Melbourne, 3205
Victoria, Australia

Nelson Canada
1120 Birchmount Road
Scarborough, Ontario
Canada M1K 5G4

International Thomson Editores
Seneca 53
Col. Polanco
11560 México, D.F., México

International Thomson Publishing GmbH
Königswinterer Strasse 418
53227 Bonn
Germany

International Thomson Publishing Asia
221 Henderson Road
#05–10 Henderson Building
Singapore 0315

International Thomson Publishing Japan
Hirakawacho Kyowa Building, 3F
2-2-1 Hirakawacho
Chiyoda-ku, Tokyo 102
Japan

All rights reserved. No part of this book may be reproduced, stored in a retrieval system, or transcribed, in any form or by any means— electronic, mechanical, photocopying, recording, or otherwise—without the prior written permission of the publisher, Brooks/Cole Publishing Company, Pacific Grove, California 93950.

Printed in the United States of America

10 9 8 7 6 5 4 3 2

Library of Congress Cataloging in Publication Data
Adler, Frederick R.
  Modeling the dynamics of life : calculus and probability for life scientists / Frederick R. Adler.
    p. cm.
  ISBN 0-534-34816-5 (alk. paper)
  1. Calculus. 2. Probabilities. 3. Life sciences—Mathematics. I. Title.
QA303.A367 1997
570′.1′185—dc21   97-38657
                                                    CIP

About the Cover

Jasper Johns, *Numbers in Color*, 1959, encaustic and newspaper on canvas, 66 1/2 × 49 1/2″, Albright-Knox Art Gallery, Buffalo, New York, Gift of Seymour H. Knox, 1959.

This book uses numbers and mathematics to model the dynamics of life much as paint and color model the numbers depicted on the cover. Jasper Johns himself best expressed how both artists and scientists use many techniques to model the world. The composer John Cage tells of watching Johns at work, "changing brushes and colors, and working everywhere at the same time rather than starting at one point." Cage asked the painter how many processes he was involved in. "He concentrated to reply and speaking sincerely said: It is all one process."

Preface

Modeling the Dynamics of Life teaches calculus, with additional chapters on probability and statistics, as a way of introducing freshman and sophomore life science majors to the insights mathematics can provide into biology. Why should there be a special book for this audience? Although the importance of quantitative skills in the life sciences is much discussed, current realities tend to conceal their vital role. Too often, biology is the natural science of last resort for students who believe "they aren't cut out for math." Most colleges and universities require little calculus for their biology majors, and those that require a full calculus course doubt its worth when students emerge unable to apply even precalculus mathematics in new contexts. Students are left with similar doubts when the techniques they learned for tests vanish as swiftly from the curriculum as from their memories. Students, biology faculty, and administrators see that biology is burgeoning as a science and as a major, apparently unhindered by pervasive mathematical illiteracy.

In fact, mathematics has played an important if under-appreciated role in biology, providing the impetus for breakthroughs in areas including epidemiology, genetics, statistics, and physiology. As a theoretical biologist who uses mathematics to make sense of complex biological systems, I see this role expanding, not contracting. Although a great deal of biology can be done without any mathematics, the powerful new technologies that are transforming fields of biology—from genetics and physiology to ecology—are increasingly quantitative, as are many of the questions at the frontiers of knowledge. Along with genetics, mathematics is one of two unifying factors in the life sciences. And as biology becomes more important in society, mathematical literacy becomes as necessary for doctors, business people, lawyers, and art historians as for researchers.

Modeling

The central goal of this book is simple: to teach biology majors the mathematical ideas I use every day in my own research and in collaborations with my more empirical colleagues. These ideas are not specific techniques such as differentiation, but concepts of modeling. The skills include describing a system, translating appropriate aspects into equations, and interpreting the results in terms of the original problem. In this process, the science is central and solving the equations is in some ways the least important step.

Even students who will do little modeling on their own will be confronted by the models of others. This book teaches students how to **read** models. Like computers, mathematical models have an aura of authority to those who cannot understand them. As a modeler myself, I want my work to be read, understood, and challenged by knowledgeable scientists. This book aims to give students the skills and confidence to do so.

The Dynamics of Life

Mathematics helps unify biology by identifying dynamical principles that underlie a remarkable diversity of biological processes. This book follows three themes throughout: **growth**, **diffusion**, and **selection**. Each theme is studied in turn with three kinds of models: discrete-time dynamical systems (chapters 1–3), differential equations (chapters 4–5), and stochastic processes (chapters 6–8). The concept of diffusion is treated as a discrete-time dynamical system describing a lung (section 1.10), as a differential equation describing movement of a chemical across a membrane (section 4.2), and as a Markov chain describing the random behavior of an individual molecule (section 6.2). Addressing these themes in different ways shows students how different mathematical ideas describe and explain different facets of a biological process.

The Contents

These three themes provide a logical context to teach the standard material of a calculus text, and more. First, students review the basic functional building blocks of applied mathematics in the context of dynamics. For example, exponential functions are studied as the solutions of linear discrete-time dynamical systems (section 1.7) and linked to power functions through the idea of allometric relationships (section 1.8). Because dynamics motivates the book from the beginning, differentiation arises naturally as a way to describe instantaneous rates of change (section 2.1). The dynamical models provide an array of functions for which differentiation is more than an algebraic exercise (section 2.8). Integration is introduced by its most important application, solving differential equations (sections 4.3 and 5.3).

Additional topics, which are often covered in less detail in traditional texts, fit naturally within the framework created by modeling and dynamics. Just as students need an intuitive understanding of basic measurements, such as distance and temperature, they need an intuitive understanding of the functions that express relationships between these measurements. The book emphasizes "reasoning about functions," and includes extensive sections devoted to applications of such general theorems as the Intermediate Value Theorem (section 3.4), sketching graphs using the method of leading behavior (section 3.6), and approximating functions with the tangent line or Taylor polynomials (section 3.7). These skills translate into the ability to reason about the stability of discrete-time dynamical systems (sections 3.1 and 3.2) and differential equations (section 5.2). Dynamics provides a natural context to derive and explain Newton's method (section 3.8) and the meaning of the second derivative (section 2.7).

Teaching probability without using calculus is like studying biology without using genetics. The final three chapters teach probability and statistics from the dynamical perspective, using discrete-time dynamical systems (sections 6.2 and 6.6) and differential equations (section 7.6) to model fundamental stochastic processes such as Markov chains (section 6.2) and the Poisson process (section 7.7). The integral and the Fundamental Theorem of Calculus have one of their most important applications in the theory of probability density functions (section 6.7), reinforcing these foundational calculus concepts.

In the minds of some, statistics is the only quantitative skill needed by life scientists. *Modeling the Dynamics of Life* stresses that it is impossible to use statistics correctly and flexibly without understanding the underlying probabilistic models, such as Markov chains and the Poisson process. The book emphasizes the links between models of those processes and the fundamental statistical notions of likelihood (section 8.1), parameter estimation (section 8.2), and hypothesis testing (section 8.4). Students learn the **principles** of statistics and emerge with a set of tools that encourage understanding statistics rather than merely choosing the correct recipe.

Graphical Methods

The focus on reasoning about what models **mean** favors graphical and computer techniques, the methods professionals use when equations cannot be solved. Understanding a discrete-time dynamical system with cobwebbing (section 1.10) makes visual sense of such advanced topics as stability (section 3.1) and chaos (section 3.2). The phase-line (section 5.2) provides a visual presentation that summarizes the meaning of an autonomous differential equation, just as the phase-plane (section 5.5) summarizes a system of coupled differential equations and describes one mechanism of biological oscillation (section 5.7). Approximation methods essential in more advanced applied mathematics are introduced through the method of leading behavior (section 3.7) and its graphical interpretation.

Algorithms

There are many crucial logical procedures students will use again and again to solve problems. These are set out as approximately 30 algorithms, including:

- Finding equilibria for discrete-time dynamical systems (section 1.11)
- Applying the chain rule (section 2.10)
- Finding global maxima and minima (section 3.3)
- Applying integration by substitution (section 4.4)
- Using the method of separation of variables (section 5.3)
- Computing descriptive statistics such as the mean, median, and mode (section 6.9)
- Applying the standard normal distribution (section 7.10)
- Finding the value of r^2 from a linear regression (section 8.7)

In-depth Explorations of Particular Models

The book includes several extended explorations. Early on, the power of cobwebbing is put to work to study the phenomenon of AV block in the heart (section 1.13). The application of maximization methods to fisheries, and their interaction with dynamics, is presented in section 3.3. These methods are used to fully analyze an extension of the basic lung model to study optimal breathing patterns (section 3.9). Phase-plane methods are used to analyze the FitzHugh-Nagumo equations describing a neuron (section 5.7). These extended developments show students how to combine a set of processes into a coherent whole, and how to use models to clarify hypotheses and answer specific questions.

Problems

In addition to more routine skills problems at the end of each section, each section includes a wide variety of modeling problems to consistently emphasize the importance of interpretation. Examples include:

- Chemical absorption (introduced in exercise 1.9.5 and revisited in exercises 1.10.9 and 3.1.3)
- Mutation (exercises 1.12.4–6, returned to in exercises 6.3.10–12)
- Consequences of the different scalings of surface area and volume (exercise 1.8.8 studies heat loss and exercises 4.1.12 and 4.2.4 explore food absorption).

Each chapter includes supplementary problems, drawn from previous tests and practice tests. These introduce a variety of new applications and can be used for review and practice sessions.

Complete Solutions Manual

A complete solutions manual containing solutions to all odd- and even-numbered problems and to the supplementary problems (excluding computer problems and projects) is available.

Computer Problems

The book includes well over 100 exercises designed for computer or graphing calculator, emphasizing visualization, experimentation, and simulation. Examples include:

- Simulation of a simple linear discrete-time dynamical system describing harvesting and observation of threshold behavior to motivate the later discussion of unstable equilibria in section 3.1 (exercise 1.5.12)
- Exploration of the power of Fourier series through seeing how cosine functions can sum to an approximate square wave (exercise 1.9.13)
- Experimentation with the approximate rate of change to motivate the subsequent derivation of the derivative of cosine (exercise 2.1.14)
- Simulation of a mutation process to show how different individual samples can be from averages (exercise 6.3.13).

The computer lab is an effective alternative learning environment for students who communicate well with machines, and an excellent place for students to work with one another and with instructors.

Projects

Each chapter includes at least two projects. These are recommended for individual or group exploration. Examples include:

- Analysis of data on the different allometric relationships in male and female lizards (project 1.3)
- Study of periodic hematopoiesis with a discrete-time dynamical system (project 2.1)
- Experimentation with different numerical schemes for solving differential equations (project 4.1)
- Careful study of models of adaptation by cells (project 5.1)
- Development of the famous Luria-Delbruck fluctuation test (project 7.2).

Teaching

Although the book is designed for a full-year, introductory-level course, there are many other ways it can be used.

- One semester calculus course: Chapters 1–4 (with the possible omission of the extended models in section 1.13 and 3.9)
- One quarter calculus course: Chapters 1–4 (possible with omission of sections 1.13, 3.1, 3.2, 3.6, 3.7, 3.8, 3.9, and 4.8)
- One semester probability course for students who have had calculus: Chapters 6–8 (with background on discrete-time dynamical systems from sections 1.2, 1.5 and 1.10, and a short review of differential equations as in section 4.2)
- One quarter probability course for students who have had calculus: Chapters 6–8 (like the one semester course, but must be taken at a rather rapid pace; can work by eliminating the final three sections)
- One quarter or one semester graduate modeling course: The entire book, with an emphasis on the main modeling ideas. Introduce discrete-time dynamical systems (sections 1.2, 1.5, 1.10–12, 3.1 and 3.2), reasoning about functions (sections 3.3–3.8), one-dimensional differential equations (sections 4.1, 4.2, 5.1–5.3), two-dimensional differential equations (sections 5.4–5.7), stochastic processes (sections 6.1–6.3, 6.6), probability models (sections 7.3–7.8), and statistics (sections 8.1–8.4). Such a course should emphasize the extended modeling sections (1.13, 3.9, and 5.7) and the projects.

Why Bother?

What are the benefits of this approach? The modeling approach is naturally a **problem-solving** approach. All instructors know that students will not remember every technique they have learned. This book emphasizes understanding what a model **is** and recognizing what models **say**. To be able to **recognize** a differential equation, **interpret** the terms, and **use** the solution is far more important than knowing how to **find** the solution. These reasoning skills, in addition to familiarity with models in general, are what stay with the motivated student and are what matter most in the end.

Most important, the course is fun to teach. Leading students through an integrated course for a full year removes the pressure for instant instructor gratification. ("All of my students could take the derivatives of polynomials.") Instead, one can allow understanding to develop as concepts return for the second or third time. Students find this unsettling and yearn for instant gratification too. But with time, they accept the challenge of thinking. When they begin to apply their new powers to their own problems, when they solve a problem on the computer without being told to, or when they teach me something about biology in the context of a mathematical idea, delayed gratification starts to feel like the best possible kind.

Acknowledgments

This book would never have been written without the support of a Hughes Foundation Grant to the University of Utah, which included as part of its mission an attempt to more effectively teach mathematics to biology majors. The grant was fathered by Gordon Lark, who has been a colleague, a mentor, and a mensch throughout this project. The grant brought together a committee of faculty to guide creation of this book and course: Aaron Fogelson, David Goldenberg, Jim Keener, Mark Lewis, David Mason, Larry Okun, Hans Othmer, Jon Seger, and Ryk Ward. Each added much to this work, generously providing ideas and corrections. Particular thanks to Jon "Preface" Seger and Mark "reasoning about functions" Lewis for discussion and ideas. Frank Wattenberg, Lou Gross, and Simon Levin looked over the book and delivered much-needed advice early in the writing process. Alan Rogers kindly let me use his exercise style and Nelson Beebe helped smooth over many technical problems. Thanks to Marlene Thom for dealing with chapter re-renumbering and numerous production frustrations, Roy Neuhaus for working so hard on the cover, the Albright-Knox Art Gallery for giving permission to use the Jasper Johns painting, Carol Benedict for dealing with all the little things, Elizabeth Rammel for liking the solutions manual, Kelly Ricci and Kristin Steffeck for seeking copy-editing perfection, Kelli Radican and Frances Hill for expertise with FedEx, and Gary Ostedt for agreeing to support a Preliminary Edition, for providing bird-watching opportunities, and for making me feel important. Most of all, thanks to Don DeLand, Leslie Galen, and the entire staff at Integre for professionalism and dedication far beyond the call of duty that made the book look so good so quickly.

The Preliminary Edition benefited greatly from the comments and complaints of the students who survived the rocky first run of the course with the brightly covered photocopies: Jennifer Aiman, Ty "Captain Flail" Corbridge, Brett Doxey, Ambur Economou, Brad Hasna, Robert Kane, Laura Krause, Jennifer Layman, Eric Mortensen, Scott Nord, Kevin Rapp, Chris Reilly, Mindi Robinson, Rachael Rosenfeld, Stephanie Spindler, Mark Stevens, Sheri Williams, Richard Wood, and Gentry Yost. Ranging from those with an excess of comments to those who suffered in eloquent silence, they helped give this book whatever value it might have as a teaching tool. These students, along with the veterans of later drafts, corrected many errors and cheerfully pointed out pedagogical shortcomings. My teaching assistants, Stephen Proulx, Peter Spiro, Vicky Solomon, Colonel Tim Lewis, Kristina Bogar, Eric Marland, and Steve Parrish were always ready to stand behind me and tell me what I was doing wrong.

Many reviewers were instrumental in making this book comprehensible. Particular thanks to the class testers: Joe Mahaffy, Kathleen Crowe, Rollie Lamberson, Daniel Bentil, and the student reviewers at San Diego State. The following reviewers were honest in their criticism and generous with their expertise, and constantly provided needed perspective and ideas: Dr. Lawrence Marx, University of California-Davis; Dr. Paul Van Steenberghe, University of Maine-Orono; Dr.

Gail Johnston, Iowa State University; Dr. Catherine Roberts, Northern Arizona University; Dr. Steven Baer, Arizona State University; Dr. Daniel Bentil, University of Vermont; Dr. Roland Lamberson, Humboldt State University; Dr. Bill Hintzman, San Diego State University; Dr. Jim Powell, Utah State University; Dr. Joe Mahaffy, San Diego State University; Dr. Steve Smith, Truman State University; Dr. Dave Carlson, San Diego State University; Dr. Fabio A. Milner, Purdue University; and Dr. Kathleen Crowe, Humboldt State University. I failed to follow their advice only when blinded by overweaning pride, misguided philosophy, or general sloth. The remaining errors and oversights are entirely my own.

In addition to her extraordinary sartorial advice and culinary support, I thank Anne Collopy for her inspirational example of writing with clear transitions, extended metaphors, and elegant sentence structure. And for filling the work-free interstices of life with the same.

Frederick R. Adler
Salt Lake City, Utah

Contents

Chapter 1 **Introduction to Discrete-Time Dynamical Systems**   **1**

     1.1   Biology and Dynamics   2
     1.2   Updating Functions: Describing Growth   6
     1.3   Units and Dimensions of Measurements and Functions   14
     1.4   Linear Functions and Their Graphs   22
     1.5   Finding Solutions: Describing the Dynamics   32
     1.6   Combining and Manipulating Functions   39
     1.7   Expressing Solutions with Exponential Functions   49
     1.8   Power Functions and Allometry   59
     1.9   Oscillations and Trigonometry   69
     1.10   Modeling and Graphical Analysis of Updating Functions   78
     1.11   Equilibria   86
     1.12   An Example of Nonlinear Dynamics   95
     1.13   Excitable Systems: The Heart   106

Chapter 2 **Limits and Derivatives**   **118**

     2.1   Introduction to Derivatives   119
     2.2   Limits   129
     2.3   Continuity   138
     2.4   Computing Derivatives: Linear and Quadratic Functions   148
     2.5   Derivatives of Sums, Powers, and Polynomials   156
     2.6   Derivatives of Products and Quotients   168
     2.7   The Second Derivative, Curvature, and Acceleration   176

2.8 Derivatives of Exponential and Logarithmic Functions 184

2.9 Derivatives of Trigonometric Functions 190

2.10 The Chain Rule 199

Chapter 3 Applications of Derivatives and Dynamical Systems 212

3.1 Stability and the Derivative 213

3.2 More Complex Dynamics 224

3.3 Maximization 237

3.4 Reasoning About Functions 248

3.5 Limits at Infinity 257

3.6 Leading Behavior and L'Hôpital's Rule 268

3.7 Approximating Functions with Lines and Polynomials 280

3.8 Newton's Method 289

3.9 Panting and Deep Breathing 299

Chapter 4 Differential Equations, Integrals, and Their Applications 310

4.1 Differential Equations 311

4.2 Basic Differential Equations 322

4.3 Solving Pure-Time Differential Equations 330

4.4 Integration of Special Functions and Integration by Substitution 340

4.5 Integrals and Sums 348

4.6 Definite and Indefinite Integrals 356

4.7 Applications of Integrals 366

4.8 Improper Integrals 374

Chapter 5 Analysis of Differential Equations 386

5.1 Equilibria and Display of Autonomous Differential Equations 387

5.2 Stable and Unstable Equilibria 395

5.3 Solving Autonomous Differential Equations 402

- 5.4 Two-Dimensional Differential Equations 409
- 5.5 The Phase Plane 417
- 5.6 Solutions in the Phase Plane 425
- 5.7 The Dynamics of a Neuron 435

Chapter 6 Probability Theory and Descriptive Statistics 447

- 6.1 Introduction to Probabilistic Models 448
- 6.2 Stochastic Models of Diffusion 454
- 6.3 Stochastic Models of Genetics 462
- 6.4 Probability Theory 468
- 6.5 Conditional Probability 474
- 6.6 Independence and Markov Chains 480
- 6.7 Displaying Probabilities 486
- 6.8 Random Variables 497
- 6.9 Descriptive Statistics 506
- 6.10 Descriptive Statistics for Spread 515

Chapter 7 Probability Models 529

- 7.1 Joint Distributions 530
- 7.2 Covariance and Correlation 538
- 7.3 Sums and Products of Random Variables 547
- 7.4 The Binomial Distribution 557
- 7.5 Applications of the Binomial Distribution 567
- 7.6 Waiting Times: Geometric and Exponential Distributions 574
- 7.7 The Poisson Distribution 584
- 7.8 The Normal Distribution 592
- 7.9 Applying the Normal Approximation 602

Chapter 8 Introduction to Statistical Reasoning 618

- 8.1 Statistics: Estimating Parameters 619
- 8.2 Confidence Limits 627
- 8.3 Estimating the Mean 636

- 8.4 Hypothesis Testing 647
- 8.5 Hypothesis Testing: Normal Theory 654
- 8.6 Comparing Experiments: Unpaired Data 663
- 8.7 Regression 671

Answers 685

Index 777

List of Algorithms

Algorithm 1.1   Procedure for converting between units   16
Algorithm 1.2   Finding the equation of a line from data   28
Algorithm 1.3   Finding the inverse of a function   45
Algorithm 1.4   The horizontal line test   46
Algorithm 1.5   Solving for equilibria   92
Algorithm 2.1   Using the chain rule   200
Algorithm 3.1   Finding global maxima and minima   240
Algorithm 3.2   Finding global maxima and minima with second derivative   242
Algorithm 3.3   The method of matched leading behaviors   274
Algorithm 4.1   Euler's method for solving a differential equation   317
Algorithm 4.2   Integration by substitution   342
Algorithm 4.3   Computing a definite integral by substitution   368
Algorithm 5.1   Finding equilibria of an autonomous differential equation   389
Algorithm 5.2   Separation of variables   404
Algorithm 5.3   Finding the nullclines and equilibria of a pair of autonomous differential equations   420
Algorithm 6.1   Finding the median of a discrete random variable   507
Algorithm 6.2   Finding the median of a continuous random variable   507
Algorithm 6.3   Finding the mode of a random variable   510
Algorithm 6.4   Finding the geometric mean   513
Algorithm 7.1   Finding the correlation of two discrete random variables   543
Algorithm 7.2   Converting into a problem about the standard normal distribution   604
Algorithm 7.3   Applying the normal approximation to the binomial distribution   609

Algorithm 7.4	Applying the normal approximation to the Poisson distribution	611
Algorithm 8.1	Monte Carlo method for finding confidence limits	632
Algorithm 8.2	Finding confidence limits with the normal approximation: Binomial case	645
Algorithm 8.3	Comparing hypotheses with the method of support	652
Algorithm 8.4	Hypothesis testing with the normal distribution	657
Algorithm 8.5	Comparing the means of two experiments	665
Algorithm 8.6	Comparing proportions from two experiments	667
Algorithm 8.7	Comparing experiments with the method of support	669
Algorithm 8.8	Calculation of the coefficient of determination	674

Chapter 1

Introduction to Discrete-Time Dynamical Systems

This chapter introduces the main tools needed to use mathematics to study biology: **functions** and **modeling**. Biological phenomena are described by **measurements**, a set of numerical values with units (such as degrees or centimeters). **Relations** between these measurements are described by functions, which take one value as an input and return another as an output. We review the important functions used to describe biological systems: linear, trigonometric, exponential, and power functions.

Modeling is the art of taking a description of a biological phenomenon and converting it into mathematical form. Living things are characterized by change. One goal of modeling is to quantify these **dynamics** with an appropriate function. By using our understanding of the system and carefully following how a set of basic measurements change step by step, we will learn to derive an **updating function** that summarizes the change. We will follow this process to derive models of bacterial population growth, gas exchange in the lung, and genetic change in a population of competing bacteria. These models describe the biological process, and developing them will make our thinking more precise. We will develop a set of algebraic and graphical tools to deduce the dynamics that result from a particular updating function.

Throughout this chapter, keep the following questions in mind.

- What biological process are we trying to describe?
- What biological questions do we seek to answer?
- What are the basic measurements and their units?
- What are the relations between the basic measurements?
- What do the results mean biologically?

1.1 Biology and Dynamics

Living systems, from cells to organisms to ecosystems, are characterized by change and dynamics. Living things grow, maintain themselves, and reproduce. Even remaining the same requires dynamic responses to a changing environment. Understanding the mechanisms behind these dynamics and deducing their consequences is crucial to understanding biology. This book uses a dynamical approach to address questions about biology.

This dynamical approach is necessarily mathematical because describing dynamics requires quantifying measurements. What is changing? How fast is it changing? What is it changing into?

In this book, we use the language of mathematics to describe the working of living systems quantitatively and develop the mathematical tools needed to compute how they change. From measurements describing the initial state of a system and a set of rules describing how change occurs, we attempt to predict what will happen to the system. For example, given the position and velocity of a planet (the initial state) and the laws of gravitation and inertia (the set of rules), Isaac Newton invented the mathematical methods of calculus to predict its position at any future time. This example illustrates the approach of **applied mathematics**, *the use of mathematics to answer scientific questions* (Figure 1.1). Applied mathematics begins with scientific observations and questions, perhaps about the position of a planet, which are then quantified into a **model**. When possible, mathematical methods are developed to answer the question. In other cases, computers are used to **simulate** the process and find answers in particular cases.

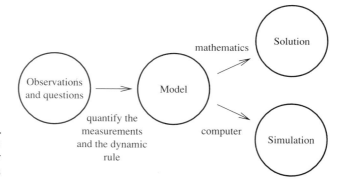

Figure 1.1
The workings of applied mathematics: The use of mathematics to answer scientific questions

The steps in applied mathematics

Step	Definition
Quantify the basic measurements.	The numerical values that describe the system
Describe a **dynamic rule**.	A description of how the basic measurements change
Develop a **model**.	A mathematical translation of the observations
Find a **solution**.	Use of mathematical methods to predict behavior
Write a **simulation**.	Use of a computer to predict behavior

This book is organized around three basic biological processes: **growth**, **maintenance**, and **replication**. Mathematical methods have contributed significantly to the understanding of each of these three processes. After briefly describing these successes, we outline the different types of models and mathematics used in this book.

Growth: Models of Malaria

Early in the 20th century, Sir Ronald Ross discovered that malaria is transmitted by certain types of mosquitos. Because the disease was (and is) difficult to treat, one promising strategy for control seemed to be reduction of the number of mosquitos. Many people believed that all mosquitos would have to be killed to eradicate the disease. Because killing every mosquito seemed impossible, it was feared that malaria might be impossible to control.

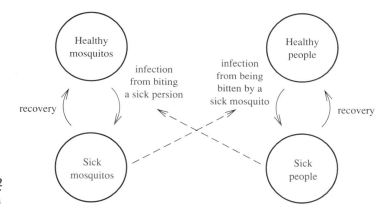

Figure 1.2
The dynamics of malaria

Sir Ross decided to use mathematics to convince people that mosquito control could be effective. The problem can be formulated dynamically as a problem in population growth. His first step was to **quantify the basic measurements**, in this case, the number of people and mosquitos with and without malaria. The **dynamical rule** describes how these numbers change. Sir Ross assumed that an uninfected person becomes infected when bitten by an infected mosquito and that an uninfected mosquito becomes infected when it bites an infected person (Figure 1.2). From these assumptions, he built a **mathematical model** describing the population dynamics of malaria. With this model, he proved that the disease *could* be eradicated without killing every mosquito (a simple version of this model is studied in Section 4.2). We see evidence of this eradication today in the United States, where malaria has been virtually eliminated although mosquitos that are capable of transmitting the disease persist in many regions.

Many dynamic biological processes other than population dynamics are forms of growth. Growth in size is ubiquitous. One might use measurements of size (such as weight, height, or stomach volume) and a rule describing change in size (increase in weight caused by large stomach volume) to predict the size of an organism over time. For example, one organism might add a constant amount to its weight every day while another might add a constant fraction to its weight every day.

Organisms can also grow in complexity. For example, a tree can add branches as well as size. The quantitative description of the system might include the number of branches, their ages, sizes, and pattern. The dynamic rule might provide the number of new branches produced each day, the probability that a given branch divides during the next month, or the rate at which new branches are formed. From the description and rule, we could compute the number of branches as a function of age.

Maintenance: Models of Neurons

Neurons are cells that transmit information throughout the brain and body. Even the simplest neuron faces a challenging task. It must be able to amplify an appropriate incoming stimulus, transmit it to neighboring neurons, and then turn off and be ready for the next stimulus. This task is not as simple as it may seem. If we imagine the stimulus to be an input of electrical charge, a plausible sounding rule is "if electrical charge is raised above a certain level, increase it further." Such a rule works well for the first stimulus but does not provide a way for the cell to turn itself off. How does a neuron maintain functionality?

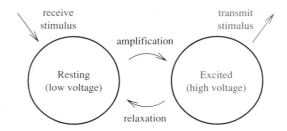

Figure 1.3
The dynamics of a neuron

In the early 1950s, Hodgkin and Huxley used their own measurements of neurons to develop a mathematical model of dynamics to explain the behavior of neurons. The idea, explained in detail in Section 4.7 is that the neuron has fast and slow mechanisms to open and close specialized ion channels in response to electrical charge (Figure 1.3). Hodgkin and Huxley measured the dynamic behavior of these channels and showed mathematically that their mechanism explained many aspects of the functioning of neurons. They received the Nobel Prize in Physiology and Medicine for this work in 1963 and, perhaps more impressively, developed a model that is still used to study neurons and other types of cells.

In general, maintenance of biological systems depends on preserving the distinction between inside and outside, while maintaining flows of necessary materials from outside to inside and vice versa. The neuron maintains itself at a different electrical potential from the surrounding tissue to be able to respond, while remaining ready to exchange ions with the outside to create the response. As applied mathematicians, we **quantify the basic measurements**, the concentrations of various substances inside and outside the cell. The **dynamical rules** express how concentrations change, generally as a function of properties of the cell membrane. Most commonly, the rule describes the process of **diffusion**, movement of materials from regions of high concentration to regions of low concentration.

Replication: Models of Genetics

Although Mendel's work on genetics from the 1860s was rediscovered around 1900, many biologists in the 1920s remained unconvinced of his proposed mechanism of genetic transmission. In particular, it was unclear whether Darwin's theory of evolution by natural selection was consistent with this, or any other, proposed mechanism.

Working independently, biologists Sir Ronald Fisher, J.B.S. Haldane, and Sewall Wright developed mathematical models of the dynamics of evolution in natural populations. These scientists quantified the basic measurement, in this case the number of individuals with a particular allele (a version of a gene). Their dynamical rules described how many individuals in a subsequent generation would have a particular allele as a function of numerous factors, including **selection** (differential success of particular types in reproducing) and **genetic drift** (the workings of chance). They showed that Mendel's ideas were indeed consistent with observations of evolution. This work established Mendelian genetics, which provide the foundation for the breakthroughs in genetics that continue today. We study some of the basic ideas in Sections 1.12, 4.2, and 6.3.

Types of Dynamical Systems

We study each of the three processes—growth, maintenance, and replication—with three types of dynamical system, termed **discrete-time**, **continuous-time**, and **probabilistic**. The first two types are **deterministic**, meaning that the dynamics includes no chance factors. In this case, the values of the basic measurements can be predicted exactly at all future times. Probabilistic dynamical systems include chance factors, and the values of the basic measurements can be predicted only on average.

Discrete-Time Dynamical Systems

Discrete-time dynamical systems describe a sequence of measurements made at equally spaced intervals (Figure 1.4a). These dynamical systems are mathematically described by a rule that gives the value at one time as a function of the value at the previous time. For example, a discrete-time dynamical system describing population growth is a rule that gives the population in a given year as a function of the population in the previous year. A discrete-time dynamical system describing the concentration of oxygen in the lung is a rule that gives the concentration of oxygen in a lung after one breath as a function of the concentration after the previous breath. A discrete-time dynamical system describing the spread of a mutant allele is a rule that gives the number of mutant alleles in one generation as a function of the number in the previous generation. Mathematical analysis of the rule

Figure 1.4

The measurements described by three types of dynamical systems

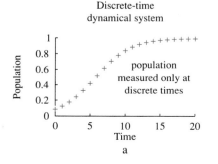
a

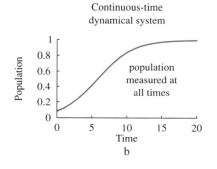
b

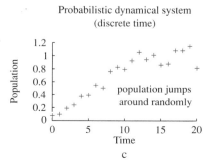
c

can make various scientific predictions, such as the maximum population size, the average concentration of oxygen in the lung, or the final number of mutant alleles. The study of these systems requires the mathematical methods of **differential calculus**, the material in the first three chapters of this book.

Continuous-Time Dynamical Systems

Continuous-time dynamical systems, often known as **differential equations**, describe measurements that are collected continuously (Figure 1.4b). A differential equation consists of a rule that gives the **instantaneous rate of change** of a set of measurements. The miracle of differential equations is that such information about a system at one time is sufficient to predict the state of a system at all future times. For example, a continuous-time dynamical system describing the growth of a population is a rule that gives the rate of change of population size as a function of the population size itself. The study of these systems requires the mathematical methods of **integral calculus**, the material in Chapters 4 and 5 of this book.

Probabilistic Dynamical Systems

Probabilistic dynamical systems describe measurements, in either discrete or continuous time, that are affected by chance. In the discrete-time case, data are collected at equally spaced time intervals (Figure 1.4c). The rule indicating how the measurements at one time depend on measurements at the previous time includes random factors. Rather than knowing with certainty the next measurements, we know only a set of possible outcomes and their associated probabilities and can therefore predict the outcome only in a probabilistic or statistical sense. For example, a probabilistic dynamical system describing population growth is a rule that gives the **probability** that a population has a particular size in a given year as a function of the population in the previous year. The study of such systems requires the mathematical methods of **probability theory**, the material in the final three chapters of this book.

1.2 Updating Functions: Describing Growth

Suppose we collect data on how much bacterial cultures grow in one hour, or how much trees grow in one year. How can we predict what will happen in the long run? In this section, we begin addressing this problem. We follow the basic steps of applied mathematics: **quantifying the basic measurement** and describing the **dynamic rule** with the **updating function**. The **updating function** is the mathematical way to describe how the basic measurement changes. Studying this function requires a review of the terminology and graphs used to describe functions in general.

A Model Population: Reproduction of Bacteria

Suppose several bacterial cultures with different initial population sizes are grown in controlled conditions for one hour and then carefully counted. The population size acts as the basic measurement. The following table presents the results for six colonies.

1.2 Updating Functions: Describing Growth

Growth of six bacterial colonies

Colony	Old population (b_{old})	New population (b_{new})	Ratio of new to old population
1	0.47×10^6	0.95×10^6	2.02
2	3.3×10^6	6.4×10^6	1.94
3	0.73×10^6	1.5×10^6	2.05
4	2.8×10^6	5.6×10^6	2.00
5	1.5×10^6	3.1×10^6	2.07
6	0.62×10^6	1.2×10^6	1.94

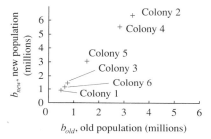

Figure 1.5
Data describing the growth of six bacterial populations

In each case, the population approximately doubled in size. For example, the first colony began with 0.47×10^6 bacteria and ended with 0.95×10^6, slightly more than twice as many. The data are graphed in Figure 1.5, where we have plotted the first measurement, b_{old}, along the horizontal axis and the second measurement, b_{new}, along the vertical axis.

How can we convert these observations into a dynamic rule? Because our measurements were made at discrete times rather than continuously, we use a **discrete-time dynamical system**. The dynamic rule for a discrete-time dynamical system is a **function** giving the measurement at the end of the experiment in terms of the measurement at the beginning. Informally, a **function** is a rule that assigns exactly one output to every possible input. In this case, the function is the rule that assigns as output the measurement at the end of the experiment to the input of the measurement at the beginning. It thus summarizes what happened, the dynamics. We call this function the **updating function** (Figure 1.6).

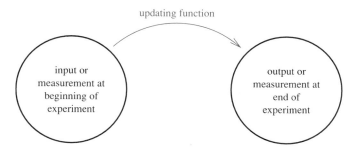

Figure 1.6
The updating function

There are several ways to write the updating function. Denoting the population at the beginning of the experiment by b_{old} and that at the end by b_{new}, we can write the updating function **as a rule**:

$$b_{new} = 2.0 b_{old} \tag{1.1}$$

This equation says that the new population can be computed by doubling the old population. Alternatively, we can write the updating function as a **function** f with formula

$$f(b_{old}) = 2.0 b_{old} \tag{1.2}$$

This function describes what happened in our experiment; the input was doubled. We can then write the dynamic rule in the **general form for an updating function** as

$$b_{new} = f(b_{old}) \tag{1.3}$$

For example, if $b_{old} = 1.2 \times 10^6$, we compute that

$$b_{new} = f(1.2 \times 10^6) = 2 \cdot (1.2 \times 10^6) = 2.4 \times 10^6$$

Our updating function **predicts** that a bacterial culture starting with a population of 1.2 million will end up with a population of 2.4 million.

It may look strange to write a function without the traditional x as input and y as output. Because biology is complicated and involves many factors, we must choose variable names that help us remember what the letters stand for. Here, for example, b_{old} represents the old *b*acterial population.

A Model Organism: A Growing Tree

Suppose you measure the heights of several trees in one year and then again the next year. Denoting the old height by h_{old} and the new height by h_{new}, we might find the data in the following table.

Growth of several trees

Tree	Old height (h_{old}) (m)	New height (h_{new}) (m)	Change in height (m)
1	23.1	24.1	1.0
2	18.7	19.8	1.1
3	20.6	21.5	0.9
4	16.0	17.0	1.0
5	32.5	33.6	1.1
6	19.8	20.6	0.8

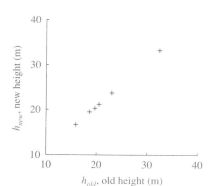

Figure 1.7
Data describing the growth of six trees

These data are graphed in Figure 1.7. Like the bacteria, the trees appear to follow a simple rule, increasing in height by about 1.0 m each year.

How can we express the updating function describing tree growth? Our updating function must express the fact that the new height can be found by adding about 1.0 m to the old height, or express the rule

$$h_{new} = h_{old} + 1.0 \tag{1.4}$$

This updating function, denoted by g to distinguish it from the updating function for the bacterial population, has formula

$$g(h_{old}) = h_{old} + 1 \tag{1.5}$$

In the general form for an updating function, we write

$$h_{new} = g(h_{old}) \tag{1.6}$$

to summarize the data in the table. For example, if $h_{old} = 12.2$, the updating function predicts a new height of

$$h_{new} = g(12.2) = 12.2 + 1 = 13.2$$

In both examples, the data points do not exactly match the updating function. The updating function captures the major trend in the data while ignoring the noise. Including only the trend corresponds to the use of a **deterministic** dynamical system to describe the behavior. To include the noise, we must use a **probabilistic** dynamical system. We will address the problem of **finding** an updating function that captures the major trends in the data when we study the technique of data-fitting called the method of least squares.

Functions: Terminology and Graphs

The analysis of the data in the preceding two subsections involved two fundamental mathematical objects: numbers and functions. Numbers describe measurements and **functions** describe **relations** between measurements. For example, the bacterial population growth example includes two measurements, the numbers denoted by b_{old} and b_{new}. The **relation** between these measurements is expressed by the updating function f. To prepare to study these functions in detail, we review the language and graphical techniques mathematicians and applied mathematicians use to describe and depict functions.

A function is a mathematical object that takes something (such as a number) as input, performs an operation on it, and returns a new object (such as another number) as output. The input is called the **argument** (or sometimes the **independent variable**) and the output is called the **value** (or sometimes the **dependent variable**) (Figure 1.8).

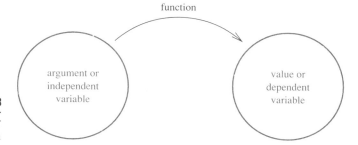

Figure 1.8

The basic terminology for describing a function

In the population growth example, the argument of the updating function f is b_{old}, the population of bacteria at the beginning of the experiment. The value of the function is b_{new}, the population at the end of the experiment. The updating function describes the relation between these two measurements: the output is double the input.

The set of things that a function can accept as inputs is called the **domain**. The updating function f takes bacterial population sizes as inputs. Because negative populations do not make sense, the domain of the function consists of all positive numbers and 0. We write that

$$f \text{ is defined on the domain } b_{old} \geq 0$$

Besides negative numbers, fruits, bacteria, and updating functions lie outside the domain of f.

The set of things a function can return as outputs is called the **range**. Because the updating function f returns population sizes as output, the range of f also consists of all positive numbers and 0. We write that

f has range $b_{new} \geq 0$

Consider the table of data shown on the left. These data describe a relation between two observations: an identification of the species and the number of legs. We could express this as a function L (to represent legs). According to the table,

$$L(\text{ant}) = 6, \quad L(\text{crab}) = 10$$

and so on. The domain of this function is "types of animals," and the range is the nonnegative integers (0, 1, 2, 3, and so on). We can plot this information as shown in Figure 1.9. To show the **relation** between the two numbers, we plot the input ("animal") along the horizontal axis and the output ("number of legs") on the vertical axis.

Animal	Number of legs
Ant	6
Crab	10
Duck	2
Fish	0
Human being	2
Mouse	4
Spider	8

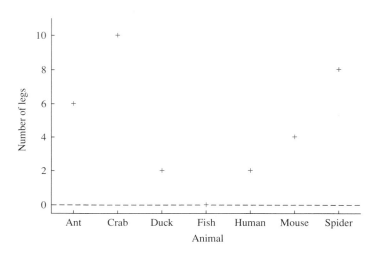

Figure 1.9
Numbers of legs on various organisms plotted on a graph

As with our updating functions f and g, some of the most important mathematical functions have domains and ranges consisting of numbers. Such functions can be described in three ways: numerically (by means of a table), as a rule, or as a graph. The bacterial population growth data are described by the function f defined by

$$f(b_{old}) = 2.0 b_{old}$$

The graph corresponding to this function is shown in Figure 1.10. Graphs are drawn using **Cartesian coordinates**, which use two perpendicular number lines to describe two numbers. The axes are labeled with both units and values. **Never draw a graph without labeling the axes.**

To graph an unknown function, it is easiest to start by plugging in some representative arguments. For example, plugging in the values 0, 1, 2 and 3, we find that

$$f(0) = 0, \quad f(1) = 2, \quad f(2) = 4, \quad \text{and} \quad f(3) = 6$$

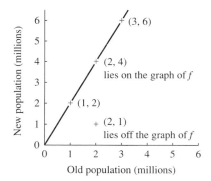

Figure 1.10
Updating function for the bacterial population

The fact that $f(2) = 4$ is plotted as the **ordered pair** (2, 4) by moving 2 units along the horizontal axis and 4 units along the vertical axis. Plotting the points (0, 0), (1, 2), and (3, 6) in the same way (Figure 1.10), we see that the points lie on the line drawn connecting them. The point (2, 1), plotted by moving 2 units along the horizontal axis and 1 unit along the vertical axis, does not lie on the graph of f because $f(2) \neq 1$.

It is important to realize that the graph of a function is *not* the function, just as the spot labeled 2 on the number line is not the number 2 and a photograph of a dog is not a dog. The graph is a depiction of the function.

Biological data are often presented as functions of time. One of our goals is to use the updating function to compute bacterial populations or tree sizes as functions of time. Results, quite different from those that would be produced by the updating function f, are presented in the following table. These data are graphed in Figure 1.11. Because the data points rise during the first ten time intervals, we can see (more easily from the graph than the table) that the bacterial population grew steadily for the first 10 hours. After that, the data points become lower, meaning that the population declined thereafter. The maximum size of the population is at time 10. This graph presents the results of an experiment and can be used to understand the results even without a mathematical formula. As biologists or applied mathematicians, we must be able to translate pictorial information into words that communicate key observations to colleagues and the public.

The population of bacteria in a culture

Time (h)	Population (millions)	Time (h)	Population (millions)
0	0.86	13	5.66
1	1.22	14	5.29
2	1.69	15	4.86
3	2.28	16	4.41
4	2.98	17	3.95
5	3.74	18	3.50
6	4.49	19	3.07
7	5.17	20	2.67
8	5.69	21	2.30
9	6.03	22	1.96
10	6.17	23	1.67
11	6.13	24	1.41
12	5.95	25	1.18

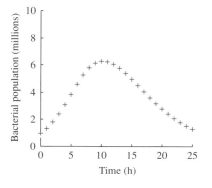

Figure 1.11

The population of bacteria in a culture

Conversely, it can be useful to draw a graph of a function from a verbal description. Suppose we are told that a population increases between time 0 and time 5, decreases nearly to 0 by time 12, increases to a maximum at time 20, and goes extinct at time 30. Our graph translates this information into pictorial form. Because we were not given exact values, our graph is not exact. Instead, it gives a **qualitative picture** of the results (Figure 1.12).

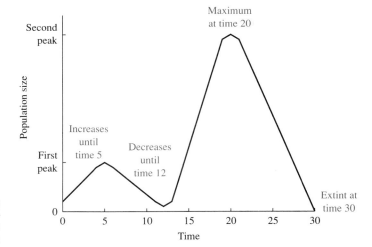

Figure 1.12
A bacterial population plotted from a verbal description

SUMMARY

Starting from data on the growth of bacterial populations and tree height, we defined the **updating function**, the **dynamical rule** that tells how the population or organism grew from one time to the next. In general, a function describes a mathematical operation that accepts an **argument** as input and returns a **value** as output. The **domain** consists of all possible arguments and the **range** consists of all possible values. A function can be presented with a table of data, a mathematical formula, or a graph. Graphs are presented in **Cartesian coordinates** by plotting the argument on the carefully labeled horizontal axis and the value on the equally carefully labeled vertical axis. The ability to translate back and forth between graphs and verbal descriptions is a key skill of all scientists.

1.2 EXERCISES

1. Compute the values of the following functions at the points indicated.
   a. $f(x) = x^2$ at $x = 0$, $x = 1$, and $x = 4$
   b. $g(y) = 3.22y$ at $y = 0.29$, $y = -1.18$, and $y = 7.32$
   c. $h(z) = 3.22 + z$ at $z = 0.29$, $z = -1.18$, and $z = 7.32$

2. Plot the values computed in Exercise 1, and sketch a graph of the function on the domain suggested.
   a. $f(x)$ for $0 \leq x \leq 5$
   b. $g(y)$ for $-2 \leq y \leq 8$
   c. $h(z)$ for $-2 \leq z \leq 8$

3. a. Which of the points $(-2, 2)$, $(-2, 2)$, $(-2, 4)$, and $(4, -2)$ lie on the graph of the function $f(x) = x^2$?
   b. Which of the points $(1.23, 3.96)$, $(1.23, 4.55)$, and $(3.96, 1.23)$ lie on the graph of the function $g(y) = 3.22y$?
   c. Which of the points $(1.23, 3.96)$, $(1.23, 4.55)$, and $(3.96, 1.23)$ lie on the graph of the function $h(z) = 3.22 + z$?

4. Plot on the same graph the data (Figure 1.5) and the updating function f (Equation 1.2 and Figure 1.10) describing the

bacterial population. Which data points lie exactly on the graph of the function? Do the same for the data (Figure 1.7) and updating function g (Equation 1.5) describing tree size.

5. The following tables display data from three experiments: cell volume after 10 min in a watery bath, fish length after 1 week in a chilly tank, and gnat population size after 3 days without food.

Cell volume (μm^3)

Old (v_{old})	New (v_{new})
1220	1830
1860	2790
1080	1620
1640	2460
1540	2310
1900	2850

Fish length (cm)

Old (l_{old})	New (l_{new})
13.1	11.4
18.2	16.5
17.3	15.6
16.0	14.3
20.5	18.8
14.7	13.0

Gnat number

Old (n_{old})	New (n_{new})
1.2×10^3	6.0×10^2
2.4×10^3	1.2×10^3
1.6×10^3	8.0×10^2
2.0×10^3	1.0×10^3
1.4×10^3	7.0×10^2
8.0×10^2	4.0×10^2

a. For each experiment, graph the new value as a function of the old value.
b. For each experiment, write the updating function.
c. What would the new cell volume be if the old volume were 1420 μm^3?
d. What would the new fish length be if the old length were 1.5 cm? Does this make sense? How might you modify your updating function?

6. Consider the following data describing the growth of a tadpole.

Age (days)	Length (cm)	Mass (g)
0.5	1.5	1.5
1.0	3.0	3.0
1.5	4.5	6.0
2.0	6.0	12.0
2.5	7.5	24.0
3.0	9.0	48.0

a. Graph length as a function of age. Make sure to label your axes.
b. Graph mass as a function of age.
c. Graph mass as a function of length.

d. Graph length as a function of mass. How does this graph compare with the one in part **c**?

7. The data in Exercise 6 can be thought of as defining several updating functions. For example, between the first measurement (on day 0.5) and the second (on day 1.0), the length increases by 1.5 cm. Between the second measurement (on day 1.0) and the third (on day 1.5), the length again increases by 1.5 cm.
a. Graph the length at the second measurement as a function of length at the first. Graph the length at the third measurement as a function of length at the second. Do the same for the rest of the measurements.
b. Do the same for mass, graphing mass at each measure after the first as a function of the mass at the previous measurement.
c. Do the same for age, graphing age at each measure after the first as a function of the age at the previous measurement.
d. Write updating functions for length, mass, and age.

8. Describe what is happening in the three graphs shown. The first plots cell volume against time in days; the second, the population of Pacific Salmon against time in years; and the last, the average height of a population of trees plotted against age in years.

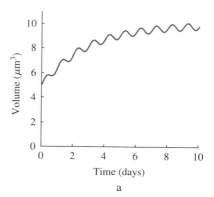
a

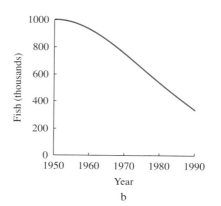
b

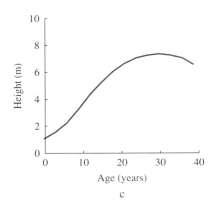

c

9. a. Draw a graph describing a population of birds that begins at a large value, decreases to a tiny value, and then increases again to an intermediate value.
   b. Draw a graph describing the amount of DNA in an experiment where the quantity increases rapidly from a very small value and then levels out at a large value before declining rapidly to 0.
   c. Draw a graph of body temperature that oscillates between high values during the day and low values at night.

10. Suppose students are permitted to take a test again and again until they get a perfect score of 100. We wish to write an updating function describing these dynamics.

   a. In words, what is the argument of this updating function?
   b. What is the value?
   c. What are the domain and range of the updating function?
   d. What value do you expect if the argument is 100?
   e. Sketch a possible graph of this updating function.
   f. Use your graph to describe how a student will do after three tries if he starts with a score of 20.

11. COMPUTER: Use your graphing calculator or computer to plot the following functions. How would you describe them in words?
   a. $x^2 e^{-x}$ for $0 \leq x \leq 10$
   b. $1.5 + e^{-0.1x} \sin(x)$ for $0 \leq x \leq 20$
   c. $\sin(x) + \cos(2.2x)$ for $0 \leq x \leq 20$

12. COMPUTER: Use your computer to plot the function

$$h(x) = e^{-x^2} - e^{-1000(x-0.13)^2} - 0.2$$

for values of x between -10 and 10.
   a. How would you describe the result in words?
   b. Zoom in on the graph by changing the range to find all points where the value of the function is 0. For example, one such value is between 1 and 2. Plot the function again for x between 1 and 2 to zoom in.
   c. If you found only two such intersections, zoom in on the region between 0 and 1 to try to find two more.

1.3 Units and Dimensions of Measurements and Functions

Suppose we wish to study the growth of a bacterial population in terms of mass rather than number, or the sizes of trees in terms of volume rather than height. We cannot simply change the letters in our updating functions. Unlike the numbers and functions studied in many mathematics courses, the measurements and relations used by scientists and applied mathematicians have **units** and **dimensions**. Measurements of number, mass, height, and volume are fundamentally different from one another and are said to have different **dimensions**. Measurements of height in inches or height in centimeters describe the same thing but are presented in different **units**. When we wish to express a measurement or relation in different units, we must use appropriate **conversion factors**. When we wish to express a measurement or relation in different dimensions, we must **translate** with a **fundamental relation**. In this section, we use conversion factors and fundamental relations to look at bacterial populations and tree sizes in different units and dimensions.

Converting Between Units

The updating function $b_{new} = 2.0 b_{old}$ (Equation 1.1) expresses a relation between numbers of bacteria. The updating function $h_{new} = h_{old} + 1$ m (Equation 1.4) expresses a relation between heights in meters. This difference is expressed by the **units** appearing in the equations: bacteria in the first and meters in the second. In the sciences, numbers represent measurements and measurements have units. In mathematics, numbers are "pure" and often lack units.

The equation

$$2 + 2 = 2$$

looks hopelessly wrong. But

$$2\,Na^+ + 2\,Cl^- = 2\,NaCl$$

is a standard formula from chemistry. The difference is that the terms in the second equation have explicit units: ions of sodium, ions of chlorine, and molecules of salt. Similarly, although it is absurd to write

$$1 = 2.54$$

it is nearly true that

$$1\text{ in.} = 2.54\text{ cm}$$

Numbers with units behave very differently from pure numbers.

Often, data are presented with more than one unit. To compare measurements, we must be able to convert between different units. Suppose we want to know how many centimeters make up a mile. We can do this in steps, first changing miles to feet, then feet to inches, and then inches to centimeters. To convert between units, first **write down the basic identities**:

$$5280\text{ ft} = 1\text{ mile}$$
$$12\text{ in.} = 1\text{ ft}$$
$$2.54\text{ cm} = 1\text{ in.}$$

These identities define how many centimeters are in an inch, how many inches are in a foot, etc. Next, **divide** to find three **conversion factors**:

$$1 = 5280\frac{\text{ft}}{\text{mile}}$$
$$1 = 12\frac{\text{in.}}{\text{ft}}$$
$$1 = 2.54\frac{\text{cm}}{\text{in.}}$$

Units are manipulated in the same way as the numerators and denominators of fractions. Finally, **multiply** the original measurement by the conversion factors (which are just fancy ways to write the number 1). In this case,

$$1\text{ mile} = 1\text{ mile} \times 1 \times 1 \times 1$$
$$= 1\text{ mile} \times 5280\frac{\text{ft}}{\text{mile}} \times 12\frac{\text{in.}}{\text{ft}} \times 2.54\frac{\text{cm}}{\text{in.}}$$
$$= 160{,}934\text{ cm}$$

The units cancel in the same way as the numerators and denominators of fractions. In chemistry, this is often called the **factor-label method**.

Algorithm 1.1 (The procedure for converting between units)[1]

1. Write the basic identities that relate the original units to the new units.
2. Divide the basic identities to create conversion factors equal to 1, placing unwanted units in the denominator.
3. Multiply the original measurement by the appropriate conversion factors.

In the example, it would be equally true that

$$1 = \frac{1}{5280}\frac{\text{mile}}{\text{ft}}$$

$$1 = \frac{1}{12}\frac{\text{ft}}{\text{in.}}$$

$$1 = \frac{1}{2.54}\frac{\text{in.}}{\text{cm}}$$

Multiplying by these conversion factors,

$$1 \text{ mile} = 1 \text{ mile} \times 1 \times 1 \times 1$$
$$= 1 \text{ mile} \times \frac{1}{5280}\frac{\text{mile}}{\text{ft}} \times \frac{1}{12}\frac{\text{ft}}{\text{in.}} \times \frac{1}{2.54}\frac{\text{in.}}{\text{cm}}$$
$$= \frac{1}{160{,}934}\frac{\text{mile}^2}{\text{cm}}$$

The units did not cancel. Even though the result is *true*, miles and centimeters are left over in a rather inconvenient way. The trick to getting unit conversions to work is making sure that unwanted units cancel.

Translating Between Dimensions

Miles and centimeters measure the same stuff, length, with different rulers. Miles and grams measure completely different quantities. **Dimensions** describe the underlying quantities. **Units** are a particular standard for measurement. Measurements in miles and centimeters share the same dimension and can be **converted** from one to the other. Measurements with different dimensions cannot. The dimensions and sample units of some common biological measurements are listed in Table 1.1.

Table 1.1 Some Common Quantities, Their Dimensions, and Sample Units

Quantity	Dimensions	Sample units
Length	Length	Inch, meter, micron
Duration	Time	Second, minute, day
Mass	Mass	Gram, kilogram
Area	Length^2	Square meter, acre
Volume	Length^3	Liter, cubic foot, gallon
Speed	Length/time	Meter/second, mph
Acceleration	Length/time^2	Meter/second^2
Force	Mass $\times$ length/time^2	Dynes, pounds
Density	Mass/length^3	Grams/liter

---

[1] This book contains many algorithms, procedures or recipes for solving particular problems. As with a recipe, following an algorithm without thinking about the steps can lead to disaster.

1.3 Units and Dimensions of Measurements and Functions

Suppose we want to measure bacterial populations in grams (mass) or tree sizes in cubic meters (volume). We cannot apply an identity such as 1 in. = 2.54 cm because we are translating between dimensions rather than converting between units. Instead of **identities**, we use **fundamental relations** among measurements with different dimensions (Table 1.2).

Table 1.2 Some Fundamental Relations

Relation	Variables	Formula
Geometric relations		
Volume of a sphere	V = volume r = radius	$V = \dfrac{4\pi}{3} r^3$
Surface area of a sphere	S = surface area r = radius	$S = 4\pi r^2$
Area of a circle	A = area r = radius	$A = \pi r^2$
Perimeter of a circle	P = perimeter r = radius	$P = 2\pi r$
Volume, area, and thickness	V = volume A = area T = thickness	$V = AT$
Relations involving mass		
Total number and mass	m = total mass μ = mass per individual b = number of individuals	$m = \mu b$
Mass, density, and volume	M = total mass ρ = density V = volume	$M = \rho V$

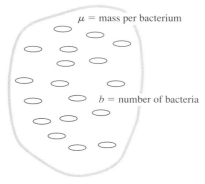

total mass = mass per bacterium × number of bacteria $m = \mu b$

Figure 1.13

Fundamental relation between mass and number

The fundamental relation between number and total mass is

$$\text{total mass} = \text{mass per bacterium} \times \text{number of bacteria} \quad (1.7)$$

(Figure 1.13). Let m represent the total mass, μ the mass per bacterium, and b the number of bacteria.[2] The fundamental relation can be rewritten in mathematical symbols as

$$m = \mu b \quad (1.8)$$

Like numbers, variables representing measurements have both dimensions and units. The variable m has units of grams, μ has units of grams, and b has units of bacteria. For example, if

$$b = 2.0 \times 10^5 \quad \text{and} \quad \mu = 3.1 \times 10^{-9}\,\text{g}$$

---

[2]This book uses many Greek letters to represent various measurements. The Greek alphabet and pronunciations of the letters are tabulated in the inside back cover.

then
$$m = (3.1 \times 10^{-9}) \cdot (2.0 \times 10^5) = 6.2 \times 10^{-4} \text{g}$$

Computing the volume of a tree from its height requires a fundamental relation between height and volume. Suppose first that trees are perfect spheres (slightly more realistic shapes are considered in Exercise 5 at the end of this section). The fundamental relation between height and volume comes from geometry. The volume V of a sphere with radius r is

$$V = \frac{4\pi}{3} r^3 \qquad (1.9)$$

where r has units of meters and V has units of cubic meters, or m³. Looking at Figure 1.14, we see that the height is twice the radius, or $h = 2r$. Therefore,

$$r = \frac{h}{2}$$

Substituting into the formula for the volume of a sphere, we have

$$V = \frac{4\pi}{3} \left(\frac{h}{2}\right)^3$$
$$= \frac{4\pi}{3} \frac{h^3}{2^3}$$
$$= \frac{\pi}{6} h^3$$

Figure 1.14

Height and radius of a spherical tree

The first tree measured in the growth experiment increased from height 23.1 m to height 24.1 m. Therefore, the volume began at

$$V_{old} = \frac{\pi}{6} h_{old}^3$$
$$= \frac{\pi}{6} 23.1^3$$
$$\approx 6450 \text{ m}^3$$

The next year, the volume is

$$V_{new} = \frac{\pi}{6} h_{new}^3$$
$$= \frac{\pi}{6} 24.1^3$$
$$\approx 7330 \text{ m}^3$$

The volume increases by much more than the height.

To compute the mass of our tree from its volume, we use the important fundamental relation

$$\text{mass} = \text{density} \times \text{volume} \qquad (1.10)$$

If we denote the density by ρ ("rho") and the mass by M, the fundamental relation can be rewritten in mathematical symbols as

$$M = \rho V \qquad (1.11)$$

For example, suppose the density of wood is 0.8 g/cm³ (slightly less than the density of water). Because we measured tree size in cubic meters, we must first

convert the density into grams per cubic meter. From the basic identity

$$1 \text{ m} = 100 \text{ cm}$$

we find the conversion factor

$$1 = 100 \frac{\text{cm}}{\text{m}}$$

Therefore, we can convert the density into the appropriate units with

$$0.8 \frac{\text{g}}{\text{cm}^3} = 0.8 \frac{\text{g}}{\text{cm}^3} \times 100 \frac{\text{cm}}{\text{m}} \times 100 \frac{\text{cm}}{\text{m}} \times 100 \frac{\text{cm}}{\text{m}}$$
$$= 8.0 \times 10^5 \frac{\text{g}}{\text{m}^3}$$

The mass of a tree with height 23.1 m is

$$M_{old} = 8.0 \times 10^5 \frac{\text{g}}{\text{m}^3} \cdot 6450 \text{ m}^3 \approx 5.16 \times 10^9 \text{ g}$$

and the mass after a year, when the height is 24.1 m, is

$$M_{new} = 8.0 \times 10^5 \frac{\text{g}}{\text{m}^3} \cdot 7330 \text{ m}^3 \approx 5.86 \times 10^9 \text{ g}$$

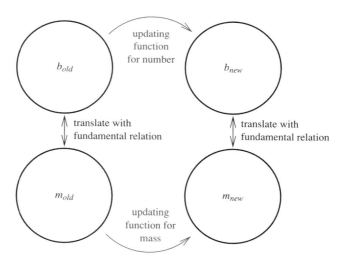

Figure 1.15
Finding the updating function for bacteria in terms of mass

As with numbers and variables, the functions used in applied mathematics have dimensions and units. For example, the updating function $f(b_{old}) = 2.0 b_{old}$ can accept as input only positive numbers with units of bacteria. However, we can translate updating functions into different dimensions. For example, suppose we wish to study the updating function for a bacterial population in terms of mass rather than number. At the beginning, the mass, denoted by m_{old}, is

$$m_{old} = \rho b_{old}$$

and the updated mass, m_{new}, is

$$m_{new} = \rho b_{new}$$

The updating function says that $b_{new} = 2.0 b_{old}$. Our goal is to find a formula for m_{new} in terms of m_{old}. We need to solve for b_{old} in terms of m_{old} and b_{new} in terms of m_{new}, or

$$b_{old} = \frac{m_{old}}{\rho} \quad \text{and} \quad b_{new} = \frac{m_{new}}{\rho}$$

Therefore,

$$b_{new} = 2.0 b_{old} \quad \text{original updating function}$$

$$\frac{m_{new}}{\rho} = 2 \frac{m_{old}}{\rho} \quad \text{translating in terms of mass}$$

$$m_{new} = 2.0 m_{old} \quad \text{solving for } m_{new}$$

The new updating function doubles its input like the original updating function, but takes as its input mass in grams rather than numbers of bacteria (Figure 1.15).

Checking: Dimensions and Estimation

It is crucial to check the dimensions and units of any unfamiliar equation you encounter. In the equation

$$\text{mass} = \text{density} \times \text{volume}$$

(Equation 1.10), the dimensions of M are mass, the dimensions of V are length^3, and the dimensions of ρ are mass/length^3. Rewriting in dimensions,

$$\text{mass} = \frac{\text{mass}}{\text{length}^3} \times \text{length}^3$$

The length^3 terms cancel, and the dimensions of the two sides match, as they must. This procedure is called **dimensional analysis**. Many errors can be nipped in the bud by checking dimensions. An equation with inconsistent dimensions is not merely incorrect, it is nonsensical.

Just as it is essential to check the dimensions and units of equations, it is essential to check the plausibility of the numerical results of calculations. For example, suppose you wanted to figure out how many tons all the people in the United States weigh. Each person weighs on average about 100 lb (counting children) or 1/20 ton per person. If there are about 260 million people, they should weigh a net amount of around 13 million tons (using the fundamental relation that total mass is equal to mass per individual times the number of individuals). If you had worked this out with a more complicated set of measurements and found a more precise answer of 14.7 million tons, everything is probably all right. If the complicated method gave an answer of 1.47 million tons, it needs to be checked.

How much area does a colony of 2.0×10^5 bacteria take up on a petri dish? One method is to use our computation of the mass (6.2×10^{-4} g), convert to volume, and find the area by dividing by the thickness. Assuming that bacteria have approximately the density of water, the volume is

$$V = \frac{M}{\rho} = \frac{6.2 \times 10^{-4} \text{ g}}{1.0 \times 10^{-12} \text{ g}/\mu\text{m}^3}$$

$$= 6.2 \times 10^8 \ \mu\text{m}^3$$

The next **fundamental relation**, which translates between volume and area, is

$$\text{volume} = \text{area} \times \text{thickness}$$

so that

$$\text{area} = \frac{\text{volume}}{\text{thickness}}$$

Estimating the thickness of the colony to be about 20 μm (roughly the thickness of a cell),

$$\text{area} \approx \frac{6.2 \times 10^8 \ \mu m^3}{20 \ \mu m}$$

$$\approx 3 \times 10^7 \ \mu m^2$$

This sounds rather large. To convert to centimeters, we use the basic identity

$$1 \ \mu m = 10^{-4} \ cm$$

so that the conversion factor is

$$1 = 10^{-4} \ \frac{cm}{\mu m}$$

Multiplying, we get

$$3 \times 10^7 \ \mu m^2 = 3 \times 10^7 \ \mu m^2 \times 10^{-4} \ \frac{cm}{\mu m} \times 10^{-4} \ \frac{cm}{\mu m}$$

$$= 0.3 \ cm^2$$

To find the radius, we use the fundamental geometric relation between the area A and radius r for a circle:

$$A = \pi r^2$$

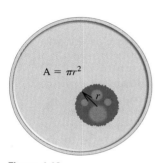

Figure 1.16
A bacterial colony on a petri dish

(Figure 1.16). The radius r of this colony satisfies

$$\pi r^2 \approx 0.3$$

Solving for r,

$$r \approx \sqrt{\frac{0.3}{\pi}} \approx 0.31 \ cm$$

In fact, this colony is pretty small.

SUMMARY   Understanding scientific equations and formulas requires understanding the **units** and **dimensions** of the measurements and variables. Dimensions describe the underlying quantities and tell what sort of thing is being measured. Units express numerical values based on a particular scale. Converting among units can be done by starting with **basic identities**, deriving **conversion factors** and multiplying. Translating between measurements with different dimensions requires using **fundamental relations**, such as those between mass and volume or volume and radius. All such relations, and every scientific formula, should be checked for consistency with **dimensional analysis**. Using basic identities and fundamental relations, we can compute useful estimates of quantities, often without using a calculator. Checking results for plausibility can help locate mistakes.

1.3 EXERCISES

1. a. What are the dimensions of pressure (force per unit area)?
    b. What are the dimensions of rate of spread for a colony of bacteria growing on a plate?
    c. What are the dimensions of rate of spread for a colony of bacteria growing in a beaker?
    d. The force of gravity between two objects is equal to Gm_1m_2/r^2, where m_1 and m_2 are the masses of the two objects and r is the distance between them. What are the dimensions of the gravitational constant G?

2. Check whether the following formulas are dimensionally consistent.
    a. Distance = rate × time
    b. Force = mass × acceleration
    c. rate of chemical production = $\dfrac{kS}{1+cS}$
    where S has dimensions of mass, k has dimensions of 1/time, and c has dimensions of 1/mass

3. Convert the following into the specified units.
    a. Find 3.4 lb in grams.
    b. Find 1 yard in millimeters.
    c. Find 60 yr in hours.
    d. Find 65 mph in centimeters per second.
    e. Find 2.3 g/cm^3 in pounds per cubic foot.

4. Find the mass of the following objects in grams (the density of water is 1.0 g/cm^3).
    a. A cube of water 40 cm on a side.
    b. A coral colony consisting of 3200 individuals each weighing 0.45 g.
    c. A spherical cow with radius 1.3 m and density 1.3 times that of water.
    d. A circular colony of mold with radius 2.4 cm and density of 0.0023 g/cm^2.

5. Find the masses of trees of height 23.1 m and 24.1 m under the following shape assumptions.
    a. A tree is a perfect cylinder with radius 0.5 m no matter what the height. (Recall that the volume of a cylinder of height h and radius r is $\pi h r^2$.)
    b. A tree is a perfect cylinder with radius equal to 0.1 times the height.
    c. A tree looks like the tree in part **a**, but half the height is a cylindrical trunk and the other half is a spherical blob on top.

6. Recall the updating functions $b_{new} = 2.0 b_{old}$ for a bacterial population.
    a. Write an updating function for the total volume of bacteria (suppose each bacterium takes up $10^4 \mu m^3$).
    b. Write an updating function for the total area taken up by the bacteria (suppose the thickness is 20 μm).

7. Recall the equation $h_{new} = h_{old} + 1.0$ for tree height.
    a. Write an updating function for the total volume of the cylindrical trees in Exercise 5a.
    b. Write an updating function for the total volume of a spherical tree (this is kind of tricky).

8. Estimate the following and compute them exactly (when possible).
    a. The density of water in pounds per cubic inch.
    b. The weight of the earth in tons. Assume that the earth is a sphere with radius 4000 miles and density 5 times that of water.
    c. The number of gallons of water needed to give a rectangular garden that is 4 ft × 5 ft an amount of water equivalent to 1 in. of rain.
    d. The speed of light in centimeters per nanosecond (ns) (10^{-9} seconds), starting from the speed of 186,000 miles/s. A fairly fast computer takes about 20 ns per operation. How big could a computer be before communication time began interfering with computations? In other words, how far does light travel in the time required by one operation?

9. (From Paulos, 1988): Estimate the speed that your hair grows in miles per hour.

10. A cell is roughly a sphere 10 μm in radius.
    a. Using the fact that the density of a cell is approximately the density of water, and that water weighs 1 g/cm^3, estimate the number of cells in your body.
    b. Try to estimate your volume in cubic meters by imagining you are shaped like a board. Pretending that cells are cubes 20 μm on a side, what do you estimate the number of cells to be?
    c. The nematode *C. elegans* is a cylinder about 1 mm long and 0.1 mm in diameter, consisting of about 1000 cells. Are these cells about the same size as the ones in your body?

1.4 Linear Functions and Their Graphs

The updating functions for bacteria and for growing trees were developed from a particular set of data. How can we predict what will happen with different starting values? To answer this question, we need a systematic way to find the formula of an updating function from a table of data. In the cases we have studied so far, the graphs of the data lie on straight lines. We will see how to summarize these data with a **linear function**, placing particular emphasis on its **slope**. Using this formula, we can **interpolate** from known values to make predictions about the results of additional experiments.

Proportional Relations

In the bacterial updating function $b_{new} = 2.0 b_{old}$ (Equation 1.1), the new bacterial population can be found by multiplying the old population by the constant value 2.0. In the fundamental relation between mass and volume $M = \rho V$ (Equation 1.11), mass can be found by multiplying volume by the constant value ρ. These relations are called **proportional relations**. The ratio (proportion) of bacterial populations is

$$\frac{b_{new}}{b_{old}} = 2.0$$

and that of mass to volume is

$$\frac{\text{mass}}{\text{volume}} = \text{density} = \rho$$

When the ratio is constant, the value is called the **constant of proportionality**.

The proportion in a proportional relation is the **slope** of the graph. Slope is defined as "rise over run," meaning

$$\text{slope} = \frac{\text{change in output}}{\text{change in input}}$$

For example, consider the data points graphed in Figure 1.17a. At the lower point, $b_{old} = 1.0 \times 10^6$ and $b_{new} = 2.0 \times 10^6$. At the upper point, $b_{old} = 3.0 \times 10^6$ and $b_{new} = 6.0 \times 10^6$. The change in the output is the difference between the two values of b_{new}, or

$$\text{change in output} = 6.0 \times 10^6 - 2.0 \times 10^6 = 4.0 \times 10^6$$

The change in the input is the difference between the two values of b_{old}, or

$$\text{change in input} = 3.0 \times 10^6 - 1.0 \times 10^6 = 2.0 \times 10^6$$

The slope is

$$\text{slope} = \frac{\text{change in output}}{\text{change in input}}$$
$$= \frac{4.0 \times 10^6}{2.0 \times 10^6} = 2.0$$

equal to the constant of proportionality.

Similarly, suppose that $\rho = 5.0$ g/cm^3. A first object with volume $V_1 = 1.0$ cm^3 has mass $M_1 = 5.0$ g. A second object with volume $V_2 = 4.0$ cm^3 has mass $M_2 = 20.0$ g (Figure 1.17b). We then find

$$\text{change in output} = 20.0 - 5.0 = 15.0 \text{ g}$$
$$\text{change in input} = 4.0 - 1.0 = 3.0 \text{ cm}^3$$

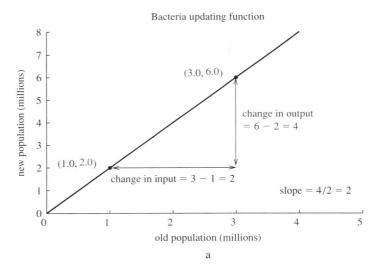

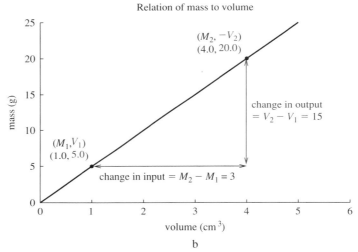

Figure 1.17
Two proportional relations and their slopes

The slope is then

$$\text{slope} = \frac{\text{change in output}}{\text{change in input}} = \frac{15.0 \text{ g}}{3.0 \text{ cm}^3} = 5.0 \ \frac{\text{g}}{\text{cm}^3}.$$

Again, the **slope** is equal to the **constant of proportionality** that multiplies the input.

In general, suppose we denote two points on the graph by (V_1, M_1) and (V_2, M_2) (Figure 1.18). Then

$$\text{slope} = \frac{\text{change in output}}{\text{change in input}} = \frac{M_2 - M_1}{V_2 - V_1} \tag{1.12}$$

The change in the output M is often written with the shorthand

$$\Delta M = M_2 - M_1$$

1.4 Linear Functions and Their Graphs

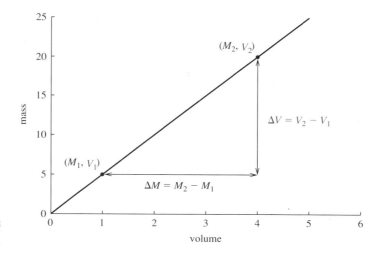

Figure 1.18
Density, slope, and Δ notation

where Δ (the Greek letter "delta") means "change in." The denominator, the amount by which the input changes, can be written, with the same shorthand, as

$$\Delta V = V_2 - V_1$$

The slope is ΔM divided by ΔV, or

$$\text{slope} = \frac{\Delta M}{\Delta V} \tag{1.13}$$

This notation will prove useful later when we study derivatives.

The Equation of a Line

Proportional relations are described by functions that perform a single operation on their input: multiplication by a constant. We have seen that the graphs of such functions are lines with slope equal to the constant of proportionality. Furthermore, these lines pass through the point $(0, 0)$ because an input of 0 produces an output of 0.

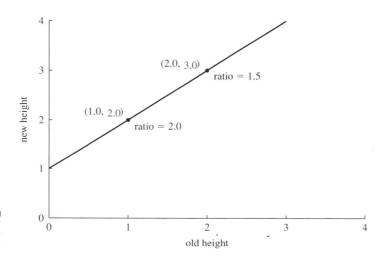

Figure 1.19
A linear function that is not a proportional relation

Many functions other than those describing proportional relations have straight-line graphs. The graph of the updating function for a growing tree certainly looks straight (Figure 1.19). The relation between h_{old} and h_{new} is not a proportional relation, however. For example, when $h_{old} = 1$, $h_{new} = h_{old} + 1 = 2$. The new height is 2 times the old height. However, when $h_{old} = 2$, $h_{new} = h_{old} + 1 = 3$. At this point, the new height is only 1.5 times the old height. The proportion of h_{new} to h_{old} is not a constant.

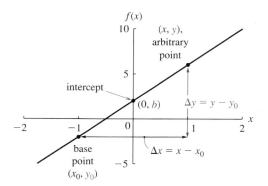

Figure 1.20
The elements of the general linear graph

A **straight line** is characterized by **constant slope**, like a constant grade on a ski slope. Suppose that the function $f(x)$ has a straight-line graph and that $f(x_0) = y_0$. The point (x_0, y_0), which lies on the graph of the function (Figure 1.20), is called the **base point**. The slope between (x_0, y_0) and an arbitrary point (x, y) on the line is

$$\text{slope} = \frac{\Delta y}{\Delta x} = \frac{y - y_0}{x - x_0}$$

Because the slope between any two points on the graph is constant,

$$\frac{y - y_0}{x - x_0} = m$$

for some fixed value of m. Multiplying both sides by $(x - x_0)$, we find

$$y - y_0 = m(x - x_0)$$

or

$$y = m(x - x_0) + y_0 \tag{1.14}$$

This equation gives the **point-slope form** for a line, expressing the vertical height y as a function of the horizontal position x, in terms of the point (x_0, y_0) and the slope m. As a check, if we plug in $x = x_0$, we find that $y = m(x_0 - x_0) + y_0 = y_0$.
In terms of the function f,

$$\text{slope} = \frac{f(x) - f(x_0)}{x - x_0} = m$$

Solving for the function $f(x)$, we find the **point-slope form for a linear function**,

$$f(x) = m(x - x_0) + f(x_0) \tag{1.15}$$

Alternatively, we can multiply out the terms on the right-hand side of the point-slope formula, finding

$$y = mx + (y_0 - mx_0)$$

Setting $b = y_0 - mx_0$, we get

$$y = mx + b \quad (1.16)$$

called the **slope-intercept form for a line**. The letter b represents the point where the graph crosses the y-axis and is called the **y-intercept** (Figure 1.20).

A function that has a formula

$$f(x) = mx + b$$

is called a **linear function**. The graph of this linear function is a line with slope m and y-intercept b. Algebraically, $b = f(0)$. When the y-intercept is 0, the function f describes a proportional relation. The updating function for the bacterial population,

$$f(b_{old}) = 2.0 b_{old}$$

is thus a linear function with a slope of 2.0 and a y-intercept of 0 (and is therefore a proportional relation). The updating function for the growing tree,

$$g(h_{old}) = h_{old} + 1.0$$

is a linear function with slope 1 and y-intercept 1.0. The y-intercept gives the height at age 1 yr (starting from height 0).

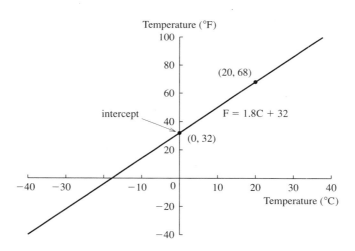

Figure 1.21
The relation between Fahrenheit and Celsius temperatures

Another important linear function gives the basic conversion between temperature in Fahrenheit and Celsius (Figure 1.21). Recall that

$$F = 1.8C + 32 \quad (1.17)$$

where C represents temperature in Celsius and F represents temperature in Fahrenheit. Unlike nearly all other unit conversions, this formula does not express a proportional relation. The y-intercept of 32 indicates that $0°C$ corresponds to $32°F$ rather than to $0°F$. The slope, however, describes the number of degrees Fahrenheit per degree Celsius as in an ordinary conversion. For example, $20°C$ corresponds

to 68°F. The change in °F (the output) is
$$\Delta F = 68°F - 32°F = 36°F$$
and the change in °C (the input) is
$$\Delta C = 20°C - 0°C = 20°C$$
Therefore,
$$\text{slope} = \frac{36°F}{20°C} = 1.8 \frac{°F}{°C}$$
meaning that it takes 1.8°F to make up 1.0°C.

Finding Equations and Graphing Lines

Pet store owners can be plagued by parasites. Suppose the employees spend a week observing populations of mites on several lizards and find the following data:

Populations of mites on lizards

Old number (x_{old})	New number (x_{new})
20	70
30	90
40	110
50	130

We are asked to estimate the number of mites we would find in a week on a lizard that has 45 mites today. To do so, we must first find the equation for the updating function and then substitute $x_{old} = 45$. We find the equation in two steps: First we find the slope, and then we plug one data point into the point-slope form for a line.

■ **Algorithm 1.2** (Finding the equation of a line from data)

1. Pick two data points.

2. Find the slope as the change in output divided by the change in input.

3. Find the equation by substituting one point into the point-slope form for a line (Equation 1.14).

4. If needed, convert into slope-intercept form.

We can follow this algorithm to find the equation of the updating function describing our data.

1. Pick the first and last data points.

2. The slope m is
$$m = \frac{\Delta x_{new}}{\Delta x_{old}}$$
$$= \frac{130 - 70}{50 - 20}$$
$$= \frac{60}{30} = 2.0$$

3. Using the point $(20, 70)$ as (x_0, y_0) in the point-slope form for a line, the equation is

$$x_{new} = m(x_{old} - x_0) + y_0$$
$$= 2.0(x_{old} - 20) + 70$$

4. We can multiply this out to find the slope-intercept form,

$$x_{new} = 2.0 x_{old} + 30$$

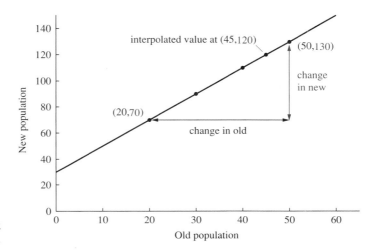

Figure 1.22
Updating function for a mite population

The updating function h describing these mites is therefore

$$h(x_{old}) = 2.0 x_{old} + 30 \quad (1.18)$$

To predict the number of mites we will find in a week on a lizard that has 45 mites today, substitute $x_{old} = 45$ into the updating function to find

$$h(x_{old}) = x_{new} = 2.0 \cdot 45 + 30 = 120$$

We have used the formula for the updating function to **interpolate** a prediction between known values.

What is the meaning of the y-intercept 30? It is the number of mites we would find on a lizard that started out with no mites. These mites probably arrived from other lizards. What is the meaning of the slope 2.0? It is the number of additional mites we would find after a week if we added 1 mite at the beginning. For example, $h(21) = 72$, two more than $h(20)$. Each mite leaves 2 progeny, just like our bacteria. One possible interpretation of the updating function is that every mite already on the lizard produces 2 babies and dies, while 30 new mites arrive from elsewhere.

In this example, we used a set of numerical data to plot a graph and derive an algebraic formula. In other cases, we are given a formula and need to produce a graph. The easiest way to plot the graph of a linear function from its formula is to substitute two plausible values of the input into the equation, graph the points, and connect them with a line. For example, suppose we wish to plot the linear function $F(x)$ given by

$$F(x) = -2x + 30$$

Plugging in $x = 0$ gives $F(0) = 30$, and $x = 10$ gives $F(10) = 10$. The graph of the line connects the points $(0, 30)$ and $(10, 10)$ (Figure 1.23). This line goes down with a negative slope of -2.

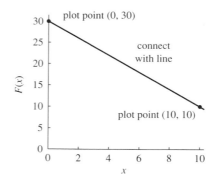

Figure 1.23
Graphing a line from its equation

Because the graph goes down, we say that F is a **decreasing function**. A larger input produces a smaller output. Linear functions with negative slopes are decreasing functions. In contrast, a positive slope corresponds to an **increasing function**, such as the updating function for mites. Larger inputs produce larger outputs. A slope of exactly 0 corresponds to a function with equation

$$f(x) = 0 \cdot x + b = b$$

Such a function always takes on the constant value b, the y-intercept, and has as graph a horizontal line (Figure 1.24).

Slope	Graph	Function
Positive	Goes up	Increasing
Negative	Goes down	Decreasing
Zero	Flat	Constant

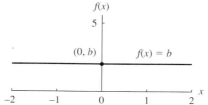

Figure 1.24
A linear function with slope equal to 0

SUMMARY The graphs of the updating functions for bacterial populations and tree heights are straight lines. We derived the link between straight lines and **linear functions**. A **proportional relation** is a special type of linear function in which the ratio of the output to the input is always the same. This constant ratio is the **slope** of the graph of the relation. Lines other than proportional relations can be expressed in **point-slope form** or **slope-intercept form**. The slope can be found as the change in output divided by the change in input. Equations of linear functions can be used to **interpolate**, or estimate outputs from untested inputs.

1.4 EXERCISES

1. Unit conversions are proportional relations. Find the slope and graph the relations between the following units.
   a. Changing inches to centimeters.
   b. Changing centimeters to inches.
   c. Changing grams to pounds.
   d. Changing pounds to grams.

2. Find the slope of the graph of each of the following fundamental relations, and sketch a graph of each.
   a. Volume = area × thickness. Think of volume as a function of area, and set thickness to 1.0 cm.
   b. Volume = area × thickness. Think of volume as a function of thickness, and set area to 7.0 cm².
   c. Total mass = mass per bacterium × number of bacteria. Think of total mass as a function of the number of bacteria, and set the mass per bacterium to 5.0×10^{-9} grams.
   d. Total mass = mass per bacterium × number of bacteria. Think of total mass as a function of mass per bacterium, and set the total number to 10^6.

3. A ski slope has a slope of −0.2. You start at an altitude of 10,000 feet.
   a. What will be your altitude when you have gone 2000 feet horizontally?
   b. The ski run ends at an altitude of 8000 feet. How far will you have gone horizontally?
   c. Write the equation giving altitude as a function of horizontal distance moved.

4. Find the slopes and y-intercepts of the following lines. Graph the lines.
   a. The line $f(x) = 2(x − 1) + 3$.
   b. A line passing through the point $(1, 6)$ with slope −2.
   c. A line passing through the points $(1, 6)$ and $(4, 3)$.

5. Suppose $h(y) = −5(y + 7) + 8$.
   a. Identify the components of point-slope form.
   b. Use them to sketch a graph of the function.
   c. Find another point that lies on the graph of the line.
   d. Use this point to compute the slope, and indicate how you did this on your graph.
   e. Write the line in slope-intercept form, and mark the intercept on your graph.

6. Consider the data in the following table (adapted from H. C. F. Godfray, 1994). It describes the number of wasps that can develop inside caterpillars of different weights.

Weight of caterpillar (g)	Number of wasps
0.5	80
1.0	115
1.5	150
2.0	175

   a. Graph these data.
   b. Find the equation of the line connecting the first two points.
   c. Which point does not lie on the line?
   d. How many wasps would you expect from a caterpillar weighing 0.72 g? Plot this point on your graph.
   e. How many wasps would you expect from a caterpillar weighing 0.0 g? Does this make sense? Plot this point on your graph.
   f. How many wasps would you expect from a caterpillar weighing 2.72 g? Plot this point on your graph.

7. Compute the equation of the updating function for mites on lizards using a different pair of rows from those used in the text (rather than the first and last).

8. Find and graph the equations of the lines giving M, V, and G as functions of a from the following table describing a growing plant. Write them in both point-slope and slope-intercept form. What do the y-intercepts mean? Use interpolation to find the mass, volume, and glucose production on day 1.75.

Age (days)	Mass (g)	Volume (cm³)	Glucose production (mg)
0.5	2.5	5.1	0.0
1.0	4.0	6.2	3.4
1.5	5.5	7.3	6.8
2.0	7.0	8.4	10.2
2.5	8.5	9.5	13.6
3.0	10.0	10.6	17.0

9. Consider the following data describing the levels of a medication in the blood of two patients over the course of several days.

Day	Medication level in patient 1 (mg/L)	Medication level in patient 2 (mg/L)
0	5.0	1.0
1	3.5	1.5
2	2.75	1.75
3	2.375	1.875

   a. Use these data to graph three points on the updating function for each patient. Make sure to label your axes with good names for the measurements.
   b. Find the equation of the updating function for the first patient.
   c. Find the equation of the updating function for the second patient.

d. Is it surprising that the two patients have the same updating function even though one has decreasing levels of the medication and the other has increasing levels?

10. The world record times for various races are decreasing at roughly linear rates (adapted from McWhirter and McWhirter, 1990).
  a. The men's Olympic record for the 1500-m race was 3:36.8 in 1972 and 3:35.9 in 1988. Find and graph the line connecting these. (Don't forget to convert everything into seconds.)
  b. The women's Olympic record for the 1500-m race was 4:01.4 in 1972 and 3:53.9 in 1988. Find and graph the line connecting these.
  c. If things continue at this rate, when will women be running this race faster than men? (Set the two speeds equal and solve for the date.)

11. COMPUTER: Try Exercise 10 on the computer. Compute the year when the times will reach 0. Give your best guess of the times in the year 1900.

12. COMPUTER: Graph the ratio of temperature measured in Fahrenheit to temperature measured in Celsius for $-273 \leq C < 200$. What happens near $C = 0$? What happens for large and small values of C? How would the results differ if the 0 for Fahrenheit were changed to match that of Celsius?

1.5 Finding Solutions: Describing the Dynamics

We have carefully defined updating functions that describe what happens from the beginning to the end of a single experiment. What happens if we run the same experiment repeatedly? Both the bacterial population and the tree will grow, but how? A description of what happens to a population as a function of time is called a **solution** of the discrete-time dynamical system. If we can find a **formula** for the solution, we can predict sizes after 2, 3, or 100 time units.

Population Growth of Bacteria

Our updating functions give the result at the end of an experiment as a function of the measurement at the beginning. What if we were to continue the experiment? A bacterial population growing according to $b_{new} = 2.0 b_{old}$ (Equation 1.2) would double again. A tree growing according to $h_{new} = h_{old} + 1.0$ (Equation 1.5) would add another meter to its height. An infested lizard would become even more infested.

We cannot use the b_{old} and b_{new} notation to describe the behavior of a system left to run for many time steps. Instead, we **index** the measurements by the time t. For the bacteria, let b_0 denote the population at the beginning of the experiment (the original b_{old}), b_1 the population after 1 h (the original b_{new}), b_2 the population after 2 h, and so on. In general, we define

$$b_t = \text{population after } t \text{ h} \quad (1.19)$$

Our updating function tells us that $b_1 = f(b_0)$, $b_2 = f(b_1)$ or, in general,

$$b_{t+1} = f(b_t) = 2.0 b_t \quad (1.20)$$

for any $t \geq 0$ (we never defined b_t for $t < 0$) (Figure 1.25). Henceforth, we will use this notation to define updating functions.

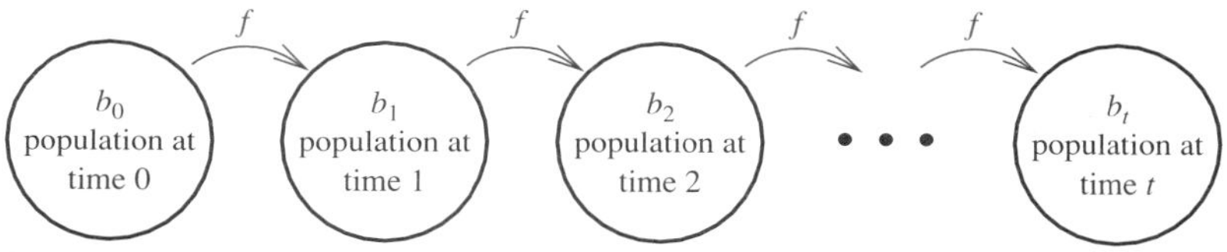

Figure 1.25
The repeated action of the bacterial updating function

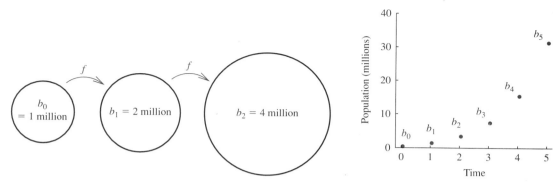

Figure 1.26
A solution: bacterial population size as a function of time

Our goal is to solve for b_t and h_t. First, we must know where we started. Without such information, it is impossible to answer a question like, "Where are you after driving 5 miles south?" The starting value is known as the **initial condition**. Suppose we begin with 1 million bacteria, an initial condition of $b_0 = 1.0 \times 10^6$. Then

$$b_1 = 2.0 b_0 = 2.0 \cdot 1.0 \times 10^6 = 2.0 \times 10^6$$
$$b_2 = 2.0 b_1 = 2.0 \cdot 2.0 \times 10^6 = 4.0 \times 10^6$$
$$b_3 = 2.0 b_2 = 2.0 \cdot 4.0 \times 10^6 = 8.0 \times 10^6$$

(Figure 1.26). Each hour, the population doubles. Examining these results, we notice that

$$b_1 = 2.0 \cdot 1.0 \times 10^6$$
$$b_2 = 2.0^2 \cdot 1.0 \times 10^6$$
$$b_3 = 2.0^3 \cdot 1.0 \times 10^6$$

After 3 h, the population has doubled three times and is $2.0^3 = 8.0$ times the original population. After t h, the population will have doubled t times, and have reached the size

$$b_t = 2.0^t \cdot 1.0 \times 10^6 \qquad (1.21)$$

This formula is the **solution** of the discrete-time dynamical system. It predicts the population after t h of reproduction for any value of t. For example, we can compute

$$b_8 = 2.0^8 \cdot 1.0 \times 10^6 = 2.56 \times 10^8$$

without ever computing b_3, b_4, or other intermediate values.

We graph the solution by plotting the time t on the horizontal axis and the number of bacteria after t h (b_t) on the vertical axis (Figure 1.26). The graph consists only of a discrete set of points describing the hourly measurements, hence the name "discrete-time dynamical system."

Suppose we started the system with a different initial condition, $b_0 = 3.0 \times 10^5$. We can find subsequent values by repeatedly applying the updating function,

$$b_1 = 2.0 \cdot 3.0 \times 10^5 = 6.0 \times 10^5$$
$$b_2 = 2.0 \cdot 6.0 \times 10^5 = 1.2 \times 10^6$$
$$b_3 = 2.0 \cdot 1.2 \times 10^6 = 2.4 \times 10^6$$

If we look for the pattern in this case, we find

$$b_1 = 2.0 \cdot 3.0 \times 10^5$$
$$b_2 = 2.0^2 \cdot 3.0 \times 10^5$$
$$b_3 = 2.0^3 \cdot 3.0 \times 10^5$$

After t h, the population will have doubled t times, as before, and will have reached the size

$$b_t = 2.0^t \cdot 3.0 \times 10^5$$

The solution is different (but related) when we use a different initial condition (Figure 1.27). Although the two solutions get further and further apart, the ratio always remains the same (Exercise 3).

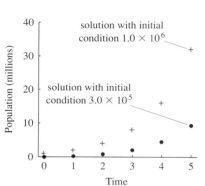

Figure 1.27
Solutions starting from two different initial conditions

Tree Height

With our new notation, let h_t represent the height of a tree after t yr of growth. The updating function g is

$$h_{t+1} = g(h_t) = h_t + 1.0$$

We can use the method of finding solutions for several steps and looking for the pattern to find h_t, tree height after t yr. Suppose the tree begins with a height of $h_0 = 10.0$ m. Then

$$h_1 = h_0 + 1.0 = 11.0 \text{ m}$$
$$h_2 = h_1 + 1.0 = 12.0 \text{ m}$$
$$h_3 = h_2 + 1.0 = 13.0 \text{ m}$$

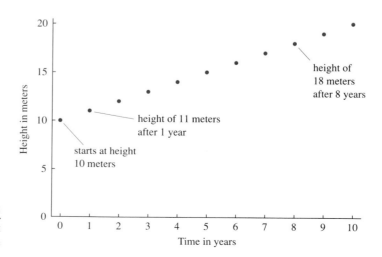

Figure 1.28
A solution for tree height as a function of time

Each year, the height of the tree increases by 1.0 m. After 3 yr, the height is 3.0 m greater than the original height. After t yr the tree has added 1.0 m to its height t times, meaning that the height will have increased by a total of t m. Therefore the solution is

$$h_t = 10.0 + t \text{ m} \qquad (1.22)$$

This formula predicts the height after t yr of growth for any t (Figure 1.28). For example, we can find that

$$h_8 = 10.0 + 8.0 = 18.0 \text{ m}$$

without computing h_3, h_4, or other intermediate values.

If the tree began at the smaller size of 2.0 m, the size for the first few years would be

$$h_1 = h_0 + 1.0 = 3.0 \text{ m}$$
$$h_2 = h_1 + 1.0 = 4.0 \text{ m}$$
$$h_3 = h_2 + 1.0 = 5.0 \text{ m}$$

Again, the tree adds t m of height in t yr, so the height is

$$h_t = 2.0 + t \text{ m}$$

The solution with this smaller initial condition is always exactly 8.0 m below the solution found before (Figure 1.29).

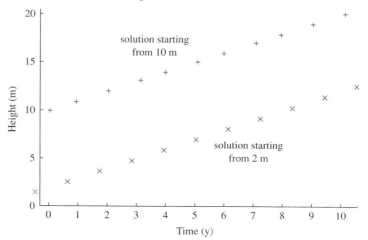

Figure 1.29
Two solutions for tree height as functions of time

Finding Solutions in More Complicated Cases

Is it always possible to guess the formula for a solution in this way? Recall the updating function

$$x_{t+1} = 2x_t + 30$$

for lizard mites, translated into our new notation. If we started our lizard with $x_0 = 10$ mites, we compute

$$x_1 = 2.0x_0 + 30 = 50$$
$$x_2 = 2.0x_1 + 30 = 130$$
$$x_3 = 2.0x_2 + 30 = 290$$

The pattern is not so obvious in this case. There is a pattern, which is a good challenge to find (Exercise 11). We will soon meet dynamical systems for which there is *no* pattern, hard as that might be to imagine.

Sometimes finding a formula is hard even though we can see the pattern. Consider the updating function based on Exercise 9 in Section 1.4. Each day, a patient is given a dose of medication sufficient to raise its concentration in the bloodstream by 1.0 mg/L. During the next 24 h, however, half the medication is used up by the body. Let M_t denote the concentration at time t. The updating function is

$$M_{t+1} = 0.5 M_t + 1.0$$

Like the lizard updating function, this is constructed from two pieces. The first piece, $0.5M_t$, indicates that only half the old medication *remains* the next day. The second piece indicates that 1.0 mg/L of medication is *added* each day (Figure 1.30).

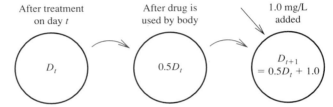

Figure 1.30
The dynamics of medication concentration in the blood

Suppose we begin from an initial condition of $M_0 = 5.0$ mg/L. Then

$$M_1 = 0.5 \cdot 5.0 + 1.0 = 3.5$$
$$M_2 = 0.5 \cdot 3.5 + 1.0 = 2.75$$
$$M_3 = 0.5 \cdot 2.75 + 1.0 = 2.375$$
$$M_4 = 0.5 \cdot 2.375 + 1.0 = 2.1875$$

If we plot these results, we notice that the values are getting closer and closer to 2.0 (Figure 1.31). More careful examination indicates that the results move exactly

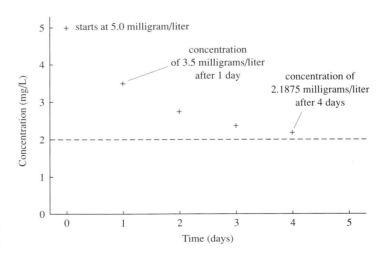

Figure 1.31
The solution of medication concentration as a function of time

halfway toward 2.0 in each step. In particular,

$$M_0 - 2.0 = 5.0 - 2.0 = 3.0$$
$$M_1 - 2.0 = 3.5 - 2.0 = 1.5 = 0.5 \cdot 3.0$$
$$M_2 - 2.0 = 2.75 - 2.0 = 0.75 = 0.5 \cdot 1.5$$
$$M_3 - 2.0 = 2.375 - 2.0 = 0.375 = 0.5 \cdot 0.75$$
$$M_4 - 2.0 = 2.1875 - 2.0 = 0.1875 = 0.5 \cdot 0.375$$

Can we convert these observations into the formula for a solution? If we write the concentration as 2.0 plus the difference,

$$M_0 = 2.0 + 3.0$$
$$M_1 = 2.0 + 0.5 \cdot 3.0$$
$$M_2 = 2.0 + 0.5^2 \cdot 3.0$$
$$M_3 = 2.0 + 0.5^3 \cdot 3.0$$

we might see that

$$M_t = 2.0 + 0.5^t \cdot 3.0$$

Finding patterns in this way and translating them into formulas can be tricky. It is much more important to be able to **describe** the behavior of solutions with a graph or in words. In this case, our calculations quickly revealed that the solution moved closer and closer to 2.0.

Similarly, if we begin with an initial concentration of $M_0 = 1.0$ mg/L, then

$$M_1 = 0.5 \cdot 1.0 + 1.0 = 1.5$$
$$M_2 = 0.5 \cdot 1.5 + 1.0 = 1.75$$
$$M_3 = 0.5 \cdot 1.75 + 1.0 = 1.875$$
$$M_4 = 0.5 \cdot 1.875 + 1.0 = 1.9375$$

Again, the values get closer and closer to 2.0 (Figure 1.32). Again, the difference from 2.0 is reduced by a factor of 0.5 each day,

$$M_0 - 2.0 = 1.0 - 2.0 = -1.0$$
$$M_1 - 2.0 = 1.5 - 2.0 = -0.5$$
$$M_2 - 2.0 = 1.75 - 2.0 = -0.25$$
$$M_3 - 2.0 = 1.875 - 2.0 = -0.125$$
$$M_4 - 2.0 = 1.9375 - 2.0 = -0.0625$$

However, unlike bacterial populations (Figure 1.27) and tree size (Figure 1.29), the graphs of solutions starting from different initial conditions have very different forms.

However, we can find the formula using the same idea as before. If we write

$$M_0 = 2.0 - 1.0$$
$$M_1 = 2.0 - 0.5 \cdot 1.0$$
$$M_2 = 2.0 - 0.5^2 \cdot 1.0$$
$$M_3 = 2.0 - 0.5^3 \cdot 1.0$$

we can see that

$$M_t = 2.0 - 0.5^t \cdot 1.0$$

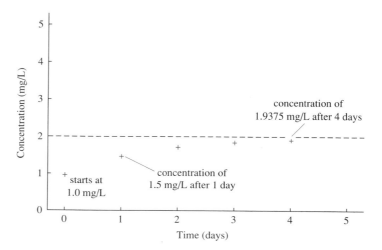

Figure 1.32
Another solution of medication concentration as a function of time

---

SUMMARY By examining the pattern, we found and graphed the **solutions** of the discrete-time dynamical systems for bacterial population growth, tree height, and medication concentration. In each case, the solution gives the value of the measurement as a function of time and depends on the starting point, or **initial condition**. Solutions with different initial conditions may look similar, as for the bacterial and tree cases, or quite different, as for medication concentration.

1.5 EXERCISES

1. Using the solution of the bacterial updating function (Equation 1.21), find the following.
   a. The bacterial population after 20 h.
   b. The bacterial population after 40 h.
   c. If each bacterium is a cube about 2.0×10^{-3} cm on a side, how much volume will the population take up at each time? How much would the population weigh if bacteria have about the density of water?

2. Using the solution of the tree height updating function (Equation 1.22), find the following.
   a. Tree height after 20 yr.
   b. Tree height after 100 yr.
   c. Suppose trees are cylinders with radius 0.5 m no matter what the height (as in Exercise 5a in Section 1.3). Graph the volume as a function of time up to time $t = 10$.
   d. Suppose instead that a tree is a perfect cylinder with radius equal to 0.1 times the height (as in Exercise 5b in Section 1.3). Graph the volume as a function of time up to time $t = 10$. Why does the volume increase faster than the height?

3. Suppose two bacterial populations follow the updating function $b_{t+1} = 2.0 b_t$, but the first starts with initial condition $b_0 = 1.0 \times 10^6$ and the second starts with initial condition $b_0 = 3.0 \times 10^5$.
   a. Find the difference between the two populations as a function of time.
   b. Find the ratio of the two populations as a function of time.
   c. Why does the ratio remain the same although the difference increases?

4. Suppose two trees follow the updating function $h_{t+1} = h_t + 1.0$, but the first starts with initial condition $h_0 = 10.0$ m and the second starts with initial condition $h_0 = 2.0$ m.
   a. Find the difference between the two heights as a function of time.
   b. Find the ratio of the two heights as a function of time.
   c. Why does the ratio change although the difference remains the same?

5. Find and graph the solutions of the following discrete-time dynamical systems (given in Exercise 5 in Section 1.2) for five steps starting from the given initial condition.
   a. $v_{t+1} = 1.5 v_t$, starting from $v_0 = 1220 \ \mu\text{m}^3$
   b. $l_{t+1} = l_t - 1.7$, starting from $l_0 = 13.1$ cm
   c. $n_{t+1} = 0.5 n_t$, starting from $n_0 = 1200$

6. Find and graph the first five steps of the solution of the medication updating function, $M_{t+1} = 0.5M_t + 1.0$, starting from the initial condition $M_0 = 18.0$. Write the equation for the solution.

7. Suppose a population of bacteria doubles every hour, but that 1.0×10^6 individuals are removed to be converted into valuable biological by-products. Suppose the population begins with $b_0 = 2.0 \times 10^6$ bacteria.
   a. Find the population after 1, 2, and 3 h.
   b. How many bacteria were harvested?
   c. Write the updating function.
   d. Suppose you waited to harvest bacteria until the end of 3 h. How many could you remove and still match the population b_3 found in part **a**? Where did all the extra bacteria come from?

8. A superior mutant gene occurs in a fraction 0.0001 of a population (one in 10,000 organisms have it). Suppose the fraction increases by 10% in each generation.
   a. Write the updating function for the fraction of organisms with the gene.
   b. When will the fraction reach 1.0?
   c. When does the updating function stop making sense?

9. The Weber-Fechner law describes how human beings perceive differences. Suppose, for example, that a person first hears a tone with a frequency of 400 Hz (hertz, or cycles per second). She is then tested with higher tones until she can hear the difference. The ratio between these values describes how well this person can hear differences.
   a. Suppose the next tone she can distinguish has a frequency of 404 Hz. What is the ratio?
   b. According to the Weber-Fechner law, the next higher tone will be greater than 404 Hz by the same ratio. Find this tone.
   c. Write the updating function for this person. Find the fifth tone she can distinguish.
   d. Suppose the experiment is repeated on a musician, and she manages to distinguish 400.5 Hz from 400 Hz. What is the fifth tone she can distinguish?

10. Consider the updating function

$$f(x) = \frac{x}{1 + x}$$

   a. Starting from an initial condition of $x_0 = 1$, compute x_1, x_2, x_3, and x_4.
   b. What is the pattern?
   c. Try the same from the initial condition $x_0 = 2$.

11. Try to find the pattern in the number of mites on a lizard starting with $x_0 = 10$ and following the updating function $x_{t+1} = 2x_t + 30$.
   HINT: Either look at the increase in the number of mites each week or add 30 to the number of mites.

12. **COMPUTER**: Use your computer (it may have a special feature for this) to find and graph the first 10 points on the solutions of the following discrete-time dynamical systems. The first two describe populations with reproduction and immigration of 100 individuals per generation, and the last two describe populations that have 100 individuals harvested or removed from each generation.
   a. $b_{t+1} = 0.5b_t + 100$ starting from $b_0 = 100$
   b. $b_{t+1} = 1.5b_t + 100$ starting from $b_0 = 100$
   c. $b_{t+1} = 1.5b_t - 100$ starting from $b_0 = 201$
   d. $b_{t+1} = 1.5b_t - 100$ starting from $b_0 = 199$
   e. What happens if you run the last one for 15 steps? What is wrong with the model?

1.6 Combining and Manipulating Functions

Many of the processes we have studied involve combining different functions. For example, in building the updating function $M_{t+1} = 0.5M_t + 1.0$ describing medication concentrations, we had to *add* the medication remaining after one day ($0.5M_t$) to the medication dose each day (1.0). Building solutions required applying the same updating function again and again. Because functions are the building blocks of models, we need to know how and when to combine them in different ways. In this section, we will study **addition** and **composition of functions**. Finally, a builder must be able to work backward and take things apart. We can figure out yesterday's measurement from today's measurement by computing the **inverse** of the updating function.

Adding Functions

Recall that the updating function describing medication concentrations combines two separate processes, use of medication by the body and addition of medication by the doctor (Figure 1.30). The updating function is a **sum** of terms describing these two processes,

$$M_{t+1} = 0.5M_t + 1.0$$

We **add** these terms because each contributes separately to the final result.

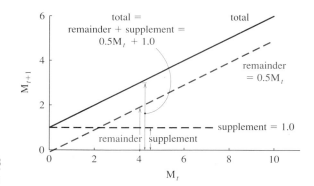

Figure 1.33
Adding two simple functions

Geometrically, we can graph each of the pieces and add them in the same way (Figure 1.33). The height of the graph of the sum of two functions is the height of the first plus the height of the second. More formally, we can define a new function, $f + g$, as the sum of the functions f and g.

Definition 1.1 The sum, $f + g$, of the functions f and g is the function that takes on the value

$$(f + g)(x) = f(x) + g(x)$$

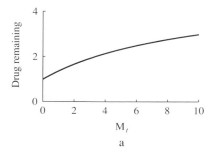
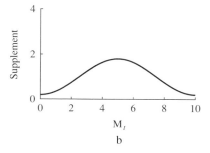

Figure 1.34
More complicated functions for medication remaining and medication given

This procedure can be used graphically to add much more complicated functions. Suppose, for example, that the amount of medication remaining in the body follows the curve in Figure 1.34a, and that the doctor decides to supplement according to the graph in Figure 1.34b. The updating function is again the sum of these two processes. Proceeding through the graph step by step as in Figure 1.33, we can sketch the combined effect of these two processes (Figure 1.35).

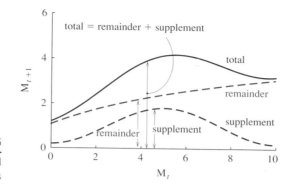

Figure 1.35

Adding two more complicated functions

Composition of Functions

We found the solutions of the bacterial and tree updating functions without using functional notation. The step-by-step calculation of values, however, proceeded by applying the updating function again and again. We will now write these results explicitly in terms of the updating function. For the bacteria, with updating function f, we have

$$b_1 = f(b_0) \tag{1.23}$$

and

$$b_2 = f(b_1) \tag{1.24}$$

These equations express the fact that the updating function **updates** the population for one time unit.

Could we find b_2 from b_0 without ever computing the intermediate value b_1? We can combine our previous two equations to write

$$\begin{aligned} b_2 &= f(b_1) && \text{definition of the updating function} \\ &= f[f(b_0)] && \text{plugging in } b_1 = f(b_0) \\ &= (f \circ f)(b_0) && \text{definition of functional composition} \end{aligned} \tag{1.25}$$

The last two equations are read "f of f of b_0". The operator "$\circ$" indicates **composition** of the function f with itself (Figure 1.36).

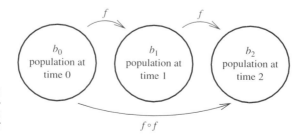

Figure 1.36

The composition of an updating function with itself

Before applying this notation, we define the composition of functions in general.

42 Chapter 1 Introduction to Discrete-Time Dynamical Systems

■ **Definition 1.2** The composition $f \circ g$ of functions f and g is a function defined by

$$(f \circ g)(x) = f[g(x)] \quad (1.26)$$

We say "f composed with g evaluated at x" or "f of g of x." The function f is called the **outer function**, and g is called the **inner function**. ■

How do we compute the composition of two functions? Suppose

$$F(x) = -2x + 1$$
$$G(x) = 5x + 3$$

with domains consisting of all numbers. To find the composition $F \circ G$, plug the definition of the **inner function** G into the formula for the **outer function** F, or

$(F \circ G)(x) = F(G(x))$	the definition
$= F(5x + 3)$	write the formula for the inner function G
$= -2(5x + 3) + 1$	plug the formula for G into the outer function F
$= -10x - 5$	simplify the expression

We find the composition $G \circ F$ by following the same steps, or

$(G \circ F)(x) = G(F(x))$	the definition
$= G(-2x + 1)$	write the formula for the inner function F
$= 5(-2x + 1) + 3$	plug the formula for F into the outer function G
$= -10x + 8$	simplify the expression

The key step is substituting the output of the inner function as x into the outer function.

This example illustrates an important point about the composition of functions: The answer is generally different when the functions are composed in a different order. If $F \circ G = G \circ F$, we say that the two functions **commute**. When the two compositions do not match, we say that the two functions do not commute. Without a good reason, never assume that two functions commute. If you think of functions as operations, this should make sense. Cleaning a wound and then stitching it closed produces a quite different result from stitching it closed and then cleaning it.

As another example, suppose that $G(x) = 5x + 3$ as before, but that $F(x) = x^2$. Then

$$(F \circ G)(x) = F(G(x))$$
$$= F(5x + 3)$$
$$= (5x + 3)^2$$
$$= 25x^2 + 30x + 9$$

and

$$(G \circ F)(x) = G(F(x))$$
$$= G(x^2)$$
$$= 5(x^2) + 3$$
$$= 5x^2 + 3$$

The results are quite different. It is not easy to find interesting functions that commute.

We can also compose functions that have units and dimensions. Suppose that F takes a radius r as input and returns the volume of a sphere with that radius as output. Then F has the formula

$$F(r) = \frac{4\pi}{3} r^3$$

(Equation 1.9). Suppose G takes a volume V as input and returns the mass of an object with that volume as output, according to mass = density × volume, or

$$G(V) = 5.0 \, \frac{\text{g}}{\text{cm}^3} V$$

(Equation 1.11 with $\rho = 5.0$ g/cm^3). The composition $G \circ F$ takes radius as input and returns mass as output in a single step (Figure 1.37). The composition is

$$(G \circ F)(r) = G(F(r))$$
$$= G\left(\frac{4\pi}{3} r^3\right)$$
$$= 5.0 \frac{4\pi}{3} r^3$$

For example, we could find the mass of an object with radius 2.5 cm in two steps by finding volume and then mass with the steps

$$V = F(2.5) = \frac{4\pi}{3} 2.5^3 = 65.45 \text{ cm}^3$$
$$M = G(65.45) = 5.0 \cdot 65.45 = 327.25 \text{ g}$$

Alternatively, we could find the mass in a single step by composing the functions G and F,

$$M = (G \circ F)(2.5) = 5.0 \frac{4\pi}{3} 2.5^3 = 327.25 \text{ g}$$

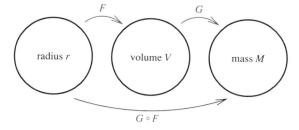

Figure 1.37

The composition of two functions with units

What about the units? F takes as input things with dimensions of length and returns as output things with dimensions of volume. G accepts an input with dimensions of volume and returns an output with dimensions of mass. Because G takes as input precisely what F provides as output, the composition makes sense.

What if we tried to compute $F \circ G$? F cannot accept an input with dimensions of mass, which are the only outputs that G can return. It is impossible to compute the volume of a sphere with a radius of 4.3 g. This composition is nonsense.

We can now apply these methods to compose the updating function f with itself. The function $f \circ f$ takes the population size at time 0 as input and returns the population size 2 h later as output. The composition is the updating function for an experiment run twice as long. We can compute $f \circ f$ with the steps

$$(f \circ f)(b_0) = f[f(b_0)]$$
$$= f(2.0 b_0)$$
$$= 4.0 b_0$$

After 2 h, the population is 4 times larger, having doubled twice. In this case, composition of f with itself looks like multiplication. Do not get confused; this is the case *only* for an updating function expressing a proportional relation.

The composition of the mite updating function $h(x_t) = 2x_t + 30$ with itself gives

$$(h \circ h)(x_0) = h[h(x_0)]$$
$$= h(2x_0 + 30)$$
$$= 2(2x_0 + 30) + 30$$
$$= 4x_0 + 90$$

Testing, we get

$$(h \circ h)(10) = 4 \cdot 10 + 90 = 130$$

This function gives the number of mites after 2 weeks, skipping over the intermediate value of 50 mites after 1 week.

Inverse Functions: Looking Backward

Suppose a lizard ends up with 52 mites. How many did it have the week before? This is a kind of "inverse" problem: Starting from where you ended up, you want to try to end up where you started. Fortunately, these problems can be expressed more clearly in equations than in words. "Word problems" can be expressed in equations by translating them into biological terms. What was x_t if x_{t+1} is 52 (Figure 1.38)? We want to solve the equation $h(x_t) = 52$ for x_t, or

$$2.0 x_t + 30 = 52$$

The two sides of the equation say the same thing in two ways. The right-hand side gives our measured value, 52. The left-hand side gives the measured value as a function of the unknown x_t. We can solve for x_t as follows:

$$2.0 x_t = 52 - 30 = 22 \quad \text{subtract 30 from both sides}$$
$$x_t = \frac{22}{2.0} = 11 \quad \text{divide both sides by 2}$$

Figure 1.38
Going backward with the mite population

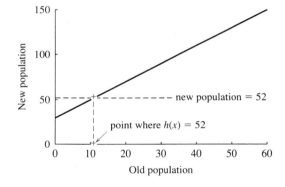

We can check this answer by substituting, finding

$$h(11) = 2 \cdot 11 + 30 = 52$$

Our updating function h gives the new population in terms of the old population. How can we find a **function** that gives the old population in terms of the new population? Such a function would help us solve the inverse problem in general. The trick is to follow the same steps we used above to find x_t when $x_{t+1} = 52$ but without substituting $x_{t+1} = 52$:

$$2.0x_t + 30 = x_{t+1} \qquad \text{the original equation}$$
$$2.0x_t = x_{t+1} - 30 \qquad \text{subtract 30 from both sides}$$
$$x_t = \frac{x_{t+1} - 30}{2.0} \qquad \text{divide both sides by 2}$$

The function h^{-1}, read "h inverse," defined by

$$h^{-1}(x_{t+1}) = \frac{x_{t+1} - 30}{2.0} = 0.5x_{t+1} - 15$$

is the **inverse** of h; the function that precisely undoes what h did in the first place. Whereas h takes the old population as input and returns the new population as output, h^{-1} takes the *new* population as input and returns the *old* population as output (Figure 1.39). Using the formula for the inverse function, we can quickly compute the old number of mites if the new number is 52 by computing $h^{-1}(52)$, or

$$h^{-1}(52) = \frac{52 - 30}{2} = 11$$

Similarly,

$$h^{-1}(120) = \frac{120 - 30}{2} = 45$$

The general definition of an inverse function does not require that the original function be linear.

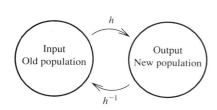

Figure 1.39

The action of a function and its inverse

■ **Definition 1.3** The function f^{-1} is the inverse of f if

$$f[f^{-1}(x)] = x$$
$$f^{-1}[f(x)] = x$$

Each of f and f^{-1} undoes the action of the other.

The following algorithm describes the steps for computing the inverse of the function $y = f(x)$. Unlike most other algorithms in this book, this one is not guaranteed to work.

■ **Algorithm 1.3** **(Finding the inverse of a function)**

1. Write the equation $y = f(x)$.
2. Solve for x in terms of y.
3. The inverse function is the operation done to y.

It may look odd to have a function defined in terms of y. *Do not* change the letters around to make it look normal: In applied mathematics different letters stand for different things and resent having their names switched as much as we do.

This algorithm may fail in two ways. First, not all functions have inverses. An operation can be undone only if you can deduce the input from the output. If any particular output is associated with more than one input, there is no way to tell where you started solely on the basis of where you end up. Consider the data in the following table, graphed in Figure 1.40.

Data that cannot be inverted

Old mass	New mass	Old mass	New mass
1.0	7.0	9.0	20.0
2.0	12.0	10.0	18.0
3.0	16.0	11.0	15.0
4.0	19.0	12.0	12.0
5.0	22.0	13.0	9.0
6.0	23.0	14.0	6.0
7.0	23.0	15.0	3.0
8.0	22.0	16.0	1.0

Suppose you were told that the mass at the end of an experiment was 12.0 g. Old masses of 2.0 g and 12.0 g both produce a new mass of 12.0 g. You cannot tell whether the input was 2.0 or 12.0; therefore, this function has no inverse.

■ **Algorithm 1.4** (The horizontal line test)

A function has no inverse if it takes on the same value twice. This can be established by graphing the function and checking whether the graph crosses some horizontal line two or more times.

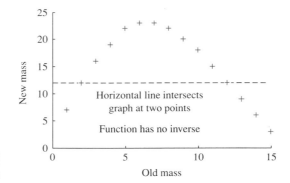

Figure 1.40

A function with no inverse

One can think of functions without inverses as losing information over the course of the experiment: Things that start out *different* end up the *same*. Graphically, a function has no inverse if some horizontal line intersects its graph in more than one place (Figure 1.40). The horizontal line at 12.0 intersects the curve at the two points (2.0, 12.0) and (12.0, 12.0).

Also, Algorithm 1.3 for computing the inverse might fail because the algebra is impossible. For example, consider the function

$$f(x) = x^5 + x + 1$$

The graph is shown in Figure 1.41; it satisfies the horizontal line test. We can try to find the inverse $f^{-1}(y)$ as follows:

1. Set $y = x^5 + x + 1$.
2. Try to solve for x. Even with the cleverest algebraic tricks, this is impossible.
3. Give up.

In mathematical modeling, however, it is often more important to know that something (such as the inverse) exists, than to be able to write a formula.

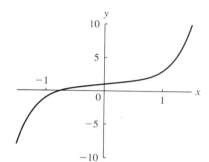

Figure 1.41
A function with an inverse that is impossible to compute

SUMMARY

Analyzing discrete-time dynamical systems requires combining functions. When two components go into building an updating function, they can be combined with **addition of functions**. Solutions found by repeatedly applying the updating function can be defined with the mathematical procedure of **functional composition**. In this method of combining functions, the output of the **inner function** is used as the input of the **outer function**. Many functions do not **commute**, meaning that combining the functions in a different order gives a different result. Functions with units can be composed only when the output of the inner function has the appropriate units to act as the input of the outer function. Composing an updating function with itself can be used to find values after two or more time steps. Finally, we can figure out where an experiment started using where it ended by computing the **inverse function**, which undoes the action of the function. Functions without inverses can be thought of as losing information.

1.6 EXERCISES

1. Find and graph the following sums.
   a. Find $(f + g)(x)$ if $f(x) = 3 + 2x$ and $g(x) = 4 + 5x$.
   b. Find $(f + g)(x)$ if $f(x) = x^2$ and $g(x) = x + 1$.
   c. Find $(F + H)(y)$ if $F(y) = y^2 - y$ and $H(y) = y/3$.

2. Find the following compositions.
   a. Find $(f \circ g)(x)$ and $(g \circ f)(x)$ if $f(x) = 3 + 2x$ and $g(x) = 4 + 5x$.
   b. Find $(f \circ g)(x)$ and $(g \circ f)(x)$ if $f(x) = x^2$ and $g(x) = x + 1$.
   c. Find $(F \circ H)(y)$ and $(H \circ F)(y)$ if $F(y) = y^2 - y$ and $H(y) = y/3$.

3. Compose each of the updating functions from Exercise 5 in Section 1.5 with itself.

4. Consider the updating function
$$f(x) = \frac{x}{1+x}$$
used in Exercise 10 in Section 1.5.
   a. Find the composition $f \circ f$ (remember to put things over a common denominator).
   b. Find the inverse of f.
   c. Show that $(f \circ f^{-1})(y) = y$ and that $(f^{-1} \circ f)(x) = x$.

5. A colony of 10,000 flat, square organisms begins growing in culture. Each organism has a density of 0.4 g/cm² and sides with length r.
   a. Sketch a picture of this situation.
   b. Denote the area of each organism by A and find A as a function of r.
   c. Denote the mass of each organism by M and find M as a function of A.
   d. Denote the total mass of all the organisms by T and find T as a function of M.
   e. Find T as a function of r.
   f. Suppose the energy consumption C of the whole colony is given by
$$C = 2.5 \times 10^2 \frac{T}{1000 + T}$$
   Find C as a function of r.
   g. What happens to C when r gets large? What might this say about the health of the colony?

6. Insects are grown at different temperatures T (°C). Suppose the length is $L(T) = 10.0 - 0.01(T - 25)^2$ cm.
   a. If the volume is $V(L) = 0.34L^3$, find volume as a function of temperature.
   b. If the mass is $M(V) = \frac{3.4V}{1.0 + 0.1V}$, find mass as a function of temperature.
   c. Compute L, V, and M for $T = 10$, $T = 25$, and $T = 40$.

7. Think of two daily activities that commute (things turn out the same in whatever order you do them) and two that do not. Answers referring to driving to and from work are not allowed.

8. Find the inverses of each of the functions in Exercise 2 when they exist, and use functional composition to check that they are correct.

9. Consider the function $h(x) = 3x + 2$.
   a. Graph this function. Indicate where $h(x) = 8$.
   b. Find another point on this line and write the equation of the line in point-slope form.
   c. Show that this function does not describe a proportional relation.
   d. Find its inverse. Use it to find where $h(x) = 8$.
   e. Graph the inverse.

10. Find and graph the inverses of the following functions. Use the inverse to solve the requested equation.
    a. $b_{t+1} = 2b_t$ (Equation 1.1). Find b_t if $b_{t+1} = 4.6 \times 10^7$.
    b. $h_{t+1} = h_t + 1$ (Equation 1.4). Find h_t if $h_{t+1} = 45.6$ m.
    c. $F = 1.8C + 32$ (Celsius to Fahrenheit conversion). Find C if $F = 72$.

11. Graph M, V, and G as functions of a from the following table. Which of these functions have inverses? Which measurements could we use to figure out the age of an organism?

Age (days)	Mass (g)	Volume (cm³)	Glucose production (mg)
0.5	1.5	5.1	0.0
1.0	3.0	6.2	3.4
1.5	4.3	7.2	6.8
2.0	5.1	8.1	8.2
2.5	5.6	8.9	9.4
3.0	5.6	9.6	8.2

12. **COMPUTER:** Compose the medication updating function $M_{t+1} = 0.5M_t + 1.0$ with itself 10 times. Plot the resulting function. Use this composition to find the concentration after 10 days starting from concentrations of 1.0 mg/L, 5.0 mg/L, and 18.0 mg/L. If the goal is to reach a stable concentration of 2.0 mg/L, do you think this is a good therapy?

13. **COMPUTER:** Have your graphics calculator or computer plot the following functions for $-2 \leq x \leq 2$. Do they have inverses?
    a. $f(x) = 2x$
    b. $g(x) = 36 - 20x$
    c. $h(x) = x^3 + 2x$
    d. $F(x) = x^3 - 2x$
    Have your computer try to find the formula for the inverses of these functions and plot the results. Does the machine always succeed in finding an inverse when there is one? Does it sometimes find an inverse when there is none?

14. **COMPUTER:** Use your computer to find and plot the following functional compositions.
    a. $(f \circ g)(x)$ and $(g \circ f)(x)$ if $f(x) = \sin(x)$ and $g(x) = x^2$
    b. $(f \circ g)(x)$ and $(g \circ f)(x)$ if $f(x) = e^x$ and $g(x) = x^2$
    c. $(f \circ g)(x)$ and $(g \circ f)(x)$ if $f(x) = e^x$ and $g(x) = \sin(x)$

1.7 Expressing Solutions with Exponential Functions

The solution associated with the bacterial updating function $b_{t+1} = 2.0b_t$ (a linear updating function) is

$$b_t = 2.0^t \cdot 1.0 \times 10^6$$

when $b_0 = 1.0 \times 10^6$. As a function of t, the solution is an example of an **exponential function**. Suppose we wish to find out how long it will take the population to reach 10^8. Solving for t requires being able to convert this function into a standard form with the base e and to work with the inverse of the exponential function, the **natural logarithm**. In particular, we will study the **laws of exponents** and the **laws of logarithms**. More generally, what happens to the updating function and solution if some of the bacteria die during the course of each hour? We will see that the solution is again an exponential function but with base equal to the **per capita reproduction** of the bacteria.

Bacterial Population Growth in General

The bacteria studied hitherto have doubled in number each hour. Each bacterium divided once, and both "daughter" bacteria survived. Suppose that only a fraction σ ("sigma") of the daughters survive. Instead of 2.0 offspring per bacterium, we find an average of 2σ offspring (Figure 1.42). For example, if only 75% of offspring survive the transition ($\sigma = 0.75$), there are an average of only 1.5 surviving offspring per parent. Let

$$r = 2\sigma \qquad (1.27)$$

The new **parameter**, r, which represents the number of new bacteria produced per bacterium, is called the **per capita reproduction**. A parameter is a value represented by a letter that remains constant in a particular problem but may change to a different value in a different problem. A **variable**, such as b_t, changes in the course of a particular problem. Scientifically, variables describe the measurements that change during an experiment and parameters describe the conditions that are supposed to remain constant.

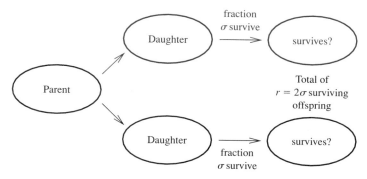

Figure 1.42
Bacterial population growth with reproduction and mortality

In terms of the parameter r, the updating function can be written

$$b_{t+1} = rb_t \qquad (1.28)$$

This fundamental equation of population biology says that the population at time $t + 1$ is equal to the per capita reproduction (the number of new bacteria per old bacterium) times the population at time t (the number of old bacteria), or

$$\text{new population} = \text{per capita reproduction} \times \text{old population}$$

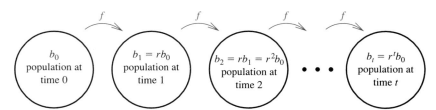

Figure 1.43
Bacterial population growth

If $\sigma = 0.75$, then $r = 2 \cdot 0.75 = 1.5$; this population increases by 50% each hour. In contrast, if $\sigma = 0.25$, then $r = 2 \cdot 0.25 = 0.5$; this population decreases by 50% each hour. Starting from a population of b_0 bacteria, we can apply the updating function repeatedly to derive a solution in general, much as we did in Section 1.6 with the value $r = 2$ (Figure 1.43). We find

$$b_1 = rb_0$$
$$b_2 = rb_1 = r^2 b_0$$
$$b_3 = rb_2 = r^3 b_0$$

Each hour, the initial population b_0 is multiplied by the per capita reproduction r. After t hours, the initial population b_0 has been multiplied by t factors of r. Therefore,

$$b_t = r^t b_0 \tag{1.29}$$

How do these solutions behave for different values of the per capita reproduction r? Results with four values of r starting from $b_0 = 1.0 \times 10^6$ are

t	$r = 2.0$	$r = 1.5$	$r = 1.0$	$r = 0.5$
0	1.0×10^6	1.0×10^6	1.0×10^6	1.0×10^6
1	2.0×10^6	1.5×10^6	1.0×10^6	5.0×10^5
2	4.0×10^6	2.25×10^6	1.0×10^6	2.50×10^5
3	8.0×10^6	3.37×10^6	1.0×10^6	1.25×10^5
4	1.6×10^7	5.06×10^6	1.0×10^6	6.25×10^4
5	3.2×10^7	7.59×10^6	1.0×10^6	3.12×10^4
6	6.4×10^7	1.14×10^7	1.0×10^6	1.56×10^4
7	1.28×10^8	1.71×10^7	1.0×10^6	7.81×10^3
8	2.56×10^8	2.56×10^7	1.0×10^6	3.91×10^3

In the first two columns, $r > 1$ and the population increases each hour (Figure 1.44a and b). In the third column, $r = 1$ and the population remains the same hour after hour (Figure 1.44c). In the final column, $r < 1$ and the population decreases each hour (Figure 1.44d). We summarize these observations in the following table.

1.7 Expressing Solutions with Exponential Functions

Value of r	Behavior of population
$r > 1$	Population increases
$r = 1$	Population remains constant
$r < 1$	Population decreases

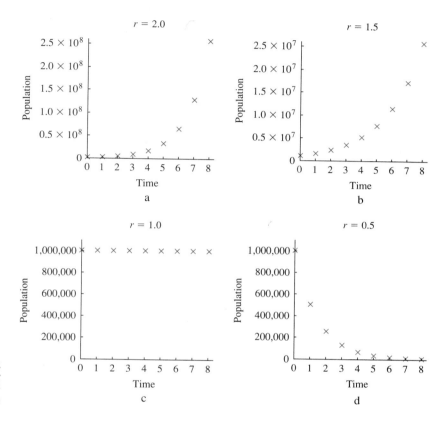

Figure 1.44
Growing, constant, and shrinking bacterial populations

Laws of Exponents and Logs

The solution $b_t = r^t b_0$ is presented in **exponential notation**. For any positive number a, the **exponential function to the base a** is written

$$f(x) = a^x \qquad (1.30)$$

and we say "a to the xth power." This function takes x as input and returns x factors of a multiplied together. The notation generalizes that used in equations such as

$$a^2 = a \cdot a$$

The key to using exponential functions is knowing the **laws of exponents**, summarized in the table. The table also includes examples using $a = 2$ that can help you to remember when to add and when to multiply.

Laws of exponents

	General formula	Example with $a = 2$, $x = 2$, and $y = 3$
Law 1	$a^x \cdot a^y = a^{x+y}$	$2^2 \cdot 2^3 = 2^5 = 32$
Law 2	$(a^x)^y = a^{xy}$	$(2^2)^3 = 2^6 = 64$
Law 3	$a^{-x} = 1/a^x$	$2^{-2} = 1/2^2 = 1/4$
Law 4	$a^y/a^x = a^{y-x}$	$2^3/2^2 = 2^{3-2} = 2$
Law 5	$a^1 = a$	$2^1 = 2$
Law 6	$a^0 = 1$	$2^0 = 1$

The exponential function is defined for all values of x, including negative numbers and fractions. What does it mean to multiply half an a or -3 as? These expressions must be computed with the laws of exponents. To compute $a^{0.5}$, we raise this unknown quantity to the 2nd power (square it), and use Law 2 to find

$$(a^{0.5})^2 = a^{0.5 \cdot 2} = a^1 = a$$

Therefore, a to the 0.5th power is the number that, when squared, gives back a. In other words, a to the 0.5 power is the square root of a. For example

$$2^{0.5} = \sqrt{2} = 1.41421$$

To compute a^{-3}, apply Law 3 to find

$$a^{-3} = \frac{1}{a^3}$$

For example

$$2^{-3} = \frac{1}{2^3} = \frac{1}{8} = 0.125$$

For reasons that will make sense only with a bit of calculus, the base a most commonly used throughout the sciences is the irrational number

$$e = 2.718281828459\ldots$$

The function

$$e^x$$

pronounced "e to the x," is called the **exponential function to the base e**, or simply the **exponential function** (Figure 1.45). The domain of this function consists of all numbers, and the range is all **positive numbers**. Starting at values less than 1 for negative x, the graph increases through 1 at $x = 0$ and increases faster and faster thereafter.

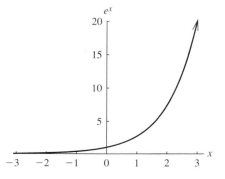

Figure 1.45
Graph of the exponential function

The graph of the exponential function crosses every positive horizontal line only once. Knowing the output of the exponential function provides enough information to find the input, so the exponential function has an inverse. The inverse function is called the **natural logarithm** (or natural log) (Figure 1.46). The natural log of x is written $\ln(x)$ or $\ln x$. From the definition of the inverse (Definition 1.3),

$$\ln e^x = x \tag{1.31}$$
$$e^{\ln x} = x \tag{1.32}$$

The graph of the natural logarithm increases from "negative infinity" near $x = 0$, through 0 at $x = 1$, and rises more and more slowly as x becomes larger (Figure 1.47).

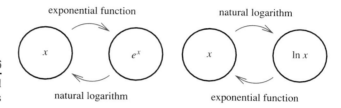

Figure 1.46
The exponential function and natural logarithm are inverses

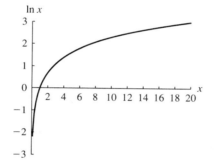

Figure 1.47
Graph of the natural logarithm

Because the exponential function outputs only positive numbers, these are the only numbers the natural log can accept as input. The key to understanding natural logarithms is knowing the laws of logarithms, described in the table.

In some disciplines, people use the **exponential function with base 10**, or

$$f(x) = 10^x$$

Its inverse is the **logarithm to the base 10**, written

$$\log_{10} x$$

We say "log base 10 of x." Just as $\ln x = y$ implies that $x = e^y$,

$$\log_{10} x = y$$

implies that

$$x = 10^y$$

For example, if $\log_{10} x = 2.3$, $x = 10^{2.3} \approx 199.5$. In most ways, the exponential function with base 10 and the log base 10 work much like the exponential function

The laws of logarithms

Law 1	$\ln(xy) = \ln x + \ln y$
Law 2	$\ln(x^y) = y \ln x$
Law 3	$\ln(1/x) = -\ln x$
Law 4	$\ln(x/y) = \ln x - \ln y$
Law 5	$\ln(e) = 1$
Law 6	$\ln(1) = 0$

with base e and and the natural logarithm. All laws of exponents and logs are the same except for Law 5, which becomes

Law 5 of exponents:	$10^1 = 10$
Law 5 of logarithms:	$\log_{10}(10) = 1$

The base e is more convenient for studying dynamics with calculus.

Expressing Results with Exponentials

How can we use the laws of exponentials and logs to express

$$b_t = r^t b_0$$

(Equation 1.29) in terms of the exponential function with base e? Because the exponential function and the natural logarithm are inverses, we can write

$$r = e^{\ln r}$$

Then, using Law 2 of exponents,

$$r^t = (e^{\ln r})^t$$
$$= e^{(\ln r)t}$$

Therefore, the *general solution* for the discrete-time dynamical system

$$b_{t+1} = r b_t$$

with initial condition b_0 can be written in exponential notation as

$$b_t = b_0 e^{(\ln r)t} \tag{1.33}$$

For example, consider the case $r = 2.0$ and $b_0 = 1.0 \times 10^6$. Because $\ln(2.0) = 0.6931$, the solution is

$$b_t = 1.0 \times 10^6 e^{(0.6931)t}$$

When using exponential notation, it is conventional to move the constant factor b_0 (here equal to 1.0×10^6) to the front. What is the value of rewriting the solution in this way?

Exponential notation makes it easier to answer questions about when a population will reach a particular value. When will this population reach 1.0×10^8? We set the formula for b_t equal to 1.0×10^8 and solve for t with the steps

$1.0 \times 10^6 e^{(0.6931)t} = 1.0 \times 10^8$	equation for t
$e^{(0.6931)t} = 100$	divide by 1.0×10^6
$0.6931 t = \ln(100)$	take the natural log
$t = \dfrac{\ln(100)}{0.6931} = 6.64$	solve for t

The population will pass 100 million between hours 6 and 7 (Figure 1.48a). The key step uses the natural log, the inverse of the exponential function, to remove the variable t from the exponent.

Similarly, we can ask how long it will take a population with $r < 1$ to decrease to some specified value. How long will it take a population starting from

1.7 Expressing Solutions with Exponential Functions

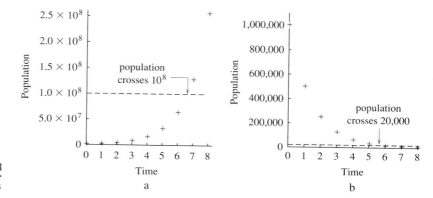

Figure 1.48
Using solutions to find times

$b_0 = 1.0 \times 10^6$ to decrease to 2.0×10^4 if $r = 0.5$? We find that $\ln(0.5) = -0.6931$, so

$$b_t = 1.0 \times 10^6 e^{-0.6931t}$$

Solving as before,

$$
\begin{aligned}
1.0 \times 10^6 e^{(-0.6931)t} &= 2.0 \times 10^4 && \text{equation for } t \\
e^{(-0.6931)t} &= 0.02 && \text{divide by } 1.0 \times 10^6 \\
-0.6931 t &= \ln(0.02) && \text{take the natural log} \\
t &= \frac{\ln(0.02)}{-0.6931} = 5.64 && \text{solve for } t
\end{aligned}
$$

All the negative signs cancel, and we see that this population will pass 2.0×10^4 between hours 5 and 6 (Figure 1.48b).

Throughout the sciences, many measurements other than population sizes are described by exponential functions. In such cases, we write the measurement S as a function of t as

$$S(t) = S(0) e^{\alpha t} \tag{1.34}$$

The parameter $S(0)$ represents the value of the measurement at time $t = 0$. The parameter α describes how the measurement changes. When $\alpha > 0$ the function is increasing (Figure 1.49a and b). When $\alpha < 0$ the function is decreasing (Figure 1.49c and d). The function increases most quickly with large positive values of α and decreases most quickly with large negative values of α.

One important number describing such measurements is the **doubling time** (Figure 1.50a). When $\alpha > 0$, the measurement is increasing and we can ask how long it will take to double. In general, we solve for the doubling time t_d with the steps:

$$
\begin{aligned}
S(t_d) = S(0) e^{\alpha t_d} &= 2S(0) && \text{equation for } t_d \\
e^{\alpha t_d} &= 2 && \text{divide by } S(0) \\
\alpha t_d &= \ln(2) && \text{take the natural log} \\
t_d &= \frac{\ln(2)}{\alpha} = \frac{0.6931}{\alpha} && \text{solve for } t_d
\end{aligned}
$$

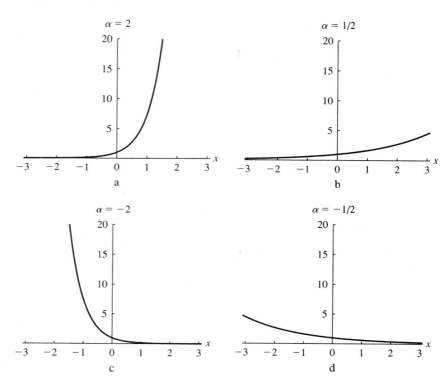

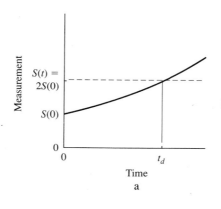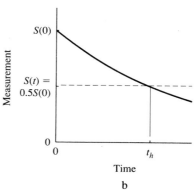

Figure 1.49 The exponential function with different parameters α

Figure 1.50 Doubling times and half-lives

Therefore, the **general formula for the doubling time** is

$$t_d = \frac{0.6931}{\alpha} \tag{1.35}$$

The doubling time becomes smaller as α becomes larger.

Suppose that

$$S(t) = 150.0e^{1.2t}$$

with t measured in hours. Then $\alpha = 1.2$ per hour, and the doubling time is

$$t_d = \frac{0.6931}{1.2} = 0.5776 \text{ h}$$

Checking, we get
$$S(0.5776) = 150.0e^{1.2 \cdot 0.5776} = 75.0$$

Larger values of α produce smaller doubling times because the measurement is growing more quickly.

When $\alpha < 0$, the measurement is decreasing, and we can ask how long it will take to become half as large. This time, denoted t_h, is called the **half-life** (Figure 1.50b) and can be found with the following steps:

$$S(t_h) = S(0)e^{\alpha t_h} = 0.5S(0) \quad \text{equation for } t_h$$
$$e^{\alpha t_h} = 0.5 \quad \text{divide by } S(0)$$
$$\alpha t_h = \ln(0.5) \quad \text{take the natural log}$$
$$t_h = \frac{\ln(0.5)}{\alpha} = -\frac{0.6931}{\alpha} \quad \text{solve for } t_h$$

Therefore, the **general formula for the half-life** is

$$t_h = -\frac{0.6931}{\alpha} \tag{1.36}$$

The half-life becomes smaller when α grows larger in absolute value. Remember to apply this equation only when $\alpha < 0$.

If a measurement follows the equation

$$M(t) = 240.0e^{-2.3t}$$

with t measured in seconds, then $\alpha = -2.3$ per second, and the half-life is

$$t_h = \frac{-0.6931}{-2.3} = 0.4347 \text{ s}$$

Conversely, if we are told the initial value and the half-life of some measurement, we can find the formula. Suppose $t_h = 6.8$ yr. Because

$$t_h = -\frac{0.6931}{\alpha}$$

we can solve for α as

$$\alpha = -\frac{0.6931}{t_h} = -\frac{0.6931}{6.8} = -0.1019$$

If $V(0) = 23.1$, then

$$V(t) = 23.1e^{-0.1019t}$$

In summary,

Finding doubling time or half-life from a formula	
Grows according to $S(t) = S(0)e^{\alpha t}$ with $\alpha > 0$.	Doubling time is $t_d = 0.6931/\alpha$.
Declines according to $S(t) = S(0)e^{\alpha t}$ with $\alpha < 0$.	Half-life is $t_h = -0.6931/\alpha$.

Finding formula from doubling time or half-life	
Doubling time is t_d.	Grows according to $S(t) = S(0)e^{\alpha t}$ with $\alpha = 0.6931/t_d$.
Half-life is t_h.	Declines according to $S(t) = S(0)e^{\alpha t}$ with $\alpha = -0.6931/t_h$.

58 Chapter 1 Introduction to Discrete-Time Dynamical Systems

SUMMARY We generalized the updating function for bacterial population growth to compute when populations would reach particular values. If some offspring die, the updating function can be written in terms of the **per capita reproduction r**. A population grows if $r > 1$ and declines if $r < 1$. The solution can be expressed as an **exponential function to base r**. For convenience, exponential functions are usually expressed to the base e, often called the **exponential function**. Using the laws of exponents, any exponential function can be expressed to the base e. The inverse of the exponential function is the **natural logarithm**, or natural log. This function can be used to solve equations involving the exponential function, including finding **doubling times** and **half-lives**.

1.7 Exercises

1. Use the laws of exponents to rewrite the following if possible. If no law of exponents applies, say so. Compute the value of your original expression and the rewritten version to check.
   a. e^0
   b. $(e^{4.5})^{5.5}$
   c. $e^{3.5} + e^{6.5}$
   d. $e^{-3.5}$
   e. $e^{4.5} \cdot e^{5.5}$
   f. $e^{4.5} + e^{5.5}$
   g. $(e^{3.5})^{6.5}$
   h. $e^{-4.5}$
   i. $e^{6.5} \cdot e^{3.5}$
   j. e^1

2. Use your calculator to compute the following. Does this give you any idea why e is special?
   a. $2^{0.001}$
   b. $10^{0.001}$
   c. $0.5^{0.001}$
   d. $e^{0.001}$

3. Use the laws of logs to rewrite the following if possible. If no law of logs applies or the quantity is not defined, say so. Compute the value of your original expression and rewritten version to check.
   a. $\ln(1)$
   b. $\ln(-6.5)$
   c. $\ln(4.5) + \ln(5.5)$
   d. $\ln(-4.5)$
   e. $\ln(3.5 + 6.5)$
   f. $\ln(1/4.5)$
   g. $\ln(4.5^{5.5})$
   h. $\ln(3.5) + \ln(6.5)$
   i. $\ln(4.5 + 5.5)$
   j. $\ln(3.5^{6.5})$

4. Find and graph the updating function $b_{t+1} = rb_t$ in the following cases. Find and graph the solution up to $t = 5$.
   a. $r = 1.5$ and $b_0 = 1.0 \times 10^6$
   b. $r = 0.7$ and $b_0 = 5.0 \times 10^5$
   c. $r = 0.8$ and $b_0 = 4.0 \times 10^6$

5. Find the times when the given population sizes are reached in the following situations. Make sure to use the exponential function to base e.
   a. If $\sigma = 0.75$ in Equation 1.27 and $b_0 = 1.0 \times 10^6$, when will the population reach 8.0×10^6?
   b. If $\sigma = 0.35$ in Equation 1.27 and $b_0 = 1.0 \times 10^6$, when will the population reach 8.0×10^4?
   c. If $\sigma = 0.4$ in Equation 1.27 and $b_0 = 1.0 \times 10^6$, when will the population reach 8.0×10^6? Does your answer look strange?

6. Suppose cell volume follows the updating function $v_{t+1} = 1.5v_t$, fish length follows the updating function $l_{t+1} = l_t - 1.7$, and gnat population follows the updating function $n_{t+1} = 0.5n_t$ (Exercise 5 in Section 1.2). How long will it take for the following to occur?
   a. A cell with initial volume of 1350 μm^3 to reach 3250 μm^3.
   b. A fish of initial length 15.2 cm to reach 2.5 cm.
   c. A gnat population of initial number 5.5×10^5 to reach 1.52×10^3.

7. Suppose the size of an organism at time t is given by
$$S(t) = S_0 e^{\alpha t}$$
where S_0 is the initial size.
   a. How long does it take for the organism to double in size if $S_0 = 1.0$ cm and $\alpha = 1.0$ per day?
   b. How long does it take for the organism to double in size if $S_0 = 2.0$ cm and $\alpha = 1.0$ per day?
   c. How long does it take for the organism to double in size if $S_0 = 2.0$ cm and $\alpha = 0.1$ per hour?

8. The amount of carbon 14 (C^{14}) left t years after the death of an organism is given by
$$Q(t) = Q_0 e^{-0.000122t}$$
where Q_0 is the amount left at the time of death. Suppose $Q_0 = 6.0 \times 10^{10}$ C^{14} atoms per gram.
   a. How much is left after 50,000 yr?
   b. How much is left after 100,000 yr?
   c. Find the half-life of C^{14}.

9. Suppose a population has a doubling time of 24 yr.
   a. Find the equation for population size $P(t)$ as a function of time if $P(0) = 5.0 \times 10^9$.
   b. What is the population in 48 yr?

10. Suppose a population is dying with a half-life of 43 h.
    a. Find the equation for population size $P(t)$ as a function of time if $P(0) = 1600$.
    b. Find the population in 86 h.
    c. How long will it take to reach 200?

11. Use the laws of exponents or logs to rewrite the following if possible. If no law applies, say so. Compute the value of your original expression and the rewritten version to check.
    a. 10^0
    b. $(10^{4.5})^{5.5}$
    c. $10^{3.5} + 10^{6.5}$
    d. $10^{-3.5}$
    e. $10^{4.5} \cdot 10^{5.5}$
    f. 10^1
    g. $\log_{10}(1)$
    h. $\log_{10}(-6.5)$
    i. $\log_{10}(4.5) + \ln(5.5)$
    j. $\log_{10}(3.5 + 6.5)$
    k. $\log_{10}(1/4.5)$

12. Suppose the size of an organism follows the equation
    $$S(t) = 2.34 \times 10^{0.5t}$$
    a. Find the size at time 2.
    b. Find the doubling time without converting to base e.
    c. Write the formula for $S(t)$ in terms of the exponential function.
    d. Find the doubling time.

13. **COMPUTER:** Use your computer to find the following. Plot the graphs to check.
    a. The doubling time of $S_1(t) = 3.4e^{0.2t}$.
    b. The doubling time of $S_2(t) = 0.2e^{3.4t}$.
    c. The half-life of $H_1(t) = 3.4e^{-0.2t}$.
    d. The half-life of $H_2(t) = 0.2e^{-3.4t}$.

14. **COMPUTER:** Have your computer solve for the times when the following hold. Plot the graphs to check your answer.
    a. $S_1(t) = S_2(t)$ with S_1 and S_2 from Exercise 13.
    b. $H_1(t) = 2H_2(t)$ with H_1 and H_2 from Exercise 13.
    c. $H_1(t) = 0.5H_2(t)$ with H_1 and H_2 from Exercise 13.

15. **COMPUTER:** Use your computer to plot the following functions.
    a. $\ln(x)$ for $10 \leq x \leq 100{,}000$
    b. $\ln[\ln(x)]$ for $10 \leq x \leq 100{,}000$
    c. $\ln\{\ln[\ln(x)]\}$ for $10 \leq x \leq 100{,}000$
    d. e^x for $0 \leq x \leq 2$
    e. e^{e^x} for $0 \leq x \leq 2$
    f. $e^{e^{e^x}}$ for $0 \leq x \leq 2$. Will your machine let you do it? Can you compute the value of $e^{e^{e^2}}$?

1.8 Power Functions and Allometry

Two measurements that grow exponentially are related to each other by a **power relation**; each can be written as a power of the other. We will derive this power relation from the underlying exponential model and show how the power relation can be graphed as a linear relation on a **log–log plot**. Power relations describe many measurements in biology, including **allometries** between the sizes of different body parts. Understanding these allometries casts light on important principles of organismal design and function.

Power Relations and Exponential Growth

Suppose that the antler size $A(t)$ in centimeters of an elk increases with age in years during the first 5 yr of growth according to the exponential function

$$A(t) = 53.2e^{0.17t}$$

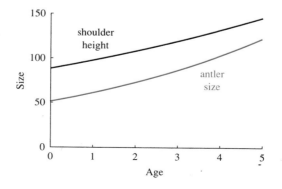

Figure 1.51
Antler size and shoulder height of an elk

and that the shoulder height $L(t)$ increases according to

$$L(t) = 88.5e^{0.1t}$$

The antlers start rather small but grow faster than the shoulder height (Figure 1.51).

If we measured the antler sizes and shoulder heights of elk at different ages, we would find the following data:

Age	Shoulder height	Antler size
0	88.5	53.2
1	97.8	63.3
2	108.0	75.0
3	119.3	88.7
4	132.0	105.3
5	145.8	124.8

If we plot antler size and shoulder height against each other on a graph (Figure 1.52), we find that the points lie on a curve that looks like the graph of some function. The curvature indicates that the antlers grow faster than the rest of the elk. How can we find this function? What sort of function is it?

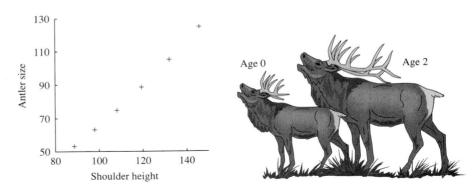

Figure 1.52
Antler size plotted against shoulder height of an elk

To find antler size as a function of shoulder height, we first find the time t as a function of shoulder height and then substitute into the formula for antler size. To solve for t in terms of L,

$$L = 88.5e^{0.1t} \qquad \text{the original equation}$$

$$\frac{L}{88.5} = e^{0.1t} \qquad \text{divide both sides by 88.5}$$

$$\ln\left(\frac{L}{88.5}\right) = 0.1t \qquad \text{take the natural log}$$

$$10\ln\left(\frac{L}{88.5}\right) = t \qquad \text{finish solving for } t$$

We next substitute this formula into the equation for antler size $A(t)$. Using the laws of logs and exponents,

$$A = 53.2e^{0.17t} \qquad \text{the original equation}$$
$$= 53.2e^{1.7\ln(L/88.5)} \qquad \text{plug in formula for } t$$
$$= 53.2e^{\ln(L/88.5)^{1.7}} \qquad \text{Law 2 of logs}$$
$$= 53.2\left(\frac{L}{88.5}\right)^{1.7} \qquad \text{exponents and logs are inverses}$$
$$= \frac{53.2}{88.5^{1.7}}L^{1.7} \qquad \text{place constants at front}$$
$$= 0.026L^{1.7} \qquad \text{evaluate the numbers}$$

Although both antler size and shoulder height are exponential functions of time, antler size is a **power function** of shoulder height. This means that antler size can be found by raising shoulder height to the 1.7 power and then multiplying by 0.026. Such a power relation is called an **allometry**.

We can check this relation directly. Using the laws of exponents,

$$0.026L(t)^{1.7} = 0.026(88.5e^{0.1t})^{1.7} \qquad \begin{array}{l}\text{take } L \text{ to the 1.7 power and}\\ \text{multiply by 0.026}\end{array}$$
$$= 0.026 \cdot 88.5^{1.7}(e^{0.1t})^{1.7} \qquad \text{Law 1 of exponents}$$
$$= 53.2e^{0.17t} \qquad \text{Law 2 of exponents}$$
$$= A(t) \qquad \text{it checks!}$$

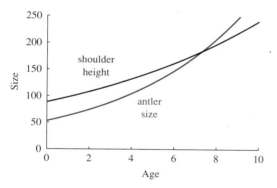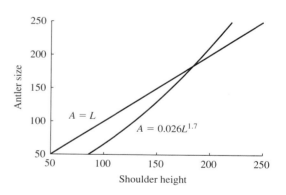

Figure 1.53
Finding when the antlers would be bigger than the shoulders

What would happen if our elk continued to grow for more than 5 yr? When does our model predict that the antlers will be broader than the shoulder height? We can solve this in two different ways (Figure 1.53). Starting with our original exponential growth equation, we could solve

$$A(t) = L(t)$$

for t by following the steps

$$53.2e^{0.17t} = 88.5e^{0.1t} \qquad \text{the original equation}$$
$$\frac{e^{0.17t}}{e^{0.1t}} = \frac{88.5}{53.2} = 1.66 \qquad \text{move all } t\text{s to one side}$$
$$e^{0.17t} \cdot e^{-0.1t} = 1.66 \qquad \text{Law 3 of exponents}$$
$$e^{0.17t - 0.1t} = 1.66 \qquad \text{Law 1 of exponents}$$

$$e^{0.07t} = 1.66 \qquad \text{do subtraction inside exponent}$$

$$0.07t = \ln(1.66) \qquad \text{take the log to eliminate the exponential function}$$

$$t = \frac{\ln(1.66)}{0.07} = 7.27 \qquad \text{solve for } t$$

Checking, we get

$$A(7.27) = 53.2e^{0.17(7.27)} = 183.1$$
$$L(7.27) = 88.5e^{0.1(7.27)} = 183.1$$

If the elk continued to grow for much more than 7 yr, or reached a shoulder height of 183 cm, its antlers would be bigger than the rest of its body.

Alternatively, we could use the allometric power relation between the two measures to solve for L by following the steps

$$L = 0.026L^{1.7} \qquad \text{the equation } L = A$$

$$\frac{L}{L^{1.7}} = 0.026 \qquad \text{place all } L\text{s on one side}$$

$$L \cdot L^{-1.7} = 0.026 \qquad \text{Law 3 of exponents}$$

$$L^{-0.7} = 0.026 \qquad \text{Law 1 of exponents}$$

$$-0.7 \ln(L) = \ln(0.026) \qquad \text{take log and apply Law 2 of logs}$$

$$\ln(L) = \frac{\ln(0.026)}{-0.7} = 5.21 \qquad \text{divide by } -0.7$$

$$L = e^{5.21} = 183.8 \qquad \text{exponentiate to solve for } L$$

As before, the shoulder height and antler height would match if both reached 183 cm (the difference between the answers results from rounding error).

Power relations between body parts or other quantities that grow exponentially is a general phenomenon. Because exponential growth is a fundamental biological principle, such power relations, known as **allometries**, occur throughout biology.

Power Relations and Lines

These calculations have been rather involved. Taking natural logarithms can greatly simplify the study of power relations. Consider the elk size data after taking natural logarithms:

Age	Log shoulder height	Log antler size
0	4.48	3.97
1	4.58	4.14
2	4.68	4.31
3	4.78	4.48
4	4.88	4.65
5	4.98	4.82

The log sizes increase **linearly** over time (Figure 1.54). Graphs of the logarithm of the measurement are called **semilog plots** because we have taken the logarithm of one measurement (the size) but not the other (the time). These plots transform graphs of exponential functions into graphs of linear functions.

1.8 Power Functions and Allometry

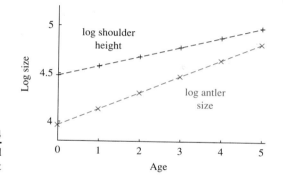

Figure 1.54
A semilog plot of antler size and shoulder height

We can find the formulas for the lines on the semilog plot by taking natural logarithms of the original exponential equations $A(t) = 53.2e^{0.17t}$ and $L(t) = 88.5e^{0.1t}$, finding

$$\ln[A(t)] = \ln(53.2e^{0.17t})$$
$$= \ln(53.2) + \ln(e^{0.17t})$$
$$= 3.97 + 0.17t$$
$$\ln[L(t)] = \ln(88.5e^{0.1t})$$
$$= \ln(88.5) + \ln(e^{0.1t})$$
$$= 4.48 + 0.1t$$

The slope of the line is equal to the constant in the exponent, and the y-intercept is equal to the logarithm of the multiplicative constant.

What is the relation between $\ln(A)$ and $\ln(L)$? We can use the same method as above, solving for t in terms of $\ln(L)$ and plugging into the equation for A. In this case

$\ln(L) = 4.48 + 0.1t$	the original equation
$\ln(L) - 4.48 = 0.1t$	subtract 4.48 from both sides
$10\ln(L) - 44.8 = t$	solve for t

We can now substitute this much simpler formula into the equation for $\ln(A)$:

$\ln(A) = 3.97 + 0.17t$	the original equation
$= 3.97 + 0.17[10\ln(L) - 44.8]$	plug in for t
$= 1.7\ln(L) - 3.64$	simplify the algebra

The **power relation** between L and A becomes a **linear relation** when we take logarithms of both measurements (Figure 1.55). This graph is called a **log–log plot** because we have taken the log of *both* measurements.

Does this match the power function we found above? Exponentiating, we find

$A = e^{1.7\ln(L) - 3.64}$	exponentiate both sides
$= e^{1.7\ln(L)}e^{-3.64}$	Law 1 of exponents
$= e^{\ln(L^{1.7})}e^{-3.64}$	Law 2 of logs
$= L^{1.7}e^{-3.64} = 0.026L^{1.7}$	cancel exponential and log

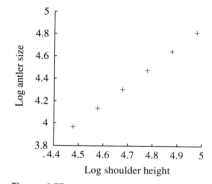

Figure 1.55
A log–log plot of antler size against shoulder height

In general, suppose that two measurements grow exponentially according to

$$A(t) = A(0)e^{\alpha t}$$
$$B(t) = B(0)e^{\beta t}$$

where $A(0)$ and $B(0)$ are known initial sizes (Figure 1.56a). Taking natural logarithms, we get

$$\ln(A) = \ln[A(0)] + \alpha t$$
$$\ln(B) = \ln[B(0)] + \beta t$$

The relation of the logarithm of each measurement to the time is linear (Figure 1.56b).

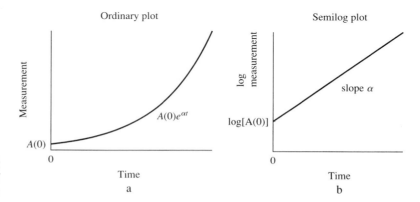

Figure 1.56
Ordinary and semilog plots of exponentially growing measurements

A log–log plot of $\ln(B)$ against $\ln(A)$ follows a line. To find the formula, we solve the equation in B for t,

$$\ln(B) = \ln[B(0)] + \beta t \qquad \text{the original equation}$$
$$\ln(B) - \ln[B(0)] = \beta t \qquad \text{subtract } \ln[B(0)] \text{ from both sides}$$
$$\frac{\ln(B) - \ln[B(0)]}{\beta} = t \qquad \text{solve for } t$$

We then plug this formula into the equation for $\ln(A)$,

$$\ln(A) = \ln[A(0)] + \alpha t \qquad \text{the original equation}$$
$$\ln(A) = \ln[A(0)] + \alpha \frac{\ln(B) - \ln[B(0)]}{\beta} \qquad \text{plug in for } t \qquad (1.37)$$
$$\ln(A) = \frac{\beta \ln[A(0)] - \alpha \ln[B(0)]}{\beta} + \frac{\alpha}{\beta} \ln(B) \qquad \text{simplify the algebra}$$

The linear relation of the logarithms of the measurements has slope equal to the ratio of the parameters in the exponents.

There are four distinct cases:

1. If $\beta > \alpha$, then B grows faster than A, the power in the power relation is greater than 1, and the plot of the power relation curves upward (Figure 1.57a).

2. If $\beta < \alpha$, then B grows more slowly than A, the power in the power relation is less than 1, and the plot of the power relation curves downward (Figure 1.57b).

3. If $\beta = \alpha$, both measurements grow at the same rate and the power relation is linear (Figure 1.57c).

4. If β and α have different signs, one increases and the other decreases. The power in the power relation is negative, and the plot of the power relation is decreasing (Figure 1.57d).

The *shapes* of these curves do not depend on the constants $A(0)$ and $B(0)$. In a log–log plot, the graphs of power relations are always linear. In these coordinates, a *slope* greater than 1 indicates that the second measurement increases more quickly than the first. A negative slope indicates that one measurement is growing and the other is shrinking.

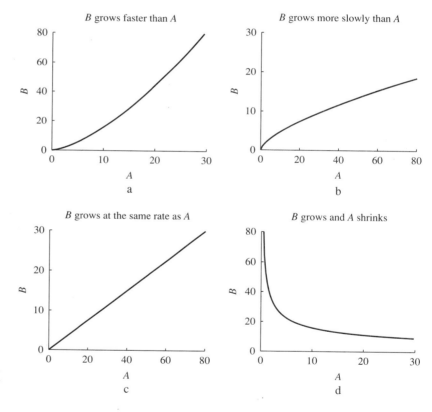

Figure 1.57
Four types of power relation

Power Relations in Biology: Shape and Flight

The curved power relation between antler size and shoulder height in elk results from a change in shape during growth. Properties of objects that change in size but not in shape also follow power relations. For example, let L represent the length of an object, A its surface area, and V its volume. For objects of fixed shape,

$$A = c_A L^2 \quad \text{area is proportional to the square of length}$$
$$V = c_V L^3 \quad \text{volume is proportional to the cube of length}$$

The constants c_A and c_V depend on the shape of the object. For example, if the object is a cube with sides of length L, then

$$A = 6L^2$$
$$V = L^3$$

If the object is a sphere with radius L, then

$$A = 4\pi L^2$$
$$V = \frac{4\pi}{3} L^3$$

These simple relations have important biological consequences. Processes that involve exchange with the outside world take place on the surface, while internal processes take place throughout the entire volume. A warm-blooded animal loses heat through its surface but generates heat throughout its entire volume. Does a large animal have to work harder to stay warm in a cold place? Suppose we compare a small animal with one twice as long. The larger animal has a surface area 4 times as large and a volume 8 times as large. Although it loses 4 times as much energy, it can produce 8 times as much heat with no more effort per unit volume. According to this logic, large animals should be easier to heat than small animals. Many species become larger in colder climates, consistent with this pattern (Figure 1.58).

Figure 1.58
Surface area and volume of a sphere

Surface area = $4\pi r^2$ = 1256.6
Volume = $4/3 \pi r^3$ = 4188.8
Ratio = 3.33

Surface area = $4\pi r^2$ = 5026.5
Volume = $4/3 \pi r^3$ = 33510.3
Ratio = 6.67

Similar power relations are important for flight. The lift F generated by a wing is proportional to the surface area of the wing A, which is itself proportional to the square of its length, or

$$F = c_L A = c_L c_A L^2$$

The constant c_L, which describes lift per unit area, depends on the curvature, material, and speed of the wing. This lift needs to raise an object with mass M. According to the fundamental relation, mass = density × volume,

$$M = \rho V = \rho c_V L^3$$

The density ρ depends on what the bird is made of.

Approximate parameters for a chickadee are given in the table. The lift is

$$F = 0.5 \cdot 0.2 \cdot 15^2 = 22.5 \text{ g}$$

Parameter	Value
L	15
c_L	0.5
c_A	0.2
ρ	0.1
c_V	0.04

and the mass is

$$M = 0.1 \cdot 0.04 \cdot 15^3 = 13.5 \text{ g}$$

The bird has some excess lift. Suppose an evil scientist creates a giant chickadee that is 50 cm long (Figure 1.59). The bird could lift

$$F = 0.5 \cdot 0.2 \cdot 50^2 = 250 \text{ g}$$

but its mass would be

$$M = 0.1 \cdot 0.04 \cdot 50^3 = 500 \text{ g}$$

This giant chickadee could not get off the ground (Figure 1.60).

Figure 1.59
Small and large chickadees

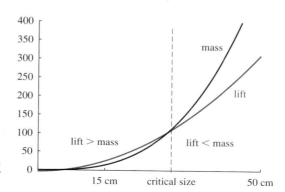

Figure 1.60
Wing area, mass, and lift of a bird

How could it cope? It could give up flying altogether, like an emu; increase its wing area, like a vulture or heron; or fly faster, like a duck or goose. Exercise 7 explores further the power relation between speed and lift.

SUMMARY

If two measurements grow exponentially, each can be written as a **power function** of the other. This sort of relation is called an **allometry**. When the logarithm of such a measurement is plotted against the time, called a **semilog** plot, the graph is linear. When the logarithms of two such measurements are plotted against each other, called a **log–log** plot, the graph is again linear, with slope equal to the power in the power relation. Power relations and allometries also arise from the geometric constraints. Organisms with the same shapes but different sizes generally have very different properties.

1.8 Exercises

1. a. Find the ratio of antler size to shoulder height at ages 0 through 5.
    b. How can you tell this is not a proportional relation?
    c. What would the antler size be at age 5 if the proportions matched those at age 0? How would this elk differ?
    d. What would the antler size be at age 0 if the proportions matched those at age 5? How would this elk differ?

2. The growth of a fly in an egg can be described allometrically (H. F. Nijhout and D. E. Wheeler, 1996). During growth, two **imaginal disks** (the first later becomes the wing, and the second becomes the haltere) expand according to
    $$S_1(t) = 0.007 e^{1.0t}$$
    $$S_2(t) = 0.007 e^{0.4t}$$
    where size is measured in cubic millimeters and time is measured in days. Development takes about 5 days.
    a. Plot both measurements as functions of time.
    b. Plot the second as a function of the first.
    c. How long will it take each to double?
    d. When will the first measurement be exactly double the second? How big will each be?
    e. Sketch a semilog plot of each measurement.
    f. Find the power relation giving S_2 in terms of S_1.
    g. Plot a log–log graph of $\ln(S_2)$ against $\ln(S_1)$.
    h. Find the power relation giving S_1 in terms of S_2.
    i. Plot a log–log graph of $\ln(S_1)$ against $\ln(S_2)$.

3. While the imaginal disks are growing (Exercise 2) the yolk of the egg is shrinking according to
    $$Y(t) = 4.0 e^{-1.2t}$$
    a. Plot yolk size as a function of time.
    b. Sketch a semilog plot of Y.
    c. Find the power relation giving S_1 in terms of Y.
    d. Plot S_1 as a function of the yolk size.
    e. Plot a log–log graph of $\ln(S_1)$ against $\ln(Y)$.
    f. Find the half-life of Y.
    g. Find the power relation giving Y in terms of S_1.
    h. Plot a log–log graph of $\ln(Y)$ against $\ln(S_1)$.

4. Find the equation of the following linear relations for elk by using the data points.
    a. $\ln(L)$ as a function of t using the data in the text at ages 0 and 5.
    b. $\ln(A)$ as a function of $\ln(L)$ using the data in the text at ages 0 and 5.
    c. $\ln(A)$ as a function of $\ln(L)$ using ages 1 and 2.

5. It has been found that the population density Y of large animals (in number per square kilometer) is related to their mass X in kilograms by the power relation
    $$Y = 36 X^{-1.14}$$
    (Peters, 1983).

    a. What would the population density be of a 1 kg animal? A 10 kg animal? A 100 kg animal?
    b. Plot this relation.
    c. Find the relation of $\ln(Y)$ and $\ln(X)$.
    d. Give a log–log plot.
    e. If humans weigh an average of 100 kg, what should the population density be? The United States is about 9.2×10^6 km². How much higher is the actual human population density than what you would expect?
    f. How small an animal should match our population density?

6. How much would the 50-cm-long chickadee have to change the following in order to be able to fly? How might the bird achieve the change?
    a. Its density.
    b. Its wing area.
    c. Wing lift per unit wing area.

7. The lift also has a power relation with velocity, following the equation
    $$F = c_F v^2$$
    Suppose a chickadee flies at 40 km/h.
    a. What are the units of c_F?
    b. If a chickadee flew twice as fast, how much more weight could it carry?
    c. How fast would a 50-cm-long chickadee have to go to be able to get off the ground?

8. Heat is lost through the surface but is generated throughout the entire volume of a warm-blooded animal. Suppose
    $$\text{heat lost per time} = h_A A$$
    and
    $$\text{heat generated per time} = h_V V$$
    where A is surface area in square centimeters, V is volume in cubic centimeters
    $$h_A = 5.0 \times 10^{-6} \frac{\text{Kcal}}{\text{cm}^2 \text{sec}} \quad \text{and} \quad h_V = 2.0 \times 10^{-7} \frac{\text{Kcal}}{\text{cm}^3 \text{sec}}$$
    a. Check that the units make sense.
    b. Find the energy balance of spherical cows with radius 50 cm and 100 cm.
    c. What is the smallest spherical cow that could maintain its temperature?
    d. Find the energy balance of cubical cows with sides of length 100 cm and sides of length 200 cm.
    e. What is the smallest cubical cow that could maintain its temperature?

9. The rate A (in cubic centimeters per second) at which liquid flows through a pipe has a power relation with the radius r

of the pipe,
$$A = cr^4$$

a. What are the units of c?
b. How much more flow do you get by tripling the radius of a blood vessel?
c. If a vessel branches into a bunch that are 0.5 as large, how many do you need to prevent back-up? Compare the total cross-sectional area of the original vessel and the full set of smaller vessels.

10. The strength of a tree trunk is proportional to its cross-sectional area. This means that the weight W it can carry follows the relation
$$W = c_w A$$
where A represents cross-sectional area. The tree shapes we will study are based on those in Exercise 5 in Section 1.3. We measure weight in kilograms and area in square meters. Suppose the density of the tree is 500 kg/m³.
a. Find the units of c_w, and suppose it has a value of 1.0×10^4.
b. If a tree is a perfect cylinder with a radius of 0.5 m, how tall can it be?
c. If a tree is a perfect cylinder with a radius equal to 0.1 times its height, how tall can it be?

11. **COMPUTER:** An alternative growth model for the fly in Exercise 2 is the **Gompertz model**. With the parameters used in that problem, growth would follow the equations
$$S_1(t) = 0.016 e^{-(e^{-1.2t}/1.2)}$$
$$S_2(t) = 0.0097 e^{-(e^{-1.2t}/3.0)}$$
$$Y(t) = 4.0 e^{-t}$$

a. Check that S_1 and S_2 start at the same size as in Exercise 2.
b. Check that S_1 and S_2 start growing in a similar way by comparing the two curves.
c. What is the final size of S_1 and S_2?
d. Plot semilog graphs of S_1 and S_2. What sort of functions do they follow?
e. Plot S_1 against S_2 on regular and log–log graphs. Do S_1 and S_2 still follow the same allometric relationship?

1.9 Oscillations and Trigonometry

We have used linear, exponential, and power functions to describe several types of relations between measurements. Important and flexible as they are, these functions cannot describe **oscillations**. Heartbeats and breathing are examples of biological oscillations. The daily and seasonal cycles imposed by the movements of the earth drive sleep–wake cycles, seasonal population cycles, and the tides. The **trigonometric functions** are the basic tools we use to describe simple **sinusoidal oscillations**. Four numbers are needed to describe such oscillations with the **cosine** function: the **average**, the **amplitude**, the **period**, and the **phase**.

Sine and Cosine: A Review

The trigonometric functions have two interpretations, geometric and dynamic. Geometrically, the trigonometric functions are used to compute angles and distances. After a brief review of the key facts about the **sine** and **cosine** functions, we will see how to use them to study biological oscillations.

In applied mathematics, angles are measured in **radians**; 2π radians corresponds to 360°, or one complete revolution (Figure 1.61). There is thus a **basic identity** between radians and degrees, given by

$$2\pi \text{ radians} = 360° \quad (1.38)$$

From this, we derive the **conversion factors**

$$1 = \frac{2\pi \text{ radians}}{360°} = \frac{\pi \text{ radians}}{180°}$$

$$1 = \frac{360°}{2\pi \text{ radians}} = \frac{180°}{\pi \text{ radians}}$$

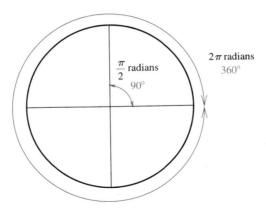

Figure 1.61
Degrees and radians

To find 60° in radians, we convert

$$60° = 60° \times 1 = 60° \times \frac{\pi \text{ radians}}{180°} = \frac{\pi}{3} \text{ radians}$$

Similarly, to find 1.0 radians in degrees, we convert

$$1.0 \text{ radians} = 1.0 \text{ radians} \times 1 = 1.0 \text{ radians} \times \frac{180°}{\pi \text{ radians}} \approx 57.3°$$

The **sine** and **cosine** functions take angles as inputs and return numbers between -1 and 1 as outputs. We write $\sin(\theta)$ and $\cos(\theta)$ to denote these functions, where the Greek letter θ ("theta") is often used for angles. Values of these functions for representative inputs are given in the following table.

Radians	Degrees	$\cos(\theta)$	$\sin(\theta)$	Radians	Degrees	$\cos(\theta)$	$\sin(\theta)$
0	$0°$	1	0	π	$180°$	-1	0
$\frac{\pi}{6}$	$30°$	$\frac{\sqrt{3}}{2}$	$\frac{1}{2}$	$\frac{7\pi}{6}$	$210°$	$-\frac{\sqrt{3}}{2}$	$-\frac{1}{2}$
$\frac{\pi}{4}$	$45°$	$\frac{\sqrt{2}}{2}$	$\frac{\sqrt{2}}{2}$	$\frac{5\pi}{4}$	$225°$	$-\frac{\sqrt{2}}{2}$	$-\frac{\sqrt{2}}{2}$
$\frac{\pi}{3}$	$60°$	$\frac{1}{2}$	$\frac{\sqrt{3}}{2}$	$\frac{4\pi}{3}$	$240°$	$-\frac{1}{2}$	$-\frac{\sqrt{3}}{2}$
$\frac{\pi}{2}$	$90°$	0	1	$\frac{3\pi}{2}$	$270°$	0	-1
$\frac{2\pi}{3}$	$120°$	$-\frac{1}{2}$	$\frac{\sqrt{3}}{2}$	$\frac{5\pi}{3}$	$300°$	$\frac{1}{2}$	$-\frac{\sqrt{3}}{2}$
$\frac{3\pi}{4}$	$135°$	$-\frac{\sqrt{2}}{2}$	$\frac{\sqrt{2}}{2}$	$\frac{7\pi}{4}$	$315°$	$\frac{\sqrt{2}}{2}$	$-\frac{\sqrt{2}}{2}$
$\frac{5\pi}{6}$	$150°$	$-\frac{\sqrt{3}}{2}$	$\frac{1}{2}$	$\frac{11\pi}{6}$	$330°$	$\frac{\sqrt{3}}{2}$	$-\frac{1}{2}$
π	$180°$	-1	0	2π	$360°$	1	0

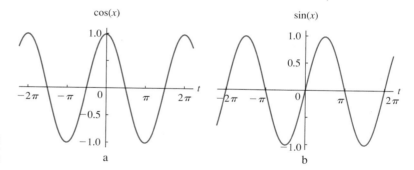

Figure 1.62
Graphs of the cosine and sine functions

Both the sine and cosine functions repeat every 2π radians (Figure 1.62). The value 2π is called the **period** of the oscillation. This means that adding or subtracting multiples of 2π from the argument does not change the value, so

$$\cos(\theta) = \cos(\theta + 2\pi) = \cos(\theta + 4\pi) = \cos(\theta + 2n\pi)$$
$$\cos(\theta) = \cos(\theta - 2\pi) = \cos(\theta - 4\pi) = \cos(\theta - 2n\pi)$$

for any value of θ and any integer n. For example,

$$\cos\left(\frac{\pi}{4}\right) = \cos\left(\frac{\pi}{4} + 2\pi\right) = \cos\left(\frac{\pi}{4} + 4\pi\right) = \frac{\sqrt{2}}{2}$$

$$\cos\left(\frac{\pi}{4}\right) = \cos\left(\frac{\pi}{4} - 2\pi\right) = \cos\left(\frac{\pi}{4} - 4\pi\right) = \frac{\sqrt{2}}{2}$$

The graphs of sine and cosine have the same shape but are shifted from each other by $\pi/2$ radians. In equations,

$$\sin(\theta) = \cos\left(\theta - \frac{\pi}{2}\right)$$

For example,

$$\sin\left(\frac{2\pi}{3}\right) = \cos\left(\frac{2\pi}{3} - \frac{\pi}{2}\right) = \cos\left(\frac{\pi}{6}\right) = \frac{\sqrt{3}}{2}$$

Because we can compute the sine function in terms of the cosine function, we will use the cosine alone to describe oscillations.

Describing Oscillations with the Cosine

Oscillations that are shaped like the graph of the sine or cosine function are called **sinusoidal**. Four numbers are needed to describe an oscillation with the cosine function: the **average**, the **amplitude**, the **period**, and the **phase**, ϕ ("phi") (Figure 1.63).

- The *average* is the middle value on the curve.
- The *amplitude* is the difference between the maximum and the average (or the minimum and the average).
- The *period* is the time between successive peaks.
- The *phase* is the time of the first peak.

The cosine function has average 0, amplitude 1, period 2π, and phase 0. The sine function has average 0, amplitude 1, period 2π, and phase $\pi/2$ (Figure 1.64).

72 Chapter 1 Introduction to Discrete-Time Dynamical Systems

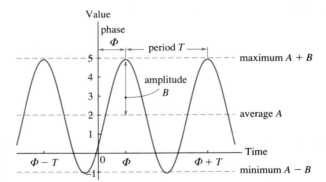

Figure 1.63
The four numbers that describe a sinusoidal oscillation

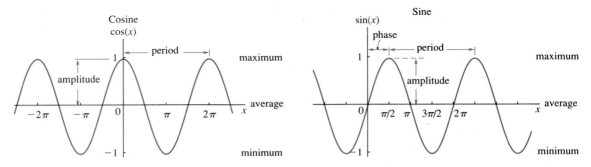

Figure 1.64
The average, amplitude, period, and phase of the cosine and sine

How can we build the oscillation pictured in Figure 1.63 from the cosine function alone? Suppose we wish to build a function with an average of 3.0, an amplitude of 2.0, a period of 4.0, and a phase of 1.0. We can construct the formula in steps. First, to raise the *average* from 0 to 3.0, we *add* 3.0 to the function, making

$$f(t) = 3.0 + \cos(t)$$

(Figure 1.65a). To increase the amplitude by a factor of 2.0, we *multiply* the oscillation (the cosine) by 2.0. The function is now

$$f(t) = 3.0 + 2.0\cos(t)$$

(Figure 1.65b).

Next, we wish to decrease the period from 2π to 4.0. We do this by multiplying t by $2\pi/4.0$. Our function is now

$$f(t) = 3.0 + 2.0\cos\left[\frac{2\pi}{4.0}t\right]$$

(Figure 1.65c). Finally, we need to shift the curve over so that the first peak is at 1.0 instead of 0.0. We do this by subtracting 1.0 from t, arriving at

$$f(t) = 3.0 + 2.0\cos\left[\frac{2\pi}{4.0}(t - 1.0)\right]$$

(Figure 1.65d). Notice where each piece of this newly built function appears.

1.9 Oscillations and Trigonometry

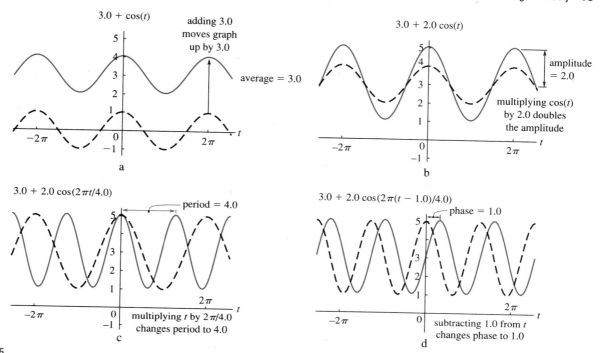

Figure 1.65

Building a function with different average, amplitude, period, and phase

More generally, a sinusoidal oscillation with average A, amplitude B, period T, and phase ϕ can be described as a function of time t with the formula

$$f(t) = A + B \cos\left[\frac{2\pi}{T}(t - \phi)\right] \quad (1.39)$$

This function has a maximum at $t = \phi$, a minimum at $t = \phi + T/2$, and averages at $t = \phi + T/4$ and $t = \phi + 3T/4$. Thereafter, it repeats every period T (Figure 1.66). Using these values, sinusoidal oscillations can be sketched from their equations. For example, suppose we wish to plot

$$f(t) = -2.0 + 0.4 \cos\left[\frac{2\pi}{10.0}(t - 7.0)\right]$$

We can see that the average is -2.0 and the amplitude is 0.4, so the maximum is at -1.6 and the minimum at -2.4. The first maximum occurs at $t = 7.0$

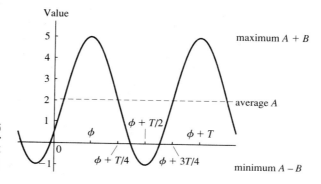

Figure 1.66

The guideposts for plotting $f(t) = A + B \cos\left[\frac{2\pi}{T}(t - \phi)\right]$

(Figure 1.67). The average occurs one-fourth of the way through the first period, or after another 2.5 time units at $t = 9.5$. The minimum occurs halfway through the first period at $t = 12.0$, the next average at $t = 14.5$, and the cycle repeats at $t = 17.0$.

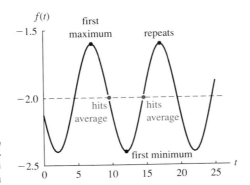

Figure 1.67

Graphing a sinusoidal oscillation from its equation

Women have two cycles affecting body temperature: a daily and a monthly rhythm (Figure 1.68). The key facts about these two cycles are given in the following table:

	Minimum (°C)	Maximum (°C)	Average (°C)	Time of maximum	Period
Daily cycle	36.5	37.1	36.8	2:00 P.M.	24 h
Monthly cycle	36.6	37.0	36.8	Day 16	28 days

Assuming these cycles are sinusoidal, we can use this information to describe these cycles with the cosine function.

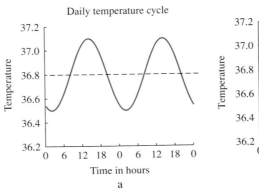
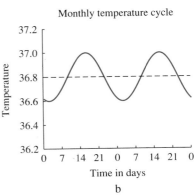

Figure 1.68

Women's daily and monthly temperature cycles

The amplitude of a cycle is

$$\text{amplitude} = \text{maximum} - \text{average} \tag{1.40}$$

For the daily cycle, the amplitude is

$$\text{daily cycle amplitude} = 37.1 - 36.8 = 0.3$$

For the monthly cycle, the amplitude is

$$\text{monthly cycle amplitude} = 37.0 - 36.8 = 0.2$$

The phase depends on the time chosen as the starting time. We define the daily cycle to begin at midnight and the monthly cycle to begin at menstruation. The maximum of the daily cycle occurs 14 h after the start and that of the monthly cycle 16 days after the start. The oscillations can be described by the fundamental formula (Equation 1.39). For the daily cycle, with t measured in hours, the formula $P_d(t)$ is

$$P_d(t) = 36.8 + 0.3 \cos\left[\frac{2\pi(t-14)}{24}\right]$$

For the monthly cycle, with t measured in days, the formula $P_m(t)$ is

$$P_m(t) = 36.8 + 0.2 \cos\left[\frac{2\pi(t-16)}{28}\right]$$

More Complicated Shapes

Real oscillations are not perfectly sinusoidal. Nonetheless, the cosine function is useful for describing many complicated oscillations. A powerful theory, which is beyond the scope of this book, called the **Fourier series**, shows how almost any oscillation can be written as the sum of many cosine functions with different amplitudes, periods, and phases (Exercise 13).

As an illustration, we will combine the daily and monthly temperature cycles. To do so, we must write both cycles in the same time units, days. In days, the period of the daily cycle is 1.0 days and the phase (the time of the maximum) is

$$\text{phase in days} = \frac{14 \text{ h}}{24 \text{ h}} \cdot \frac{1.0 \text{ days}}{24 \text{ h}} = 0.583 \text{ days}$$

The equation for the daily cycle, with t measured in days, is

$$P_d(t) = 36.8 + 0.3 \cos[2\pi(t-0.583)]$$

To figure out how the daily and monthly cycles combine, we cannot simply add them together because

$$P_d(t) + P_m(t) = 36.8 + 0.3 \cos[2\pi(t-0.583)] + 36.8 + 0.2 \cos\left[\frac{2\pi(t-16)}{28}\right]$$

$$= 73.6 + 0.3 \cos[2\pi(t-0.583)] + 0.2 \cos\left[\frac{2\pi(t-16)}{28}\right]$$

which has an average of 73.6.

To keep the average at the appropriate value of 36.8, we add only the two cosine terms to the average, getting for the combined cycle the formula

$$P_t(t) = 36.8 + 0.2 \cos\left[\frac{2\pi(t-16)}{28}\right] + 0.3 \cos[2\pi(t-0.583)]$$

(Figure 1.69). In the course of one month, there is a single slow cycle, with the 28 daily cycles superimposed. The temperature takes on a maximum at 2:00 P.M. on the 16th day of the cycle. The temperature can be found by adding the sum of the amplitudes of the daily and monthly cycles to the overall average, or

$$\text{overall maximum} = 36.8 + (0.2 + 0.3) = 37.3$$

The overall minimum occurs at 2:00 A.M. on the second day of the cycle, with temperature found by subtracting the sum of the amplitudes of the daily and monthly cycles from the overall average, or

$$\text{overall minimum} = 36.8 - (0.2 + 0.3) = 36.3$$

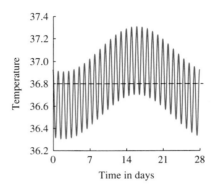

Figure 1.69
The combined effect of the daily and monthly temperature cycles

SUMMARY

Sinusoidal oscillations can be described mathematically with the **cosine** function. Four factors change the shape of the graph: the **average** (the middle value), the **amplitude** (the distance from the middle to the minimum or maximum), the **period** (the time between successive maxima), and the **phase** (the time of the first maximum). Oscillations with more complicated shapes can be described by adding appropriate cosine functions.

1.9 EXERCISES

1. Convert the following angles from degrees to radians or vice versa.
   a. $30°$
   b. $330°$
   c. $1°$
   d. $-30°$
   e. 2.0 radians
   f. $\pi/5$ radians
   g. $-\pi/5$ radians
   h. 30 radians

2. There are $360°$ in a full circle because the ancient Babylonians were fond of the number 60 and its multiples. An attempt was made to introduce another system for measuring angles, called "grads," where 400 grads make up a full circle (or 100 grads make up $90°$). Although found on many calculators, these units are almost never used. Write the basic identities between degrees and grads and radians and grads, and use them to make the following conversions.
   a. $180°$ into grads
   b. $60°$ into grads
   c. $\pi/3$ radians into grads
   d. 3.0 radians into grads
   e. 150 grads into degrees
   f. 250 grads into radians

3. Use the table in this section or a calculator to find the values of sine and cosine for the following inputs (all in radians), and plot them on a graph.
   a. $\pi/2$
   b. $3\pi/4$
   c. $\pi/9$
   d. 5.0
   e. -2.0
   f. 3.2
   g. -3.2

4. The other trigonometric functions (tangent, cotangent, secant, and cosecant) are defined in terms of sine and cosine by

$$\tan(x) = \frac{\sin(x)}{\cos(x)}$$

$$\cot(x) = \frac{\cos(x)}{\sin(x)}$$

$$\sec(x) = \frac{1}{\cos(x)}$$

$$\csc(x) = \frac{1}{\sin(x)}$$

Calculate the value of each of these functions at the points in Exercise 3.

5. Sketch graphs of the following for the domain -2π to 2π.
   a. $\tan(x)$ **b.** $\cot(x)$
   c. $\sec(x)$ **d.** $\csc(x)$

6. Find the average, minimum, maximum, amplitude, period, and phase from the graphs of the following oscillations.

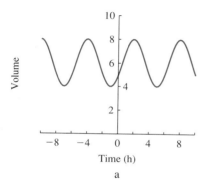

a

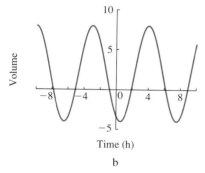

b

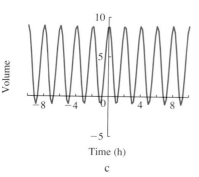

c

7. Graph the following functions. Give the average, maximum, minimum, amplitude, period, and phase of each and mark them on your graph.
   a. $f(x) = 3.0 + 4.0 \cos\left(\dfrac{2\pi(x - 1.0)}{5.0}\right)$.
   b. $g(t) = 4.0 + 3.0 \cos[2\pi(t - 5.0)]$. Is there a simpler way to write this function?
   c. $h(z) = 1.0 + 5.0 \cos\left(\dfrac{2\pi(z - 3.0)}{4.0}\right)$.

8. The following are alternative ways to write formulas for sinusoidal oscillations. Convert them to the standard form and sketch a graph.
   a. $f(t) = 2.0 - 1.0 \cos(t)$, which has negative amplitude.
   b. $g(t) = 2.0 + 1.0 \sin(t)$, which uses sine instead of cosine.
   c. $h(t) = 2.0 + 1.0 \cos(2\pi t - 3.0)$, in which the factor of 2π does not multiply the 3.0.

9. Sleepiness has two cycles, a circadian rhythm with a period of approximately 24 h and an ultradian rhythm with a period of approximately 4 h. Both have phase 0 and average 0, but the amplitude of the circadian rhythm is 1.0 sleepiness units and of the ultradian is 0.4 sleepiness units.
   a. Find the formula and sketch the graph of sleepiness over the course of a day due to the circadian rhythm.
   b. Find the formula and sketch the graph of sleepiness over the course of a day due to the ultradian rhythm.
   c. Sketch the graph of the two cycles combined.
   d. At what time of day are you sleepiest?
   e. At what time of day are you least sleepy?

10. Use the formula for the monthly temperature cycle $P_t(t)$,

$$P_t(t) = 36.8 + 0.2 \cos\left[\frac{2\pi(t - 16)}{28}\right]$$
$$+ 0.3 \cos[2\pi(t - 0.583)]$$

to compute the temperature at midnight, 2:00 A.M., noon, and 2:00 P.M. on days 0, 2, 14, and 16 of the cycle. Plot them on a graph and compare with Figure 1.69.

11. Functions that describe oscillations can be combined with other functions to describe different phenomena. Plot a graph of the function $f(x) = x + \cos(x)$ with the following steps:
   a. Plot graphs of x and $\cos(x)$ for $0 \leq x \leq 4\pi$.
   b. Add the graphs by moving the cosine graph up by an amount x at evenly spaced values of x (0, $\pi/2$, π, and so forth).
   c. Compute the exact values of $f(x)$ at your evenly spaced values, and plot these on your graph.
   d. Can you imagine a biological process that might produce such a graph?

12. Oscillations can be combined with exponential growth to describe other phenomena. Plot a graph of the function $g(x) = e^x \cdot \cos(2\pi x)$ with the following steps:
   a. Plot graphs of e^x and $\cos(2\pi x)$ for $0 \leq x \leq 2$.

b. Multiply the graphs by amplifying the cosine graph by an amount e^x at evenly spaced values of x (0, 1/4, 1/2, and so forth).
c. Compute the exact values of $g(x)$ at your evenly spaced values, and plot these on your graph.
d. Can you imagine a biological process that might produce such a graph?

13. **COMPUTER:** Consider the following functions.

$$f_1(x) = \cos\left(x - \frac{\pi}{2}\right)$$

$$f_3(x) = \frac{\cos\left(3x - \frac{\pi}{2}\right)}{3}$$

$$f_5(x) = \frac{\cos\left(5x - \frac{\pi}{2}\right)}{5}$$

$$f_7(x) = \frac{\cos\left(7x - \frac{\pi}{2}\right)}{7}$$

a. Plot them all on one graph.
b. Plot the sum $f_1(x) + f_3(x)$
c. Plot the sum $f_1(x) + f_3(x) + f_5(x)$
d. Plot the sum $f_1(x) + f_3(x) + f_5(x) + f_7(x)$
e. What does this sum look like?
f. Try to guess the pattern, and add $f_9(x)$ and $f_{11}(x)$ to your graph. This is an example of a **Fourier series**, a sum of cosine functions that add up to a **square wave** that jumps between values of -1 and 1.

14. **COMPUTER:** Use your computer to graph solutions of discrete-time dynamical systems with the following updating functions. Try three initial conditions for each. Can you make any sense of what happens? Why don't the solutions follow a sinusoidal oscillation?
a. $x_{t+1} = \cos(x_t)$
b. $y_{t+1} = \sin(y_t)$
c. $z_{t+1} = \sin(z_t) + \cos(z_t)$

1.10 Modeling and Graphical Analysis of Updating Functions

One of the most important skills for a biologist is to be able to translate facts about a biological process into a mathematical model. One of the most important biological processes is the exchange of materials between an organism and its environment. By following the amount of a chemical step by step through the breathing process, we can derive an updating function that models this process for a simplified lung. This updating function, which describes how the **ambient air** mixes with internal air, takes the form of a **weighted average**. Because this discrete-time dynamical system is difficult to solve, we develop a graphical technique called **cobwebbing** to help us sketch solutions.

A Model of the Lungs

Consider a simplified breathing process. Suppose a lung has a volume of 2.0 L when full. With each breath, 1.5 L of the air are exhaled and replaced by 1.5 L of outside (or **ambient**) air. After exhalation, the volume of the lung is 0.5 L, and after inhalation it returns to 2.0 L (Figure 1.70).

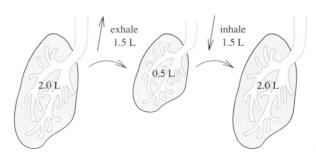

Figure 1.70
Gas exchange in the lung: volume

1.10 Modeling and Graphical Analysis of Updating Functions

Suppose further that the lung contains a particular chemical with concentration 2.0 mmol (millimoles)/L before exhalation. (A mole is a convenient chemical unit indicating 6.02×10^{23} molecules, and a millimole is 6.02×10^{20} molecules). The ambient air, which is inhaled, has a chemical concentration of 5.0 mmol/L. What is the chemical concentration after one breath?

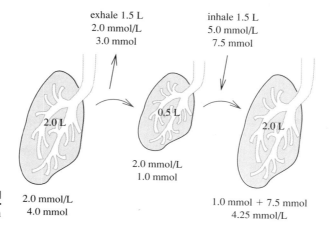

Figure 1.71
Gas exchange in the lung: concentration

We must track three quantities through the steps: the volume (Figure 1.70), the total amount of chemical, and the chemical concentration (Figure 1.71). To find the total amount from the concentration, we use the fundamental relation

$$\text{total amount} = \text{concentration} \times \text{volume} \qquad (1.41)$$

Conversely, to find the concentration from the total amount, we rearrange the fundamental relation

$$\text{concentration} = \frac{\text{total amount}}{\text{volume}} \qquad (1.42)$$

Step	Volume (L)	Total chemical (mmol)	Concentration (mmol/L)	What we did
Air in lung before breath	2.0	4.0	2.0	Multiplied volume of lung (2.0) by concentration (2.0) to get 4.0.
Air exhaled	1.5	3.0	2.0	Multiplied volume exhaled (1.5) by concentration (2.0) to get 3.0.
Air in lung after exhalation	0.5	1.0	2.0	Multiplied volume remaining (0.5) by concentration (2.0) to get 1.0.
Air inhaled	1.5	7.5	5.0	Multiplied volume inhaled (1.5) by ambient concentration (5.0) to get 7.5.
Air in lung after breath	2.0	8.5	4.25	Found total by adding $1.0 + 7.5 = 8.5$, and divided by volume (2.0) to get 4.25.

Three basic assumptions underlie our reasoning:

1. Air breathed out has a concentration equal to that of the whole lung.
2. The total amount after breathing out is the total amount before breathing out minus the amount breathed out.
3. The total amount after breathing in is the total amount before breathing in plus the amount breathed in.

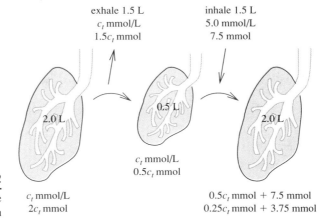

Figure 1.72
Gas exchange in the lung: finding the updating function

Breathing creates a discrete-time dynamical system. The original concentration of 2.0 mmol/L is updated to 4.25 mmol/L after a breath. To write the updating function, we must figure out the concentration after a breath, c_{t+1}, as a function of the concentration before the breath, c_t. We follow the same steps but replace 2.0 with c_t (Figure 1.72).

Step	Volume (L)	Total chemical (mmol)	Concentration (mmol/L)	What we did
Air in lung before breath	2.0	$2.0c_t$	c_t	Multiplied volume of lung (2.0) by concentration (c_t) to get $2.0c_t$.
Air exhaled	1.5	$1.5c_t$	c_t	Multiplied volume exhaled (1.5) by concentration (c_t) to get $1.5c_t$.
Air in lung after exhalation	0.5	$0.5c_t$	c_t	Multiplied volume remaining (0.5) by concentration (c_t) to get $0.5c_t$.
Air inhaled	1.5	7.5	5.0	Multiplied volume inhaled (1.5) by ambient concentration (5.0) to get 7.5.
Air in lung after breath	2.0	$7.5 + 0.5c_t$	$3.75 + 0.25c_t$	Found total by adding $7.5 + 0.5c_t$ and divided by volume (2.0) to get $3.75 + 0.25c_t$.

1.10 Modeling and Graphical Analysis of Updating Functions

The updating function is therefore

$$c_{t+1} = 3.75 + 0.25 c_t \tag{1.43}$$

(Figure 1.73). This equation summarizes the calculations in the table; for example, an input of $c_t = 2.0$ gives $c_{t+1} = 4.25$. The graph of the updating function is a line with y-intercept 3.75 and slope 0.25. We graph it by connecting the y-intercept $(0, 3.75)$ with another point, such as the value at $c_t = 10.0$:

$$c_{t+1} = 3.75 + 0.25 \cdot 10 = 6.25$$

(Figure 1.73).

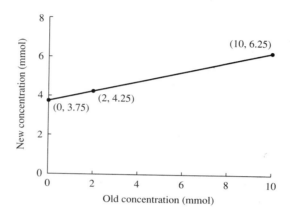

Figure 1.73
Updating function for the lung model

The Lung Updating Function in General

In the previous subsection, we assumed that the lung had a volume of 2.0 L, that 1.5 L of air were exhaled and inhaled, and that the ambient concentration of chemical was 5.0 mmol/L. Suppose, more generally, that the lung has a volume of V L, that W L of air are exhaled and inhaled with each breath, and that the ambient concentration of chemical is γ ("gamma"). Our goal, as before, is to find the updating function giving c_{t+1} as a function of c_t. We do so by following the same steps as before (Figure 1.74).

Step	Volume (L)	Total chemical (mmol)	Concentration (mmol/L)	What we did
Air in lung before breath	V	$c_t V$	c_t	Multiplied volume of lung (V) by concentration (c_t) to get $c_t V$.
Air exhaled	W	$c_t W$	c_t	Multiplied volume exhaled (W) by concentration (c_t) to get $c_t W$.
Air in lung after exhalation	$V - W$	$c_t(V - W)$	c_t	Multiplied volume remaining ($V - W$) by concentration (c_t) to get $c_t(V - W)$.
Air inhaled	W	γW	γ	Multiplied volume inhaled (W) by ambient concentration (γ) to get γW.
Air in lung after breath	V	$c_t(V - W) + \gamma W$	$\dfrac{c_t(V - W) + \gamma W}{V}$	Found total by adding $c_t(V - W)$ to γW and divided by volume (V).

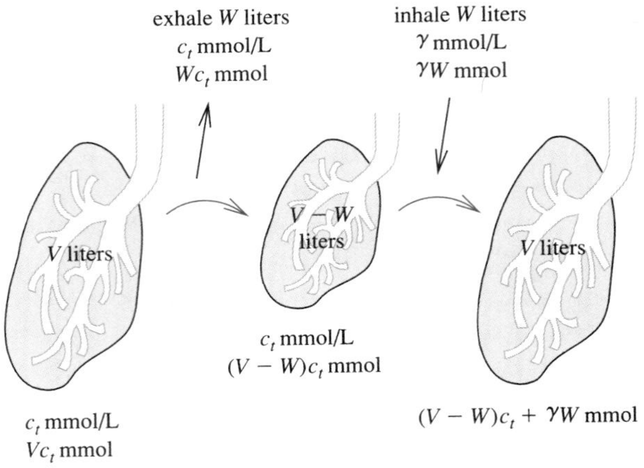

82 Chapter 1 Introduction to Discrete-Time Dynamical Systems

Figure 1.74
Gas exchange in the lung: general case

The new concentration appears at the bottom of the fourth column of the table, giving the updating function

$$c_{t+1} = \frac{c_t(V - W) + \gamma W}{V}$$

This equation can be simplified by multiplying out the first term and dividing out the V,

$$c_{t+1} = \frac{c_t(V - W) + \gamma W}{V}$$

$$= \frac{c_t V - c_t W + \gamma W}{V}$$

$$= c_t - c_t \frac{W}{V} + \gamma \frac{W}{V}$$

The two values W and V appear only as the ratio W/V, which is the fraction of the total volume exchanged with each breath. For example, when $W = 1.5$ L and $V = 2.0$ L, $W/V = 0.75$, meaning that 75% of air is exhaled with each breath. We define a new parameter,

$$q = \frac{W}{V} = \text{fraction of air exchanged}$$

to represent this quantity. We can then write the updating function as

$$c_{t+1} = c_t - c_t q + \gamma q$$

or

$$c_{t+1} = (1 - q)c_t + q\gamma \tag{1.44}$$

In our original example, $q = 0.75$ and $\gamma = 5.0$. The general equation matches our original updating function because

$$c_{t+1} = (1 - 0.75)c_t + 0.75 \cdot 5.0 = 0.25c_t + 3.75$$

The right-hand side of this equation is a **weighted average**. After a breath, the air in the lung is a mix of old air and ambient air (Figure 1.75). The fraction $1 - q$ is old air, and the remaining fraction q is ambient air. If $q = 0.5$, half the air in the lung after a breath came from outside, and c_{t+1} is the average of the old concentration and the ambient concentration. If q is small, little of the internal

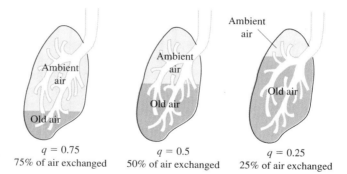

Figure 1.75
Effects of different values of q

$q = 0.75$
75% of air exchanged

$q = 0.5$
50% of air exchanged

$q = 0.25$
25% of air exchanged

air is replaced with ambient air and c_{t+1} is close to c_t. If q is near 1, most of the internal air is replaced with ambient air. The air in the lung then resembles ambient air and c_{t+1} is close to the ambient concentration γ.

Cobwebbing: A Graphical Solution Technique

What will happen to the concentration in the lung after many breaths? It is tricky (but possible) to write a solution, as we did with bacterial and tree growth. This solution would give us a formula for the concentration c_2 after two breaths, c_3 after three breaths, and c_t after t breaths in terms of the initial concentration c_0. Alternatively, there is a **graphical** method to figure out what solutions look like. The technique is called **cobwebbing** (Figure 1.76).

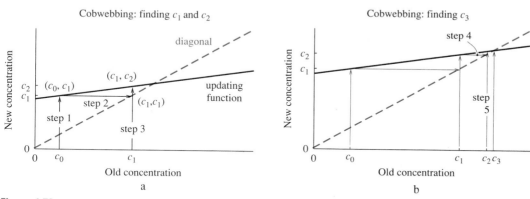

Figure 1.76
Cobwebbing

There are two steps. First, we must remember the meaning of the updating function. To find c_1 from c_0, we use the fact that c_1 is the result of applying the updating function to c_0 and that c_1 is therefore the point on the graph of the updating function directly above c_0 (step 1 in Figure 1.76a). Similarly, c_2 is the point on the graph of the updating function directly above c_1, and so on.

The missing step is getting c_1 off of the vertical axis and onto the horizontal axis. The trick is to reflect it off the diagonal line $c_{t+1} = c_t$. Move the point (c_0, c_1) horizontally until it intersects the diagonal line $c_{t+1} = c_t$. Moving a point horizontally does not change the vertical coordinate. The intersection with

84 Chapter 1 Introduction to Discrete-Time Dynamical Systems

$c_{t+1} = c_t$ takes place at the point (c_1, c_1) (step 2 in Figure 1.76a). The point $(c_1, 0)$ lies directly below.

What have we done? Starting from the initial value c_0, plotted on the horizontal axis, we used the updating function to find c_1 on the vertical axis and the reflecting trick to project c_1 onto the horizontal axis. We then can find c_2 by moving vertically to the graph of the updating function (step 3 in Figure 1.76a). To find c_3, we move horizontally to the diagonal to reach the point (c_2, c_2), and then vertically to the point (c_2, c_3) (steps 4 and 5 in Figure 1.76b).

Having found c_1, c_2, and c_3 on our cobwebbing graph, we would like to translate this into a graph of the solution that shows the concentration as a function of time. In Figure 1.76, we began at $c_0 = 1$ mmol/L. This is plotted as the point $(0, c_0) = (0, 1)$ in the solution (Figure 1.77). The value c_1, approximately 4.0, is plotted as the point $(1, c_1)$ in the solution. The values of c_2 and c_3 increase slightly, but less so, and are thus plotted on the graph. Without plugging numbers into the formula, we have used the **graph** of the updating function to figure out that the solution increases more and more slowly as time passes.

Similarly, we can find how the concentration would behave over time if we started from the different initial condition $c_0 = 8.0$ (Figure 1.78). In this case, the solution decreases quickly at first and then more and more slowly. If the initial concentration is less than the ambient concentration, the concentration increases (Figures 1.76 and 1.77). If the initial concentration is greater than the ambient concentration, the concentration decreases (Figure 1.78).

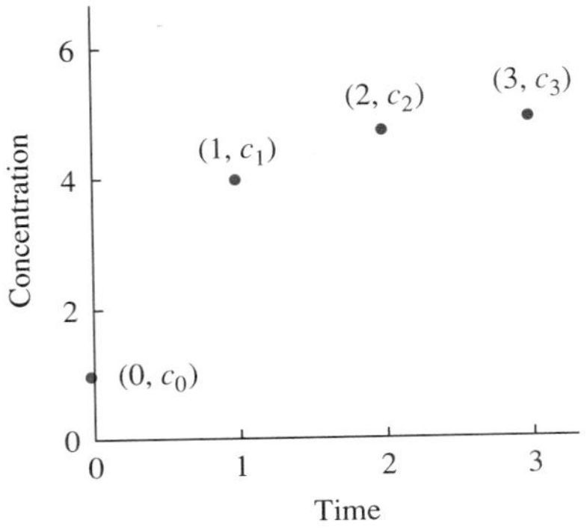

Figure 1.77
The solution of the lung model

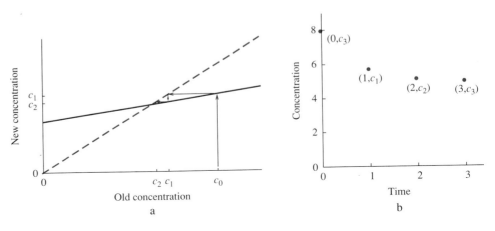

Figure 1.78
Cobwebbing and solution with a diferent initial condition

SUMMARY

This section has taken two steps in mathematical modeling. Starting from an understanding of how a lung exchanges air, we derived an updating function for the chemical concentration. The updating function can be described as a **weighted average** of the internal concentration and the **ambient concentration**. Next, we used this updating function to understand the consequences. Even without finding a formula for the solution, we used the graph of the updating function to estimate a solution with the method of **cobwebbing**. The cobwebbing diagrams show that an initial concentration lower than the ambient concentration increases, and one greater than the ambient concentration decreases.

1.10 EXERCISES

1. Suppose that the volume of the lung is 2.0 L, the amount breathed in and out is 0.5 L, and $\gamma = 5.0$ mmol/L. If $c_t = 1.0$ mmol/L, find
   a. the amount of chemical in the lung.
   b. the amount of chemical breathed out.
   c. the amount of chemical in the lung after breathing out.
   d. the amount of chemical breathed in.
   e. the amount of chemical in the lung after breathing in.
   f. the concentration of chemical in the lung after breathing in.
   g. Compare this result with the result of using Equation 1.44 (make sure to compute q).

2. Redo the calculations in Exercise 1 for the following two situations.
   a. $V = 1.0$ L, $W = 0.1$ L, $\gamma = 8.0$ mmol/L, $c_t = 4.0$ mmol/L
   b. $V = 1.0$ L, $W = 0.9$ L, $\gamma = 5.0$ mmol/L, $c_t = 9.0$ mmol/L

3. Find the average score on the following tests (find the total number of points scored and divide by the total number of students). Try to write this in the form of a weighted average.
   a. 50 students score 60, 50 score 90.
   b. 70 students score 60, 30 score 90.
   c. 20 students score 60, 80 score 90.
   d. 20 students score 60, 60 score 70, 20 score 80.

4. Find and graph the updating function in the following cases. Try cobwebbing for three steps starting from the points indicated in the earlier problems. Sketch the solutions.
   a. The situation in Exercise 1.
   b. The situation in Exercise 2a.
   c. The situation in Exercise 2b.
   d. The situation in Exercise 2b but with $W = 1.0$ L.

5. Our model of chemical dynamics in the lung ignored any absorption of the chemical by the body. The following steps outline construction of one type of model that includes this important factor.
   a. Suppose the lung in Exercise 1 absorbs $A = 1.0$ mmol of the chemical right before breathing out. Find the total amount and concentration right after absorption, after breathing out, and again after breathing in.
   b. Try the same steps starting from c_t rather than 1.0 mmol.
   c. Write the updating function in this case.
   d. For what values of c_t does this derivation fail to make sense?

6. Another way to incorporate absorption is for the body to absorb a particular fraction of the chemical. The following steps outline construction of this alternative model.
   a. Suppose the lung in Exercise 1 absorbs 20% of the chemical in the lung just before breathing out. Find the total amount and concentration right after absorption, after breathing out, and again after breathing in.
   b. Try the same steps starting from c_t rather than 1.0 mmol.
   c. Write the updating function in this case.
   d. Does this updating function always make sense? Why?

7. Try the following cobwebbing exercises based on the growth of bacterial populations.
   a. Suppose $b_{t+1} = 2.0b_t$ and $b_0 = 1.0 \times 10^6$. Graph the updating function.
   b. Use your graph of the updating function to find the point (b_0, b_1).
   c. Reflect this point off the diagonal to find the point (b_1, b_1).
   d. Use the graph of the updating function to find (b_1, b_2).
   e. Reflect this point off the diagonal to find the point (b_2, b_2).
   f. Use the graph of the updating function to find (b_2, b_3).
   g. Sketch the solution as a function of time.

8. Follow the same steps as in Exercise 7 for the updating function $b_{t+1} = 0.5b_t$ with $b_0 = 1.0 \times 10^6$.

9. Bacterial populations that have $r < 1$ but that are supplemented by immigration have an updating function much like that of the lung. Suppose a population of bacteria has per capita reproduction $r = 0.6$, and that 1.0×10^6 bacteria are added in each generation.
   a. Starting from 3.0×10^6 bacteria, find the number after reproduction.
   b. Find the number after the new bacteria are added.
   c. Find the updating function.
   d. What values of q and γ would have produced the same updating function?

10. Lakes receive water from streams each year, and they lose water to outflowing streams and evaporation. The salt level in a lake depends greatly on the ratio of outflow to evaporation. Consider a lake that receives 100 gal of water per year with salinity of 1 part per thousand. The lake contains 10,000 gal of water and starts with salinity 2 parts per thousand.
    a. Suppose there is no evaporation and that 100 gal of water flow out each year. Find the total salt before the inflow, total water, total salt, and salt concentration after inflow and total water, total salt, and salt concentration after outflow.
    b. Find the updating function in this case.
    c. Do parts **a** and **b** assuming that 50 gal flow out and 50 gal evaporate (no salt is removed by evaporation).
    d. Do parts **a** and **b** assuming that all 100 gal evaporate.
    e. What do you think will happen to the salinity in each of these lakes?

11. **COMPUTER:** Use your computer to plot solutions for 20 steps of the lung model

$$c_{t+1} = (1-q)c_t + q\gamma$$

for the following values of q, γ, and c_0. Explain what is going on in each graph.
   a. $q = 0.9$, $\gamma = 5.0$, $c_0 = 1.0$
   b. $q = 0.9$, $\gamma = 5.0$, $c_0 = 9.0$
   c. $q = 0.5$, $\gamma = 5.0$, $c_0 = 1.0$
   d. $q = 0.5$, $\gamma = 5.0$, $c_0 = 9.0$
   e. $q = 0.1$, $\gamma = 5.0$, $c_0 = 1.0$
   f. $q = 0.1$, $\gamma = 5.0$, $c_0 = 9.0$
   g. $q = 0.01$, $\gamma = 5.0$, $c_0 = 1.0$
   h. $q = 0.01$, $\gamma = 5.0$, $c_0 = 9.0$

12. **COMPUTER:** If your computer has a feature for cobwebbing, produce cobwebbing diagrams for Exercise 11. Label the values of the solution at each point. Otherwise, use your computer to graph the updating function and plot the first few steps of the solution by hand on a cobwebbing diagram, being careful to label each point.

1.11 Equilibria

Quantitative understanding of a biological system with a discrete-time dynamical system requires three steps: quantifying the basic measurements, expressing the dynamical rule as an updating function, and analyzing and interpreting the results. Sometimes we can find a formula for the solution of a discrete-time dynamical system starting from any initial condition. When this is impossible (or very difficult), we need not give up. In Section 1.10, we derived the graphical method of **cobwebbing** to understand solutions. Another powerful technique for understanding the behavior of dynamical systems without a great deal of algebra involves finding special points called **equilibria**, points where the updating function does nothing.

Graphical Approach

Consider two new updating functions describing a population of plants (Figure 1.79a) and a population of birds (Figure 1.79b). Each graph includes the diagonal line used in cobwebbing. A great deal of information can be read directly from this graph.

If we begin cobwebbing to describe the plant population from an initial condition where the graph of the updating function lies *above* the diagonal, the population increases (Figure 1.80a). In contrast, if we begin cobwebbing from an initial condition where the graph of the updating function lies *below* the diagonal, the population decreases (Figure 1.80b). The plant population will thus increase if the initial condition lies below the crossing point and decrease if it lies above.

Similarly, the updating function for the bird population lies below the diagonal for initial conditions less than the first crossing and the population decreases (Figure 1.81a). The updating function is above the diagonal for initial conditions between the crossings, and the population increases (Figure 1.81b). Finally, the updating function is again below the diagonal for initial conditions greater than the second crossing, and the population decreases (Figure 1.81c).

What happens when the updating function crosses the diagonal? At these points, the population neither increases nor decreases; it remains the same. Such a point is called an **equilibrium**.

■ **Definition 1.4**   A point x^* is called an equilibrium of the updating function f if $f(x^*) = x^*$.

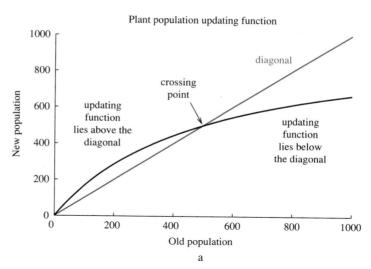

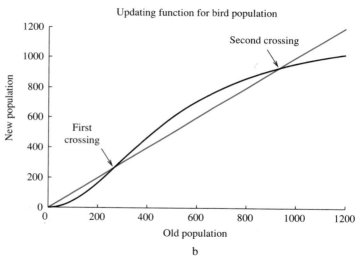

Figure 1.79
Updating functions for two populations

This definition says that the updating function does precisely nothing to the particular input x^*. As we have seen, these points can be found graphically by looking for intersections of the graph of the updating function with the diagonal line.

When there is more than one equilibrium, they are called **equilibria**. The plant population has two equilibria, one at $P = 0$ and one at the crossing point. If we start cobwebbing from an initial condition exactly equal to an equilibrium, not much happens. The cobweb goes up to the crossing point and gets stuck there (Figure 1.82a). The solution is a flat series of dots (Figure 1.82b).

Each of the three intersections of the bird population updating function with the diagonal is also an equilibrium. Why does this graphical method work? The diagonal is the graph of a line with slope 1 and y-intercept 0 and is described by the equation

$$B_{t+1} = B_t$$

The diagonal line can therefore be thought of as the graph of the "do nothing"

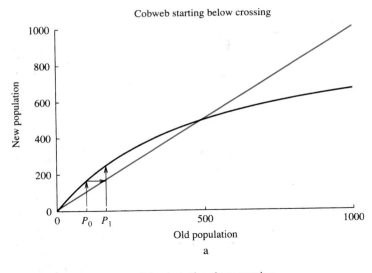
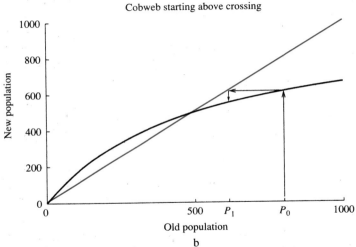

Figure 1.80
Behavior of plant updating function starting from two initial conditions.

updating function. The intersections of the actual updating function with the "do nothing" updating function are points where the updating function "does nothing." These are the equilibria.

Algebraic Approach

When we know the formula for the updating function, we can solve algebraically for the equilibria. The first step is to write the equation that expresses the fact that an equilibrium is unchanged by the updating function. Recall the updating function for a lung (Equation 1.43) that exchanges 75% of its contents for ambient air with a concentration of 5.0 mmol/L,

$$c_{t+1} = 3.75 + 0.25c_t$$

Let the new variable c^* stand for an equilibrium. The equation for equilibrium

1.11 Equilibria

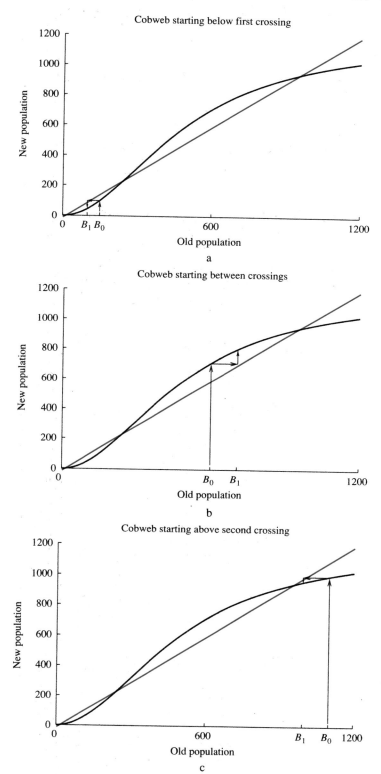

Figure 1.81
Behavior of bird updating function starting from three initial conditions

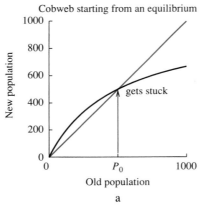
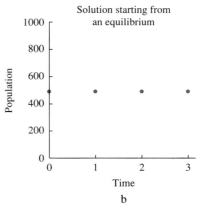

Figure 1.82
Cobweb and solution of plant updating function starting from an equilibrium

Figure 1.83
Equilibrium of the lung updating function

says that c^* is unchanged by the updating function, or

$$c^* = 3.75 + 0.25c^*$$

The solutions of this equation are equilibria (Figure 1.83).
  The next step is to use algebra to find the solution. In this case,

$$c^* = 3.75 + 0.25c^* \quad \text{the original equation}$$
$$c^* - 0.25c^* = 3.75 \quad \text{subtract } 0.25c^* \text{ to get unknowns on one side}$$
$$0.75c^* = 3.75 \quad \text{do the subtraction}$$
$$c^* = \frac{3.75}{0.75} = 5.0 \quad \text{divide by 0.75}$$

The equilibrium value is 5.0 mmol/L. We can check this by plugging $c_t = 5.0$ into the updating function, finding that

$$c_{t+1} = 3.75 + 0.25 \cdot 5.0 = 5.0$$

A concentration of 5.0 is indeed unchanged by the breathing process.
  We can follow the same steps to find the equilibria for the bacterial population's updating function,

$$b_{t+1} = 2b_t$$

which says that a bacterial population doubles each hour. The first step is to write the equation for equilibria,

$$b^* = 2b^*$$

We then solve this equation:

$$b^* = 2b^* \quad \text{the original equation}$$
$$b^* - b^* = 2b^* - b^* \quad \text{subtract } b^* \text{ from both sides}$$
$$0 = b^* \quad \text{do the subtraction}$$

Consistent with our picture (Figure 1.84), the only equilibrium is at $b_t = 0$. The only number that remains the same after doubling is 0.

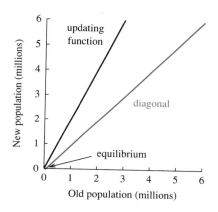

Figure 1.84
Equilibrium of the bacterial updating function

Two situations without equilibria are illustrated in Figure 1.85. Figure 1.85a shows the updating function for a growing tree (Equation 1.4):

$$h_{t+1} = h_t + 1.0$$

To solve for the equilibria, we might try

$$h^* = h^* + 1 \quad \text{the equation for the equilibrium}$$
$$h^* - h^* = 1 \quad \text{subtract } h^* \text{ to get unknowns on one side}$$
$$0 = 1 \quad \text{do the subtraction}$$

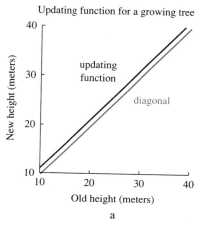
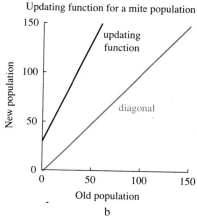

Figure 1.85
Two updating functions without equilibria

This looks bad. It means that the graph of the updating function and the graph of the diagonal do not intersect because they are parallel lines. Something that grows 1.0 m/yr has no equilibrium.

Figure 1.85b shows the updating function for a mite population (Equation 1.18):

$$x_{t+1} = 2x_t + 30$$

To solve for the equilibria, try

$$\begin{aligned}
x^* &= 2x^* + 30 & &\text{the equation for the equilibrium} \\
x^* - 2x^* &= 30 & &\text{subtract } 2x^* \text{ to get unknowns on one side} \\
-x^* &= 30 & &\text{do the subtraction} \\
x^* &= -30 & &\text{divide both sides by } -1
\end{aligned}$$

This looks like nonsense. It means that although there is a *mathematical* equilibrium, there is no *biological* equilibrium. If we extend the graph to include biologically meaningless negative values, we see that the graph of the updating function does intersect the diagonal. Because the updating function lies above the diagonal for all positive inputs, any positive population of mites will grow.

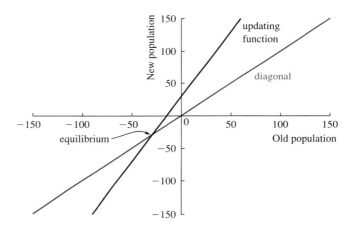

Figure 1.86
Extending the updating function for mites into the negative zone

Algebra Involving Parameters

We have seen that studying the general form of an updating function with parameters instead of numbers can sometimes simplify both the algebra and the understanding. The same is often true when solving for equilibria. When we work with parameters, however, we must be more careful with some of the algebra.

■ **Algorithm 1.5** (Solving for equilibria)

1. Write the equation for the equilibrium.

2. Use subtraction to move all the terms to one side, leaving 0 on the other.

3. Factor.

4. Set each factor equal to 0 and solve for the equilibria.

5. Meditate on the results.

As always, we begin by setting up the problem. The next three steps give a safe method to do the algebra. The final step is perhaps the most important; a result is worthwhile only if it makes sense.

The general updating function for the lung model is (Equation 1.44)

$$c_{t+1} = (1-q)c_t + q\gamma$$

The algorithm for finding equilibria is

$c^* = (1-q)c^* + q\gamma$	the equation for the equilibrium
$c^* - (1-q)c^* - q\gamma = 0$	move everything to one side
$qc^* - q\gamma = 0$	do the subtraction
$q(c^* - \gamma) = 0$	factor out the common factor, q
$q = 0$ or $c^* - \gamma = 0$	set both factors to 0
$q = 0$ or $c^* = \gamma$	solve each term

The key algebraic step comes after factoring. Remember that the product of two terms (such as q and $c^* - \gamma$) can equal 0 only if one of the terms is equal to 0.

What do these results mean? The first case, $q = 0$, occurs when no air is exchanged. Because there is no expression for c^* in this case, *any* value of c_t is an equilibrium. This make sense because a lung that is doing nothing (exchanging no air) is, in fact, at equilibrium. The second case is perhaps more interesting. It says that the equilibrium value of the concentration is equal to the ambient concentration. Exchanging air with the outside world has no effect when the inside and the outside match.

In general, doing the calculation shows that the fact that the equilibrium of 5.0 found earlier must match the ambient concentration of 5.0. It has also revealed another possibility: A lung that does not breathe does not change.

We can apply Algorithm 1.5 to the general bacterial updating function (Equation 1.28) written in terms of the per capita reproduction r,

$$b_{t+1} = rb_t$$

with the five steps

$b^* = rb^*$	the equation for the equilibrium
$b^* - rb^* = 0$	move everything to one side
$b^*(1 - r) = 0$	factor out the common factor of b^*
$b^* = 0$ or $1 - r = 0$	set both factors to 0
$b^* = 0$ or $r = 1$	solve each term

Again, there are two possibilities. The first matches what we found earlier: A population of 0 is at equilibrium. This makes sense because an extinct population cannot reproduce. The second is new. If the per capita reproduction r is exactly 1, any value of b_t is an equilibrium. In this case, every bacterium exactly replaces itself. The population will remain the same no matter what its size, even though the bacteria are not "doing nothing."

SUMMARY

By further examining the diagrams used for cobwebbing, we have seen that intersections of the graph of the updating function with the diagonal line play a special role. These **equilibria** are points that are unchanged by the updating function, and they separate the regions where the updating function makes things larger from regions where it makes things smaller. Algebraically, we find equilibria by solving the equation that expresses the fact that the updating function leaves its input unchanged. With a little extra care, we can solve for equilibria in general, without substituting numerical values for the parameters. Solving the equations in this way can help elucidate the underlying biological process.

1.11 EXERCISES

1. Graph the following updating functions. Identify equilibria and the regions where the updating function lies above the diagonal.
   a. $c_{t+1} = 0.5c_t + 8.0$, for $0 \leq c_t \leq 30$
   b. $b_{t+1} = 3b_t$, for $0 \leq b_t \leq 10$
   c. $b_{t+1} = 0.3b_t$, for $0 \leq b_t \leq 10$

2. Find the equilibria of the following updating functions. Cobweb for two steps starting from the point labeled x_0 and sketch the solution.

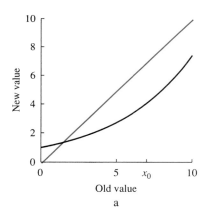

a

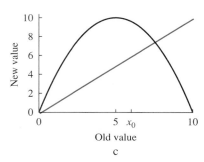

c

3. Find the lung updating function with the following parameter values and compute the equilibrium. Check that it matches the formula $c^* = \gamma$.
   a. $V = 2.0$ L, $W = 0.5$ L, $\gamma = 5.0$ mmol/L
   b. $V = 1.0$ L, $W = 0.1$ L, $\gamma = 8.0$ mmol/L
   c. $V = 1.0$ L, $W = 0.9$ L, $\gamma = 5.0$ mmol/L

4. Graph the following updating functions and find their equilibria both graphically and algebraically.
   a. The lung in Exercise 5 in Section 1.10 with updating function $c_{t+1} = 0.75c_t + 0.875$.
   b. The lung in Exercise 6 in Section 1.10 with updating function $c_{t+1} = 0.6c_t + 1.25$.
   c. The bacteria in Exercise 9 in Section 1.10 with updating function $b_{t+1} = 0.6b_t + 1.0 \times 10^6$.

5. Consider the model of a medication presented in Section 1.5,

$$M_{t+1} = 0.5M_t + 1.0$$

a. Graph this updating function and cobweb starting from $M_0 = 1.0$ and $M_0 = 5.0$.
b. Solve for the equilibrium.
c. Why does this model looks so much like the lung model?

6. Graph the following updating functions and find their equilibria both graphically and algebraically.
   a. The lake in Exercise 10b in Section 1.10 with updating function $s_{t+1} = 0.99s_t + 0.0099$.
   b. The lake in Exercise 10c in Section 1.10 with updating function $s_{t+1} = 0.995s_t + 0.00995$.

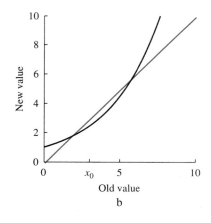

b

c. The lake in Exercise 10d in Section 1.10 with updating function $s_{t+1} = s_t + 0.01$.
d. Do these results make sense?

7. Assume that a lab is growing a valuable culture of bacteria described by the updating function
$$b_{t+1} = rb_t - h$$
This equation says that the bacteria have per capita reproduction r but that h are harvested each generation. Consider the case $r = 1.5$ and $h = 1.0 \times 10^6$ bacteria.
   a. Graph the updating function and cobweb starting from 3.0×10^6 and 1.0×10^6.
   b. Find the equilibrium.
   c. What is different from the lung model?

8. Consider again bacteria described by the updating function from Exercise 7:
$$b_{t+1} = rb_t - h$$
Without setting r and h to particular values, try the following:
   a. Find the equilibrium.
   b. Compare with the result from Exercise 7.
   c. Does your result make sense when $h = 0$?
   d. Does your result make sense when $r < 1$?

9. Recall the model of a lung with absorption in Exercise 5 in Section 1.10. The following steps outline construction of a general model that includes this important factor. The lung has volume V and begins with a concentration c_t. A volume W is exhaled and inhaled, and the ambient concentration is γ. An amount A of chemical is absorbed right before breathing out. All amounts are measured in millimoles and volumes in liters.
   a. Find the total amount and concentration of chemical before absorption, after absorption, after breathing out, and after breathing in.
   b. Write the updating function in this case.
   c. Set $A = 0$ and check that this matches the standard updating function for a lung, $c_{t+1} = (1-q)c_t + q\gamma$.
   d. Find the equilibrium.
   e. Does your result make sense?

10. Recall the model of a lung with absorption in Exercise 6 in Section 1.10. The following steps outline construction of a general model that includes this important factor. The lung has volume V and begins with a concentration c_t. A volume W is exhaled and inhaled, and the ambient concentration is γ. A fraction α ("alpha") of chemical is absorbed right before breathing out. All amounts are measured in millimoles and volumes in liters.
   a. Find the total amount and concentration of chemical before absorption, after absorption, after breathing out, and after breathing in.
   b. Write the updating function in this case.
   c. Set $\alpha = 0$ and check that this matches the standard updating function for a lung, $c_{t+1} = (1-q)c_t + q\gamma$.
   d. Find the equilibrium.
   e. Does your result make sense?

11. **COMPUTER:** Use your computer to graph the trigonometric updating functions from Exercise 14 in Section 1.9.
   a. $x_{t+1} = \cos(x_t)$
   b. $y_{t+1} = \sin(y_t)$
   c. $z_{t+1} = \sin(z_t) + \cos(z_t)$
   Find the equilibria of each, and produce cobweb diagrams starting from three different initial conditions (as in Exercise 14 in Section 1.9). Do the new diagrams help make sense of the solutions?

12. **COMPUTER:** Consider the updating function
$$x_{t+1} = e^{ax_t}$$
for the following values of the parameter a. Use your computer to graph the function and the diagonal to look for equilibria. Cobweb starting from $x_0 = 1$ in each case.
   a. $a = 0.3$
   b. $a = 0.4$
   c. $a = 1/e$

1.12 An Example of Nonlinear Dynamics

The updating functions we have studied in detail, for bacterial populations, tree height, mite populations, and the lung (Equations 1.1, 1.4, 1.18, and 1.44) have linear graphs. We will see that a model of two competing bacterial populations leads naturally to an updating function that is not linear. Such a **nonlinear dynamical system** can have much more complicated behavior than a linear system. For example, it can have more than one equilibrium. By comparing the two equilibria we will find in this model of competition, we will catch a glimpse of an important theme treated in Sections 3.1 and 3.2, the **stability** of equilibria.

A Model of Selection

Our original model of bacterial growth followed the population of a single type of bacteria, denoted by b_t at time t. Suppose that a mutant type with population m_t appears and begin reproducing. If the original (or **wild**) type has a per capita reproduction of 1.5 and the mutant type has a per capita reproduction of 2.0, the two populations will have updating functions

$$b_{t+1} = 1.5 b_t \quad \text{updating function for wild type} \tag{1.45}$$
$$m_{t+1} = 2.0 m_t \quad \text{updating function for mutants}$$

The per capita reproduction of the mutant type is greater than that of the wild type. The mutant type might be be better able to survive in a particular laboratory. Over time, we would expect the population to include a larger and larger proportion of mutant bacteria (Figure 1.87). The establishment of this mutant is an example of **selection**. Selection occurs when the frequency of a gene (the mutation) changes over time.

Imagine observing this mixed population for many hours. Counting all the bacteria each hour would be impossible. Nonetheless, we could track the mutant invasion by taking a sample and measuring the fraction of the mutant type by counting or using a specific stain. If this fraction becomes larger and larger, we would know that the mutant type was taking over.

How can we model the dynamics of the fraction? The vital first step is to **define a new variable**. In this case, we set p_t to be the fraction that are of the mutant type at time t. Then

$$\begin{aligned} p_t &= \frac{\text{number of mutants}}{\text{total number}} \\ &= \frac{\text{number of mutants}}{\text{number of mutants} + \text{number of wild type}} \\ &= \frac{m_t}{m_t + b_t} \end{aligned} \tag{1.46}$$

For example, if $m_t = 2.0 \times 10^5$ and $b_t = 3.0 \times 10^6$, there are a total of 3.2×10^6 bacteria. The fraction that is of the mutant type is

$$\begin{aligned} p_t &= \frac{2.0 \times 10^5}{2.0 \times 10^5 + 3.0 \times 10^6} \\ &= \frac{2.0 \times 10^5}{3.2 \times 10^6} = 0.0625 \end{aligned}$$

What is the fraction of the wild type? It is the number of wild-type bacteria divided by the total number of bacteria, or

$$\begin{aligned} \text{fraction of wild type} &= \frac{\text{number of wild type}}{\text{total number}} \\ &= \frac{\text{number of wild type}}{\text{number of mutants} + \text{number of wild type}} \\ &= \frac{b_t}{m_t + b_t} \end{aligned}$$

(Figure 1.88). If $m_t = 2.0 \times 10^5$ and $b_t = 3.0 \times 10^6$, then

$$\text{fraction of wild type} = \frac{3.0 \times 10^6}{2.0 \times 10^5 + 3.0 \times 10^6} = \frac{3.0 \times 10^6}{3.2 \times 10^6} = 0.9375$$

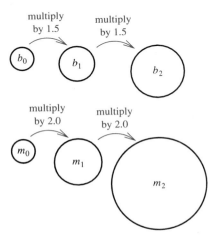

Figure 1.87
An invasion by mutant bacteria

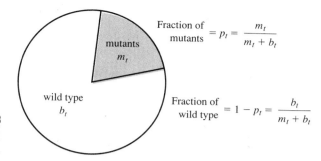

Figure 1.88
The fraction of mutant and wild-type bacteria

We do not need to give a new name to this fraction because we know that

fraction of mutants + fraction of wild type = 1

Because all bacteria are of these two types, the fractions must add up to 1. In our example, the fraction of mutants is 0.0625 and the fraction of wild type is 0.9375, which do indeed add up to 1. Solving for the fraction of wild type,

fraction of wild type = 1 − fraction of mutants
= $1 - p_t$

Putting this together with our original calculation, we find

$$\frac{b_t}{m_t + b_t} = 1 - p_t \tag{1.47}$$

Our goal is to find p_{t+1}, the fraction of the mutant type after 1 h. In our example with $m_t = 2.0 \times 10^5$ and $b_t = 3.0 \times 10^6$, the updated populations are

$$m_{t+1} = 2.0 m_t = 4.0 \times 10^5$$
$$b_{t+1} = 1.5 b_t = 4.5 \times 10^6$$

The new fraction of the mutant type, p_{t+1}, is then

$$p_{t+1} = \frac{4.0 \times 10^5}{4.0 \times 10^5 + 4.5 \times 10^6} = 0.082$$

As expected, the fraction has increased.

We can follow these same steps to find the updating function for p_t. By definition,

$$p_{t+1} = \frac{m_{t+1}}{m_{t+1} + b_{t+1}}$$

Using the updating functions for the two types (Equation 1.45), we find

$$p_{t+1} = \frac{2.0 m_t}{2.0 m_t + 1.5 b_t} \tag{1.48}$$

Although mathematically correct, this is not a satisfactory updating function. We have supposed that the actual values of m_t and b_t are impossible to measure. The updating function must give p_{t+1} in terms of p_t, which we can measure by sampling.

We can do this by using an algebraic trick: dividing the numerator and denominator by the same thing. Because the definition for p_t has the total population $m_t + b_t$ in the denominator, we divide it into the numerator and denominator, finding

$$p_{t+1} = \frac{2.0\dfrac{m_t}{m_t + b_t}}{2.0\dfrac{m_t}{m_t + b_t} + 1.5\dfrac{b_t}{m_t + b_t}}$$

Using the definition of p_t (Equation 1.46) and the formula for the fraction of wild type $1 - p_t$ (Equation 1.47), we can rewrite this as

$$p_{t+1} = \frac{2.0 p_t}{2.0 p_t + 1.5(1 - p_t)} \qquad (1.49)$$

This is the updating function we sought, giving a formula for the fraction at time $t + 1$ in terms of the fraction at time t. For example, if $p_t = 0.062$, the updating function tells us that

$$p_{t+1} = \frac{2.0 \cdot 0.062}{2.0 \cdot 0.062 + 1.5(1 - 0.062)} = 0.082$$

This matches the answer we found before but is based only on **measurable quantities**.

This calculation illustrates one of the great strengths of mathematical modeling. Our **derivation** of this measurable updating function used the values m_t and b_t, which are impossible to measure. But just because things cannot be measured does not mean they do not exist. These numbers do exist, and they can be worked with mathematically. One can think of a mathematical model as a way to "do the impossible."

The updating function for the fraction (Equation 1.49) is not linear because it involves division along with multiplication by constants and addition. The graph of the function is curved (Figure 1.89). We plotted it by substituting representative values for a fraction, which must lie between 0 and 1. The points $(0, 0)$, $(0.5, 0.57)$, and $(1, 1)$ lie on the graph, and we connected them with a smooth curve.

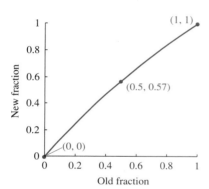

Figure 1.89
Graph of updating function from the competition model

Equilibria in the General Case

As before, we can gain a better understanding of this process by studying the general case. Suppose that the mutant type of bacteria has per capita reproduction s and the wild type has per capita reproduction r (as in the updating function in Section 1.2). The updating functions are then

$$\begin{aligned} m_{t+1} &= s m_t \\ b_{t+1} &= r b_t \end{aligned} \qquad (1.50)$$

We can follow the steps above to derive the updating function for the fraction:

$$p_{t+1} = \frac{m_{t+1}}{m_{t+1} + b_{t+1}}$$

$$= \frac{s m_t}{s m_t + r b_t}$$

$$= \frac{s\dfrac{m_t}{m_t + b_t}}{s\dfrac{m_t}{m_t + b_t} + r\dfrac{b_t}{m_t + b_t}}$$

$$= \frac{sp_t}{sp_t + r(1 - p_t)}$$

which gives the general form

$$p_{t+1} = \frac{sp_t}{sp_t + r(1 - p_t)} \tag{1.51}$$

Our previous derivation considered the case $s = 2.0$ and $r = 1.5$. With these parameters, the general form for competition of bacteria is

$$p_{t+1} = \frac{2.0 p_t}{2.0 p_t + 1.5(1 - p_t)}$$

matching what we found before.

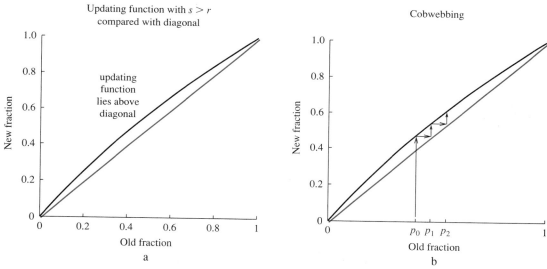

Figure 1.90
Behavior of competition equation when mutants have an advantage

When $s > r$, the graph of the updating function lies above the diagonal except at the intersection points $p_t = 0$ and $p_t = 1$ (Figure 1.90a). This means that any value of p_t between 0 and 1 will be increased by the updating function, consistent with the higher per capita reproduction of the mutants. The cobwebbing moves up, indicating this increase (Figure 1.90b).

What happens if the per capita reproduction of the wild-type exceeds that of the mutants? With $r = 2.0$ and $s = 1.5$, the updating function is

$$p_{t+1} = \frac{1.5 p_t}{1.5 p_t + 2.0(1 - p_t)}$$

The three points $(0, 0)$, $(0.5, 0.43)$, and $(1, 1)$ lie on the graph, which itself lies below the diagonal (Figure 1.91a). Values of p_t between 0 and 1 are decreased by the updating function as shown by the decreasing cobweb (Figure 1.91b). This is consistent with the lower per capita reproduction of the mutants.

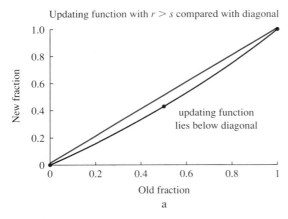
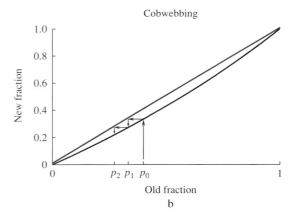

Figure 1.91

Graph of updating functions with different parameter values

Finally, what happens if the two types grow at exactly the same rate? If $r = s$, the updating function simplifies to

$$p_{t+1} = \frac{sp_t}{sp_t + r(1 - p_t)}$$
$$= \frac{sp_t}{sp_t + s(1 - p_t)}$$
$$= \frac{sp_t}{sp_t + s - sp_t}$$
$$= \frac{sp_t}{s}$$
$$= p_t .$$

In this case, the updating function is precisely the "do nothing" function. The fraction never changes, no matter what it is. When both types reproduce at the same rate, the fraction of the mutant neither increases nor decreases and every value of p_t is an equilibrium. This makes biological sense; there is no selection when the two types have the same growth rate. This does not say that the *total number* of bacteria is unchanged; if $r = s = 2.0$, the total number will double each hour. The *fraction* of mutants remains the same.

We can use the five steps of Algorithm 1.5 to find the equilibria:

$p^* = \dfrac{sp^*}{sp^* + r(1 - p^*)}$	the equation for the equilibrium
$\left[sp^* + r(1 - p^*)\right] p^* = sp^*$	multiply both sides by the denominator
$\left[sp^* + r(1 - p^*)\right] p^* - sp^* = 0$	move everything to one side
$(sp^* + r - rp^*)p^* - sp^* = 0$	multiply out inner term
$sp^{*2} + rp^* - rp^{*2} - sp^* = 0$	multiply out all terms
$sp^{*2} - sp^* + rp^* - rp^{*2} = 0$	collect like terms
$(s - r)p^*(p^* - 1) = 0$	factor
$s - r = 0$ or $p^* = 0$ or $p^* = 1$	set each factor to 0
$s = r$ or $p^* = 0$ or $p^* = 1$	solve each term

Factoring involves some tricky algebra, which is worth working through yourself.

1.12 An Example of Nonlinear Dynamics

What do these three equilibria mean? If $s = r$, the updating function does nothing and every value of p_t is an equilibrium. Otherwise, the equilibria are at $p_t = 0$ and $p_t = 1$. When $p_t = 0$, the population consists entirely of the wild type. Because our model includes no mutation or immigration, there is nowhere for the mutant type to arise. Similarly, when $p_t = 1$, the population consists entirely of the mutant type and the wild type will never arise. These equilibria correspond to the extinction equilibrium for a population of one type of bacteria. At $p_t = 0$ the mutants are extinct, and at $p_t = 1$ the wild-type bacteria are extinct.

Figure 1.92

Solution of the competition model starting near $p_0 = 0$

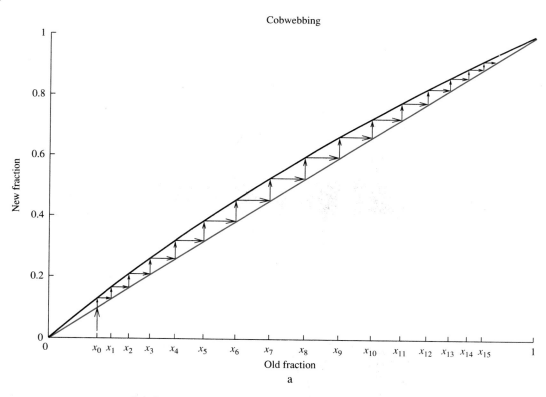
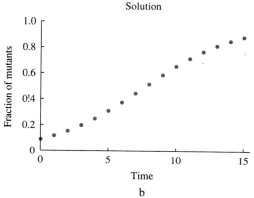

Stable and Unstable Equilibria

Figure 1.92 shows many steps of cobwebbing the updating function with $s = 2.0$ and $r = 1.5$, starting near the equilibrium $p^* = 0$. The solution moves slowly away from 0, swiftly through the halfway point at $p_t = 0.5$, and then slowly approaches the other equilibrium at $p_t = 1$.

If we started *exactly* at $p_0 = 0$, the solution would remain at $p_t = 0$ for all times t. Similarly, if we started *exactly* at $p_t = 1$, the solution would remain at $p_t = 1$ for all times t. The two equilibria behave quite differently, however, if our starting point is nearby. A solution starting *near* $p_0 = 0$ moves steadily *away* from the equilibrium (Figure 1.92). A solution starting *near* $p_0 = 1$ moves *toward* the equilibrium (Figure 1.93).

Figure 1.93

Solution of the competition model starting near $p_0 = 1$

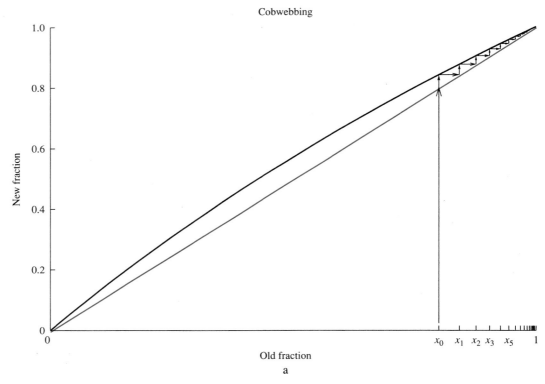

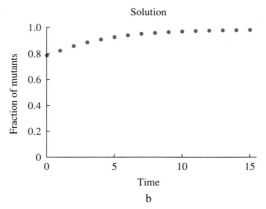

This situation is analogous to keeping a ball from rolling around on a surface. If we place it on the bottom of a small depression, it will remain at an equilibrium (Figure 1.94). If it is moved slightly away from this equilibrium, it will come back, much like the solution that starts near the equilibrium $p^* = 1$ (Figure 1.93). An equilibrium with this property is called **stable**. Similarly, if we place the ball exactly on top of a small hill, it will remain there (Figure 1.94). However, if it is moved even slightly away from this equilibrium, it will roll farther and farther away, much like the solution that starts near the equilibrium $p^* = 0$ (Figure 1.92). An equilibrium with this property is called **unstable**.

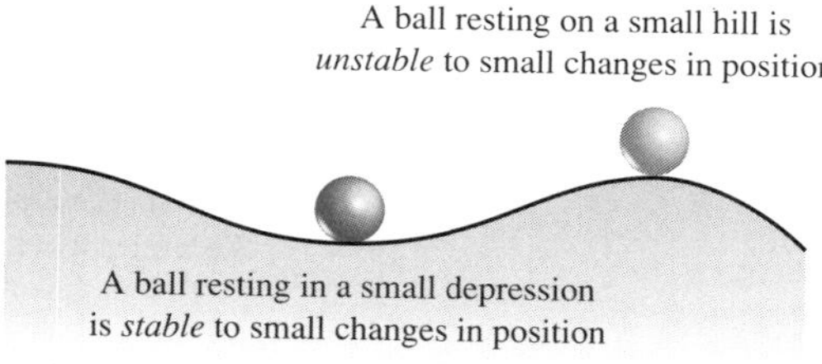

Figure 1.94
Stable and unstable resting points for a ball

A ball resting on a small hill is *unstable* to small changes in position

A ball resting in a small depression is *stable* to small changes in position

This leads us to an informal definition of stable and unstable equilibria.

■ **Definition 1.5** An equilibrium is **stable** if solutions that begin near that equilibrium stay near, or approach, that equilibrium. An equilibrium is **unstable** if solutions that begin near that equilibrium move away from that equilibrium. ■

We will soon derive powerful methods to analyze updating functions and determine whether their equilibria are stable or unstable. Because these techniques require the **derivative**, a central idea from calculus, we first study the foundational notions of limits and rate of change.

SUMMARY

As an example of a **nonlinear dynamical system**, an updating function with a curved graph, we derived the equation for the fraction of mutants invading a population of wild-type bacteria. This dynamical system, unlike the linear ones studied hitherto, has two equilibria. One of these equilibria is **unstable**; solutions starting nearby move farther and farther away. The other is **stable**; solutions starting nearby move closer and closer to the equilibrium.

1.12 EXERCISES

1. Find p_t, m_{t+1}, b_{t+1}, and p_{t+1} in the following situations.
    a. $s = 1.2, r = 2.0, m_t = 1.2 \times 10^5, b_t = 3.5 \times 10^6$
    b. $s = 1.2, r = 2.0, m_t = 1.2 \times 10^5, b_t = 1.5 \times 10^6$
    c. $s = 1.8, r = 0.8, m_t = 1.2 \times 10^5, b_t = 3.5 \times 10^6$
    d. $s = 0.3, r = 0.5, m_t = 1.2 \times 10^5, b_t = 3.5 \times 10^6$
    e. $s = 1.8, r = 1.8, m_t = 1.2 \times 10^5, b_t = 3.5 \times 10^6$

2. Find and graph the updating functions for the following cases (used in Exercise 1). Find the equilibria. Cobweb starting from $p_0 = 0.1$ and $p_0 = 0.9$. Which equilibria seem to be stable in each case?
    a. $s = 1.2, r = 2.0$ b. $s = 1.8, r = 0.8$
    c. $s = 0.3, r = 0.5$. What is the difference from part **a**?
    d. $s = 1.8, r = 1.8$

3. For each of the following updating functions (from Figure 1.79), indicate which of the equilibria are stable and which are unstable.

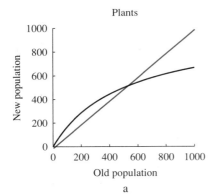

a

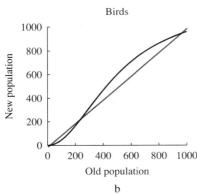

b

4. One evolutionary force we ignored in this section is mutation. We begin with a model that considers only mutation without reproduction. Suppose that each hour 20% of wild-type bacteria transform into mutants and that 10% of mutants transform back into wild type ("revert"). Suppose that we begin with 1.0×10^6 wild type and 1.0×10^5 mutants. (It might help to draw a picture illustrating this process.)
   a. Find the number of wild-type bacteria that mutate and the number of mutants that revert.
   b. Find the number of wild-type bacteria and the number of mutants after mutation and reversion.
   c. Find the total number of bacteria before and after mutation. Why is it the same?
   d. Find the fraction of mutants before and after mutation.

5. We will now find the updating function describing the fraction of mutants and wild type of the bacteria in Exercise 4. Suppose that 20% mutate and 10% revert, but that the numbers of wild type and mutants are b_t and m_t, respectively.
   a. Find the number of wild-type bacteria that mutate and the number of mutants that revert.
   b. Find the number of wild-type bacteria and the number of mutants after mutation and reversion.
   c. Find the total number of bacteria before and after mutation.
   d. Find the fraction of mutants before and after mutation. (Divide m_{t+1} by $b_{t+1} + m_{t+1}$ to find p_{t+1}, and use the fact that $b_{t+1} + m_{t+1} = b_t + m_t$ when there is no reproduction.)
   e. Graph the updating function and find the equilibrium fraction of mutants.
   f. Cobweb starting from the initial condition in Exercise 4. Is the equilibrium stable? What will happen to the fraction of mutants?

6. Follow the steps in Exercise 5 to find the updating function when the fraction of wild type that mutate is μ ("mu") and the fraction of mutants that revert is ν ("nu").
   a. Find the number of wild-type bacteria that mutate and the number of mutants that revert.
   b. Find the number of wild-type bacteria and the number of mutants after mutation and reversion.
   c. Find the total number of bacteria before and after mutation.
   d. Find the fraction of mutants before and after mutation.
   e. Find the equilibrium fraction of mutants.
   f. Under what condition is the equilibrium fraction exactly 0.5?

7. The model describing the dynamics of the concentration of medication in the bloodstream,
$$M_{t+1} = 0.5 M_t + 1.0$$
becomes nonlinear if the fraction of medication used by the body is a function of the concentration. In the basic model, half is used no matter how much there is. Suppose instead that the fraction used is a **decreasing function** of the concentration.
   a. Write the basic updating function as

   new concentration
   = old concentration − fraction used
   × old concentration + supplement

   and figure out what each term is.
   b. Consider the expression
   $$\text{fraction used} = \frac{0.5}{1.0 + 0.1 M_t}$$
   Is this fraction larger or smaller than the basic value of 0.5? What happens to the fraction used as M_t becomes large? What does this mean in biological terms?
   c. Substitute this expression to write the updating function.
   d. Solve for the equilibrium. Can you explain why it is larger?

8. We will now generalize the situation in Exercise 7. Suppose that
$$\text{fraction used} = \frac{0.5}{1.0 + \alpha M_t}$$
for some parameter α.
   a. Write the updating function.
   b. Solve for the equilibrium.
   c. Plug in values to sketch a graph of the equilibrium as a function of α. What happens when $\alpha > 0.5$? Can you explain this in biological terms?

d. Cobweb starting from $M_0 = 1.0$ for $\alpha = 0.1$ and $\alpha = 1.0$. Does your cobweb diagram with $\alpha = 1.0$ remind you of another model without an equilibrium?

9. The model of selection studied in this section is similar in many ways to a model of migration. The next three problems study the following situation. Suppose two nearby islands have populations of butterflies, with x_t on the first island and y_t on the second. Each year, 20% of the butterflies from the first island fly to the second and 30% of the butterflies from the second fly to the first. We wish to find an updating function for the *fraction* of the butterflies on the first island.

a. Suppose there are 100 butterflies on each island at time $t = 0$. How many are on each island at $t = 1$? At $t = 2$?
b. Find an equation for x_{t+1} in terms of x_t and y_t.
c. Find an equation for y_{t+1} in terms of x_t and y_t.
d. Divide both sides of the updating function for x_t by $x_{t+1} + y_{t+1}$ to find an updating function for the fraction p_t on the first island.
e. Compare the results of this updating function with the results from part **a** (start from the initial condition $p_0 = 0.5$).
f. Sketch the updating function and cobweb starting from $p_0 = 0.1$.
g. Find the equilibrium. Does the value make sense?

10. Consider again the situation in Exercise 9, but suppose that the butterflies that begin the year on the first island reproduce after migration (making one additional butterfly). Those that begin the year on the second island do not reproduce. Do parts **a** through **f** of Exercise 9 for this situation.

11. Consider again the situation in Exercise 9, but suppose the butterflies that do not migrate reproduce (making one additional butterfly each) and those that do migrate fail to reproduce. Do parts **a** through **f** of Exercise 9 for this situation.

12. A widely used nonlinear model of competition is the "logistic" model. In this model, bacteria produce fewer offspring in large populations, unlike the updating function $b_{t+1} = rb_t$, which describes a situation in which bacteria produce r offspring whatever the population size. Try the following steps in general, but if you get stuck, try with $r = 1.5$ and $K = 1.0 \times 10^8$ bacteria.
a. Suppose the per capita production of bacteria is r when the bacterial population is near 0 and decreases linearly as the population gets larger. Suppose per capita reproduction reaches 0 when the bacterial population has size K; K is known as the **carrying capacity**. Graph per capita reproduction as a function of population size. Find the equation of this line.
b. The updated population is equal to the per capita reproduction times the population size. Use this to write the updating function.
c. Graph the updating function. The function is decreasing when b_t is large. What does this mean?
d. The graph of the updating function takes on negative values for $b_t > K$. Does this make sense? What is a more reasonable model when $b_t > K$?

13. COMPUTER: Consider the updating function

$$f(x) = rx(1 - x)$$

as in Exercise 12 but with $K = 1$. Plot the updating function, and have your computer find solutions for 50 steps starting from $x_0 = 0.3$ for the following values of r:
a. Some value of r between 0 and 1. What is the only equilibrium?
b. Some value of r between 1 and 2. Where are the equilibria? Which one seems to be stable?
c. Some value of r between 2 and 3. Where are the equilibria? Which one seems to be stable?
d. Try several values of r between 3 and 4. What is happening to the solution? Is there any stable equilibrium?
e. The solution is **chaotic** when $r = 4$. One property of chaos is **sensitive dependence on initial conditions**. Compare a solution starting at $x_0 = 0.3$ with one starting at $x_0 = 0.30001$. Even though they start off very close, they soon separate and become completely different. Why might this be a problem for a scientific experiment?

14. COMPUTER: Consider the equation describing the dynamics of selection (Equation 1.51)

$$p_{t+1} = \frac{sp_t}{sp_t + r(1 - p_t)}$$

Suppose you have two bacteria cultures, 1 and 2. In culture 1, the mutant does better than the wild type, and in culture 2 the wild type does better. In particular, suppose that $s = 2.0$ and $r = 0.3$ in culture 1, and that $s = 0.6$ and $r = 2.0$ in culture 2. Define updating functions f_1 and f_2 to describe the dynamics in the two cultures.

a. Graph the functions f_1 and f_2 along with the identity function. Find the first five values of solutions starting from $p_0 = 0.02$ and $p_0 = 0.98$ in each culture. Explain in words what each solution shows and why.
b. Suppose you change the experiment. Begin by taking a population with a fraction p_0 of mutants. Split this population in half, and place one half in culture 1 and the other half in culture 2. Let the bacteria reproduce once in each culture, and then mix them together. Split the mix in half and repeat the process. The updating function is

$$f(p) = \frac{f_1(p) + f_2(p)}{2}$$

Can you derive this? Plot this updating function along with the identity function. Have your computer find the equilibria, and label them on your graph. Do they make sense?
c. Use cobwebbing to figure out which equilibria are stable.
d. Find one solution starting from $p_0 = 0.001$ and another starting from $p_0 = 0.999$. Are these results consistent with the stability of the equilibria? Explain in words why the solutions behave as they do. Why don't they move toward $p = 0.5$?

1.13 Excitable Systems: The Heart

We can use cobwebbing and equilibria to study a simplified model of the heart. Our goal is to understand how simple changes in the parameters of a heart can produce heartbeat patterns called **second-degree block** or the **Wenckebach phenomenon**. These syndromes cause people's hearts either to beat half as often as they should or to beat normally for a while, skip a beat, and return to beating normally. We will see that we can understand these problems, and the normal heartbeat, as related phenomena.

A Simple Heart

Figure 1.95 shows the basic apparatus for beating of the heart. The sinoatrial (SA) node is the pacemaker, which sends regular signals to the atrioventrical (AV) node. The AV node then tells the heart to beat if conditions are suitable.

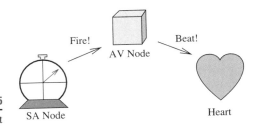

Figure 1.95
A mathematician's version of the heart

The AV node can be thought of as keeping track of the condition of the heart using an electrical potential. Denote the potential after it responds to a signal from the SA node as V_t. Two processes go into updating this potential. During the time τ between signals from the SA node, the electrical potential of the AV node decays exponentially at rate α. Setting $\hat{V}_t$ to be the potential of the AV node just before it receives the next signal from the SA node, we have

$$\hat{V}_t = e^{-\alpha\tau} V_t \qquad (1.52)$$

(Figure 1.96). Whether the heart beats depends on the response of the AV node when a signal arrives (Figure 1.97). If the potential $\hat{V}_t$ is too high, the heart has not had enough time to recover from the last beat and the AV node ignores the signal. Otherwise, the AV node accepts the signal, tells the heart to beat, and increases its potential by u.

Let V_c be the threshold potential. If $\hat{V}_t > V_c$, the heart is not ready to beat and

$$V_{t+1} = \hat{V}_t \qquad \text{if } \hat{V}_t > V_c \qquad (1.53)$$

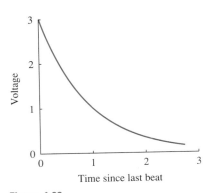

Figure 1.96
The exponential decay of voltage between beats

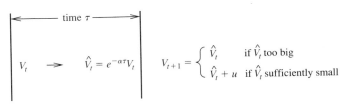

Figure 1.97
Schematic diagram of the potential of the AV node

$$V_{t+1} = \begin{cases} \hat{V}_t & \text{if } \hat{V}_t \text{ too big} \\ \hat{V}_t + u & \text{if } \hat{V}_t \text{ sufficiently small} \end{cases}$$

If $\hat{V}_t \leq V_c$, the AV node responds and tells the heart to beat and

$$V_{t+1} = \hat{V}_t + u \qquad \text{if } \hat{V}_t \leq V_c \tag{1.54}$$

To translate this description into a discrete-time dynamical system, we must write V_{t+1} entirely in terms of V_t, eliminating the $\hat{V}_t$ terms. Because $\hat{V}_t = e^{-\alpha \tau} V_t$, the two cases can be summarized as

$$V_{t+1} = \begin{cases} e^{-\alpha \tau} V_t & \text{if } e^{-\alpha \tau} V_t > V_c \\ e^{-\alpha \tau} V_t + u & \text{if } e^{-\alpha \tau} V_t \leq V_c \end{cases} \tag{1.55}$$

This updating function is graphed in Figure 1.98a.

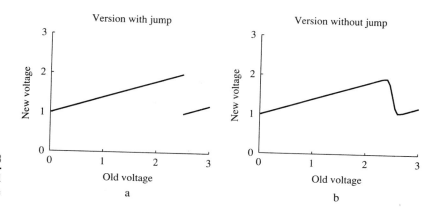

Figure 1.98
The updating function for the potential of the AV node

This function, unlike those we have studied hitherto, has a jump (where $e^{-\alpha \tau} V_t = V_c$). This jump reflects the sharp response threshold. In the real heart, the threshold is not precise and the two branches of the updating function are connected (Figure 1.98b).

We can use the graphical method for finding equilibrium as intersections with the diagonal to study this updating function. Each piece of the updating function is a straight line with slope $e^{-\alpha \tau} < 1$. There are two possible pictures: Either the upper branch of the updating function crosses the diagonal at an equilibrium (Figure 1.99a) or the diagonal sneaks through the gap between the two branches and there is no equilibrium (Figure 1.99b).

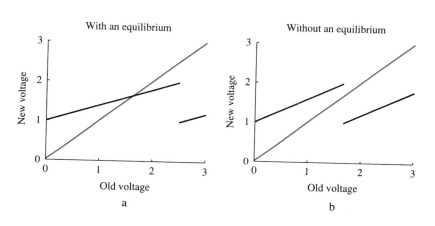

Figure 1.99
The heart updating function with and without an equilibrium

What does equilibrium mean? We sometimes think of an equilibrium as a point where nothing happens. In the present case, an equilibrium represents a value of the potential where the decay (by a factor of $e^{-\alpha\tau}$) is exactly balanced by the response to the signal (an increase of u). This means that the heart will beat steadily.

What are the algebraic conditions for an equilibrium? An equilibrium is a value of V_t that solves $V_{t+1} = V_t$. The heart must proceed through the cycle

$$V_t \stackrel{\text{decay}}{\to} \hat{V}_t = e^{-\alpha\tau}V_t \stackrel{\text{signal}}{\to} e^{-\alpha\tau}V_t + u = V_t$$

and end up where it started. Setting V^* to be the equilibrium, we can solve

$$V^* = e^{-\alpha\tau}V^* + u$$

to find

$$V^* = \frac{u}{1 - e^{-\alpha\tau}} \qquad (1.56)$$

This equilibrium is correct only if the heart is indeed ready to beat when the next signal comes, or if

$$e^{-\alpha\tau}V^* = e^{-\alpha\tau}\frac{u}{1 - e^{-\alpha\tau}} \leq V_c \qquad (1.57)$$

For example, suppose that $u = 1$, $V_c = 1$, and $\tau = 1$. If $\alpha = \ln(3) = 1.099$, then $e^{-\alpha\tau} = 1/3$. The equilibrium is

$$V^* = \frac{1}{1 - 1/3} = 1.5$$

This equilibrium is correct because $e^{-\alpha\tau}V^* = 0.5$ is less than $V_c = 1$. The heart will beat every time, with voltage decaying from 1.5 to 0.5 between beats and increasing back to 1.5 on the beat (Figure 1.100).

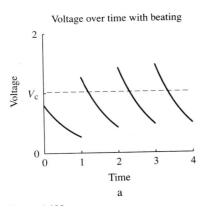
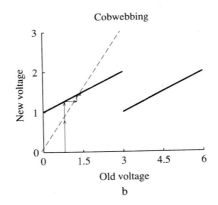
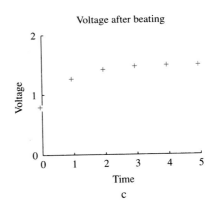

Figure 1.100

The behavior of a heart with an equilibrium

What happens if α, the recovery rate, becomes smaller? If $\alpha = \ln(1.5) = 0.405$, then $e^{-\alpha\tau} = 2/3$. The equilibrium is

$$V^* = \frac{1}{1 - 2/3} = 3.0$$

This equilibrium is correct only if $e^{-\alpha\tau}V^* = 2.0$ is less than $V_c = 1.0$. Because it is not, the heart cannot beat every time. If α is too small, the AV node recovers too slowly from one signal to be ready to respond to the next. Similarly, if the time τ between beats is decreased by too much, the heart might not have time to recover and there will be no equilibrium. The AV node cannot respond to every signal when signals from the SA node arrive too frequently. The more complicated dynamics that result are our next topic.

Second Degree Block

When the heart fails to beat in response to every signal from the SA node, the condition is called **second degree block**. In one type, called **2:1 AV block**, the heart beats only with every other stimulus. In another, called the **Wenckebach phenomenon**, the heart beats normally for a while, skips a beat, and then resumes normal beating and repeats the cycle. Our model of the heart can help us understand these phenomena.

Graphically, 2:1 AV block corresponds to the situation in Figure 1.101. There is no equilibrium. The potential of the AV node alternates between a high value and a low value. When high, the potential does not decay sufficiently to respond to the next signal. After another cycle (time τ), however, the potential has reached a low enough value to respond.

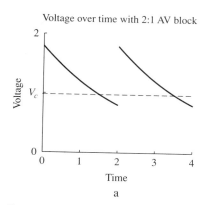
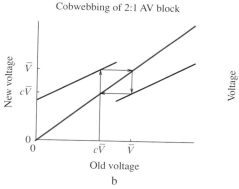
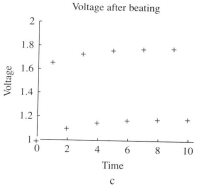

Figure 1.101
The dynamics of 2:1 AV block

To find the conditions for 2:1 AV block, we use techniques similar to those used to find an equilibrium. For convenience we substitute the new parameter c for $e^{-\alpha\tau}$ in this subsection and the next. A value of c near 1 means that the potential decays very little, and a value of c near 0 means that the potential decays a great deal. Suppose the potential is $\bar{V}$ just after a beat. If the node responds to the second signal but not the first, we have

$$\bar{V} \xrightarrow{\text{decay}} c\bar{V} \xrightarrow{\text{signal}} c\bar{V} \xrightarrow{\text{decay}} c^2\bar{V} \xrightarrow{\text{signal}} c^2\bar{V} + u \quad (1.58)$$

If the potential after these two full cycles comes back exactly to where it started, the heart beats with every other signal, producing 2:1 AV block. The updated potential after two cycles matches the original potential if

$$c^2\bar{V} + u = \bar{V}$$
$$c\bar{V} > V_c \quad (1.59)$$
$$c^2\bar{V} < V_c$$

110 Chapter 1 Introduction to Discrete-Time Dynamical Systems

which has solution

$$\bar{V} = \frac{u}{1-c^2} \quad (1.60)$$

if the inequalities are satisfied. Note the similarity to the equation for an equilibrium for this model (Equation 1.56).

In Figure 1.101, we have set $u = V_c = 1.0$ and $c = 2/3$, corresponding to the second case considered above. We have seen that there is no equilibrium. Equation 1.60 implies that $\bar{V} = 1.8$. We can follow the dynamics through a complete cycle, finding

$$\bar{V} = 1.8 \xrightarrow{\text{decay}} c\bar{V} = 1.2 \xrightarrow{\text{signal}} c\bar{V} = 1.2 \xrightarrow{\text{decay}} c^2\bar{V} = 0.8 \xrightarrow{\text{signal}} c^2\bar{V} + u = 1.8$$

The AV node does not respond to the first signal because $1.2 > V_c = 1.0$, but it is ready to respond to the second and return to its original potential.

The Wenckebach Phenomenon

With the parameter values $u = V_c = 1$, we can compute the conditions on c for the existence of an equilibrium. The equation for an equilibrium is

$$V^* = \frac{1}{1-c}$$

requiring that $cV^* \leq 1$ (Equation 1.57). The equilibrium at V^* exists only if

$$\frac{c}{1-c} \leq 1$$

We can solve for c, finding

$$c \leq 1 - c$$
$$2c \leq 1$$
$$c \leq 0.5$$

The value $c = 1/3$ that produced an equilibrium and normal beating (Figure 1.100) is well below this value. The value $c = 2/3$ that produced 2:1 AV block (Figure 1.101) is well above this value. What happens if c is only slightly above 0.5 and the heart can nearly recover?

Figure 1.102

The Wenckebach phenomenon

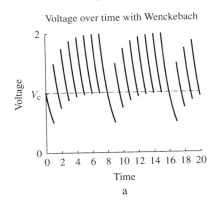

a

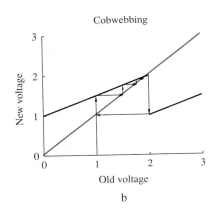

b

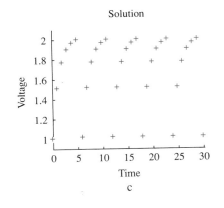

c

Figure 1.102 shows the behavior of the system when $c = 0.5001$, a hair above the threshold for existence of an equilibrium. The heart beats about 8 times, building up to a higher and higher potential. Eventually, the potential becomes too high, the AV node cannot recover, and the heart fails to beat. After this rest, the potential drops and the process begins again. This is the Wenckebach phenomenon.

Actual measurements of the Wenckebach phenomenon correspond in part to this model but show that the heart beats a bit more slowly before missing a beat. Why might this be the case? Our model assumes that the SA node sends out precise pulses at precise times. If the signals from the SA node take a little while to build up, an AV node at low potential will respond right at the beginning of a signal from the SA node. An AV node close to the threshold will be slower and might respond near the end of the signal from the SA node, delaying the heart beat slightly. This slowing indicates that the AV node will soon exceed the threshold and that the heart will miss a beat.

SUMMARY

A simplified model of the heart includes two phases, decay of potential in the AV node (recovery from the last beat) and response to a rhythmic signal from the SA node. We derived conditions for the heart to beat properly with each signal and showed that if the recovery time is not long enough, two types of **second degree block** can result. In the first, **2:1 AV block**, the heart beats with every other signal. In the second, the **Wenckebach phenomenon**, the heart misses a beat every once in a while.

1.13 EXERCISES

1. For the following circumstances, compute $\hat{V}_t$ and V_{t+1} and state whether the heart will beat.
   a. $V_c = 20.0$ mV, $u = 10.0$ mV, $c = 0.5$, $V_t = 30.0$ mV
   b. $V_c = 20.0$ mV, $u = 10.0$ mV, $c = 0.6$, $V_t = 30.0$ mV
   c. $V_c = 20.0$ mV, $u = 10.0$ mV, $c = 0.7$, $V_t = 30.0$ mV
   d. $V_c = 20.0$ mV, $u = 10.0$ mV, $c = 0.8$, $V_t = 30.0$ mV

2. Describe the long-term dynamics for each case in Exercise 1. Find which cases will beat every time, which display 2:1 AV block and which show some sort of Wenckebach phenomenon.

3. For the circumstances of Exercise 1 (except for the values of c), state whether the heart will beat every time with the following values of α and τ.

   a. $\alpha = 1.0$, $\tau = 1.0$
   b. $\alpha = 1.0$, $\tau = 0.5$
   c. $\alpha = 2.0$, $\tau = 0.5$
   d. $\alpha = 0.5$, $\tau = 0.5$

4. **COMPUTER:** Study the dynamics of Exercise 3 for values of c ranging from 0.4 to 1.0. Are there any cases where the behavior looks like neither 2:1 AV block nor the Wenckebach phenomenon? How would you describe these behaviors?

5. **COMPUTER:** What happens to the dynamics of the example illustrated in Figure 1.102 if c is made even closer to 0.5? What does it look like on a cobwebbing diagram? If $c = 0.5000000000001$, do you think it would be possible to distinguish the Wenckebach phenomenon from normal beating? Is it?

Supplementary Problems for Chapter 1

EXERCISE 1
Suppose you have a culture of bacteria, where the density of each bacterium is 2.0 g/cm³.

a. If each bacterium is 5 μm × 5 μm × 20 μm in size, find the number of bacteria if their total mass is 30 g. Recall that 1 μm = 10^{-6} m.

112 Chapter 1 Introduction to Discrete-Time Dynamical Systems

c. Suppose you learn that the sizes of bacteria range from $4\,\mu m \times 5\,\mu m \times 15\,\mu m$ to $5\,\mu m \times 6\,\mu m \times 25\,\mu m$. What is the range of the possible number of bacteria making up the total mass of 30 g?

EXERCISE 2
Consider the functions $f(x) = e^{-2x}$ and $g(x) = x^3 + 1$.
   a. Find the inverses of f and g, and use these to find when $f(x) = 2$ and when $g(x) = 2$.
   b. Find $f \circ g$ and $g \circ f$ and evaluate each at $x = 2$.
   c. Find the inverse of $g \circ f$. What is the domain of this function?

EXERCISE 3
Suppose the number of bacteria in culture is a linear function of time.
   a. If there are 2.0×10^8 bacteria in your lab at 5 P.M. on Tuesday and 5.0×10^8 bacteria the next morning at 9 A.M., find the equation of the line describing the number of bacteria in your culture as a function of time.
   b. At what time will your culture have 1.1×10^9 bacteria?
   c. The lab across the hall has a bacterial culture where the number of bacteria is a linear function of time. If they have 2.0×10^8 bacteria at 5 P.M. on Tuesday and 3.4×10^8 bacteria the next morning at 9 A.M., when will your culture have twice as many bacteria as theirs?

EXERCISE 4
The lab upstairs has a culture of a new kind of bacteria in which each individual takes 2 h to split into three bacteria. Suppose that these bacteria never die and that all offspring are viable.
   a. Write the updating function describing this system.
   b. Suppose there are 2.0×10^7 bacteria at 9 A.M. How many will there be at 5 P.M.?
   c. Write an equation for how many bacteria there are as a function of how long the culture has been running.
   d. When will this population reach 10^9?

EXERCISE 5
The number of bacteria in a lab downstairs is found to be

Time (t)	Number (b_t) (millions)
0.0 h	1.5
1.0 h	3.0
2.0 h	4.5
3.0 h	5.0
4.0 h	7.5
5.0 h	9.0

   a. Graph these points.
   b. Find the line connecting them and the time t^* at which the value does not lie on the line.
   c. Find the equation of the line, and use it to find what the value at t^* would have to be to lie on the line.
   d. How many bacteria would you expect at time 7.0 h?

EXERCISE 6
The number of bacteria in another lab downstairs follows the updating function

$$b_{t+1} = \begin{cases} 2.0 b_t & b_t \leq 1.0 \\ -0.5(b_t - 1.0) + 2.0 & b_t > 1.0 \end{cases}$$

where t is measured in hours and b_t in millions of bacteria.
   a. Graph this updating function. For what values of b_t does it make sense?
   b. Find the equilibrium.
   c. Cobweb starting from $b_0 = 0.4$ million bacteria. What do you think happens to this population?

EXERCISE 7
Convert the following angles from degrees to radians, and find the sine and cosine of each. Plot the related point both on a circle and on a graph of the sine or cosine.
   a. $\theta = 60°$
   b. $\theta = -60°$
   c. $\theta = 110°$
   d. $\theta = -190°$
   e. $\theta = 1160°$

EXERCISE 8
Suppose the temperature H of a bird follows the equation

$$H = 38.0 + 3.0 \cos\left[\frac{2\pi(t - 0.4)}{1.2}\right]$$

where t is measured in days and H is measured in °C.
   a. Sketch a graph of the temperature of this bird.
   b. Write the equation if the period changes to 1.1 days. Sketch a graph.
   c. Write the equation if the amplitude increases to 3.5°. Sketch a graph.
   d. Write the equation if the average decreases to 37.5°. Sketch a graph.

EXERCISE 9
The butterflies on a particular island are not doing well. Each autumn, every butterfly produces, on average, 1.2 eggs and then dies. Half these eggs survive the winter and produce new butterflies by late summer. At this time, 1000 butterflies arrive from the mainland to escape overcrowding.
   a. Write the updating function for the population on this island.
   b. Graph the updating function and cobweb starting from 1000.
   c. Find the equilibrium number of butterflies.

EXERCISE 10
A culture of bacteria has mass 3.0×10^{-3} g and consists of spherical cells of mass 2.0×10^{-10} g and density 1.5 g/cm^3.
   a. How many bacteria are in the culture?
   b. What is the radius of each bacterium?
   c. If the bacteria were mashed into mush, how much volume would they occupy?

EXERCISE 11
A person develops a small liver tumor. It grows according to

$$S(t) = S(0) e^{\alpha t}$$

where $S(0) = 1.0$ g and $\alpha = 0.1$ per day. At time $t = 30$ days, the tumor is detected and treatment begins. The size of the tumor then decreases linearly with slope of -0.4 g/day.
  a. Sketch a graph of the size of the tumor over time.
  b. When will the tumor disappear completely?

EXERCISE 12
Two similar objects are left to cool for 1 h. One starts at 80°C and cools to 70°C; the other starts at 60°C and cools to 55°C. Suppose the general updating function for cooling objects is linear.
  a. Find the updating function. Find the temperature of the first object after 2 h. Find the temperature after 1 h of an object that starts at 20°C.
  b. Graph the updating function and cobweb starting from 80°C.
  c. Find the equilibrium. Explain in words what the equilibrium means.

EXERCISE 13
A culture of bacteria increases in area by 10% each hour. Suppose the area is 2.0 cm² at 2:00 P.M.
  a. What will the area be at 5:00 P.M.?
  b. What was the area at 1:00 P.M.?
  c. If all bacteria are the same size and each adult produces two offspring each hour, what fraction of offspring must survive?
  d. If the culture medium is only 10 cm² in size, when will it be full?
  e. Write the relevant updating function and cobweb starting from the initial condition 2.0 cm.

EXERCISE 14
Candidates Dewey and Howe are competing for fickle voters. Each of the 100,000 people registered to vote in the election will vote for one of these two candidates. Each week, some voters switch their allegiance. Of Dewey's supporters, 20% switch to Howe each week. Howe's supporters are more likely to switch when Dewey is doing well: The fraction switching from Howe to Dewey is proportional to Dewey's percentage of the vote—none switch if Dewey commands 0% of the vote, and 50% switch if Dewey commands 100% of the vote. Suppose Howe starts with 90% of the vote.
  a. Find the number of votes Dewey and Howe have after a week.
  b. Find Dewey's percentage after a week.
  c. Find the updating function describing Dewey's percentage.
  d. Graph the updating function and find the equilibrium or equilibria.
  e. Who will win the election?

EXERCISE 15
A certain bacterial population exhibits the following odd behavior: If the population is less than 1.5×10^8 in a given generation, each bacterium produces two offspring. If the population is greater than or equal to 1.5×10^8 in a given generation, it will be exactly 1.0×10^8 in the next.

  a. Graph the updating function.
  b. Cobweb starting from an initial population of 10^7.
  c. Graph a solution starting from an initial population of 10^7.
  d. Find the equilibrium or equilibria of this population.

EXERCISE 16
An organism is breathing a chemical that modifies the depth of its breaths. In particular, suppose that the fraction q of air exchanged is given by

$$q = \frac{c_t}{c_t + \gamma}$$

where γ is the ambient concentration and c_t is the concentration in the lung. After a breath, a fraction q of the air came from outside, and a fraction $1 - q$ remained inside. Suppose $\gamma = 0.5$ mol/L.
  a. Describe in words the breathing of this organism.
  b. Find the updating function for the concentration in the lung.
  c. Find the equilibrium or equilibria.

EXERCISE 17
Lint is building up in a dryer. With each use, the old amount of lint x_t is divided by $1 + x_t$ and 0.5 lintons (the units of lint) are added.
  a. Find and graph the updating function.
  b. Cobweb starting from $x_0 = 0$. Graph the associated solution.
  c. Find the equilibrium or equilibria.

EXERCISE 18
Suppose people in a bank are waiting in two lines. Each minute several things happen: Some people are served, some people join the lines, and some people switch lines. Suppose that each minute 1/10 of the people in the first line are served and 3/10 of the people in the second line are served. Suppose the number of people who join each line is equal to 1/10 of the total number of people in both lines, and that 1/10 of the people in each line switch to the other.
  a. Suppose there are 100 people in each line at the beginning of a minute. How many people are in each line at the end of the minute?
  b. Write an updating function for the number of people in the first line and another updating function for the number of people in the second.
  c. Write an updating function for the fraction of people in the first line.

EXERCISE 19
A gambler faces off against a very poor casino. She begins with $1000 and the casino with $11,000. In each round, the gambler loses 10% of her current funds to the casino and the casino loses 2% of its current funds to the gambler.
  a. Find the amount of money each has after one round.
  b. Find an updating function for the amount of money the gambler has and another for the amount of money the casino has.

c. Find the updating function for the fraction p of money the gambler has.
d. Find the equilibrium fraction of the money held by the gambler.
e. Using the fact that the total amount of money is constant, find the equilibrium amount of money held by the gambler.

EXERCISE 20
Let V represent the volume of a lung and c the concentration of some chemical inside the lung. Suppose the internal surface area is proportional to volume, and a lung with volume 400 cm^3 has a surface area of 100 cm^2. The lung absorbs the chemical at a rate per unit surface area given by

$$R = \alpha \frac{c}{4.0 \times 10^{-2} \text{ mol} + c}$$

Time is measured in seconds, surface area in square centimeters and volume in cubic centimeters. The parameter α takes on the value 6.0 in the appropriate units.
a. Find surface area as a function of volume. Make sure your dimensions make sense.
b. What are the units of R? What must be the units of α?
c. Suppose that $c = 1.0 \times 10^{-2}$ mol and $V = 400$ cm^3. Find the total amount of chemical absorbed.
d. Suppose that $c = 1.0 \times 10^{-2}$ mol. Find the total chemical absorbed as a function of V.

EXERCISE 21
Suppose a person's head diameter D and height H grow according to

$$D(t) = 10.0 e^{0.03t}$$
$$H(t) = 50.0 e^{0.09t}$$

during the first 15 years of life.
a. Find D and H at $t = 0$, $t = 7.5$, and $t = 15$.
b. Sketch graphs of these two measurements as functions of time.
c. Sketch semilog graphs of these two measurements as functions of time.
d. Graph height as a function of head diameter.
e. Graph the logarithm of height as a function of the logarithm of head diameter.
f. Find the doubling time of each measurement.

EXERCISE 22
On another planet, people have three hands and like to compute tripling times instead of doubling times.
a. Suppose a population follows the equation $b(t) = 3.0 \times 10^3 e^{0.333t}$, where t is measured in hours. Find the tripling time.
b. Suppose a population has a tripling time of 33 h. Find the equation for population size $b(t)$ if $b(0) = 3.0 \times 10^3$.

EXERCISE 23
A Texas millionaire (with $1,000,001 in assets in 1995) got rich by clever investments. She managed to earn 10% interest per year for the last 20 yr, and she plans to do the same in the future.
a. How much did she have in 1975?
b. When will she have $5,000,001?
c. Write and graph the updating function.
d. Write and graph the solution.

EXERCISE 24
A major university hires a famous Texas millionaire to manage its endowment. The millionaire decides to follow this plan each year:

- Spend 25% of all funds above $100 million on university operations.
- Invest the remainder at 10% interest.
- Collect $50 million in donations from wealthy alumni.

a. Suppose the endowment has $340 million to start. How much will it have after spending on university operations? After collecting interest on the remainder? After the donations roll in?
b. Find the updating function.
c. Graph the updating function and cobweb starting from $340 million.

EXERCISE 25
Another major university hires a different famous Texas millionaire (see Exercise 23) to manage its endowment. This millionaire starts with $340 million, brings back $355 million the next year, and claims to be able to guarantee a linear increase in funds thereafter.
a. How much money will this university have after 8 yr?
b. Graph the endowment as a function of time.
c. Write the updating function, graph, and cobweb starting from $340 million.
d. Which university do you think will do better in the long run? Which Texan would you hire?

EXERCISE 26
A recent model of plant growth finds that some grow according to the rule

$$F_{t+1} = F_t \frac{F_t}{b + \sqrt{F_t}}$$

where b represents the competitive pressure from other plants and F is the size of the plants. Plants are measured once per year. Suppose that $b = 2$.
a. Find the equilibria (try guessing).
b. Sketch the updating function.
c. Guess the stability of the equilibria. What would happen the next year to a plant that started with a size 0.2 above the positive equilibrium?
d. Suppose a population started out with a mixture of plants of size 1 and plants of size 9. What would happen?
e. What would you do if b increased over time according to $b = 2 + t$? What questions might you ask? How might you answer them?

EXERCISE 27
A heart receives a signal to beat every second. If the voltage when the signal arrives is below 50 mV, the heart beats and increases its voltage by 30 mV. If the voltage when the signal arrives is greater than 50 mV, the heart does not beat and the voltage does not change. The voltage of the heart decreases by 25% between beats in either case.

a. Suppose the voltage of the heart is 40 mV right after one signal arrives. What is the voltage before the next signal? Will the heart beat?
b. Graph the updating function for the voltage of this heart.
c. Will this heart exhibit normal beating or some sort of AV block?

Projects for Chapter 1

PROJECT 1
Combine the model of competition from Section 1.12 with the model of mutation and reversion discussed in Exercises 4–5 in Section 1.12. That is, assume that wild-type bacteria have per capita reproduction r, mutants have per capita reproduction s, a fraction μ of the offspring of the wild type mutate into the mutant type, and a fraction ν of the offspring of the mutant type revert. First, set $b_t = 4.0 \times 10^6$, $m_t = 2.0 \times 10^5$, $\mu = 0.2$, $\nu = 0.1$, $r = 1.5$, and $s = 2.0$.

a. How many wild-type individuals will there be after reproduction and before mutation? How many of these will mutate? How many will remain wild type?
b. How many mutant individuals will there be after reproduction and before mutation? How many of these will revert? How many will not?
c. Find the total number of wild-type bacteria after reproduction, mutation, and reversion.
d. Find the total number of mutants after reproduction, mutation, and reversion.
e. Find the total number of bacteria after reproduction, mutation, and reversion. Why is it different from the initial number?
f. Find the fraction of mutants to begin with, and the fraction after reproduction, mutation, and reversion.

Now, treat b_t and m_t as variables. The following steps will help you find the updating function.

i. Use the steps in parts **a–f** to find m_{t+1} in terms of b_t and m_t.
ii. Find the total number of bacteria, $b_{t+1} + m_{t+1}$, in terms of b_t and m_t.
iii. Divide your equation for m_{t+1} by $b_{t+1} + m_{t+1}$ to find an expression for p_{t+1} in terms of b_t and m_t.
iv. Divide the numerator and denominator by $m_t + b_t$ as in the derivation of Equation 1.49, and write the updating function in terms of p_t.
v. Find the equilibrium of this updating function.

Finally, we can do this in general by treating r, s, μ, and ν as parameters. Do the above steps exactly to find the updating function. After doing so, study the following special cases. In each case, explain your answer.

g. $r = s$ (no selection) and mutation in only one direction ($\mu = 0.0$ and $\nu > 0$). Find the updating function and the equilibrium. What does this mean?
h. $r = s$ and $\mu = \nu$. Find the updating function and the equilibrium. What does this mean?
i. $r > s$, $\nu = 0$, and $\mu > 0$. This means that the wild-type bacteria have a reproductive advantage but keep mutating. Find the updating function and the equilibrium. The result is called mutation–selection balance. Can you guess why?

PROJECT 2
Suppose a measurement follows

$$f(t) = \cos\left(\frac{2\pi t}{T}\right)$$

where the period T is unknown. You guess that the period is 1.0 and figure you can get a good fix on the behavior of the measurement by checking every 1.0 time steps (once per period). Try the following for these values of T: 1.0, 0.6, 0.601, 0.602, 0.603, and 0.618. Experimenting with other values of T is highly recommended.

a. Graph the actual data as a continuous function of time, $0 \le t \le 100$.
b. Graph what you find by plotting the data every one time unit.
c. Are you able to figure out what is going on?

PROJECT 3
The following data (kindly provided by J. Ragsdale) describe the growth of male and female lizards as both juveniles and adults. The standard measurement of overall body size is called "snout–vent length," abbreviated SVL. Allometric relations between this value, head width (HW), and abdomen length (A) were measured. Denote measurements for juvenile males with a subscript m, adult males with a subscript M, juvenile females with a subscript f, and adult females with a subscript F. The allometric relations are

$$\log(\mathrm{HW}_m) = 0.78\log(\mathrm{SVL}_m) - 0.35$$
$$\log(\mathrm{HW}_M) = 0.85\log(\mathrm{SVL}_M) - 0.47$$
$$\log(\mathrm{HW}_f) = 0.75\log(\mathrm{SVL}_f) - 0.295$$
$$\log(\mathrm{HW}_F) = 0.66\log(\mathrm{SVL}_F) - 0.14$$
$$\log(\mathrm{A}_m) = 1.05\log(\mathrm{SVL}_m) - 0.41$$
$$\log(\mathrm{A}_M) = 1.06\log(\mathrm{SVL}_M) - 0.425$$

$$\log(A_f) = 1.03 \log(SVL_f) - 0.38$$
$$\log(A_F) = 1.24 \log(SVL_F) - 0.74$$

Both males and females switch from juvenile to adult when $\log(SVL) = 1.72$.

  a. Plot the allometric relation for head width for juvenile and adult males on the same log–log graph.
  b. Do the same for female head width, male abdomen length, and female abdomen length.
  c. Replot each relation in the original measurements (with the logs). Remember that these are logs base 10.
  d. What aspect of male allometry changes when they switch from juvenile to adult? Why might they do this?
  e. What aspects of female allometry change when they switch from juvenile to adult? Why might they do this?
  f. Sketch small juvenile, small adult, and large adult lizards of each sex.

Bibliography for Chapter 1

Section 1.1

Anderson, R. M., and R. M. May. *Infectious Diseases of Humans*. Oxford University Press, Oxford, England, 1992.

Hodgkin, A. L., and A. F. Huxley. A quantitative description of membrane current and its application to conduction and excitation in nerve. *Journal of Physiology* 117:500–544, 1952. (Reprinted in *Bulletin of Mathematical Biology* 52:25–71, 1990.)

Provine, W. B. *The Origins of Theoretical Population Genetics*. University of Chicago Press, Chicago, 1971.

Rinzel, J. Electrical excitability of cells, theory and experiment: review of the Hodgkin-Huxley foundation and an update. *Bulletin of Mathematical Biology* 52:5–23, 1990.

Ross, R. *The Prevention of Malaria*. Murray, London, England, 1911.

Section 1.3

Hecht, E. *Physics*. Brooks/Cole, Pacific Grove, Calif., 1994.

Paulos, J. A. *Innumeracy*. Hill and Wang, New York, 1988.

Section 1.4

Godfray, H. C. F. *Parasitoids*. Princeton University Press, Princeton, N.J., 1994.

McWhirter, N., and R. McWhirter. *Guinness Book of World Records 1990*. Sterling Publishing Co., New York, 1990.

Section 1.8

Gould, S. J. The origin and function of "bizarre" structures: antler size and skull size in the "Irish elk," *Megaloceros giganteus*. *Evolution* 28:191–220, 1974.

McNeill Alexander, R. *Optima for Animals*. Princeton University Press, Princeton, N.J., 1996.

Nijhout, H. F., and D. E. Wheeler. Growth models of complex allometries in holometabolous insects. *American Naturalist* 148:40–56, 1996.

Peters, R. H. *The Ecological Implications of Body Size*. Cambridge University Press, Cambridge, England, 1983.

Vogel, S. *Life in Moving Fluids*. Princeton University Press, Princeton, N.J., 1989.

Section 1.9

Stampi, C. *Why We Nap*. Birkhauser, Boston, 1990.

Section 1.10

Hoppensteadt, F. C., and C. S. Peskin. *Mathematics in Medicine and the Life Sciences.* Springer-Verlag, New York, 1992.

Section 1.12

Hartl, D. L., and A. G. Clark. *Principles of Population Genetics.* Sinauer Associates, Sunderland, Mass., 1989.

Section 1.13

Glass, L., and M. C. Mackey. *From Clocks to Chaos: The Rhythms of Life.* Princeton University Press, Princeton, N.J., 1988.

Keener, J. P. On cardiac arrhythmias: AV conduction block. *Journal of Mathematical Biology* 12:215–225, 1981.

Chapter 2

Limits and Derivatives

We have used discrete-time dynamical systems to describe how biological processes change when measurements are made at discrete intervals. In Chapter 2, we will learn how to describe change in a value that is measured **continuously**. This description is based on the two central ideas of differential calculus, the **limit** and the **instantaneous rate of change**, or **derivative**. We will find a **geometric** interpretation of the derivative as the **slope of the tangent line**, which will help us to graph and analyze complicated functions.

Achieving these goals requires both understanding the idea of the derivative and having the tools to **compute** the derivatives of a wide range of functions. Complicated functions are built by combining simple pieces, and their derivatives are computed with a set of rules for combining the derivatives of sums, powers, products, quotients, and compositions. Starting with the derivatives of linear, exponential, logarithmic, and trigonometric functions, we will be able to differentiate pretty much any function we can write. The ability to compute the derivative, combined with its dual interpretation as the instantaneous rate of change and as the slope, will open up a dazzling array of applications.

2.1 Introduction to Derivatives

Discrete-time dynamical systems are a powerful tool for describing the dynamics of biological systems when change can be accurately described by measurements made at **discrete times**. To understand other systems fully, however, measurements must be made at all times, or **continuously**. We have only to think of the growth of a plant or the motion of an animal to realize that some change is best described by a continuous set of measurements.

In this section, we develop the tools needed to describe measurements that change continuously. We will switch from thinking about what *happens* to what *is happening* to a measurement. The central idea is the **instantaneous rate of change**, or the **derivative**. Graphically, the derivative is equal to the slope of the *tangent line* to a curve.

These two ideas of the derivative, as the instantaneous rate of change of a measurement and as the slope of a curve, are the keys to appreciating the many applications of **calculus**. Understanding that the derivative must be computed as a **limit** is the key to using calculus correctly.

The Average Rate of Change

Suppose we measured a bacterial population continuously and found that the population size $b(t)$ followed the equation

$$b(t) = 2.0^t$$

(Figure 2.1). How can we completely describe the growth of this population?

If we checked the population size every hour, we would find the data shown in the following table (Figure 2.2).

Population measured every hour

t	$b(t)$	Change, $\Delta b = b(t) - b(t-1)$
0	1.0	—
1	2.0	1.0
2	4.0	2.0
3	8.0	4.0
4	10.0	8.0
5	16.0	16.0

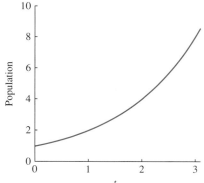

Figure 2.1

A bacterial population measured continuously

If this were the whole story, we could describe the growth with the updating function

$$b_{t+1} = 2.0 b_t$$

To begin working toward a description of how the population is *changing* at any given time, we have added to the table a column Δb, denoting the change in the population between times $t - 1$ and t. Recall that Δ (the Greek letter "delta") means "change in" (Section 1.4). For example, the bacterial population increased by 2.0 million between $t = 1.0$ and $t = 2.0$. The change in population is not defined at $t = 0$ because there was no previous measurement.

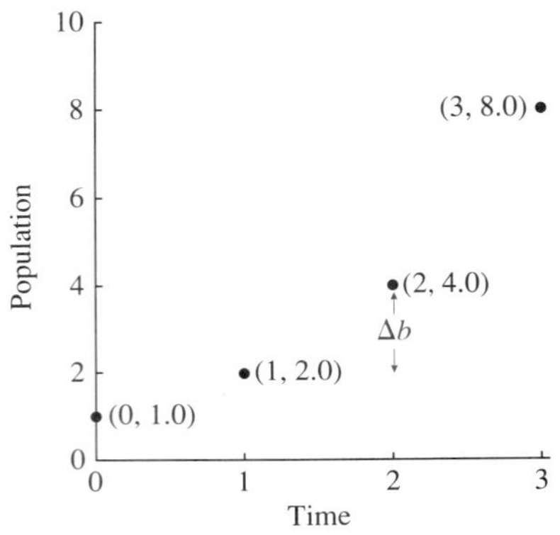

120 Chapter 2 Limits and Derivatives

Figure 2.2
Hourly measurement of a bacterial population

For more accuracy, we could check the population every half-hour (Figure 2.3a).

t	$b(t)$	Change, $\Delta b = b(t) - b(t - 0.5)$	Average rate of change, $\dfrac{\Delta b}{\Delta t} = \dfrac{b(t) - b(t - 0.5)}{\Delta t}$
0.0000	1.0000	—	—
0.5000	1.4142	0.4142	0.8284
1.0000	2.0000	0.5858	1.1716
1.5000	2.8284	0.8284	1.6568
2.0000	4.0000	1.1716	2.3431
2.5000	5.6569	1.6569	3.3137
3.0000	8.0000	2.3431	4.6863

We have added another column to the table, indicating the **average rate of change** of the population, defined as

$$\text{average rate of change} = \frac{\text{change in population}}{\text{change in time}} = \frac{\Delta b}{\Delta t}$$

Figure 2.3
Hourly and half-hourly measurements of a bacterial population

The average rate of change is an indication of how fast the population is changing. Between $t = 1.0$ and $t = 1.5$, the population grew by 0.8284 million. Because this change took only $\Delta t = 0.5$ h, the average rate of change is 1.6568 million bacteria per hour.

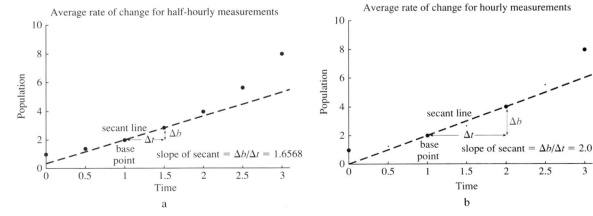

There is an important graphical interpretation of the average rate of change, as the slope of the line connecting the two points on a graph (Section 1.4). This line is called a **secant line**, and the point (1, 2.0) is called the **base point**. The average rate of change between $t = 1.0$ and $t = 1.5$ corresponds to the slope of the line connecting the data points at these two times (Figure 2.3a).

Using our less accurate hourly measurements, the average rate of change between times $t = 1.0$ and $t = 2.0$ is 2.0 million bacteria per hour, a larger value than the average rate between $t = 1.0$ and $t = 1.5$. The two rates fail to match because they cover different time intervals. The average rate between $t = 1.0$ and $t = 2.0$ is larger because it includes the more rapid change that occurred after $t = 1.5$ (Figure 2.3b).

To find an accurate estimate of the rate of change exactly at $t = 1.0$, we should look only at the change near that point. The following table shows the results when measurements are taken every 0.1 h.

t	$b(t)$	Change, $\Delta b = b(t) - b(t - 0.1)$	Average rate of change, $\dfrac{\Delta b}{\Delta t}$
0.5000	1.4142	0.0947	0.9471
0.6000	1.5157	0.1015	1.0150
0.7000	1.6245	0.1088	1.0879
0.8000	1.7411	0.1166	1.1660
0.9000	1.8661	0.1250	1.2496
1.0000	2.0000	0.1339	1.3393
1.1000	2.1435	0.1435	1.4355
1.2000	2.2974	0.1538	1.5385
1.3000	2.4623	0.1649	1.6489
1.4000	2.6390	0.1767	1.7673
1.5000	2.8284	0.1894	1.8941

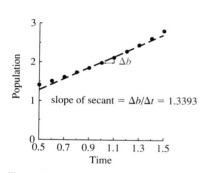

Figure 2.4

Measurement of a bacterial population every 0.1 h

Again, as measurements become more frequent, the average rate of change becomes smaller; we have included only change near $t = 1.0$ (Figure 2.4).

Instantaneous Rates of Change

As the time Δt between measurements becomes smaller, the average rate of change between times $t = 1.0$ and $t = 1.0 + \Delta t$ becomes a better description of what is happening exactly at $t = 1.0$. Average rates of change over smaller and smaller intervals are computed in the following table.

Δt	$1.0 + \Delta t$	$b(1.0 + \Delta t)$	Δb	$\dfrac{\Delta b}{\Delta t}$
1.0	2.0	4.0000	2.0000	2.0000
0.5	1.5	2.8284	0.8284	1.6568
0.1	1.1	2.1435	0.1435	1.4354
0.01	1.01	2.0139	0.0139	1.3911
0.001	1.001	2.00139	0.00139	1.3868
0.0001	1.0001	2.000139	0.000139	1.3863

As Δt becomes smaller, there is less time for anything to happen, and the *change* in population Δb also becomes small. However, the average *rate* of change does not become tiny because the change takes place over shorter time intervals. In fact, the average rate of change seems to approach a value of about 1.386. We would like to define this value as the **instantaneous rate of change**, the rate of change *exactly* at $t = 1.0$.

What do we mean by instantaneous rate of change? The most familiar instantaneous rate of change is speed, the rate of change of position, as measured by a speedometer. But how does a speedometer *measure* this rate? One method is to measure position at one time and again at a later time, and to divide the distance moved by the time elapsed:

$$\text{average rate of change} = \frac{\text{change in position}}{\text{change in time}}$$

Again, this value tells only what happened *on average* during the interval.

Suppose a car is moving at a constant speed of 20.0 m/s. It moves 20.0 m in 1.0 s, for an average rate of change of 20.0 m/s (Figure 2.5a). It moves only 2.0 m in 0.1 s, again with an average rate of change of 20.0 m/s (Figure 2.5b). The distance moved becomes smaller as the time between checks becomes smaller. The average rate of change of position remains the same. If we continued to make more and more accurate measurements, assessing the position of the car every Δt s, we would find the data listed in the table at left.

Δt	Distance moved	Average rate of change
1.0	20.0	20.0
0.1	2.0	20.0
0.01	0.2	20.0
0.001	0.02	20.0

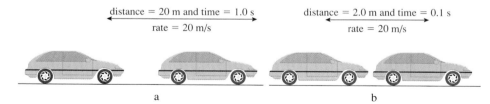

Figure 2.5
A car moving at constant speed travels a shorter distance in a shorter time

The average rate of change is always 20.0 because the car is moving at a constant speed. This constant speed must be the instantaneous rate of change.

With the bacterial population, it is not as easy to guess the instantaneous rate of change. Ideally, we would compute the instantaneous rate of change by picking $\Delta t = 0$, the smallest possible value, resulting in

Δt	$1.0 + \Delta t$	$b(1.0 + \Delta t)$	Δb	$\dfrac{\Delta b}{\Delta t}$
0.0	1.0	2.0	0.0	$\dfrac{0.0}{0.0}$

Dividing by 0 is a mathematical misdemeanor; dividing 0 by 0 is a mathematical felony. Not only have we failed to find the answer, but we have violated a major mathematical law.

Choosing a smaller and smaller value of Δt corresponds to picking the second point on the curve closer and closer to the base point (Figure 2.6). With $\Delta t = 0$, the second point lies right on top of the base point. It is impossible to draw a unique line through a single point. In fact, there are an infinite number of possible lines through this point (Figure 2.7). Only one of these lines, the **tangent line**, is close to the secant lines in Figure 2.6. This line does touch the curve in only a single point, but does so by just kissing the side of it rather than rudely crossing straight through. In fact, if we zoom in on the tangent line it looks more and more similar to the curve (Figure 2.8).

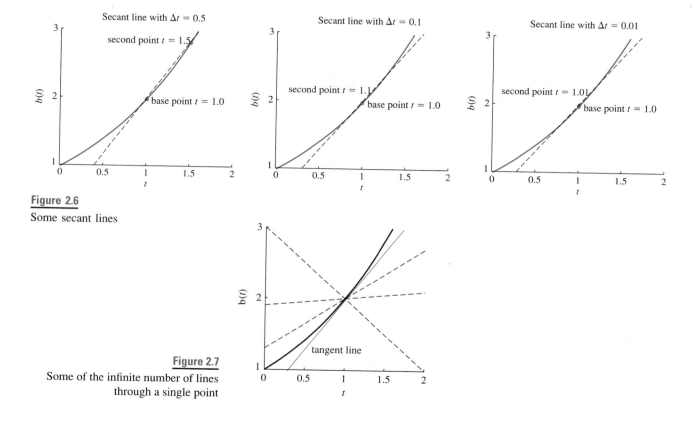

Figure 2.6
Some secant lines

Figure 2.7
Some of the infinite number of lines through a single point

Figure 2.8
Zooming in on the tangent line

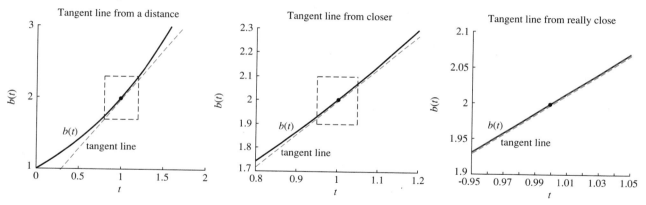

We thus find the instantaneous rate of change by computing the average rate of change with smaller and smaller values of Δt, but without ever reaching $\Delta t = 0$. Graphically, this corresponds to moving the second point closer and closer to the base point and seeing that the secant line gets closer and closer to the tangent line. Just as the slope of the secant line is equal to the average rate of change, the slope of the tangent line is equal to the instantaneous rate of change.

Limits, Derivatives, and Differential Equations

We now need a way to work with the idea of "smaller and smaller" or "closer and closer." The mathematical notion is called the **limit**. The expression

$$\lim_{\Delta t \to 0} \frac{\Delta b}{\Delta t}$$

pronounced "the limit as delta t approaches 0 of delta b over delta t," means what we *should* get by plugging in $\Delta t = 0$, but *cannot* get exactly because we cannot divide 0 by 0. Near $t = 1.0$, we found that the limit seems to have a value of about 1.386. We will study the limit more carefully in Section 2.2.

Using the limit, we can define the derivative.

■ **Definition 2.1**   The instantaneous rate of change of a function $f(t)$, called the **derivative of f**, is computed as

$$\lim_{\Delta t \to 0} \frac{\Delta f}{\Delta t}$$
■

The derivative measures how fast a measurement is changing at a particular instant.
There are two notations for the derivative:

$$\text{The derivative of } f \text{ at } t = \begin{cases} \dfrac{df}{dt} & \text{differential notation} \\ f'(t) & \text{prime notation} \end{cases}$$

In differential notation, the d's represent small versions of the letter Δ. Prime notation defines a new function $f'(t)$ that outputs the rate of change of $f(t)$ at any input time t. **Differential notation** is convenient for writing differential equations and the complicated expressions needed to calculate derivatives. **Prime notation** is convenient for analyzing discrete-time dynamical systems and for describing rates of change.

Geometrically, the slope of the tangent closely matches the slope of the curve. We use this observation to define the slope of a curve.

■ **Definition 2.2**   The slope of the graph of a function is equal to the slope of the tangent line to the graph, which is itself equal to the derivative of the function. ■

These two ways of thinking about the derivative are linked in Figure 2.9.

We can also use the point-slope form of a line (Section 1.4) to write the *equation* of the tangent line. Suppose the function $f(x)$ has derivative equal to $f'(a)$ at the base point a. The tangent line passes through the point $(a, f(a))$ with slope $f'(a)$. The tangent line, which we can think of as the **tangent line approximation**, $\hat{f}$, has formula

$$\hat{f}(x) = f'(a)(x - a) + f(a) \tag{2.1}$$

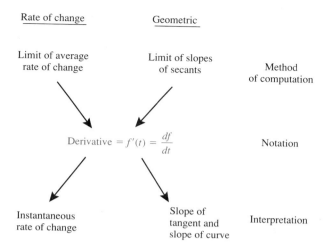

Figure 2.9
The interrelated meanings of the derivative

(Figure 2.10). For example, the slope of the line tangent to the function $b(t) = 2.0^t$ is 1.386 at the base point $t = 1.0$. Therefore, the tangent line approximation is

$$\hat{b}(t) = 1.386(t - 1.0) + b(1.0) \quad \text{the general formula with } a = 1.0$$
$$= 1.386(t - 1.0) + 2.0 \quad \text{evaluate } b(1.0) = 2.0$$
$$= 1.386t + 0.614 \quad \text{write in slope-intercept form}$$

The tangent line is a good approximation because it lies so close to the curve (Figure 2.8). In some sense, the central trick in calculus is realizing that from up close, "nice" curves look like lines (we will soon see examples of "nasty" curves that do not).

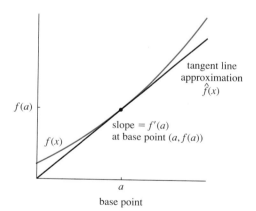

Figure 2.10
Approximating a function with the tangent line

Using the derivative, we can find the instantaneous rate of change of the bacterial population at $t = 1.0$, or indeed at any time. However, this description does not tell us the *rule* followed by the population. We can use the derivative along with our intuition about the behavior of growing populations to derive such a rule. Instead of finding an updating function, a formula for the new population as a function of the old population, we will derive a **differential equation**, a formula for the **rate of change** of the population.

The following data estimate the instantaneous rate of change with the average rate of change by using a fairly small value of $\Delta t = 0.1$. In other words, we estimate that the derivative db/dt at each time t is approximately

$$\frac{db}{dt} \approx \frac{\Delta b}{\Delta t} = \frac{b(t) - b(t-0.1)}{0.1}$$

t	$b(t)$	Change, Δb	Average rate of change, $\frac{\Delta b}{\Delta t} \approx \frac{db}{dt}$	Per capita rate of change $\frac{\frac{db}{dt}}{b(t)}$
0.00	1.0	—	—	—
0.1000	1.0718	0.0718	0.7177	0.6697
0.2000	1.1487	0.0769	0.7692	0.6697
0.3000	1.2311	0.0824	0.8245	0.6697
0.4000	1.3195	0.0884	0.8836	0.6697
0.5000	1.4142	0.0947	0.9471	0.6697
0.6000	1.5157	0.1015	1.0150	0.6697
0.7000	1.6245	0.1088	1.0879	0.6697
0.8000	1.7411	0.1166	1.1660	0.6697
0.9000	1.8661	0.1250	1.2496	0.6697
1.0000	2.0000	0.1339	1.3393	0.6697
1.1000	2.1435	0.1435	1.4355	0.6697
1.2000	2.2974	0.1538	1.5385	0.6697
1.3000	2.4623	0.1649	1.6489	0.6697
1.4000	2.6390	0.1767	1.7673	0.6697
1.5000	2.8284	0.1894	1.8941	0.6697
1.6000	3.0314	0.2030	2.0301	0.6697
1.7000	3.2490	0.2176	2.1758	0.6697
1.8000	3.4822	0.2332	2.3319	0.6697
1.9000	3.7321	0.2499	2.4993	0.6697
2.0000	4.0000	0.2679	2.6787	0.6697

We have added another new column to the table, the **per capita rate of change**, defined as

$$\text{per capita rate of change} = \frac{\text{instantaneous rate of change}}{\text{population}}$$

When studying the updating function for a growing bacterial population, we found that the **per capita reproduction** provided a useful description. Similarly, the per capita rate of change, or per capita *rate* of reproduction, provides a useful description of a continuous set of measurements of population size.

Mathematically, we have

$$\text{per capita rate of reproduction} = \frac{\frac{db}{dt}}{b(t)} \approx 0.6697$$

Solving for the derivative db/dt, we get

$$\frac{db}{dt} = 0.6697 b(t) \tag{2.2}$$

This is a **differential equation**. A differential equation sets the derivative of some function (usually written in differential notation) equal to some combination of measurements not involving derivatives. In this case, the rate of change of the population is proportional to the population size itself. Of course, this differential equation is not exactly correct because we estimated the instantaneous rate of change as the average rate of change with $\Delta t = 0.1$. Exercise 8 follows the same steps to find a more accurate differential equation by using smaller values of Δt.

Differential equations and updating functions are two types of rule describing how a measurement changes. How do they differ? An updating function, such as $b_{t+1} = 2.0 b_t$, gives the new value of the population (at time $t + 1$) as a function of the previous value (at time t). A differential equation instead gives the **rate of change** of the population as a function of the population. Unlike the updating function, it does not involve the population at two different times (e.g., b_{t+1} and b_t). The differential equation relates two pieces of information about the population at one time, the population size $b(t)$ and the rate of change db/dt (Figure 2.11).

Consider again the car moving at a constant speed of 20.0 m/s. The instantaneous rate of change of position, or derivative, is equal to 20.0 m/s. If we denote the position at time t by $P(t)$, we can describe the movement of the car with the differential equation

$$\frac{dP}{dt} = 20.0$$

This differential equation says mathematically that the rate of change of position is exactly 20.0 m/s at all times.

Differential equations are the single most powerful tool in applied mathematics. Their discovery by Sir Isaac Newton, the great English physicist, is perhaps the greatest insight in the history of science. If differential equations seem hard to understand, remember that it took hundreds of years of hard work by the world's leading mathematicians and scientists to realize their importance. The detailed study of differential equations in this book begins in Chapter 4, after we practice working with derivatives and learn some of their many applications.

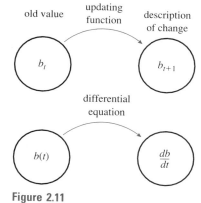

Figure 2.11
The difference between updating functions and differential equations

SUMMARY

To describe how a measurement changes, we began by defining the **average rate of change** of a population, the change in population divided by the change in time. The average rate of change corresponds graphically to the slope of the **secant line** connecting two data points. To find the **instantaneous rate of change**, or **derivative**, we estimated the average rate of change over shorter and shorter intervals. Because it is impossible to use an interval of length zero, we found the **limit** as the interval approaches zero. Graphically, the instantaneous rate of change corresponds to the slope of the **tangent line**, which is defined to be equal to the slope of the curve itself. In some cases, we can write a **differential equation**, a formula for the derivative.

2.1 EXERCISES

1. Suppose that a population of bacteria is described by the formula
$$b(t) = 1.5^t$$
where the time t is measured in hours.
   a. Graph the data.
   b. Find the population at times 0, 1, 2, and 3.

c. Find the average rate of change of the population between times 0 and 1, and indicate it on a graph.
d. Find the average rate of change of the population between times 1 and 2, and indicate it on a graph.
e. Find the population at times 0, 0.5, 1.0, 1.5, and 2.0.
f. Find the average rate of change of the population between times 0 and 0.5, and indicate it on a graph.
g. Find the average rate of change of the population between times 0.5 and 1, and indicate it on a graph.

2. For a population of bacteria that follows the formula

$$b(t) = 1.5^t$$

find the following.
a. The average rate of change between times 0 and 0.1.
b. The average rate of change between times 0 and 0.01.
c. The average rate of change between times 0 and 0.001.
d. The average rate of change between times 0 and 0.0001.
e. What do you think the limit is?
f. Write the formula and graph the tangent line.

3. Suppose a population follows the formula

$$b(t) = 2.0^t$$

Use as base point (0.0, 1.0).
a. Graph the secant line with second point (1.0, 2.0), and find its slope.
b. Do the same with second point $(0.1, 2.0^{0.1})$.
c. Do the same with second point $(0.01, 2.0^{0.01})$.
d. Try to guess the slope of the tangent line at base point (0.0, 1.0).
e. Write the equation of the tangent line.

4. Suppose a population follows the formula

$$h(t) = 5.0t^2$$

Use as base point (1.0, 5.0).
a. Graph the secant line with second point (2.0, h(2.0)), and find its slope.
b. Do the same with second point (1.1, h(1.1)).
c. Do the same with second point (1.01, h(1.01)).
d. Write an equation for the slope of the secant line connecting the points (1.0, h(1.0)) and (1.0 + Δt, h(1.0 + Δt)).
e. What does this look like when Δt is small?

5. Suppose a bacterial population follows the equation $b(t) = 1.0 + 2.0t$, where t is measured in hours and b is measured in millions.
a. Graph the population as a function of time.
b. Find the average rate of change during the first hour.
c. Find the average rate of change during the second hour.
d. Find the average rate of change during the first half-hour.
e. Find the average rate of change during the second half-hour.
f. Find the average rate of change during the first Δt hours, and try to guess the limit as Δt becomes smaller and smaller.

g. Find the average rate of change between time 1.0 and 1.0 + Δt hours, and try to guess the limit as Δt becomes smaller and smaller.

6. For the following bacterial populations, find the average rate of change during the first hour and during the first and second half-hours. Graph the data and the secant lines associated with the average rates of change.
a. $b(t) = 3.0 \times 2.0^t$
b. $b(t) = 0.3 \times 2.0^t$
c. $b(t) = 3.0 \times 1.5^t$
d. $b(t) = 3.0 \times 0.5^t$

7. The distance a rock falls is given by $p(t) = 5t^2$, where t is measured in seconds and distance is measured in meters.
a. Graph the position of the rock as a function of time.
b. Find the average rate of change during the first second.
c. Find the average rate of change during the second second.
d. Find the average rate of change during the first half-second.
e. Find the average rate of change during the second half-second.
f. Find the average rate of change during the first Δt seconds, and try to guess the limit as Δt becomes smaller and smaller. Write the formula, and graph the tangent line at $t = 0.0$.
g. Find the average rate of change between time 1.0 and 1.0 + Δt seconds, and try to guess the limit as Δt becomes smaller and smaller. Write the formula, and graph the tangent line at $t = 1.0$.

8. Follow the steps in the text used to derive the approximate differential Equation 2.2:

$$\frac{db}{dt} = 0.6697 b(t)$$

with the following values of Δt.
a. $\Delta t = 1.0$
b. $\Delta t = 0.5$
c. $\Delta t = 0.01$
d. $\Delta t = 0.001$

9. Consider the following data on the height of a tree.

Age	Height (m)	Age	Height (m)
0	10.11		
1	11.18	6	18.27
2	12.40	7	20.17
3	13.74	8	22.01
4	15.01	9	24.45
5	16.61	10	26.85

a. Estimate the rate of change of height at each age.
b. Graph the rate of change of height as a function of age.
c. Find and graph the rate of change of height divided by the height as a function of age.
d. Use these results to describe the growth of this tree.

10. For the function $b(t) = 2.0^t$, we found that the tangent line with base point $t = 1.0$ has slope 1.386 and formula $\hat{b}(t) = 2.0 + 1.386(t - 1.0)$.
   a. Find the equation of a line with slope 2.0 passing through the base point, calling it $\tilde{b}(t)$.
   b. Graph $b(t)$, $\hat{b}(t)$, and $\tilde{b}(t)$.
   c. Compare $b(1.1)$, $\hat{b}(1.1)$, and $\tilde{b}(1.1)$. Is the tangent line a better approximation to the value of b? Why?
   d. Compare $b(1.9)$, $\hat{b}(1.9)$, and $\tilde{b}(1.9)$. Which line is a better approximation to the value of b?

11. The procedure banks use to compute continuously compounded interest is similar to the process we used to derive a differential equation. Suppose several banks claim to be giving 5% annual interest and that you have $1000 to deposit.
   a. How much would you have after a year from a bank that has no compounding?
   b. A bank that compounds twice yearly really gives 2.5% interest twice. How much would you have after a year from this bank?
   c. A bank that compounds monthly really gives 5/12% interest each month. How much would you have after a year from this bank?
   d. How much would you have after a year from a bank that compounds daily?
   e. Write a limit that expresses the amount of money you would get from a bank that compounds continuously, and try to guess the answer.

12. List five different names or formulas for the instantaneous rate of change of the function $g(t)$.

13. **COMPUTER:** Consider the function
$$f(x) = \sqrt{1 - x^2}$$
defined for $-1 \leq x \leq 1$. This is the equation for a semicircle. The tangent line at the base point $(\sqrt{2}/2, \sqrt{2}/2)$ has slope -1. Graph this tangent line. Now zoom in on the base point. Does the circle look more and more like the tangent line? How far to you need to go before the circle looks flat? Would a tiny insect be able to tell that his world is curved?

14. **COMPUTER:** Suppose a bacterial population oscillates with the formula
$$b(t) = 2.0 + \cos(t)$$
   a. Graph this function.
   b. Find and graph the function that gives the rate of change between times t and $t + 1$ as a function of t for $0 \leq t \leq 10$.
   c. Find and graph the function that gives the rate of change between times t and $t + 0.1$ as a function of t for $0 \leq t \leq 10$.
   d. Try the same with smaller values of Δt. Do you have any idea what the limit might be?

15. **COMPUTER:** Follow the steps in Exercise 14 for a bacterial population that follows the formula
$$b(t) = e^{-0.1t}[2.0 + \cos(t)]$$

2.2 Limits

The derivative, the mathematical version of the instantaneous rate of change and the slope of a curve, includes a **limit** in its definition. We now study the mathematical and scientific basis of this fundamental idea. Understanding the useful properties of limits enables us to calculate limits of polynomials and rational functions. We study three extensions of the basic idea. **Left-hand limits** and **right-hand limits** are used for functions with domains lying entirely to the left or right of a point of interest. **Infinite limits** are used for measurements or functions that become extremely large.

Limits of Functions

We begin by formalizing the steps we used to compute the instantaneous rate of change of a population, following the law

$$b(t) = 2.0^t$$

at $t = 1.0$. First, we found the change in the population, Δb, between times 1.0 and $1.0 + \Delta t$ as

$$\Delta b = b(1.0 + \Delta t) - b(1.0)$$
$$= 2^{1.0+\Delta t} - 2.0$$

The average rate of change, the change in the population divided by the change in time, is then

$$\frac{\Delta b}{\Delta t} = \frac{2^{1.0+\Delta t} - 2.0}{\Delta t}$$

The instantaneous rate of change (or derivative) is

$$\frac{db}{dt} = b'(1.0) = \lim_{\Delta t \to 0} \frac{2^{1.0+\Delta t} - 2.0}{\Delta t}$$

As Δt becomes smaller, the average rate of change gets closer to a value near 1.386. We could not take the smallest possible value, $\Delta t = 0$, because that would have led to division of 0 by 0.

We can think of the average rate of change

$$\frac{\Delta b}{\Delta t} = \frac{2^{1.0+\Delta t} - 2.0}{\Delta t}$$

as a **function** of Δt that is defined at all points except $\Delta t = 0$ (Figure 2.12). We guessed that the limit is 1.386 because values get closer and closer to 1.386. But what does it mean for a function to get "closer and closer" to a "limit"? What, indeed, does it mean to be "close"?

As is often the case, this mathematical question can be understood by thinking of it *scientifically*. What does it mean *scientifically* for one measurement to be close to another?

Two measurements are close when they cannot be distinguished by a precise measuring device.

Consider the measurement of temperature. With a crude measuring device, such as waving your hand around in the air, it might be difficult to distinguish temperatures of 20°C and 25°C. In science, if you cannot *measure* a difference, it does not exist. Without a more precise device, the two temperatures might as well be the same. With an ordinary thermometer, we can distinguish 20°C and 25°C, but we may not be able to distinguish 20°C and 20.5°C. Two temperatures are *exactly equal* if no thermometer, no matter how precise, can distinguish them.

We can translate this scientific idea into the mathematical idea of a limit. Think of our average rate of change as a function of Δt (Figure 2.12). Measurements, however, are uncertain. If we can measure the rate of change with an accuracy of only 0.1, we do not need a very small value of Δt to be within 0.1 of the limit (Figure 2.13a). With a more precise measurement of the average rate of change with an accuracy of 0.01, we need a much smaller Δt to be within measurement accuracy of the limit (Figure 2.13b). No matter how precisely we can measure rate of change, however, we can always pick a sufficiently small Δt to be within measurement accuracy of the limit.

Consider finding the instantaneous rate of change of a falling rock whose position follows the quadratic function

$$y(t) = 10.0t^2$$

(Figure 2.14). We want to find the rate of change at $t = 1.0$. The change Δy between times 1.0 and $1.0 + \Delta t$ is

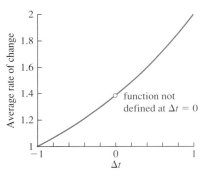

Figure 2.12

The limit of the average rate of change

2.2 Limits

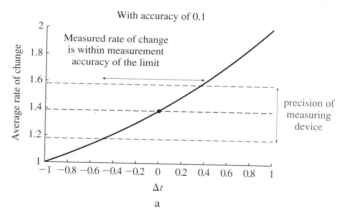

Figure 2.13
An experimental output approaching a limit

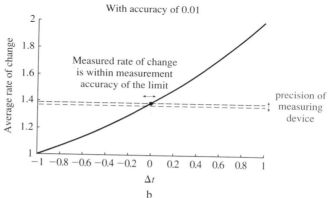

Figure 2.14
The position of a falling rock and the average rate of change

$$\Delta y = y(1.0 + \Delta t) - y(1.0)$$
$$= 10.0(1.0 + \Delta t)^2 - 10.0$$
$$= 10.0(1.0 + 2.0\Delta t + \Delta t^2) - 10.0$$
$$= 20.0\Delta t + 10.0\Delta t^2$$

Therefore,

$$\text{average rate of change} = \frac{\Delta y}{\Delta t}$$
$$= \frac{20.0\Delta t + 10.0\Delta t^2}{\Delta t}$$

(Figure 2.15). As long as $\Delta t \neq 0$, we can divide out the Δt terms, so that

$$\text{average rate of change} = 20.0 + 10.0\Delta t$$

To find the exact rate of change, we must take the limit as $\Delta t \to 0$. For small values of Δt, the average rate of change gets as close as we might wish to 20.0. The limit, as well as the velocity, is exactly 20.0.

Left-Hand and Right-Hand Limits   Sometimes a function is defined only on one side of a point. For example, the value of the function shown in Figure 2.16a seems to get closer and closer to 2.0 even

Figure 2.15

The average rate of change for the falling rock

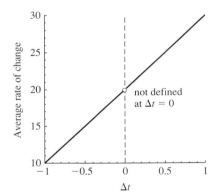

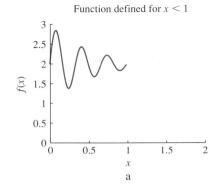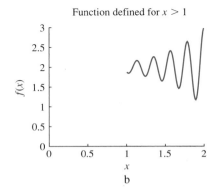

Figure 2.16

Functions with left-hand and right-hand limits

though it is defined only for $x < 1$. If we wish to find the limit as x approaches 1, we must plug in values of x that lie in the domain. We write

$$\lim_{x \to 1^-} f(x) = 2$$

where the superscript minus sign on 1 indicates that x approaches 1 from *below*. This is called the **left-hand limit**, and the notation is pronounced "the limit of $f(x)$ as x approaches 1 from below."

Similarly, if the domain of f contains only values of $x > 1$ (Figure 2.16b), we write

$$\lim_{x \to 1^+} f(x) = 2$$

The superscript plus sign on 1 indicates that x approaches 1 from *above*. This is called the **right-hand limit**, and the notation is pronounced "the limit of $f(x)$ as x approaches 1 from above."

Left- and right-hand limits are most often used to describe functions and measurements that have only positive numbers in their domains, (e.g., $\sqrt{x}$, $\ln x$, and population sizes). Suppose we were asked to find

$$\lim_{x \to 0^+} \sqrt{x}$$

We must compute the right-hand limit because we cannot take the square root of a negative number. Because the square root of 0 is 0 and the graph gets closer and closer to 0, the limit itself is 0.

Consider the **signum** function, defined by

$$\begin{cases} \text{signum}(x) = -1 & \text{if } x < 0 \\ \text{signum}(x) = 0 & \text{if } x = 0 \\ \text{signum}(x) = 1 & \text{if } x > 0 \end{cases}$$

This function gives the "sign" of a number, telling whether it is negative, 0, or positive (Figure 2.17). The limit as x approaches 0 is not the value signum(0) = 0 because the function does not get close to 0. The left-hand limit is -1, and the right-hand limit is 1. This is an example of a function whose left-hand and right-hand limits match neither each other nor the value of the function.

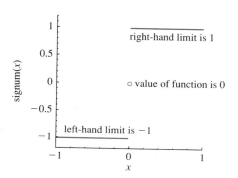

Figure 2.17
The signum function

Properties of Limits

Limits have many properties that simplify their computation, and the basic idea is important. First, we find the limits of some simple functions, and then we build limits of more complicated functions by combining these simple pieces with rules that tell how limits add, multiply, and divide.

The simple functions we begin with are constant functions, with formula

$$f(x) = c$$

and the **identity function**, with formula

$$f(x) = x$$

■ **THEOREM 2.1** (Limits of Basic Functions)

a. Suppose $f(x) = c$ for all x. Then, for any value of a,

$$\lim_{x \to a} f(x) = c$$

b. Suppose $f(x) = x$ for all x. Then, for any value of a,

$$\lim_{x \to a} f(x) = a$$

Second, we can combine limits by adding, multiplying, and dividing (as long as we avoid dividing by 0).

■ **THEOREM 2.2** (Rules for Combining Limits)

Suppose $f(x)$ and $g(x)$ are functions with well-defined limits at $x = a$.

a. The limit of the sum is the sum of the limits, or
$$\lim_{x \to a} [f(x) + g(x)] = \lim_{x \to a} f(x) + \lim_{x \to a} g(x)$$

b. The limit of the product is the product of the limits, or
$$\lim_{x \to a} f(x)g(x) = \left[\lim_{x \to a} f(x)\right] \cdot \left[\lim_{x \to a} g(x)\right]$$

c. The limit of the product of a constant and a function is the product of the constant and the limit of the function:
$$\lim_{x \to a} [cf(x)] = c \cdot \lim_{x \to a} f(x)$$

d. Suppose $\lim_{x \to a} g(x) \neq 0$. Then the limit of the quotient is the quotient of the limits, or
$$\lim_{x \to a} \left[\frac{f(x)}{g(x)}\right] = \frac{\lim_{x \to a} f(x)}{\lim_{x \to a} g(x)}$$

For example, Theorem 2.2 implies that
$$\lim_{x \to 7} 5 = 5$$
and
$$\lim_{x \to 7} x = 7$$
(Figure 2.18).

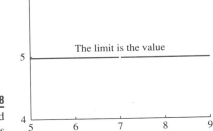
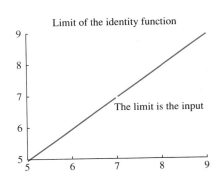

Figure 2.18
Limits of the constant and identity functions

Together, these theorems imply that we can find the limit of any **polynomial** by **plugging in the value**. Remember that a polynomial is a function that can be written as a sum of constants multiplied by powers of x. For example,
$$\lim_{x \to 5} 2x^2 + 3x = 2 \cdot 5^2 + 3 \cdot 5 = 65$$

For the average rate of change of a falling rock, we write
$$\lim_{\Delta t \to 0} 20.0 + 10.0\Delta t = 20.0 + 10.0 \cdot 0 = 20.0$$

More generally, these theorems make it easy to compute the limit of a **rational function**, defined as the **ratio of polynomials**. For instance,
$$\lim_{x \to 5} \frac{2x^2 + 3x}{2x + 1} = \frac{2 \cdot 5^2 + 3 \cdot 5}{2 \cdot 5 + 1} = \frac{65}{11} = 5.91$$

Left- and right-hand limits share the properties of ordinary limits summarized in Theorems 2.1 and 2.2: They add, multiply, and divide. For example, we can move constants outside of limits to find that

$$\lim_{x \to 0^+} \left(3\sqrt{x} + 2\right) = \lim_{x \to 0^+} 3\sqrt{x} + 2$$
$$= 3 \lim_{x \to 0^+} \sqrt{x} + 2$$
$$= 3 \cdot 0 + 2 = 2$$

Infinite Limits

One sometimes hears it said that a function gets "infinitely large" near some input. What might such a statement mean in scientific terms? As with ordinary limits, we can clarify this idea by thinking of measurements. It is impossible to measure a value of infinity. Instead, we encounter values that exceed the capacity of our measuring devices. A thermometer would explode if exposed to a temperature above its capacity. A value is **infinitely large** if exceeds the capacity of *every possible* measuring device. We say "the limit of $f(x)$ as x approaches a is equal to infinity" and write

$$\lim_{x \to a} f(x) = \infty$$

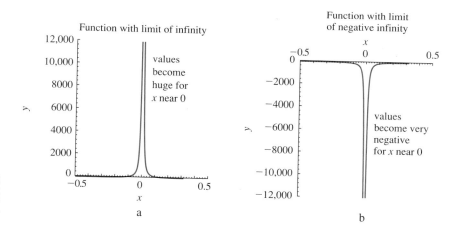

Figure 2.19
Functions with limits of infinity and negative infinity

Consider the function

$$f(x) = \frac{1}{x^2}.$$

(Figure 2.19a). If we plug in values of x closer and closer to 0, $f(x)$ becomes larger and larger. In fact, almost any calculator will give an error message if x is too small (some calculators are smart enough to say "infinity"). In such a case,

$$\lim_{x \to 0} f(x) = \infty$$

Similarly, a limit is equal to **negative infinity**, written $-\infty$, when the output becomes smaller than any given value. We write

$$\lim_{x \to a} f(x) = -\infty$$

136 Chapter 2 Limits and Derivatives

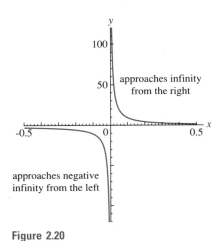

Figure 2.20
A function with a left-hand limit of $-\infty$ and a right-hand limit of ∞

The function

$$g(x) = \frac{-1}{x^2}$$

has a limit of negative infinity as x approaches 0 (Figure 2.19b).
Consider the function

$$h(x) = \frac{1}{x}$$

(Figure 2.20). Suppose we wish to find the limit as x approaches 0. This function shoots off to both positive and negative infinity near $x = 0$. The left-hand and right-hand limits help us to separate these two types of behavior. As we move toward 0 from the left (negative x), the values of the function get smaller. As we move toward 0 from the right (positive x), the values of the function get larger.

During our study of the natural logarithm, we indicated that $\ln(x)$ is not defined for negative values of x and that the graph of $\ln(x)$ rises from negative infinity near $x = 0$ (Figure 2.21). We can now describe this in terms of the limit as

$$\lim_{x \to 0^+} \ln(x) = -\infty$$

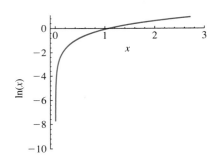

Figure 2.21
The natural logarithm approaching negative infinity

SUMMARY

Starting with scientific reasoning, we have developed an intuitive idea of the limit of a function. A function $f(x)$ approaches the limit L as x approaches a if the values of f become very close to L when x is very close to a. We defined **left-hand limits** and **right-hand limits** for situations in which functions or measurements are meaningful on only one side of a point. Limits have the useful property that they can be added, multiplied, and divided. Finally, we defined limits of **infinity** and **negative infinity**, based on the idea that a measurement is effectively infinite when it exceeds the capacity of any measuring device.

2.2 EXERCISES

1. Using a computer or calculator, estimate the following limits. Sketch the function.

   a. $\lim_{x \to 0} (1 + x)^{1/x}$

   b. $\lim_{x \to 0} \frac{\sin x}{x}$

   c. $\lim_{x \to 0} \frac{1 - \cos x}{x}$

   d. $\lim_{x \to 0^+} x^x$

   e. $\lim_{x \to 0^+} x \ln x$

   f. $\lim_{x \to 0^+} \frac{\ln(x)}{x}$

   g. $\lim_{x \to 1^-} \frac{\ln(1 - x)}{x}$

   h. $\lim_{x \to 1^+} \sqrt{\ln(x)}$

2. From the following pictures, find the left-hand and right-hand limits as x approaches 1.

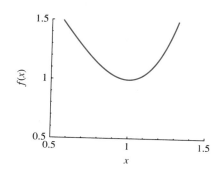

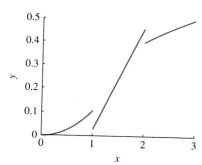

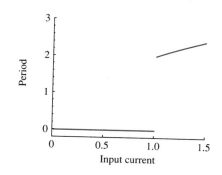

3. Use your calculator to check that if $g(x) = (e^x - 1)/x$, then
$$\lim_{x \to 0} g(x) = 1$$
Then find the following, being sure to state which theorem you used.
   a. $\lim_{x \to 0} 3g(x)$
   b. $\lim_{x \to 0} g(x) + 3$
   c. $\lim_{x \to 0} xg(x)$
   d. $\lim_{x \to 0} \dfrac{x}{g(x)}$
   e. $\lim_{x \to 0} \dfrac{g(x)}{x}$

4. Find the following limits.
   a. $\lim_{x \to 0} 5x + 7$
   b. $\lim_{x \to 1} 5x + 7$
   c. $\lim_{x \to 1} 5x^2 + 7x + 3$
   d. $\lim_{x \to 1} \dfrac{5x^2 + 7x + 3}{5x + 7}$
   e. $\lim_{x \to 1} \dfrac{5x + 7}{5x^2 + 7x + 3}$

5. Find the average rate of change of the following functions as a function of Δx. Find the limit. Graph the function, indicate the rate of change on your graph, and graph the average rate of change as a function of Δx.
   a. $f(x) = 5x + 7$ near $x = 0$
   b. $f(x) = 5x + 7$ near $x = 1$
   c. $f(x) = 5x^2$ near $x = 0$
   d. $f(x) = 5x^2$ near $x = 1$
   e. $f(x) = 5x^2 + 7x + 3$ near $x = 1$

6. Graph the following functions, and explain why the limit does not exist.
   a. The Heaviside function, defined by
   $$\begin{cases} H(x) = 0 & \text{if } x < 0 \\ H(x) = 1 & \text{if } x \geq 0 \end{cases}$$
   Try to find $\lim_{x \to 0} H(x)$.
   b. A function $j(x)$ is defined by
   $$\begin{cases} j(x) = 0 & \text{if } 0 < x < 0.5 \text{ or } 1.5 < x < 2.0 \\ j(x) = 1 & \text{if } 0.5 < x < 0.75 \text{ or } 1.25 < x < 1.5 \\ j(x) = 0 & \text{if } 0.75 < x < 0.875 \\ & \quad \text{or } 1.125 < x < 1.25 \end{cases}$$
   and so forth. Try to find $\lim_{x \to 1} j(x)$.
   c. $\lim_{x \to 0} \sin\left(\dfrac{2\pi}{x}\right)$.

7. Suppose that the volume $V(T)$ (in cubic centimeters), and hardness $H(T)$ of an object follow the equations
$$V(T) = \dfrac{1 + T^2}{1 + 2T}$$
$$H(T) = 0.1\left(8 + \dfrac{1}{1+T}\right)^2$$
where T is measured in Kelvins (so that $T = 0$ is absolute 0). We are interested in the values of these measurements at $T = 0$, but we cannot measure it because it is impossible to reach absolute 0.

a. Find the limits of $V(T)$ and $H(T)$.
b. If we could lower temperature to 2 K, how close would we be to the limit for each?
c. If we could lower temperature to 1 K, how close would we be to the limit for each? Do the same for 0.1 K.

8. Consider the functions
$$f_1(x) = \sqrt{x}$$
$$f_2(x) = x$$
$$f_3(x) = x^2$$
defined for $x \geq 0$.
a. For each function, how close must the input be to 0 for the output to be within 0.1 of the limit?
b. To be within 0.001 of the limit?
c. Which function would you say approaches 0 the fastest?

9. Consider the functions
$$g_1(x) = \frac{1}{\sqrt{x}}$$
$$g_2(x) = \frac{1}{x}$$
$$g_3(x) = \frac{1}{x^2}$$
defined for $x \geq 0$.
a. For each function, how close must the input be to 0 for the output to be greater than 100?
b. To be greater than 10^4?
c. Which function would you say approaches infinity the fastest?

10. Suppose that
$$\lim_{x \to 1} f(x) = 0$$
a. Find a function $f(x)$ such that
$$\lim_{x \to 1} \frac{1}{f(x)} = \infty$$
b. Find a function $f(x)$ such that
$$\lim_{x \to 1} \frac{1}{f(x)} = -\infty$$
c. Find a function $f(x)$ such that
$$\lim_{x \to 1^+} \frac{1}{f(x)} = \infty \quad \text{and} \quad \lim_{x \to 1^-} \frac{1}{f(x)} = -\infty$$
d. Find a function $f(x)$ such that
$$\lim_{x \to 1^+} \frac{1}{f(x)} = -\infty \quad \text{and} \quad \lim_{x \to 1^-} \frac{1}{f(x)} = \infty$$
e. Are there any other possibilities?

11. **COMPUTER**: There are various ways to find the smallest and largest numbers your calculator or computer can handle.
a. Try doubling some number until you get an overflow.
b. Try halving some number until you get an underflow.
c. Compute $1.0 + \epsilon$ for smaller and smaller values of ϵ. At what point does your calculator return 1.0 as the answer?
d. Compare the answers of parts **b** and **c**. Which value do you think is more important in computation?

2.3 Continuity

Even with the useful properties of limits, computing them by graphing can be rather slow. When we are finding the limit of some function $f(x)$ as x approaches a, we would like simply to compute $f(a)$. In this section, we learn the conditions under which this method works. The chief new tool is the **continuous function**, the mathematically precise definition of a function without "jumps." The definition of the continuous function depends on the limit, and continuous functions share the useful properties of limits. Scientifically, continuous functions represent relations between measurements where a small change in the input produces a small change in the output. Functions that are not continuous do occur, and they pose challenges for scientists.

Continuous Functions

We can find
$$\lim_{x \to 0} 2x + 1 = 1$$

in two ways. We can plug in values of x close to 0 and see that the results get close to 1. Alternatively, we could break the function into component parts and use the properties of limits. But why not simply compute $2 \cdot 0 + 1 = 1$ to find the limit?

We can do just this when the function is **continuous**. What is a continuous function? Our intuitive notion of "continuous" is "connected," without jumps. Figure 2.22 illustrates functions with and without jumps. The graph of a continuous function can be drawn without lifting one's pencil.

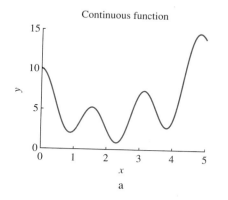
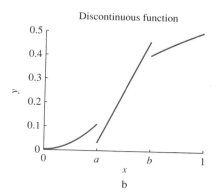

Figure 2.22
Continuous and discontinuous functions

The mathematical definition is supposed to capture this intuitive idea.

■ **Definition 2.3**   A function f is **continuous** at a point a if

$$\lim_{x \to a} f(x) = f(a)$$

Otherwise, we say the function is **discontinuous** at the point a. ■

This definition requires that the left- and right-hand limits are equal. As the graph approaches a particular point in the domain, the value of a continuous function gets closer and closer to the value of the function at that point. Otherwise, the graph would have a jump and could not be drawn without lifting a pencil.

Continuity is defined at *points*. A function can be continuous at some points and discontinuous at others. The points a and b are points of discontinuity of the function in Figure 2.22b; the function is continuous at all other points. When a function is said to be continuous without specifying a particular point, the function is continuous at *all* points of its domain.

■ **Definition 2.4**   A function is **continuous** if it is continuous at every point in its domain. ■

What sorts of functions are not continuous? We have met examples of several types. Some functions have jumps—for example, the signum function, which is equal to -1 for negative arguments, 0 when the argument is 0, and 1 for positive arguments (Figure 2.23). At the point $x = 0$, the value of the function is 0; the left-hand limit is -1, and the right-hand limit is 1. Because the left-hand limit and the right-hand limit do not match, the limit does not exist. Furthermore, neither of these limits matches the value of the function.

Functions with limits of infinity or negative infinity are not defined at all points and cannot be graphed without a jump (strictly speaking, these functions can be

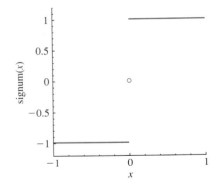

Figure 2.23

The signum function is not continuous at 0

continuous at all points in their domain). In Figure 2.24a, the limit is infinity from both below and above. However, the function cannot be evaluated at $x = 0$ and has a jump. In Figure 2.24b, the limit is negative infinity from below and positive infinity from above. This function has an infinitely large jump.

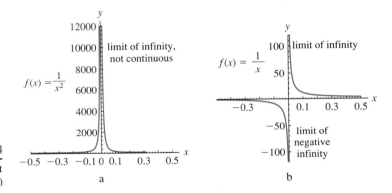

Figure 2.24

Two functions that are not defined at $x = 0$

Continuous functions are useful because we can find limits by plugging in. How can we recognize a continuous function? Like limits, continuous functions can be combined in many useful ways, including addition, multiplication, division, and composition. Mathematically, we build the edifice of continuous functions on a stable foundation of two theorems: one giving the basic continuous functions and the other saying how they can be combined.

■ **THEOREM 2.3** (The Basic Continuous Functions)

    a. The constant function $f(x) = c$ is continuous.

    b. The identity function $f(x) = x$ is continuous.

    c. The exponential function $f(x) = e^x$ is continuous.

    d. The logarithmic function $f(x) = \ln(x)$ is continuous for $x > 0$.

    e. The cosine function $f(x) = \cos x$ is continuous. ■

■ **THEOREM 2.4** (Combining Continuous Functions)

Suppose $f(x)$ and $g(x)$ are continuous at $x = a$.

a. The sum $f(x) + g(x)$ is continuous at $x = a$.

b. The product $f(x)g(x)$ is continuous at $x = a$.

c. The quotient $f(x)/g(x)$ is continuous at $x = a$ if $g(a) \neq 0$.

d. If $f(x)$ is continuous at $x = g(a)$, then the composition $(f \circ g)(x)$ is continuous at $x = a$. ■

The rules for recognizing continuous functions are summarized in three tables: the basic continuous functions (Table 2.1), the rules for combining (Table 2.2), and the three trouble signs (Table 2.3).

Table 2.1 Basic Continuous Functions

Type of function	Example	Where continuous
Linear	$f(x) = mx + b$	All x
Polynomial	$f(x) = ax^3 + bx^2 + cx + d$	All x
Exponential	$f(x) = e^{ax}$	All x
Cosine	$f(x) = \cos(x)$	All x
Logarithm	$f(x) = \ln(x)$	$x > 0$

Table 2.2 Continuous Combinations of Continuous Functions $f(x)$ and $g(x)$

Type of combination	Formula	Where continuous	Example
Product	$f(x)g(x)$	Where defined	$(2x + 1)e^x$
Sum	$f(x) + g(x)$	Where defined	$2x + 1 + e^x$
Composition	$(f \circ g)(x)$	Where defined	e^{2x+1}
Quotient	$\dfrac{f(x)}{g(x)}$	Where $g(x) \neq 0$	$\dfrac{e^x}{2x+1}$ if $2x + 1 \neq 0$

Table 2.3 The Three Trouble Signs: Points Where a Function May Be Undefined or Discontinuous

Division by 0
Logs of 0
Points where the definition of the function changes

Suppose we want to find

$$\lim_{x \to 0} f(x)$$

where

$$f(x) = e^{x^2 + 3}$$

If the function is continuous at $x = 0$, we can evaluate the limit by evaluating $f(0)$. Is the function continuous at $x = 0$? This function is built up as a **composition** of the polynomial function $x^2 + 3$ and the exponential function. Both pieces are continuous (Table 2.1), and the composition of these pieces is also continuous (Table 2.2). Therefore,

$$\lim_{x \to 0} e^{x^2 + 3} = e^{0^2 + 3} = e^3 = 20.08$$

If we want to find

$$\lim_{x \to 0} g(x)$$

where

$$g(x) = \frac{f(x)}{h(x)} = \frac{e^{x^2+3}}{2 + x + \ln(x+1)}$$

we must take more care in taking apart the function. First, it is the quotient of the previous function, $f(x) = e^{x^2+3}$, which is continuous, and a new function $h(x) = 2 + x + \ln(x+1)$. The function $h(x)$ is the sum of a continuous linear function $2 + x$ and the natural log, which is continuous when its argument is positive. The quotient will then be continuous at 0 as long as $h(0) \neq 0$. But

$$h(0) = 2 + 0 + \ln(0 + 1) = 2 + 0 + 0 = 2$$

Therefore,

$$\lim_{x \to 0} \frac{e^{x^2+3}}{2 + x + \ln(x+1)} = \frac{e^{0^2+3}}{2} = \frac{e^3}{2} = 10.04$$

Similarly, the function

$$P_d(t) = 36.8 + 0.3 \cos\left[\frac{2\pi(t-14)}{24}\right]$$

is continuous because it is built from the linear function

$$\frac{2\pi(t-14)}{24}$$

composed with the cosine function, then composed with another linear function

$$g(c) = 36.8 + 0.3c$$

Division by 0 can be disguised. Suppose we are asked to find

$$\lim_{x \to 1^-} F(x) = \lim_{x \to 1^-} (1-x)^{-3}$$

Negative exponents can be written in the denominator (Law 3 of exponents), so

$$F(x) = \frac{1}{(1-x)^3}$$

Because the denominator is equal to 0 at $x = 1$, this function may not be continuous. If we try to evaluate $F(1)$, we find ourselves dividing by 0. In fact, the limit is equal to infinity.

Functions defined in pieces look like the signum function (Figure 2.23) or the functions used to describe the heart (Section 1.13). An example is the updating function

$$V_{t+1} = \begin{cases} 3.0V_t & \text{if } V_t > 2.0 \\ 2.0V_t + 1.0 & \text{if } V_t \leq 2.0 \end{cases} \quad (2.3)$$

The only way to check whether this updating function is continuous is to check whether the two parts match up at $V_t = 2$, the point where the definition changes. As shown in Figure 2.25, the value at $V_t = 2$ jumps from 6.0 to 5.0. Because these do not match, the function is not continuous. In contrast, the function

$$V_{t+1} = \begin{cases} 3.0V_t & \text{if } V_t > 2.0 \\ 2.0V_t + 2.0 & \text{if } V_t \leq 2.0 \end{cases}$$

is continuous because the two pieces match (Figure 2.25b).

2.3 Continuity

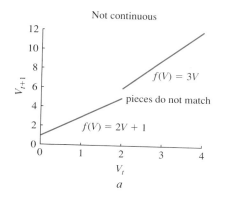
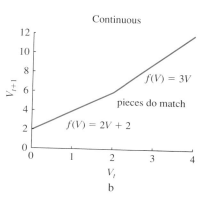

Figure 2.25
Two functions that are defined in pieces

Input and Output Tolerances

In applied mathematics, functions describe relationships between measurements. Continuous functions therefore represent a special sort of relationship between measurements, characterized by input and output tolerances.

Recall the updating function

$$b_{t+1} = 2.0 b_t$$

describing a bacterial population (Equation 1.1). Suppose we want a population of 2.0×10^6 at time $t = 1$. To hit 2.0×10^t exactly, we require $b_0 = 1.0 \times 10^6$. How close must b_0 be to 1.0×10^6 for b_1 to be within 0.1×10^6 of our target (Figure 2.26a)? We require

$$1.9 \times 10^6 \leq b_1 \leq 2.1 \times 10^6$$
$$1.9 \times 10^6 \leq 2.0 b_0 \leq 2.1 \times 10^6$$
$$0.95 \times 10^6 \leq b_0 \leq 1.05 \times 10^6$$

As long as we can guarantee that b_0 is within 0.05×10^6 of 1.0×10^6, our output is within our tolerance. A small error in the input produces a small (but somewhat larger) error in the output. In other words, the output tolerance of 0.1×10^6 translates into an input tolerance of 0.05×10^6.

Similarly, consider the updating function

$$M_{t+1} = 0.5 M_t + S$$

Figure 2.26
Tolerances and continuous functions

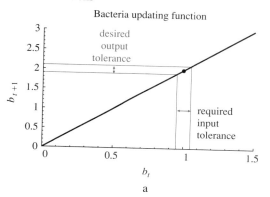
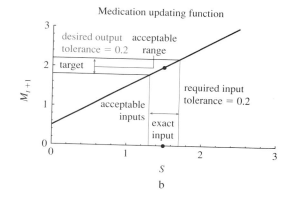

for the concentration of medication in the blood (see Section 1.5), modified so that the dosage is a variable S rather than the constant 1.0. Suppose the concentration on day t is $M_t = 1.0$ and we wish to hit $M_t = 2.0$. What value of S should be used? The value of M_{t+1} is

$$M_{t+1} = 0.5 \cdot 1.0 + S = 0.5 + S$$

Therefore, to hit $M_{t+1} = 2.0$ exactly, we require $S = 1.5$. Suppose we really need be only within 0.2 of the exact target. Because M_{t+1} is a continuous function of S, there should also be a tolerance around $S = 1.5$. For M_{t+1} to be between 1.8 and 2.2,

$$1.8 \leq 0.5 + S \leq 2.2$$
$$1.3 \leq S \leq 1.7$$

(Figure 2.26b). If the input S is within 0.2 of 1.5, the output M_{t+1} will be within the desired tolerance of 2.0. If we needed to hit 2.0 exactly, we would need to use a method that offers better control.

In contrast, if the updating function is discontinuous, a tiny error in input can produce a huge error in output. Suppose we wish to have a voltage $V_1 = 5.0$ from the updating function

$$V_{t+1} = \begin{cases} 3.0V_t & \text{if } V_t > 2.0 \\ 2.0V_t + 1.0 & \text{if } V_t \leq 2.0 \end{cases}$$

(Equation 2.3). We could hit $V_1 = 5.0$ exactly with $V_0 = 2.0$. But if the input is even a tiny bit too high, the output will be very different. If $V_0 = 2.001$, then $V_1 = 6.003$ (Figure 2.27). If our output tolerance is 0.1, meaning that only values between 4.9 and 5.1 are satisfactory, there is no input tolerance at all.

Relations described by functions that are not continuous are difficult to deal with experimentally. Because no measurement can be controlled exactly, we hope that small errors in one measurement will not result in large errors in another. Systems with threshold behavior, or that display hysteresis (as in the Hysteresis section), can include discontinuous relations and must be treated with caution.

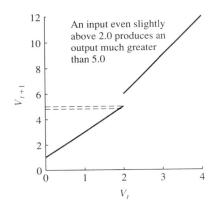

Figure 2.27
A discontinuous function with no input tolerance

Hysteresis

Different left-hand and right-hand limits arise in biological systems that display **hysteresis** (Figure 2.28). In the experiment shown, a neuron is stimulated with different levels of electric current. Low levels of current generally have no effect; large levels tend to induce an oscillation. If the current starts at a low level and increases steadily, the neuron switches from no oscillation (designated as period 0 on the graph) to oscillation with period 2.0 when the input current crosses 3.0. If the current starts at a high value and decreases steadily, the neuron switches from an oscillation of period 1 to no oscillation when the input current crosses 2.0. In both cases, the jump indicates that the left-hand and right-hand limits do not match.

More oddly, the two graphs jump at different points. If we wished to *measure* the left-hand limit, we would slowly increase the current toward the jump, finding the results in Figure 2.28a. If we wished to measure the right-hand limit, we would

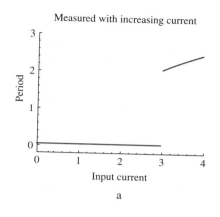
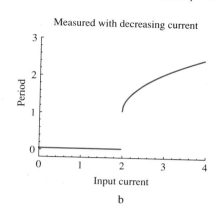

Figure 2.28 Hysteresis

slowly decrease the current toward the jump, finding the results in Figure 2.28b. However, both the position and the size of the jump depend on which way we change the current.

Why do the graphs fail to match? The idea is related to the idea of equilibria. When the current increases, the equilibrium without oscillation eventually becomes **unstable** and the cell begins to oscillate. When the current decreases, the oscillation itself becomes unstable and the cell stops oscillating. For input currents between 2.0 and 3.0, *both* the oscillation and the lack of oscillation are stable and the neuron's behavior depends on what it was last doing. Hysteresis is a simple form of memory, because knowing the input current at one time is insufficient to describe the behavior of the system.

If the device starts off tilted to the left, the ball is on the left when device is flat

If the device starts off tilted to the right, the ball is on the right when device is flat

Figure 2.29 Hysteresis and stability

Hysteresis can be pictured with a physical device. Figure 2.29 compares two experiments on a device that has two possible stable resting places for a ball. In the top panel, the device begins tilted to the left, with the ball resting in the left depression, and is slowly rotated toward the right. When the device is flat, the ball is still on the left side. As the device is tilted to the right, the ball rolls into the right depression. In the bottom panel, the device begins tilted to the right, with the ball resting in the right depression, and is slowly rotated toward the left. When the device is flat, the ball is still on the right side. The position of the ball is an example of hysteresis (Figure 2.30). If you were asked to guess the position of the ball when the device is flat, you could say only that it depended on how it got there.

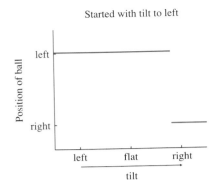
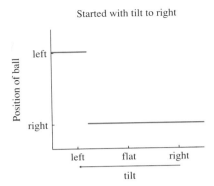

Figure 2.30
The position of a ball as a function of tilt

SUMMARY

Many important mathematical and scientific functions are **continuous**, meaning that they have graphs with no jumps. In these cases, limits can be computed by evaluating the function. Many continuous functions are built from four basic pieces: polynomials (including linear functions), exponentials, logarithms, and cosines, combined with addition, multiplication, division (except by 0), and functional composition. Most functions encountered in biology are continuous unless they exhibit one of the three warning flags: division by 0, taking the logarithm of 0, or definition in parts. Continuous functions describe scientific measurements when small changes in input produce only small changes in output. With a discontinuous function, even a tiny error in input can produce a large jump in output. We examined discontinuous functions created by **hysteresis**, where the results of an experiment depend on the order in which conditions are tested.

2.3 EXERCISES

1. Describe how the following functions are built from the basic continuous functions. Where might they be discontinuous?
    a. $l(t) = 5t + 6$
    b. $p(x) = x^5 + 6x^3 + 7$
    c. $f(x) = \dfrac{e^x}{x + 1}$
    d. $h(y) = y^2 \ln(y - 1)$
    e. $g(z) = \dfrac{\ln(z - 1)}{z^2}$
    f. $F(t) = \cos(e^{t^2})$
    g. $\sin(x)$ [recall that $\sin(x) = \cos(x - \pi/2)$]
    h. $a(t) = t^2$ if $t > 0$ and 0 otherwise
    i. $r(w) = (1 - w)^{-4}$

2. Using the functions in Exercise 1, find the following limits by plugging in (if possible). Compute the value of the function 0.1 and 0.01 above and below the limiting argument to see if your answer is correct.

    a. $\lim\limits_{t \to 5} l(t)$
    b. $\lim\limits_{x \to 2} p(x)$
    c. $\lim\limits_{x \to 0} f(x)$
    d. $\lim\limits_{y \to 1} h(y)$
    e. $\lim\limits_{y \to 2} h(y)$
    f. $\lim\limits_{z \to 2} g(z)$
    g. $\lim\limits_{z \to 0} g(z)$
    h. $\lim\limits_{t \to 2} F(t)$
    i. $\lim\limits_{t \to 0} a(t)$
    j. $\lim\limits_{t \to 3} a(t)$
    k. $\lim\limits_{w \to 1} r(w)$

3. Find the accuracy of input necessary to achieve the desired output accuracy.
    a. Suppose the mass of an object as a function of volume is given by $M = \rho V$ (Equation 1.11). If $\rho = 2.0$ g/cm³, how close must V be to 2.5 cm³ for M to be within 0.2 g of 5.0 g?
    b. The area of a disk as a function of radius is given by $A = \pi r^2$. How close must r be to 2.0 cm to guarantee an area within 0.5 cm² of 4π?

c. The flow rate F through a pipe is proportional to the fourth power of the radius, or
$$F(r) = ar^4$$
Suppose $a = 1.0$/cm s. How close must r be to 1.0 cm to guarantee a flow within 5% of 1 cm^3/s?

d. Consider an organism growing according to $S(t) = S(0)e^{\alpha t}$. Suppose $\alpha = 0.001$/s, and $S(0) = 1.0$ mm. At time 1000 s, $S(t) = 2.71828$ mm. How close must t be to 1000 s to guarantee a size within 0.1 mm of 2.71828 mm?

4. Suppose a population of bacteria obeys the updating function
$$b_{t+1} = 2.0 b_t$$
and we wish to have a population within 1.0×10^8 of 1.0×10^9 at $t = 10$.

a. What values of b_9 produce a result within the desired tolerance? What is the input tolerance?
b. What values of b_5 produce a result within the desired tolerance? What is the input tolerance?
c. What values of b_0 produce a result within the desired tolerance? What is the input tolerance?
d. Why is it harder and harder to hit the target from farther away?

5. Suppose the amount of toxin in a culture declines according to
$$T_{t+1} = 0.5 T_t$$
and we wish to have a concentration within 0.02 of 0.5 g/L at $t = 10$.

a. What values of T_9 produce a result within the desired tolerance? What is the input tolerance?
b. What values of T_5 produce a result within the desired tolerance? What is the input tolerance?
c. What values of T_0 produce a result within the desired tolerance? What is the input tolerance?
d. Why is it easier and easier to hit the target from farther away?

6. The following functions do odd things near $x = 0$. Try to figure out what they do and graph the function.
a. $f(x) = (e^x - 1)/x$.
b. $g(x) = e^{-1/x}$.
c. $h(x) = e^{-1/x^2}$. Compute $h(0.5)$, $h(0.3)$, and $h(0.1)$.

7. Suppose a neuron has the following response to inputs. If it receives a voltage input V greater than or equal to a threshold of V_0, it outputs a voltage of kV for some constant k. If it receives an input less than the threshold value of V_0, it outputs a fixed voltage V^*. Suppose that $k = 2.0$, $V_0 = 50$, and $V^* = 80$.
a. Write this as a function defined in pieces.
b. Graph output as a function of input.
c. What would V^* have to be for the function to be continuous? Graph the resulting function.
d. What would k have to be to make the function continuous? Graph the resulting function.

8. We can build a continuous approximation of signum (the function giving the sign of a number) as follows.
a. Graph a continuous function that is -1 for $x \le -0.1$, 1 for $x \ge 0.1$, and linear for $-0.1 < x < 0.1$.
b. Find the formula for this function.
c. Do the same, but replace 0.1 by the small number ϵ.
d. What happens as ϵ approaches 0?
e. If you want to hit an output within 0.1 of 0, how hard would it be? How hard would it be for signum itself?

9. A child is swinging on a swing that makes a horrible screeching noise. Starting from when the swing is furthest back, the pitch of the screeching noise increases as it swings forward and then decreases as it swings back.
a. Draw a graph of the pitch as a function of position without hysteresis.
b. Draw a graph with hysteresis. Which graph seems more likely?
c. Imagine the sound each graph describes. Which is more irritating?

10. Little Billy walks due east to school, but he must cross from the south side to the north side of the street. Because he is a careful child, he crosses at great speed at the first possible opportunity.
a. Graph Little Billy's latitude as a function of distance from home on the way to school.
b. Graph Little Billy's latitude as a function of distance from home on the way home.
c. Is this an example of hysteresis?

11. **COMPUTER:** Consider a population with
$$\text{per capita reproduction} = r \frac{x_t}{1.0 + x_t^2}$$
a. Write the updating function.
b. Set $r = 3$ and $x_0 = 2$. Compute the solution until it gets close to an equilibrium, and record this value.
c. Decrease r to 2.9. Use the equilibrium found with $r = 3$ as an initial condition to compute the solution until it approaches an equilibrium. Record this value.
d. Continue decreasing r down to $r = 1$, using the last equilibrium as the initial condition. What happens when r crosses 2?
e. If this were a real population and r were a measure of the quality of a habitat, how would you interpret this behavior?
f. Follow the same procedure, but start with $r = 1$ and increase r to 3.
g. Can you explain what is going on?

12. **COMPUTER:** Consider again the basic updating function for medication,

$$M_{t+1} = 0.5M_t + 1.0$$

Figure out how to run this system backward. What happens if you start from an initial condition of 2.1? An initial condition of 1.9? What does this have to do with input and output tolerances?

13. **COMPUTER:** Consider again the updating function for mites on lizards,

$$x_{t+1} = 2.0M_t + 30$$

Figure out how to run this system backward. What happens if you start from an initial condition of 200? If you found a lizard with 200 mites, how many mites did it have 4 weeks ago?

2.4 Computing Derivatives: Linear and Quadratic Functions

We now begin building a set of techniques to compute derivatives. First, we learn to recognize those functions, called **differentiable functions**, that have derivatives. Second, we use the definition of the derivative and find the derivatives of two simple functions, a **linear function** and a **quadratic function**. Along the way, we will see that the derivative, thought of as the slope of a graph, can help us reason more effectively about the measurements that graphs depict. We can tell whether a function is increasing or decreasing simply by checking whether the derivative is positive or negative.

Differentiable Functions

Not all functions have derivatives. A function that has a well-defined derivative is called **differentiable**. The function shown in Figure 2.31 is not differentiable at three points in three different ways. At a, the function is not continuous; there is no way to draw a tangent. In a graph of position against time, a point of discontinuity indicates that the object jumped instantly from one place to another. The idea of speed or instantaneous rate of change makes no sense. At b, there is a corner; at such a point, the object has instantly changed direction and has no well-defined speed. At c, the tangent line is vertical; vertical lines have "infinite slope," meaning, loosely speaking, that the object has infinite speed at this time.

The three ways a function can fail to be differentiable
The graph has a jump
The graph has a corner
The graph is vertical

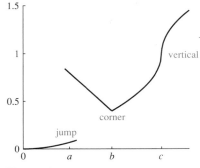

Figure 2.31
The graph of a function that is not differentiable at $x = a$, $x = b$, and $x = c$

Differentiability, like continuity, is a property of a function at a point. If we say that a function is differentiable without mentioning a particular point, the function is differentiable on its entire domain. The function shown in Figure 2.31 is differentiable everywhere except at points a, b, and c.

Linear Functions

What is the slope of the graph of the function $f(x) = 2x + 1$ at $x = 1$? We now have two ways to answer this question. Because f is a linear function (written

2.4 Computing Derivatives: Linear and Quadratic Functions

here in slope-intercept form), we know that the slope is the factor 2 multiplying x. Alternatively, we could compute the slope by computing the derivative, which requires finding the slope of a secant line and taking the limit as Δx approaches 0. These two methods should give the same answer.

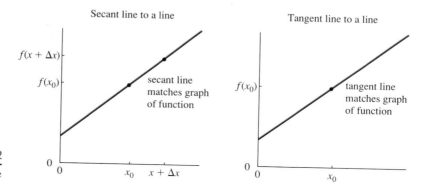

Figure 2.32
The secant and tangent lines for a line

The secant line to $f(x) = 2x + 1$ connects two points on the graph of the function. Because the graph of the function is straight, the graph of the secant line matches the graph of the function (Figure 2.32). With base point $x_0 = 1.0$ and $\Delta x = 0.5$,

$$\begin{aligned}
\text{slope of secant} &= \frac{\Delta f}{\Delta x} \\
&= \frac{f(x_0 + \Delta x) - f(x_0)}{\Delta x} \\
&= \frac{f(1.0 + 0.5) - f(1.0)}{0.5} \\
&= \frac{f(1.5) - f(1.0)}{0.5} \\
&= \frac{4.0 - 3.0}{0.5} \\
&= 2.0
\end{aligned}$$

The slope of the secant matches the slope of the function. In general, with base point x_0 and second point $x_0 + \Delta x$,

$$\begin{aligned}
\text{slope of secant} &= \frac{\Delta f}{\Delta x} \\
&= \frac{f(x_0 + \Delta x) - f(x_0)}{\Delta x} \\
&= \frac{2(x_0 + \Delta x) + 1 - (2x_0 + 1)}{\Delta x} \\
&= \frac{2\Delta x}{\Delta x} \\
&= 2
\end{aligned}$$

as long as $\Delta x \neq 0$. Every secant line has slope 2 and coincides exactly with the graph of the original function.

The slope of the tangent is the limit as Δx approaches 0 of the slope of the secant. The slope of the secant is the **constant** function 2. This function is **continuous** (Theorem 2.3a), and we can find the limit by evaluating the function at $\Delta x = 0$, so

$$\text{slope of tangent} = \lim_{\Delta x \to 0} 2 = 2$$

The derivative does match the usual idea of the slope of a line.

In general, we can follow the same steps with the linear function $f(x) = mx + b$. The secant line connecting points x_0 and $x_0 + \Delta x$ has slope

$$\begin{aligned}
\text{slope of secant} &= \frac{\Delta f}{\Delta x} \\
&= \frac{f(x_0 + \Delta x) - f(x_0)}{\Delta x} \\
&= \frac{[m(x_0 + \Delta x) + b] - (mx_0 + b)}{\Delta x} \\
&= \frac{m \Delta x}{\Delta x} \\
&= m
\end{aligned}$$

as long as $\Delta x \neq 0$. The derivative of a linear function, the limit of m as Δx approaches 0, is equal to the slope m of the line.

In differential and prime notation,

$$\frac{df}{dx} = m \quad \text{differential notation}$$

$$f'(x_0) = m \quad \text{prime notation}$$

Sometimes we use a different version of differential notation and write

$$\frac{d(mx + b)}{dx} = m$$

or

$$\frac{d}{dx}(mx + b) = m$$

One special case is worthy of note: The function

$$f(x) = b$$

the constant function, is a linear function with slope 0. According to the general formula for the derivative of a linear function,

$$\frac{df}{dx} = 0$$

when $f(x)$ is constant. The rate of change of something that does not change is 0 (Figure 2.33).

Lines can have positive, negative, or zero slope, corresponding to increasing, decreasing, and constant functions (Section 1.4). Because the derivative is equal to the slope, this correspondence holds in general. If the derivative is positive, the rate of change of the function is positive and the function is **increasing** (Figure 2.34a). If the derivative is negative, the function is **decreasing** (Figure 2.34b). If the derivative is exactly 0, the function is neither increasing nor decreasing (Figure 2.34c) at that point. Points where the derivative is 0, called **critical points**, will prove extremely useful for finding maxima and minima of functions (Section 3.3).

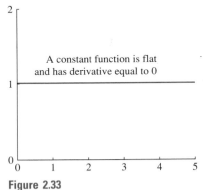

Figure 2.33
The slope of a constant function is 0

2.4 Computing Derivatives: Linear and Quadratic Functions

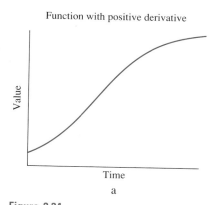
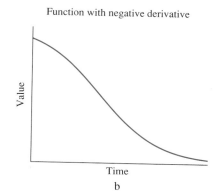
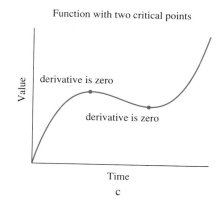

Figure 2.34
Functions with positive and negative derivatives

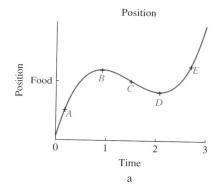
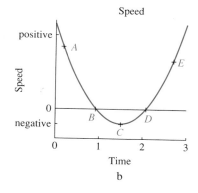

Figure 2.35
A function and its derivative

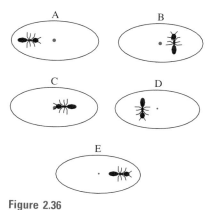

Figure 2.36
The experiences of an ant

Furthermore, we know that lines with large slopes are steep. Large values of the derivative are therefore associated with points where the graph of the function is steep. The derivative of the function plotted in Figure 2.35a is shown in Figure 2.35b. Steeply increasing portions of the graph correspond to large positive values of the derivative; steeply decreasing portions of the graph correspond to large negative values of the derivative. The derivative passes through 0 wherever the graph is flat.

Suppose Figure 2.35a represents the position of an ant as a function of time. There is a bit of food at position C. At point A the ant is moving forward quickly toward the food. At B, the ant has overshot the food and stopped. At C the ant is walking slowly through the food in a reverse direction. By D the ant has passed the food and stopped, preparatory to turning around and walking quickly away at E (Figure 2.36).

A Quadratic Function

Consider a rock dropped from a building in a constant gravitational field. The rock falls a distance y (measured in meters) in time t (measured in seconds), where $y(t)$ satisfies

$$y(t) = 5t^2$$

152 Chapter 2 Limits and Derivatives

What is the velocity of the rock after 1.0 s? (We use **velocity** as a more precise term for speed.) The velocity is the rate of change of position. The average velocity between $t = 1.0$ and $t = 1.0 + \Delta t$ is found by dividing the distance traveled by the time elapsed, or

$$\begin{aligned}
\text{average velocity} &= \frac{\text{distance}}{\text{time}} \\
&= \frac{\Delta y}{\Delta t} \\
&= \frac{y(1.0 + \Delta t) - y(1.0)}{\Delta t} \\
&= \frac{5(1.0 + \Delta t)^2 - 5}{\Delta t} \\
&= \frac{(5 + 10\Delta t + 5\Delta t^2) - 5}{\Delta t} \\
&= \frac{10\Delta t + 5\Delta t^2}{\Delta t} \\
&= 10 + 5\Delta t
\end{aligned}$$

as long as $\Delta t \neq 0$. To compute the derivative, we must take the limit of this function as Δt approaches 0. The average velocity is a linear function of Δt and is therefore continuous (Table 2.1), and so we can find the limit by evaluating at $\Delta t = 0$,

$$\begin{aligned}
\text{velocity} &= \lim_{\Delta t \to 0} (10 + 5\Delta t) \\
&= 10 + 5 \cdot 0 = 10
\end{aligned}$$

What are the units of the velocity? The change in distance, Δy, is measured in meters, and the change in time, Δt, is measured in seconds. The velocity at $t = 1.0$ is therefore 10 m/s.

We can follow this procedure to find the exact velocity at any time t. The velocity is a function of time because it is constantly changing (Figure 2.37). The average velocity between times t and $t + \Delta t$ is

$$\text{average velocity} = \frac{\Delta y}{\Delta t}$$

Figure 2.37

The slope of a quadratic function is constantly changing

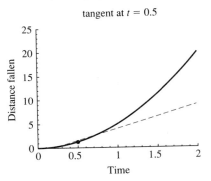
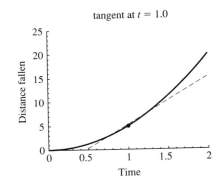
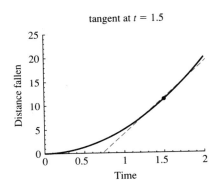

$$= \frac{y(t+\Delta t) - y(t)}{\Delta t}$$

$$= \frac{5(t+\Delta t)^2 - 5t^2}{\Delta t}$$

$$= \frac{5(t^2 + 2t\Delta t + \Delta t^2) - 5t^2}{\Delta t}$$

$$= \frac{(5t^2 + 10t\Delta t + 5\Delta t^2) - 5t^2}{\Delta t}$$

$$= \frac{10t\Delta t + 5\Delta t^2}{\Delta t}$$

$$= 10t + 5\Delta t$$

as long as $\Delta t \neq 0$. Taking the limit by evaluating this continuous function at $\Delta t = 0$, we find

$$\text{velocity} = \lim_{\Delta t \to 0} 10t + 5\Delta t$$
$$= 10t + 5 \cdot 0 = 10t$$

In differential and prime notation, we have

$$\frac{dy}{dt} = 10t \quad \text{differential notation}$$

$$y'(t) = 10t \quad \text{prime notation}$$

$$\frac{d}{dt}(5t^2) = 10t \quad \text{alternative differential notation}$$

The function $y'(t)$ gives the velocity as a function of time (Figure 2.38). Does the result make sense? At time $t = 0$, the velocity is 0. As time passes, the velocity increases linearly, consistent with the behavior of a falling object.

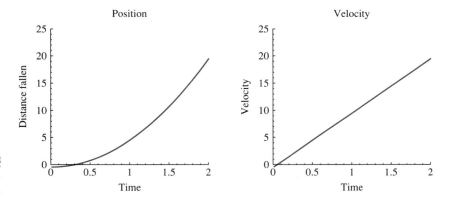

Figure 2.38
Position and velocity as functions of time

When written in differential notation, the equation

$$\frac{dy}{dt} = 10t$$

is a **differential equation**, giving a formula for the rate of change as a function of time. In this case, we know the **solution**, $y(t) = 5t^2$, of the differential equation because we began with a formula for $y(t)$ itself.

SUMMARY

A function might fail to have a derivative at a particular point in three ways: if the graph has a jump, a corner, or is vertical. The graphs of linear and quadratic functions have none of these problems, and we can compute their derivatives directly from the definition. The derivative of a linear function is equal to the usual idea of slope, and we saw in general that the derivative is positive when the function is increasing and negative when the function is decreasing. Points where the function is neither increasing nor decreasing are characterized by a derivative of 0 and are called **critical points**. The derivative of a quadratic function is a linear function.

2.4 EXERCISES

1. On the following two graphs, identify points where
   a. the function is not continuous,
   b. the function is not differentiable (and say why),
   c. the derivative is 0.

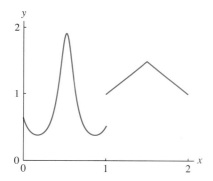

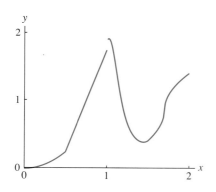

2. Find the derivatives of the following functions. Write your answers in both differential and prime notation.
   a. $M(x) = 0.5x + 2$
   b. $L(t) = 2t + 30$
   c. $g(y) = -3y + 5$
   d. $Q(z) = -3.5 \times 10^8$

3. Find the slope of the secant line and use the definition of the derivative to find the slope of the following lines. Graph the lines. Make sure that the units make sense.
   a. $M = 3a + 1$ where a is age in days and M is mass in grams. Find the secant at $a = 1$ with $\Delta a = 0.5$ and then with $\Delta a = -0.5$.
   b. $V = 2.2a + 4$ where a is age in days and V is volume in cubic centimeters. Find the secant at $a = 2$ and $\Delta a = 0.5$ and then with $\Delta a = -0.5$.
   c. $G = -8a + 34$ where a is age in days and G is glucose production in grams. Find the secant at $a = 3$ and $\Delta a = 0.5$ and then with $\Delta a = -0.5$.

4. Graph the positions of a bear and a hiker as functions of time from the following descriptions.
   a. A bear sets off in pursuit of a hiker. Both move at constant speed, but the bear is faster and eventually catches the hiker.
   b. A bear sets off in pursuit of a hiker. Both increase speed until the bear catches the hiker.
   c. The bear increases speed, and the hiker steadily slows down until the bear catches the hiker.
   d. The bear runs at constant speed, and the hiker steadily runs faster until the bear gives up and stops. The hiker slows down and stops soon after that.

5. On the figures, label the following points and sketch the derivative.

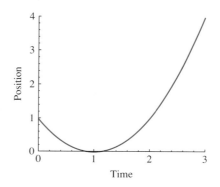

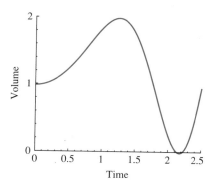

a. Two points where the derivative is positive.
b. Two points where the derivative is negative.
c. The point with maximum derivative.
d. The point with minimum (most negative) derivative.
e. Points with derivative of 0.

6. On the figures, identify which of the curves is a graph of the derivative of the other.

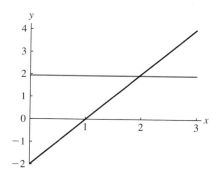

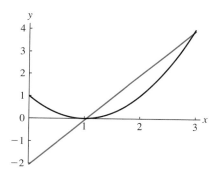

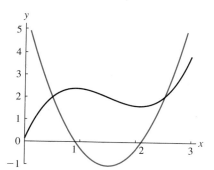

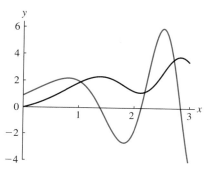

7. Sketch the function $f(x) = 4 - x^2$ for $0 \leq x \leq 2$. Identify $x, x + \Delta x, f(x), f(x + \Delta x)$, and Δf for the following values of x and Δx. Draw the secant line and compute its slope.
   a. $x = 1.0, \Delta x = 0.5$
   b. $x = 1.0, \Delta x = -0.5$
   c. Find the derivative at $x = 1$.
   d. $x = 0.5, \Delta x = 0.1$
   e. $x = 0.5, \Delta x = -0.1$
   f. Find the derivative at $x = 0.5$.
   g. Write the equation of the tangent line with base point $x = 0.5$.
   h. Find $f'(x)$.

8. Consider a falling rock described by
$$M(t) = 2.5t^2$$
with $a = 2.5$ m/s^2, sort of like on the moon. We wish to find the velocity at $t = 10$.
   a. Find the average velocity with $\Delta t = 5$.
   b. Find the average velocity with $\Delta t = 2$.
   c. Find the average velocity with $\Delta t = 1$.
   d. Find the velocity at $t = 10$ by taking the limit.
   e. Find the derivative $M'(t)$.

9. We will try to compute the derivative of the function $f(x) = |x|$ (the absolute value) at $x = -1$, $x = 0$, and $x = 1$.
   a. Graph the function.
   b. Compute the slope of secant lines at $x = -1, x = 0$, and $x = 1$ using $\Delta x = 0.5$ and $\Delta x = -0.5$.

c. Find the slope of the graph of the function at $x = -1$ and $x = 1$.
d. Graph the derivative.

10. We will try to compute the derivative of the function $f(x) = \sqrt{x}$ at $x = 0$.
    a. Before computing the derivative, sketch a graph of the function.
    b. Find the slope of secants at $x = 0$ with $\Delta x = 1.0$, $\Delta x = 0.1$, $\Delta x = 0.01$, and $\Delta x = 0.0001$.
    c. What is the limit of the slopes?
    d. What is the tangent line?
    e. Was your original graph correct? If not, sketch a new graph of the function.

11. **COMPUTER:** Have your computer plot the following for the function
    $$y(t) = 10t^{2.5}$$
    Use base point $t = 1$.
    a. The secant line with $\Delta t = 1$
    b. The secant line with $\Delta t = -1$
    c. The secant line with $\Delta t = 0.1$
    d. The secant line with $\Delta t = -0.1$

e. Use smaller and smaller values of Δt and try to estimate the slope of the tangent. Graph it, then zoom in on your graph. Does the tangent look right?

12. **COMPUTER:** The functions we have seen that are not differentiable fail only at a few points. Surprisingly, it is possible to find continuous curves that have a corner at every point. These curves are called fractals. The following construction creates a fractal called Koch's snowflake.
    a. Draw an equilateral triangle.
    b. Take the middle third of each side and expand it as shown.
    c. Do the same for the middle third of each straight piece.
    d. Repeat the process for as long as you can.

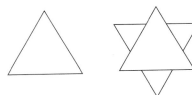

If this process is continued for an infinite number of steps, the resulting curve has a corner at every point.

2.5 Derivatives of Sums, Powers, and Polynomials

In Section 2.4, we used the definitions of the derivative and the limit to compute some derivatives. Like limits and continuous functions, derivatives of complicated functions can be computed easily when the functions have been built from simpler component parts. If we have one set of rules for computing derivatives of these components and another set for putting these derivatives together, we will be able to find the derivative of almost any function. Sections 2.5 to 2.10 are dedicated to finding these rules. We derive ways to compute derivatives of the most important components used to build biological functions: power functions, exponential functions, and trigonometric functions. We then combine these with the **sum rule**, the **product rule**, the **quotient rule**, and the **chain rule** (for computing derivatives when functions have been combined with functional composition).

This section presents our first rule for putting derivatives together, the **sum rule**, which says that the derivative of the sum is the sum of the derivatives. We also learn the **power rule** for computing derivatives of one important set of building blocks, the power functions. In combination with one more general rule, the **constant product rule**, these are the tools we need to compute the derivative of any **polynomial**.

The Sum Rule

Consider the function
$$s(t) = 5t^2 + 15t$$

This is the sum of two functions we studied in Section 2.4: the quadratic function $y(t) = 5t^2$ and a linear function, $l(t) = 15t$. The derivatives of the pieces are

$$y'(t) = 10t$$
$$l'(t) = 15$$

How can we use our knowledge of the derivatives of the building blocks $y(t)$ and $l(t)$ to find the derivative of the sum $s(t)$? The technique is provided by the following theorem, which says that **the derivative of the sum** of two functions is **the sum of the derivatives**.

■ **THEOREM 2.5** (The Sum Rule for Derivatives)

Suppose

$$s(x) = f(x) + g(x)$$

where f and g are both differentiable functions. Then

$$s'(x) = f'(x) + g'(x) \qquad \text{prime notation}$$

$$\frac{ds}{dx} = \frac{df}{dx} + \frac{dg}{dx} \qquad \text{differential notation}$$

Applying the sum rule to our example with $y(t) = 5t^2$ and $l(t) = 15t$, we get

$$s'(t) = y'(t) + l'(t) = 10t + 15$$

The functions and their derivatives are shown in Figures 2.39, 2.40, and 2.41.

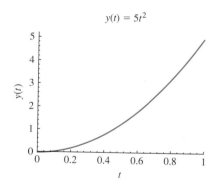
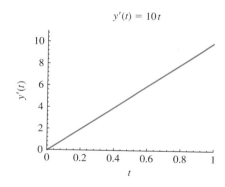

Figure 2.39
Graph and derivative of $y(t)$

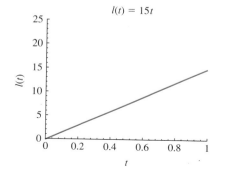
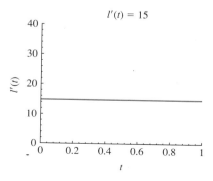

Figure 2.40
Graph and derivative of $l(t)$

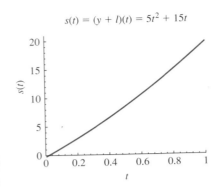
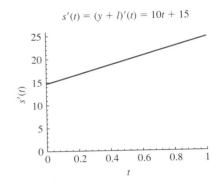

Figure 2.41
Graph and derivative of
$s(t) = y(t) + l(t)$

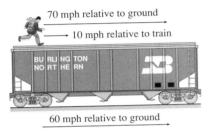

Figure 2.42
Velocities add

Why does the sum rule work? The derivative gives the velocity of a moving object. Suppose a train has a speed of 60 mph, and a person is running forward on the train at a speed of 10 mph. Relative to the ground, the person is moving at $60 + 10 = 70$ mph, the sum of the two velocities (Figure 2.42). Similarly, if the person is running backward on the train at a velocity of 10 mph, her velocity relative to the ground is $60 - 10 = 50$ mph. Velocities add (and subtract). Derivatives, the mathematical version of velocity, also add.

There is another geometric way to understand the sum rule. Suppose we measure the height of a tree as the sum of the trunk length and the crown height (Figure 2.43). The increase in trunk height, $T(t)$, between times t and $t + \Delta t$ is

$$\Delta T = T(t + \Delta t) - T(t)$$

and the increase in crown height, $C(t)$, between times t and $t + \Delta t$ is

$$\Delta C = C(t + \Delta t) - C(t)$$

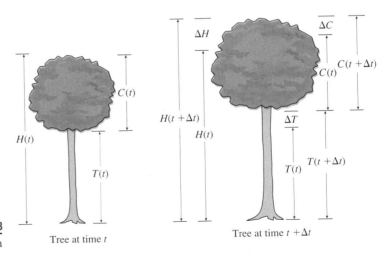

Figure 2.43
The height of a tree as a sum

The total increase in height, $H(t)$, is

$$\Delta H = \Delta T + \Delta C$$

the sum of the changes in trunk and crown. Therefore,

$$\frac{dH}{dt} = \lim_{\Delta t \to 0} \frac{\Delta H}{\Delta t}$$

$$= \lim_{\Delta t \to 0} \frac{\Delta T + \Delta C}{\Delta t}$$

$$= \lim_{\Delta t \to 0} \frac{\Delta T}{\Delta t} + \lim_{\Delta t \to 0} \frac{\Delta C}{\Delta t}$$

$$= \frac{dT}{dt} + \frac{dC}{dt}$$

The total rate of change of the tree's height is the sum of the rate of change of trunk height and the rate of change of crown height.

One useful example concerns adding a constant to a function. We measured the position of our falling rock from the point where we dropped it, finding

$$y(t) = 5t^2$$

Suppose we measure the position of the rock from a point 5 m higher instead. The distance fallen is then

$$s(t) = 5t^2 + 5$$

(Figure 2.44a). In differential notation, the velocity of the rock is

$$\frac{ds}{dt} = \frac{d}{dt}(5t^2 + 5)$$

$$= \frac{d(5t^2)}{dt} + \frac{d}{dt}5$$

$$= 10t + 0 = 10t$$

because the derivative of the constant 5 is equal to 0. The derivative of $s(t)$ exactly matches that of $y(t)$ itself (Figure 2.44b). This makes physical sense because velocity does not depend on where we begin measuring distance. Graphically, two curves differing only by a constant have the same slope at each point. Mathematically, adding a constant to a function does not change the derivative.

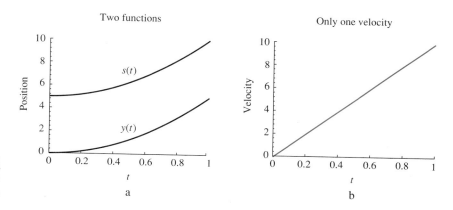

Figure 2.44 Two functions with the same derivative

We summarize the sum rule, along with the special cases involving constants, in the following tables.

Prime notation

Rule	Function	Derivative
Sum rule	$f(x) + g(x)$	$f'(x) + g'(x)$
Constant sum rule	$f(x) + c$	$f'(x)$

Differential notation

Rule	Function	Derivative
Sum rule	$\dfrac{d(f+g)}{dx}$	$\dfrac{df}{dx} + \dfrac{dg}{dx}$
Constant sum rule	$\dfrac{d(f+c)}{dx}$	$\dfrac{df}{dx}$

Derivatives of Basic Power Functions

The simplest power functions have the form

$$f(x) = x^n$$

where n is a positive integer $(1, 2, 3, \ldots)$. Is there a simple formula to find the derivative of these power functions?

There are two ways to compute the derivative of $f(x) = x^n$. One uses the **binomial theorem**, and the other uses **mathematical induction**. We use the binomial theorem to derive the result in the text, leaving mathematical induction for later (Exercise 12 in Section 2.6). Recall how we found the derivative of the quadratic $f(x) = x^2$ (where we use h instead of Δx to make the calculation more readable),

$$\begin{aligned}
\frac{dx^2}{dx} &= \lim_{h \to 0} \frac{f(x+h) - f(x)}{h} \\
&= \lim_{h \to 0} \frac{(x+h)^2 - x^2}{h} \\
&= \lim_{h \to 0} \frac{(x^2 + 2xh + h^2) - x^2}{h} \\
&= \lim_{h \to 0} \frac{2xh + h^2}{h} \\
&= \lim_{h \to 0} 2x + h \\
&= 2x
\end{aligned}$$

We computed the limit by evaluating the continuous function $2x + h$ at $h = 0$.

When $n = 3$, we expand $(x + h)^3$, finding

$$\begin{aligned}
\frac{dx^3}{dx} &= \lim_{h \to 0} \frac{(x+h)^3 - x^3}{h} \\
&= \lim_{h \to 0} \frac{(x^3 + 3x^2h + 3xh^2 + h^3) - x^3}{h} \\
&= \lim_{h \to 0} \frac{3x^2h + 3xh^2 + h^3}{h}
\end{aligned}$$

2.5 Derivatives of Sums, Powers, and Polynomials

$$= \lim_{h \to 0} 3x^2 + 3xh + h^2$$
$$= 3x^2$$

We used the expansion

$$(x + h)^3 = x^3 + 3x^2h + 3xh^2 + h^3$$

finding that the only term that appears in the derivative is the second. The first term canceled when we subtracted $f(x) = x^3$, and the terms with h^2 and h^3 disappear because they approach 0 even after being divided by h.

We see the same pattern when studying higher powers. In general, the **binomial theorem**, or **binomial formula**, says that

$$(x + h)^n = x^n + nx^{n-1}h + \frac{n(n-1)}{2}x^{n-2}h^2 + \cdots + nxh^{n-1} + h^n \quad (2.4)$$

The terms indicated by the dots have somewhat complicated formulas; they involve increasing powers of h. Substituting into the definition of the derivative, we get

$$\frac{df}{dx} = \lim_{h \to 0} \frac{(x+h)^n - x^n}{h}$$

$$= \lim_{h \to 0} \frac{\left[x^n + nx^{n-1}h + \frac{n(n-1)}{2}x^{n-2}h^2 + \cdots + nxh^{n-1} + h^n\right] - x^n}{h}$$

$$= \lim_{h \to 0} \frac{nx^{n-1}h + \frac{n(n-1)}{2}x^{n-2}h^2 + \cdots + nxh^{n-1} + h^n}{h}$$

$$= \lim_{h \to 0} nx^{n-1} + \frac{n(n-1)}{2}x^{n-2}h + \cdots + nxh^{n-2} + h^{n-1}$$

$$= nx^{n-1}$$

When we evaluate this continuous function at $h = 0$, all terms except the first are equal to 0. Therefore, all terms indicated by dots disappear.

We have derived the **power rule**:

■ **THEOREM 2.6** (The Power Rule for Derivatives: Integer Case)

Suppose

$$f(x) = x^n$$

when n is a positive integer. Then

$$f'(x) = nx^{n-1} \quad \text{prime notation}$$

$$\frac{df}{dx} = nx^{n-1} \quad \text{differential notation}$$

■

For the first few values of n, see the table. In words, to find the derivative of a power function, multiply by the power out front and then decrease the power by 1.

With $n = 1$, the power function is a line with constant slope of 1 (Figure 2.45a). With $n = 2$, the power function is a parabola, and the derivative is a line with slope 2 (Figure 2.45b). The function changes from decreasing to increasing at $x = 0$, indicated where the derivative crosses from negative to positive values. With $n = 3$, the power function is always increasing but has a critical point at $x = 0$ (Figure 2.45c).

Function	Derivative
x^1	1
x^2	$2x$
x^3	$3x^2$
x^4	$4x^3$

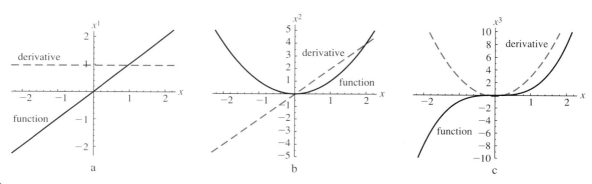

Figure 2.45
The first few power functions and their derivatives

The power rule does work when the power is $n = 0$. Recall that $x^0 = 1$, a constant, so that the derivative should be 0. According to the power rule

$$\frac{d}{dx}x^0 = 0x^{-1} = 0$$

Negative and Fractional Powers

Although the binomial theorem cannot be used to derive it, the power rule also works for negative and fractional powers.

■ **THEOREM 2.7** (The Power Rule for Derivatives)

Suppose

$$f(x) = x^p$$

defined for $x > 0$, where p is any number. Then

$$f'(x) = px^{p-1} \quad \text{prime notation}$$

$$\frac{df}{dx} = px^{p-1} \quad \text{differential notation}$$ ■

Function	Derivative
x^{-1}	$-x^{-2}$
x^{-2}	$-2x^{-3}$
x^{-3}	$-3x^{-4}$

For the first few negative powers, see the table. To find the derivative, multiply out front by the power and decrease the power by 1. Be careful to *subtract* 1 from the power even when it is negative. The function and derivative with power equal to -1 are shown in Figure 2.46. The derivative is always negative, indicating that the function is decreasing (except at $x = 0$, where it is not defined). Near $x = 0$, the slope approaches negative infinity.

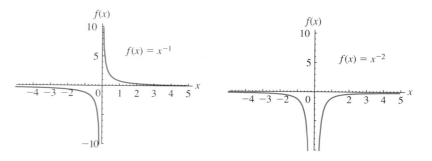

Figure 2.46
Graph of power function with a negative power

The generalized power rule can be used to find the derivative of $\sqrt{x} = x^{1/2}$:

$$\frac{d}{dx} x^{1/2} = \frac{1}{2} x^{(1/2)-1}$$
$$= \frac{1}{2} x^{-1/2}$$
$$= \frac{1}{2\sqrt{x}}$$

Near $x = 0$, the slope becomes infinite. For large x, the slope becomes small (Figure 2.47).

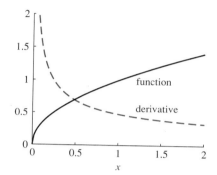

Figure 2.47
The square root function and its derivative

We can use the generalized power rule to interpret power relations between measurements. For example, the surface area S is proportional to the volume V raised to the $2/3$ power (see Section 1.8), or

$$S = c_s V^{2/3}$$

where the constant c_s depends on the shape of the object. If we imagine increasing the volume of an object, the surface area will increase, with

$$\frac{dS}{dV} = \frac{2}{3} c_s V^{-1/3}$$

Surface area does increase, but at a **decreasing rate** (Figure 2.48). In words, as an object gets larger, the surface area increases more slowly than the volume.

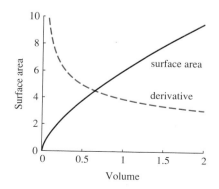

Figure 2.48
The derivative of the power relation between surface area and volume

Derivatives of Polynomials

A **polynomial** is a function built from power functions with positive integer values of n that are multiplied by constants and added. For example, the function

$$p(x) = 2x^3 - 5x^2 + 7x - 8$$

is a polynomial constructed by multiplying x^3 by 2, subtracting 5 times x^2, adding 7 times x^1, and subtracting 8. The **degree** of a polynomial is the highest power that appears in the formula, in this case, 3. Using the power rule, the sum rule, and a new rule, the **constant product rule**, we can find the derivative of any polynomial.

The constant product rule says constants come "outside" of derivatives. For example, we know that if $f(x) = x^2$, then $f'(x) = 2x$. The new function built by multiplying $f(x)$ by the constant 5, then, has derivative equal to 5 times the derivative of $f(x)$, or

$$\frac{d}{dx} 5x^2 = 5 \frac{d}{dx} x^2 = 5 \cdot 2x = 10x$$

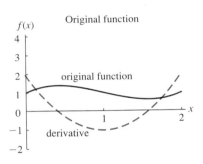

Figure 2.49
The constant product rule for derivatives

In general, this works. For example, doubling a function doubles the derivative (Figure 2.49). We formalize this observation as the constant product rule.

■ **THEOREM 2.8** **(The Constant Product Rule for Derivatives)**

If c is a constant and $f(x)$ is a differentiable function, then

$$\frac{d}{dx} cf(x) = c \frac{d}{dx} f(x)$$ ■

Using the power rule, the constant product rule, and the sum rule, we can find the derivative of the polynomial $p(x)$ as

$$\begin{aligned}
\frac{dp}{dx} &= \frac{d(2x^3)}{dx} + \frac{d(-5x^2)}{dx} + \frac{d(7x)}{dx} + \frac{d(-8)}{dx} & \text{sum rule} \\
&= 2\frac{dx^3}{dx} - 5\frac{d(x^2)}{dx} + 7\frac{dx}{dx} - 8\frac{d1}{dx} & \text{constant product rule} \\
&= 2 \cdot 3x^2 - 5 \cdot 2x + 7 \cdot 1 & \text{power rule} \\
&= 6x^2 - 10x + 7 & \text{the answer}
\end{aligned}$$

The derivative of a polynomial is always a polynomial. Because the power rule reduces the power by 1, the degree of the derivative is always 1 less than the

2.5 Derivatives of Sums, Powers, and Polynomials

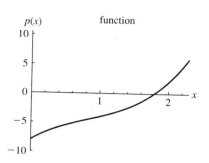
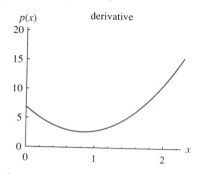

Figure 2.50
The polynomial $p(x)$ with its derivative

degree of the original polynomial. The function and its derivative are plotted in Figure 2.50.

Polynomials are convenient to work with mathematically and describe many biological processes. Consider the following model of offspring production. Let N be the number of eggs a bird lays. Suppose that each chick survives with probability

$$P(N) = 1 - 0.1N$$

(Figure 2.51a). The more offspring a parent has, the smaller chance each one has to survive. If the bird lays 1 egg, the chick has a 90% chance of surviving. If the bird lays 5 eggs, each chick has only a 50% chance of surviving. If the bird lays 10 eggs, none of the chicks survive. Mathematically, the total number of offspring that survive, $S(N)$, is the product of the number of eggs and the probability of survival, or

$$S(N) = NP(N) = N(1 - 0.1N) = N - 0.1N^2$$

(Figure 2.51b). For example, when $N = 1$, 0.9 offspring survive on average. When $N = 5$, 50% of the 5 chicks survive, producing 2.5 offspring on average. When $N = 10$, none of the chicks survive.

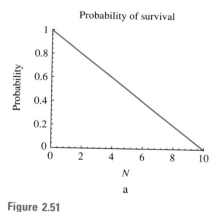
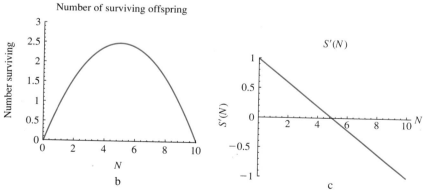

Figure 2.51
A model of offspring number and survival

The formula for $S(N)$ is a polynomial. We use the constant product and power rules to find

$$\frac{dS}{dN} = \frac{d(N - 0.1N^2)}{dN}$$

$$= \frac{dN}{dN} - 0.1\frac{dN^2}{dN}$$

$$= 1 - 0.1(2N)$$
$$= 1 - 0.2N$$

At $N = 0$, the derivative is positive, meaning that a higher value of N produces more surviving offspring. At $N = 5$, the derivative is 0, after which the derivative becomes negative. The negative derivative indicates that the number of surviving offspring *decreases* when the number of eggs becomes too large.

SUMMARY

By combining three rules, the **sum rule**, the **power rule**, and the **constant product rule**, we can find the derivative of any polynomial. The sum rule states that the derivative of the sum of two functions is equal to the sum of the derivatives. This corresponds to the physical fact that velocities add. As a special case, we found the **constant sum rule**, which states that adding a constant to a function does not change the derivative. The power rule gives the formula for the derivative of the power function x^p as px^{p-1} for any power. The constant product rule says that constants come "outside" the derivative; multiplying a function by a constant multiplies the derivative by the same constant. The power rule works with negative and fractional powers and can be used to study power relations.

2.5 EXERCISES

1. Find the derivatives of the following functions with the sum rule.
   a. $f(x) = 5x^2 + 7$
   b. $g(z) = 5z^2 + 7z$
   c. $h(y) = 5y^2 + 5y^2$

2. One car is towing another using a rigid 50-ft pole. The car starts from a stop, slowly speeds up, cruises for a while, and then stops abruptly. Sketch the positions and speeds of the two cars as functions of time.

3. A passenger is traveling on a luxury train moving west at 80 mph.
   a. The passenger starts running east at 10 mph. What is her velocity relative to the ground?
   b. While running, the passenger flips a roll over her shoulder (west) at 25 mph. What is the velocity of the roll relative to the train? What is the velocity of the roll relative to the ground?
   c. A roll weevil jumps east off the roll at 5 mph. What is the velocity of the weevil relative to the ground?

4. The above-ground volume of a plant is growing according to
   $$V_a(t) = 3.0t + 20.0$$
   and the below-ground volume according to
   $$V_b(t) = -1.0t + 40.0$$
   where t is measured in days and volumes are measured in cubic centimeters.
   a. Sketch the plant at $t = 0$ and $t = 10$.
   b. Compute the derivative of $V_a(t)$. What are the units?
   c. Compute the derivative of $V_b(t)$.
   d. Find the function $V(t) = V_a(t) + V_b(t)$ giving the total volume of the plant.
   e. Find the derivative of $V(t)$ and compare with the sum of the results from parts **a** and **b**.
   f. Describe in words what this plant is up to.

5. Suppose two populations of bacteria follow the differential equations
   $$\frac{db_1}{dt} = 2b_1$$
   $$\frac{db_2}{dt} = 2b_2$$
   The total population is $b(t) = b_1(t) + b_2(t)$.
   a. Find a differential equation for b.
   b. Suppose instead that
   $$\frac{db_2}{dt} = 3b_2$$
   Write a differential equation for b. Why is it impossible to find a formula for the rate of change of b in terms of b itself?

6. Find the derivatives of the following functions. Justify each step.
   a. $f(x) = 3x^2 + 3x + 1$
   b. $s(x) = 1 - x + x^2 - x^3 + x^4$
   c. $g(z) = 3z^3 + 2z^2$
   d. $h(y) = y^{10} - y^9$
   e. $p(x) = 1 + x + \dfrac{x^2}{2} + \dfrac{x^3}{6} + \dfrac{x^4}{24}$

7. Find the derivatives of the following functions.
   a. In its early phase, the number of AIDS cases in the United States grew approximately according to a cubic equation, $A(t) = 175t^3$, where t is measured in years since the beginning of the epidemic in 1972. Find and interpret the derivative.
   b. $F(r) = 1.5r^4$, where F represents the flow in cubic centimeters per second through a pipe of radius r.
   c. The area of a circle as a function of radius is $A(r) = \pi r^2$, with area measured in square centimeters and radius measured in centimeters. Find the derivative of area with respect to radius. On a geometric diagram, illustrate the area corresponding to $\Delta A = A(r + \Delta r) - A(r)$. What is a geometric interpretation of the derivative? Do the units make sense?
   d. The volume of a sphere as a function of radius is $V(r) = (4/3)\pi r^3$. Find the derivative of volume with respect to radius. On a geometric diagram, illustrate the volume corresponding to $\Delta V = V(r + \Delta r) - V(r)$. What is a geometric interpretation of this derivative? Do the units make sense?

8. Use the binomial formula to compute the following.
   a. $(x + 1)^2$
   b. $(x + 3)^2$
   c. $(2x + 1)^2$
   d. $(x + 1)^3$
   e. $(x + 2)^3$
   f. $(2x + 1)^3$

9. Find the derivatives of the following functions.
   a. x^5
   b. x^{-5}
   c. $x^{0.2}$
   d. $x^{-0.2}$
   e. x^e
   f. x^{-e}
   g. $x^{1/e}$
   h. $x^{-1/e}$

10. In Section 1.8, we found that antler size A and shoulder height L are related by the power relation
    $$A = 0.026L^{1.7}$$
    a. Thinking of A as a function of L, find $\dfrac{dA}{dL}$.
    b. Sketch the derivative.
    c. Solve for L in terms of A.
    d. Find and sketch $\dfrac{dL}{dA}$. Why does it look so different from $\dfrac{dA}{dL}$?

11. Write the power relation in Exercise 10 in terms of $\ln(A)$ and $\ln(L)$.
    a. Solve for $\ln(A)$ in terms of $\ln(L)$.
    b. Find the derivative $\dfrac{d \ln(A)}{d \ln(L)}$.
    c. Solve for $\ln(L)$ in terms of $\ln(A)$.
    d. Find the derivative $\dfrac{d \ln(L)}{d \ln(A)}$.
    e. Do the graphs of these derivatives look similar? How could you tell from the derivative that antlers were growing faster than shoulder height?

12. Find functions with the following as their derivatives.
    a. 2
    b. $2x$
    c. $15x^{14}$
    d. x^{14}
    e. $-x^{-2}$
    f. $3x^{-4}$
    g. Can you think of a function with derivative equal to x^{-1}?

13. Suppose the size of an organism follows the differential equation
    $$\dfrac{dS}{dt} = 12 \dfrac{S(t)}{t}$$
    a. Describe in words the growth of this organism. How does it differ from one following
    $$\dfrac{dS}{dt} = 12S(t)$$
    b. Try to guess a solution (try different power functions for $S(t)$).
    c. Suppose the organism grows twice as fast, according to
    $$\dfrac{dS}{dt} = 24 \dfrac{S(t)}{t}$$
    Try to guess a solution. Is this organism more than twice as big at $t = 1$? How about at $t = 2$?

14. **COMPUTER:** Consider the polynomials
    $$p_0(x) = 1$$
    $$p_1(x) = 1 + x$$
    $$p_2(x) = 1 + x + \dfrac{x^2}{2}$$
    $$p_3(x) = 1 + x + \dfrac{x^2}{2} + \dfrac{x^3}{6}$$
    To find the fourth polynomial in the series, add a term equal to the last term of $p_3(x)$ multiplied by $x/4$. To find the fifth polynomial in the series, add a term equal to the last term of $p_4(x)$ multiplied by $x/5$, and so forth.
    a. Find the first six polynomials in this series.
    b. Plot them all for $-3 \leq x \leq 3$.
    c. Take the derivative of $p_6(x)$. What is it equal to? Compare the graph of $p_6(x)$ with the graph of its derivative.
    d. Can you guess what function these polynomials are becoming more and more close to?

2.6 Derivatives of Products and Quotients

In this section, we find the derivatives of the products and quotients of known functions by using the **product rule** and **quotient rule** for derivatives. These two rules allow us to study the derivatives of **rational functions**, functions that are the ratios of two polynomials.

The Product Rule

Tempting as it might sound, the derivative of a product is *not* the product of the derivatives. Finding the derivative of a product is a bit more complicated. Like the derivative of a sum, there is a useful geometric argument to help us to derive and understand the formula.

Suppose the density, $\rho(t)$, and the volume, $V(t)$, of an object are functions of time. The mass, $M(t)$, the product of the density and volume, is also a function of time and has the formula

$$M(t) = \rho(t)V(t)$$

What is the derivative of the mass? For example, if both the density and volume are increasing, the mass is also increasing. But if the density increases while the volume decreases, we cannot tell whether mass increases without more information.

The mass, $M(t)$, can be represented geometrically as the area of a rectangle with base $\rho(t)$ and height $V(t)$ (Figure 2.52). Between the times t and $t + \Delta t$, the density changes from $\rho(t)$ to $\rho(t) + \Delta\rho$ and the volume changes from $V(t)$ to $V(t) + \Delta V$. The mass at time $t + \Delta t$ is then

$$M(t + \Delta t) = [\rho(t) + \Delta\rho][V(t) + \Delta V]$$
$$= \rho(t)V(t) + \rho(t)\Delta V + V(t)\Delta\rho + \Delta\rho\Delta V$$
$$= M(t) + \rho(t)\Delta V + V(t)\Delta\rho + \Delta\rho\Delta V$$

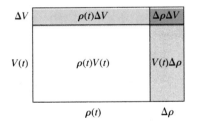

Figure 2.52
The product rule

The change in mass, ΔM, is

$$\Delta M = M(t + \Delta t) - M(t)$$
$$= \rho(t)\Delta V + V(t)\Delta\rho + \Delta\rho\Delta V$$

Each component of this algebraic computation corresponds to one of the regions in Figure 2.52.

How does this help us compute the derivative of the product? To find the derivative of M with respect to t we divide ΔM by Δt and take the limit. ΔM divided by Δt is

$$\frac{\Delta M}{\Delta t} = \frac{\rho(t)\Delta V + V(t)\Delta\rho + \Delta\rho\Delta V}{\Delta t}$$
$$= \rho(t)\frac{\Delta V}{\Delta t} + V(t)\frac{\Delta\rho}{\Delta t} + \frac{\Delta\rho\Delta V}{\Delta t}$$

Taking the limit, we get

$$\lim_{\Delta t \to 0}\frac{\Delta M}{\Delta t} = \rho(t)\lim_{\Delta t \to 0}\frac{\Delta V}{\Delta t} + V(t)\lim_{\Delta t \to 0}\frac{\Delta\rho}{\Delta t} + \lim_{\Delta t \to 0}\frac{\Delta\rho\Delta V}{\Delta t}$$

where we used the properties of limits (Theorems 2.1 and 2.2) to break up the sum and take constants outside the limits. The first two terms on the right-hand side are

$$\rho(t) \lim_{\Delta t \to 0} \frac{\Delta V}{\Delta t} = \rho(t)V'(t)$$

$$V(t) \lim_{\Delta t \to 0} \frac{\Delta \rho}{\Delta t} = V(t)\rho'(t)$$

by the definition of the derivative. The final piece can be analyzed as a product,

$$\lim_{\Delta t \to 0} \frac{\Delta \rho \Delta V}{\Delta t} = \lim_{\Delta t \to 0} \Delta \rho \lim_{\Delta t \to 0} \frac{\Delta V}{\Delta t}$$

However,

$$\lim_{\Delta t \to 0} \Delta \rho = 0$$

because ρ is a continuous function. Therefore,

$$\lim_{\Delta t \to 0} \frac{\Delta \rho \Delta V}{\Delta t} = 0 \cdot V'(t) = 0$$

Geometrically, the region labeled $\Delta \rho \Delta V$ becomes very small as Δt approaches 0 and disappears in the limit. We make this calculation precise in the following theorem.

■ **THEOREM 2.9** (The Product Rule for Derivatives)

Suppose

$$p(x) = f(x)g(x)$$

where f and g are both differentiable. Then

$$p'(x) = f'(x)g(x) + g'(x)f(x) \quad \text{prime notation}$$

$$\frac{dp}{dx} = \frac{df}{dx}g(x) + f(x)\frac{dg}{dx} \quad \text{differential notation} \quad ■$$

Special Cases and Examples

The constant product rule is a special case of the product rule. Suppose we *multiply* a function by a constant. For example, if we measure distance fallen in centimeters rather than meters, the distance fallen will be 100 times the original distance $y(t)$, or

$$p(t) = 100y(t)$$

This function is a product of the constant function 100 and the function $y(t)$. The velocity is then

$$\frac{dp}{dt} = \frac{d(100y)}{dt} \quad \text{definition of } p(t)$$

$$= 100\frac{dy}{dt} + y(t)\frac{d\,100}{dt} \quad \text{the product rule with } f(t) = 100 \text{ and } g(t) = y(t)$$

$$= 100\frac{dy}{dt} + y(t) \cdot 0 = 100\frac{dy}{dt} \quad \text{the derivative of the constant function } f(t) = 100 \text{ is } 0$$

In centimeters per second, the velocity is 100 times the velocity in meters per second. Mathematically, multiplying a function by a constant multiplies the derivative by the same constant.

We summarize these rules in the following tables.

Prime notation

Rule	Function	Derivative
Product rule	$f(x)g(x)$	$f(x)g'(x) + g(x)f'(x)$
Constant product rule	$cf(x)$	$cf'(x)$

Differential notation

Rule	Function	Derivative
Product rule	$\dfrac{d(fg)}{dx}$	$f(x)\dfrac{dg}{dx} + g(x)\dfrac{df}{dx}$
Constant product rule	$\dfrac{d(cf)}{dx}$	$c\dfrac{df}{dx}$

To find the derivative of

$$p(x) = (x - 1)(x + 1)$$

think of $p(x)$ as the product of $f(x) = x - 1$ and $g(x) = x + 1$. Both $f(x)$ and $g(x)$ are linear functions with slope 1; therefore

$$f'(x) = 1$$
$$g'(x) = 1$$

Then, by the product rule

$$p'(x) = (fg)'(x) = (x - 1) \cdot 1 + (x + 1) \cdot 1 = 2x$$

Suppose the volume V of a plant is increasing according to

$$V(t) = 100.0 + 12.0t$$

where t is measured in hours and V is measured in cubic centimeters, and that the density is decreasing according to

$$\rho(t) = 0.8 - 0.05t$$

where ρ is measured in grams per cubic centimeter. Is the mass increasing or decreasing? We can use the product rule to find out:

$$M'(t) = \rho(t)V'(t) + V(t)\rho'(t)$$
$$= (0.8 - 0.05t)12.0 + (100.0 + 12.0t)(-0.05)$$
$$= 9.6 - 0.6t + -5.0 - 0.6t = 4.6 - 1.2t$$

At $t = 0$, the derivative is positive and the mass is increasing. By $t = 4$, the decrease in density has overwhelmed the increase in volume and the mass is decreasing (Figures 2.53 and 2.54).

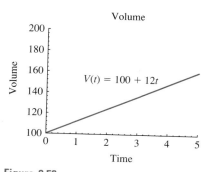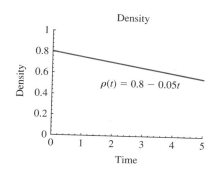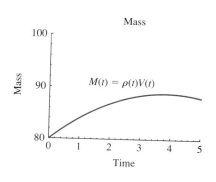

Figure 2.53
The volume, density, and mass of a plant

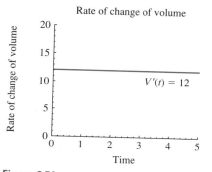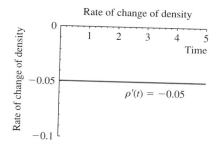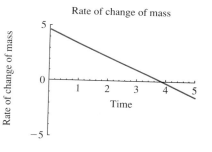

Figure 2.54
The rates of change of volume, density, and mass of a plant

The Quotient Rule

Suppose we are interested in finding the derivative of a function $w(x)$ defined as the quotient of $u(x)$ and $v(x)$,

$$w(x) = \frac{u(x)}{v(x)}$$

With a bit of a trick, we can find $w'(x)$ from the product rule. First, multiply both sides by $v(x)$,

$$v(x)w(x) = u(x)$$

Now take the derivative of both sides using the product rule,

$$v(x)w'(x) + v'(x)w(x) = u'(x)$$

This can be thought of as an equation for $w'(x)$, the derivative of the quotient, and can be solved for $w'(x)$ as

$$v(x)w'(x) = u'(x) - v'(x)w(x)$$

$$w'(x) = \frac{u'(x) - v'(x)w(x)}{v(x)}$$

To write the answer entirely in terms of the component functions $u(x)$ and $v(x)$, substitute in $w(x) = u(x)/v(x)$ to find

$$w'(x) = \frac{u'(x) - v'(x)u(x)/v(x)}{v(x)}$$

$$= \frac{u'(x)v(x) - v'(x)u(x)}{[v(x)]^2}$$

where the last step involves putting the result over a common denominator. We summarize this result in the following theorem.

THEOREM 2.10 (The Quotient Rule for Derivatives)

Suppose

$$w(x) = \frac{u(x)}{v(x)}$$

where $u(x)$ and $v(x)$ are differentiable and $v(x) \neq 0$. Then

$$w'(x) = \frac{u'(x)v(x) - v'(x)u(x)}{[v(x)]^2} \quad \text{prime notation}$$

$$\frac{dw}{dx} = \frac{v(x)\dfrac{du}{dx} - u(x)\dfrac{dv}{dx}}{[v(x)]^2} \quad \text{differential notation}$$

Consider the updating function for the fraction of invading mutant bacteria, p_t,

$$p_{t+1} = \frac{2.0 p_t}{2.0 p_t + 1.5(1 - p_t)}$$

(Equation 1.49). In functional form, this can be written as

$$f(p) = \frac{2.0p}{2.0p + 1.5(1-p)}$$

The numerator is $u(p) = 2.0p$ and the denominator is $v(p) = 2.0p + 1.5(1 - p)$, with derivatives

$$\frac{du}{dp} = 2.0$$

$$\frac{dv}{dp} = 2.0 - 1.5 = 0.5$$

Therefore, the derivative is

$$\frac{df}{dp} = \frac{v(p)\dfrac{du}{dp} - u(p)\dfrac{dv}{dp}}{v(p)^2}$$

$$= \frac{2.0p + 1.5(1-p)\dfrac{d(2.0p)}{dp} - 2.0p\dfrac{d[2.0p + 1.5(1-p)]}{dp}}{[2.0p + 1.5(1-p)]^2}$$

$$= \frac{2.0p + 1.5(1-p) \cdot 2.0 - 2.0p \cdot 0.5}{[2.0p + 1.5(1-p)]^2}$$

$$= \frac{3.0}{[2.0p + 1.5(1-p)]^2}$$

At $p = 0$, the derivative is

$$\frac{3.0}{[2.0 \cdot 0 + 1.5(1 - 0)]^2} = 1.333$$

and at $p = 1$, the derivative is

$$\frac{3.0}{[2.0 \cdot 1 + 1.5(1 - 1)]^2} = 0.75$$

Compared to the diagonal line, which has slope 1, the graph of this function starts out steep and ends up rather flat (Figure 2.55).

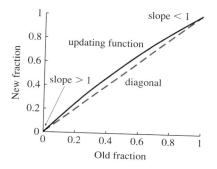
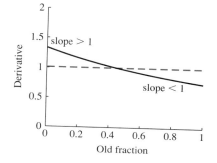

Figure 2.55
An updating function and its derivative

An important family of functions used to describe biological processes is the set of **Hill functions**, which are of the form

$$h(x) = \frac{x^n}{1 + x^n} \tag{2.5}$$

where n can be any positive number. We will compute the derivatives of these functions with $n = 1$ and $n = 2$. With $n = 1$, $u(x) = x$ and $v(x) = 1 + x$, so

$$\frac{dh}{dx} = \frac{(1 + x)\frac{dx}{dx} - x\frac{d(1 + x)}{dx}}{(1 + x)^2}$$

$$= \frac{(1 + x) - x}{(1 + x)^2}$$

$$= \frac{1}{(1 + x)^2}$$

This derivative is always positive and takes on the value $h'(0) = 1$ at $x = 0$. When $n = 2$, $u(x) = x^2$ and $v(x) = 1 + x^2$, so

$$\frac{dh}{dx} = \frac{(1 + x^2)\frac{d(x^2)}{dx} - x^2\frac{d(1 + x^2)}{dx}}{(1 + x^2)^2}$$

$$= \frac{(1 + x^2)2x - x^2 \cdot 2x}{(1 + x^2)^2}$$

$$= \frac{2x}{(1 + x^2)^2}$$

This derivative too is always positive, but it takes on the value $h'(0) = 0$ at $x = 0$. Both Hill functions are increasing but with different shapes (Figure 2.56). These functions are useful for describing the response to a stimulus (Exercise 11).

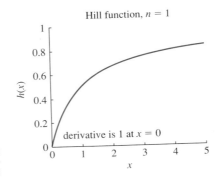
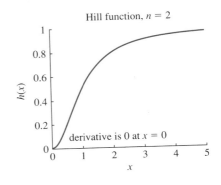

Figure 2.56
Two Hill functions

SUMMARY

The **product rule** states that the derivative of the product of two functions is equal to the first function times the derivative of the second plus the second function times the derivative of the first. The constant product is a special case of the product rule. We also derived the **quotient rule** for differentiating quotients. With these rules, we can find derivatives of **rational functions**, the ratios of polynomials, such as the **Hill functions**.

2.6 EXERCISES

1. Find the derivatives of the following functions.
   a. $f(x) = (2x + 3)(-3x + 2)$
   b. $g(z) = (5z - 3)(z + 2)$
   c. $h(x) = (x + 2)(2x + 3)(-3x + 2)$ (Apply the product rule twice.)
   d. $r(y) = (5y - 3)(y^2 - 1)$
   e. $s(t) = (t^2 + 2)(3t^2 - 1)$

2. Suppose $p(x) = f(x)g(x)$. Test the *incorrect* formula
$$p'(x) = f'(x)g'(x)$$
on the following functions.
   a. $f(x) = x, g(x) = x$
   b. $f(x) = 1, g(x) = x$
   c. $f(x) = 1$, any function $g(x)$

3. Suppose the population P of a city is
$$P(t) = 2.0 \times 10^6 + 2.0 \times 10^4 t$$
where t is measured in years. Suppose the average weight per person, $W(t)$, in kilograms declines according to
$$W(t) = 80 - 0.5t$$
   a. What is the total weight of all the people in the city as a function of time?
   b. Sketch a graph of this function.
   c. Compute and sketch the derivative.
   d. When is the derivative equal to 0?
   e. Find the population, the average weight, and the total weight at this time.

4. For the situation described in Exercise 3, draw a geometric diagram illustrating the product rule.
   a. Draw a rectangle illustrating the total weight at time $t = 0$.
   b. With $t = 0$ and $\Delta t = 4$, find ΔP and ΔW, and draw these on your diagram.
   c. Label each region of your diagram, and find its area.

5. Recall the plant in Exercise 4 in Section 2.5, with above-ground and below-ground volumes
$$V_a(t) = 3.0t + 20.0$$
$$V_b(t) = -1.0t + 40.0$$
Suppose the density, ρ_b, below ground is increasing according to
$$\rho_b(t) = 1.8 + 0.02t$$
and the density, ρ_a, above ground is decreasing according to
$$\rho_a(t) = 1.2 - 0.01t$$
where time is measured in days and density is measured in grams per cubic centimeter.
   a. Find the total mass below ground as a function of time, and compute the derivative.
   b. Find the total mass above ground as a function of time, and compute the derivative.
   c. Find the total mass (both above and below ground), and compute the derivative.

6. Find the derivatives of the following functions. Justify the steps.
   a. $f(x) = \dfrac{x^2}{1 + 2x^3}$
   b. $g(z) = \dfrac{1 + z^2}{(1 + 2z)^3}$
   c. $h(z) = \dfrac{1 + 2z^3}{(1 + z)^2}$
   d. $F(y) = \dfrac{1 + y^2}{2y^3}$

7. Compute the derivative of $f(x) = x/(1 + x)$ in three ways. State which rules you use at each step.
   a. Use the quotient rule directly.
   b. Rewrite the function as
   $$f(x) = \dfrac{1}{1 + \dfrac{1}{x}}$$
   and use the quotient rule.
   c. Now show you can rewrite the function as
   $$f(x) = 1 - \dfrac{1}{1 + x}$$
   and use the constant sum and quotient rules.

8. Check the following.
   a. Use the quotient rule to find the derivative of $v(x)/v(x)$. Does the result make sense?
   b. Suppose $f(x)$ is a positive increasing function (both $f(x) > 0$ and $f'(x) > 0$). Find the derivative of $1/f(x)$. Is it positive or negative? Sketch a graph of $f(x)$ and $1/f(x)$.

9. Suppose that the fraction of chicks that survive, $P(N)$, as a function of the number, N, of eggs laid is given by the following forms (variants of the model studied in Section 2.5 under Derivatives of Polynomials). The total number of offspring that survive is $S(N) = NP(N)$. Find the number of surviving offspring when the bird lays 1, 5, or 10 eggs. Find $S'(N)$. Sketch a graph of $S(N)$. What do you think is the best strategy for each bird?
   a. $P(N) = 1 - 0.08N$
   b. $P(N) = \dfrac{1}{1 + 0.5N}$
   c. $P(N) = \dfrac{1}{1 + 0.1N^2}$

10. Suppose that the mass, $M(t)$, of a cell is $M(t) = 1 + t^3$ and the density, $\rho(t)$, of a cell is $\rho(t) = 1 + t$.
    a. Find the volume as a function of time.
    b. Find the derivative of the volume.
    c. At what times is the volume increasing?

11. The following steps should help you to figure out what happens to the Hill function,
    $$h_n(x) = \dfrac{x^n}{1 + x^n}$$
    for large values of n.
    a. Compute $h_n(0)$, $h_n(1)$, and $h_n(2)$ for $n = 2$, $n = 5$, and $n = 10$.
    b. Compute the derivative at $x = 0$, $x = 1$, and $x = 2$ for $n = 2$, $n = 5$, and $n = 10$.
    c. Use this information to sketch graphs.
    d. If you think of $h_n(x)$ as representing a response to a stimulus of strength x, describe the responses for small and large n. How do they differ?

12. Deriving the power rule with mathematical induction: The idea behind mathematical induction is to show first that a formula is true for $n = 1$, and then that if it is true for some particular n, it must then also be true for $n + 1$. First, check that the power rule is true for $n = 1$. Then, assuming it is true for n, find
    $$\dfrac{d(x^{n+1})}{dx}$$
    using the product rule, and check that it too satisfies the power rule.

13. **COMPUTER:** Consider the functions
    $$g_n(x) = 1 + x + x^2 + \cdots + x^n$$
    for various values of n. We will compare these functions with
    $$g(x) = \dfrac{1}{1 - x}$$
    a. Plot $g_1(x)$, $g_3(x)$, $g_5(x)$, and $g(x)$ on the intervals $0 \le x \le 0.5$ and $0 \le x \le 0.9$.
    b. Take the derivative of $g(x)$. Can you see how the derivative is related to $g(x)$ itself? In other words, what function could you apply to the formula for $g(x)$ to get the formula for $g'(x)$?
    c. Apply the same function to $g_1(x)$, $g_3(x)$, and $g_5(x)$. Can you see why these functions are good approximations to $g(x)$? Can you see why these approximations are best for small values of x? What happens to the approximations for x near 1?

14. **COMPUTER:** Consider the function
    $$r(x) = \dfrac{u(x)}{v(x)} = \dfrac{1 + x}{2 + x^2 + x^3}$$
    a. Make one graph of $u(x)$ and $v(x)$ for $0 \le x \le 1$, and another of $r(x)$. Could you have guessed the shape of $r(x)$ from looking at the graphs of $u(x)$ and $v(x)$?
    b. What happens at the critical point?
    c. Find the exact location of the critical point $x = x_c$.
    d. Compare $\dfrac{u'(x_c)}{u(x_c)}$ with $\dfrac{v'(x_c)}{v(x_c)}$. Why are they equal?

2.7 The Second Derivative, Curvature, and Acceleration

The sign of the derivative can be used to figure out when a function is increasing or decreasing. Much more information can be gleaned about a function and its graph from the derivative of the derivative, or the **second derivative**. In particular, we will see that the second derivative tells whether the graph of a function curves upward (**concave up**) or downward (**concave down**). Furthermore, just as the first derivative of position is **velocity**, the second derivative of position is **acceleration**.

The Second Derivative

Consider the two graphs in Figure 2.57. In the first, the slope becomes steeper, indicating that the measurement is increasing faster and faster. In the second, the slope becomes smaller. Although this measurement increases, it does so at a decreasing rate. These differences are clear on graphs of the derivatives of the two functions. Each derivative is **positive** because the functions are increasing. The derivative of the first function is **increasing** (Figure 2.58a), and the derivative of the second function is **decreasing** (Figure 2.58b).

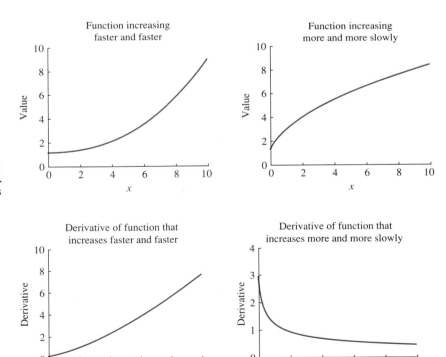

Figure 2.57 Two increasing functions

Figure 2.58 Derivatives of two increasing functions

The derivative of any function is positive when the function is increasing and negative when the function is decreasing. We can apply this observation to

2.7 The Second Derivative, Curvature, and Acceleration

the derivative itself. Because the derivative of the first function is increasing, the **derivative of the derivative** must be positive. Similarly, because the derivative of the second function is decreasing, the **derivative of the derivative** must be negative (Figure 2.59).

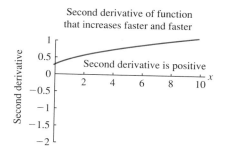

Figure 2.59
Second derivatives of two increasing functions

The derivative of the derivative is called the **second derivative**. For clarity, the derivative itself is often called the **first derivative**. We write the second derivative of f in prime notation as

$$\text{the second derivative of } f = f''(x) \tag{2.6}$$

The two $'$ symbols represent taking two derivatives. In differential notation,

$$\text{the second derivative of } f = \frac{d^2 f}{dx^2} \tag{2.7}$$

This notation may look odd. Think of the object d/dx as a sort of function (called an **operator** by mathematicians) that takes one function as input and returns another function as output. This operator returns as output the derivative of its input. To find the second derivative, apply this operator twice. More generally, the result of taking n derivatives is written

$$\text{the } n\text{th derivative of } f = f^{(n)}(x) \quad \text{prime notation}$$
$$= \frac{d^n f}{dx^n} \quad \text{differential notation}$$

Figure 2.60
A function, its derivative, and its second derivative

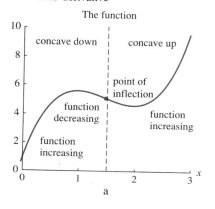
a

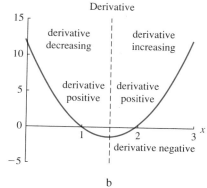
b

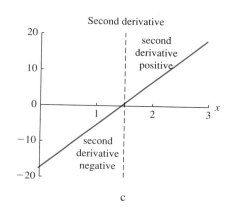
c

The second derivative has a general interpretation in terms of the **curvature** of the graph. Consider the function shown in Figure 2.60a. There are many ways to describe this graph. First of all, the graph of the function itself is positive, meaning that the associated measurement takes on only positive values. Furthermore, the function is increasing for $x < 1$, decreasing for $1 < x < 2$, and increasing for $x > 2$. The derivative, therefore, is positive for $x < 1$, negative for $1 < x < 2$, and positive for $x > 2$ (Figure 2.60b).

With more careful examination, we can see that the graph breaks into two regions of **curvature**. Between $x = 0$ and $x = 1.5$, the graph curves downward; this portion of the graph is said to be **concave down**. The slope of the curve is a **decreasing** function in this region (Figure 2.60b), implying that the **second derivative is negative** (Figure 2.60c). Between $x = 1.5$ and $x = 3$, the graph curves upward, like a bowl. This portion of the graph is said to be **concave up**. In this region, the slope of the curve becomes steeper and steeper, the derivative is increasing, and the **second derivative is positive** (Figure 2.60c).

At a point where the *second derivative* is 0, the curvature of the graph can change. Such a point is called a **point of inflection**. In Figure 2.60a, the graph has a point of inflection at $x = 1.5$, where the function switches from concave down to concave up and the second derivative switches from negative to positive. Neither the function itself nor the derivative changes sign at this point. In fact, points of inflection can occur when the derivative is positive, negative, or 0.

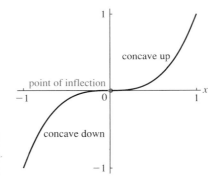

Figure 2.61

The function $C(x) = x^3$ has a point of inflection at a critical point

Consider the power function $C(x) = x^3$ (Figure 2.61). Using the power rule, we find that $C'(x) = 3x^2$ and $C''(x) = 6x$. Therefore, this function has both derivative and second derivative equal to 0 at the same point, $x = 0$. In other words, it has both a critical point and a point of inflection at $x = 0$. If this graph described a moving object, the object would move forward for $x < 0$, come to a temporary stop at $x = 0$, and then move forward again for $x > 0$.

Using the Second Derivative for Graphing

When we can compute the derivative and second derivative of a function, we can use these tools to sketch graphs quickly. As a test, we check the second derivative of a linear function,

$$f(x) = 2x + 1$$

2.7 The Second Derivative, Curvature, and Acceleration

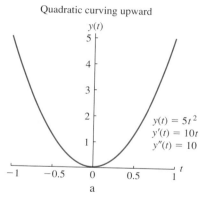
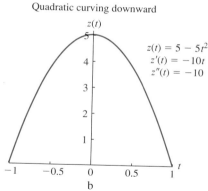

Figure 2.62
Concave up and concave down quadratic functions

The first derivative is $f'(x) = 2$, a constant, implying that the second derivative is 0, consistent with the fact that a line is straight (has no curvature).

Consider now the quadratic function

$$y(t) = 5t^2$$

The first derivative is $y'(t) = 10t$, computed with the constant product and power rules. Because this is a linear function, the second derivative is $y''(t) = 10$, which is a positive constant (Figure 2.62a). The quadratic function

$$z(t) = 5 - 5t^2$$

has first derivative $z'(t) = -10t$ and second derivative $z''(t) = -10$, a negative constant (Figure 2.62b).

A quadratic function with a *positive* coefficient on the quadratic term has a positive second derivative and is concave up at all points. A quadratic function with a *negative* coefficient on the quadratic term has a negative second derivative and is concave down at all points. The quadratic function

$$f(x) = 3x^2 - 6x + 5$$

which has a positive coefficient 3 in front of x^2, must be concave up. The derivative is

$$f'(x) = 6x - 6$$

There is a critical point at $x = 1$. Because the graph of this function is an upward-pointing parabola, this critical point must be the bottom of the bowl. Using the fact that $f(1) = 2$, we can easily sketch a graph of this function (Figure 2.63).

The second derivative helps to categorize all **power functions**. Consider the power function

$$g(x) = x^p$$

According to the power rule, the first and second derivatives are

$$g'(x) = px^{p-1}$$
$$g''(x) = p(p-1)x^{p-2}$$

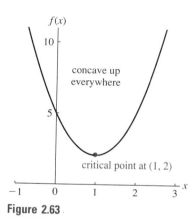

Figure 2.63
Graphing a quadratic using the first and second derivatives

By substituting various values of p, we can create the following table and graph the three possibilities (Figure 2.64). This table and the graphs apply only for $x > 0$, the values for which power functions are most often applied.

Behavior of power function $g(x) = x^p$ for $x > 0$

Power	First derivative	Second derivative
$p > 1$	Positive	Positive
$0 < p < 1$	Positive	Negative
$p < 0$	Negative	Positive

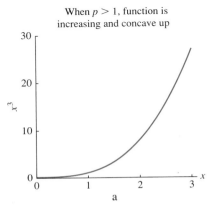
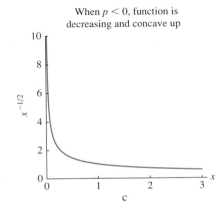

Figure 2.64
Graphs of the basic power functions

The cubic power function $g(x) = x^3$ has first derivative $g'(x) = 3x^2$ and second derivative $g''(x) = 6x$. For $x > 0$, both are positive (Figure 2.64a). This function is increasing and concave up on this domain, consistent with the fact that the power is greater than 1.

The square root function $g(x) = \sqrt{x}$ is the power function

$$g(x) = x^{1/2}$$

The first and second derivatives are

$$g'(x) = \frac{1}{2}x^{-1/2}$$

$$g''(x) = -\frac{1}{2} \cdot \frac{1}{2}x^{-3/2} = -\frac{1}{4}x^{-3/2}$$

In accordance with the table, the second derivative is negative and the function is concave down (Figure 2.64b). Finally, the function

$$g(x) = x^{-1/2}$$

has first and second derivatives

$$g'(x) = -\frac{1}{2}x^{-3/2}$$

$$g''(x) = \frac{1}{2} \cdot \frac{3}{2}x^{-5/2} = \frac{3}{4}x^{-5/2}$$

This decreasing function is concave up.

2.7 The Second Derivative, Curvature, and Acceleration

Suppose we wish to study the polynomial

$$p(x) = 2x^3 - 7x^2 + 5x + 2$$

What does the graph look like? Does $p(x)$ take on negative values for $x > 0$? To begin, we find the derivatives

$$p'(x) = 6x^2 - 14x + 5$$
$$p''(x) = 12x - 14$$

The point of inflection occurs where $12x - 14 = 0$, or at $x = 1.167$. The graph is concave down for $x < 1.167$ and concave up for $x > 1.167$. The critical points occur at solutions of $p'(x) = 0$. Applying the quadratic formula, we find that these occur at

$$\frac{7 - \sqrt{19}}{6} \approx 0.440, \qquad \frac{7 + \sqrt{19}}{6} \approx 1.893$$

At these points,

$$p(0.440) = 3.105, \qquad p(1.893) = -0.052$$

Figure 2.65 combines this information in a single graph. This function does indeed take on negative values, but only right near the critical point at $x = 1.893$.

The guideposts for reading and interpreting a graph are summarized in the following table.

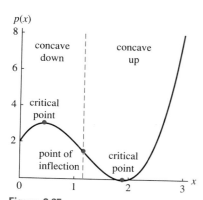

Figure 2.65

Graphing a polynomial with the first and second derivatives

What the graph does	What the derivatives do
Jump	Function discontinuous, derivative not defined
Corner	Derivative not defined
Graph vertical	Derivative not defined (infinite)
Graph increasing	Derivative positive
Graph decreasing	Derivative negative
Graph horizontal	Critical point, derivative equal to 0
Graph concave up	Second derivative positive
Graph concave down	Second derivative negative
Graph switches curvature	Point of inflection, second derivative equal to 0

Acceleration

On a graph of position against time, the second derivative has an important physical interpretation as the **acceleration**. Suppose the position of an object is $y(t)$. The derivative dy/dt is the velocity, and the second derivative is the rate of change of velocity. Formally, we have

$$\frac{d^2y}{dt^2} = \text{acceleration}$$

(Differential notation is usually used in physical applications.) A positive acceleration indicates that an object is speeding up, and a negative acceleration that it is slowing down.

A rock that has fallen a distance

$$y(t) = 5.0t^2$$

in time t has second derivative, and acceleration, of

$$\frac{d^2y}{dt^2} = 10.0 \qquad (2.8)$$

This acceleration is positive because we are measuring how far the rock has fallen, and the downward speed is increasing. Equation 2.8, the fundamental type of differential equation studied in physics, says that acceleration is proportional to force (here the force of gravity). In fact, we can rewrite Newton's famous law

$$F = ma$$

where F is the force, m the mass of the object, and a the resulting acceleration, as

$$a = \frac{d^2y}{dt^2} = \frac{F}{m}$$

If we know the force, F, and the mass, m, we have a differential equation for the position, y.

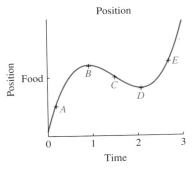 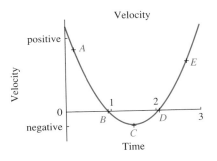 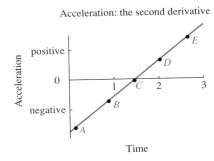

Figure 2.66
The position, velocity, and acceleration of an ant

Consider again the ant pictured in Figures 2.35 and 2.36. The position, velocity, and acceleration are shown in Figure 2.66. This ant is slowing down, or decelerating, until point C and then speeds up afterward. Deceleration includes both slowing down in the forward direction and speeding up in the negative direction. Physically, the ant acts as if there is a force pushing it to the left before it reaches C. After that time, the ant accelerates. Acceleration includes any change tending to move the ant more toward the right.

SUMMARY We have seen that the **second derivative**, defined as the derivative of the derivative, is positive when the graph of a function is **concave up** and negative when the graph of a function is **concave down**. A point where a function changes curvature is called a **point of inflection**. Using the second derivative, it is easy to graph quadratic functions, power functions, and complicated polynomials. Physically, the second derivative of the position is the **acceleration**, the tool used to write many physical laws as differential equations.

2.7 EXERCISES

1. Draw graphs of functions with the following properties.
   a. A function with a positive, increasing derivative.
   b. A function with a positive, decreasing derivative.
   c. A function with a negative, increasing (becoming less negative) derivative.
   d. A function with a negative, decreasing (becoming more negative) derivative.

2. On the figures, label the following.
   a. Critical points.
   b. One point with a positive derivative.
   c. One point with a negative derivative.
   d. One point with a positive second derivative.
   e. One point with a negative second derivative.
   f. One point of inflection.

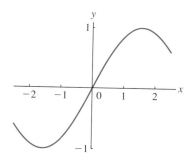

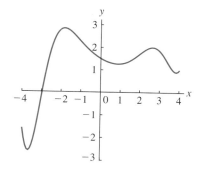

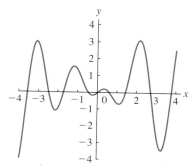

3. On the first curve in the figure for Exercise 2a, find one point with negative *third* derivative. On the second curve in the figure for Exercise 2b, find one point with positive *third* derivative.
   (**HINT:** Think about points of inflection.)

4. Find the first and second derivatives of the following functions. Justify each step.
   a. $f(x) = 3x^2 + 3x + 1$
   b. $s(x) = 1 - x + x^2 - x^3 + x^4$
   c. $g(z) = 3z^3 + 2z^2$
   d. $h(y) = y^{10} - y^9$
   e. $p(x) = 1 + x + \frac{x^2}{2} + \frac{x^3}{6} + \frac{x^4}{24}$

5. Find the first and second derivatives of the following functions, and use them to sketch a graph.
   a. $f(x) = x^{-3}$ for $x > 0$
   b. $g(z) = z + \frac{1}{z}$ for $z > 0$
   c. $h(x) = (1 - x)(2 - x)(3 - x)$
   d. $M(t) = \frac{t}{1+t}$ for $t > 0$

6. Graph the following functions on the given domains. Find the first and second derivatives. Make sure that the function is increasing when the derivative is positive and that the function is concave up when the second derivative is positive.
   a. $f(x) = 2x^3 + 1$ for $-5 \leq x \leq 5$
   b. $f(x) = \frac{1}{x^2}$ for $0 < x \leq 2$
   c. $f(x) = 10x^2 - 50x$ on $-5 \leq x \leq 5$
   d. $f(x) = x - x^2$ on $0 \leq x \leq 1$

7. Find the following.
   a. Find the tenth derivative of x^9 (think first).
   b. Find the first six derivatives of x^5.
   c. Is the eighth derivative of $p(x) = 7x^8 - 8x^7 - 5x^6 + 6x^5 - 4x^3$ positive or negative?

8. Suppose an object follows the equation
$$p(t) = 4.9t^2 - 8.0t + 10.0$$
   a. Find the velocity and the acceleration of this object.
   b. Describe in words what is happening.
   c. Graph the object's position for $t = 0$ to $t = 3$.

9. Sketch a differentiable updating function that has three equilibria. How many points of inflection does it have? Why must it have at least one?

10. The following are extreme cases where the tangent line is either vertical or horizontal.
    a. Consider the function $f(x) = x^2$. Find the slopes of three secant lines near the base point $(0, 0)$, and guess the slope of the tangent.
    b. Find the equation of the tangent line. Is it a good approximation?

c. Consider the function $g(x) = \sqrt{x}$. Find the slopes of three secant lines near the base point $(0,0)$, and guess the slope of the tangent.
d. What is the tangent line? Sketch a graph of this function and the tangent line.

11. Consider the function
$$M(t) = \frac{t}{1+t}$$
We can approximate this function at $t = 0$ with the tangent line, which matches the value of the function and its first derivative. A better approximation uses the parabola that matches the value of the function and its first and second derivatives.
a. Find the tangent line.
b. Find the second derivative and evaluate at $t = 0$.
c. Add a quadratic term to the formula of your tangent line to match the second derivative.
d. Sketch a graph of the function, its tangent line, and the approximating quadratic for $0 < t < 1$.

12. **COMPUTER:** In our universe, acceleration due to gravity is pretty much constant, so that position follows a differential equation rather like
$$\frac{d^2p}{dt^2} = -g$$
when g points in the downward direction. One can imagine a universe where gravity changes over time, making objects accelerate according to
$$\frac{d^2p}{dt^2} = -gt^n$$
for some power n. Set $g = 10$ meters per second.
a. Find a solution of the normal differential equation.
b. Find solutions of the modified differential equation for different values of n. Would objects fall faster or slower in the imagined universe?

13. **COMPUTER:** Have your computer find all critical points and points of inflection of the function
$$P(x) = 8x^5 - 18x^4 - x^3 + 18x^2 - 7x$$
Show that these match what you see on a graph of the function.

2.8 Derivatives of Exponential and Logarithmic Functions

Using the sum, product, power, and quotient rules, we can differentiate polynomials and ratios of polynomials. To be able to differentiate all biologically important functions, however, we need three more building blocks: exponential, logarithmic, and trigonometric functions. In this section we find and apply the derivatives of the exponential and logarithmic functions.

The Exponential Function

Suppose we wish to find the derivative of the function $b(t) = 2^t$. None of our rules tells us how to find a formula for the derivative. The power rule looks promising, but it applies only to functions of the form t^n, not when t is in the exponent. The best way to find the derivative of an unfamiliar function is to return to its definition (Definition 2.1):
$$\frac{db}{dt} = \lim_{h \to 0} \frac{b(t+h) - b(t)}{h}$$

The derivative of the function 2^t is

$$\frac{d\,2^t}{dt} = \lim_{h \to 0} \frac{2^{t+h} - 2^t}{h} \quad \text{definition of the derivative}$$
$$= \lim_{h \to 0} \frac{2^t 2^h - 2^t}{h} \quad \text{Law 1 of exponents}$$

2.8 Derivatives of Exponential and Logarithmic Functions

$$= \lim_{h \to 0} \frac{2^t(2^h - 1)}{h} \quad \text{factor out } 2^t$$

$$= 2^t \left(\lim_{h \to 0} \frac{2^h - 1}{h} \right) \quad \text{pull the constant out of the limit}$$

The quantity 2^t acts as a constant because the limit depends on h, not on t. The derivative is the product of two factors, the function 2^t and the *number*

$$\lim_{h \to 0} \frac{2^h - 1}{h}$$

This is the same limit we studied in Section 2.1. There we estimated the limit to be 0.6697 by evaluating the expression with $h = 0.1$. Smaller values of h produce results close to 0.693. (Exercise 8 in Section 2.1). Based on this more accurate estimate, we find

$$\lim_{h \to 0} \frac{2^h - 1}{h} = 0.693$$

Therefore,

$$\frac{d(2^t)}{dt} = 2^t \lim_{h \to 0} \frac{2^h - 1}{h} = 0.693 \times 2^t$$

or

$$\frac{db}{dt} = 0.693 \, b(t)$$

The derivative of this function is 0.693 times the function itself (Figure 2.67). The graph of the derivative looks like the graph of the function, but it is shifted down by a constant factor.

What happens when we try to find the derivative of the exponential function itself? We can follow the same steps to compute

$$\frac{de^x}{dx} = \lim_{h \to 0} \frac{e^{x+h} - e^x}{h} \quad \text{definition of the derivative}$$

$$= \lim_{h \to 0} \frac{e^x e^h - e^x}{h} \quad \text{Law 1 of exponents}$$

$$= \lim_{h \to 0} \frac{e^x(e^h - 1)}{h} \quad \text{factor out } e^x$$

$$= e^x \lim_{h \to 0} \frac{e^h - 1}{h} \quad \text{pull the constant out of the limit}$$

Again, we can factor out e^x, leaving an unknown term,

$$\lim_{h \to 0} \frac{e^h - 1}{h}$$

We can guess the limit by evaluating the function for smaller and smaller values of h (as in Exercise 3 in Section 2.2).

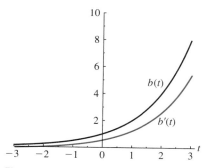

Figure 2.67
The function 2^t and its derivative

h	e^h	$e^h - 1$	$\dfrac{e^h - 1}{h}$
1.0	2.718	1.718	1.718
0.1	1.105	0.105	1.052
0.01	1.010	0.010	1.005
0.001	1.001	0.001	1.0005

The limit seems to be 1.0. In fact, the number e is *defined* to be the number for which the limit is 1. Mathematically, we have the following definition.

■ **Definition 2.5** The number e is the number for which

$$\lim_{h \to 0} \frac{e^h - 1}{h} = 1$$

■

The number e is equal to $2.7182818284590\ldots$, rather than a familiar number, such as 2 or 3.

Therefore,

$$\frac{de^x}{dx} = e^x$$

The exponential function is its own derivative. At every point on its graph, the slope of the curve is equal to the height of the curve (Figure 2.68).

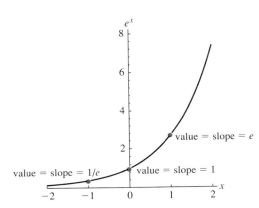

Figure 2.68
The function e^x and its derivative

Are there any other functions with this remarkable property? The function $g(x) = 2e^x$ has the derivative

$$\frac{d}{dx} 2e^x = 2 \frac{de^x}{dx} \qquad \text{the constant product rule}$$
$$= 2e^x \qquad \text{derivative of the exponential function}$$

Any constant multiple of the exponential function is its own derivative, but these functions are the *only* functions with this property.

Every function of the form $b(t) = Ke^t$ is the solution of the differential equation

$$\frac{db}{dt} = b$$

which says that the rate of change of a population is equal to the population's size. When we study differential equations in more detail, we will find that the constant K appearing in front of the exponential depends on the **initial condition**, as in the solution of a discrete-time dynamical system.

Using the derivative of the exponential function and the basic rules for differentiation, we can find the derivatives of more complicated functions. Consider the function $F(x) = xe^x$ (Figure 2.69). This is a *product* of a power function and

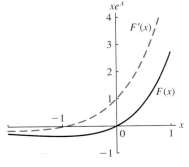

Figure 2.69
The function $F(x) = xe^x$ and its derivative

the exponential function. Therefore,

$$\frac{dF}{dx} = x\frac{de^x}{dx} + e^x\frac{dx}{dx} \quad \text{product rule}$$
$$= xe^x + e^x \quad \text{derivative of exponential and power rule}$$
$$= (1+x)e^x \quad \text{factoring}$$

The Natural Logarithm

The formula for the derivative of the inverse of the exponential function, the natural logarithm, is also simple

$$\frac{d\ln(x)}{dx} = \frac{1}{x}$$

(Figure 2.70). Because $\ln(x)$ is defined only for $x > 0$, the derivative is also defined only on this domain. We will use the fact that the natural log and exponential functions are inverses to derive this formula in Section 2.10. The derivative of the natural logarithm is the **power function** with a power of -1. We can therefore find the second derivative with the power rule, computing

$$\frac{d^2\ln(x)}{dx^2} = -\frac{1}{x^2}$$

The graph of the natural logarithm is concave down, increasing more and more slowly.

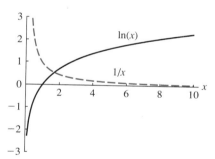

Figure 2.70
The natural logarithm and its derivative

The derivative of the function $g(x) = x\ln(x)$ is

$$\frac{dg}{dx} = x\frac{d\ln(x)}{dx} + \ln(x)\frac{dx}{dx} \quad \text{product rule}$$
$$= x \cdot \frac{1}{x} + \ln(x) \cdot 1 \quad \text{derivative of log and power rule}$$
$$= 1 + \ln(x) \quad \text{simplifying}$$

Similarly, we can find the derivative of the function $\ln(2x)$ using the laws of logs,

$$\frac{d\ln(2x)}{dx} = \frac{d[\ln(2) + \ln(x)]}{dx} \quad \text{Law 1 of logs}$$
$$= \frac{d\ln(2)}{dx} + \frac{d\ln(x)}{dx} \quad \text{sum rule}$$
$$= 0 + \frac{1}{x} = \frac{1}{x} \quad \text{constant sum rule and derivative of log}$$

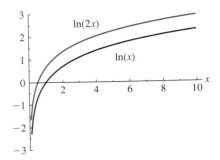

Figure 2.71
The graphs of $\ln(x)$ and $\ln(2x)$

The derivative of $\ln(2x)$ matches the derivative of $\ln(x)$ because the graphs of the two functions differ by a constant (Figure 2.71).

Applications

An important family of functions often used in probability theory are the **gamma distributions**, with formulas

$$g(x) = x^n e^{-x}$$

for various powers n. We study some of their properties in Section 7.8 under The Central Limit Theorem for Sums. What do their graphs look like? Because we do not yet know the derivative of e^{-x}, we use a law of exponents to rewrite $g(x)$ as the quotient

$$g(x) = \frac{x^n}{e^x}$$

With $n = 1$, the gamma distribution has derivative

$$g'(x) = \frac{e^x - xe^x}{e^{2x}} = \frac{1-x}{e^x}$$

where we used a law of exponents to cancel e^x from the top and bottom. This derivative is positive for $x < 1$ and negative for $x > 1$. Furthermore,

$$g''(x) = \frac{-e^x - (1-x)e^x}{e^{2x}} = \frac{x-2}{e^x}$$

The second derivative is negative when $x < 2$ and positive for $x > 2$. Using the facts that $g(0) = 0$ and $g(x) > 0$ when $x > 0$, we can draw an accurate graph of this function (Figure 2.72a).

With $n = 2$, the gamma distribution is

$$g(x) = \frac{x^2}{e^x}$$

The derivative and second derivative are

$$g'(x) = \frac{2xe^x - x^2 e^x}{e^{2x}} = \frac{2x - x^2}{e^x}$$

$$g''(x) = \frac{(2-2x)e^x - (2x - x^2)e^x}{e^{2x}} = \frac{2 - 4x + x^2}{e^x}$$

The derivative is positive for $x < 2$ and negative for $x > 2$. There are two points of inflection, which we find by solving

$$2 - 4x + x^2 = 0$$

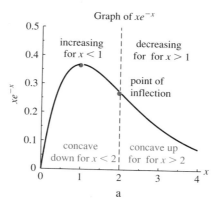
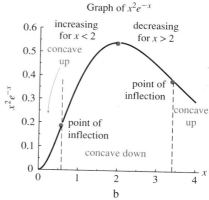

Figure 2.72
Graphs of two gamma distributions

for the roots

$$x = 2 - \sqrt{2} \approx 0.586, \qquad x = 2 + \sqrt{2} \approx 3.414$$

The second derivative is negative only between these roots. The graph incorporating all this information is shown in Figure 2.72b.

SUMMARY We added the derivatives of the exponential function and the natural log to our collection of building blocks. With these pieces and the basic rules for derivatives, we can find derivatives of many more complicated functions, including the gamma distributions.

2.8 EXERCISES

1. Suppose $f(x) = 5^x$.
   a. Write the definition of the derivative of this function.
   b. Simplify with a law of exponents and factoring.
   c. Estimate the limit by substituting small values of h.
   d. Graph the function and its derivative.
   e. Try exponentiating the limit to figure out what it is.

2. We can use two methods to try to find the derivative of e^{2x}.
   a. Work from the definition of the derivative (as in Exercise 1) and substitute small values of h to guess the limit.
   b. Show that e^{2x} is equal to the product of e^x with e^x, and use the product rule to find the derivative.
   c. Examine the limit in part **a**. Define a new variable $\Delta y = 2h$, and try to evaluate the limit.

3. Compute the derivatives of the following functions. Find the value and slope of the function at $x = 0$ and $x = 1$. Sketch a graph of the function for $0 \le x \le 1$.
   a. $f(x) = x^2 e^x$
   b. $g(x) = \dfrac{e^x}{x+1}$
   c. $F(x) = x^2 + e^x$
   d. $H(x) = \dfrac{e^x}{10 + e^x}$. Sketch the graph for $0 \le x \le 5$.

4. Consider the differential equation describing a bacterial population,

$$\frac{db}{dt} = b$$

   a. Use your knowledge of the exponential function to find a solution.
   b. Suppose you know that $b(0) = 100$. Can you guess a solution that matches this value?
   c. Graph the solution, including marking the slope at times $t = 0$ and $t = 1$.

5. a. Write the definition of the derivative for $\ln(x)$ at $x = 1$.
   b. Substitute some small values of h to guess the limit.
   c. Does your answer match the value of the derivative according to the formula?

6. a. Write the definition of the derivative for $\ln(x)$ at $x = 2$.
   b. Use a law of logs to rewrite your answer.
   c. Let $\Delta y = h/2$, and rewrite the limit in terms of Δy.
   d. Use the answer of Exercise 5 to find the values.
   e. Follow steps **a–d** to find the derivative at $x = a$.

7. Find the derivatives of the following functions.
   a. $x^2 \ln(x)$.
   b. $e^x \ln(x)$.
   c. $\ln(x^2)$ (use a law of logs).
   d. $\ln(x) - \dfrac{1}{x}$.

8. Show that the following are solutions of the given differential equation.
   a. $\dfrac{dL}{dt} = \dfrac{1}{t}$ has solution $L(t) = \ln(5t)$.
   b. $\dfrac{db}{dt} = e^t$ has solution $b(t) = e^t$.
   c. $\dfrac{db}{dt} = -b(t)$ has solution $b(t) = 1/e^t$.
   d. $\dfrac{dx}{dt} = e^{-x}$ has solution $x(t) = \ln(t)$.

9. Find and graph the tangent line to the exponential function at the following base points. What is the y-intercept?
   a. $x = 0$
   b. $x = 2$
   c. $x = -2$
   d. Do the same for the general base point $x = x_0$. For what value of x_0 is the y-intercept equal to 0? Why this value of x_0? (Think about the fact that the slope of the exponential function is always equal to the value of the function.)

10. Polynomials form a useful set of functions, in part because the derivative of a polynomial is another polynomial. Another set of functions with this useful property is called the **generalized polynomials**, formed as products of polynomials and exponential functions. Consider all functions of the form

    $$h(x) = (ax + b)e^x$$

    for various values of a and b. We call these **generalized first-order polynomials**.

    a. Show that the first and second derivatives of $h(x)$ are generalized first-order polynomials.
    b. Set $a = b = 1$. Find the first and second derivatives. Try to see the pattern, and find the tenth derivative. Where is $h(x) = 0$? Where is $h'(x) = 0$? Where is $h''(x) = 0$?
    c. Find a generalized first-order polynomial such that $h(1) = 0$. Where is the point of inflection?
    d. Show that the critical point of a generalized first-order polynomial is always 1 less than the point where $h(x) = 0$ and that the point of inflection is always 1 less than the critical point.

11. **COMPUTER:** Consider again the function

    $$g(x) = \dfrac{x^n}{e^x}$$

    which describes the gamma distribution.
    a. Find the critical point in general.
    b. Find the point or points of inflection. What happens when $n < 1$?
    c. Graph this function for $n = 0.5$, $n = 2$, and $n = 5$. What happens for very large values of n?

12. **COMPUTER:** Consider the generalized polynomial

    $$R(x) = x^4 + (88x^3 - 76x^2 - 65x + 25)e^x$$

    As defined in Exercise 10, a generalized polynomial is formed by multiplying and adding polynomials and exponential functions.
    a. Find all critical points and points of inflection of $R(x)$ for $-1 \le x \le 1$.
    b. Find the fifth derivative of $R(x)$. Does it look any simpler than $R(x)$ itself? Compare with what happens when you take many derivatives of either a polynomial or the exponential function.

2.9 Derivatives of Trigonometric Functions

The last group of special functions important in biology are the trigonometric functions. In this section, we compute the derivatives of these special functions. Like the derivative of the exponential function, the derivatives of sine and cosine have special properties that link them to solutions of important differential equations.

Sine and Cosine

When a new function cannot be built from existing pieces with products, sums, and quotients, we must compute the derivative from the definition. For the sine and cosine functions,

2.9 Derivatives of Trigonometric Functions

$$\frac{d \sin(x)}{dx} = \lim_{h \to 0} \frac{\sin(x+h) - \sin(x)}{h}$$

$$\frac{d \cos(x)}{dx} = \lim_{h \to 0} \frac{\cos(x+h) - \cos(x)}{h}$$

These formulas do us little good unless we can expand the expressions for $\sin(x+h)$ and $\cos(x+h)$. When finding the derivative of the exponential function, we used a law of exponents to rewrite e^{x+h} as $e^x e^h$. Trigonometric functions have similarly useful sum laws,

$$\sin(x+h) = \sin(x)\cos(h) + \cos(x)\sin(h)$$
$$\cos(x+h) = \cos(x)\cos(h) - \sin(x)\sin(h)$$

We can use the angle addition formula for the sine function to try to find the derivative:

$$\lim_{h \to 0} \frac{\sin(x+h) - \sin(x)}{h}$$

$$= \lim_{h \to 0} \frac{\sin(x)\cos(h) + \cos(x)\sin(h) - \sin(x)}{h} \quad \text{apply sum law for sine}$$

$$= \lim_{h \to 0} \frac{\sin(x)[\cos(h) - 1] + \cos(x)\sin(h)}{h} \quad \text{combine terms involving } \sin(x)$$

$$= \lim_{h \to 0} \frac{\sin(x)[\cos(h) - 1]}{h} + \lim_{h \to 0} \frac{\cos(x)\sin(h)}{h} \quad \text{break up limit}$$

$$= \sin(x) \lim_{h \to 0} \frac{\cos(h) - 1}{h} + \cos(x) \lim_{h \to 0} \frac{\sin(h)}{h} \quad \text{pull terms without } h\text{s outside}$$

Although this might not look much better than the limit we started with, it has an important simplification, much like that found with the exponential function. All terms involving x have come outside the limits. If we could figure out the numerical values of

$$\lim_{h \to 0} \frac{\cos(h) - 1}{h}$$

and

$$\lim_{h \to 0} \frac{\sin(h)}{h}$$

we would be done. Testing with a calculator gives the table. It seems as though the first limit is 0 and the second is 1.

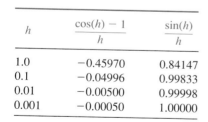

h	$\frac{\cos(h)-1}{h}$	$\frac{\sin(h)}{h}$
1.0	-0.45970	0.84147
0.1	-0.04996	0.99833
0.01	-0.00500	0.99998
0.001	-0.00050	1.00000

We can see why this occurs from a geometric diagram of sine and cosine. First, we show that

$$\lim_{h \to 0} \frac{\sin(h)}{h} = 1$$

In Figure 2.73, the length of the arc is equal to the angle in radians and is greater than $\sin(h)$, the length of the vertical line segment. Therefore,

$$\frac{\sin(h)}{h} \leq 1$$

Furthermore, the length of the line segment tangent to the circle is equal to $\tan(h)$ (Exercise 3), which is greater than the length of the arc, so $\tan(h) > h$ or

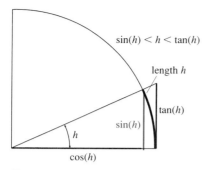

Figure 2.73
Geometric components of sine

$$\frac{\sin(h)}{\cos(h)} > h$$

implying that

$$\frac{\sin(h)}{h} > \cos(h)$$

Therefore,

$$\lim_{h \to 0} \cos(h) \leq \lim_{h \to 0} \frac{\sin(h)}{h} \leq \lim_{h \to 0} 1$$

Because $\cos(h)$ is a continuous function, $\lim_{h \to 0} \cos(h) = \cos(0) = 1$. Furthermore, the limit of the constant 1 is also 1. Therefore,

$$1 \leq \lim_{h \to 0} \frac{\sin(h)}{h} \leq 1$$

implying that

$$\lim_{h \to 0} \frac{\sin(h)}{h} = 1$$

as our computer experiment indicated.

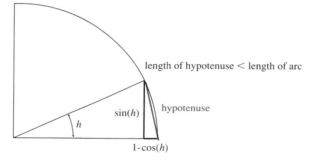

Figure 2.74
The geometry of the limit of $\dfrac{\cos(h) - 1}{h}$

To find the other limit

$$\lim_{h \to 0} \frac{\cos(h) - 1}{h}$$

we can use the fact that $\sin(h)$ and $1 - \cos(h)$ are two sides of a right triangle with hypotenuse of length less than the length of the arc (Figure 2.74). By the Pythagorean theorem, the square of the length of the hypotenuse is the sum of the squares of the sides,

$$\sin^2(h) + [1 - \cos(h)]^2 < h^2 \qquad \text{hypotenuse is shorter than arc}$$
$$\sin^2(h) + 1 - 2\cos(h) + \cos^2(h) < h^2 \qquad \text{expand the quadratic}$$
$$2 - 2\cos(h) < h^2 \qquad \text{use fact that } \sin^2(h) + \cos^2(h) = 1$$
$$1 - \cos(h) < \frac{h^2}{2} \qquad \text{divide by 2}$$

So

$$\lim_{h \to 0} \frac{\cos(h) - 1}{h} < \lim_{h \to 0} \frac{h}{2} = 0$$

This limit, in accord with computer calculations, is 0.

2.9 Derivatives of Trigonometric Functions

Putting this all together, we have found that

$$\frac{d \sin(x)}{dx} = \sin(x) \lim_{h \to 0} \frac{\cos(h) - 1}{h} + \cos(x) \lim_{h \to 0} \frac{\sin(h)}{h}$$

$$= \sin(x) \cdot 0 + \cos(x) \cdot 1 = \cos(x)$$

Similarly, the derivative of $\cos(x)$ is

$$\lim_{h \to 0} \frac{\cos(x + h) - \cos(x)}{h}$$

$$= \lim_{h \to 0} \frac{\cos(x)\cos(h) - \sin(x)\sin(h) - \cos(x)}{h} \quad \text{apply sum law for cosine}$$

$$= \lim_{h \to 0} \frac{\cos(x)(\cos(h) - 1) - \sin(x)\sin(h)}{h} \quad \text{combine terms involving } \cos(x)$$

$$= \lim_{h \to 0} \frac{\cos(x)(\cos(h) - 1)}{h} - \lim_{h \to 0} \frac{\sin(x)\sin(h)}{h} \quad \text{break up limit}$$

$$= \cos(x) \lim_{h \to 0} \frac{(\cos(h) - 1)}{h} - \sin(x) \lim_{h \to 0} \frac{\sin(h)}{h} \quad \text{pull terms without } h\text{s out}$$

$$= \cos(x) \cdot 0 - \sin(x) \cdot 1 \quad \text{use the limits found earlier}$$

$$= -\sin(x) \quad \text{multiply out}$$

In summary,

$$\frac{d \sin(x)}{dx} = \cos(x)$$

$$\frac{d \cos(x)}{dx} = -\sin(x)$$

At $x = 0$, the cosine is flat but beginning to decrease, meaning that the derivative must begin at 0 and become negative. At $x = 0$, sine is increasing, meaning the derivative should take on a positive value at that point (Figure 2.75a). To recall where the negative sign goes, remember the graphs of sine and cosine (Figure 2.75b).

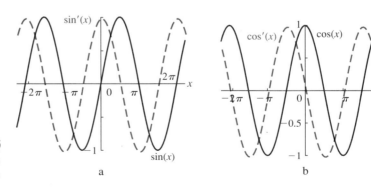

Figure 2.75
The graphs and derivatives of sine and cosine

What is the second derivative of each of these functions?

$$\frac{d^2 \sin(x)}{dx^2} = \frac{d \cos(x)}{dx} = -\sin(x)$$

$$\frac{d^2 \cos(x)}{dx^2} = -\frac{d \sin(x)}{dx} = -\cos(x)$$

Each of these functions is equal to the **negative** of its second derivative. These functions are concave down when positive and concave up when negative. Furthermore, both have points of inflection every time they cross 0. This property may remind you of the fact that the exponential function is its own derivative. There is a deep connection between these two types of functions that requires the more advanced topic of **complex numbers** to understand.

We can use rules for differentiation to find the derivatives of more complex functions. For example, is the function

$$F(x) = x + \sin(x)$$

an increasing function? We can check by computing the derivative

$$\frac{d\,[x + \sin(x)]}{dx} = \frac{dx}{dx} + \frac{d\,\sin(x)}{dx} \quad \text{sum rule}$$

$$= 1 + \cos(x) \quad \text{power rule and derivative of sine}$$

Because $\cos(x)$ is never smaller than -1, the derivative is never negative and this function is increasing. The function has a derivative critical points where $\cos(x) = -1$, such as $x = -\pi$ and $x = \pi$ (Figure 2.76).

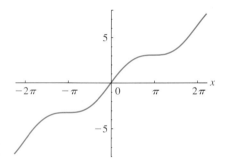

Figure 2.76
The graph of $x + \sin(x)$

Other Trigonometric Functions

We can use the quotient rule to find the derivatives of the other trigonometric functions. For $\tan(\theta)$,

$$\frac{d}{d\theta}\tan(\theta) = \frac{d}{d\theta}\frac{\sin(\theta)}{\cos(\theta)}$$

$$= \frac{\cos(\theta)\,[d\,\sin(\theta)/d\theta] - \sin(\theta)\,[d\,\cos(\theta)/d\theta]}{\cos^2\theta} \quad \text{quotient rule}$$

$$= \frac{\cos(\theta)\cos(\theta) + \sin(\theta)\sin(\theta)}{\cos^2(\theta)} \quad \text{derivatives of sine and cosine}$$

$$= \frac{1}{\cos^2\theta} \quad \sin^2(\theta) + \cos^2(\theta) = 1$$

$$= \sec^2\theta \quad \text{definition of the secant function}$$

On a graph of $\tan(\theta)$, the function has a slope of 1 at $\theta = 0$ and is always increasing except at points where it is not defined (Figure 2.77).

2.9 Derivatives of Trigonometric Functions

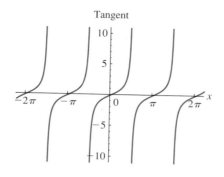
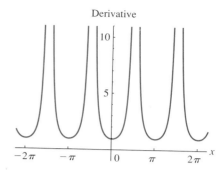

Figure 2.77
The tangent function and its derivative

Similarly (Exercise 2),

$$\cot(x) = -\csc^2(x)$$
$$\sec(x) = \tan(x)\sec(x)$$
$$\csc(x) = -\cot(x)\csc(x)$$

Although these functions also have simple derivatives, they do not have the remarkable properties of sine and cosine themselves.

Applications

We have studied the behavior of a rock falling in a constant gravitational field. Consider now an object attached to a perfect spring (Figure 2.78). The spring produces an outward force when compressed and an inward force when stretched. Assume that

$$\text{Force} = -\text{amount of stretch}$$

The negative sign indicates that the force acts opposite the direction of stretch (Figure 2.78). For example, if the spring is stretched to the right, it produces a force to the left.

Newton's law says that acceleration, the second derivative of position, is proportional to force according to

$$F = ma$$

where m is the mass of the object. Suppose the mass of the object is 1.0 (we examine other cases in Section 2.10). Then,

$$a = \frac{d^2 x}{dt^2} = -x$$

Applies no force when not stretched

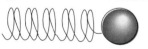

Applies outward force when compressed

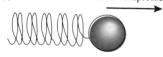

Applies inward force when stretched

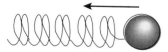

Figure 2.78
A perfect spring

The solution of this differential equation is a function that has second derivative equal to the negative of itself. We have just met two such functions: sine and cosine. The position of an object attached to such an ideal spring (without friction) will follow exactly a sinusoidal oscillation. This spring is called the "simple harmonic oscillator" in physics and is one of the reasons why the sinusoidal functions are so important (in the same way that the differential equation $\frac{db}{dt} = b$ is one of the reasons why exponential functions are so important).

For example, suppose the spring is stretched 1 unit and then the object is released. The subsequent movement follows the function

$$p(t) = \cos(t)$$

We can check that $p(0) = 1$ and that $\dfrac{dp}{dt} = -\sin(0) = 0$, meaning that the object does begin at rest at a position of 1. Furthermore,

$$\frac{d^2 p}{dt^2} = -\cos(t) = -p(t)$$

(Figure 2.79a). If the object is released when the spring is *compressed* by 1 unit, the solution is

$$p(t) = -\cos(t)$$

(Figure 2.79b). Finally, if the object is released when the spring is stretched outward by 0.2 units, the solution is

$$p(t) = 0.2 \cos(t)$$

(Figure 2.79c).

Figure 2.79

Three solutions of the spring equation

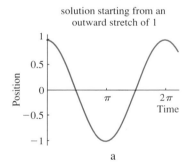
a

solution starting from an outward stretch of 1

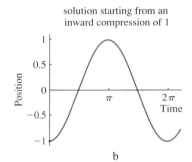
b

solution starting from an inward compression of 1

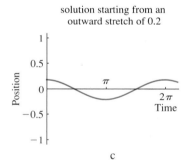
c

solution starting from an outward stretch of 0.2

SUMMARY To complete our collection of building blocks, we derived the derivatives of the trigonometric functions.

Function	Derivative
$\sin(x)$	$\cos(x)$
$\cos(x)$	$-\sin(x)$
$\tan(x)$	$\sec^2(x)$
$\cot(x)$	$-\csc^2(x)$
$\sec(x)$	$\tan(x)\sec(x)$
$\csc(x)$	$-\cot(x)\csc(x)$

The sinusoidal functions provide the solutions of the differential equation describing a spring, the **simple harmonic oscillator**. They describe the position of an object that tends to move back to a resting state with force proportional to the displacement.

2.9 EXERCISES

1. Find the derivatives of the following functions.
   a. $x^2 \sin(x)$
   b. $x^2 \cos(x)$
   c. $\sin(\theta) + \cos(\theta)$
   d. $\sin(\theta)\cos(\theta)$
   e. $\dfrac{\sin(\theta)}{1 + \cos(\theta)}$

2. Check the derivatives of the other trigonometric functions.
   a. $\sec(\theta)$
   b. $\csc(\theta)$
   c. $\cot(\theta)$

3. Show that the length of the tangent line segment in Figure 2.73 is equal to the tangent of the angle.

4. Compute the derivatives of the following functions. Find the value of the function and the slope at $x = 0$, $x = \pi/2$, and $x = \pi$. Sketch a graph of the function for $0 \le x \le \pi$.
   a. $a(x) = 3x + \cos(x)$
   b. $b(y) = y^2 + \cos(y)$
   c. $c(z) = \dfrac{\sin(z)}{e^z}$

5. Compute the following in two ways: (1) by using the product or quotient rule and then simplifying with trigonometric identities, and (2) by simplifying first and then taking the derivative.
   a. $\tan(\theta)\cos(\theta)$
   b. $\dfrac{\cot(\theta)}{\csc(\theta)}$
   c. $\dfrac{\tan(\theta)}{\sin(\theta)}$

6. Use addition formulas to find the derivatives of the following.
   a. $\cos(2\theta)$. Try to simplify the answer in terms of $\sin(2\theta)$.
   b. $\sin(2\theta)$. Try to simplify the answer in terms of $\cos(2\theta)$.
   c. $\cos(\theta + \phi)$. Try to simplify the answer in terms of $\sin(\theta + \phi)$ (think of ϕ as a constant).
   d. $\sin(\theta + \phi)$. Try to simplify the answer in terms of $\cos(\theta + \phi)$.

7. Use derivatives to sketch graphs of the following functions. How would you describe the behavior in words?
   a. $r(t) = t \cos(t)$ for $0 \le t \le 4\pi$
   b. $s(t) = e^{0.2t} \cos(t)$ for $0 \le t \le 40$
   c. $p(t) = e^t [1 + 0.2 \cos(t)]$ for $0 \le t \le 40$

8. Show that the following are solutions of the given differential equation.
   a. $\dfrac{ds}{dt} = \cos(t)$ has solution $s(t) = \sin(t) + 4$.
   b. $\dfrac{dy}{dt} = 2t \sin(t) + t^2 \cos(t)$ has solution $y(t) = t^2 \sin(t)$.
   c. $\dfrac{d^2 f}{dt^2} = -f(t) - 2\sin(t)$ has solution $f(t) = t \cos(t)$.
   d. $\dfrac{d^4 g}{dt^4} = g(t)$ has solution $g(t) = \cos(t) + 2\sin(t) - 3e^t$.
   e. $\dfrac{d^{40} h}{dt^{40}} = h(t)$ has solution $h(t) = \cos(t) + 2\sin(t) - 3e^t$.

9. Find and graph the following tangent lines. Which look like the best approximations? Compute the second derivative at the base point and try to explain why.
   a. Tangent to $\sin(x)$ at $x = 0$
   b. Tangent to $\cos(x)$ at $x = 0$
   c. Tangent to $\tan(x)$ at $x = 0$
   d. Tangent to $\sin(x)$ at $x = \pi/4$
   e. Tangent to $\cos(x)$ at $x = \pi/4$
   f. Tangent to $\tan(x)$ at $x = \pi/4$

10. Blood flow is pulsatile. Suppose the blood flow along the artery of a whale is
    $$F(t) = 212.0 \cos(t)$$
    where F is measured in cm^3/s and t is measured in seconds.
    a. Find the average flow and the amplitude.
    b. Find the period of this flow. How many heartbeats does this whale have per minute?
    c. When is the flow 0? What is happening at these times?
    d. Find the rate of change of the flow. What does it mean when this is 0? What is the flow at these times?

11. The spring we studied had no **friction**. Friction acts as a force much like the spring but is proportional to velocity rather than displacement. One possible equation describing this is
    $$\dfrac{d^2 x}{dt^2} = -2x - 2\dfrac{dx}{dt}$$
    a. Explain each term in this equation.
    b. Show that $x(t) = e^{-t} \cos(t)$ is a solution of this differential equation.
    c. Graph the solution and explain what is happening. Is friction strong in this system?

12. In London, the number of hours of daylight follows roughly
    $$L(t) = 12.0 - 4.5 \cos(t)$$
    where t represents time measured in units, where 2π corresponds to 1 year, and where the shortest day is December 21. A plant puts out leaves in the spring in response to the *change* in day length.
    a. Find the value of t producing the longest day. What day is this? How many hours of daylight?
    b. Find the smallest value of t producing a day of average length. What day is this? How many hours of daylight?
    c. Find the rate of change of day length. When is this equal to 0?
    d. At what time of year would it be easiest for the plant to detect changes in day length?

13. **COMPUTER**: This problem requires a computer program with a built-in ability to solve differential equations. The spring equation
    $$\dfrac{d^2 y}{dt^2} = -y(t)$$

is only an approximation to the behavior of a pendulum, which is in fact better described by the equation

$$\frac{d^2y}{dt^2} = -\sin[y(t)]$$

It is impossible to write a solution to this equation.
a. Starting from $y(0) = 0.1$ and $dy/dt = 0$ at $t = 0$, the solution of the spring equation is $y(t) = 0.1\cos(t)$. Compare this with the solution of the pendulum equation for one period (from $t = 0$ to $t = 2\pi$). Graph the two solutions.
b. Do the same starting from $y(0) = 0.2$.
c. Do the same starting from $y(0) = 0.5$.
d. Do the same starting from $y(0) = 1.0$.
e. Do the same starting from $y(0) = 1.5$.
f. How long does it take the pendulum to swing all the way back in each case? Does the period of a pendulum depend on the amplitude?

2.10 The Chain Rule

Composition is the final, and perhaps the most important, way to combine functions. If we know the derivatives of the components parts, the **chain rule** gives the formula for the derivative of the composition. Using the chain rule (along with the sum, product, power, and quotient rules and the derivatives of special functions), we can find the derivative of *any* function that can be built from polynomial, exponential, and trigonometric functions. We can also apply the chain rule to compute the derivatives of **inverse functions**.

The Derivative of a Composite Function

Consider the function

$$F(x) = e^{-x^2}$$

which we use later to describe the **normal distribution**. This function is the **composition** of the exponential function $f(g) = e^g$ with the quadratic function $g(x) = -x^2$, or

$$F(x) = (f \circ g)(x)$$

(Figure 2.80). Similarly, the function

$$H(y) = \frac{1}{1 + y^2}$$

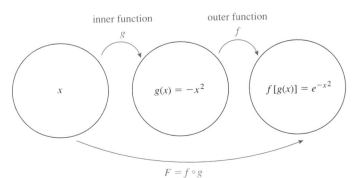

Figure 2.80
$F(x)$ written as a composition of functions

is the composition of the reciprocal function $r(p) = 1/p$ and the polynomial $p(y) = 1 + y^2$, or

$$H(y) = (r \circ p)(y)$$

(Figure 2.81). In each case, we know the derivative of the component functions. Can we use this information to find the derivative of the composition?

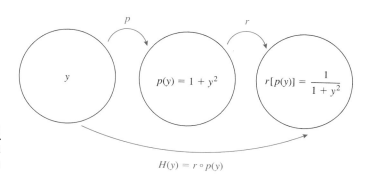

Figure 2.81
$H(y)$ written as a composition of functions

We need to first recognize a function as a composition. To determine how a function was built, think of how you would compute the value on a calculator. To calculate $F(0.5)$, find the negative of the square of 0.5 (the **inner function** g) as -0.25, and then exponentiate that number (the **outer function** f) to find the result $e^{-0.25} = 0.779$. To calculate $H(1)$, evaluate the polynomial $1 + y^2$ at $y = 1$ (the inner function p) to find 2, and then take the reciprocal of that (the outer function r) to find the result 0.5.

The chain rule tells us how to compute the derivative by combining information about the inner and outer functions and their derivatives.

■ **THEOREM 2.11** **(The Chain Rule for Derivatives)**

Suppose

$$h(x) = (f \circ g)(x)$$

where f and g are both differentiable functions. Then

$$h'(x) = f'[g(x)]g'(x) \quad \text{prime notation}$$

$$\frac{dh}{dx} = \frac{df}{dg}\frac{dg}{dx} \quad \text{differential notation} \quad ■$$

The expression $f'[g(x)]$ means the derivative of the function f evaluated at the point $g(x)$. The expression

$$\frac{df}{dg}$$

refers to the same quantity, but means the derivative of f thought of as a function of g, again evaluated at the point $g(x)$.

How do we *use* the chain rule? The following algorithm gives the necessary steps.

Algorithm 2.1 (Using the chain rule)

1. Write the function as a composition.
2. Take the derivatives of the component pieces.
3. Multiply the derivatives together.
4. Put everything in terms of the original variable. ∎

We apply this method to the two functions F and H defined above. To find the derivative of $F(x)$,

1. Write $F(x) = f[g(x)]$, where
$$f(g) = e^g$$
$$g(x) = -x^2$$

2. Find the derivative of each component function.
$$f'(g) = e^g$$
$$g'(x) = -2x$$

3. The derivative is the product
$$F'(x) = e^g \cdot (-2x)$$

4. Substitute the definition of the inner function $g(x)$ to write the answer in terms of x.
$$F'(x) = e^g \cdot (-2x) = e^{-x^2} \cdot (-2x) = -2xe^{-x^2}$$

It can also help to organize this information into a table as follows:

The derivative of $F(x) = (f \circ g)(x) = e^{-x^2}$

Function	Derivative (prime notation)	Derivative (differential notation)
$g(x) = -x^2$	$g'(x) = -2x$	$\dfrac{dg}{dx} = -2x$
$f(g) = e^g$	$f'(g) = e^g$	$\dfrac{df}{dg} = e^g$
$F(x) = f[g(x)]$	$F'(x) = -2xe^g = -2xe^{-x^2}$	$\dfrac{dF}{dx} = -2xe^g = -2xe^{-x^2}$

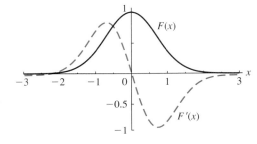

Figure 2.82

The function $F(x) = e^{-x^2}$ and its derivative

2.10 The Chain Rule

To find the derivative of the function H defined by

$$H(y) = \frac{1}{1+y^2}$$

we follow the same steps.

1. Write $H(y) = (r \circ p)(y)$ where $p(y) = 1 + y^2$ and $r(p) = 1/p$.
2. Find the derivatives of r and p,

$$p'(y) = 2y$$
$$r'(p) = -\frac{1}{p^2}$$

3. The derivative is the product

$$H'(y) = -\left(\frac{1}{p^2}\right) 2y$$

4. Substitute the definition of the inner function p to write the answer in terms of y,

$$H'(y) = -\frac{2y}{(1+y^2)^2}$$

This calculation can be summarized in a table (prime notation only):

The derivative of $H(y) = (r \circ p)(y) = \dfrac{1}{1+y^2}$

Function	Derivative
$p(y) = 1 + y^2$	$p'(y) = 2y$
$r(p) = \dfrac{1}{p}$	$r'(p) = -\dfrac{1}{p^2}$
$H(y) = r[p(y)]$	$H'(y) = -\dfrac{2y}{p^2} = -\dfrac{2y}{(1+y^2)^2}$

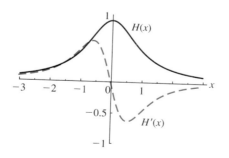

Figure 2.83
The function $H(y) = 1/(1+y^2)$ and its derivative

For more complicated functions, the chain rule can be applied several times. Some functions must be broken into three or more component pieces. The version of the chain rule for triple compositions is

$$(f \circ g \circ h)'(x) = f'\{g[h(x)]\}\, g'[h(x)]\, h'(x) \tag{2.9}$$

or, in differential notation,

$$\frac{d(f \circ g \circ h)}{dx} = \frac{df}{dg}\frac{dg}{dh}\frac{dh}{dx} \qquad (2.10)$$

Because the chain rule follows the same pattern, we use Algorithm 2.1 to find derivatives.

Consider the function

$$F(x) = \sin\left[(1-x)^4 + 1\right]$$

1. Write as the composition $F(x) = f\{g\,[h(x)]\}$, where

$$h(x) = 1 - x$$
$$g(h) = h^4 + 1$$
$$f(g) = \sin(g)$$

(Figure 2.84).

2. Take the derivatives of the components:

$$h'(x) = -1$$
$$g'(h) = 4h^3$$
$$f'(g) = \cos(g)$$

3. Multiply the derivatives together:

$$F'(x) = \cos(g) \cdot 4h^3 \cdot (-1).$$

4. Substitute the definitions of the functions $h(x)$ and $g(h)$ to write the answer in terms of x:

$$\begin{aligned}F'(x) &= \cos(h^4 + 1) \cdot 4h^3 \cdot (-1) \\ &= \cos\left[(1-x)^4 + 1\right] \cdot 4(1-x)^3 \cdot (-1) \\ &= -4\cos\left[(1-x)^4 + 1\right](1-x)^3\end{aligned}$$

In tabular form,

The derivative of $F(x) = f\{g[h(x)]\} = \sin[(1-x)^4 + 1]$

Function	Derivative
$h(x) = 1 - x$	$h'(x) = -1$
$g(h) = h^4 + 1$	$g'(h) = 4h^3$
$f(g) = \sin(g)$	$f'(g) = \cos(g)$
$F(x) = f\{g[h(x)]\}$	$F'(x) = -1 \cdot 4h^3 \cos(g) = -4(1-x)^3 \cos[(1-x)^4 + 1]$

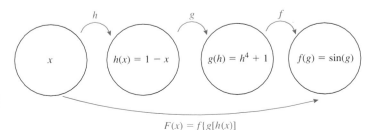

Figure 2.84

The function $F(x) = \sin[(1-x)^4 + 1]$ written as a composition

Derivatives of Inverse Functions

Many important functions, such as the natural logarithm, are defined as the **inverses** of other functions. The inverse f^{-1} of the function $f(x)$ is defined in terms of functional composition as

$$(f \circ f^{-1})(x) = x$$

(Definition 1.3). We can use the chain rule to find the derivative. However, in this case we use the derivative of the composition (which has the simple formula x) to find the derivative of the component part f^{-1}.

We can take the derivatives of both sides,

$$(f \circ f^{-1})'(x) = f'\left[f^{-1}(x)\right](f^{-1})'(x) = 1$$

We can solve this equation for $(f^{-1})'(x)$, the derivative of the inverse, finding

$$(f^{-1})'(x) = \frac{1}{f'\left[f^{-1}(x)\right]}$$

■ **THEOREM 2.12** (The Derivative of an Inverse Function)

Suppose f is a differentiable function with inverse f^{-1} and that

$$f'\left[f^{-1}(x)\right] \neq 0$$

Then

$$(f^{-1})'(x) = \frac{1}{f'\left[f^{-1}(x)\right]}$$ ∎

We can use this theorem to check the derivative of $\ln(x)$. Setting $f(x) = e^x$,

$$f'(x) = e^x$$
$$f^{-1}(x) = \ln(x)$$

Therefore,

$$(f^{-1})'(x) = \frac{1}{f'\left[f^{-1}(x)\right]}$$
$$= \frac{1}{e^{f^{-1}(x)}}$$
$$= \frac{1}{e^{\ln(x)}}$$
$$= \frac{1}{x}$$

The simple formula for the derivative of the natural logarithm can be found from the simple formula for the derivative of the exponential function.

We can also check the derivative of $\sqrt{x}$. Setting $S(x) = x^2$,

$$S'(x) = 2x$$
$$S^{-1}(x) = \sqrt{x}$$

Therefore

$$(f^{-1})'(x) = \frac{1}{f'\left[f^{-1}(x)\right]}$$

$$= \frac{1}{2f^{-1}(x)}$$

$$= \frac{1}{2\sqrt{(x)}}$$

matching the power rule for fractional powers (Theorem 2.7).

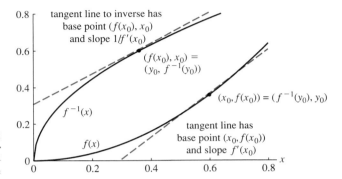

Figure 2.85

The tangent lines to a function and its inverse

This theorem has a useful geometric interpretation (Figure 2.85). The graph of the inverse function is the mirror image of the graph of the function itself. Therefore, the tangent line to the inverse function is the mirror image of the tangent line to the function. The tangent to the inverse at the point $[f(x_0), x_0]$, the reflection of the point $[x_0, f(x_0)]$, will have slope

$$\text{slope of tangent to inverse at } [f(x_0), x_0]$$
$$= \frac{1}{\text{slope of tangent to function at } [x_0, f(x_0)]}$$

Therefore

$$(f^{-1})'[f(x_0)] = \frac{1}{f'(x_0)}.$$

With this geometric interpretation, we can find the slope of a curve for which we have no formula. When studying the inverse, we found that the function

$$f(x) = x^5 + x + 1$$

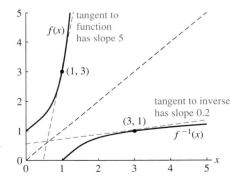

Figure 2.86

The tangent lines to an uncomputable inverse

has an inverse even though we cannot find a formula for it. Suppose, nonetheless, that we wish to find the slope of the inverse at the point (3, 1) (Figure 2.86). This point lies on the inverse because $f(1) = 3$. Furthermore,

$$f'(x) = 5x^4 + 1$$

so $f'(1) = 5$. Therefore, the slope of the inverse is the reciprocal, or

$$(f^{-1})'(3) = \frac{1}{5}$$

Applications

We can use the chain rule to find the derivative of $b(t) = 2.0^t$, the function we first studied in Section 2.1. We will finally be able to explain that factor of 0.693. The trick is to write the function in terms of the exponential function as

$$2.0^t = \left(e^{\ln(2.0)}\right)^t \quad \text{exponential and natural log are inverses}$$
$$= e^{\ln(2.0)t} \quad \text{Law 2 of exponents}$$

The derivative of this composition can be found by following the chain rule, Algorithm 2.1.

1. $b(t)$ is a composition of $f[g(t)]$, where

$$f(g) = e^g$$
$$g(x) = \ln(2.0)t$$

2. The derivatives of the components are

$$f'(g) = e^g$$
$$g'(t) = \ln(2.0)$$

3. Using the chain rule, the derivative of b is

$$b'(t) = e^g \ln(2.0)$$

4. Putting back in terms of t, we get

$$b'(t) = e^{\ln(2.0)t} \ln(2.0) \quad \text{substitute definition of } g$$
$$= (e^{\ln(2.0)})^t \ln(2.0) \quad \text{Law 2 of exponents}$$
$$= 2.0^t \ln(2.0) \quad \text{exponential and log are inverses}$$
$$= \ln(2.0) 2.0^t = 0.693 \cdot 2.0^t \quad \text{reorder and evaluate}$$

The mysterious factor 0.693, which cropped up in Section 2.1, is the natural logarithm of 2.0. Perhaps better than any other calculation, this one shows the pivotal role of the number e and the exponential function. Although the function 2.0^t does not contain any reference to e, a natural log is generated spontaneously by the derivative.

Exponential measurements are generally written in the form

$$M(t) = M(0)e^{\alpha t}$$

(Section 1.8). The constant $M(0)$ represents the value of the measurement at $t = 0$. The parameter α determines whether and how fast the measurement is increasing as a function of time. If α is negative, the measurement is decreasing and decreases faster the smaller α is. If α is positive, the measurement is increasing and increases faster the larger α is.

We can find the derivative of $M(t)$ using the chain rule:

1. $M(t)$ is the composition of $f[g(t)]$ where
$$f(g) = M(0)e^g$$
$$g(t) = \alpha t$$

2. The derivatives of the components are
$$f'(g) = M(0)e^g$$
$$g'(t) = \alpha$$

3. Using the chain rule, we find that the derivative of M is
$$M'(t) = M(0)e^g \alpha$$

4. Putting back in terms of t, we get
$$\begin{aligned} M'(t) &= M(0)e^{\alpha t} \alpha & \text{substitute definition of } g \\ &= \alpha M(0) e^{\alpha t} & \text{reorder} \\ &= \alpha M(t) & \text{rewrite in terms of } M(t) \end{aligned}$$

The derivative of the general exponential function can be found by multiplying the original function by the parameter α that appears in the exponent.

Rule for Finding Derivatives of the General Exponential Function

The derivative of any function of the form
$$f(x) = ce^{\alpha x}$$
is
$$\frac{df}{dx} = c\alpha e^{\alpha x} = \alpha f(x)$$ ∎

We can use this rule to find a solution of the differential equation
$$\frac{db}{dt} = 0.5b$$
which says that the rate of change of a population is 0.5 times the population size. One solution is
$$b(t) = e^{0.5t}$$

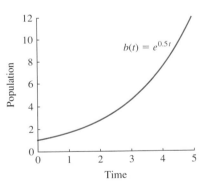

Figure 2.87
The growth of a population obeying $db/dt = 0.5b$.

(Figure 2.87). This population grows more slowly than one following the exponential function itself.

The same type of rule works for oscillations written in terms of the general cosine function: The constant comes out in front of the function.

Rule for Finding Derivatives of the General Trigonometric Function

The derivative of any function of the form
$$f(x) = A \cos\left[\frac{2\pi(x-\phi)}{T}\right]$$
is
$$f'(x) = \frac{-2\pi A}{T} \sin\left[\frac{2\pi(x-\phi)}{T}\right]$$ ∎

We can show this with the chain rule again:

1. $f(x)$ is the composition of $h[g(x)]$, where

$$h(g) = A\cos(g)$$
$$g(x) = \frac{2\pi(x - \phi)}{T}$$

2. The derivatives of the components are

$$h'(g) = -A\sin(g)$$
$$g'(x) = \frac{2\pi}{T}$$

3. Using the chain rule, we find that the derivative of f is

$$h'(g) = -A\frac{2\pi}{T}\sin(g)$$

4. Putting back in terms of x, we get

$$h'(x) = \frac{-2\pi A}{T}\sin\left[\frac{2\pi(x - \phi)}{T}\right]$$

This rule can be applied to the daily and monthly temperature cycles introduced in Section 1.9 (written in units of days),

$$P_d(t) = 36.8 + 0.3\cos[2\pi(t - 0.583)]$$
$$P_m(t) = 36.8 + 0.2\cos\left[\frac{2\pi(t - 16)}{28}\right]$$

Using the rule for finding the derivative of a general trigonometric function, along with the constant sum rule, we get

$$\frac{dP_d}{dt} = -0.3 \cdot 2\pi\sin[2\pi(t - 0.583)]$$
$$\frac{dP_d}{dt} = -0.2 \cdot \frac{2\pi}{28}\sin\left[\frac{2\pi(t - 16)}{28}\right]$$

The scales on the graphs of the two derivatives are very different (compare Figures 2.88 and 2.89). The daily oscillation has a much greater rate of change than the monthly one, nearly 2° per day, with the other reaching only 0.05° per day. Biologically, this corresponds to the fact that the daily oscillation is much faster. Mathematically, this results from the factor of 28 dividing the amplitude of the derivative of the monthly rhythm.

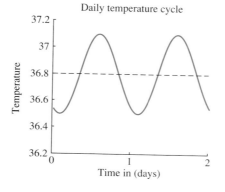
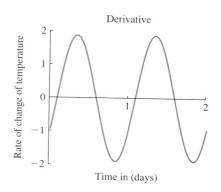

Figure 2.88
The daily temperature cycle and its derivative

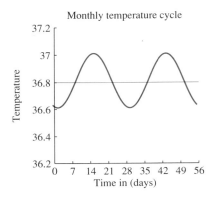
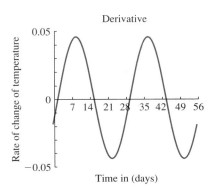

Figure 2.89

The monthly temperature cycle and its derivative

SUMMARY We derived the **chain rule** for differentiating the composition of two functions. Almost any complicated function can be differentiated with the four-step algorithm that consists of (1) breaking the function into components, (2) taking the derivative of each component, (3) multiplying the derivatives together, and (4) rewriting in terms of the original variable. The same technique works for compositions of more than two functions. The chain rule can be used to derive the formulas for the derivatives of inverse, exponential, and trigonometric functions.

2.10 EXERCISES

1. Use the chain rule to compute the following derivatives.
   a. $g(x) = (1 + 3x)^2$
   b. $F(z) = \left[1 + \dfrac{2}{1+z}\right]^3$
   c. $f_1(t) = (1 + t)^{30}$
   d. $f_2(t) = (1 + t^2)^{15}$
   e. $h(x) = (1 + 2x)^3$
   f. $r(x) = \dfrac{(1 + 3x)^2}{(1 + 2x)^3}$

2. Compute the following derivatives.
   a. $f(x) = e^{-3x}$
   b. $g(y) = \ln(1 + y)$
   c. $h(x) = 2^x 3^x$
   d. $F(z) = \dfrac{e^z}{z}$
   e. $G(x) = 8e^{x^2}$
   f. $s(w) = 4.2\sqrt{1 + e^w}$
   g. $p(x) = \dfrac{1 - e^x}{1 + e^x}$
   h. $q(y) = y^y$

   (**HINT:** Rewrite part **h** with the exponential function.)

3. Take the derivatives of the following functions in two ways: first using the chain rule, and then using a law of logs.
   a. $\ln(x^3)$
   b. $\ln(3x)$

4. Suppose the volume V of a cell is

$$V(t) = 10.0t^2$$

where V is measured in microns and t represents time in hours. Suppose the mass as a function of volume is

$$M(V) = \dfrac{10V}{1000 + V}$$

where M is measured in micrograms.
   a. Sketch a graph of mass as a function of volume over the first 24 h.
   b. Find the density as a function of volume. Describe the behavior of this function. Does it make biological sense?
   c. Use the chain rule to find the derivative of mass with respect to time.
   d. Evaluate the derivative at $t = 0$, $t = 12$, and $t = 24$. Does this fit your graph?

5. Test the chain rule on the following, where $I(x)$ represents the identity function defined by $I(x) = x$. Explain your results.
   a. $F(x) = I[f(x)]$
   b. $G(x) = g[I(x)]$

6. Consider finding the derivative of functions with the form

$$F(x) = (1 + 2x)^n$$

with and without using the chain rule. Try the following with $n = 2$, $n = 3$, and $n = 4$.
   a. Expand by multiplying out the function and taking the derivative of the whole expression.

b. Take the derivative with the chain rule and compare.
c. Which method would you rather use with $n = 50$? What is the result?

7. We will compare the chain rule with the quotient rule. Find the derivative of

$$H(y) = \frac{1}{1 + y^3}$$

in the following ways.
a. Use the chain rule.
b. Use the quotient rule.
c. Which method do you prefer?

8. Consider the function

$$p(t) = \cos\left(\frac{2\pi t}{T}\right)$$

where T is a constant.
a. Find the first and second derivatives.
b. Show that $p(t)$ obeys an equation like the spring equation,

$$\frac{d^2 x}{dt^2} = -x$$

c. What happens when $T > 2\pi$? What does this mean physically about the spring and the oscillation?

9. In Exercise 7 in Section 1.9, we graphed the following functions and found their amplitudes, periods, and phases.

$$f(x) = 3.0 + 4.0 \cos\left[\frac{2\pi(x - 1.0)}{5.0}\right]$$

$$g(t) = 4.0 + 3.0 \cos\left[2\pi(x - 5.0)\right]$$

$$h(z) = 1.0 + 5.0 \cos\left[\frac{2\pi(x - 3.0)}{4.0}\right]$$

Find and graph their derivatives.

10. Consider the sum of the temperature cycles $P_d(t) + P_m(t)$ (introduced in Section 1.9).
a. Find the derivative.
b. Sketch a graph of the derivative over 1 month.
c. If you measured only the derivative, which oscillation would you see?

11. Suppose a bacterial population follows the equation

$$\frac{db}{dt} = 0.1b[1 + \cos(0.8t)]$$

a. Describe in words what is happening.
b. Show that $b(t) = e^{0.1t + 0.125 \sin(0.8t)}$ is a solution.
c. Graph this solution for $0 \le t \le 20$.
d. Consider the more general equation

$$\frac{db}{dt} = 0.1b[1 + A\cos(0.8t)]$$

where A is a constant. How does this equation differ when A is large? When A is small?

e. Show that $b(t) = e^{0.1t + 0.125A \sin(0.8t)}$ is a solution.
f. Graph this solution for $A = 0$, $A = 0.5$, and $A = 2.0$.

12. The equation for the top half of a circle is

$$f(x) = \sqrt{1 - x^2}$$

a. Draw a picture of this function complete with a radius connecting the point $(0, 0)$ to the point $(0.5, \sqrt{3}/2)$ and a tangent at $(0.5, \sqrt{3}/2)$.
b. Find the slope of the radius.
c. Use the fact that a line perpendicular to a line with slope m has slope $-1/m$ to find the slope of the tangent.
d. Do the same for the point $(x_0, \sqrt{1 - x_0^2})$.
e. Write the formula for the derivative of $f(x)$ at x_0. Test your formula at $x_0 = 0$, $x_0 = \sqrt{2}/2$, and $x_0 = -\sqrt{2}/2$.
f. Compare with what you find with the chain rule.

13. COMPUTER: Consider the family of functions

$$h(t) = \cos(3.0t) + 1.5 \cos(3.6t) + 2.0 \cos(vt)$$

for various values of v ranging from 2.0 to 3.0. Graph them. Why do they look so weird? When do they look least weird?

14. COMPUTER: Find the derivatives of the following functions. Where are the critical points and points of inflection? What happens as more and more cosines are piled up? Explain this in terms of the updating function

$$x_{t+1} = \cos(x_t)$$

a. $\cos[\cos(x)]$
b. $\cos\{\cos[\cos(x)]\}$
c. $\cos(\cos\{\cos[\cos(x)]\})$
d. $\cos[\cos(\cos\{\cos[\cos(x)]\})]$

15. COMPUTER: An object attached to a spring with friction of strength α (a generalized version of Exercise 11 in Section 2.9) that oscillates with period T can have solution

$$x(t) = e^{-\alpha t} \cos\left(\frac{2\pi t}{T}\right)$$

a. Take the first and second derivatives.
b. Show that $x(t)$ is a solution of

$$\frac{d^2 x}{dt^2} = -\left[\left(\frac{2\pi}{T}\right)^2 + \alpha^2\right]x - 2\alpha \frac{dx}{dt}$$

c. What is the strength of the spring?
d. Find two different values of spring strength and friction that produce the same period T.

Supplementary Problems for Chapter 2

EXERCISE 1
Find the derivatives of the following functions. Note any points where the derivative does not exist.
  a. $F(y) = y^4 + 5y^2 - 1$
  b. $H(c) = \dfrac{c^2}{1 + 2c}$
  c. $G(r) = (4r^2 + 3r - 1)(2r + 2)$
  d. $a(x) = 4x^7 + 7x^4 - 28$
  e. $f(x) = \dfrac{x}{1 + \sqrt{x}}$ for $x \geq 0$
  f. $g(y) = e^{3y^3 + 2y^2 + y}$
  g. $h(z) = \dfrac{z}{1 + \ln(z^2)}$ for $z \geq 0$
  h. For $y \geq 0$, $b(y) = \dfrac{1}{y^{0.75}}$
  i. For $z \geq 0$, $c(z) = \dfrac{z}{(1 + z)(2 + z)}$

EXERCISE 2
Find the derivatives of the following functions.
  a. $g(x) = (4 + 5x^2)^6$
  b. $s(t) = \ln(2t^3)$
  c. $p(t) = t^2 e^{2t}$
  d. $s(x) = e^{-3x+1} + 5\ln(3x)$
  e. $c(x) = (1 + 2/x)^5$

EXERCISE 3
Find the derivatives of the following functions. Find the equation of the tangent line at the specified point.
  a. $f(x) = xe^{-x}$. Find the tangent line at $x = 0$.
  b. $g(t) = t/(1 + 2t^2)$. Find the tangent line at $t = 1$.
  c. $h(x) = \ln(1 + 2x^2)$. Find the tangent line at $x = 0$.

EXERCISE 4
Find all critical points and points of inflection of the following. Sketch graphs.
  a. $f(x) = e^{-x^3}$
  b. $g(x) = e^{-x^4}$
  c. $h(y) = \sin(y^2)$
  d. $F(c) = e^{-2c} + e^c$

EXERCISE 5
A population has size

$$N(t) = 1000 + 10t^2$$

where t is measured in years.
  a. What units should follow the 1000 and the 10?
  b. Graph the population between $t = 0$ and $t = 10$.
  c. Find the population after 7 yr and 8 yr. Find the approximate growth rate between these two times. Where does this approximate growth rate appear on your graph?
  d. Find and graph the derivative $N'(t)$.
  e. Find the per capita rate of growth. Is it increasing?

EXERCISE 6
Find the limits of the following functions or say why you cannot.
  a. $\lim\limits_{x \to 0} 1 + x^2$
  b. $\lim\limits_{x \to -2} 1 + x^2$
  c. $\lim\limits_{x \to 1} \dfrac{1}{x^2 - 1}$
  d. $\lim\limits_{x \to 0} \dfrac{e^x}{1 + e^{2x}}$
  e. $\lim\limits_{x \to 1^+} \dfrac{1}{x - 1}$
  f. $\lim\limits_{x \to 1^-} \dfrac{1}{x - 1}$

EXERCISE 7
Suppose the volume of a cell is described by the function $V(t) = 2 - 2t + t^2$, where t is measured in minutes and V is measured in thousands of cubic microns.
  a. Graph the secant line from time $t = 2$ to $t = 2.5$.
  b. Find the equation of this line.
  c. Find the value of the derivative at $t = 2$, and express it in both differential and prime notation. Don't forget the units.
  d. What is happening to cell volume at time $t = 2$?
  e. Graph the derivative of V as a function of time.

EXERCISE 8
Suppose a machine is invented to measure the amount of knowledge in a student's head in units called "factoids." One student is measured at $F(t) = t^3 - 6t^2 + 9t$ factoids at time t, where t is measured in weeks.
  a. Find the rate at which the student is gaining (or losing) knowledge as a function of time (be sure to give the units).
  b. During what time between $t = 0$ and $t = 11$ is the student losing knowledge?
  c. Sketch a graph of the function $F(t)$.
  d. What sort of class might produce this educational result?

EXERCISE 9
Write the tangent line approximation for the following functions, and estimate the requested values.
  a. $f(x) = \dfrac{1 + x}{1 + e^{3x}}$. Estimate $f(-0.03)$.
  b. $g(y) = (1 + 2y)^4 \ln(y)$. Estimate $g(1.02)$.
  c. $h(z) = 2 + z^2 - \dfrac{3}{z^2}$. Estimate $h(3.04)$.

EXERCISE 10
The following measurements are made of a plant's height in centimeters and its rate of growth:

Day	Height	Growth rate
1	8.0	5.2
2	15.0	9.4
3	28.0	17.8
4	53.0	34.6

a. Graph these data. Make sure to give units.
b. Write the equation of a secant line connecting two of these data points, and use it to guess the height on day 5. Why did you pick the points you did?
c. Write the equation of a tangent line, and use it to guess the height on day 5.
d. Which guess do you think is better?
e. How do you think the growth rate might have been measured?

EXERCISE 11
Suppose the fraction of mutants in a population follows the updating function

$$p_{t+1} = \frac{2.0 p_t}{2.0 p_t + 1.5(1 - p_t)}$$

Suppose you wish the fraction at time $t = 1$ to be close to $p_1 = 0.4$.

a. What would p_0 have to be to hit 0.4 exactly?
b. How close would p_0 have to be to this value to produce p_1 within 0.1 of the target?
c. What happens to the input tolerance as the output tolerance becomes smaller (as p_1 is required to be closer and closer to 0.4)?

EXERCISE 12
Suppose the position of an object attached to a spring is described by

$$x(t) = 2.0 \cos\left(\frac{2\pi t}{3.2}\right)$$

a. Find the derivative of $x(t)$.
b. Find the second derivative of $x(t)$. What does this mean physically?
c. What are the position and velocity at $t = 0$?
d. What are the position and velocity at $t = 1.6$?
e. What are the position and velocity at $t = 3.2$?
f. What differential equation does this object follow?

EXERCISE 13
Suppose a system exhibits hysteresis. As temperature, T, increases, the voltage response follows

$$V_i(T) = \begin{cases} T & \text{if } T \leq 50 \\ 100 & \text{if } 50 < T \leq 100 \end{cases}$$

As temperature decreases, however, the voltage response follows

$$V_d(T) = \begin{cases} T & \text{if } T \leq 20 \\ 100 & \text{if } 20 < T \leq 100 \end{cases}$$

a. Graph these functions.
b. Find the left- and right-hand limits of V_i and V_d as T approaches 50.
c. Find the left- and right-hand limits of V_i and V_d as T approaches 20.

Project for Chapter 2

PROJECT 1
Periodic hematopoiesis is a disease characterized by large oscillations in the red blood cell count. Red blood cells are generated by a feedback mechanism that approximately obeys the following equation:

$$x_{t+1} = \frac{\tau}{1 + \gamma \tau} F(x_t) + \frac{1}{1 + \gamma \tau} x_t$$

when x_t is the number of red blood cells on day t.
The function $F(x_t)$ describes production as a function of number of cells and takes the form of a Hill function,

$$F(x) = F_0 \frac{\theta^n}{\theta^n + x^n}$$

The terms are described in the table.

a. Graph and explain $F(x)$ with these parameter values. Does this sort of feedback system make sense?
b. Use a computer to find the equilibrium with healthy parameters.
c. Graph and cobweb the updating function. If you start near the equilibrium, do values remain nearby? What would a solution look like?

Parameter	Meaning	Normal value
x_t	Number of cells at time t (billions)	About 330
τ	Time for cell development	5.7 days
γ	Fraction of cells that die each day	0.0231
F_0	Maximum production of cell	76.2 billion
θ	Value where cell production is halved	247. billion
n	Shape parameter of Hill function	7.6

d. Certain autoimmune diseases increase γ, the death rate of cells. Study what happens to the equilibrium and the solution as γ increases. Explain your results in biological terms.
e. Explore what would happen if the value of n were decreased. Does the equilibrium become more sensitive to small changes in γ?

Chapter 3

Applications of Derivatives and Dynamical Systems

We have developed techniques for computing derivatives of many functions, and we have interpreted the derivative as either the rate of change of a measurement or the slope of the graph of a function. The derivative has a remarkable range of applications for understanding the behavior of biological measurements, with the general theme of **reasoning about functions**. To begin, we use the derivative to characterize the **stability** of equilibria of discrete-time dynamical systems. Next, we use the derivative as a tool to find **maxima** and **minima** of functions. When combined with appropriate models of a biological process, these methods can be used to solve **optimization** problems.

Next, we develop tools for reasoning about continuous and differentiable functions with a minimum of calculation. On the basis of general properties of functions, we show that certain equations have solutions or that a maximum exists, and we deduce more specific conclusions with a system of simple comparisons called the **method of leading behavior**. We use the tangent line approximation to solve impossible equations with **Newton's method**, and we extend the theme of approximating complicated functions with simple functions to **Taylor polynomials**.

3.1 Stability and the Derivative

The slope of a curve has the remarkable power to identify the **stability** of equilibria of discrete-time dynamical systems. By examining the graphs and cobwebbing the updating functions we have studied, we develop a method for evaluating the stability of an equilibrium with the **derivative of the updating function**. This method will explain why some solutions, like that for a growing bacterial population, move away from equilibrium, while others, like that of the lung model, move toward equilibrium.

Motivation

Figures 3.1–3.5 review the cobwebbing diagrams for five of the updating functions we have studied, and the following list gives a review of the terminology of discrete-time dynamical systems.

Figure 3.1
Bacterial population model $b_{t+1} = 2b_t$

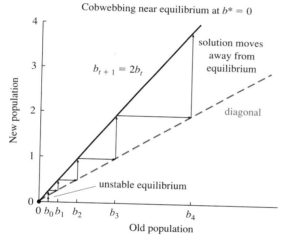
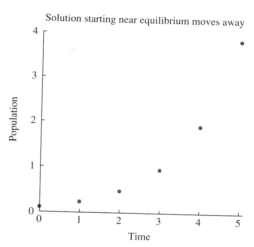

Figure 3.2
Bacterial population model $b_{t+1} = 0.5b_t$

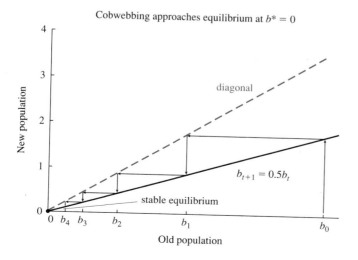
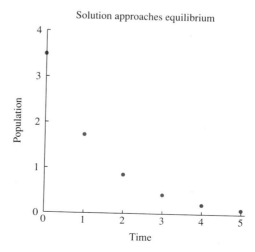

Figure 3.3

Lung model $c_{t+1} = 0.5c_t + 2.5$

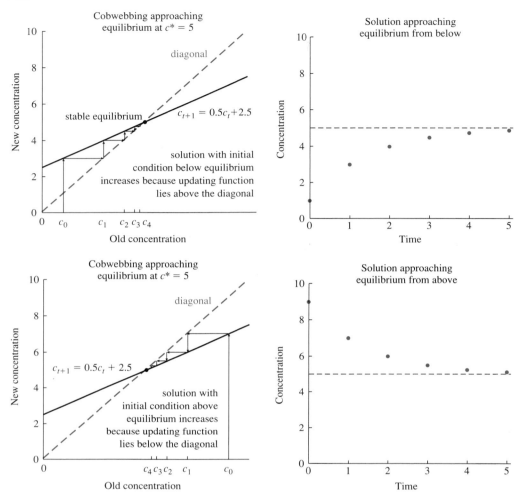

Figure 3.4

Bacterial competition model
$p_{t+1} = 2.0p_t / [2.0p_t + 1.5(1 - p_t)]$

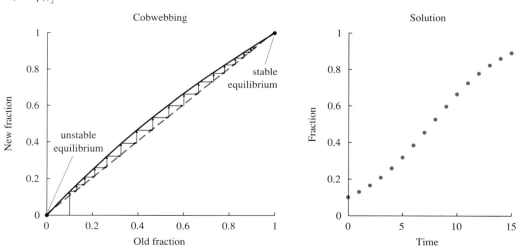

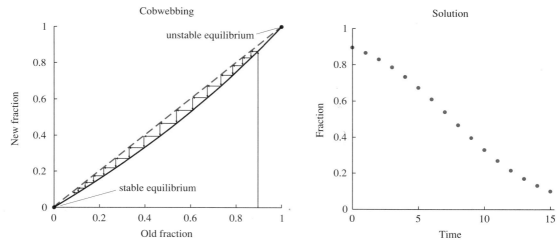

Figure 3.5

Bacterial competition model
$p_{t+1} = 1.5p_t / [1.5p_t + 2.0(1 - p_t)]$

- An *updating function*, or discrete-time dynamical system, is a rule that takes a measurement at one time step as input and returns the measurement at the next time step as output.
- *Cobwebbing* is a graphical method for finding solutions of discrete-time dynamical systems. Starting from an initial condition, the first vertical line gives the value after one time step. The horizontal line reflects this point onto the diagonal, after which steps are repeated.
- A *solution* is a graph of the values of the measurement as a function of time.
- An *equilibrium* is a point where the updating function leaves the value unchanged (Definition 1.4). Graphically, the equilibrium is a point where the updating function crosses the diagonal line, the "do nothing" updating function.
- An equilibrium is *stable* if solutions that start near the equilibrium get closer to the equilibrium and *unstable* if solutions that start near the equilibrium move away from the equilibrium.

The figures show four stable equilibria: $b^* = 0$ in Figure 3.2, $c^* = 5$ in Figure 3.3, $p^* = 1$ in Figure 3.4, and $p^* = 0$ in Figure 3.5. There are three unstable equilibria: $b^* = 0$ in Figure 3.1, $p^* = 0$ in Figure 3.4, and $p^* = 1$ in Figure 3.5. What is the pattern? How can we recognize which equilibria are stable?

Think about cobwebbing near a stable equilibrium, such as the one in Figure 3.3. If we start slightly above the equilibrium, the solution *decreases* because the graph of the updating function lies *below* the diagonal. Similarly, if we start slightly below the equilibrium, the solution *increases* because the graph of the updating function lies *above* the diagonal. In other words, if the graph of the updating function crosses from above the diagonal to below the diagonal, the equilibrium is stable. Does this work for the other stable equilibria? It does if we imagine extending the graphs beyond the biologically meaningful realm. For example, in Figure 3.4, the updating function intersects the diagonal from above near $p^* = 1$. If we extended the curve beyond $p^* = 1$, it would cross from above to below.

Unstable equilibria are the opposite; points where the updating function crosses the diagonal from below to above. In this case, a solution that starts *below* the equilibrium will *decrease* further, moving away from the equilibrium.

A solution that starts *above* the equilibrium will *increase* further, again moving away from the equilibrium. For example, at the unstable equilibrium at $p^* = 0$ in Figure 3.4, the updating function would cross from below to above if we extended it below $p^* = 0$.

We can summarize our observations with the following condition.

Tentative Condition I for an Equilibrium to Be Stable or Unstable

- An equilibrium is stable if the graph of the updating function crosses the diagonal from above to below.
- An equilibrium is unstable if the graph of the updating function crosses the diagonal from below to above. (See Figure 3.6.)

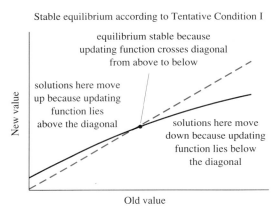

Figure 3.6
Tentative Condition I

Stability and the Slope of the Updating Function

When does the updating function cross from below to above the diagonal? The diagonal is the graph of $y = x$, a line with a slope equal to 1. The graph of the updating function will cross from below to above when its slope is *greater than 1*. Conversely, a curve will cross from above to below when the slope is *less than 1*. We can therefore translate Tentative Condition I into a condition about the **derivative of the updating function**.

Tentative Condition II for an Equilibrium to Be Stable or Unstable

- An equilibrium is stable if the derivative of the updating function is less than 1 at the equilibrium.
- An equilibrium is unstable if the derivative of the updating function is greater than 1 at the equilibrium. (See Figure 3.7.)

A highly nonlinear discrete-time dynamical system can have many equilibria. With the aid of Tentative Conditions I and II, we can recognize which are stable and unstable just by examining the slope of the updating function (Figure 3.8). We will see in Section 3.2, however, that the dynamics can be much more complicated when the updating function has a negative slope.

Tentative Condition II says nothing about what happens when the slope is exactly 1. Figure 3.9 shows an equilibrium where the updating function does not

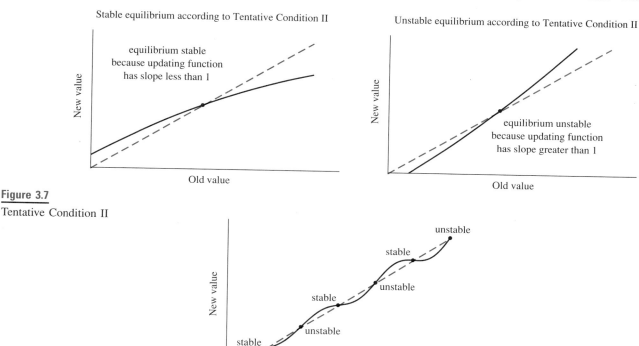

Figure 3.7
Tentative Condition II

Figure 3.8
Recognizing the stability of a dynamical system with many equilibria

cross the diagonal but remains below the diagonal both to the left and to the right of the equilibrium. A solution that starts from any point below the equilibrium will decrease and move away from the equilibrium, as in the unstable case. A solution that starts from any point above the equilibrium will decrease and move toward the equilibrium, as in the stable case. This sort of half-stable equilibrium is possible because the diagonal is tangent to the updating function. It touches the curve but does not cross it.

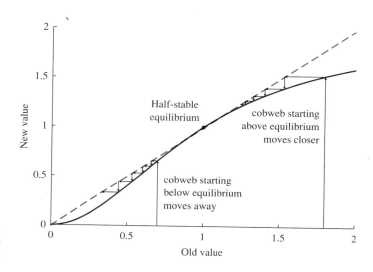

Figure 3.9
An unusual equilibrium

Evaluating Stability with the Derivative

We can test Tentative Condition II by computing the derivatives of the updating functions pictured in Figures 3.1–3.5. In Figure 3.1, the discrete-time dynamical system is

$$b_{t+1} = 2.0 b_t$$

with updating function

$$f(b) = 2.0b$$

The only equilibrium is $b^* = 0$, corresponding to extinction. The updating function is a line with slope 2.0. Because this slope is greater than 1, the equilibrium is unstable.

In Figure 3.2, the updating function is

$$b_{t+1} = 0.5 b_t$$

or

$$f(b) = 0.5b$$

Again, the only equilibrium is $b^* = 0$. This updating function, however, has a slope of 0.5, less than 1. This equilibrium is stable. These results make biological sense because the first population is doubling in each generation and growing away from extinction, while the second population is being halved in each generation and moving toward extinction.

The results with the lung model,

$$c_{t+1} = 0.5 c_t + 2.5$$

are a little less obvious (Figure 3.3). Once again, the updating function is linear with a slope of 0.5, which is less than 1. Therefore, the single equilibrium must be stable. Combined with the result (from Section 1.11) that the equilibrium is always equal to the ambient concentration, we thus know that the concentration in the lung (at least in the absence of absorption) will always get closer and closer to the ambient concentration.

What happens when the updating function is nonlinear? Figure 3.4 uses the bacterial competition equation

$$p_{t+1} = \frac{2.0 p_t}{2.0 p_t + 1.5(1 - p_t)}$$

Recall that p_t represents the proportion of mutant bacteria at time t, 2.0 is the per capita reproduction of mutant bacteria, and 1.5 is the per capita reproduction of wild-type bacteria. There are two equilibria, at $p^* = 0$ and $p^* = 1$ (Section 1.12). The first corresponds to extinction of the mutant and the second, to extinction of the wild type.

By writing the updating function as

$$f(p) = \frac{2.0p}{2.0p + 1.5(1 - p)}$$

we computed the derivative in Section 2.6, finding

$$f'(p) = \frac{3.0}{[2.0p + 1.5(1 - p)]^2}$$

At the equilibrium $p = 0$, the derivative is 1.333, which is greater than 1. This equilibrium is unstable. At the equilibrium $p = 1$, the derivative is 0.75, which is less than 1. This equilibrium is stable (Figure 3.10).

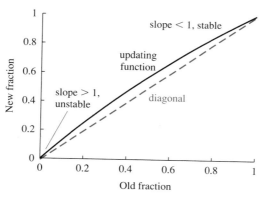
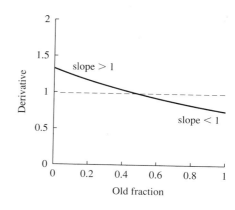

Figure 3.10

The updating function for the fraction of mutants and its derivative

When $s = 1.5$ and $r = 2.0$, the updating function is

$$f(p) = \frac{1.5p}{1.5p + 2.0(1-p)}$$

The derivative (Exercise 5) is

$$f'(p) = \frac{3.0}{[1.5p + 2.0(1-p)]^2}$$

With these parameter values, $f'(0) = 0.75 < 1$ and $f'(1) = 1.333 > 1$. The $p = 0$ equilibrium is stable and the $p = 1$ equilibrium is unstable. When the wild type have an advantage ($r > s$), mutants cannot invade.

Consider the updating function for medication concentration in the bloodstream,

$$M_{t+1} = 0.5 M_t + 1$$

describing a patient who is administered 1 unit of medication each day, but uses up half the previous amount (Section 1.5). The equilibrium concentration is 2.0. Because the updating function is linear with slope $0.5 < 1$, this equilibrium is stable (Figure 3.11a). What happens if we run this system backward in time (as in Section 1.6)? The backward updating function is the inverse of the forward updating function, which is found by solving for M_t:

$$M_t = 2.0(M_{t+1} - 1) = 2.0 M_{t+1} - 2.0$$

The backward updating function is

$$B(M) = 2.0 M - 2.0$$

It shares the equilibrium at 2.0 (Figure 3.11b). However, because the backward updating function is the mirror image, the slope at the equilibrium is the reciprocal of 0.5, or 2.0. An equilibrium that is stable when time runs forward is *unstable* when time runs backward.

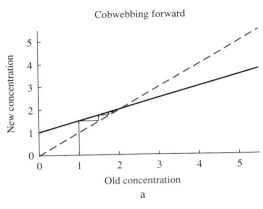
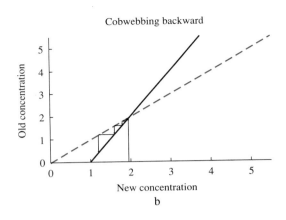

Figure 3.11
An updating function and its inverse

As a final example, consider a population in which the per capita reproduction is a decreasing function of the population size, x, according to

$$\text{per capita reproduction} = \frac{2.0}{1 + 0.001x}$$

The per capita reproduction is 2.0 when $x = 0$ and decreases as x becomes larger (Figure 3.12a). For example, when $x = 500$,

$$\text{per capita reproduction} = \frac{2.0}{1 + 0.001 \cdot 500} = 1.33$$

We use the fact that the updated population is the per capita reproduction (offspring per individual) times the old population (the number of individuals), to find that the updating function for this population is

$$x_{t+1} = (\text{per capita reproduction}) \cdot x_t$$
$$= \frac{2x_t}{1 + 0.001x_t}$$

(Figure 3.12b).

Figure 3.12
An updating function with reduced per capita reproduction

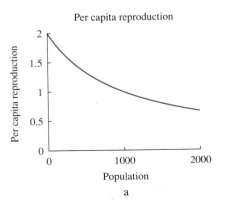
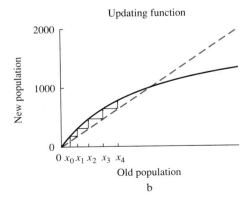

To find the equilibria, we follow Algorithm 1.5:

$$x^* = \frac{2x^*}{1 + 0.001x^*} \quad \text{the original equation}$$

$$x^* - \frac{2x^*}{1 + 0.001x^*} = 0 \quad \text{subtract to get unknowns on one side}$$

$$(1 + 0.001x^*)x^* - 2x^* = 0 \quad \text{multiply both sides by } 1 + 0.001x^*$$

$$x^*\left(1 + 0.0001x^* - 2\right) = 0 \quad \text{factor}$$

$$x^* = 0 \text{ or } 1 + 0.001x^* - 2 = 0 \quad \text{set both factors to 0}$$

$$x^* = 0 \text{ or } x^* = 1000 \quad \text{solve the second equation}$$

The equilibrium $x^* = 0$ corresponds to extinction. The equilibrium $x^* = 1000$ is the point where per capita reproduction has been so reduced by crowding that the population can only break even.

The equilibrium at $x^* = 0$ appears to be unstable, and the equilibrium at $x^* = 1000$ appears to be stable (Figure 3.12b). In accord with Tentative Condition I, the graph of the updating function crosses the diagonal from below to above at $x^* = 0$ and from above to below at $x^* = 1000$. We therefore expect that the derivative of the updating function is greater than 1 at $x^* = 0$ and less than 1 at $x^* = 1000$.

In functional notation, the updating function is

$$f(x) = \frac{2x}{1 + 0.001x}$$

The derivative of the updating function, found with the quotient rule, is

$$f'(x) = \frac{(1 + 0.001x)\frac{d(2x)}{dx} - 2x\frac{d(1 + 0.001x)}{dx}}{(1 + 0.001x)^2}$$

$$= \frac{2(1 + 0.001x) - 2x \cdot 0.001}{(1 + 0.001x)^2}$$

$$= \frac{2}{(1 + 0.001x)^2}$$

Evaluating at the equilibria, we find

$$f'(0) = 2 \quad \text{and} \quad f'(1000) = 0.5$$

As indicated by the graph, the equilibrium at $x = 0$ is unstable and the equilibrium at $x = 1000$ is stable. If reduced to a low level, this population would increase back toward the stable equilibrium at $x^* = 1000$. If artificially increased to a high level, the population would decrease toward the stable equilibrium at $x^* = 1000$.

SUMMARY

We used cobwebbing and logic to derive two tentative conditions for the stability of equilibria. When the updating function crosses the diagonal from below to above, initial conditions both above and below the equilibrium are pushed away, making the equilibrium unstable. For the updating function to cross from below to above, the slope at the equilibrium must be greater than 1. The opposite occurs at stable equilibria, where the updating function crosses from above to below and has slope less than 1. In the special case that the slope is exactly equal to 1, the equilibrium might be neither stable nor unstable.

3.1 EXERCISES

1. Apply Tentative Condition I to the equilibria of the following curves (studied in Figure 1.79 and Exercise 3 in Section 1.12). Check by cobwebbing.

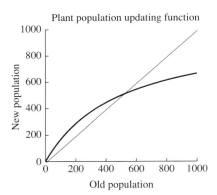

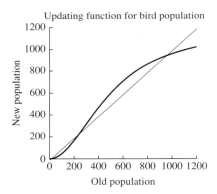

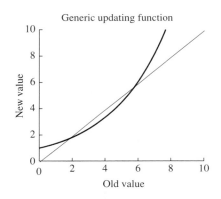

2. Recall the updating function describing mites on lizards, given by
$$x_{t+1} = 2x_t + 30$$
   a. Find the equilibrium.
   b. Graph the updating function, making sure to include the equilibrium on your graph although it makes no biological sense.
   c. Apply Tentative Condition I to this equilibrium.
   d. Apply Tentative Condition II to this equilibrium.
   e. Explain why your result is consistent with the behavior of solutions.

3. Find the equilibria, sketch graphs, and check Tentative Conditions I and II for the following linear updating functions.
   a. $c_{t+1} = 0.75c_t + 0.875$ (Exercise 5 in Section 1.10).
   b. $c_{t+1} = 0.6c_t + 1.25$ (Exercise 6 in Section 1.10).
   c. $s_{t+1} = 0.99s_t + 0.0099$ (Exercise 10b in Section 1.10).
   d. $b_{t+1} = 1.5b_t - 10^6$ (Exercise 7 in Section 1.11).

4. Consider the nonlinear updating function describing the dynamics of medication concentration due to use by the body and external supplementation:
$$M_{t+1} = M_t - \frac{0.5}{1.0 + 0.1M_t} M_t + 1.0$$
(studied in Exercise 7 in Section 1.12). The equilibrium concentration of medication is 2.5.
   a. Sketch the updating function.
   b. Find the slope of the updating function at the equilibrium.
   c. Explain in biological terms why this equilibrium is stable.

5. Consider the updating function for a bacterial population (Equation 1.49 in Section 1.12) given by
$$f(p) = \frac{sp}{sp + r(1-p)}$$
where s is the per capita reproduction of the mutant and r is the per capita reproduction of the wild type. Find the derivative in the following cases, and evaluate at the equilibria $p = 0$ and $p = 1$. Are the equilibria stable?
   a. $s = 1.5, r = 2.0$
   b. $s = 1.5, r = 1.5$
   c. Try this calculation in general (without substituting numerical values). Can both equilibria be stable?

6. Consider again the updating function
$$f(x) = \frac{2x}{1 + 0.001x}$$
illustrated in Figure 3.12.

a. Find and sketch the inverse of this function. Why does it look so different from the original updating function? What happens if $x_{t+1} = 2500$?
b. Find the equilibria.
c. Find their stability.

7. The unusual equilibrium in Figure 3.9 has an updating function that lies below the diagonal both to the left and to the right of the equilibrium.
   a. Draw an updating function that lies above the diagonal both to the left and to the right of an equilibrium.
   b. Cobweb starting from points to the left and to the right of the equilibrium.
   c. Describe the stability of this equilibrium.

8. There are two other ways that an updating function can be tangent to the diagonal at an equilibrium.
   a. Draw a case where the updating function is tangent to the diagonal at an equilibrium but crosses from below to above. Show by cobwebbing that the equilibrium is unstable. What is the second derivative at the equilibrium?
   b. Draw a case where the updating function is tangent to the diagonal at an equilibrium but crosses from above to below. Show by cobwebbing that the equilibrium is stable.

9. Another peculiarity of the function shown in Figure 3.9 is that slight changes in the graph can produce big changes in the number of equilibria.
   a. Move the curve slightly down (while keeping the diagonal in the same place). How many equilibria are there now? What happens when you cobweb starting from a point at the right-hand edge of the figure?
   b. Move the curve slightly up (again keeping the diagonal in the same place). How many equilibria are there? Describe their stability.
   c. Imagine starting the curve from the position in part **a** and moving it gradually up to the position in part **b**. What happens to the number and stability of equilibria along the way? This change is known as a **bifurcation**.

10. Sketch an updating function that has a corner at an equilibrium. Draw a case where the equilibrium is stable, one where it is unstable, and one where it is neither.

11. Consider a population with

$$\text{per capita reproduction} = r\frac{x_t}{1.0 + x_t^2}$$

as in Exercise 11 in Section 2.3.
   a. Write the updating function.
   b. Find the equilibria.
   c. Graph the updating function with $r = 1$, $r = 1.9$, $r = 2$, $r = 2.1$, and $r = 3$.
   d. Identify and indicate which are stable and which are unstable.
   e. Describe in words how each of these populations would behave. Can you imagine circumstances that would produce this effect?

12. Consider the updating function for a heart studied in Section 1.13:

$$V_{t+1} = \begin{cases} cV_t & \text{if } cV_t > V_c \\ cV_t + u & \text{if } cV_t \le V_c \end{cases}$$

In each case, sketch the updating function. Why is the equilibrium always stable when it exists?
   a. $V_c = 20.0$ mV, $u = 10.0$ mV, $c = 0.5$
   b. $V_c = 20.0$ mV, $u = 10.0$ mV, $c = 0.6$
   c. $V_c = 20.0$ mV, $u = 10.0$ mV, $c = 0.7$
   d. $V_c = 20.0$ mV, $u = 10.0$ mV, $c = 0.8$

13. **COMPUTER:** Consider the updating function found in Exercise 14 in Section 2.1. In that exercise, there were two cultures, 1 and 2. In culture 1 the mutant does better than the wild type, and in culture 2 the wild type does better. In particular, suppose that $s = 2.0$ and $r = 0.3$ in culture 1 and that $s = 0.6$ and $r = 2.0$ in culture 2. Define updating functions f_1 and f_2 to describe the dynamics in the two cultures. The overall updating function after mixing equal amounts from the two is

$$f(p) = \frac{f_1(p) + f_2(p)}{2}$$

   a. Write the updating function explicitly.
   b. Find the equilibria.
   c. Find the derivative of the updating function.
   d. Evaluate the stability of the equilibrium.

14. **COMPUTER:** Consider again the discrete-time dynamical systems with the following updating functions (Exercise 11 in Section 1.11). Check the stability of the equilibria.
   a. $x_{t+1} = \cos(x)$
   b. $y_{t+1} = \sin(y)$
   c. $z_{t+1} = \sin(z) + \cos(z)$

15. **COMPUTER:** Consider the updating function

$$z_{t+1} = a^{z_t}$$

We will study this for different values of a.
   a. Follow the dynamics starting from initial condition $z_0 = 1.0$ for $a = 1.0, 1.1, 1.2, 1.3, 1.4$. Keep running the system until it seems to reach an equilibrium.
   b. Do the same as part **a**, but increase a slowly past a critical value of about 1.4446679. Solutions should creep up for a while and then blow up very quickly. Graph the updating function for values above and below the critical value, and try to explain why this happens.
   c. At the critical value, the slope of the updating function is 1 at the equilibrium. Show that this occurs with $a = e^{1/e}$. What is the equilibrium?

3.2 More Complex Dynamics

The derivative or slope of an updating function determines whether an equilibrium is stable or unstable. Tentative Conditions I and II tell how to assess stability by checking whether the slope is greater than or less than 1. We have yet, however, to consider cases where the slope of the updating function is **negative** at an equilibrium. By studying several such cases, we will see that rather exotic behaviors are possible. Using the idea of **qualitative dynamical systems**, where we approximate a nonlinear discrete-time dynamical system with its **tangent line approximation** at an equilibrium, we will find a general condition that includes these new cases.

The Logistic Dynamical System

In Section 3.1, we looked at a model where large population size reduced per capita reproduction according to

$$\text{per capita reproduction} = \frac{2.0}{1 + 0.001x}$$

(Figure 3.12). Another widely studied model is the **logistic dynamical system** (introduced in Exercise 12 in Section 1.12). In this model, the per capita reproduction of a population decreases linearly with population size according to

$$\text{per capita reproduction} = r\left(1 - \frac{N}{K}\right)$$

where N represents population size. The parameter r is the greatest possible reproduction, and K is the maximum possible population. For populations greater than K, the per capita reproduction is assumed to be 0. We use the fact that the new population is the per capita reproduction times the old population, to find that the updating function for the logistic dynamical system is

$$N_{t+1} = r\left(1 - \frac{N_t}{K}\right)N_t$$

To make the algebra simpler, we set K equal to 1. This does not mean that the maximum population is 1 (which would be absurd), but rather that the actual population is measured as the fraction of the maximum possible population. If K is 1000, a population of $N_t = 500$ corresponds to the fraction $x_t = 0.5$. We can write the updating function for the fraction as

$$x_{t+1} = f(x_t) = rx_t(1 - x_t) \tag{3.1}$$

(Exercise 8). Graphs and cobwebs of this dynamical system for several values of r are shown in Figure 3.13.

To understand these diagrams, we begin by finding the equilibria by setting $x_t = x_{t+1} = x^*$ and solving with Algorithm 1.5.

$$x^* = rx^*(1 - x^*) \quad \text{the original equation}$$
$$x^* - rx^*(1 - x^*) = 0 \quad \text{place unknowns on one side}$$
$$x^*\left[1 - r(1 - x^*)\right] = 0 \quad \text{factor}$$
$$x^* = 0 \text{ or } 1 - r(1 - x^*) = 0 \quad \text{set each factor equal to 0}$$
$$x^* = 0 \text{ or } x^* = 1 - \frac{1}{r} \quad \text{solve each piece}$$

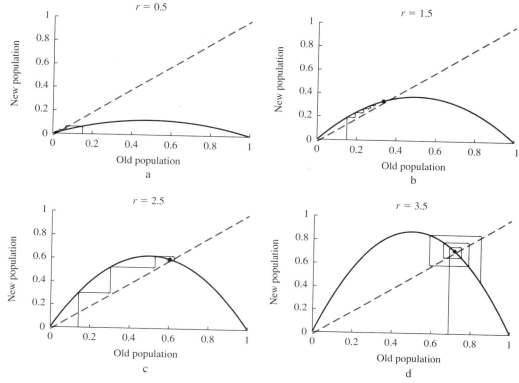

Figure 3.13
The behavior of the logistic dynamical system

Do these solutions make sense? The first, $x^* = 0$, is the extinction equilibrium that shows up in all population models without immigration. The second equilibrium depends on the maximum per capita reproduction r. If $r < 1$, this equilibrium is negative and biologically impossible. A population with a *maximum* per capita reproduction less than 1 cannot replace itself and will go extinct. For larger values of r, the equilibrium becomes larger, consistent with the fact that a population with higher potential reproduction grows to a larger value.

We will examine four cases: $r = 0.5$, $r = 1.5$, $r = 2.5$, and $r = 3.5$. The equilibria and their behavior, deduced from cobwebbing, are summarized in the following table:

r	x^*	Stability
0.5	0	Stable
1.5	0	Unstable
1.5	$1 - \frac{1}{1.5} = 0.333$	Stable
2.5	0	Unstable
2.5	$1 - \frac{1}{2.5} = 0.600$	Stable (oscillates)
3.5	0	Unstable
3.5	$1 - \frac{1}{3.5} \approx 0.714$	Unstable (oscillates)

The results match Tentative Condition I pretty well until $r = 3.5$ (Figure 3.13d). The updating function crosses the diagonal from above to below at the positive equilibrium, so we expect this equilibrium to be stable. However, a solution starting near the positive equilibrium $x^* = 0.714$ moves away, and does so by jumping back and forth. Closer examination of the case with $r = 2.5$ (Figure 3.13c) reveals that solutions jump back and forth as they approach the equilibrium at $x^* = 0.6$. What is going on?

Qualitative Dynamical Systems

We have seen that an equilibrium whose updating has a positive slope is stable precisely when that slope is less than 1. These results match the behavior of the bacterial growth model,

$$b_{t+1} = rb_t$$

at the $b^* = 0$ equilibrium. This simpler linear system is stable precisely when the slope r is less than 1.

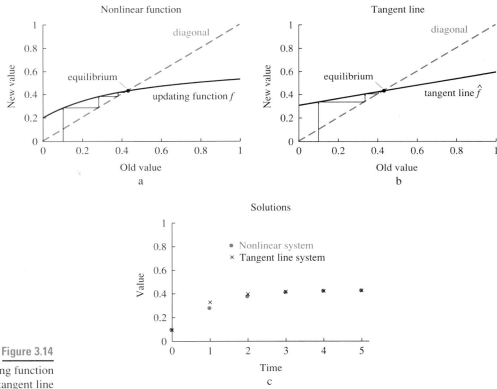

Figure 3.14
Comparing an updating function with its tangent line

What happens if $r < 0$? Of course, this does not make biological sense but will provide insight into what happens near equilibria with negative slopes, as found in Figure 3.13c and d. Why should this work? Remember the central trick in calculus: replacing problems about curves with simpler problems about lines. Consider a nonlinear updating function $x_{t+1} = f(x_t)$ and its tangent line at an equilibrium x^* (Figure 3.14). Because the graph of $\hat{f}$ is the tangent line at the equilibrium x^* of the original system, the two systems share the equilibrium at x^* and have the same slope. Furthermore, solutions remain very close to each other (Figure 3.14c). According to Tentative Condition II, the equilibrium of $\hat{f}$ should be stable precisely when the equilibrium of f is stable (Figure 3.14). Starting from a point near x^*, the solutions are nearly identical because the tangent line lies so close to the curve (Figure 3.15).

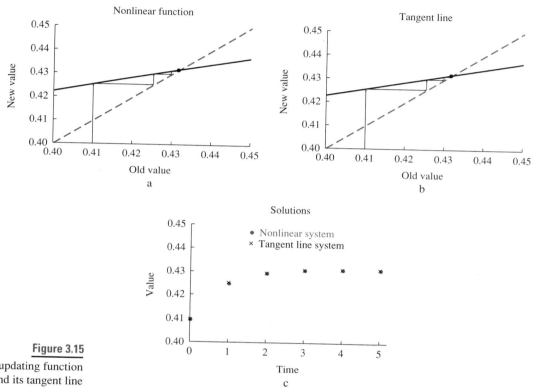

Figure 3.15
Zooming in on an updating function and its tangent line

Strictly speaking, the original updating function and the tangent line are *not equal*. The exact values in the solution will therefore also be different. When we ask about stability, however, we are concerned with general behavior rather than with exact values. Studying general aspects of dynamical systems without requiring exact measurements is the realm of **qualitative dynamical systems**. The approach is essential in biology, where measurements often include a great deal of noise. For example, we neither know nor believe that the updating function for the lung model (Section 1.10) is *exactly* linear. Nonetheless, we would like to use the linear model to make predictions. Some predictions are *qualitative*, verbal descriptions of behavior, such as "the concentration of chemical in the lung will approach an equilibrium." Others are *quantitative*, numerical descriptions of behavior, such as "the concentration of chemical in the lung will reach a concentration of 5.23 mmol/L after seven breaths."

The qualitative theory of dynamical systems requires comparing the dynamics produced by similar updating functions. If two nearly indistinguishable updating functions produce qualitatively different dynamics, we are in trouble. Such situations do occur. In this section, we restrict our attention to cases where similar updating functions produce qualitatively similar dynamics. Other cases are treated in more advanced texts on **bifurcation theory**.

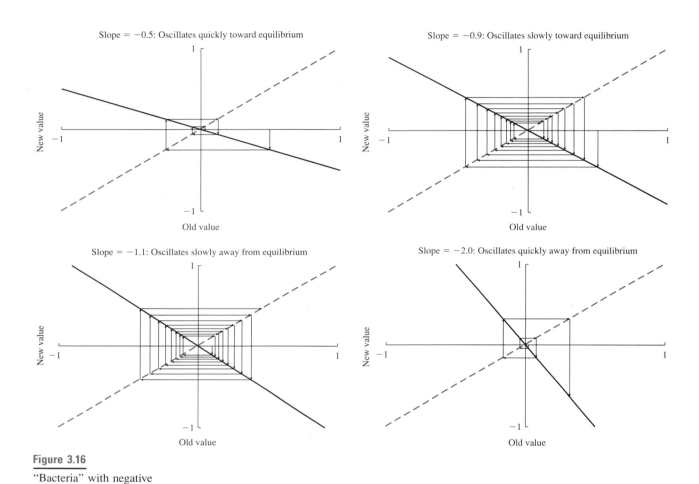

Figure 3.16
"Bacteria" with negative per capita reproduction

Using this philosophy, we can try to figure out what happens at an equilibrium where the slope is negative by studying the bacterial growth model,

$$b_{t+1} = rb_t$$

with unrealistic negative values for the per capita reproduction r. Cobwebs with several negative values of r are presented in Figure 3.16. In each case, the solution jumps between positive and negative values because the old value is multiplied by a negative number. When that negative number is between -1 and 0, the solution gets closer and closer to 0 (Figure 3.17a and b). When that negative number is less than -1, the solution moves further and further from 0 (Figure 3.17c and d).

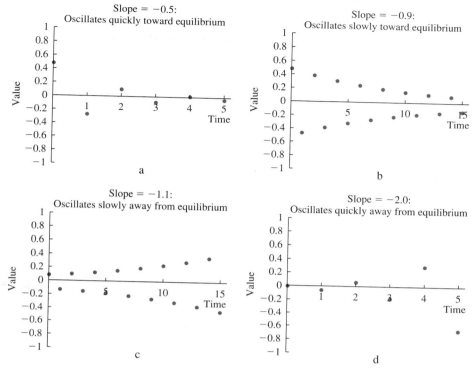

Figure 3.17
Solutions for bacteria with negative per capita reproduction

For example, starting from an initial condition of $b_0 = 0.5$ with $r = -0.5$, the solution is

$$b_1 = -0.25$$
$$b_2 = 0.125$$
$$b_3 = -0.0625$$
$$b_4 = 0.03125$$

The absolute value decreases while the sign switches back and forth. Starting from an initial condition of $b_0 = 0.02$ with $r = -2.0$, the solution is

$$b_1 = -0.04$$
$$b_2 = 0.08$$
$$b_3 = -0.16$$
$$b_4 = 0.32$$

Again, the sign switches back and forth, but the absolute value now increases.

The behavior for positive and negative values of r is summarized as follows:

Behavior of solutions of $b_{t+1} = rb_t$

r	Stability	Behavior
$r > 1$	Unstable	Moves away from equilibrium
$0 < r < 1$	Stable	Moves toward equilibrium
$-1 < r < 0$	Stable	Oscillates toward equilibrium
$r < -1$	Unstable	Oscillates away from equilibrium

An equilibrium is stable if r is less than 1 in *absolute value* and unstable if r is greater than 1 in *absolute value*.

Because the behavior of a nonlinear dynamical system resembles that of its tangent line, we can extend these results to the following theorem.

■ **THEOREM 3.1** (Stability Theorem for Discrete-Time Dynamical Systems)

Suppose that the discrete-time dynamical system

$$x_{t+1} = f(x_t)$$

has an equilibrium at x^*. Let $f'(x)$ be the derivative of f with respect to x. The equilibrium at x^* is stable if

$$|f'(x^*)| < 1$$

and unstable if

$$|f'(x^*)| > 1$$ ■

The proof depends on the Mean Value Theorem (presented in Section 3.4).

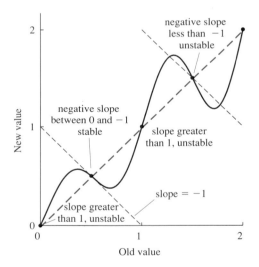

Figure 3.18
Applying the stability theorem for discrete-time dynamical systems

Using this method, we can read off the stability of equilibria of more complicated updating functions (Figure 3.18). By including lines with slope -1 on our graph in addition to the diagonal, we can see that the equilibrium at $(0.5, 0.5)$ is stable because the slope is between 0 and -1, and the equilibrium at $(1.5, 1.5)$ is stable because the slope is less than -1.

Analysis of the Logistic Dynamical System

Do these results help make sense of the behavior of the logistic dynamical system? Because r is a constant, the derivative of the updating function is

$$f'(x) = \frac{d}{dx} rx(1-x)$$

$$= r\left[\frac{dx}{dx}(1-x) + \frac{d(1-x)}{dx}x\right]$$
$$= r[(1-x) - x]$$
$$= r(1-2x)$$

Next, we evaluate the derivative at the two equilibria. At $x^* = 0$,

$$f'(0) = r - 2r \cdot 0 = r$$

According to the stability theorem for discrete-time dynamical systems, the equilibrium at 0 is stable if $r < 1$ and unstable if $r > 1$. When $r > 1$, the derivative at the positive equilibrium $x^* = 1 - (1/r)$ is

$$f'\left(1 - \frac{1}{r}\right) = r - 2r\left(1 - \frac{1}{r}\right)$$
$$= r - 2r + \frac{2r}{r}$$
$$= r - 2r + 2$$
$$= 2 - r$$

The results are summarized in the following table:

r	x^*	$f'(x^*)$	Stability
0.5	0	0.5	Stable
1.5	0	1.5	Unstable
1.5	0.333	0.5	Stable
2.5	0	2.5	Unstable
2.5	0.600	−0.5	Stable (oscillates)
3.5	0	3.5	Unstable
3.5	0.714	−1.5	Unstable (oscillates)

(See Figure 3.19.) When $r = 3.5$, neither equilibrium is stable. This population has nowhere to settle down and will continue jumping around indefinitely. Some of the interesting dynamics that can result are shown in Figure 3.20. The solutions of this simple discrete-time dynamical system for $3.5 < r \leq 4$ are extremely complicated, including examples of **chaos** (Exercise 12).

A model commonly used for reproduction in fisheries is the **Ricker model**, defined by

$$x_{t+1} = \lambda x_t e^{-x_t}$$

In this model, the maximum per capita reproduction is usually written as λ rather than r. The per capita reproduction declines exponentially from a maximum of λ as the population becomes larger. As with the logistic dynamical system, reproduction decreases rapidly, but this model is a bit more realistic because per capita reproduction never reaches 0 (Figure 3.21). This updating function is the constant λ multiplied by one of the gamma distributions of Section 2.8 (see Figure 3.22). If we write the updating function as

$$R(x) = \lambda x e^{-x}$$

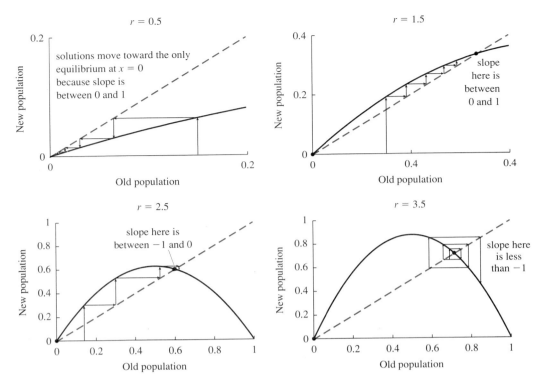

Figure 3.19
The behavior of the logistic dynamical system explained

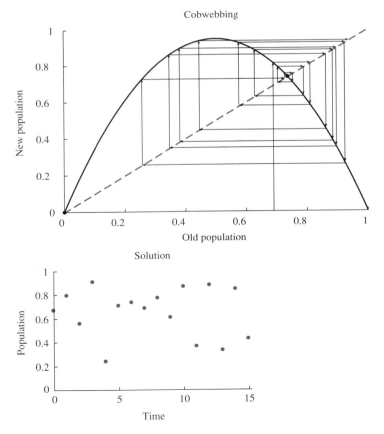

Figure 3.20
The long-term behavior of the logistic dynamical system with $r = 3.8$

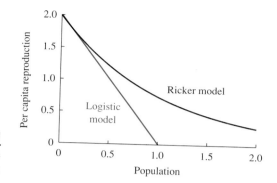

Figure 3.21

The per capita reproduction in the Ricker and logistic models

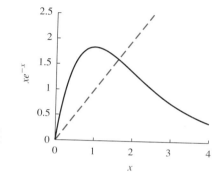

Figure 3.22

The Ricker updating function with $\lambda = 5.0$

the first derivative is

$$R'(x) = \lambda(1 - x)e^{-x}$$

Where are the equilibria of the Ricker model? We solve:

$$x^* = \lambda x^* e^{-x^*}$$
$$x^* - \lambda x^* e^{-x^*} = 0$$
$$x^*(1 - \lambda e^{-x^*}) = 0$$
$$x^* = 0 \text{ or } 1 - \lambda e^{-x^*} = 0$$
$$x^* = 0 \text{ or } e^{x^*} = \lambda$$
$$x^* = 0 \text{ or } x^* = \ln(\lambda)$$

At the equilibrium $x = 0$, $R'(0) = \lambda$. This equilibrium is stable when the maximum per capita reproduction $\lambda < 1$ (Figure 3.23a) and unstable otherwise. The equilibrium $x^* = \ln(\lambda)$ is positive when $\lambda > 1$. At this point, the derivative of the updating function is

$$R'(\ln(\lambda)) = \lambda(1 - \ln(\lambda))e^{-\ln(\lambda)}$$
$$= \lambda(1 - \ln(\lambda))\frac{1}{\lambda}$$
$$= 1 - \ln(\lambda)$$

Results for various values of λ are summarized in the following table:

234 Chapter 3 Applications of Derivatives and Dynamical Systems

λ	x^*	$f'(x^*)$	Stability
0.5	0	0.5	Stable
2.0	0	2.0	Unstable
2.0	0.693	0.307	Stable
4.0	0	4.0	Unstable
4.0	1.386	−0.386	Stable (oscillates)
8.0	0	8.0	Unstable
8.0	2.079	−0.079	Unstable (oscillates)

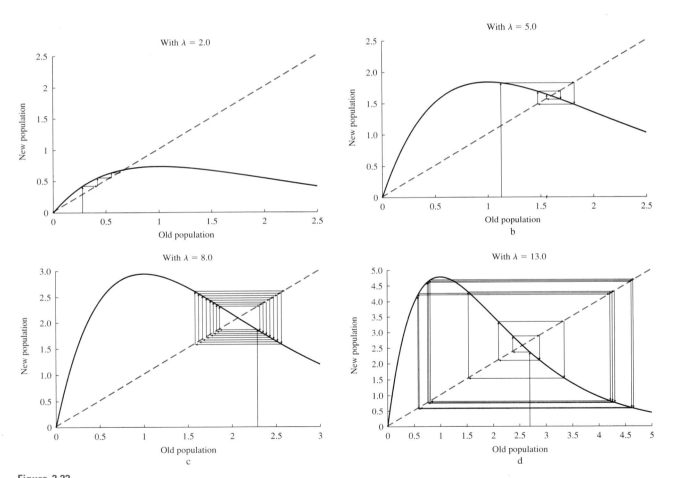

Figure 3.23

The dynamics of the Ricker model with various values of λ

As long as $\ln(\lambda) < 2$, or $\lambda < e^2 \approx 7.39$, the positive equilibrium is stable (Figure 3.23b). Like the logistic model, this updating function produces unstable dynamics when the maximum possible per capita reproduction is large (Figure 3.23c). In this case, fish can produce so many offspring that they destroy their resource base and induce a population crash.

SUMMARY

The idea of **qualitative dynamical systems** is to think of behavior in general terms without depending on quantitative details about the updating function. As an example, we compare the dynamics generated by a nonlinear updating function with those generated by the tangent line at the equilibrium. Because stability of equilibria can be understood by studying linear discrete-time dynamical systems, we examined linear systems with negative slopes, finding that such systems oscillate. Based on these results, we proposed the **stability theorem for discrete-time dynamical systems**: An equilibrium is stable if the *absolute value* of the slope is less than 1. We applied this condition to the **logistic dynamical system** and the **Ricker model**, and we found that there are parameter values with no stable equilibrium.

3.2 EXERCISES

1. Draw the tangent line approximating the given system at the specified equilibrium, and compare the cobweb diagrams.
   a. The bacterial competition equation (Equation 1.51 in Section 1.12)
   $$p_{t+1} = \frac{1.5 p_t}{1.5 p_t + 2.0(1 - p_t)}$$
   with $s = 1.5$ and $r = 2.0$ at the equilibrium $p^* = 0$.
   b. As in part **a**, but at the equilibrium $p^* = 1$.
   c. $x_{t+1} = 1.5 x_t(1 - x_t)$ at the equilibrium $x^* = 0$.
   d. $x_{t+1} = 1.5 x_t(1 - x_t)$ at the equilibrium $x^* = 1/3$.
   e. $c_{t+1} = 3.75 + 0.25 c_t$ at the equilibrium $c_t^* = 5.0$.

2. Starting from the given initial condition, find the solution for five steps of each of the following.
   a. $y_{t+1} = 1.2 y_t$ with $y_0 = 2.0$.
   b. $y_{t+1} = -1.2 y_t$ with $y_0 = 2.0$.
   c. How long will it take for the values to exceed 100?
   d. $y_{t+1} = 0.8 y_t$ with $y_0 = 2.0$.
   e. $y_{t+1} = -0.8 y_t$ with $y_0 = 2.0$.
   f. How long will it take before the values are less than 0.2?

3. Consider the linear updating function
   $$y_{t+1} = 1.0 + m(y_t - 1.0)$$
   a. Find the equilibrium.
   b. Graph and cobweb with $m = 0.1$.
   c. Graph and cobweb with $m = 0.9$.
   d. Graph and cobweb with $m = 1.5$.
   e. Graph and cobweb with $m = -0.5$.
   f. Graph and cobweb with $m = -1.5$.
   g. Compare your results with the stability condition.

4. Consider the logistic dynamical system with $r = 2$.
   a. Find the equilibria.
   b. Find the derivative of the updating function.
   c. Show that the positive equilibrium is a critical point.
   d. What does this say about the stability of the equilibrium? (Think about a linear dynamical system with the same slope. How quickly do solutions approach the equilibrium?) Draw a cobweb diagram to illustrate these results.

5. Expanding oscillations can be a result of improperly tuned feedback systems. Suppose that a thermostat is supposed to keep a room at 20°C. You come in one morning and find that the temperature is 21°C. To correct this, you move the thermostat down by 1°C to 19°C. But the next day the temperature has dropped to 18°C.
   a. Suppose the temperature produced is a linear function of the temperature on the thermostat. Find this function.
   b. You continue to respond to a temperature x°C above 20 by setting the thermostat to $20 - x$°C, and to a temperature x°C below 20 by setting the thermostat to $20 + x$°C. Find the temperature for the next few days.
   c. Denote the temperature on day t by T_t. Find a formula for the thermostat setting z_t in response.
   d. Use the answer to part **a** to find T_{t+1}. Write the updating function.
   e. Use the stability condition to describe what will happen in this room.
   f. If the thermostat were controlled by computer, how might you deal with this problem?

6. Suppose a plant population obeys the following rule: If there are n seeds (say 20), each sprouts and grows to a size

$s = 100/n$ (so $s = 5$ when $n = 20$). Each plant produces $s - 1$ seeds (because it must use one unit of energy to survive).

a. Starting from $n = 20$, find the total number of seeds for the next few years (if the number of seeds per plant is a fraction, don't worry. Think of it as an average).
b. Write the updating function.
c. Find the equilibrium.
d. Graph the updating function and cobweb.
e. How does this result relate to the stability condition?

7. Repeat the analysis of Exercise 6 for the following.
   a. A plant uses 0.5 units of energy to survive, so one of size s makes $s - 0.5$ seeds.
   b. A plant uses 2.0 units of energy to survive, so one of size s makes $s - 2.0$ seeds.

8. We can change from total population N_t in the dynamical system,
$$N_{t+1} = rN_t\left(1 - \frac{N_t}{K}\right)$$
to the fraction of the maximum possible population by careful change of variables.
   a. Define $x_t = N_t/K$. Find x_{t+1} in terms of N_{t+1}.
   b. Solve for N_t in terms of x_t and N_{t+1} in terms of x_{t+1}.
   c. Substitute x for N everywhere in the original equation.
   d. Solve for x_{t+1}.

9. The logistic model quantifies a competitive interaction wherein per capita reproduction is a decreasing function of population size. In some situations, per capita reproduction is enhanced by population size. In one simple model, per capita growth increases linearly as a function of population size. Consider the case
$$\text{per capita reproduction} = 0.5 + 0.5b_t$$
where population size is measured in millions.
   a. Write the updating function.
   b. Find the equilibria.
   c. Graph the updating function and cobweb. Which equilibrium is stable?
   d. Explain what this population is doing.

10. Find the following for the Ricker model.
    a. The value of λ where the positive equilibrium switches from having a positive to a negative slope.
    b. Graph and cobweb this case.
    c. Graph and cobweb for one value of λ between 1 and the value you found in part **a**.
    d. Graph and cobweb for one value of λ between the value you found in part **a** and $\lambda = e^2$.

11. Consider a modified version of the logistic dynamical system,
$$x_{t+1} = rx_t(1 - x_t^2)$$
    a. Sketch the updating function with $r = 2$.
    b. Find the equilibria.
    c. Find the derivative of the updating function at the equilibria.
    d. For what values of r is the $x = 0$ equilibrium stable? For what values of r is the positive equilibrium stable?

12. **COMPUTER:** Study the behavior of the logistic dynamical system, first for values of r near 3.0 and then for values between 3.5 and 4.0. Try the following with five values of r near 3.0 (such as 2.9, 2.99, 3.0, 3.01, and 3.1) and ten values of r between 3.5 and 4.0.
    a. Use your computer to find solutions for 100 steps.
    b. Look at the last 50 or so points on the solution and try to describe what is going on.
    c. The case with $r = 4.0$ is rather famously chaotic. One of the properties of chaotic systems is sensitivity to initial conditions. Run the system for 100 steps from one initial condition, and then run it again from another initial condition that is very close to the first. If you compare your two solutions, they should be similar for a while but eventually become completely different. What if a real system had this property?

13. **COMPUTER:** Consider an even more general version of the logistic dynamical system,
$$x_{t+1} = rx_t(1 - x_t^p)$$
    a. Sketch the updating function for several values of $p > 2$ (such as $p = 3$, $p = 4$, and $p = 10$).
    b. Find the equilibria.
    c. Find the derivative of the updating function at the equilibria.
    d. For what values of r is the $x = 0$ equilibrium stable? For what values of r is the positive equilibrium stable?
    e. What happens to solutions when r is too large?

14. **COMPUTER:** Consider a population following
$$x_{t+1} = \lambda x_t^2 e^{-x_t}$$
    a. Graph the updating function for $\lambda = 1$, $\lambda = 2.6$, $\lambda = e$, $\lambda = 2.8$, $\lambda = 3.6$, $\lambda = e^2/2$, $\lambda = 3.8$, $\lambda = 6.6$, $\lambda = e^3/3$, $\lambda = 6.8$, and $\lambda = 10.0$.
    b. Find the equilibria for these values of λ (the equation cannot be solved in general, but your computer should have a routine for solving; or you can just guess). Make sure you find them all.
    c. Find the derivative of the updating function at each equilibrium.
    d. Find which of the equilibria are stable.
    e. When all the positive equilibria are unstable, how might this model behave differently from the Ricker model with $\lambda > e^2$? Can you explain why?

3.3 Maximization

A bee arrives at a flower. The more time she spends eating nectar, the more slowly the nectar comes out. But she knows that she must fly a long distance to find the next flower. When should she give up and leave? On a larger scale, a fisherman must decide how many fish to catch. The more he catches, the more he gets that year, but the more he depletes the fishery for next year. How many fish should he catch to catch the most fish in the long run? In fisheries jargon, what is the **maximum sustained yield** from the fish population?

Both of these are problems in **optimization**, finding the best solution to a problem. In this section, we learn how to use the derivative to find **optima**. When the problem has been set up correctly, these best solutions are points where a function takes on a **minimum** or **maximum value**. In particular, **critical points** where the derivative is equal to 0 are candidates for being minima or maxima, which can be distinguished with the **second derivative**. We will use these techniques to figure out the optimal behaviors for the bee and the fisherman.

Minima and Maxima

The graphs of the functions

$$f(x) = xe^{-x}$$
$$g(x) = x(1-x)$$

rise to a peak, known as a **maximum of the function** (Figure 3.24). How do we find where the maximum occurs? A function has a maximum where it switches from increasing to decreasing, just as a trail reaches a peak when it switches from ascending to descending. Mathematically, a maximum occurs when the derivative switches from positive to negative.

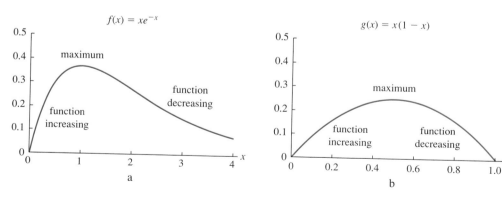

Figure 3.24
Two functions with peaks

What are the derivatives of our two functions? For the first, we have

$$\frac{df}{dx} = \frac{d}{dx} xe^{-x}$$
$$= \frac{dx}{dx} e^{-x} + \frac{d(e^{-x})}{dx} x$$

$$= e^{-x} + -e^{-x}x$$
$$= (1-x)e^{-x}$$

(Figure 3.25a). The derivative is positive for $x < 1$, negative for $x > 1$, and 0 for $x = 1$. The point where $f'(x) = 0$ is called a **critical point** (Section 2.4); in this case it is the maximum.

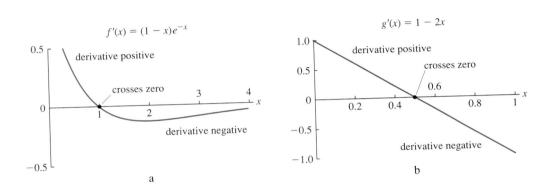

Figure 3.25
The derivatives of two functions with peaks

Similarly, for the second function, we have

$$\frac{dg}{dx} = \frac{d}{dx}(x - x^2) = 1 - 2x$$

(Figure 3.25b). The derivative is exactly 0 when

$$1 - 2x = 0$$
$$x = 0.5$$

Because $f'(x) > 0$ for $x < 0.5$ and $f'(x) < 0$ for $x > 0.5$, this critical point is also a maximum.

Now consider the function

$$h(x) = x^3 - x$$

for $-1 < x < 1$. Does this function have a maximum? To begin, take the derivative:

$$\frac{dh}{dx} = 3x^2 - 1$$

Next, set the derivative equal to 0 to find critical points:

$$3x^2 - 1 = 0$$

Therefore, we have

$$3x^2 = 1$$
$$x^2 = \frac{1}{3}$$
$$x = \pm\sqrt{\frac{1}{3}}$$

These points are candidates for where the function changes from increasing to decreasing. How can we tell what is happening at these points? One simple method is to evaluate the function, finding

$$h\left(\sqrt{\frac{1}{3}}\right) = -0.385$$

$$h\left(-\sqrt{\frac{1}{3}}\right) = 0.385$$

The point $x = \sqrt{1/3}$ is a minimum, and the point $x = -\sqrt{1/3}$ is a maximum (Figure 3.26).

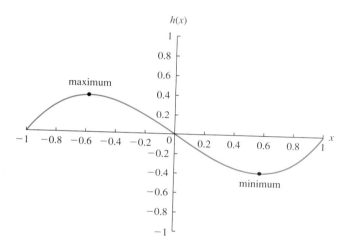

Figure 3.26
A function with a minimum and a maximum: $h(x) = x^3 - x$

To develop a general method for finding maxima and minima of functions, we must classify the types of maxima and minima. Maxima are classified as **global** or **local**. The global maximum of a function f is the largest value taken on by the function. Similarly, the global minimum is the smallest value taken on by the function. A local maximum is a "peak" where the function takes on its largest value in a region of the domain (Figure 3.27). The peak of the tallest mountain in Utah is a local maximum but not a global one. A function may have many local maxima with different values, just as a mountain range has many peaks with different heights. Local and global minima are defined in the same way.

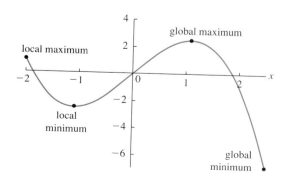

Figure 3.27
Local and global maxima

A special kind of maximum occurs at the boundary of the domain. The function in Figure 3.27 has a local maximum at the left-hand edge of its domain and a global minimum at the right. Maxima and minima at the boundaries can also be analyzed with the derivative. A function with negative derivative is decreasing and has a local maximum at the left-hand boundary and a local minimum at the right. The function in Figure 3.27 matches these criteria. A function with positive derivative is increasing and has a local minimum at the left-hand boundary and a local maximum at the right.

Algorithm 3.1 (Finding global maxima and minima)

1. Compute the value of the function at the endpoints and anywhere the function is not differentiable.

2. Find all critical points.

3. Compute the value of the function at all critical points.

4. The largest of the numbers found in steps 1 and 3 is the global maximum. The smallest of the numbers found in steps 1 and 3 is the global minimum.

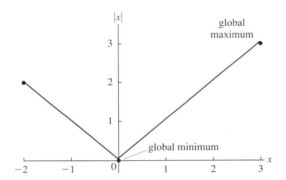

Figure 3.28
The minimum of the absolute value

We must check separately all points where the function is not differentiable because these points can be maxima or minima without being critical points. Consider the absolute value function $|x|$ defined for $-2 \leq x \leq 3$ (Figure 3.28). Following the algorithm, we have

1. The endpoints are -2 and 3, where $|-2| = 2$, $|3| = 3$. The function is not differentiable at $x = 0$ (because of the corner), where $|0| = 0$.

2. This function has no critical points.

4. The largest of these values is 3, so the global maximum is at the right-hand endpoint. The smallest is 0, so $x = 0$ is the global minimum.

To find the global minima and maxima of the differentiable function $h(x) = x^3 - x$ for $-1 \leq x \leq 1$ (Figure 3.26), we follow the steps:

1. $h(-1) = 0$ and $h(1) = 0$.

2. We found the critical points earlier at $x = \sqrt{1/3}$ and $x = -\sqrt{1/3}$.

3. $h(\sqrt{1/3}) = -0.385$, and $h(-\sqrt{1/3}) = 0.385$.

4. The largest of these values is 0.385, so the global maximum occurs at $x = -\sqrt{1/3}$. The smallest is -0.385, so the global minimum occurs at $x = \sqrt{1/3}$.

(See Figure 3.29a–c.) If we extend the domain to include $-1 \leq x \leq 2$, the global maximum moves to $x = 2$ (Figure 3.29d). The critical point $x = -\sqrt{1/3}$ is still a **local maximum**.

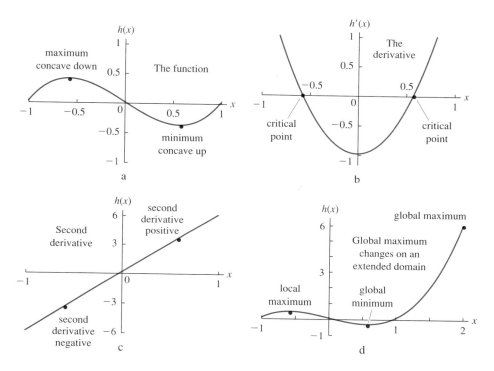

Figure 3.29

The function $h(x) = x^3 - x$, its derivative and second derivative

The second derivative provides a method to check whether a critical point is a maximum or a minimum. On the graph of $h(x) = x^3 - x$, we can see that a critical point is a local maximum when the graph is **concave down** and is a local minimum when the graph is **concave up**. The second derivative of $h(x) = x^3 - x$ is

$$\frac{d^2}{dx^2}x^3 - x = \frac{d}{dx}3x^2 - 1 = 6x$$

This function is negative when $x < 0$, so the critical point $x = -\sqrt{1/3}$ is a local maximum, and the critical point $x = \sqrt{1/3}$ is a local minimum (Figure 3.29c). Similarly, the second derivative of $f(x) = xe^{-x}$ is

$$\frac{d^2 f}{dx^2} = \frac{d}{dx}(1-x)e^{-x} \qquad \text{take derivative of derivative}$$

$$= \frac{d(1-x)}{dx}e^{-x} + \frac{d(e^{-x})}{dx}(1-x) \qquad \text{product rule}$$

$$= -e^{-x} + -e^{-x}(1-x) \qquad \text{linear function and exponential rules}$$
$$= (x-2)e^{-x} \qquad \text{factoring}$$

At the critical point $x = 1$,
$$f''(1) = (1-2)e^{-1} = -e^{-1} < 0$$

Because the second derivative is negative at the critical point, the function switches from increasing to decreasing at $x = 1$. There is a local maximum at $x = 1$. Finally, the second derivative of $g(x) = x(1-x)$ is

$$\frac{d^2g}{dx^2}x(1-x) = \frac{d}{dx}1 - 2x = -2$$

Because the second derivative is negative, the critical point at $x = 1/2$ is a local maximum (Figure 3.24b).

We can formalize these observations into another algorithm.

■ **Algorithm 3.2** (Finding local maxima and minima with the second derivative)

1. Find all critical points.

2. Compute the sign of the second derivative at all critical points.

3. Critical points where the second derivative is positive correspond to local minima, and critical points where the second derivative is negative correspond to local maxima. ■

If the second derivative is 0 at a critical point, we cannot tell whether the critical point is a local minimum, a local maximum, or neither. The function $C(x) = x^3$ has both a critical point and a point of inflection at $x = 0$ because

$$C'(x) = 3x^2$$
$$C''(x) = 6x$$

(Figure 3.30a). Because the derivative is never negative, the critical point is neither a minimum nor a maximum. However, the quartic function $Q(x) = x^4$ has a local minimum at $x = 0$ even though the second derivative is 0 (Figure 3.30b).

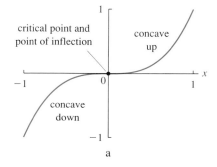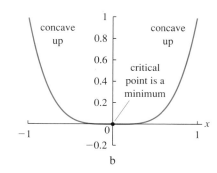

Figure 3.30
Two functions for which the second derivative is 0 at a critical point

Maximizing the Rate of Food Intake

Consider the bee described at the beginning of this section. After she finds a flower, she sucks up nectar at a slower and slower rate as the flower is depleted. However, she does not want to leave too soon because she must fly some distance to find the next flower. When should she give up and leave? If she stays a long time at each flower, she will get most of the nectar from each flower but will visit few flowers. If she stays a short time at each flower, she gets to skim the best nectar off the top but spends most of her time flying to new flowers (Figure 3.31).

Figure 3.31

The bee's maximization problem

First, we must formulate this as a *maximization problem*. What is the bee trying to achieve? Her job is to bring back as much nectar as possible over the course of a day. To do so, she should maximize the *rate per visit*, which includes the travel time τ between flowers. Suppose $F(t)$ is the amount of nectar collected by a bee that stays on a flower for time t (Figure 3.32). The rate is

$$\text{rate at which nectar is collected} = R(t)$$
$$= \frac{\text{food per visit}}{\text{time per visit}}$$
$$= \frac{F(t)}{t + \tau}$$

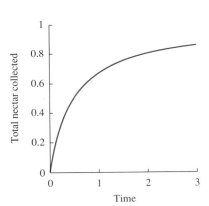

Figure 3.32

Nectar collected as a function of time

As a particular case, assume that the travel time τ is equal to 1.0 minutes, and that

$$F(t) = \frac{t}{t + 0.5}$$

Then

$$R(t) = \frac{t}{(t + 0.5)(t + 1.0)}$$

The derivative can be found with the quotient rule and a lot of algebra:

$$\frac{dR}{dt} = \frac{(t + 0.5)(t + 1.0) - t\,[(t + 0.5) + (t + 1.0)]}{[(t + 0.5)(t + 1.0)]^2}$$
$$= \frac{0.5 - t^2}{[(t + 0.5)(t + 1.0)]^2}$$

This derivative is 0 when the numerator is 0, or when

$$0.5 - t^2 = 0$$

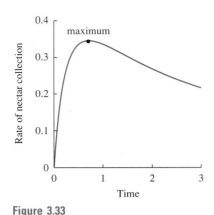

Figure 3.33

The average rate of return for the bee

which has positive solution $t = 0.707$. The numerator is positive for $t < 0.707$ and negative for $t > 0.707$, implying that this value is a maximum.

By looking at the problem in general, without substituting a specific functional form for $F(t)$ or a value of τ, we can find a valuable graphical method to solve the bee's problem. Again, we differentiate $R(t)$ with the quotient rule, finding

$$\frac{dR}{dt} = \frac{(t + \tau)F'(t) - F(t)}{(t + \tau)^2}$$

Because the denominator is positive, the critical points occur where

$$(t + \tau)F'(t) = F(t)$$

or

$$F'(t) = \frac{F(t)}{t + \tau} = R(t)$$

We will show in Exercise 8 that the solution of this equation is a local maximum as long as $F(t)$ is concave down. This fundamental equation says that the bee should leave when the derivative of F, the instantaneous rate of food collection, is equal to the average rate. This is called the **Marginal Value Theorem** and is a powerful tool in both ecology and economics. The idea is simple; leave when you can do better elsewhere. Graphically, the slope of the food collection curve at the critical point is equal to the slope of the line connecting that point with a point at negative τ (Figure 3.34a). From the graph, we can see that the optimal time to remain becomes shorter when the travel time between flowers is shorter (Figure 3.34b).

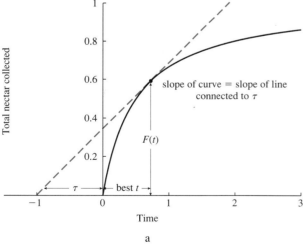

a

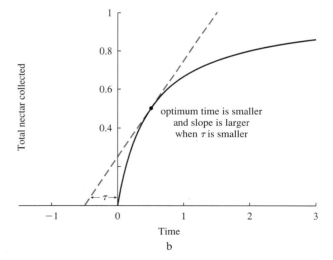

b

Figure 3.34

The Marginal Value Theorem

Maximizing Fish Harvest

Consider a variant of the logistic dynamical system (Equation 3.1) that includes harvesting,

$$N_{t+1} = 2.5N_t(1 - N_t) - hN_t \tag{3.2}$$

(Figure 3.35). The population of fish at the beginning of one fishing season is denoted by N_t, and N_{t+1} denotes the population at the beginning of the next. The population is measured as the fraction of the maximum possible population size. The term $-hN_t$ is the harvest, where h is called the "harvesting effort." Harvesting effort depends on the number of ships, the number of fishing days, the quality of the fishing vessels, and many other factors. Total harvest is the product of harvesting effort and population size. In this case, the harvest is computed by subtracting a factor h times the number of fish before reproduction.

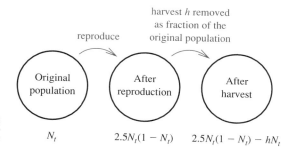

Figure 3.35
The dynamics of a simple fishery

What harvesting effort brings in the maximum long-term harvest? No harvesting effort ($h = 0$) brings in nothing. An enormous harvesting effort might bring in many fish in the short term but end up depleting the population. We suspect that an intermediate harvesting effort will maximize the long-term harvest.

The long-term behavior of this system is described by a stable equilibrium. What are the equilibria? The equilibrium value N^*, found with Algorithm 1.5, is

$$N^* = 2.5N^*(1 - N^*) - hN^* \quad \text{the original equation}$$

$$N^* - 2.5N^*(1 - N^*) + hN^* = 0 \quad \text{move everything to one side}$$

$$N^*\left[1 - 2.5(1 - N^*) + h\right] = 0 \quad \text{factor}$$

$$N^* = 0 \text{ or } 1 - 2.5(1 - N^*) + h = 0 \quad \text{set each piece equal to 0}$$

$$N^* = 0 \text{ or } N^* = 1 - \frac{1 + h}{2.5} \quad \text{do the algebra}$$

We have the usual extinction equilibrium as well as a second equilibrium that is positive only if $(1+h)/2.5 < 1$. If we choose h larger than 1.5 the only equilibrium is $N^* = 0$ and the population goes extinct. The peculiar possibility that $h > 1$ is because our model measures the population before reproduction and collects the harvest after reproduction.

Suppose we have chosen a harvesting effort h and that the population has reached the positive equilibrium N^* (conditions for stability are derived in Exercise 11). The equilibrium harvest, denoted $P(h)$, is the product of the harvesting effort h and the population size N^*, so we have

$$P(h) = hN^* = h\left(1 - \frac{1 + h}{2.5}\right) \tag{3.3}$$

(Figure 3.36). With Algorithm 3.1, we can find the value of h that maximizes harvest by checking the endpoints and locating critical points. The endpoints are $h = 0$ and $h = 1.5$ (above which the population goes extinct), where $P(0) = 0$ and $P(1.5) = 0$, as we suspected. To find critical points, we solve

$$P'(h) = 1 - \frac{1 + 2h}{2.5} = 0$$

for h, finding

$$h = \frac{1.5}{2} = 0.75$$

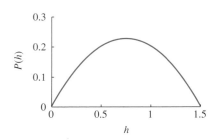

Figure 3.36
Long-term harvest as a function of harvesting effort with $r = 2.5$

On the basis of our knowledge of the graph of the function (Figure 3.36) we know this is a maximum. We would do best by letting the fish reproduce and then collecting a harvest equal to 75% of the population before reproduction. The payoff $P(h)$ is

$$P(0.75) = 0.75 \left(1 - \frac{1 + 0.75}{2.5} \right) = 0.225$$

A larger harvest would deplete the population. With $h = 1.0$,

$$P(1.0) = 1.0 \left(1 - \frac{1 + 1.0}{2.5} \right) = 0.2$$

which is smaller because the equilibrium is only 0.2. In contrast, a smaller harvest is inefficient. With $h = 0.5$,

$$P(0.5) = 0.5 \left(1 - \frac{1 + 0.5}{2.5} \right) = 0.2$$

which is smaller than the maximum even though the equilibrium population is larger.

SUMMARY

We have seen how to use the derivative to find **maxima** and **minima** of functions by locating **critical points** where the derivative is 0. To find **global maxima** and **minima**, values at critical points must be compared with values at the endpoints, where maxima and minima often occur. The second derivative is positive at a local minimum and negative at a local maximum. We first applied this method to find the optimal length of time a bee should spending harvesting nectar from a flower. The solution is an example of the **Marginal Value Theorem**, which states that the best time to leave occurs when the rate of collecting resources falls below the average rate. We next applied the methods of maximization to a discrete-time dynamical system with harvesting to find the optimal way to harvest a fish population.

3.3 EXERCISES

1. Find the critical points of the following functions.
   a. $a(x) = \dfrac{x}{1 + x}$
   b. $f(x) = 1 + 2x - 2x^2$
   c. $c(w) = w^3 - 3w$
   d. $g(y) = \dfrac{y}{1 + y^2}$
   e. $h(z) = e^{z^2}$
   f. $c(\theta) = \cos(2\pi\theta)$

2. Find the global minimum and maximum of the following functions on the given interval. Don't forget to check the endpoints.
   a. $a(x) = \dfrac{x}{1 + x}$ for $0 \leq x \leq 1$
   b. $f(x) = 1 + 2x - 2x^2$ for $0 \leq x \leq 1$
   c. $f(x) = 1 + 2x - 2x^2$ for $0 \leq x \leq 2$
   d. $c(w) = w^3 - 3w$ for $-2 \leq x \leq 2$

e. $g(y) = \dfrac{y}{1 + y^2}$ for $0 \leq x \leq 2$

f. $h(z) = e^{z^2}$ for $0 \leq x \leq 1$

g. $F(x) = |1 - x|$ for $0 \leq x \leq 3$ (This function cannot be differentiated at $x = 1$.)

3. Find the second derivative at the critical points found in Exercise 1. Classify the critical points as minima and maxima. Use the second derivative to draw an accurate graph of the function for the ranges given in Exercise 2.

4. Suppose $f(x)$ is a positive function with a maximum at x^*. Consider the function $g(x) = 1/f(x)$.
   a. Show that g has a critical point at x^*.
   b. Compute the second derivative at this point.
   c. Can you explain this result?

5. Consider the bee faced by the problem in the subsection entitled "Maximizing the Rate of Food Intake." Find the optimal strategies for the following travel times τ, and draw the associated diagram. For each value of τ, find the equation of the tangent line at the optimal t and prove that it goes through the point $(-\tau, 0)$.
   a. $\tau = 2.0$
   b. $\tau = 0.5$
   c. $\tau = 0.1$
   d. Find the solution in general (without plugging in a value for τ).
   e. What is the limit as τ approaches 0? Does this answer make sense?

6. Suppose now that the total food collected by a bee follows
   $$F(t) = \dfrac{t}{c + t}$$
   where c is some parameter. If $\tau = 1.0$, find the optimal departure time for the following circumstances. Sketch the associated graph.
   a. $c = 2.0$
   b. $c = 1.0$
   c. $c = 0.1$
   d. Find the solution in general (without plugging in a value for c).
   e. What does the parameter c mean biologically? (Think about how long it takes the bee to collect half the nectar.) Explain in words why the bee leaves sooner when c is smaller.

7. Mathematical models can help us to estimate values that are difficult to measure. Consider again a bee sucking nectar from a flower, with
   $$F(t) = \dfrac{t}{0.5 + t}$$
   We also measure that the bee remains on the flower a length of time t. Estimate the travel time τ assuming that the bee understands the Marginal Value Theorem.
   a. $t = 1.0$
   b. $t = 0.1$

c. $t = 4.0$

d. Find the solution in general (without plugging in a value for t).

8. We never showed that the value found in computing the optimal t with the Marginal Value Theorem is in fact a maximum.
   a. Find the second derivative of $R(t)$.
   b. Evaluate at the point where $F'(t) = F(t)/(t + \tau)$, and show that it is negative if $F(t)$ is concave down.
   c. Draw a figure illustrating a case where $F(t)$ is concave up. What does this mean biologically? What is the optimal solution?

9. Animals have more to worry about than finding food. One theory assumes that they try to maximize the ratio of food collected to predation risk. Suppose that different flowers with nectar of quality n attract $P(n)$ predators. For example, flowers with higher quality nectar (large values of n) might attract more predators (large value of $P(n)$). Bees must decide which flowers to visit.
   a. What function are the bees trying to maximize?
   b. Take the derivative and find the condition for the maximum.
   c. Find a graphical interpretation of this condition.
   d. Suppose that $P(n) = 1 + n^2$. Find the optimal n for the bees.
   e. Suppose that $P(n) = 1 + n$. Find the optimal n for the bees, and draw the associated graph. Does this make sense? Why is the result so different?

10. Follow the steps to find the maximum harvest from a population following
    $$N_{t+1} = 2.0 N_t (1 - N_t) - h N_t$$
    a. Find the equilibrium population as a function of h. What is the largest h consistent with a positive equilibrium?
    b. Find the equilibrium harvest as a function of h.
    c. Find the harvesting effort that maximizes harvest.
    d. Find the maximum harvest.

11. Find the conditions for stability of the equilibrium of Equation 3.2 (page 244). Show that the equilibrium N^* is stable when h is set to the value that maximizes the long-term harvest. Graph the updating function and cobweb.

12. Calculate the maximum long-term harvest for an alternative model of competition with updating function
    $$N_{t+1} = \dfrac{r N_t}{1 + c N_t} - h N_t$$
    Do so with $r = 2.5$ and $c = 1$ and then with $r = 1.5$ and $c = 1$.
    a. Find the equilibrium as a function of h.
    b. What is the largest value of h consistent with a positive equilibrium?
    c. Find the harvest level giving the maximum long-term harvest.

d. Sketch a graph of $P(h)$ and compute the value at the maximum.
e. Why are the answers so different for different values of r?

13. The model of fish harvesting includes nothing about harvesting cost. Suppose that the population follows
$$N_{t+1} = 2.5N_t(1 - N_t) - hN_t$$
as in the text, but that the payoff is
$$P(h) = hN^* - ch$$
where c is the cost per unit effort of harvesting.
a. Find the optimal harvest when $c = 0.1$.
b. Do the same with $c = 0.2$, $c = 0.5$, and $c = 1.0$.
c. Do your answers all make sense? What should a fisherman do if c becomes too large?

14. **COMPUTER**: Suppose a population follows the updating function
$$N_{t+1} = 2.5N_t(1 - N_t) - hN_t$$
but can be harvested only every second year. This means that harvest alternates between the chosen value h and 0.
a. Find the 2-yr updating function.
b. Find the optimal harvest.
c. Compare with the results in the text. Is it better to harvest less often?

3.4 Reasoning About Functions

Continuous and differentiable functions have many useful properties we can use to reason about the biological processes they describe. We will use the **Intermediate Value Theorem** to show, without solving any equations, that a discrete-time dynamical system has an equilibrium, the **Extreme Value Theorem** to show, without computing any derivatives, that a function has a maximum, and the **Mean Value Theorem** to find, without taking any limits, the value of a derivative.

Continuous Functions: The Intermediate Value Theorem

Consider a model for chemical concentration in the lung that includes absorption:
$$c_{t+1} = (1-q)(1-\alpha(c_t))c_t + q\gamma \tag{3.4}$$

In this updating function c_t represents the concentration before a breath, c_{t+1} the concentration before the next breath, q the fraction of air exchanged, γ the concentration of chemical in the ambient air, and $\alpha(c_t)$ the fraction of chemical *absorbed* as a function of the chemical concentration in the lung (Figure 3.37a). This equation subtracts $\alpha(c_t)$ from the basic updating function for the lung,
$$c_{t+1} = (1-q)c_t + q\gamma$$
(Equation 1.44). Suppose that $\alpha(c_t)$ is
$$\alpha(c_t) = 0.5(1 - e^{-0.5c_t})$$
(Figure 3.37b). Absorption is equal to 0 when $c_t = 0$ because there is nothing to absorb, and the fraction absorbed increases to 0.5 as the concentration becomes larger. What happens to the concentration in this lung? Does it still have an equilibrium?

We begin by trying to find an equilibrium,

$c^* = (1-q)[1 - 0.5(1 - e^{-5c^*})]c^* + q\gamma$ the original equation
$0 = (1-q)[1 - 0.5(1 - e^{-5c^*})]c^* + q\gamma - c^*$ place unknowns on one side

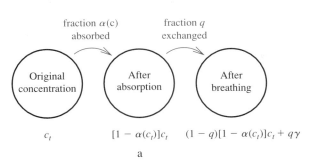
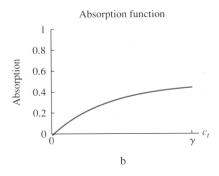

Figure 3.37
Dynamics of a lung with absorption

This equation cannot be factored. Is there any way to establish whether there is an equilibrium and to get some idea where it is?

We can answer these questions without doing any algebra by reasoning about the general process of breathing. One process, breathing in, adds chemical to the lungs, and two processes, breathing out and absorption, remove chemical from the lungs. Without absorption, we found that the equilibrium concentration is γ, the ambient concentration. With absorption, then, we expect the equilibrium to be decreased below γ. We can test this intuition by reasoning about the updating function.

Suppose the lung starts out with $c_t = 0$, the lowest possible concentration. Then

$$c_{t+1} = (1-q)[1-\alpha(0)] \cdot 0 + q\gamma = q\gamma$$

As long as $q > 0$ (meaning that some air is exchanged) and $\gamma > 0$ (meaning that some chemical is available), $c_{t+1} = q\gamma > 0$. In other words, the amount of chemical has *increased*. Conversely, suppose the lung starts out with a concentration of $c_t = \gamma$. We expect that this concentration will decrease because of absorption. Substituting into the updating function, we have

$c_{t+1} = (1-q)[1-\alpha(\gamma)]\gamma + q\gamma$	updating function with $c_t = \gamma$
$= (1-q)\gamma + q\gamma - (1-q)\alpha(\gamma)\gamma$	separate out term with α
$= \gamma - (1-q)\alpha(\gamma)\gamma$	sum of first two terms is γ
$< \gamma$	as long as $\alpha(\gamma) > 0$ and $q < 1$

Absorption reduces the concentration below the ambient concentration, the equilibrium value without absorption.

The graph of the updating function, therefore, lies above the diagonal at $c = 0$ and below it at $c = \gamma$ (Figure 3.38). Furthermore, the updating function is *continuous* because it is built by combining continuous linear and exponential functions using only multiplication and composition (Section 2.3). The graph of a continuous function has "no jumps," meaning that it is impossible to draw a graph connecting these two points without crossing the diagonal. Such a crossing point is an equilibrium (Section 1.11).

We have found, without solving any equations, that this updating function *must* have an equilibrium between 0 and γ, in accord with our biological intuition. Furthermore, the updating function must cross the diagonal from above to below. However, we cannot be sure that the equilibrium is stable because the updating function could be decreasing steeply at the equilibrium (Exercise 13).

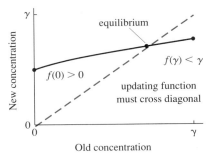

Figure 3.38
Reasoning about the equilibrium of the lung updating function with absorption

How do we prove mathematically that the updating function must cross the diagonal? This result follows from the **Intermediate Value Theorem** for continuous functions.

■ **THEOREM 3.2** (Intermediate Value Theorem)

If $f(x)$ is continuous for $a \leq x \leq b$ and $f(a) < c < f(b)$, then there is some x between a and b such that $f(x) = c$. ■

The proof of this simple theorem is quite subtle, requiring deep facts about real numbers and continuity. The idea of the Intermediate Value Theorem is shown in Figure 3.39. The theorem guarantees that there is *at least* one crossing point, but there may be more.

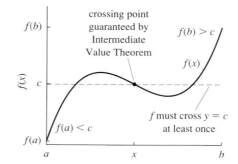

Figure 3.39
The Intermediate Value Theorem

Physically, the theorem says that if you grew from 2 ft to 6 ft in height, you must have been exactly 4 ft tall at some time; if you accelerate from 0 to 60 mph, you must have been going exactly 31.4159 mph at some time (Figure 3.40).

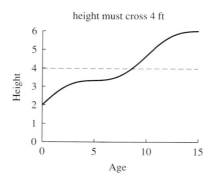 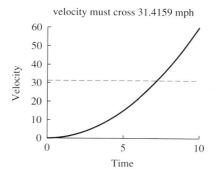

Figure 3.40
The Intermediate Value Theorem applied to height and speed

We need to make a transformation to apply this theorem to the equilibria of the lung updating function with absorption. An equilibrium is a point where the *change in concentration* is equal to 0 (Figure 3.41). The change in concentration Δc is

$$\Delta c = c_{t+1} - c_t$$

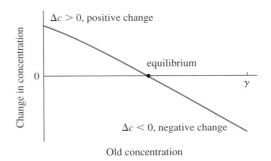

Figure 3.41
Reasoning about the equilibrium of the lung updating function using the Intermediate Value Theorem

At $c_t = 0$, $c_{t+1} > c_t$ and $\Delta c > 0$. At $c_t = \gamma$, $c_{t+1} < \gamma$ and $\Delta c < 0$. The Intermediate Value Theorem guarantees that there must be some point where $\Delta c = 0$ (Figure 3.41). This point is the equilibrium.

Maximization: The Extreme Value Theorem

Suppose that the per capita reproduction of a population of fish is

$$\text{per capita reproduction} = 2.5e^{-N_t}$$

and that a factor h times the prereproductive population is harvested each year (modified from Equation 3.2 and Exercise 12 in Section 3.3). The updating function for the population is

$$N_{t+1} = 2.5N_t e^{-N_t} - hN_t \tag{3.5}$$

We want to find the harvesting effort h that maximizes long-term harvest. Recall the steps we used in Maximizing Fish Harvest: First find the equilibrium N^* as a function of h, then find the fish harvested as $P(h) = hN^*$, and then compute the maximum of $P(h)$ by differentiating.

To find the equilibria, follow the usual steps:

$N^* = 2.5N^* e^{-N^*} - hN^*$ the original equation

$N^* - 2.5N^* e^{-N^*} + hN^* = 0$ move everything to one side

$N^*(1 - 2.5e^{-N^*} + h) = 0$ factor

$N^* = 0$ or $(1 - 2.5e^{-N^*} + h) = 0$ set each piece equal to 0

Solving the second part requires a bit of algebra:

$1 - 2.5e^{-N^*} + h = 0$ original equation

$1 + h = 2.5e^{-N^*}$ move unknowns to one side

$\dfrac{1+h}{2.5} = e^{-N^*}$ divide by 2.5

$\ln\left(\dfrac{1+h}{2.5}\right) = -N^*$ take the natural logarithm

$N^* = -\ln\left(\dfrac{1+h}{2.5}\right)$ solve for N^*

$N^* = \ln\left(\dfrac{2.5}{1+h}\right)$ use Law 3 of logs

Is this value positive? Recall that $\ln(x) > 0$ if $x > 1$. Therefore N^* is positive if

$$\frac{2.5}{1+h} > 1$$
$$2.5 > 1 + h$$
$$1.5 > h$$

If $h = 1.5$ the equilibrium is 0.

The harvest $P(h)$ is the factor h times the total population N^*, so

$$P(h) = hN^* = h \ln\left(\frac{2.5}{1+h}\right)$$

Does this equation have a maximum? We have two algorithms for finding a maximum, each requiring that we find critical points by computing the derivative and finding where it is equal to 0. In this case,

$$P'(h) = \ln\left(\frac{2.5}{1+h}\right) - \frac{h}{1+h}$$

Finding critical points requires solving the equation $P'(h) = 0$. This equation cannot be solved algebraically.

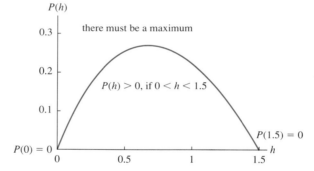

Figure 3.42
Reasoning about harvest: the Extreme Value Theorem

Nonetheless, we can still prove that this function has a maximum. We know that $P(0) = 0$ (no harvesting) and that $P(1.5) = 0$ (no fish). Furthermore, $P(h) > 0$ if $0 < h < 1.5$ because both h and N^* are positive (Figure 3.42). A function that is 0 at its endpoints and positive between them must have a maximum. Mathematically, this result follows from the **Extreme Value Theorem**.

■ **THEOREM 3.3** **(Extreme Value Theorem)**

If $f(x)$ is continuous for $a \leq x \leq b$, then there is a point c_h, $a \leq c_h \leq b$, where $f(x)$ takes on its global maximum and a point c_l, $a \leq c_l \leq b$, where $f(x)$ takes on its global minimum. ■

Again, the proof of this theorem is quite subtle. The conclusions are illustrated in Figure 3.43. The theorem does not guarantee that the maximum and minimum must occur between a and b. Either might lie at one of the endpoints (the minimum in Figure 3.43 lies at the endpoint b).

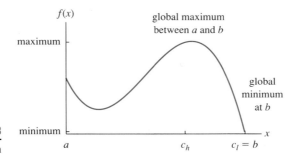

Figure 3.43
The Extreme Value Theorem

The function $P(h)$ is continuous between $h = 0$ and $h = 1.5$ because it involves no division or logs of 0. Therefore, the Extreme Value Theorem guarantees that it has a maximum. We know that the maximum does not lie at the endpoints because $P(h)$ takes on positive values, larger than the value at each endpoint, for all h between 0 and 1.5. The maximum guaranteed by the Extreme Value Theorem must occur for $0 < h < 1.5$. The minimum guaranteed by the Extreme Value Theorem is shared by the two endpoints.

Why must the function in the Extreme Value Theorem be continuous? Consider the function $f(x) = 1/x$, defined first for x between -1 and 1, but not for $x = 0$, where the function is not continuous (Figure 3.44a). This function has a right-hand limit of infinity and a left-hand limit of negative infinity at $x = 0$, and it has neither a maximum nor a minimum value. Even if we quarantine the trouble point $x = 0$ to the end of the interval, the Extreme Value Theorem breaks down. If we define the function only for $0 < x \leq 1$, $f(x)$ is perfectly continuous. However, because it is not defined at the endpoint it fails to have a maximum (Figure 3.44b).

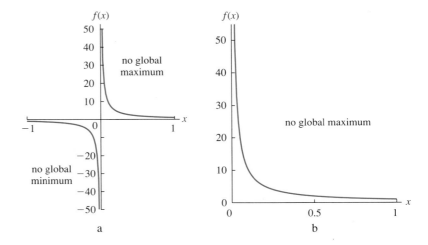

Figure 3.44
The function $f(x) = 1/x$ fails to satisfy the conditions for the Extreme Value Theorem

Rolle's Theorem and the Mean Value Theorem

The Intermediate Value Theorem and the Extreme Value Theorem guarantee that a continuous function must take on particular values. **Rolle's theorem** and the **Mean Value Theorem** guarantee that the *derivative* takes on particular values.

Rolle's theorem is closely related to the Extreme Value Theorem. The maximum of $P(h)$ is a critical point, or point with derivative 0 (Figure 3.42). Rolle's theorem states that a differentiable function that takes on equal values at its endpoints must have a critical point between them.

■ **THEOREM 3.4** (Rolle's Theorem)

If $f(x)$ is differentiable for all x, $a \leq x \leq b$ and $f(a) = f(b)$, then there exists some c with $a < c < b$ and $f'(c) = 0$. ■

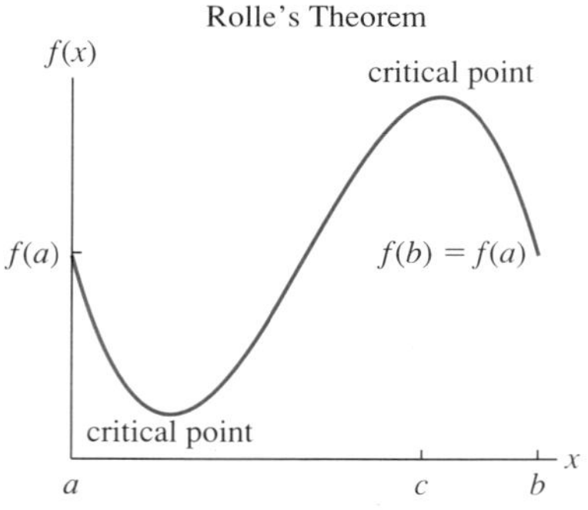

As with the Intermediate Value Theorem, Rolle's theorem guarantees only that there is at least one critical point; there may be more than one (Figure 3.45). This theorem applies directly to the harvest $P(h)$, guaranteeing the existence of a critical point for $0 < h < 1.5$. Because the function is positive, the value of the function P at the critical point must be greater than the value at the endpoints. If there is more than one critical point, the one where the function takes on its largest value is the maximum.

The proof of Rolle's theorem uses the Extreme Value Theorem to show that the function must have a minimum or maximum in the interior of the interval (unless the function is constant) as well as the fact that an interior minimum or maximum must occur at a critical point.

Figure 3.45 Rolle's theorem

The Mean Value Theorem is a tilted version of Rolle's theorem.

■ **THEOREM 3.5** (Mean Value Theorem)

If $f(x)$ is differentiable for $a \leq x \leq b$, then there exists some c such that $a < c < b$ and

$$f'(c) = \frac{f(b) - f(a)}{b - a}$$

■

This theorem says that the slope of the tangent line matches the slope of the secant at some point in the interval spanned by the secant (Figure 3.46).

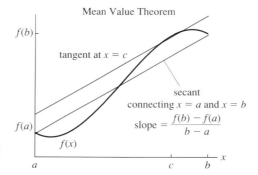

Figure 3.46 The Mean Value Theorem

The most popular application of the Mean Value Theorem involves speeds. If a car travels 140 miles in 2 h, the average rate of change over this time is 70 mph. The Mean Value Theorem guarantees that the instantaneous speed (on the speedometer) must have been exactly 70 mph at some time (Figure 3.47).

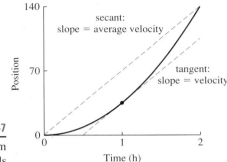
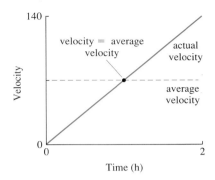

Figure 3.47
The Mean Value Theorem applied to speeds

SUMMARY

With a minimum of algebra, we can use theorems about continuous and differentiable functions to deduce mathematical conclusions. We began by showing that a version of the lung model with absorption must have an equilibrium by using the **Intermediate Value Theorem**, which guarantees that a continuous function takes on all values between those at its endpoints. We then argued that a new version of the harvesting model must have taken on a maximum value with the **Extreme Value Theorem**, which states that a continuous function that is defined on a domain including the endpoints must have a maximum and a minimum. We applied **Rolle's theorem** to the same problem, arguing that a differentiable function that takes on equal values at its endpoints must have a critical point between them. Its generalization, the **Mean Value Theorem**, states that the derivative of a differentiable function must match the slope of the secant.

3.4 EXERCISES

1. Consider the updating function given by Equation 3.4,
$$c_{t+1} = (1-q)[1 - \alpha(c_t)]c_t + q\gamma$$
   but assume that $\alpha(c_t) = 0.5$.
   a. Write the updating function.
   b. Solve for the equilibrium.
   c. Graph the updating function (pick reasonable values of q and γ).
   d. Does this match the deductions made in the text?

2. An old problem has a farmer setting off one morning at 6 A.M. to bring a crop to market, arriving in town at noon. The next day she sets off in the opposite direction at 6 A.M. and returns home along the same route, arriving once again at noon. Use the Intermediate Value Theorem to show that at some point along the path, her watch must read exactly the same time on the two days. (Graph position as a function of time on the two days and show that the curves cross.)

3. Try to apply the Intermediate Value Theorem to the following problems.
   a. The price of gasoline rises from $1.19.9 to $1.27.9. Why is it not necessarily true that the price was exactly $1.25 at some time?
   b. A cell takes up 1.5×10^{-9} mL of water in the course of an hour. Show that the cell must have taken up exactly 1.0×10^{-9} mL at some time. Is it possible that the cell took up exactly 2.0×10^{-9} mL at some time?
   c. Show that $\sin(\theta) = 0.34$ for some θ between 0 and π. Extend your argument to show that there are at least two such θ.
   d. Prove that the updating function
   $$x_{t+1} = \cos(x_t)$$
   has an equilibrium between 0 and $\pi/2$.

4. Apply the Intermediate Value Theorem to the following
$$P'(h) = \ln\left(\frac{2.5}{1+h}\right) - \frac{h}{1+h}$$
   to show that there is a critical point for P between $h = 0$ and $h = 1.5$.

5. An organism grows from 4.0 kg to 60 kg in 14 yr. Suppose mass is a differentiable function of time.
   a. Why must the mass have been exactly 10 kg at some time?

b. Why must the rate of increase have been exactly 4.0 kg/yr at some time?

c. Draw a graph of mass against time where the mass is increasing, is equal to 10.0 kg at 13 yr, and has a growth rate of exactly 4.0 kg/yr after 1 yr.

d. Draw a graph of mass against time where the organism reaches 10.0 kg at 1 yr, and has a growth rate of exactly 4.0 kg/yr at 13 yr.

6. A lung follows the updating function $c_{t+1} = f(c_t)$. We know only that neither c_{t+1} nor c_t can exceed 1 mol/L. Use the Intermediate Value Theorem to show that this discrete-time dynamical system must have an equilibrium.

7. Draw functions with the following properties.
  a. A function with a global minimum and global maximum between the endpoints.
  b. A function with a global maximum at the left endpoint and global minimum between the endpoints.
  c. A differentiable function with a global maximum at the left endpoint, a global minimum at the right endpoint, and no critical points.
  d. A function with a global maximum at the left endpoint, a global minimum at the right endpoint, and at least one critical point.
  e. A function with a global minimum and global maximum between the endpoints, but no critical points.

8. Find the points guaranteed by the Mean Value Theorem and sketch the associated graph.
  a. The slope of the function $f(x) = x^2$ must match the slope of the secant connecting $x = 0$ and $x = 1$.
  b. The slope of the function $f(x) = x^2$ must match the slope of the secant connecting $x = 0$ and $x = 2$.
  c. The slope of the function $g(x) = \sqrt{x}$ must match the slope of the secant connecting $x = 0$ and $x = 1$.
  d. The slope of the function $g(x) = \sqrt{x}$ must match the slope of the secant connecting $x = 0$ and $x = 2$.

9. Draw the positions of cars with the following properties, and apply the Mean Value Theorem.
  a. A car starts at 60 mph and slows down to 0 mph. The average speed is 20 mph after 1 h.
  b. A car starts at 60 mph, steadily slows down to 20 mph, and then speeds up to 50 mph by the end of 1 h. The average speed over the whole time is 40 mph.
  c. A car drives at a constant speed of 60 mph for 1 h.

10. There is a clever proof of the Mean Value Theorem from Rolle's theorem. The idea is to tilt the function f so that it takes on the same values at the endpoints a and b. In particular, we apply Rolle's theorem to the function

$$g(x) = f(x) - (x - a)\frac{f(b) - f(a)}{b - a}$$

Try the following first in the special case $f(x) = x^2$, $a = 1$, and $b = 2$, and then in general.

  a. Show that $g(a) = g(b)$.
  b. Apply Rolle's theorem to g.
  c. Find the derivative of f at a point where $g'(x) = 0$.

11. We can apply our tools for reasoning about functions to a model of **frequency-dependent selection**. For example, the growth rate of a type of bacteria might depend on the proportion it makes up of the population. Recall the bacteria described in Equation 1.51,

$$p_{t+1} = \frac{sp_t}{sp_t + r(1 - p_t)}$$

Suppose that the per capita growth of type a is $s(1 - \alpha p_t)$ instead of s and that the per capita growth of type b is $r[1 - \beta(1 - p_t)]$ instead of r.

  a. Graph the per capita reproduction of a as a function of p_t. Explain your graph. What might have caused it?
  b. Do the same for the per capita reproduction of b.
  c. Write down the updating function.
  d. Show that $p_t = 0$ and $p_t = 1$ are equilibria.
  e. Find the derivative of the updating function at $p = 0$ and $p = 1$.
  f. Under what conditions are both derivatives greater than 1? (Think about the case $s = r$ if you get stuck.)
  g. Sketch the graph of the updating function under these conditions. What is the meaning of the new equilibrium? Is it stable?

12. The Marginal Value Theorem discussed in Section 3.3 (Maximizing Food Intake Rate) states that the best time t for a bee to leave a patch of flowers is the solution t of the equation

$$f'(t) = \frac{f(t)}{t + \tau}$$

where τ is the travel time to the next patch and $f(t)$ is the total amount of food gathered until time t. Suppose that $\tau = 1$ and $f(t) = 1 - e^{-t}$.

  a. Sketch the associated figure (as in Figure 3.34), and estimate the solution.
  b. Use the Intermediate Value Theorem to prove that there is a solution.

13. In the absorption equation

$$c_{t+1} = (1 - q)[1 - \alpha(c_t)]c_t + q\gamma$$

consider the absorption function

$$\alpha(c_t) = \frac{c_t^n}{1 + c_t^n}$$

This is a Hill function (Equation 2.5, Figure 2.56) (page 174). Set $q = 0.5$ and $\gamma = 1.5$.

  a. Write the updating function.
  b. Show that there is always an equilibrium at $c_t = 1$ no matter what n is equal to.

c. Find the derivative of the updating function at the equilibrium.
d. Show that the equilibrium is unstable if $n \geq 10$.
e. Sketch the absorption function and the updating function with $n \geq 10$.
f. Describe the dynamics of this lung when the equilibrium is unstable. Do they make sense?

14. **COMPUTER:** Consider again the model in Exercise 13, but do not set $q = 0.5$. Study the behavior of the model for $n = 1$, $n = 5$, and $n = 15$, and for values of q between 0 and 1. When is the equilibrium stable? Can you explain in biological terms why the equilibrium is stable when $n = 15$ and q is either near 0 or near 1?

3.5 Limits at Infinity

Reasoning about functions often requires computing the behavior of the function at the endpoints of its domain. When there is no natural upper bound to the domain, we must figure out what happens to the function as its argument gets very large. To do so, we generalize the limit to include **limits at infinity** and study the behaviors of the exponential, power, and logarithmic functions, finding which approach infinity, 0, or other values. Because many biological processes involve more than one basic function, we must be able to *compare* the functions. The key tool we use is a way to formalize whether one function approaches infinity or 0 *faster* or *slower* than another. The concept of limits at infinity can be used to study **limits of sequences**, and the output of discrete-time dynamical systems, as well as to characterize stable equilibria in a new way.

The Behavior of Functions at Infinity

In Section 3.4, we studied the function
$$\alpha(c) = 0.5(1 - e^{-0.5c})$$
which describes the fraction of a chemical absorbed by a lung during each breath. The total amount absorbed with each breath is $\alpha(c)cV$, the product of the fraction absorbed, the concentration c, and the volume V. What does the graph of the function $\alpha(c)$ look like? What does the graph of the total amount absorbed look like?

The amount of a chemical or resource used as a function of the amount available is important throughout biology. For consumers, such as predators, this relation is called the **functional response**. In chemical reactions, this relation is often described by **Michaelis-Menton** or **Monod** reaction kinetics.

When using the Intermediate Value Theorem and the Extreme Value Theorem, we began the reasoning process by computing the behavior at the endpoints. Often, there is no logical upper limit for the input of a function. For example, although we expect the concentration in a lung to be less than the ambient concentration when there is absorption, some other biological process, such as release or storage, could raise concentrations well above this level. To study absorption functions in general, we must study them over their whole range of possible inputs, from 0 to infinity.

A set of possible total absorption functions are given in Table 3.1; graphs are given in Figure 3.48. Each table entry describes the total amount of chemical absorbed as a function of the concentration c. The parameter α is a measure of efficiency; small values of α produce low absorption, and large values produce high absorption. The parameters k and β describe the shape of the function.

Table 3.1 Some Different Absorption Functions

Function of c	Description	Figure
αc	Linear absorption	3.48a
$\dfrac{\alpha c}{k+c}$	Saturated absorption	3.48b
$\dfrac{\alpha c^2}{k+c^2}$	Saturated absorption with threshold	3.48c
$\dfrac{\alpha c}{e^{\beta c}}$	Saturated absorption with overcompensation I	3.48d
$\dfrac{\alpha c}{k+c^2}$	Saturated absorption with overcompensation II	3.48e
$\alpha c(1+kc)$	Enhanced absorption	3.48f

In each case, absorption is 0 when $c = 0$. The behavior of the absorption function for large values of c describes absorption at high concentrations. How do we compute and describe the functions as their arguments become large?

Recall that infinity is the mathematician's abstraction of the scientist's idea of "very large." Because infinity is not a number we can substitute into equations, we can only *approach* it. The limit (Section 2.2) tells us how a function behaves as the argument approaches some finite value. We now extend this idea to find limits as the argument x becomes very large, or "approaches infinity."

To say that a function approaches a limit L as x approaches infinity means that the value gets closer and closer to L as x gets huge (Figure 3.49a). In other words,

Figure 3.48

Various absorption functions

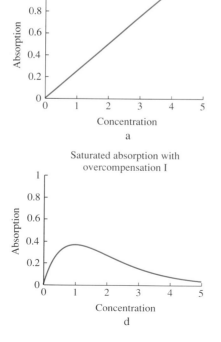
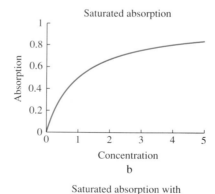
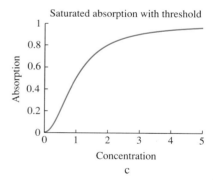
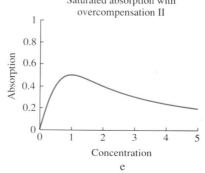
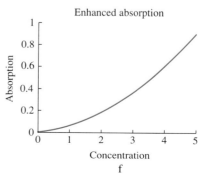

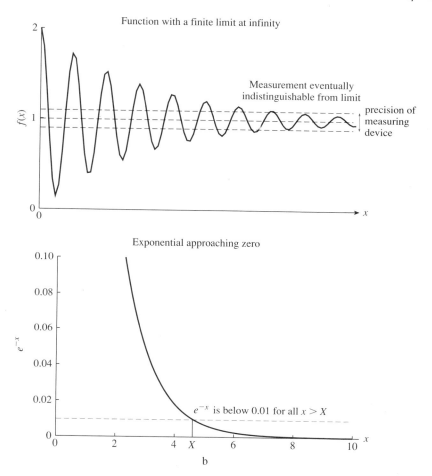

Figure 3.49
Functions with finite limits at infinity

a function f approaches the limit L as x approaches infinity if the measurement eventually becomes indistinguishable from the limit.

A function approaches infinity as x approaches infinity if the value gets larger and larger as x gets huge (Figure 3.50a). In other words, the function f approaches infinity as x approaches infinity if the output eventually overflows any given measuring device. Similarly, a function approaches negative infinity as x approaches infinity if the value gets smaller and smaller as x gets huge (Figure 3.50b).

Before studying the limits of the absorption functions shown in Figure 3.48a–f as c approaches infinity, we will learn how to compare the limits of power, logarithmic, and exponential functions.

Comparing Functions that Approach Infinity at Infinity

Many fundamental functions of biology approach infinity at infinity, including the natural logarithm, $\ln x$, the power function, x^n with positive n, and the exponential function, e^x.

To reason about more complicated functions, like those in Table 3.1, we must *compare* the behaviors of these basic functions. The graph of e^x increases very quickly, while that of $\ln(x)$ increases slowly. Is there a precise way in which the exponential function increases "faster" than the logarithmic function? What does

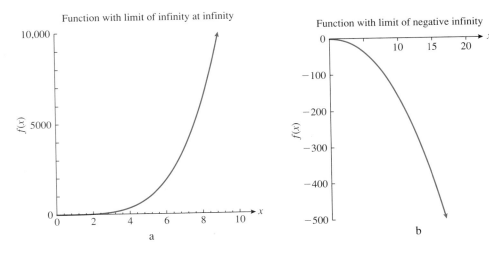

Figure 3.50
Functions with infinite limits at infinity

it mean to approach infinity "faster" or "slower"? These relations are summarized in the following definition.

■ **Definition 3.1** Suppose

$$\lim_{x \to \infty} f(x) = \infty$$
$$\lim_{x \to \infty} g(x) = \infty$$

1. The function $f(x)$ approaches infinity **faster** than $g(x)$ as x approaches infinity if

$$\lim_{x \to \infty} \frac{f(x)}{g(x)} = \infty$$

2. The function $f(x)$ approaches infinity **slower** than $g(x)$ if

$$\lim_{x \to \infty} \frac{f(x)}{g(x)} = 0$$

3. The functions $f(x)$ and $g(x)$ approach infinity at the **same rate** if

$$\lim_{x \to \infty} \frac{f(x)}{g(x)} = L$$

where L is any finite number other than 0. ■

When $f(x)$ approaches infinity faster, $f(x)$ gets farther and farther ahead (in the sense that the ratio $f(x)/g(x)$ becomes increasingly large) even though $g(x)$ increases to infinity (Figure 3.51a). When $f(x)$ approaches infinity slower, $f(x)$ falls farther and farther behind (in the sense that the ratio $f(x)/g(x)$ becomes increasingly small), even though $f(x)$ itself does increase to infinity (Figure 3.51b). When the two functions approach infinity at the same rate, the ratio $f(x)/g(x)$ becomes neither very large nor very small (Figure 3.51c).

"Faster," "slower," and "at the same rate" act like "greater than," "less than," and "equal to" for numbers. They provide a way to compare the "sizes" of functions.

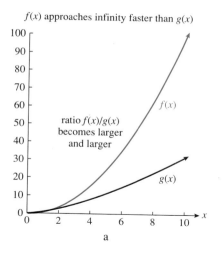
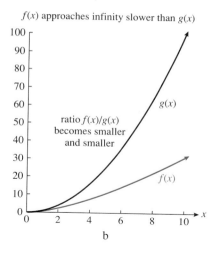
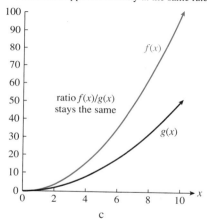

Figure 3.51
Comparing functions at infinity

Table 3.2 The Basic Functions in Increasing Order of Speed

Function	Comments
$a \ln(x)$	Goes to infinity slowly
ax^n with $n > 0$	Approaches infinity faster for larger n
$ae^{\beta x}$ with $\beta > 0$	Approaches infinity faster for larger β

The basic functions are shown in increasing order in Table 3.2 and Figure 3.52. The constant a in front of each function can be any positive number and does not change the order of the functions. Any power function, however small the power n, beats the logarithm. Any exponential function with a positive parameter β in the exponent beats any power function. To order the functions $0.1e^{2x}$, $4.5\ln(x)$, $23.2x^{0.5}$, $10.1e^{0.2x}$, $0.03x^4$ in increasing order, first spot functions of the three types: logarithmic, power, and exponential. There is only one logarithmic function, which is therefore the slowest. There are two power functions; $23.2x^{0.5}$ has the smaller power, and $0.03x^4$ has the larger power. There are two exponential functions;

$10.1e^{0.2x}$ has the smaller parameter (0.2) inside the exponent, and $0.1e^{2x}$ has the larger parameter (2) inside the exponent. The constants in front do not affect the ordering.

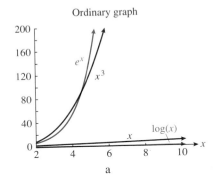
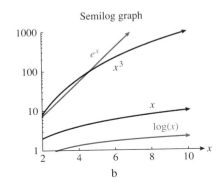

Figure 3.52
The behavior of the basic functions that approach infinity

There are two cautions regarding this method. First, comparing functions that are not logarithmic, power, or exponential functions requires different techniques (Section 3.6). Second, these results hold only for very large values of x. The power function x^2 does eventually grow faster than $1000x$ because it has a larger power (Figure 3.53). However, $x^2 < 1000x$ for $x < 1000$. If x cannot realistically take on values greater than 1000, the comparison in Table 3.2 is not relevant. When a comparison includes a large or small parameter (1000 in this case), we must first check whether the faster function becomes larger for biologically reasonable values of the argument.

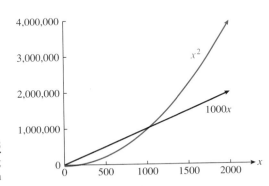

Figure 3.53
A faster function eventually overtaking a slower function

Functions Approaching 0 at Infinity

We use a similar approach to compare the rate at which functions approach a limit of 0.

■ **Definition 3.2** Suppose

$$\lim_{x \to \infty} f(x) = 0$$
$$\lim_{x \to \infty} g(x) = 0$$

1. The function $f(x)$ approaches 0 **faster** than $g(x)$ as x approaches infinity if
$$\lim_{x \to \infty} \frac{f(x)}{g(x)} = 0$$

2. The function $f(x)$ approaches 0 **slower** than $g(x)$ as x approaches infinity if
$$\lim_{x \to \infty} \frac{f(x)}{g(x)} = \infty$$

3. The function $f(x)$ approaches 0 at the **same rate** as $g(x)$ if
$$\lim_{x \to \infty} \frac{f(x)}{g(x)} = L$$

where L is any finite number other than 0.

Be careful not to confuse this with the definition for functions approaching infinity (Definition 3.1). The function $f(x)$ approaches 0 faster if it becomes *small* faster than $g(x)$ (Figure 3.54).

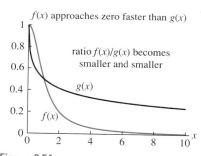
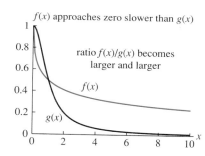
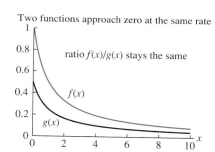

Figure 3.54
Comparing functions that approach a limit of 0 at infinity

The basic examples are reciprocals of the functions in Table 3.2 (Table 3.3 and Figure 3.55). If $f(x)$ approaches *infinity* quickly, the reciprocal $1/f(x)$ approaches 0 quickly. Again, the positive constant c does not change the ordering of the functions.

Table 3.3 The Basic Functions Approaching 0

Function	Comments
ax^{-n} with $n > 0$	Approaches 0 faster for larger n
$ae^{-\beta x}$ with $\beta > 0$	Approaches 0 faster for larger β
$ae^{-\beta x^2}$ with $\beta > 0$	Approaches 0 really fast

To order the functions $0.1e^{-2x}$, $23.2x^{-0.5}$, $10.1e^{-0.2x}$, $0.03x^{-4}$ from the one that approaches 0 fastest to the one that approaches slowest, first identify the functions as exponential and power functions. The fastest is the exponential function, which has the larger parameter, $0.1e^{-2x}$, followed by the exponential function with the smaller parameter, $10.1e^{-0.2x}$, then the power function $0.03x^{-4}$, which has the larger power, and finally the power function $23.2x^{-0.5}$, which has the smaller power.

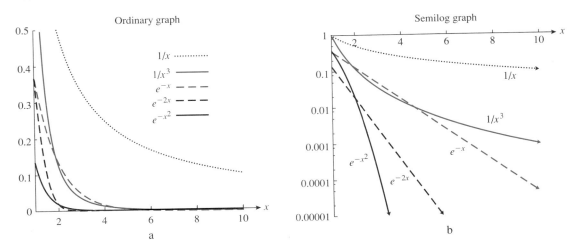

Figure 3.55
The behavior of the basic functions that approach 0

Application to Absorption Functions

How can we use these facts about the basic functions to understand the absorption functions in Table 3.1? The results are summarized in Table 3.4. We compare the numerator and denominator of the absorption function as functions of the concentration c. If the numerator grows faster than the denominator, absorption grows without bound as c gets large. If the numerator and denominator grow at the same rate, absorption approaches a constant as c gets large (Figure 3.48b and c). If the denominator grows faster than the numerator, absorption approaches 0 as c gets large (Figure 3.48d and e).

Table 3.4 Analyzing Absorption Functions

Number	Numerator	Denominator	Behavior at Infinity
3.48a	Linear	None	Approaches infinity
3.48b	Linear	Linear	Approaches constant
3.48c	Quadratic	Quadratic	Approaches constant
3.48d	Linear	Exponential	Approaches zero
3.48e	Linear	Quadratic	Approaches zero

With linear absorption (Figure 3.48a), absorption grows without bound (there is no denominator to balance the numerator). Because there are almost always limits to absorption, the saturated absorption functions (Figure 3.48b and c) provide more reasonable models. The saturated absorption form shown in Figure 3.48b, a ratio of linear functions, is among the most important in biology and is known as the **Michaelis-Menton** or **Monod** equation. In Section 3.6 we will deduce the difference in shape between the forms shown in Figures 3.48b and 3.48c and formalize the calculation of the behavior at infinity with the **method of leading behavior**.

In the forms shown in Figures 3.48d and 3.48e, the denominator grows faster than the numerator (both exponential and quadratic functions grow faster than linear functions). These functions begin at 0, increase to a maximum, and eventually decrease again to 0. This behavior is called **overcompensation** because absorption decreases when the concentration is too large.

The enhanced absorption function

$$A(c) = \alpha c(1 + kc)$$

is not a ratio of functions that we can compare. It is a quadratic polynomial that approaches infinity faster than a linear function.

Limits of Sequences

We use the same idea to define the limit of a solution of a discrete-time dynamical system. The solution of the basic bacterial updating function $b_{t+1} = rb_t$ is

$$b_t = b_0 r^t = b_0 e^{\ln(r)t} \tag{3.6}$$

This differs from an ordinary function in that populations are defined only at integer values of t. The solution is a list of numbers known as a **sequence**.

To find the limit, define the **associated function** $b(t)$ as the exponential function

$$b(t) = b_0 e^{\ln(r)t}$$

defined for all values of t. This function fills in the gaps in the sequence (Figure 3.56). If the associated function has a limit, whether 0, infinity, or some other value, the sequence will share that limit. In this case, the associated function is an exponential function. If $r > 1$, the parameter in the exponent is positive and the limit of the function, and therefore of the sequence, is infinity. If $r < 1$, the parameter in the exponent is negative and the limit of the function, and therefore of the sequence, is 0. If $r = 1$, the function has the constant value b_0 and the function and sequence share the limit b_0.

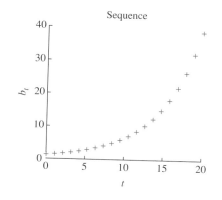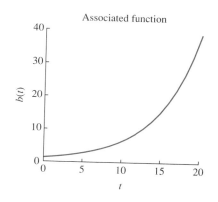

Figure 3.56
A sequence and its associated function

This definition has an important connection with the idea of **stable equilibrium**. If the sequence of points that represents a solution approaches a particular value as a limit, that limit is a stable equilibrium (Figure 3.57).

If the associated function has a limit, the sequence shares that limit. The sequence, however, may have a limit even when the associated function does not. Consider the sequence

$$a_t = \sin(2\pi t)$$

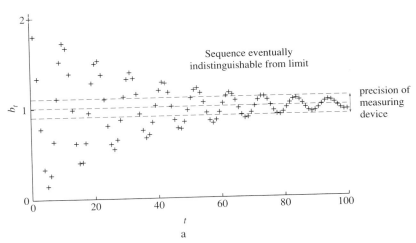

Figure 3.57

The limits correspond to a stable equilibrium of a sequence at infinity

The associated function $a(t) = \sin(2\pi t)$ has no limit because it oscillates forever. The sequence, on the other hand, takes on only the value 0 and has the limit 0 (Figure 3.58).

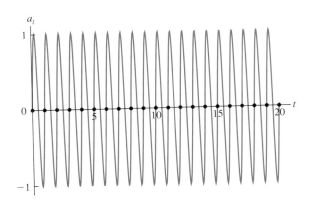

Figure 3.58

A sequence with a limit different from its associated function

---

SUMMARY In order to reason about the behavior of functions with large inputs, we defined the **limit** of a function as its argument approaches infinity. As with ordinary limits, the limit formalizes the idea that the value gets closer and closer to some particular number (if the limit is finite) or larger than any given number (if the limit is infinite). One function approaches infinity *faster* than another if the limit of the ratio is infinity. Exponential functions approach infinity faster than power functions, which in turn approach infinity faster than the logarithmic functions. Conversely, one function approaches 0 faster than another if the limit of the ratio is 0. Exponential functions with negative parameters approach 0 faster than power functions with negative powers. Using limits at infinity, we analyzed the limits of **sequences**, lists of numbers generated as solutions of discrete-time dynamical systems, by studying the **associated function**.

3.5 EXERCISES

1. Find the following limits.
   a. $\lim_{x \to \infty} x^{-0.25}$
   b. $\lim_{x \to \infty} \ln(x)$
   c. $\lim_{x \to \infty} 0.8^x$
   d. $\lim_{x \to \infty} 1 - e^{-4x}$
   e. $\lim_{x \to \infty} x^4$
   f. $\lim_{x \to \infty} 1.2^x$
   g. $\lim_{x \to \infty} e^{-x}$
   h. $\lim_{x \to \infty} x^{0.25}$

2. Put the following functions in order from the one approaching infinity fastest as x approaches infinity to the one approaching infinity slowest. Explain which rule you used to compare each pair. Compute the value of each function at $x = 1$, $x = 10$, and $x = 100$. How do these compare with the order of the functions in the limit?
   a. x^2
   b. e^{2x}
   c. $1000x$
   d. $x^{3.5}$
   e. $5e^x$
   f. $0.1x^{10}$
   g. $30 \ln x$
   h. $10x^{0.1}$

3. Put the following functions in order from the one approaching 0 fastest as x approaches infinity to the one approaching 0 slowest.
   a. x^{-2}
   b. e^{-2x}
   c. $1000/x$
   d. $x^{-3.5}$
   e. $5e^{-x}$
   f. x^{-10}
   g. $30/\ln x$
   h. $x^{-0.1}$

4. Find the derivatives of the absorption functions shown in Figure 3.48. Compute the value at $c = 0$ and the limit of the derivative as c approaches infinity. Are your results consistent with the figure?

5. The absorption functions of Figure 3.48d and e start at 0 and have limits of 0 as c approaches infinity.
   a. Deduce that each must have a maximum for c between 0 and infinity.
   b. Find the maximum using the derivative.
   c. Deduce that the function must take on a value equal to half the maximum at least twice.

6. The following are possible absorption functions. What happens to each as c approaches infinity?
   a. $\dfrac{\beta c^2}{1 + e^c}$
   b. $\dfrac{\alpha c}{\ln(1 + c)}$
   c. $\dfrac{\gamma(e^c - 1)}{e^{2c}}$
   d. $\dfrac{c^2}{1 + 10c}$

7. When $r > 1$, the bacterial population described by Equation 3.6 increases to infinity. For the given initial conditions and value of r, find how long it takes before the population exceeds the given threshold.
   a. $b_0 = 10^8$, $r = 1.1$, threshold $= 10^{10}$
   b. $b_0 = 10^8$, $r = 1.1$, threshold $= 10^{12}$
   c. $b_0 = 10^8$, $r = 1.5$, threshold $= 10^{10}$
   d. $b_0 = 10^8$, $r = 2.0$, threshold $= 10^{10}$

8. In the polymerase chain reaction (PCR) used to amplify DNA, some sequences of DNA produced are too long and others are the right length. Denote the number of overly long pieces after t generations of the process by l_t and the number of pieces of the right length by r_t. The dynamics follow approximately
$$l_{t+1} = l_t + 2$$
$$r_{t+1} = 2r_t$$
because two new overly long pieces are produced in each step, while the number of good pieces doubles.
   a. Supposing that $l_0 = 0$ and $r_0 = 2$, find expressions for l_t and r_t.
   b. Compute the fraction of pieces of the right length after 1, 5, 10, and 20 generations of the process.
   c. Why are the ratios getting smaller?
   d. How long would you wait to make sure that less than 1 in a million pieces are too long? (This can't be solved exactly; just substitute some numbers.)

9. Consider the model for medication (Section 1.5)
$$M_{t+1} = 0.5M_t + 1.0$$
starting with $M_0 = 3$.
   a. Find the equilibrium.
   b. We found the solution to be $M_t = 2.0 + 0.5^t \cdot 3.0$. Find the limit.
   c. How long will it take to be within 1% of the equilibrium?
   d. What are two ways to show that this equilibrium is stable?

10. As mentioned in the text, the amount of food a predator eats as a function of prey density is called the **functional response**. Functional response is broken into three categories:
    - Type I: linear.
    - Type II: increasing, concave down, finite limit.
    - Type III: increasing with finite limit, concave up for small prey densities, concave down for large prey densities.

    a. Sketch pictures of these three types. Which of the absorption functions do they resemble?
    b. What is the optimal prey density for a predator in each case?
    c. Suppose that the amount of food actually gathered decreases linearly with the number of prey (the prey join together and fight back). Write the equation for the optimal prey density. (Let p be the number of prey and $F(p)$ be the functional response, and subtract a constant multiple cp. The constant c represents how effectively the prey can fight.)
    d. Write down simple formulas for the type I and type II cases. Try to compute the optimal prey density. Consider both small and large values of c.
    e. Draw a picture illustrating the optimal prey density in the type II and type III cases.

11. Try to think of
    a. a biological mechanism leading to saturation.
    b. a biological mechanism leading to overcompensation.
    c. a biological mechanism leading to enhanced absorption.
12. **COMPUTER:** Use your computer to find out how large x must be before the faster function finally overtakes the slower function.

   a. $e^{0.1x}$ catches up with x^3.
   b. $0.1e^x$ catches up with x^3.
   c. $0.1e^{0.1x}$ catches up with x^3.
   d. $0.1x$ catches up with $\ln(x)$.
   e. $x^{0.1}$ catches up with $\ln(x)$.

3.6 Leading Behavior and L'Hôpital's Rule

Our previous methods, including finding limits at infinity, have given general information about the behavior of functions. In particular, we learned how to compare the ratios of different functions by describing which increased to infinity or decreased to 0 faster. We now study a much larger class of functions, **sums** of functions and the ratios of sums. The technique, called the **method of leading behavior**, consists of stripping off the largest piece of the function. By determining the leading behavior of a function at both infinity and 0, we can deduce a great deal about the *shape* of the function by using the technique of **matched leading behaviors**. When the method of leading behavior fails, **L'Hôpital's rule** provides an alternative way to compare the behavior of functions.

Leading Behavior of Functions at Infinity

Suppose we wish to describe how a function that is the sum of several functions, such as

$$f(x) = 5e^{2x} + 34e^x + 45x^5 + 56 \ln x + 10$$

behaves for large values of x. We might suspect that this function will be dominated by the fastest term, the one that becomes much bigger than all the others in the sense of Definition 3.1. The **method of leading behavior** is based on this idea.

■ **Definition 3.3** The leading behavior of a function at infinity is based on the term that approaches infinity fastest as the argument approaches infinity. We write f_∞ to represent the leading behavior of the function f at infinity. ■

The fastest term in the function $f(x)$ above is $5e^{2x}$; this term grows to infinity most quickly because it is the exponential term with the largest parameter in the exponent. Therefore, we write

$$f_\infty(x) = 5e^{2x}$$

In what sense does the leading behavior describe a function? The graph of the leading behavior looks indistinguishable from the graph of the original function when x is large, even though the two functions are quite different for small x (Figure 3.59). Mathematically, if we divide a function by its leading behavior, the limit is 1 (the function and its leading behavior approach infinity at the same rate). For example,

$$\lim_{x \to \infty} \frac{f(x)}{f_\infty(x)} = \lim_{x \to \infty} \frac{5e^{2x} + 34e^x + 45x^5 + 56 \ln x + 10}{5e^{2x}}$$

3.6 Leading Behavior and L'Hôpital's Rule

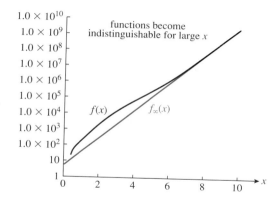

Figure 3.59
A comparison of a function and its leading behavior in a semilog plot

$$= \lim_{x \to \infty} 1 + \frac{34e^x}{5e^{2x}} + \frac{45x^5}{5e^{2x}} + \frac{56 \ln x}{5e^{2x}} + \frac{10}{5e^{2x}}$$
$$= 1 + 0 + 0 + 0 + 0 = 1$$

All terms after the first approach 0 because $5e^{2x}$ approaches infinity the fastest.

Complicated functions can be compared by comparing their leading behaviors. To find whether the function $f(x)$ increases to infinity faster than the function

$$g(x) = 23e^{2.5x} + 3e^{2x} + 2x^6$$

we need only compare the leading behavior of $f(x)$ with that of $g(x)$. The leading behavior of the function $g(x)$ is the first term (the term with the largest parameter in the exponent), so

$$g_\infty(x) = 23e^{2.5x}$$

To see whether $f(x)$ approaches infinity faster than $g(x)$, we compute the limit of the ratio of the functions as x approaches infinity. Because each function is well represented by its leading behavior, we have

$$\lim_{x \to \infty} \frac{f(x)}{g(x)} = \lim_{x \to \infty} \frac{f_\infty(x)}{g_\infty(x)}$$
$$= \lim_{x \to \infty} \frac{5e^{2x}}{23e^{2.5x}}$$
$$= 0$$

because the exponential in the denominator has a larger parameter. The function $g(x)$ approaches infinity faster than $f(x)$.

However, leading behavior gives more information than the limit. The same calculation tells us that for large x,

$$\frac{f(x)}{g(x)} \approx \frac{f_\infty(x)}{g_\infty(x)}$$
$$= \frac{5e^{2x}}{23e^{2.5x}}$$
$$= \frac{5}{23} e^{-0.5x}$$

This term not only tells us that $g(x)$ approaches infinity faster than $f(x)$, but also gives us an idea how much faster. The ratio of the functions behaves much like the ratio of the leading behaviors for large x (Figure 3.60).

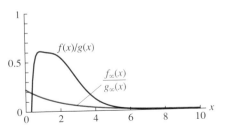

Figure 3.60

The ratio of two functions compared with the ratio of their leading behaviors

The same definition of leading behavior works for sums of functions that approach 0. Remember that the largest term is the term approaching 0 the *most slowly*, in the sense of Definition 3.2. For example, the leading behavior of the function

$$h(x) = 5e^{-2x} + 34e^{-x} + 45x^{-5}$$

is the power term $45x^{-5}$ because this term approaches 0 the most slowly and is therefore the largest for large x,

$$h_\infty(x) = 45x^{-5}$$

(Figure 3.61a). Similarly, the function

$$H(x) = 5e^{-2x} + 34e^{-x} + 45x^{-5} + 5$$

has leading behavior

$$H_\infty(x) = 5$$

because the constant 5 does not approach 0 and is therefore the largest term (Figure 3.61b).

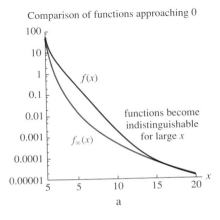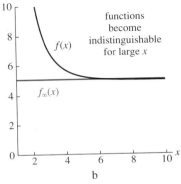

Figure 3.61

Functions that approach finite values and their leading behavior

We can apply the idea of leading behavior to supplement our reasoning about the absorption functions in Table 3.1. The absorption function with saturation

(Figure 3.48b) has the formula

$$A(c) = \frac{\alpha c}{k + c}$$

where c is the concentration and α and k are constant parameters. Both numerator and denominator increase to infinity. The leading behavior of the denominator is the larger of the two terms k (a constant) and c, which increases to infinity. Replacing the denominator by the leading behavior, we have

$$A(c) \approx \frac{\alpha c}{c} = \alpha$$

for large values of c (Figure 3.62a). In the figure, both k and α have been set to 1. Absorption saturates at α.

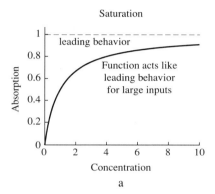
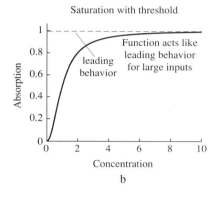
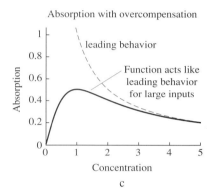

Figure 3.62
Comparing absorption functions with approximations

Similarly, the absorption function with saturation and a threshold (Figure 3.48c) has the formula

$$A(c) = \frac{\alpha c^2}{k + c^2}$$

The numerator is a single term, which is its own leading behavior. The leading behavior of the denominator is c^2 because it increases faster than the constant k. For large values of c,

$$A(c) = \frac{\alpha c^2}{k + c^2} \approx \frac{\alpha c^2}{c^2} = \alpha$$

This function also saturates at α (Figure 3.62b). In the figure, the parameters k and α have been set to 1.

The method of leading behavior tells us nothing new about the absorption function

$$A(c) = \frac{\alpha c}{e^{\beta c}}$$

We know that the limit as c approaches infinity is 0 because an exponential function grows faster than a linear function. Because both the numerator and denominator have only a single term, we cannot simplify further.

The alternative absorption function with overcompensation (Equation 3.1e),

$$A(c) = \frac{\alpha c}{k + c^2}$$

can be simplified with the method of leading behavior. In this case, the leading behavior of the denominator is again the quadratic term c^2, so that

$$A(c) = \frac{\alpha c}{k + c^2} \approx \frac{\alpha c}{c^2} = \frac{\alpha}{c}$$

Absorption decreases to 0, and does so like the function α/c (Figure 3.62c).

Leading Behavior of Functions at 0 and the Method of Matched Leading Behaviors

In each comparison of an absorption function with its leading behavior, we have done well for large values of the concentration and poorly for small values. To complete our analysis of the absorption functions, we need a better sense of what is happening near 0. Once again, the method of leading behavior can be used to identify which small terms can be ignored.

First, we define the leading behavior of a function at 0.

■ **Definition 3.4**   The leading behavior of a function at 0 is the term that is largest as the argument approaches 0. We write f_0 to represent the leading behavior of the function f at 0.

The idea of largest, the same at 0 and at infinity, is formalized in the following definition.

■ **Definition 3.5**
1. The function $f(x)$ is larger than $g(x)$ as x approaches 0 if

$$\lim_{x \to 0} \frac{f(x)}{g(x)} = \infty$$

2. The function $f(x)$ is smaller than $g(x)$ if

$$\lim_{x \to 0} \frac{f(x)}{g(x)} = 0$$

This definition really includes two cases: The functions $f(x)$ and $g(x)$ approach infinity as x approaches 0, and $f(x)$ and $g(x)$ approach 0 as x approaches 0. In the first case, all power functions of the form

$$f(x) = x^{-n}$$

with positive values of n approach infinity as x approaches 0 (from the right). The larger the value of n, the faster the function approaches infinity and thus the *larger* it is. For example, the sum

$$f(x) = x^{-1} + 5x^{-5}$$

has leading behavior

$$f_0(x) = 5x^{-5}$$

because $5x^{-5}$ approaches infinity faster than x^{-1}. The limit of the ratio of the terms is

$$\lim_{x \to 0} \frac{5x^{-5}}{x^{-1}} = \lim_{x \to 0} 5x^{-4} = \infty$$

Power functions with large negative powers do everything quickly, approaching infinity quickly at 0 and 0 quickly at infinity and are thus *largest* for small values of x and *smallest* for large values of x (Figure 3.63).

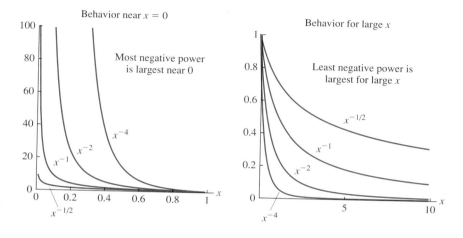

Figure 3.63
Power functions with negative powers

Similarly, if two functions $f(x)$ and $g(x)$ approach 0 as x approaches 0, the *larger* function approaches 0 more *slowly*. All power functions of the form

$$f(x) = x^n$$

for positive values of n approach 0 as x approaches 0. As before, power functions with large powers do everything quickly. The larger the power of n the *faster* the function approaches 0 for x near 0 and the faster it approaches infinity as x approaches infinity (Figure 3.64). These functions are small for small x and large for large x.

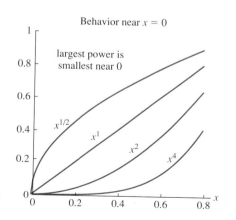
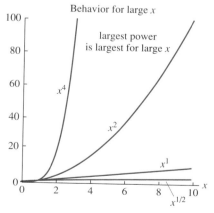

Figure 3.64
Power functions with positive powers

Consider the function

$$F(x) = 4x + x^3$$

The leading behavior near $x = 0$ is given by the first term, x, because it has the smallest power, or

$$F_0(x) = 4x$$

The leading behavior for large x, on the other hand, is given by the second term, $4x^3$, because it has the larger power, or

$$F_\infty(x) = x^3$$

(Figure 3.65).

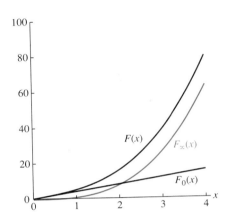

Figure 3.65
A function compared with the leading behavior at 0 and infinity

The idea of studying a function for both large and small values of x can be formalized as the **method of matched leading behaviors**.

■ **Algorithm 3.3** (The method of matched leading behaviors)

1. Find the leading behavior at 0 and at infinity.
2. Sketch graphs of each leading behavior.
3. Connect the graphs with a smooth curve. ■

For example, consider the absorption function with saturation (Equation 3.1b),

$$\frac{\alpha c}{k + c}$$

Near 0, we first simplify by finding the leading behavior of the denominator as

$$(k + c)_0 = k$$

because the constant value k is much *larger* than c for small c. Therefore,

$$\left(\frac{\alpha c}{k + c}\right)_0 = \frac{\alpha c}{k}$$

For large values of c, the denominator becomes

$$(k + c)_\infty = c$$

because the constant value k is much *smaller* than c for large c. Therefore,

$$\left(\frac{\alpha c}{k + c}\right)_\infty = \frac{\alpha c}{c} = \alpha$$

3.6 Leading Behavior and L'Hôpital's Rule

These simple linear functions can easily be connected with a smooth curve (Figure 3.66a).

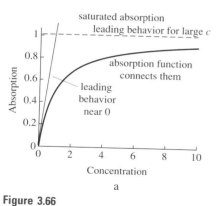
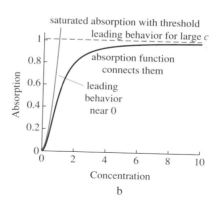
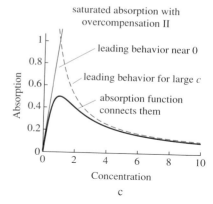

Figure 3.66

The method of matched leading behaviors

For absorption with saturation and a threshold

$$\frac{\alpha c^2}{k + c^2}$$

(Figure 3.48c), we can again use the method of matched leading behaviors to plot an accurate graph. For small values of c, the denominator has leading behavior

$$(k + c^2)_0 = k$$

because the constant value k is much larger than c^2 for small c. Therefore,

$$\left(\frac{\alpha c^2}{k + c^2}\right)_0 = \frac{\alpha c^2}{k}$$

This part of the curve looks like a parabola. For large values of c, the denominator becomes

$$(k + c^2)_\infty = c^2$$

because the constant value k is much smaller than c^2 for large c. Therefore,

$$\left(\frac{\alpha c^2}{k + c^2}\right)_\infty = \frac{\alpha c^2}{c^2} = \alpha$$

Connecting these portions of the graph with a smooth curve shows that the graph is concave up for small c and concave down for large c (Figure 3.66b).

Finally, for saturated absorption with overcompensation II (Figure 3.48e),

$$\frac{\alpha c}{k + c^2}$$

the denominator is the same, so

$$(k + c^2)_0 = k$$
$$(k + c^2)_\infty = c^2$$

Therefore,

$$\left(\frac{\alpha c}{k + c^2}\right)_0 = \frac{\alpha c}{k}$$

$$\left(\frac{\alpha c}{k+c^2}\right)_\infty = \frac{\alpha c}{c^2} = \frac{\alpha}{c}$$

This curve begins by increasing like a line with slope α/k but eventually begins decreasing like α/c. As a consequence, this curve must have a maximum (Figure 3.66c).

Sometimes the method of leading behavior does not give any useful information. For example, consider the function

$$A(c) = \frac{\alpha c}{1 + e^c}$$

For c near 0, both terms in the denominator have a limit of 1, and neither can be thrown out. All we can say is

$$A(c)_0 = \frac{\alpha c}{1 + e^c} = A(c)$$

However, if c is large, the exponential term is dominant, so

$$A(c)_\infty = \frac{\alpha c}{e^c}$$

When two terms have the same size, neither can be thrown out. To approximate functions in these circumstances, the tangent line approximation (and the more general approach of Taylor series we see in Section 3.7) can be more effective.

L'Hôpital's Rule

Suppose we wish to find

$$\lim_{x \to 0} \frac{e^{\alpha x} - 1}{x} \tag{3.7}$$

Both numerator and denominator approach 0. We cannot simplify the numerator with the method of leading behavior because both terms approach the same limit of 1 and neither is faster than the other.

There is a general and powerful rule for dealing with cases like this. Instead of comparing the functions, we compare their derivatives. This is known as **L'Hôpital's rule**. First, we define indeterminate forms:

■ **Definition 3.6**   The limit of the ratio

$$\lim_{x \to a} \frac{f(x)}{g(x)}$$

(where a could be ∞) is called an **indeterminate form** if

$$\lim_{x \to a} f(x) = \lim_{x \to a} g(x) = 0$$

or

$$\lim_{x \to a} f(x) = \lim_{x \to a} g(x) = \infty$$ ■

If we tried to substitute $x = a$ into an indeterminate form, we would be committing a mathematical felony. When we cannot recognize the functions involved or use the method of leading behavior, these limits can often be computed with the following rule.

■ **THEOREM 3.6** (L'Hôpital's Rule)

Suppose that f and g are differentiable functions and that

$$\lim_{x \to a} \frac{f(x)}{g(x)}$$

is an indeterminate form. If

$$\lim_{x \to a} \frac{f'(x)}{g'(x)} = L$$

then

$$\lim_{x \to a} \frac{f(x)}{g(x)} = L \qquad ■$$

Using this theorem we can prove that exponentials grow faster than power functions and that power functions grow faster than logarithmic functions as x approaches infinity (Table 3.2, page 261). L'Hôpital's rule says that

$$\lim_{x \to \infty} \frac{e^x}{x} = \lim_{x \to \infty} \frac{de^x/dx}{dx/dx} \qquad \text{take derivative of numerator and denominator because this is an indeterminate form}$$

$$= \lim_{x \to \infty} \frac{e^x}{1} \qquad \text{compute derivatives}$$

$$= \infty \qquad \text{the exponential function has a limit of infinity}$$

To compare the behavior of x and $\ln(x)$ at infinity using L'Hôpital's rule, we compute

$$\lim_{x \to \infty} \frac{x}{\ln x} = \lim_{x \to \infty} \frac{dx/dx}{d \ln x/dx}$$

$$= \lim_{x \to \infty} \frac{1}{1/x}$$

$$= \lim_{x \to \infty} x$$

$$= \infty$$

Some comparisons require repeated application of L'Hôpital's rule. To show that e^x increases faster than x^3, we must take the derivative three times:

$$\lim_{x \to \infty} \frac{e^x}{x^3} = \lim_{x \to \infty} \frac{e^x}{3x^2} \qquad \text{indeterminate form, take derivatives}$$

$$= \lim_{x \to \infty} \frac{e^x}{6x} \qquad \text{still indeterminate, take derivatives}$$

$$= \lim_{x \to \infty} \frac{e^x}{6} = \infty \qquad \text{limit of the exponential function is infinity}$$

L'Hôpital's rule is indispensable when the method of leading behaviors fails. We have seen that

$$\lim_{x \to 0} \frac{e^{\alpha x} - 1}{x}$$

(Equation 3.7) is indeterminate. We apply L'Hôpital's rule, finding that

$$\lim_{x \to 0} \frac{e^{\alpha x} - 1}{x} = \lim_{x \to 0} \frac{\alpha e^{\alpha x}}{1} = \alpha \qquad (3.8)$$

Why does L'Hôpital's rule work? Recall that the derivative can be used to approximate functions with the tangent line. The idea behind L'Hôpital's rule is to replace $f(x)$ by $\hat{f}(x)$ and $g(x)$ by $\hat{g}(x)$ and prove that

$$\lim_{x \to a} \frac{f(x)}{g(x)} = \lim_{x \to a} \frac{\hat{f}(x)}{\hat{g}(x)}$$

Remember that

$$\hat{f}(x) = f(a) + f'(a)(x - a)$$
$$\hat{g}(x) = g(a) + g'(a)(x - a)$$

If this is an indeterminate form of the 0/0 type, it must be true that $f(a) = g(a) = 0$. Therefore,

$$\frac{\hat{f}(x)}{\hat{g}(x)} = \frac{f'(a)}{g'(a)}$$

if $x \neq a$.

In the case

$$\lim_{x \to 0} \frac{e^{\alpha x} - 1}{x}$$

the function $f(x) = e^{\alpha x} - 1$ and $g(x) = x$. Then $f'(x) = \alpha e^{\alpha x}$ and $f'(0) = \alpha$. The tangent line approximation of f near $x = 0$ (Equation 2.1) is

$$\hat{f}(x) = f(0) + f'(0)x = 0 + \alpha x = \alpha x$$

Therefore

$$\lim_{x \to 0} \frac{e^{\alpha x} - 1}{x} = \lim_{x \to 0} \frac{\alpha x}{x} = \alpha$$

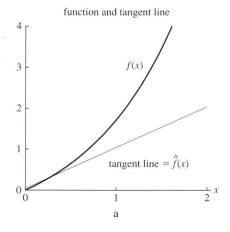
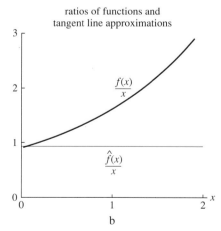

Figure 3.67
The tangent line as the leading behavior of a function

In a way, this method recalls the idea of leading behavior. The tangent line $\hat{f}$ ignores the curvy parts of f as being smaller (Figure 3.67). Better approximations include some of the curve. This idea, called **Taylor series**, will be covered in Section 3.7.

SUMMARY

We have learned two ways to compute the behavior of complicated functions. The **method of leading behavior** is a way to examine sums of functions and determine which piece increases to infinity *fastest* or to 0 *slowest*. For large values of the input, this piece is called the **leading behavior at infinity** and can be used to approximate the behavior of the function. For small values of the input, the largest piece is called the **leading behavior at 0** and can be used to approximate the behavior of the function. By graphing the leading behavior of functions at both 0 and infinity, we can use the **method of matched leading behaviors** to sketch an accurate graph. In other cases, we can evaluate **indeterminate forms** with **L'Hôpital's rule**, which says that a ratio of functions that both approach 0 or both approach infinity has the same limit as the ratio of their derivatives.

3.6 EXERCISES

1. Find the leading behavior of the following functions at 0 and infinity.
   a. $f(x) = 1 + x$
   b. $g(y) = y + y^3$
   c. $h(z) = z + e^z$
   d. $m(z) = 100z + 30z^2 + \dfrac{1}{z}$
   e. $F(a) = e^{5a} + 100a^{50} - 40e^{6a}$
   f. $G(c) = e^{-4c} + \dfrac{5}{c^2} + \dfrac{3}{c^5} + 10e^{-3c}$

2. Use the method of matched leading behaviors to sketch graphs of the following Hill functions and their variants.
   a. $h_3(x) = \dfrac{x^3}{1 + x^3}$
   b. $g_3(x) = \dfrac{x^3}{10 + x^3}$
   c. $h_{10}(x) = \dfrac{x^{10}}{1 + x^{10}}$
   d. $g_{10}(x) = \dfrac{x^{10}}{0.1 + x^{10}}$

3. Consider the absorption functions
   $$A_1(c) = \dfrac{2c^2}{1 + c}$$
   $$A_2(c) = \dfrac{c^2}{1 + 2c}$$
   $$A_3(c) = \dfrac{1 + c + c^2}{1 + c}$$
   $$A_4(c) = \dfrac{1 + c}{1 + c + c^2}$$
   $$A_5(c) = \dfrac{3c}{1 + \ln(1 + c)}$$
   $$A_6(c) = \dfrac{e^c + 1}{e^{2c} + 1}$$

   For each function, find the leading behavior of the numerator, of the denominator, and of the whole function at both 0 and infinity. Find the limit of each function at 0 and infinity, and use the method of matched leading behaviors to sketch a graph.

4. The absorption functions graphed in Figures 3.48b and c are of the form
   $$\alpha \dfrac{r(c)}{k + r(c)} \quad (3.9)$$
   for some function $r(c)$ that increases from 0 at 0 and to infinity at infinity.
   a. What is $r(c)$ in Figure 3.48b? In Figure 3.48c?
   b. Show, by finding the derivative, that the function in Equation 3.9 is increasing.
   c. Use L'Hôpital's rule to find the limit as $c \to \infty$ if $r(c) \to \infty$.
   d. Use leading behavior to describe absorption near $c = 0$.

5. Find the limits of the following functions as c approaches 0 and as c approaches infinity. Use L'Hôpital's rule, if appropriate.
   $$B_1(c) = \dfrac{2c^2}{1 + c}$$
   $$B_2(c) = \dfrac{c^2}{e^c - 1}$$
   $$B_3(c) = \dfrac{c + c^2}{\ln(1 + c)}$$
   $$B_4(c) = \dfrac{1 + c}{1 + c + c^2}$$
   $$B_5(c) = \dfrac{\ln(1 + c/2)}{\ln(1 + c)}$$
   $$B_6(c) = \dfrac{e^c - 1}{e^{2c} - 1}$$

6. Write linear approximations to the functions in the numerator and denominator of the following functions, and show that the result of applying L'Hôpital's rule matches that of comparing the linear approximations.
   a. $f(x) = \dfrac{\ln(1 + x)}{e^{2x} - 1}$ at $x = 0$

b. $f(x) = \dfrac{\ln x}{x^2 - 1}$ at $x = 1$

c. $f(x) = \dfrac{1 - 2x + x^2}{1 - 3x + 2x^2}$ at $x = 1$

7. The expression given in Equation 3.7 (page 276) is the definition of the derivative at 0 of the function $h(x) = e^{\alpha x}$.
   a. Verify this.
   b. Compute the derivative directly to find the limit.
   c. Recopy the definition of the derivative in general; show it is an indeterminate form; and apply L'Hôpital's rule.

8. Recall Equation 1.45, which describes two competing strains of bacteria, type a and type b:

   $$a_{t+1} = s a_t$$
   $$b_{t+1} = r b_t$$

   a. Supposing that $a_0 = 10^4$ bacteria per milliliter and $b_0 = 10^6$ bacteria per milliliter, find expressions for a_t and b_t.
   b. Suppose that $s > r > 1$. Which population increases faster?
   c. Compute the fraction of type a bacteria after 1, 5, 10, 20, and 50 generations of the process if $s = 1.8$ and $r = 1.4$.
   d. Use the method of leading behavior or L'Hôpital's rule to show why the fraction of type a bacteria increases to 1.
   e. Now suppose that $1 > s > r$. Which population declines faster?
   f. Compute the fraction of type a bacteria after 1, 5, 10, 20, and 50 generations of the process if $s = 0.9$ and $r = 0.7$.
   g. Use the method of leading behavior or L'Hôpital's rule to show why the fraction of type a bacteria increases to 1.

9. In Exercise 5 in Section 3.3, we computed the optimal amount of time t a bee should spend on a flower as a function of the travel time τ when resources collected follow the function

   $$F(t) = \dfrac{t}{0.5 + t}$$

   a. Find the rate of food gain as a function of τ when the bee uses the optimal strategy.
   b. Find the limit as τ approaches 0.
   c. Explain this result in words.

10. **COMPUTER:** Consider the following functions.

    $$f(c) = \dfrac{c^2}{1 + c^2}$$

    $$g(c) = \dfrac{c}{1 + c^2}$$

    Find the leading behavior of each at 0 and infinity. Suppose we approximated each by a function defined in pieces:

    $$\check{f}(c) = \begin{cases} f_0(c) & \text{if } f_0(c) < f_\infty(c) \\ f_\infty(c) & \text{if } f_0(c) > f_\infty(c) \end{cases}$$

    and similarly for $\check{g}(c)$. Plot this approximation in each case. Find and plot the ratios

    $$\dfrac{\check{f}(c)}{f(c)} \quad \text{and} \quad \dfrac{\check{g}(c)}{g(c)}$$

    When is the approximation best? When is it worst?

11. **COMPUTER:** Consider the function

    $$f(x) = \dfrac{1 + x + x^2 + x^3 + x^4}{5 + 4x + 3x^2 + 2x^3 + x^4}$$

    Find the leading behavior for large x. Next find a function that retains both the largest and second-largest terms from the numerator and denominator. How much better is this new approximation? How much improvement do you get by adding more and more terms?

3.7 Approximating Functions with Lines and Polynomials

The method of leading behavior provides a way to *approximate* complicated functions with simpler functions. In this section, we extend the related idea of the **tangent line approximation** in several ways. First, we compare the tangent line approximation with the **secant line approximation**, showing that the tangent line is the *best* linear approximation to a curve near the point of tangency, but that the secant line can be more useful over larger ranges.

The tangent line matches the value and derivative of a function at a point. More accurate approximations can be found by also matching the second, third, and higher derivatives. If we use *polynomials* to match these higher derivatives, the resulting approximation is called a **Taylor polynomial**.

The Tangent and Secant Lines

Suppose we wish to approximate the exponential function e^x near 0 with a line, perhaps in order to compare a complicated dynamical system with its linear approximation. The general formula for the tangent line, or the tangent line approximation to the function f at base point a, is

$$\hat{f}(x) = f'(a)(x - a) + f(a)$$

(Figure 3.68). The graph of the function $\hat{f}(x)$ is the tangent line to $f(x)$ at a, which matches both the value and the slope at the point of tangency.

Figure 3.68
Approximating a function with the tangent line

The tangent line to the exponential function

$$g(x) = e^x$$

at $x = 0$ matches the value $g(0) = e^0 = 1$ and the slope $g'(0) = e^0 = 1$. Hence

$$\hat{g}(x) = g(a) + g'(0)(x - 0) = 1 + x$$

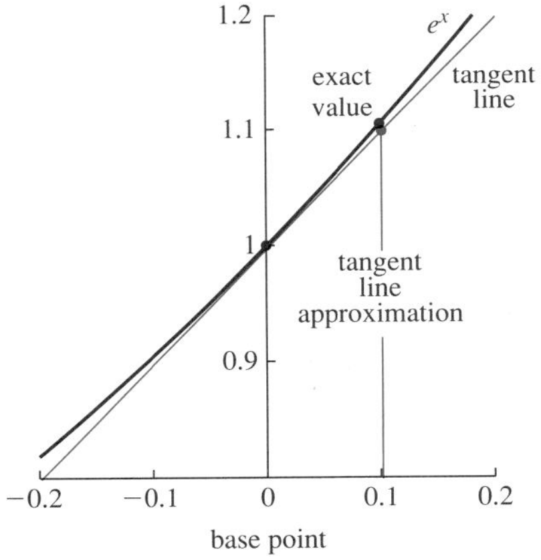

(Figure 3.69). The graph of the tangent line hugs the curve near the point of tangency and provides a good way to approximate values. For example,

$$\hat{g}(0.1) = 1.1$$

and the exact value is 1.105. Before the age of computers, this sort of approximation was indispensable. As we will soon see when we study Newton's method for solving equations (Section 3.8) and Euler's method for solving differential equations (Section 4.1), the tangent line approximation remains necessary for complicated problems.

Similarly, to estimate $\ln(0.9)$ without a calculator, note that the input value 0.9 is near 1.0. Because $\ln(1.0) = 0$, the answer is close to 0. To do better, we match both the value and the slope of the function $\ln(x)$ at $x = 1$. In this case,

$$\frac{d \ln(x)}{dx} = \frac{1}{x}$$

so the slope at $x = 1$ is 1. The tangent line is

$$\hat{\ln}(x) = 0 + 1(x - 1) = x - 1$$

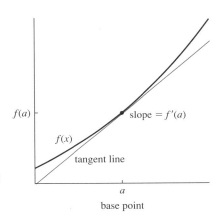

Figure 3.69
Approximating $e^{0.1}$ with the tangent line

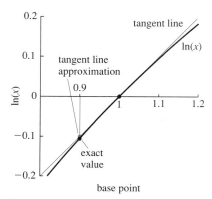

Figure 3.70

Approximating ln(0.9) with the tangent line

(Figure 3.70). Substituting $x = 0.9$, we find an approximate value of -0.1, which is close to the exact value of -0.105. $\hat{\ln}(x)$ represents the approximation of $\ln(x)$.

Recall the bacterial population growing according to

$$b(t) = 2.0^t$$

Can we find a good linear approximation to this function for times between 0 and 1? One method is to use the tangent line at $t = 0$. First, we find the derivative:

$$\begin{aligned}\frac{db}{dt} &= \frac{d}{dt} 2.0^t \\ &= \frac{d}{dt} e^{\ln(2.0)t} \\ &= \ln(2.0) e^{\ln(2.0)t} \\ &= 0.693 e^{\ln(2.0)t}\end{aligned}$$

using the rule for finding derivatives of the general exponential function (Section 2.9). The tangent line at $t = 0$ is therefore

$$\hat{b}_0(t) = b(0) + b'(0)(t - 0) = 1.0 + 0.693t$$

where the subscript 0 indicates the base point. If we are interested in approximating values near $t = 0$, the tangent line is accurate (Figure 3.71a). If instead we are interested in approximating values near $t = 1$, this tangent line is quite inaccurate, but the tangent line

$$\hat{b}_1(t) = b(1) + b'(1)(t - 1) = 2.0 + 1.386(t - 1)$$

with base point $t = 1$ is accurate (Figure 3.71b). Over the whole interval from $t = 0$ to $t = 1$, the secant line is reasonably good everywhere (Figure 3.71c). The secant line has slope

$$\text{slope of secant} = \frac{\Delta b}{\Delta t} = \frac{b(1.0) - b(0.0)}{1.0 - 0.0} = 1.0$$

and equation

$$b_s(t) = b(0) + 1.0(t - 0) = 1 + t$$

Figure 3.71

Two tangent lines and a secant line as approximations

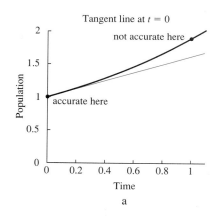

a

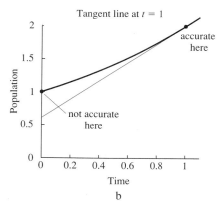

b

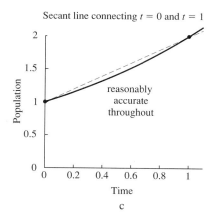

c

How can we quantify the accuracy of these alternative approximations? Results near both endpoints and at the middle of the interval are given in the following table:

t	$b(t)$	$b_0(t)$	$b_1(t)$	$b_s(t)$
0.01	1.00695	1.00693	0.62757	1.01
0.10	1.07177	1.06931	0.75244	1.10
0.50	1.41421	1.34657	1.30685	1.50
0.90	1.86607	1.62383	1.86137	1.90
0.99	1.98618	1.68622	1.98614	1.99

Each tangent line is an excellent approximation near its point of tangency. The secant line is always fairly close. This is one of the two primary strengths of the secant line, which is also called **linear interpolation**. The other is that the secant can be directly estimated from data, even when we do not know the underlying equation.

Nonetheless, the tangent line is the *best* possible approximation near the point of tangency. No other line is closer. Suppose we compare the tangent and secant lines near the base point 0. To quantify how close the estimates are to the exact values, define errors e_0 and e_s for the two lines as

$$e_0 = \hat{b}_0(t) - b(t)$$
$$e_s = \hat{b}_s(t) - b(t)$$

To check how small these are for t near 0, we divide the errors e_0 and e_s by t:

t	$b_0(t)$	$b_s(t)$	e_0	e_s	$\dfrac{e_0}{t}$	$\dfrac{e_s}{t}$
0.20	1.13863	1.20	-0.01007	0.05130	-0.05034	0.25651
0.10	1.06931	1.10	-0.00246	0.02822	-0.02459	0.28226
0.01	1.00693	1.01	-0.0000241	0.00304	-0.00241	0.30444

Both errors e_0 and e_s get smaller as t gets smaller. With the tangent line approximation, the *relative* error e_0/t also gets smaller as t gets smaller. With the secant line approximation, the relative error e_s/t remains roughly constant (Figure 3.72).

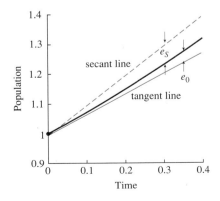

Figure 3.72
The errors associated with tangent and secant line approximations

Quadratic Approximation

The tangent and secant lines provide ways to approximate curves with lines. We can do better by approximating curves with curves. Suppose, for example, that we wish to approximate the exponential function near 0 with a quadratic function. Just as we can match the value of the function and the first derivative with the tangent line, we can match the function, the first derivative, *and* the second derivative with a quadratic. For the exponential function $g(x) = e^x$, $g'(x) = e^x$ and $g''(x) = e^x$. Therefore,

$$g(0) = 1$$
$$g'(0) = 1$$
$$g''(0) = 1$$

This *does not mean* that the quadratic approximation is $\hat{g}(x) = 1 + x + x^2$; we must be a bit more careful. Suppose the quadratic approximation is

$$\hat{g}(x) = c_0 + c_1 x + c_2 x^2$$

Then

$$\hat{g}'(x) = c_1 + 2c_2 x$$
$$\hat{g}''(x) = 2c_2$$

At $x = 0$, where we wish the values to match, we have

$$\hat{g}(0) = c_0$$
$$\hat{g}'(0) = c_1$$
$$\hat{g}''(0) = 2c_2$$

To match, we need each of these to equal the derivative of g itself, or

$$\hat{g}(0) = c_0 = g(0) = 1$$
$$\hat{g}'(0) = c_1 = g'(0) = 1$$
$$\hat{g}''(0) = 2c_2 = g''(0) = 1$$

Solving, we find $c_0 = 1$, $c_1 = 1$, and $c_2 = 1/2$. The approximating quadratic

$$\hat{g}(x) = 1 + x + \frac{x^2}{2}$$

lies extremely close to the curve (Figure 3.73).

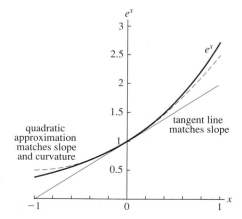

Figure 3.73
The linear and quadratic approximations of e^x

3.7 Approximating Functions with Lines and Polynomials

To find the quadratic approximation to a function $f(x)$ at the base point a in general, we again match the value, the derivative, and the second derivative at the point a. To make the calculation easier, we write the quadratic in a version of point-slope form:

$$\hat{f}(x) = c_0 + c_1(x - a) + c_2(x - a)^2$$

We want to choose the values c_0, c_1, and c_2 so that

$$\hat{f}(a) = f(a)$$
$$\hat{f}'(a) = f'(a)$$
$$\hat{f}''(a) = f''(a)$$

Then

$$\hat{f}'(x) = c_1 + 2c_2(x - a)$$
$$\hat{f}''(x) = 2c_2$$

and

$$\hat{f}(a) = c_0 + c_1(a - a) + c_2(a - a)^2 = c_0$$
$$\hat{f}'(a) = c_1 + 2c_2(a - a) = c_1$$
$$\hat{f}''(a) = 2c_2$$

Therefore,

$$c_0 = f(a)$$
$$c_1 = f'(a)$$
$$2c_2 = f''(a)$$

The approximating quadratic is therefore

$$\hat{f}(x) = f(a) + f'(a)(x - a) + \frac{f''(a)}{2}(x - a)^2$$

The first two terms in this approximate function exactly match the tangent line. The last term is an additional correction. Recall from the method of leading behavior that $(x - a)^2$ is smaller than $x - a$ when $x - a$ itself is small. This new term has little effect near a. If the second derivative is equal to 0, this term vanishes and the best approximating quadratic is the tangent line itself.

We found earlier that the function

$$h(x) = xe^{-x}$$

has derivatives

$$h'(x) = (1 - x)e^{-x}$$
$$h''(x) = (x - 2)e^{-x}$$

This function has a critical point at $x = 1$ and a point of inflection at $x = 2$. Using the formula for the approximating quadratic at three points, we find that

$$\hat{h}_0(x) = h(0) + h'(0)x + \frac{h''(0)}{2}x^2 = x - x^2$$

$$\hat{h}_1(x) = h(1) + h'(1)(x - 1) + \frac{h''(1)}{2}(x - 1)^2 = \frac{1}{e} - \frac{1}{e}(x - 1)^2$$

$$\hat{h}_2(x) = h(2) + h'(2)(x-2) + \frac{h''(2)}{2}(x-2)^2 = \frac{1}{e^2} - \frac{1}{e^2}(x-2)$$

(Figure 3.74). At the critical point, the approximating quadratic has no $x - 1$ term because $h'(1) = 0$. At the point of inflection, the quadratic term drops out, leaving us with the tangent line.

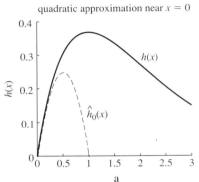
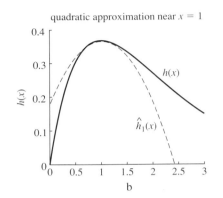
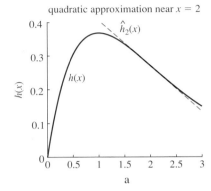

Figure 3.74
Three quadratic approximations to $h(x) = xe^{-x}$

Taylor Polynomials

The idea of matching derivatives can be extended to the third, fourth, and higher derivatives. With each added derivative, the approximation becomes more accurate but requires a polynomial of higher *degree* (largest power). The derivation of the following formula is the same as the derivation of the quadratic approximation. The approximating polynomial is called a **Taylor polynomial** of degree n.

■ **Definition 3.7** (The Taylor Polynomial)

Suppose the first n derivatives of the function f are defined at $x = a$. Then the Taylor polynomial of degree n matching the values of the first n derivatives is

$$P_n(x) = f(a) + f'(a)(x-a) + \frac{f''(a)}{2}(x-a)^2 + \cdots + \frac{f^{(i)}(a)}{i!}(x-a)^i$$
$$+ \cdots + \frac{f^{(n)}(a)}{n!}(x-a)^n$$
■

We used two new notations in this definition. First, the notation

$$f^{(i)}(x)$$

indicates the ith derivative of f. For example, we could write

$$f^{(2)}(x) = f''(x)$$

for the second derivative. Second, the terms with exclamation points are called **factorials**. The value of $i!$ is the product of i and all numbers smaller than i, or

$$i! = i \cdot (i-1) \cdot (i-2) \cdots \cdots 3 \cdot 2 \cdot 1$$

For example,

$$2! = 2 \cdot 1 = 2$$
$$3! = 3 \cdot 2 \cdot 1 = 6$$
$$4! = 4 \cdot 3 \cdot 2 \cdot 1 = 24$$
$$5! = 5 \cdot 4 \cdot 3 \cdot 2 \cdot 1 = 120$$

The values of factorials increase very quickly, which means that later terms in a Taylor polynomial become ever smaller.

These polynomials are easiest to understand from examples. Consider again the exponential function $g(x) = e^x$. Using our new notation, we have

$$g^{(i)}(x) = g(x)$$

because each derivative of the exponential function is equal to the function itself. For $a = 0$ then, the Taylor polynomial of degree 4 is

$$P_n(x) = 1 + x + \frac{1}{2}x^2 + \frac{1}{6}x^3 + \frac{1}{24}x^4$$

Figure 3.75
Three Taylor polynomial approximations to $g(x) = e^x$

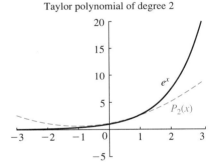
Taylor polynomial of degree 2

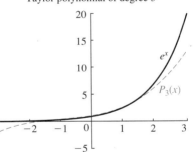
Taylor polynomial of degree 3

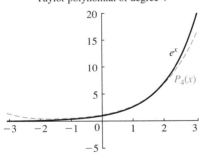
Taylor polynomial of degree 4

SUMMARY

Using lines to approximate curves is one of the central ideas in calculus. We have compared the tangent line with three other approximations. The secant line, or **linear interpolation**, has the joint virtues of using actual data and remaining fairly accurate over a broad domain. The tangent line is the **best linear approximation** near the base point. To do even better, we can use a quadratic polynomial to match both the first and second derivatives of the original function at the base point. This idea can be expanded to the **Taylor polynomial**, which is a polynomial of degree n that matches the first n derivatives of the function.

3.7 EXERCISES

1. Approximate the following values using the tangent line approximation. Make sure to identify the base point a and to write the equation for the tangent line. No calculators allowed!

a. 2.02^3
b. 3.03^2
c. $\ln(1.02)$
d. $\sqrt{4.01}$
e. $e^{\ln(2)+0.01}$
f. $\sin(0.02)$
g. $\cos(-0.02)$

2. For each function, find the tangent line approximation of the value and compare with the true value. Say which approximations are too high and which are too low. From graphs of the functions, try to explain what it is about the graph that does this.
   a. $e^{0.1}$ and $e^{-0.1}$
   b. $\ln(1.1)$ and $\ln(0.9)$
   c. 1.1^2 and 0.9^2
   d. $\sqrt{1.1}$ and $\sqrt{0.9}$

3. Consider the following table, which gives mass as a function of age.

Age (days)	Mass (g)
0.5	0.125
1.0	1.000
1.5	3.375
2.0	8.000
2.5	15.625
3.0	27.000

   The data follow the equation $M(a) = a^3$.
   a. Estimate $M(1.25)$ by using the tangent line approximation to M at $a = 1.0$ days.
   b. Estimate $M(1.25)$ by using the tangent line approximation to M at $a = 1.5$ days.
   c. Estimate $M(1.25)$ by using the secant line approximation to M connecting $a = 1.0$ and $a = 1.5$. Which approximation is closest to the exact answer?
   d. Use linear interpolation and the tangent line approximation to estimate the weight at age 1.45 days. Which is closer? Why?

4. Consider a declining population that follows the formula
   $$b(t) = 10^6 \cdot 0.5^t$$
   a. Approximate the population at times $t = 0.1$, $t = 0.5$, and $t = 0.9$ using a tangent line with base point $t = 0$. Graph the tangent line.
   b. Approximate the population at times $t = 0.1$, $t = 0.5$, and $t = 0.9$ using a tangent line with base point $t = 1$. Graph the tangent line.
   c. Approximate the population at times $t = 0.1$, $t = 0.5$, and $t = 0.9$ using a secant line connecting times $t = 0$ and $t = 1$. Graph the secant line.
   d. Which method is best for what?

5. Suppose a population follows the differential equation
   $$\frac{db}{dt} = b$$

We have seen that $b(t) = e^t$ is a solution if $b(0) = 1$. We can also estimate the answer with the tangent line approximation.
   a. Approximate the population at $t = 0.1$ using the tangent line at $t = 0$. Graph the tangent line.
   b. Approximate the population at $t = 0.2$ using the tangent line at $t = 0.1$. Instead of using the exact value of $b(0.1)$, use the approximate value. Graph this approximate tangent line.
   c. Continue in this way, step by step, until you reach $t = 1$. Compare your answer with the exact answer of e.

6. Use the tangent line approximation to evaluate the following in two ways. First, find the tangent line to the whole function using the chain rule. Second, break the calculation into two pieces by writing the function as a composition, approximate the inner function with its tangent line, and substitute this value in the tangent line of the outer function. Do your answers match?
   a. $(1 + 3 \cdot 1.01)^2$
   b. $\ln(\sqrt{0.98})$
   c. $e^{\sin(0.02)}$
   d. $\sin\{\ln[(1+0.1)^3]\}$

7. Find the quadratic approximations for the values in Exercise 1. How close are the answers to the exact value?

8. Find the third-order Taylor polynomials for the functions in Exercise 1.

9. Consider the absorption functions
   $$A_1(c) = \frac{2c^2}{1+c}$$
   $$A_2(c) = \frac{c^2}{1+2c}$$
   $$A_3(c) = \frac{1+c+c^2}{1+c}$$
   $$A_4(c) = \frac{1+c}{1+c+c^2}$$
   $$A_5(c) = \frac{3c}{1+\ln(1+c)}$$
   $$A_6(c) = \frac{e^c+1}{e^{2c}+1}$$
   from Exercise 3 in Section 3.6. Find the tangent lines and quadratic approximations at $c = 0$, and compare with the result using the method of leading behavior.

10. The tangent line approximation can be combined with the method of matched leading behaviors to sketch accurate graphs. Find the tangent line at $x = 0$ and the leading behavior for large x for the following, and use them to plot an accurate graph. How do the last two curves differ from the first two?
    a. $f(x) = 1 - e^{-x}$
    b. $g(x) = 1 - e^{-5x}$
    c. $F(x) = \dfrac{x}{1+x}$
    d. $G(x) = \dfrac{x}{0.2+x}$

11. **COMPUTER:** Find the Taylor polynomials for e^x, $\cos(x)$, and $\sin(x)$ with base point $x = 0$ up to degree 10. Can you see the pattern? Graph them on domains around 0 that get larger and larger. What happens to the approximation for values of x far from 0?

12. **COMPUTER:** Find the Taylor polynomials for $e^{-1/x}$ with base point $x = 0$ up to degree 10. Can you see the pattern? Graph the function on the domain $-1 \leq x \leq 1$. What happens to the approximation for values of x far from 0? Do the Taylor polynomials make sense?

13. **COMPUTER:** A simple equation that is impossible to solve algebraically is
$$e^x = x + 2$$

   a. Graph the two sides and convince yourself there is a solution.
   b. Replace e^x with its tangent line at $x = 0$, and try to solve for the point where the tangent line is equal to $x + 2$. This is an approximate solution. What goes wrong in this case?
   c. Replace e^x with its quadratic approximation at $x = 0$, and solve for the point where it is equal to $x + 2$.
   d. Replace e^x with its tangent line at $x = 1$ and solve.
   e. Replace e^x with its quadratic approximation at $x = 1$ and solve.
   f. How close are these solutions to the exact answer?

3.8 Newton's Method

With today's powerful calculators and computers, it might seem unnecessary to approximate a function with a tangent line. On a calculator, exponentiation is no harder than multiplying or adding. Although computing specific functional values is easy, solving equations for specific values can be difficult. We have seen how to use the Intermediate Value Theorem to show that an equation has a solution. When we cannot solve the equation with algebraic methods, **Newton's method** can be implemented on a computer to find the exact value. The method replaces the original equation with the tangent line approximation and derives a discrete-time dynamical system that converges to the solution with remarkable speed.

Finding the Equilibrium of the Lung Model with Absorption

Suppose a lung is following the equation
$$c_{t+1} = (1 - q)[1 - \alpha(c_t)]c_t + q\gamma$$

where
$$\alpha(c_t) = 0.5(1 - e^{-0.5c_t})$$

(Equation 3.4). In this updating function c_t represents the concentration, q the fraction of air exchanged, γ the concentration of chemical in the ambient air, and $\alpha(c_t)$ the fraction of chemical absorbed as a function of the chemical concentration in the lung. In Section 3.4, we used the Intermediate Value Theorem to show that this function has an equilibrium between 0 and γ. What if we wish to compute the exact value of the equilibrium with a particular set of parameter values?

If we set $q = 0.5$ and $\gamma = 5.0$, the updating function is
$$c_{t+1} = 0.5\left[1 - 0.5\left(1 - e^{-0.5c_t}\right)\right]c_t + 2.5$$

It is impossible to set $c_{t+1} = c_t = c^*$ to solve the equation

$$c^* = 0.5 \left[1 - 0.5\left(1 - e^{-0.5c^*}\right)\right] c^* + 2.5$$

algebraically for the equilibrium value because equations involving both polynomial and exponential functions cannot be solved except in unusual circumstances. How can we use a computer or calculator to find the value?

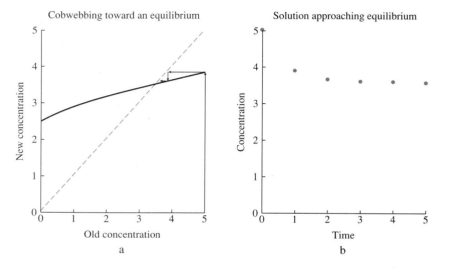

Figure 3.76
The iterative method of solving an equation

It looks as though the equilibrium we seek is stable (Figure 3.76a). A solution will approach the equilibrium (Figure 3.76b). This seems like a good way to get the computer to find the answer. The results of solving the discrete-time dynamical system starting from $c_0 = 5.0$ are given in the following table.

Iteration	Concentration	Distance from equilibrium	Factor by which distance decreased
0	5.0000000000	1.4654361738	—
1	3.8526062482	0.3180424220	0.2170291874
2	3.6034690548	0.0689052286	0.2166542067
3	3.5495215521	0.0149577259	0.2170767914
4	3.5378126599	0.0032488337	0.2172010455
5	3.5352695692	0.0007057430	0.2172296671
6	3.5347171389	0.0001533127	0.2172359626
7	3.5345971314	0.0000333052	0.2172373338
8	3.5345710613	0.0000072351	0.2172376315
9	3.5345653979	0.0000015717	0.2172376945
10	3.5345641676	0.0000003414	0.2172377007

The columns give the concentrations, the difference from the true solution (which we do not really know yet), and the ratio of the distance in the current step to the distance in the previous step. For example, the factor in the second row is

$$\frac{0.3180424220}{1.4654361738} = 0.2170291874$$

After 10 steps, the first five digits have stopped changing, meaning that we have found the equilibrium to about five decimal places of accuracy. The factor in the final column is approximately equal to the slope of the tangent at the unknown equilibrium (Exercise 13). Each step gets us almost five times closer to the answer, and we have a highly accurate answer in only 10 steps.

Suppose now we wish to find the optimal behavior for a bee (as in Section 3.4), but that the food intake follows the function

$$F(t) = 1 - e^{-t}$$

We can see from Figure 3.77 that there is an optimal behavior when the travel time τ is equal to 1. To compute it, we must solve

$$F'(t) = \frac{F(t)}{t + \tau}$$

$$e^{-t} = \frac{1 - e^{-t}}{t + 1}$$

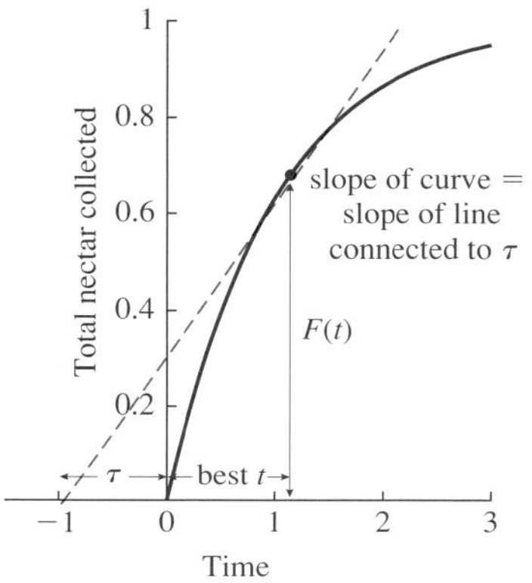

We do not know how to solve an equation like this. We can rearrange, multiplying both sides by $t + 1$ and isolating e^t:

$(t + 1)e^{-t} = 1 - e^{-t}$	multiply both sides by $t + 1$
$t + 1 = e^t - 1$	multiply both sides by e^t
$t + 2 = e^t$	isolate e^t

This equation looks simple. At $t = 0$, the right-hand side is smaller, and at $t = 2$, the right-hand side is larger. The Intermediate Value Theorem (Theorem 3.2) guarantees that these two functions cross (Figure 3.78). But this equation is not written in the form of a discrete-time dynamical system. How can we compute the solution?

Figure 3.77
Applying the Marginal Value Theorem when $F(t) = 1 - e^{-t}$

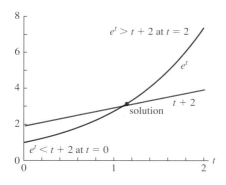

Figure 3.78
Finding the solution of an impossible equation

Newton's Method

Newton's method is a method for solving equations numerically. When it works, the method is incredibly fast, *doubling* the number of digits of accuracy with each step.

The equilibrium equation is $g(c) = c$. If we define a new function f by

$$f(c) = g(c) - c$$

the equilibrium of $g(c)$ is a point where $f(c) = 0$. This is the form of equation used by Newton's method (and is the same starting point used to apply the Intermediate Value Theorem).

Suppose that in general we want to solve the equation

$$f(x) = 0$$

If we have some idea that a solution is near the value x_0, we can replace the original equation by the approximate equation

$$\hat{f}(x) = 0$$

where $\hat{f}(x)$ is the tangent line approximation at x_0 (Figure 3.79). The equation for the tangent line $\hat{f}(x)$ is

$$\hat{f}(x) = f(x_0) + f'(x_0)(x - x_0)$$

(Equation 2.1), so our approximate equation $\hat{f}(x) = 0$ is

$$f(x_0) + f'(x_0)(x - x_0) = 0$$

As long as we can compute the derivative of $f(x)$, we can solve this equation exactly for x. Graphically, the solution of the original equation is the point where the curve intersects the horizontal axis. The solution of the approximate equation is the point where the tangent line intersects the horizontal axis.

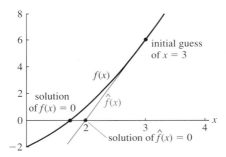

Figure 3.79
Newton's method: the first step

Suppose we wish to solve the equation

$$f(x) = x^2 - 3 = 0$$

The solution of this equation is $\sqrt{3}$, a numerical value we may not know and one that used to be hard to compute. The first step in Newton's method is to take a guess. We might begin with a rather poor guess of $x_0 = 3$. Next, we need to find the tangent line approximation by using the derivative:

$$f'(x) = 2x$$

The tangent line approximation is

$$\hat{f}(x) = f(3) + f'(3)(x - 3)$$
$$= (3^2 - 3) + 2 \cdot 3(x - 3)$$
$$= 6 + 6(x - 3)$$

The approximate equation is

$$\hat{f}(x) = 6 + 6(x - 3) = 0$$

3.8 Newton's Method

which has solution

$$6 + 6(x - 3) = 0$$
$$6 + 6x - 18 = 0$$
$$6x = 12$$
$$x = 2$$

This is closer to the right answer because 2^2 is much closer to 3.

We have replaced the difficult equation $f(x) = 0$ with the linear equation $\hat{f}(x) = 0$. It is possible to solve this linear equation for x (as long as $f'(x_0) \neq 0$). We find

$f(x) = 0$	original equation
$f(x_0) + f'(x_0)(x - x_0) = 0$	substitute tangent line approximation
$f'(x_0)(x - x_0) = -f(x_0)$	begin solving for x
$x - x_0 = \dfrac{-f(x_0)}{f'(x_0)}$	divide by $f'(x_0)$
$x = x_0 - \dfrac{f(x_0)}{f'(x_0)}$	solve for x

This value x is the point where the tangent line intersects the horizontal axis. Our hope is that this point is closer to the unknown exact answer than the original guess was. If so, we can use x as a new guess, x_1, with formula

$$x_1 = x_0 - \frac{f(x_0)}{f'(x_0)} \quad (3.10)$$

Starting from the point x_1, we can follow the same steps to find the tangent line and solve for the intersection with the horizontal axis. The new guess, x_2, will have the same formula but with x_1 substituted for x_0, or

$$x_2 = x_1 - \frac{f(x_1)}{f'(x_1)}$$

In the example with $f(x) = x^2 - 3$, our first guess was $x_0 = 3$. By finding the tangent line and solving the equation, we found $x_1 = 2$. Alternatively, we could use Equation 3.10, with $f(3) = 6$ and $f'(3) = 6$, to find

$$x_1 = 3 - \frac{f(3)}{f'(3)}$$
$$= 3 - \frac{6}{6} = 2$$

Starting from the new guess, $x_1 = 2$, we get

$$x_2 = x_1 - \frac{f(x_1)}{f'(x_1)}$$
$$= 2 - \frac{f(2)}{f'(2)}$$
$$= 2 - \frac{2^2 - 3}{2 \cdot 2}$$
$$= 2 - \frac{1}{4} = 1.75$$

(Figure 3.80). This is much closer to the exact answer because $1.75^2 = 3.0625$.

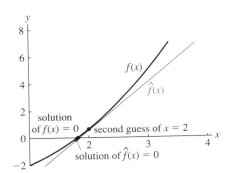

Figure 3.80
Newton's method: the second step

At each step, we applied the **Newton's method updating function**

$$x_{t+1} = x_t - \frac{f(x_t)}{f'(x_t)} \tag{3.11}$$

The algorithm for using Newton's method is to pick a good guess for x_0 and then use the Newton's method updating function until the answer converges. This unusual formula is derived by replacing the function with the tangent line at each step. Why does it approach a point where $f(x_t) = 0$? If $f(x_t) = 0$, then

$$x_{t+1} = x_t - \frac{f(x_t)}{f'(x_t)} = x_t$$

as long as $f'(x_t) = 0$. We have transformed the problem of solving the equation $f(x) = 0$ into the problem of finding the equilibrium of an updating function. We can solve this problem by repeatedly applying the updating function to some initial guess and hoping that it converges.

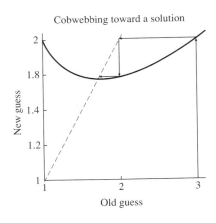
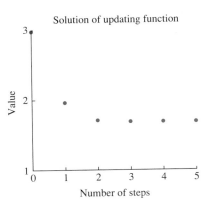

Figure 3.81
The Newton's method updating function

Iteration	Value
0	3.0000000000
1	2.0000000000
2	1.7500000000
3	1.7321428571
4	1.7320508100

With the function $f(x) = x^2 - 3$, the Newton's method updating function is

$$x_{t+1} = x_t - \frac{f(x_t)}{f'(x_t)}$$
$$= x_t - \frac{x_t^2 - 3}{2x_t}$$

The results of cobwebbing this equation and finding the solution are shown in Figure 3.81 and the table. Newton's method converges to the answer very quickly and is correct to eight decimal places after only four steps. And the calculation involved nothing more complicated than multiplication, subtraction, and division.

How do we apply Newton's method to finding the optimal behavior of a bee, the solution of

$$e^t = t + 2$$

First, we write the equation in the form $f(x) = 0$:

$$f(x) = e^x - x - 2 = 0$$

Next, we find $f'(x) = e^x - 1$. Therefore, the Newton's method updating function is

$$x_{t+1} = x_t - \frac{e^{x_t} - x_t - 2}{e^{x_t} - 1}$$

To start the algorithm, we need a guess. From our graphs (figures 3.77 and 3.78), it looks at though the solution is fairly near $x_0 = 1$. The results are at left. After only four steps, the values have converged to about seven decimal places, surely good enough for a bee.

Iteration	Value
0	1.0
1	1.163953414
2	1.146421185
3	1.146193259
4	1.146193221

In our original problem of finding the equilibrium of a complicated lung updating function, we already have an updating function whose solution converges to the equilibrium. Do we gain anything by replacing the original updating function with the Newton's method updating function?

The original lung updating function is

$$c_{t+1} = g(c_t) = 0.5 \left[1 - 0.5(1 - e^{-0.5c_t})\right] c_t + 2.5$$

To apply Newton's method, we first replace the equation for equilibrium, $g(c) = c$, with the equation $f(c) = g(c) - c = 0$, or

$$f(c) = 0.5 \left[1 - 0.5(1 - e^{-0.5c})\right] c + 2.5 - c = 0$$

To find the Newton's method updating function we must compute the derivative:

$$f'(c) = 0.25 + 0.25e^{-0.5c} - 0.125ce^{-0.5c} - 1$$

(Exercise 2). The Newton's method updating function is

$$c_{t+1} = c_t - \frac{f(c_t)}{f'(c_t)}$$

$$= c_t - \frac{0.5 \left[1 - 0.5\left(1 - e^{-0.5c_t}\right)\right] c_t + 2.5 - c_t}{0.25 + 0.25e^{-0.5c_t} - 0.125c_t e^{-0.5c_t} - 1}$$

We have replaced the original, somewhat messy, updating function with a truly huge updating function. Both share the equilibrium point we seek. Do we do better by repeatedly applying the Newton's method updating function? Starting from the same initial guess, $c_0 = 5$, we find the data in the following table.

Iteration	Value	Distance from equilibrium	Factor by which distance decreased
0	5.0000000000	1.4654361738	—
1	3.5304554457	−0.0041083804	−0.0028035205
2	3.5345638801	0.0000000539	−0.0000131336
3	3.5345638261	−2.0872 × 10^{-14}	−0.0000003868

We have a full *13 decimal places* of accuracy after only three steps. With a sufficiently powerful computer, we could *double* this number of digits with only one more step. It takes the ordinary updating function many more steps to achieve this level of accuracy.

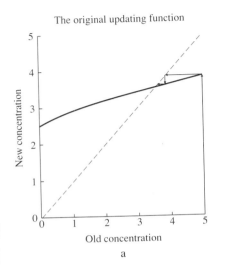

Figure 3.82
Comparison of updating functions from our original and Newton's method

Why Newton's Method Works and When It Fails

Figure 3.82 indicates why Newton's method is so fast. The two updating functions share the same equilibrium point, but the slope of the Newton's method updating function is 0 at the equilibrium. The slope of the curve at an equilibrium determines stability. If the absolute value of the slope is less than 1, the distance from the equilibrium decreases and the solution moves toward the equilibrium. If the slope is very small (near 0), the distance decreases very quickly. If the slope of an updating function is 0 at the equilibrium, the solution shoots right in toward the equilibrium (Figure 3.82b). An equilibrium where the slope is 0 is called **superstable**.

Is the slope of the updating function for Newton's method (Equation 3.11) really equal to 0 at the equilibrium? The derivative of the Newton's method updating function

$$h(x) = x - \frac{f(x)}{f'(x)}$$

can be computed with the quotient rule:

$$h'(x) = 1 - \frac{f'(x)f'(x) - f(x)f''(x)}{f'(x)^2} = \frac{f(x)f''(x)}{f'(x)^2}$$

The equilibrium of the updating function $h(x)$ is any point x^* where $f(x^*) = 0$. As long as $f'(x^*) \neq 0$,

$$h'(x^*) = \frac{f(x^*)f''(x^*)}{f'(x^*)^2} = 0$$

The equilibrium of the Newton's method updating function is superstable.

When $f(x) = x^2 - 3$, the Newton's' method updating function is

$$h(x_t) = x_t - \frac{x_t^2 - 3}{2x_t}$$

The derivative is

$$h'(x_t) = 1 - \frac{2x_t \cdot 2x_t - 2 \cdot (x_t^2 - 3)}{4x_t^2}$$

$$= 1 - \left[1 - \frac{2 \cdot (x_t^2 - 3)}{4x_t^2}\right]$$

$$= \frac{2 \cdot (x_t^2 - 3)}{4x_t^2}$$

At any point where $x_t^2 - 3 = 0$, the derivative is indeed 0 (Figure 3.81).

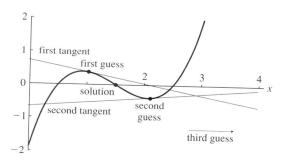

Figure 3.83
Newton's method failing miserably

As with many finely tuned machines, things can go drastically wrong with Newton's method. In Figure 3.83, our first guess was not so good and the solution eventually shot off to infinity. This picture shows geometrically why the slope $f'(x)$ must be different from 0. If we start at a point on the original curve where $f'(x) = 0$, the tangent never intersects the horizontal axis. The next step of Newton's method is not defined.

Furthermore, for functions with multiple solutions, Newton's method may not converge to the desired point. In Figure 3.83, Newton's method will eventually converge to the solution farthest to the right, which is not close to the first guess. To avoid these problems, computer algorithms usually start with the Intermediate Value Theorem to get close to a particular solution and then capitalize on the speed of Newton's method to gain accuracy.

SUMMARY

Solving equations algebraically for equilibria or other quantities is often impossible. Newton's method is a technique that can be implemented on the computer. By replacing the original equation with the tangent line approximation, we derived the **Newton's method updating function**, which converges very quickly to the solution. The awesome speed of Newton's method results from the fact that the solution is a **superstable** equilibrium, where the slope of the graph of the updating function is 0 at the equilibrium.

3.8 EXERCISES

1. Try Newton's method for two steps graphically on the figure, starting from points A, B, and C.

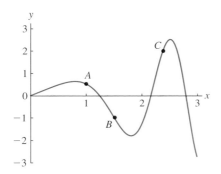

2. Compute the derivatives of the following to check the calculations in the text.
   a. $g'(c_t) = 0.25 + 0.25e^{-0.5c_t} - 0.125c_t e^{-0.5c_t}$
   b. $f'(c) = 0.25 + 0.25e^{-0.5c} - 0.125ce^{-0.5c} - 1$

3. Use Newton's method for three steps to find the following. Find and sketch the tangent line for the first step, and then find the Newton's method updating function to check your answer for the first step and to compute the next two values. Compare with the result on your calculator.
   a. $\sqrt[3]{20}$ (solve $f(x) = x^3 - 20 = 0$).
   b. A positive solution of $e^{x/2} = x + 1$ (solve $h(x) = e^{x/2} - x - 1 = 0$).
   c. The point where $\cos(x) = x$ (in radians, of course).

4. Many problems can also be solved with iteration. Compare the following updating functions with the Newton's method updating function. Show that each has the same equilibrium, and see how close you get in three steps.
   a. Updating function $x = e^x - 2$ to solve $e^x = x + 2$ (as in the text).
   b. Updating function $x = \ln(x + 2)$ to solve $e^x = x + 2$.
   c. Updating function $g(x) = x^3 + x - 20$ for Exercise 3a.
   d. Updating function $g(x) = e^{x/2} - 1$ for Exercise 3b.
   e. Updating function $g(x) = \cos(x)$ for Exercise 3c.

5. On a Newton's method diagram for solving $x(x-1)(x+1) = 0$, find all initial values from which the method fails. Which starting points converge to a negative solution? Which starting points go straight to the exact answer?

6. Suppose the total amount of nectar that comes out of a flower after time t follows the function
   $$F(t) = \frac{t^2}{1 + t^2}$$
   a. Describe how this function differs from the forms studied in the text.
   b. Find the optimum time for the bee to remain at a flower if the travel time between flowers is $\tau = 0$ (this can be done algebraically), and draw the associated diagram.
   c. Write the equation for the optimum time for the bee to remain at a flower if the travel time between flowers is $\tau = 1$.
   d. Rewrite this function in a convenient form for Newton's method.
   e. Write the Newton's method updating function, and run it for four steps starting from a reasonable guess. Draw the associated Marginal Value Theorem diagram.

7. Suppose a fish population follows the updating function
   $$N_{t+1} = rN_t e^{-N_t} - hN_t$$
   with $r = 2.5$.
   a. Find the equilibrium N^* as a function of h.
   b. Write the equation for the critical point of the payoff function $P(h) = hN^*$.
   c. Use Newton's method to find the best h.

8. Consider a variant of the medication updating function
   $$M_t = p(M_t)M_t + 1.0$$
   where the function $p(M_t)$ represents the fraction used (see Exercise 7). Suppose that
   $$p(M_t) = 0.5e^{-0.1M_t}$$
   a. Explain what is going on with this equation.
   b. Convince yourself there is an equilibrium by using the Intermediate Value Theorem.
   c. Follow the solution until it gets close to the equilibrium (to about three decimal places).
   d. Find the equilibrium with Newton's method.

9. Find the Newton's method updating function for the following. Graph the function, and try a couple of steps. Why are the functions behaving oddly?
   a. Use Newton's method to solve $x^2 = 0$ (the solution is 0).
   b. Use Newton's method to solve $\sqrt{x} = 0$ (the solution is 0). Pretend that $\sqrt{-x} = -\sqrt{x}$ on your graph.

10. For what value of r will the logistic dynamical system
    $$x_{t+1} = rx_t(1 - x_t)$$
    most rapidly converge to its equilibrium? Follow the system for three steps starting from $x_0 = 0.75$.

11. Suppose you wish to solve the equation $f(x) = e^x - x - 2 = 0$, but you do not know how to compute $f'(x)$ (this kind of problem does arise when it is impossible to compute the derivative of a complicated function).
    a. Approximate $f'(x)$ with $f(x + 1) - f(x)$. Why is this reasonable?

b. Write an approximate Newton's method updating function.
c. Use it for a few steps. Does it get close to the solution 1.146?

12. There is an alternative way to approximate the slope of the function at the value x_t:

$$f'(x_t) \approx \frac{f(x_t) - f(x_{t-1})}{x_t - x_{t-1}}$$

To get started, pick x_0 and x_1 from Exercise 11.
a. Write an approximate Newton's method updating function. Why is this different from an ordinary updating function?
b. Run it for a few steps. Does it converge faster than the method used in Exercise 11? Why?

13. Consider the updating function

$$b_{t+1} = rb_t$$

with equilibrium at $b^* = 0$.
a. Show that the solution gets closer to the equilibrium by a factor of r in each step when $0 < r < 1$.
b. Draw a picture illustrating why this is true for any linear dynamical system with a stable equilibrium.
c. Use the idea of qualitative dynamical systems to argue that a solution approaches a stable equilibrium by a factor approximately equal to the slope at that equilibrium.

14. COMPUTER: One alternative to Newton's method is called *bisection*. The method, based on the Intermediate Value Theorem, is both slower and safer than Newton's method. We will use it to solve $g(x) = e^x - x - 2 = 0$.
a. We know there is a solution between $x = 0$ and $x = 2$. Show that there is a solution between $x = 1$ and $x = 2$.
b. By computing $g(1.5)$, show there is a solution between 1.0 and 1.5.
c. Compute $g(1.25)$. There is a solution between either 1.0 and 1.25 or 1.25 and 1.5. Which is it?
d. Compute the value of g at the midpoint of the previous interval, and find an interval half as big that contains a solution.
e. Continue *bisecting* the interval until your answer is right to three decimal places.
f. About how many more steps would it take to reach six decimal places?

15. COMPUTER: Solve the equation $e^x - x - 2 = 0$ by using the quadratic approximation to the function and solving each step by using the quadratic formula. Compare how fast it converges with Newton's method. Which method do you think is better?

3.9 Panting and Deep Breathing

Why do some animals pant and others breathe slowly and deeply? Panting has the advantage of taking more breaths per second, but the breaths are shallower and leave the lung less time to absorb oxygen. Conversely, deep breathing has the advantage of giving the lung more time to absorb oxygen but at the cost of taking fewer breaths per second. We will derive updating functions describing a lung with absorption in order to generate hypotheses to explain these different breathing strategies. By writing a detailed model of the lung that includes absorption and different breathing rates, we can ask which kind of breathing maximizes the rate of oxygen absorption. We will find that the answer depends on the exact shape of the absorption function.

Breathing at Different Rates

Suppose lung expansion as a function of time behaves as in Figure 3.84. The lung receives a signal every T seconds to switch from exhaling to inhaling. The volume of air in the lung increases at a constant rate until the lung is full and then decreases at a constant rate until the next signal is received. In Figure 3.84a the signal arrives frequently and the animal breathes quickly and shallowly. In Figure 3.84b the signal arrives after a longer delay and the animal breathes slowly and deeply. In Figure 3.84c the signal does not arrive until the lung has emptied and rested.

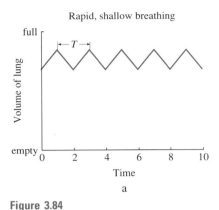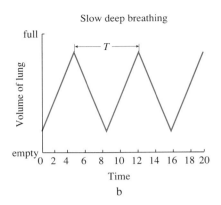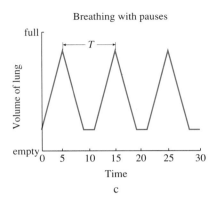

Figure 3.84

Three types of breathing

Suppose an animal is running and needs to gain oxygen as quickly as possible. Should it pant (breathe rapidly and shallowly), or should it breathe more slowly and deeply? The answer depends on how absorption depends on the internal concentration and the rate of breathing.

We can write an updating function describing the concentration c_t of oxygen in the lung:

$$c_{t+1} = (1 - q)[c_t - c_t A(T)] + q\gamma$$

As before, q represents the fraction of air exchanged and γ the ambient concentration. The function $A(T)$ denotes the fraction of oxygen absorbed as a function of time T between breaths. This is a different sort of absorption function from $\alpha(c)$, studied in Section 3.5, where absorption depended on the concentration rather than the time. If $A(T) = 0$, there is no absorption and we recover the original lung updating function (Equation 1.44).

Besides changing the fraction of oxygen absorbed, changing the breathing rate changes the fraction of air exchanged. The fraction of air exchanged is proportional to the amount of time spent inhaling (Figure 3.84). We can therefore write

$$q = rT$$

for some value of r. Suppose that $T = 1.0$ represents maximum exhalation, or the longest possible time. If $r = 1.0$ and $T = 1.0$, all the air in the lung is exchanged. For most organisms, complete exchange is impossible. The value of r, the fraction of air exchanged if breathing totally exchanges air, must be less than or equal to 1.

The updating function is then

$$c_{t+1} = (1 - rT)[c_t - c_t A(T)] + rT\gamma \tag{3.12}$$

If we knew the equation for $A(T)$, we could compare this lung with different values of T.

Deep Breathing

It is simplest to assume that the fraction of oxygen absorbed is proportional to T. If the lung has twice as much time to absorb oxygen, twice as much oxygen will be absorbed. If $A(T)$ is proportional to T, then

$$A(T) = \alpha T \quad (3.13)$$

for some value α (Figure 3.85a). If $\alpha = 1.0$ and $T = 1.0$ (maximum exchange), all the oxygen is absorbed. Because complete absorption is impossible, the value of α must be less than or equal to 1.0.

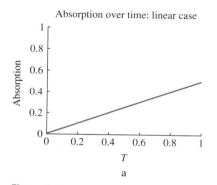
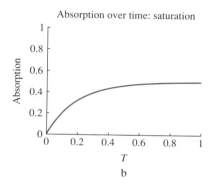
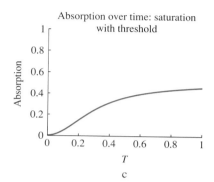

Figure 3.85

Three different absorption functions

Substituting $A(T) = \alpha T$ into the general updating function (Equation 3.12), we get

$$c_{t+1} = (1 - rT)(c_t - \alpha c_t T) + r\gamma T \quad (3.14)$$

How do we compute how much oxygen the lung absorbs as a function of T? As in the fisheries model (Section 3.3), there are two steps. First, we find the equilibrium; then, we compute how much oxygen is absorbed at this equilibrium.

The equilibrium can be found in the usual way (Section 1.11):

$$c^* = (1 - rT)(c^* - \alpha c^* T) + r\gamma T \quad \text{equation for equilibrium}$$

$$c^* - (1 - rT)(c^* - \alpha c^* T) = r\gamma T \quad \text{move all the } c^*\text{s to one side}$$

$$c^*[1 - (1 - rT)(1 - \alpha T)] = r\gamma T \quad \text{factor out } c^*$$

$$c^* = \frac{\gamma r T}{1 - (1 - rT)(1 - \alpha T)} \quad \text{solve for } c^*$$

With a bit of algebra, we can cancel a T from the numerator and denominator to find

$$c^* = \frac{\gamma r T}{1 - (1 - rT - \alpha T + r\alpha T^2)}$$

$$= \frac{\gamma r T}{rT + \alpha T - r\alpha T^2}$$

$$= \frac{\gamma r}{r + \alpha - r\alpha T} \quad (3.15)$$

How well does the lung do? We know that

$$\text{amount absorbed} = c^* A(T)$$

the equilibrium concentration times the fraction absorbed. This does not describe the *rate*, however. A lung that absorbs more oxygen over a longer time might absorb at a lower rate. The rate of absorption is

$$\text{rate of absorption} = \frac{\text{amount absorbed}}{\text{time}}$$

$$= \frac{c^* A(T)}{T}$$

Our goal is to maximize this rate. In this case, we have

$$\frac{c^* A(T)}{T} = \frac{\alpha c^* T}{T}$$

$$= \frac{\alpha \gamma r}{r + \alpha - r\alpha T} \qquad (3.16)$$

The only appearance of T in this formula is in the denominator, and the larger the value of T, the larger the function (Exercise 2). This rate takes on its maximum at $T = 1$, the maximum possible value of T (Figure 3.86). In this case, the optimal breathing rate is slow, like that of a well-conditioned runner.

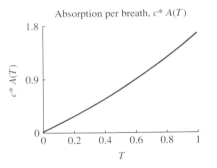
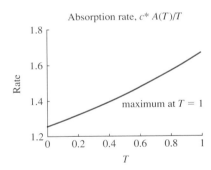

Figure 3.86
Absorption and rate of absorption as a function of T: linear case

Panting

What if absorption becomes less efficient as the length of the breath increases? One function describing saturation is

$$A(T) = \alpha(1 - e^{-kT}) \qquad (3.17)$$

In this case, long breaths might not make sense because absorption becomes less and less efficient over time (Figure 3.85b).

Substituting into the updating function (Equation 3.12), we get

$$c_{t+1} = (1 - rT)\left[c_t - \alpha c_t(1 - e^{-kT})\right] + rT\gamma \qquad (3.18)$$

We can follow the same steps as before to find the equilibrium and maximize the rate of absorption. In this case, we have

$$c^* = \frac{\gamma rT}{1 - (1 - rT)\left[1 - \alpha(1 - e^{-kT})\right]} \qquad (3.19)$$

(Exercise 3). The absorption per breath is

$$c^*A(T) = \alpha c^*(1 - e^{-kT})$$

$$= \frac{\alpha \gamma r T (1 - e^{-kT})}{1 - (1 - rT)\left[1 - \alpha(1 - e^{-kT})\right]}$$

and the rate of absorption is

$$\frac{c^*A(T)}{T} = \frac{\alpha \gamma r (1 - e^{-kT})}{1 - (1 - rT)\left[1 - \alpha(1 - e^{-kT})\right]}$$

Figure 3.87 shows the absorption and absorption rate with the parameter values $\alpha = 0.5, r = 0.5, \gamma = 5,$ and $k = 5$. Although the amount of oxygen absorbed *per breath* increases with longer breaths, the absorption *rate* decreases. The optimal value of T is $T = 0$. This organism does best by breathing as fast as it can.

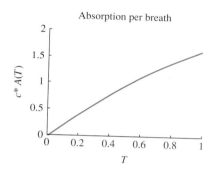
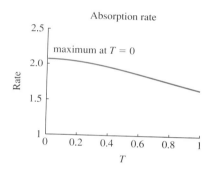

Figure 3.87
Absorption and rate as functions of T: saturating case

What if absorption takes a little time to get started? It might have a graph like that in Figure 3.85c. One function of T that produces this shape is

$$A(T) = \frac{\alpha T^2}{k + T^2} \qquad (3.20)$$

Two processes are involved: absorption saturates for large T, and absorption takes some time to get started.

We follow the same steps to find the equilibrium and maximize the rate of absorption. The equilibrium is

$$c^* = \frac{\gamma r T}{1 - (1 - rT)\left(1 - \alpha \frac{T^2}{k + T^2}\right)} \qquad (3.21)$$

(Exercise 3). The absorption is

$$c^*A(T) = \alpha c^* \frac{T^2}{k + T^2}$$

$$= \frac{\alpha \gamma r T \frac{T^2}{k + T^2}}{1 - (1 - rT)\left(1 - \alpha \frac{T^2}{k + T^2}\right)}$$

and the rate of absorption is

$$\frac{c^* A(T)}{T} = \frac{\alpha \gamma r \dfrac{T^2}{k+T^2}}{1-(1-rT)\left(1-\alpha \dfrac{T^2}{k+T^2}\right)}$$

Figure 3.88b shows the absorption and rate of absorption with the parameter values $\alpha = 0.5$, $r = 0.5$, $\gamma = 5$, and $k = 0.1$. Long breaths are not best because of saturation, and panting is not optimal because of the delay. Instead, there is an intermediate maximum. The position of this maximum depends on the parameters (the detailed shape of the graph of absorption as a function of time) and could be longer or shorter depending on the organism (Exercise 5).

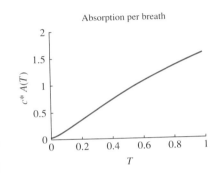
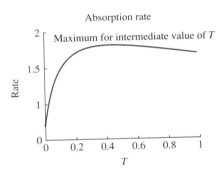

Figure 3.88
Absorption and rate as a functions of T: case with saturation and threshold

Have we explained why some animals pant and others breathe deeply? No, these models provide a *hypothesis*; animals that take long breaths should have lungs that do not saturate quickly (for example, $A(T) = \alpha T$), and animals that pant should have lungs that do saturate quickly (for example, $A(T) = \alpha(1-e^{-kT})$). The model has given us an idea what to measure next: oxygen absorption as a function of breath length. If measurements seem to be consistent with the assumptions of the model and its predictions, we could attempt to discover the physiological basis of the different absorption functions.

SUMMARY By incorporating explicit consideration of breath length into our model for the concentration of a chemical in the lung, we found optimal breathing rates. If absorption is a linear function of time, slow, deep breaths maximize the rate of oxygen uptake. If absorption saturates with time, panting can be best. If absorption begins slowly, an intermediate breathing rate is best. This provides a hypothesis to explain the different breathing behavior of organisms.

3.9 EXERCISES

1. Use a computer to reproduce all the figures in this section.
2. Show that the rate of absorption is maximized at $T = 1$ for any feasible values of α and r with the absorption function

$$A(T) = \alpha T$$

(Equation 3.13).

3. Check the formulas for the equilibria given in equations 3.19 and 3.21.
4. Use the computer to experiment with the effects of the parameter α on the absorption function of Equation 3.17. Set $r = 0.5$, $\gamma = 5.0$, and $k = 5$. Test values of α ranging from 0.1 to 1.0. Can you explain your results?

5. Show mathematically that the rate of absorption is maximized at
$$T = \sqrt{\frac{k}{1-\alpha}}$$
when the absorption function is
$$A(T) = \alpha \frac{T^2}{k+T^2}$$
(Equation 3.20). Use parameter values $\alpha = 0.5$, $r = 0.5$, $\gamma = 5.0$, and $k = 0.1$ if you get stuck. How does the optimal T change if α becomes larger? Does this make sense?

6. Try to solve for the following without substituting a particular functional form for $A(T)$.
   a. The equilibrium concentration.
   b. The equilibrium absorption per breath.
   c. The equilibrium rate of absorption.

7. Plug the forms for $A(T)$ used in the text into the expressions found in Exercise 6, and compare with the results in the text.

8. Find
$$\lim_{T \to 0} \frac{A(T)}{T}$$
for the three forms of $A(T)$ used in the text. Do the results make sense?

9. Do a complete analysis of the case
$$A(T) = \frac{\alpha T}{k+T}$$

Supplementary Problems for Chapter 3

EXERCISE 1
Between days 0 and 150 (measured from November 1), the amount of snow that falls at a certain ski resort is given by
$$S(t) = -\frac{1}{4}t^4 + 60t^3 - 4000t^2 + 96{,}000t$$
where S is measured in microns (1 μm = 10^{-4} cm).
   a. A computer wizard tells you that this function has critical points at $t = 20$, $t = 40$, and $t = 120$. Confirm this assertion.
   b. Find the global maximum and global minimum amounts of snow in feet. Remember that 1 in. = 2.54 cm.
   c. Use the second derivative test to identify the critical points as local maxima and minima.
   d. Sketch the function.

EXERCISE 2
For the functions shown:
   a. Sketch the derivative.
   b. Label local and global maxima.
   c. Label local and global minima.
   d. Find subsets of the domain with positive second derivative.

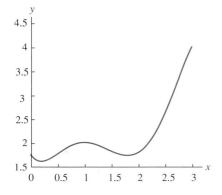

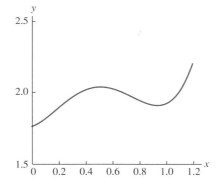

EXERCISE 3
Use the tangent line and the quadratic Taylor polynomial to find approximate values of the following. Make sure to write down the function or functions you use as well as the equation of the tangent line. Check with your calculator.
   a. $(-0.97)^3$ b. $1/(3 + 1.01^2)$ c. $e^{3(1.02)^2 + 2(1.02)}$

EXERCISE 4
Let $r(x)$ be the function giving the per capita reproduction by the formula
$$\text{per capita reproduction} = \frac{4x_t}{1 + 3x_t^2}$$
as a function of population size x.
   a. Find the population size that produces the highest per capita reproduction.
   b. Find the highest per capita reproduction.
   c. Check with the second derivative test.

EXERCISE 5
An organism is replacing 25% of the air in its lung with each breath, and the external concentration of a chemical is

$\gamma = 5.0 \times 10^{-4}$ mol/L. Suppose the body uses a fraction α of the chemical just after breathing. That is, the chemical follows the process

$$c_t \to (1-\alpha)c_t \to \text{mix of 75\% used air and 25\% ambient air}$$

a. Write the updating function for this process.
b. Find the equilibrium level in the lung as a function of α.
c. Find the amount absorbed by the body with each breath at equilibrium as a function of α.
d. Find the value of α that maximizes the amount of chemical absorbed at equilibrium.
e. Explain your result in words.

EXERCISE 6
Sketch graphs of the following functions. Find all critical points, and state whether they are minima or maxima. Find the limit of the function as $x \to \infty$.
a. $(x^2 + 2x)e^{-x}$ for positive x.
b. $\ln(x)/(1+x)$ for positive x. Do not solve for the maximum, just show that there must be one.
c. $e^{\ln[1+(x-1)^2]}$ for all x.

EXERCISE 7
Suppose the volume of a plant cell follows $V(t) = 1000(1 - e^{-t})$ μm^3 for t measured in days. Suppose the fraction of cell in a vacuole (a water-filled portion of the cell) is $H(t) = e^t/(1+e^t)$.
a. Sketch a graph of the total size of the cell as a function of time.
b. Find the volume of cell outside the vacuole.
c. Find and interpret the derivative of this function. Don't forget the units.
d. Find when the volume of the cell outside the vacuole reaches a maximum.

EXERCISE 8
During Thanksgiving dinner, the table is replenished with food every 5 min. Let F_t represent the fraction of the table laden with food:

$$F_{t+1} = F_t - \text{amount eaten} + \text{amount replenished}$$

Suppose that

$$\text{amount eaten} = \frac{bF_t}{1+F_t}$$

$$\text{amount replenished} = a(1 - F_t)$$

and that $a = 1.0$ and $b = 1.5$.
a. Explain the terms describing amount eaten and amount replenished.
b. If the table starts out empty, how much food is there after 5 min? How much is there after 10 min?
c. Use the quadratic formula to find the equilibria.
d. How much food will there be 5 min after the table is 60% full? Sketch the solution.

EXERCISE 9
Consider looking for a positive solution of the equation

$$e^x = 2x + 1$$

a. Draw a graph and pick a reasonable starting value.
b. Write the Newton's method iteration for this equation.
c. Find your next guess.
d. Show explicitly that the slope of the updating function for this iteration is 0 at the solution.

EXERCISE 10
An organism is replacing a fraction q of the air in its lung with each breath, and the external concentration of a chemical is $\gamma = 5.0 \times 10^{-4}$ mol/L. Suppose the body uses a fraction $\alpha = 1 - q$ of the chemical just after breathing. The chemical follows the process

$$c_t \xrightarrow{\text{absorption}} (1-\alpha)c_t \xrightarrow{\text{breathing}} \text{mix of used and ambient air}$$

a. Write the updating function for this process.
b. Find the equilibrium level in the lung as a function of q. Is it stable?
c. Find the amount absorbed by the body with each breath at equilibrium as a function of q.
d. Find the value of q that maximizes the amount of chemical absorbed at equilibrium.
e. Explain your result in words.

EXERCISE 11
Consider trying to solve the equation

$$\ln(x) = \frac{x}{3}$$

a. Convince yourself there is indeed a solution, and find a reasonable guess.
b. Use Newton's method to update your guess twice.
c. What would be a bad choice of an initial guess?

EXERCISE 12
Let N_t represent the difference between the sodium concentration inside and outside a cell at some time. After 1 s, the value of N_{t+1} is

$$\begin{cases} N_{t+1} = 0.5N_t & \text{if } N_t < 2 \\ N_{t+1} = 4.0N_t - 7 & \text{if } 2 < N_t < 4 \\ N_{t+1} = -0.25N_t + 10 & \text{if } 4 < N_t \end{cases}$$

a. Graph the updating function, and show that it is continuous.
b. Find the equilibria and their stability.
c. Find all initial conditions that end up at $N = 0$.

EXERCISE 13
Suppose a bee gains an amount of energy,

$$F(t) = \frac{3t}{1+t}$$

after it has been on a flower for time t but that it uses $2t$ energy units in that time (it has to struggle with the flower).
a. Find the net energy gain as a function of t.
b. Find when the net energy gain per flower is maximum.
c. Suppose the travel time between flowers is $\tau = 1$. Find the time spent on the flower that maximizes the rate of energy gain.

d. Draw a diagram illustrating the results of parts **b** and **c**. Why is the answer to part **c** smaller?

EXERCISE 14
Consider again the function for net energy gain from Exercise 13.
  a. Use the Extreme Value Theorem to show that there must be a maximum.
  b. Use the Intermediate Value Theorem to show that there must be a residence time t that maximizes the rate of energy gain.

EXERCISE 15
Find the Taylor polynomial of degree 2 approximating each of the following.
  a. $f(x) = \dfrac{1+x}{1+x^2}$ for x near 0.
  b. $f(x) = \dfrac{1+x}{1+x^2}$ for x near 1.
  c. $g(x) = \dfrac{1+x}{1+e^x}$ for x near 0.
  d. $h(x) = \dfrac{x}{2-e^x}$ for x near 0.

EXERCISE 16
Combine the Taylor polynomials from Exercise 14 with the leading behavior of the functions for large x to sketch graphs of $f(x)$, $g(x)$, and $h(x)$.

EXERCISE 17
Consider the function $F(t) = \dfrac{\ln(1+t)}{t+t^2}$.
  a. What is $\lim_{t \to 0} F(t)$?
  b. What is $\lim_{t \to 0} F'(t)$?
  c. What is $\lim_{t \to 0} F''(t)$?
  d. Sketch a graph of this function.

EXERCISE 18
A peculiar variety of bacteria enhances its own per capita reproduction. In particular, the number of offspring per bacteria increases according to the function

$$\text{per capita reproduction} = r\left(1 + \dfrac{b_t}{K}\right)$$

Suppose that $r = 0.5$ and that $K = 10^6$.
  a. Graph per capita reproduction as a function of population size.
  b. Find and graph the updating function for this population.
  c. Find the equilibria.
  d. Find the derivative of the updating function.

EXERCISE 19
A population of size x_t follows the rule

$$\text{per capita reproduction} = \dfrac{4x_t}{1 + 3x_t^2}$$

  a. Find the updating function for this population.
  b. Find the equilibrium or equilibria.
  c. What is the stability of each equilibrium?
  d. Find the equation of the tangent line at each equilibrium.
  e. What is the behavior of the approximate dynamical system defined by the tangent line at the middle equilibrium?

EXERCISE 20
A type of butterfly has two morphs, a and b. Each type reproduces annually after predation. Of type a, 20% are eaten, and 10% of type b are eaten. Each type doubles its population when it reproduces. However, the types do not breed true. Only 90% of the offspring of type a are of type a; the rest are of type b. Only 80% of the offspring of type b are of type b; the rest are of type a.
  a. Suppose there are 10,000 of each type before predation and reproduction. Find the number of each type after predation and reproduction.
  b. Find the updating function for types a and b.
  c. Find the updating function for the fraction p of type a.
  d. Find the equilibria.
  e. Find the derivative of the updating function.

EXERCISE 21
Consider a population following the updating function

$$N_{t+1} = \dfrac{rN_t}{1 + N_t^2}$$

  a. What is the per capita reproduction?
  b. Find the equilibrium as a function of r.
  c. Find the stability of the equilibrium as a function of r.
  d. Does this positive equilibrium become unstable as r becomes large?

Projects for Chapter 3

PROJECT 1
Consider the following alternative version of the Ricker model for a fishery:

$$x_{t+1} = r x_t e^{-x_{t-1}}$$

The idea is that the per capita reproduction this year (year t) is a decreasing function of the population size last year (year $t-1$).

  a. Discuss why this model might make more sense than the basic Ricker model.
  b. Find the equilibrium of this equation (set $x_{t-1} = x_t = x_{t+1} = x^*$).
  c. Use a computer to study the stability of the equilibrium. Do you have any idea why it might be different from that of the basic Ricker model?

d. Write a model where the per capita reproduction depends on the population size 2 yr ago. Find the equilibrium, and use a computer to test stability.

e. Subtract a harvest. Find the maximum sustained yield for different values of h. Is the equilibrium still stable?

f. Models with an extra delay are supposed to be simplified versions of models with two variables. The per capita reproduction is an increasing function of the amount of food available in that year, while the amount of food available is a decreasing function of the number of fish in the previous year. Try to write a pair of equations describing this situation. Try to find equilibria, and discuss whether your results make sense.

PROJECT 2

Many bees collect both pollen and nectar. Pollen is used for protein, and nectar is used for energy. Suppose the amount of nectar harvested during t s on a flower is

$$F(t) = \frac{t}{1+t}$$

and that the amount of pollen harvested during t s on a flower is

$$G(t) = \frac{t}{2+t}$$

The bee collects pollen and nectar simultaneously. Travel time between flowers is $\tau = 1.0$.

a. What is the optimal time for the bee to leave one flower for the next in order to collect nectar at the maximum rate?

b. What is the optimal time for the bee to leave one flower for the next in order to collect pollen at the maximum rate? Why are the two times different?

c. Suppose that the bee values pollen twice as much as nectar. Find a single function $V(t)$ that gives the value of resources collected by time t. What is the optimal time for the bee to leave one flower for the next? (Solving the equation requires Newton's method.)

d. Suppose that the bee values pollen k times as much as nectar. Compute the optimal time for the bee to leave one flower for the next, and graph it as a function of k. Do the values at $k = 0$ and the limit as k approaches infinity make sense?

e. Suppose that the bee collects nectar first and then switches to pollen. Assume it spends 1.0 s collecting nectar. How long should it spend collecting pollen? Now suppose it first spends 1.0 s collecting pollen. How long should it spend collecting nectar?

f. Suppose again that the bee values pollen k times as much as nectar. Experiment to try to find a solution that is best when nectar and pollen are harvested sequentially.

PROJECT 3

We know enough to study the solutions of the logistic dynamical system for values of r up to 3.5. Define the updating function to be

$$f(x) = rx(1-x)$$

a. Find the two-step updating function $g = f \circ f$.

b. Graph it along with the diagonal for values of r ranging from 2.8 to 3.6.

c. Write the equation for the equilibria of g.

d. Two of the equilibria, 0 and $x^* = 1 - (1/r)$ match those for f. Why?

e. The terms x and $x - x^*$ factor out of the equilibrium equation $g(x) - x = 0$. Why?

f. Factor $g(x) - x$ and find the other two equilibria, x_1 and x_2. For what values of r do they make sense? What do they mean? (Think about $f(x_1)$ and $f(x_2)$.)

g. For what values of r are x_1 and x_2 stable?

h. Describe the dynamics of f and g in these situations.

i. What happens for values of r just above the point where x_1 and x_2 become unstable?

Bibliography for Chapter 3

Section 3.2

Devaney, R. L. *An Introduction to Chaotic Dynamical Systems*. Benjamin/Cummings, Menlo Park, Calif., 1986.

May, R. M. Simple mathematical models with very complicated dynamics. *Nature* 261:459–467, 1976.

Section 3.3

Charnov, E. L. Optimal foraging: the marginal value theorem. *Theoretical Population Biology* 9:129–136, 1976.

Clark, C. W. *Mathematical Bioeconomics: The Optimal Management of Renewable Resources*. Wiley, New York, 1990.

Stephens, D. W., and J. R. Krebs. *Foraging Theory*. Princeton University Press, Princeton, N.J., 1986.

Section 3.5

Edelstein-Keshet, L. *Mathematical Models in Biology*. Random House, New York, 1988.

Section 3.8

Press, W. H., S. A. Teukolsy, W. T. Vetterling, and B. P. Flannery. *Numerical Recipes: The Art of Scientific Computing*. Cambridge University Press, Cambridge, England, 1992.

Section 3.9

Hoppensteadt, F. C. and C. S. Peskin. *Mathematics in Medicine and the Life Sciences*. Springer-Verlag, New York, 1992.

Chapter 4

Differential Equations, Integrals, and Their Applications

Biological systems are constantly changing. Describing this change and deducing its consequences constitute the dynamical approach to the understanding of life. In the first three chapters of this course, we used updating functions to study problems that could be treated in discrete time. Given the state of a system (such as a population size) at one time, the updating function gives the state of the system at a later time. In the course of analyzing the behavior of discrete-time dynamical systems, we developed the idea of the derivative, the slope or *instantaneous rate of change of a function*.

Using the derivative, we can study a different kind of dynamical system, the **differential equation**. The instantaneous rate of change of a measurement takes the place of the updating function and provides the information needed to deduce the future state of a system from its present state. Because the system can be measured at any time, these systems are a type of *continuous-time dynamical system*.

Differential equations were invented by Isaac Newton to study the motion of planets. They have proven to be the most powerful method for describing dynamics in all of the sciences. Like discrete-time dynamical systems, they can be applied to the three main areas of focus of this course: growth, maintenance, and replication.

4.1 Differential Equations

When we have a measurement, we can differentiate to find the rate of change (Figure 4.1). What if we know the rate of change and want to compute the measurement (Figure 4.2)? If we know, for example, that the velocity of an object is $v(t)$, a function of t, then the position, p, must satisfy the **differential equation**

$$\frac{dp}{dt} = v(t)$$

This equation relates two quantities, saying that one [the velocity, $v(t)$] is the derivative of the other [the position, $p(t)$]. Until now, we have usually thought of ourselves as knowing the position and taking the derivative to find the velocity. With a differential equation, we know the velocity and wish to find some method to compute the position.

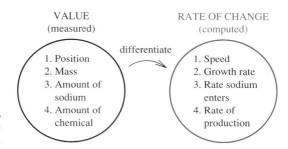

Figure 4.1
Finding the rate of change with the derivative

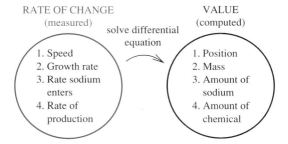

Figure 4.2
Finding a measurement from the rate of change

When might we measure the rate of change of some quantity in this way? We might want to find position or cell sodium concentration, for example. When you are lost, it is much easier to look at your car's speedometer to determine your speed than to locate landmarks and identify them on a map to determine your position. Similarly, it might be easier to measure how many ions enter and leave a cell each second by measuring changes in electrical charge than to track down and count every sodium ion floating around in a cell. In such cases, when we *measure* the rate of change, we can write what is called a **pure-time differential equation**:

derivative of unknown measurement = measured rate of change

There is another way to arrive at a differential equation describing a measurement. Based on biological principles, we might know a *rule* describing how a measurement changes, just as we did when deriving updating functions. For example, when resources are not limiting, the rate of population growth is proportional to population size. In such cases, we can write down an **autonomous differential equation**:

derivative of unknown quantity = some function of unknown quantity

Given that the unknown quantity appears on both sides of this equation, it is quite remarkable that it can be solved.

Our plan of attack in this chapter is to begin with a detailed study of pure-time differential equations, developing the method of **integration** to solve them. After studying several other applications of the integral, we return to autonomous differential equations in Chapter 5, where we show how to use both integration and graphical methods to analyze these important equations.

Differential Equations: Examples and Terminology

Suppose that $1\,\mu\text{m}^3$ of water enters a cell each second. The volume, V, of the cell is thus increasing by $1.0\,\mu\text{m}^3/\text{s}$. Mathematically, the rate of change of V is $1.0\,\mu\text{m}^3/\text{s}$, or

$$\frac{dV}{dt} = 1.0 \tag{4.1}$$

The quantity being differentiated (V in this case) is called the **state variable**. Because we *measured* the rate of change, this is a **pure-time differential equation**.

Our goal is to find the volume V of the cell as a function of time. Such a function is called a **solution**, just as with a discrete-time dynamical system. When it works, the best method for solving a differential equation is *guessing*. What function has a constant rate of change equal to 1.0? Geometrically, what function has a constant slope of 1.0? The answer is a line with slope 1. One such line is

$$V(t) = t$$

(Figure 4.3). We can check whether this guess solves the differential equation by taking the derivative,

$$\frac{dV}{dt} = \frac{dt}{dt} = 1$$

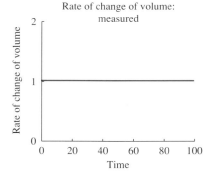
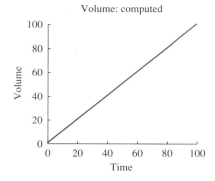

Figure 4.3
A function that solves the differential equation $dV/dt = 1$.

Suppose we know that the volume of the cell at time $t = 0$ is 300 μm^3. Although the guess $V(t) = t$ is one solution of the differential equation, it does not match this **initial condition**. As with discrete-time dynamical systems, a solution also depends on where something starts. From our knowledge of lines, we realize that any equation for $V(t)$ with the form

$$V(t) = t + c$$

has a constant slope of 1 when c is a constant. In terms of the differential equation,

$$\frac{dV}{dt} = \frac{d}{dt}(t + c) = 1$$

The value of c can be chosen to match the initial condition. In this case, because

$$V(0) = 300 = 0 + c$$

we know that $c = 300$. The solution of the differential equation

$$\frac{dV}{dt} = 1$$

with the initial condition $V(0) = 300$ is

$$V(t) = t + 300$$

(Figure 4.4). From knowledge of the initial condition and a measurement of the rate of change, we have a formula giving the volume at any time. At $t = 50$, the volume is 350 μm^3.

Figure 4.4
Volume as a function of time

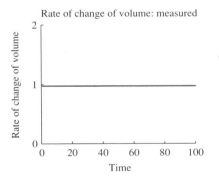
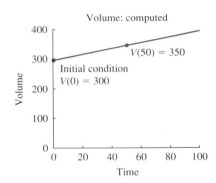

As a more complex example, suppose we measure that the rate of chemical production is e^{-t}, in units of moles per second (Figure 4.5a). The **state variable** P, the total chemical produced, follows the pure-time differential equation

$$\frac{dP}{dt} = e^{-t} \quad (4.2)$$

The rate at which the chemical is produced becomes slower and slower as time progresses. The solution for $P(t)$ (which can be found by guessing) can be sketched graphically. It must be a function that increases more and more slowly (Figure 4.5b). The solution sketched has initial condition $P(0) = 0$.

The differential equations for volume and chemical product (Equations 4.1 and 4.2) are pure-time differential equations because we *measured* the rate of change of the state variable. The equation for chemical production indicates the source for the name. The formula for the rate of change, e^{-t} in this case, depends *purely* on the *time*, t.

Suppose instead we wish to describe a phenomenon with a differential equation derived from biological principles. For a population, the simplest rule is that

Figure 4.5
Product as a function of time

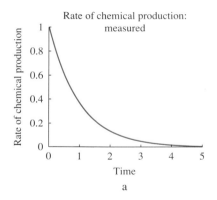
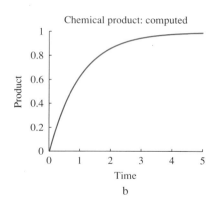

the rate of reproduction is proportional to the population size. The discrete-time dynamical system describing this process is

$$b_{t+1} = rb_t \tag{4.3}$$

(Equation 1.28), where the parameter r is the per capita reproduction (the number of offspring per bacterium). If population size can be measured *continuously* (at any time), the change in population is expressed as the *rate of change*, in this case equal to the rate of reproduction. Therefore,

$$\frac{db}{dt} = \lambda b \tag{4.4}$$

where b represents the bacterial population, and the constant of proportionality, λ, is the per capita growth rate. Although b represents a function of time, it is conventional to write just b instead of $b(t)$ on both sides of the equation. The dimensions of λ are 1/time. Because it was derived from a rule, this is an **autonomous differential equation**.

Notice how this autonomous differential equation differs from a pure-time differential equation (Equations 4.1 and 4.2). In the pure-time differential equations, the rate of change was a *measured* function of time. In this case, the rate of change of population was derived from a *rule* and is a function of the state variable population size.

Type of differential equation	Example	When used
Pure-time differential equation	$\frac{dP}{dt} = e^{-t}$	When the rate of change is a measured function of time
Autonomous differential equation	$\frac{db}{dt} = 2b$	When a rule gives the rate of change as a function of the state variable

As before, we can find the solution by guessing. Suppose $\lambda = 2$ and that the population begins at $b(0) = 1.0 \times 10^6$. The differential equation is

$$\frac{db}{dt} = 2b$$

What function has a derivative equal to double itself? Remember that the derivative of the exponential function $b(t) = b_0 e^{\alpha t}$ is

$$\frac{db}{dt} = \alpha b_0 e^{\alpha t} = \alpha b(t)$$

(Section 2.10). The guess $b(t) = 1.0 \times 10^6 e^{2t}$ is a solution of the differential equation because

$$\begin{aligned}\frac{db}{dt} &= \frac{d}{dt} 1.0 \times 10^6 e^{2t} && \text{substituting into the formula} \\ &= 1.0 \times 10^6 \frac{d}{dt} e^{2t} && \text{constant product rule} \\ &= 1.0 \times 10^6 \cdot 2 e^{2t} && \text{derivative of exponential} \\ &= 2 \cdot 1.0 \times 10^6 e^{2t} && \text{reorganizing} \\ &= 2b(t) && \text{recognizing the formula for } b(t)\end{aligned}$$

Furthermore, this formula matches the initial condition because

$$b(0) = 1.0 \times 10^6 e^{2 \cdot 0} = 1.0 \times 10^6$$

Like a population described by the discrete-time dynamical system $b_{t+1} = rb_t$, this population grows exponentially (Figure 4.6).

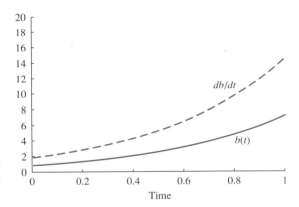

Figure 4.6
Exponential growth of a bacterial population

Euler's Method Applied to Pure-Time Differential Equations

How do we solve a differential equation if the method of guessing does not work? We can sketch solutions of pure-time differential equations using our understanding of the derivative. We will soon learn how to use *integration* to solve both pure-time differential equations and autonomous differential equations. Even that method, however, does not work for many equations. An alternative approach, called **Euler's method**, can be implemented on the computer. Like Newton's method for solving equations (Section 3.8), Euler's method works by converting the original problem into a problem about a discrete-time dynamical system. Also like Newton's method, Euler's method begins with the *tangent-line approximation*, replacing a problem about *curves* with a problem about *lines*.

Consider again the pure-time differential equation that describes the volume of a cell,

$$\frac{dV}{dt} = 1.0$$

(Equation 4.1) with initial condition $V(0) = 300$. How can we use this information to estimate the volume at time $t = 1$ if we fail to guess the solution? We have two pieces of information: the initial condition (a base point) and the differential equation (the derivative or slope), exactly the information needed to write the tangent line approximation for $V(t)$ near the base point $t = 0$ (Figure 4.7).

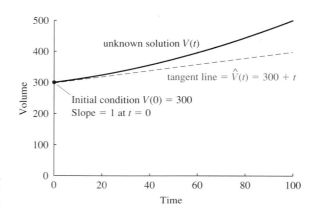

Figure 4.7

The tangent line approximation of a differential equation

In this case, the approximate function $\hat{V}(t)$ with base point 0 is

$$\begin{aligned} \hat{V}(0 + \Delta t) &= V(0) + V'(0)\Delta t && \text{equation for tangent line} \\ &= 300 + 1 \cdot \Delta t && V(0) = 300 \text{ (initial condition) and} \\ & && V'(0) = 1 \text{ (differential equation)} \\ &= 300 + \Delta t && \text{simplify} \end{aligned}$$

Although we do not know the solution, $V(t)$, we do know its tangent line. This is a graphical way of expressing the fact that a pure-time differential equation gives the rate of change of an unknown measurement.

Euler's method begins by using this tangent line to estimate the state variable at some time after $t = 0$. With $\Delta t = 1$,

$$\hat{V}(1) = 300 + 1 = 301$$

To continue, we use this new information and the differential equation to estimate the volume at $t = 2$. The tangent line with base point $t = 1$ is

$$\begin{aligned} \hat{V}(1 + \Delta t) &= V(1) + V'(1)\Delta t && \text{equation for tangent line} \\ &\approx \hat{V}(1) + V'(1)\Delta t && \text{substitute } \hat{V}(1) \text{ for unknown } V(1) \\ &= 301 + 1 \cdot \Delta t && \hat{V}(1) = 301 \text{ and } V'(1) = 1 \\ & && \text{(from the differential equation)} \\ &= 301 + \Delta t && \text{simplify} \end{aligned}$$

We estimate that

$$\hat{V}(1 + 1) = 301 + 1 \cdot 1 = 302$$

Algorithm 4.1 (Euler's method for solving a differential equation)

1. Choose a **time step** Δt (the length of time between estimated values).

2. Use the initial condition and the differential equation to find the tangent line $\hat{V}(0 + \Delta t)$ with base point $t = 0$, and use it to estimate $\hat{V}(\Delta t)$.

3. Use the estimate $\hat{V}(\Delta t)$ and the differential equation to find the tangent line $\hat{V}(t + \Delta t)$ with base point $t = \Delta t$, and estimate $\hat{V}(2\Delta t)$.

4. Repeat the previous step as many times as needed.

In our example, we chose a time step of $\Delta t = 1$.

Sometimes we can write a simple updating function to summarize the steps. Suppose we have an estimate $\hat{V}(t)$ for $V(t)$ at some time t. The tangent line approximation with base point t is

$$\begin{aligned}\hat{V}(t + \Delta t) &= V(t) + V'(t)\Delta t & \text{equation for tangent line}\\ &\approx \hat{V}(t) + V'(t)\Delta t & \text{substitute } \hat{V}(t) \text{ for unknown } V(t)\\ &= \hat{V}(t) + 1 \cdot \Delta t & V'(1) = 1 \text{ (from the differential equation)}\\ &= \hat{V}(t) + \Delta t & \text{simplify}\end{aligned}$$

Therefore,

$$\hat{V}(t + 1) = \hat{V}(t) + 1$$

This updating function says that approximately 1.0 μm^3 is added to the volume each second. We have seen this updating function as a model for a growing tree (Equation 1.4) with solution

$$\hat{V}(t) = 300 + t$$

Because the true solution is in fact a line, the results of Euler's method match the exact solution found by guessing.

The results are more interesting when we apply Euler's method to the differential equation for chemical production,

$$\frac{dP}{dt} = e^{-t}$$

Suppose the initial condition is $P(0) = 0$. Following Algorithm 4.1, we perform the following steps:

1. Pick a time step of $\Delta t = 1$.

2. The tangent line with base point $t = 0$ is

$$\begin{aligned}\hat{P}(0 + \Delta t) &= P(0) + P'(0)\Delta t & \text{equation for tangent line}\\ &= 0 + 1 \cdot \Delta t & P(0) = 0 \text{ (initial condition) and}\\ & & P'(0) = e^{-0} = 1\\ & & \text{(differential equation)}\\ &= \Delta t & \text{simplify}\end{aligned}$$

We therefore estimate that

$$\hat{P}(1) = 1$$

3. For the next step, we have

$$\hat{P}(1 + \Delta t) = P(1) + P'(1)\Delta t \quad \text{equation for tangent line}$$
$$\approx \hat{P}(1) + P'(1)\Delta t \quad \text{substitute } \hat{P}(1) \text{ for unknown } P(1)$$
$$= 1 + 0.367\Delta t \quad \hat{P}(1) = 1 \text{ and } P'(1) = e^{-1} = 0.367$$

We therefore estimate that

$$\hat{P}(2) = 1.0 + 0.367 \cdot 1.0 = 1.367$$

4. To find $\hat{P}(3)$, we have

$$\hat{P}(2 + \Delta t) = P(2) + P'(2)\Delta t$$
$$\approx \hat{P}(2) + P'(2)\Delta t$$
$$= 1.367 + e^{-2} \cdot \Delta t$$
$$= 1.367 + 0.135\Delta t$$
$$\hat{P}(3) = 1.502$$

To find $\hat{P}(4)$, we have

$$\hat{P}(3 + \Delta t) = P(3) + P'(3)\Delta t$$
$$\approx \hat{P}(3) + P'(3)\Delta t$$
$$= 1.502 + e^{-3} \cdot \Delta t$$
$$= 1.502 + 0.050\Delta t$$
$$\hat{P}(4) = 1.552$$

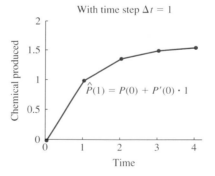
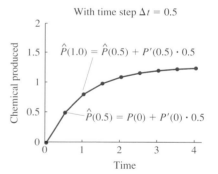

Figure 4.8
Euler's method applied to a pure-time differential equation

If we try the same method with a smaller time step of $\Delta t = 0.5$, we get the data in the table.

t	$\hat{P}(t + \Delta t)$	$t + \Delta t$	$\hat{P}(t + \Delta t)$
0.0	$P(0) + P'(0)\Delta t = 0.0 + e^{-0}\Delta t$	0.5	0.5
0.5	$\hat{P}(0.5) + P'(0.5)\Delta t = 0.5 + e^{-0.5}\Delta t$	1.0	0.803
1.0	$\hat{P}(1.0) + P'(1.0)\Delta t = 0.803 + e^{-1.0}\Delta t$	1.5	0.987
1.5	$\hat{P}(1.5) + P'(1.5)\Delta t = 0.987 + e^{-1.5}\Delta t$	2.0	1.098
2.0	$\hat{P}(2.0) + P'(2.0)\Delta t = 1.098 + e^{-2.0}\Delta t$	2.5	1.166
2.5	$\hat{P}(2.5) + P'(2.5)\Delta t = 1.166 + e^{-2.5}\Delta t$	3.0	1.207
3.0	$\hat{P}(3.0) + P'(3.0)\Delta t = 1.207 + e^{-3.0}\Delta t$	3.5	1.232
3.5	$\hat{P}(3.5) + P'(3.5)\Delta t = 1.232 + e^{-3.5}\Delta t$	4.0	1.247

The smaller time step of $\Delta t = 0.5$ gives a more accurate answer but requires more calculation (Figure 4.8).

Euler's Method Applied to an Autonomous Differential Equation

Earlier, we guessed that the solution of
$$\frac{db}{dt} = 2b$$
with $b(0) = 1.0 \times 10^6$ is
$$b(t) = 1.0 \times 10^6 e^{2t}$$
The population size at $t = 1$ is approximately 7.39×10^6.

Alternatively, we can apply Euler's method to this autonomous differential equation (Figure 4.9) to estimate $b(1)$. First, suppose we break the interval from 0 to 1 into four intervals with width $\Delta t = 0.25$.

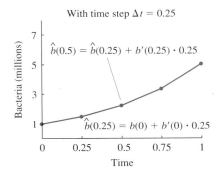

Figure 4.9
The tangent line approximation of population growth

1. Pick a time step of $\Delta t = 0.25$.
2. The tangent line with base point $t = 0$ is

   $\hat{b}(0 + \Delta t)$
   $= b(0) + b'(0)\Delta t$ equation for tangent line
   $= 1.0 \times 10^6 + 2.0 \times 10^6 \cdot \Delta t$ $b(0) = 1.0 \times 10^6$
                                                 (initial condition) and
                                                 $b'(0) = 2b(0) = 2.0 \times 10^6$
                                                 (differential equation)

   We therefore estimate that
   $$\hat{b}(0.25) = 1.0 \times 10^6 + 2.0 \times 10^6 \cdot 0.25 = 1.5 \times 10^6$$

3. To find $\hat{b}(0.5)$, we find the tangent line at $t = 0.25$:

   $\hat{b}(0.25 + \Delta t)$
   $= b(0.25) + b'(0.25)\Delta t$ equation for tangent line
   $\approx \hat{b}(0.25) + b'(0.25)\Delta t$ substitute $\hat{b}(0.25)$ for unknown $b(0.25)$

320 Chapter 4 Differential Equations, Integrals, and Their Applications

$$= \hat{b}(0.25) + 2b(0.25)\Delta t \qquad b'(0.25) = 2b(0.25) \text{ from the differential equation}$$

$$\approx \hat{b}(0.25) + 2\hat{b}(0.25)\Delta t \qquad \text{substitute } \hat{b}(0.25) \text{ for unknown } b(0.25)$$

$$= 1.5 \times 10^6 + 2 \cdot 1.5 \times 10^6 \Delta t \qquad \hat{b}(0.25) = 1.5 \times 10^6$$

We therefore estimate that

$$\hat{b}(0.5) = \hat{b}(0.25 + 0.25)$$
$$= 1.5 \times 10^6 + 2 \cdot 1.5 \times 10^6 \cdot 0.25 = 2.25 \times 10^6$$

4. To find $\hat{b}(0.75)$,

$$\hat{b}(0.5 + \Delta t) = b(0.5) + b'(0.5)\Delta t$$
$$\approx \hat{b}(0.5) + b'(0.5)\Delta t$$
$$= \hat{b}(0.5) + 2b(0.5)\Delta t$$
$$\approx \hat{b}(0.5) + 2\hat{b}(0.5)\Delta t$$
$$= 2.25 \times 10^6 + 2 \cdot 2.25 \times 10^6 \Delta t$$
$$\hat{b}(0.75) = 2.25 \times 10^6 + 2 \cdot 2.25 \times 10^6 \cdot 0.25 = 3.375 \times 10^6$$

5. To find $\hat{b}(1.0)$,

$$\hat{b}(0.75 + \Delta t) = b(0.75) + b'(0.75)\Delta t$$
$$\approx \hat{b}(0.75) + b'(0.75)\Delta t$$
$$= \hat{b}(0.75) + 2b(0.75)\Delta t$$
$$\approx \hat{b}(0.75) + 2\hat{b}(0.75)\Delta t$$
$$= 3.375 \times 10^6 + 2 \cdot 3.375 \times 10^6 \Delta t$$
$$\hat{b}(1.0) = 3.375 \times 10^6 + 2 \cdot 3.375 \times 10^6 \cdot 0.25 = 5.0625 \times 10^6$$

What if we used intervals of length $\Delta t = 0.1$ hours instead?

t	$\hat{b}(t + \Delta t)$	$t + \Delta t$	$\hat{b}(t + \Delta t)$
0.0	$b(0) + b'(0)\Delta t = 1.0 + 2 \cdot 1.0\Delta t$	0.1	1.2
0.1	$\hat{b}(0.1) + b'(0.1)\Delta t = 1.2 + 2 \cdot 1.2\Delta t$	0.2	1.44
0.2	$\hat{b}(0.2) + b'(0.2)\Delta t = 1.440 + 2 \cdot 1.440\Delta t$	0.3	1.728
0.3	$\hat{b}(0.3) + b'(0.3)\Delta t = 1.728 + 2 \cdot 1.728\Delta t$	0.4	2.074
0.4	$\hat{b}(0.4) + b'(0.4)\Delta t = 2.074 + 2 \cdot 2.074\Delta t$	0.5	2.488
0.5	$\hat{b}(0.5) + b'(0.5)\Delta t = 2.488 + 2 \cdot 2.488\Delta t$	0.6	2.986
0.6	$\hat{b}(0.6) + b'(0.6)\Delta t = 2.986 + 2 \cdot 2.986\Delta t$	0.7	3.583
0.7	$\hat{b}(0.7) + b'(0.7)\Delta t = 3.583 + 2 \cdot 3.583\Delta t$	0.8	4.300
0.8	$\hat{b}(0.8) + b'(0.8)\Delta t = 4.300 + 2 \cdot 4.300\Delta t$	0.9	5.159
0.9	$\hat{b}(0.9) + b'(0.9)\Delta t = 5.159 + 2 \cdot 5.159\Delta t$	1.0	6.192

Euler's method is more accurate with smaller time steps, but still not very close to the exact answer of 7.389.

For the many differential equations for which solutions cannot be found by guessing or other methods, computers are used to find an approximate solution using a more sophisticated version of Euler's method. The accuracy depends on the time step, the value of Δt, chosen. Larger values of Δt (e.g., 0.25) give less accurate answers but require less computation (only four steps instead of ten to reach $t = 1$). Modified versions of this technique have been developed to increase accuracy with fewer steps and to adjust the time step when solutions go astray. Research into further improving these techniques is called **numerical analysis**.

SUMMARY A **differential equation** expresses the rate of change of a quantity, the **state variable**, as a function of time or of the state variable itself. If the rate of change has been measured as a function of time, the equation is a **pure-time differential equation**. If the rate of change has been derived from a rule, the equation is an **autonomous** differential equation. A **solution** gives the value of the state variable as a function of time. The solution also depends on the the **initial condition**, the initial value of the state variable. When guessing fails, **Euler's method**, which is based on the tangent line approximation, can be used to convert any differential equation into a discrete-time dynamical system.

4.1 EXERCISES

1. For each of the measurements given in Figures 4.1 and 4.2 (page 311), give circumstances under which you could measure the following.
   a. The value but not the rate of change.
   b. The rate of change but not the value.

2. Suppose a cell starts at a volume of 600 μm^3 and loses water at a rate of 2 μm^3 per second.
   a. Write a differential equation describing this process.
   b. Find and graph the solution.
   c. When does the solution stop making sense? Can you think of a way to correct this?

3. Suppose a snail starts crawling across the sidewalk at 8:30 A.M. The velocity of the snail t min after it starts is t cm/min (it is speeding up to beat the heat).
   a. Draw a graph of the speed of the snail as a function of time.
   b. Write a differential equation for the position of the snail as a function of time (use initial condition $p(0) = 0$).
   c. What kind of differential equation is this?
   d. Try to guess the solution of this equation, using your knowledge of the derivatives of quadratic functions.
   e. Draw a graph of the position of the snail as a function of time.
   f. If the sidewalk is 0.5 m wide, how long will it take this snail to cross? How fast is it going when it reaches the other side?

4. Apply Euler's method to the differential equation in Exercise 3 using a step size of $\Delta t = 1$. Find the approximate position at $t = 5$. Graph your values and compare them with the exact solution.

5. Use the following hints to "guess" the solution of the differential equation
$$\frac{dP}{dt} = e^{-t}$$
with initial condition $P(0) = 0$.
   a. Find the derivative of e^{-t}.
   b. What do you need to multiply e^{-t} by to make the derivative equal to e^{-t}?
   c. Evaluate your answer at $t = 0$. What do you need to add to match the initial condition $P(0) = 0$?
   d. Write, graph, and check your solution.
   e. What is the limit of the solution as $t \to \infty$?

6. Consider Equation 4.2 with $P(0) = 0$ (as in Exercise 5). Use Euler's method to estimate the following.
   a. $P(0.3)$ with $\Delta t = 0.3$.
   b. $P(0.3)$ in three steps with $\Delta t = 0.1$.
   c. How do the results compare with each other and with the exact solution from Exercise 5?

7. Consider the differential equation
$$\frac{db}{dt} = -3b$$
with initial condition $b(0) = 100$.
   a. What kind of differential equation is this?
   b. Show that $b(t) = 100e^{-3t}$ is a solution.
   c. Graph your solution and its derivative. Explain what this has to do with the original differential equation.

8. Apply Euler's method to the differential equation in Exercise 7 using a step size of $\Delta t = 0.1$. Find the approximate

position at $t = 1$. Graph your values and compare them with the exact solution.

9. Find the updating function associated with Euler's method applied to the bacterial population growth model

$$\frac{db}{dt} = 2b$$

using a time step of $\Delta t = 0.1$. Use it to solve for the approximate population at time $t = 1.0$ (assume that $b(0) = 1.0 \times 10^6$).

10. Follow the steps in Exercise 9 to find the approximate population at time $t = 1.0$ using time steps of $\Delta t = 0.01$ and $\Delta t = 0.001$.

11. **COMPUTER:** Apply Euler's method to solve the differential equation

$$\frac{db}{dt} = b + t$$

with the initial condition $b(0) = 1.0$. Compare with the sum of the results of

$$\frac{db_1}{dt} = b_1$$

and

$$\frac{db_2}{dt} = t$$

Do you think there is a sum rule for differential equations? If your computer has a method for solving differential equations, find the solution and compare it with your approximate solution from Euler's method.

12. **COMPUTER:** Apply Euler's method to solve the differential equation

$$\frac{db}{dt} = b^2$$

with the initial condition $b(0) = 0.1$. Try smaller and smaller time steps. What is the largest value of t for which your solution makes sense?

4.2 Basic Differential Equations

Temperature change, chemical exchange in the lung, and selection can be simply and accurately modeled with **autonomous differential equations**. In each case, the model describes the same biological situation as a related discrete-time dynamical system, but the equation looks different. We derive these fundamental equations from biological or physical laws.

Newton's Law of Cooling

Because the state variables and measurements are continuous, differential equations are appropriate tools for studying fluid and heat transport. In some cases, it is easier to measure the amount of fluid crossing a membrane than to measure the entire quantity of fluid inside. More important, the rules describing the movement of fluids and heat can be described compactly and intuitively with differential equations.

The basic idea is expressed by **Newton's law of cooling**, which states that

The rate at which heat is lost from an object is proportional to the difference between the temperature of the object and the ambient temperature.

Denote the temperature of an object by H and the fixed ambient temperature by A. Newton's law of cooling says that

$$\frac{dH}{dt} = \alpha(A - H) \tag{4.5}$$

where α is a positive constant with dimensions of 1/time. This is an **autonomous differential equation** because it describes a **rule**, and because the rate of change is a function of the state variable H and not of the time t. Make sure to remember that H is a function of time but that α and A are constants.

If the temperature of the object is higher than the ambient temperature ($H > A$), the rate of change of temperature is negative and the object cools. If the temperature of the object is lower than the ambient temperature ($H < A$), the rate of change of temperature is positive and the object warms up (Figure 4.10). If the temperature of the object is equal to the ambient temperature, the rate of change of temperature is 0, and the temperature of the object does not change.

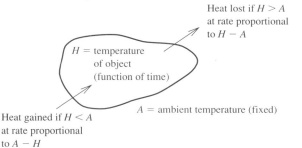

Figure 4.10
Newton's law of cooling

The coefficient α depends on the **specific heat** of the material and on its shape. Materials with high specific heat, such as water, retain heat for a long time, have low values of α, and experience low rates of heat change for a given temperature difference. Materials with low specific heat, such as metals, lose heat rapidly, have high values of α, and experience high rates of heat change for a given temperature difference. The coefficient α also depends on the object's ratio of surface area to volume. An object with a large exposed surface relative to its volume will heat or cool rapidly, as a shallow puddle freezes quickly on a cold evening. An object with a small surface area heats or cools more slowly.

Although it is possible to guess the solution (Exercise 2) or to solve the equation mathematically with the method of separation of variables (Section 5.3), we first study Newton's law of cooling using Euler's method. Suppose that α has the relatively small value of 0.1 per minute, that the ambient temperature A is $20°C$, and that the initial condition is $H(0) = 40°C$. To study how the temperature changes over the first 10 min, we choose a time step of $\Delta t = 1$, as shown in the following table.

Euler's method applied to Newton's law of cooling ($\alpha = 0.1$, $A = 20.0$, $H(0) = 40.0$)

t	$\hat{H}(t + \Delta t)$	$t + \Delta t$	$\hat{H}(t + \Delta t)$
0	$H(0) + H'(0)\Delta t = 40.0 + 0.1(20 - 40)\Delta t$	1	38.0
1	$\hat{H}(1) + H'(1)\Delta t = 38.0 + 0.1(20 - 38.0)\Delta t$	2	36.2
2	$\hat{H}(2) + H'(2)\Delta t = 36.2 + 0.1(20 - 36.2)\Delta t$	3	34.6
3	$\hat{H}(3) + H'(3)\Delta t = 34.6 + 0.1(20 - 34.6)\Delta t$	4	33.1
4	$\hat{H}(4) + H'(4)\Delta t = 33.1 + 0.1(20 - 33.1)\Delta t$	5	31.8
5	$\hat{H}(5) + H'(5)\Delta t = 31.8 + 0.1(20 - 31.8)\Delta t$	6	30.6
6	$\hat{H}(6) + H'(6)\Delta t = 30.6 + 0.1(20 - 30.6)\Delta t$	7	29.6
7	$\hat{H}(7) + H'(7)\Delta t = 29.6 + 0.1(20 - 29.6)\Delta t$	8	28.6
8	$\hat{H}(8) + H'(8)\Delta t = 28.6 + 0.1(20 - 28.6)\Delta t$	9	27.7
9	$\hat{H}(9) + H'(9)\Delta t = 27.7 + 0.1(20 - 27.7)\Delta t$	10	27.0

If we start instead at $H(0) = 0°C$, we get the data shown in the following table. In both cases, the temperature approaches the ambient temperature (Fig-

Euler's method applied to Newton's law of cooling ($\alpha = 0.1$, $A = 20.0$, $H(0) = 0.0$)

t	$\hat{H}(t + \Delta t)$	$t + \Delta t$	$\hat{H}(t + \Delta t)$
0	$H(0) + H'(0)\Delta t = 0.0 + 0.1(20 - 0.0)\Delta t$	1	2.0
1	$\hat{H}(1) + H'(1)\Delta t = 2.0 + 0.1(20 - 2.0)\Delta t$	2	3.8
2	$\hat{H}(2) + H'(2)\Delta t = 3.8 + 0.1(20 - 3.8)\Delta t$	3	5.4
3	$\hat{H}(3) + H'(3)\Delta t = 5.4 + 0.1(20 - 5.4)\Delta t$	4	6.9
4	$\hat{H}(4) + H'(4)\Delta t = 6.9 + 0.1(20 - 6.9)\Delta t$	5	8.2
5	$\hat{H}(5) + H'(5)\Delta t = 8.2 + 0.1(20 - 8.2)\Delta t$	6	9.4
6	$\hat{H}(6) + H'(6)\Delta t = 9.4 + 0.1(20 - 9.4)\Delta t$	7	10.4
7	$\hat{H}(7) + H'(7)\Delta t = 10.4 + 0.1(20 - 10.4)\Delta t$	8	11.4
8	$\hat{H}(8) + H'(8)\Delta t = 11.4 + 0.1(20 - 11.4)\Delta t$	9	12.3
9	$\hat{H}(9) + H'(9)\Delta t = 12.3 + 0.1(20 - 12.3)\Delta t$	10	13.0

ure 4.11), much the same way that the chemical concentration approaches the ambient concentration in a lung without absorption (Section 1.10). If the object had a higher value of α, for example, 0.3, the temperature would approach the ambient temperature much more quickly (Figure 4.12, Exercise 3).

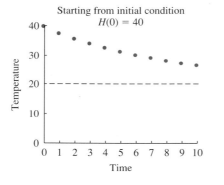
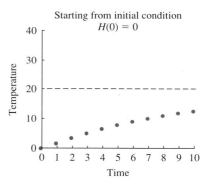

Figure 4.11
Newton's law of cooling approximated with Euler's method: small α

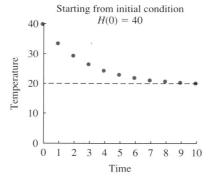
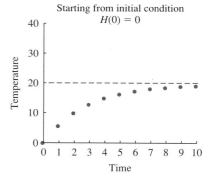

Figure 4.12
Newton's law of cooling approximated with Euler's method: large α

Diffusion Across a Membrane

The processes of heat exchange and chemical exchange have many parallels (which is remarkable given that heat is not a substance, as once thought). The underlying mechanism is that of **diffusion**. Substances wander out of a cell at a rate proportional to their concentration inside and into the cell at a rate proportional to their concentration outside. The constants of proportionality depend on the properties of the substance and the membrane separating the two regions (Figure 4.13).

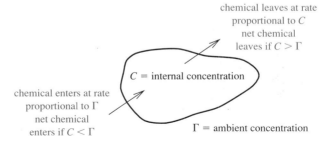

Figure 4.13
Diffusion between two regions

Denote the concentration inside the cell by C, the concentration outside by Γ ("gamma"), and the constant of proportionality by β. The rate at which the chemical leaves the cell is βC, and the rate at which it enters is $\beta \Gamma$. Therefore,

$$\frac{dC}{dt} = \text{rate at which chemical enters} - \text{rate at which chemical leaves}$$
$$= \beta\Gamma - \beta C$$
$$= \beta(\Gamma - C) \tag{4.6}$$

Except for the letters, this is exactly the same as Newton's law of cooling! The rate of change of concentration is proportional to the difference in concentration inside and outside the cell.

This model looks completely different from the updating function for the lung,

$$c_{t+1} = (1-q)c_t + q\gamma$$

(Equation 1.44), even though it describes a very similar situation. To be fair, the differential equation gives the rate of change of C, not the new value. The change in concentration for the lung described by the updating function is

$$\Delta c = c_{t+1} - c_t = -qc_t + q\gamma = q(\gamma - c_t)$$

The two models now look similar, with the fraction exchanged, q, playing the role of the rate, β.

Because this model is identical to Newton's law of cooling, their solutions must match. The concentration of chemical will get closer and closer to the ambient concentration, just as it did with the discrete-time dynamical system.

A Continuous-Time Model of Selection

Suppose we have two strains of bacteria, a and b, with rate of growth proportional to population size (Equation 4.4). If the per capita growth rate of strain a is μ and that of b is λ, they follow the equations

$$\frac{da}{dt} = \mu a$$
$$\frac{db}{dt} = \lambda b$$

If $\mu > \lambda$, type a has a higher per capita growth than b and we expect it to take over the population. However, we can only measure the *fraction* of bacteria of type a, not the total number. We would like to write a differential equation for the fraction, and we can do so by following steps much like those used to find the updating function for the fraction of bacteria competing in a discrete-time model (Section 1.12).

The fraction, p, of type a is

$$p = \frac{a}{a+b}$$

We can use the quotient rule (Theorem 2.10) to compute the derivative of p, finding

$$\frac{dp}{dt} = \frac{d}{dt}\left(\frac{a}{a+b}\right) \qquad \text{the derivative of } p$$

$$= \frac{(a+b)\frac{da}{dt} - a\frac{d(a+b)}{dt}}{(a+b)^2} \qquad \text{the quotient rule}$$

$$= \frac{a\frac{da}{dt} + b\frac{da}{dt} - a\frac{da}{dt} - a\frac{db}{dt}}{(a+b)^2} \qquad \text{the sum rule}$$

$$= \frac{b\frac{da}{dt} - a\frac{db}{dt}}{(a+b)^2} \qquad \text{cancel the } a\frac{da}{dt} \text{ terms}$$

$$= \frac{\mu ab - \lambda ab}{(a+b)^2} \qquad \text{substitute differential equations for } a \text{ and } b$$

$$= \frac{(\mu - \lambda)ab}{(a+b)^2} \qquad \text{factor}$$

We are not yet done, however. To write this as an autonomous differential equation, we must express the derivative as a function of the state variable p, not of the unmeasurable total populations a and b. Note that

$$1 - p = \frac{b}{a+b}$$

because $1 - p$ is the fraction of bacteria of type b. Then we have

$$\frac{dp}{dt} = \frac{(\mu - \lambda)ab}{(a+b)^2} \qquad \text{previous equation}$$

$$\frac{dp}{dt} = (\mu - \lambda)\left(\frac{a}{a+b}\right)\left(\frac{b}{a+b}\right) \qquad \text{factor}$$

$$= (\mu - \lambda)p(1-p) \qquad \text{write in terms of } p \text{ and } 1 - p$$

Although our derivation was similar, this differential equation looks nothing like the updating function for this system,

$$p_{t+1} = \frac{sp_t}{sp_t + r(1-p_t)}$$

(Equation 1.51). This is a key point. Never try to make a discrete-time dynamical system into a differential equation by cheerfully changing the letters. It is safer to go back to the underlying mechanism. In addition, the parameters in the two equations have different meanings. The parameters s and r in the updating function represent the per capita **reproduction**, the number of bacteria per bacterium in one generation, whereas the parameters μ and λ in the differential equation represent the per capita **growth rate**, the rate at which one bacterium produces new bacteria.

Like the updating function describing this system (Equation 1.51), this differential equation is **nonlinear** because the state variable p appears inside a nonlinear function, in this case a quadratic. As before, nonlinear equations are generally difficult to solve. In this case, however, the structure of the system makes it possible to find a solution. We begin by solving each of the differential equations for a and b. Suppose the initial number of type a bacteria is a_0 and the initial number of type b is b_0 (even though these values cannot be measured). The solutions are

$$a(t) = a_0 e^{\mu t}$$
$$b(t) = b_0 e^{\lambda t}$$

(Figure 4.14). We can check these directly:

$$\frac{da}{dt} = \frac{d}{dt} a_0 e^{\mu t} \quad \text{taking the derivative}$$
$$= a_0 \frac{d}{dt} e^{\mu t} \quad \text{the constant product rule}$$
$$= a_0 \mu e^{\mu t} \quad \text{the derivative of an exponential function}$$
$$= \mu a(t) \quad \text{recognize the formula for } a$$

Checking the solution for $b(t)$ is similar.

We can now find the equation for the fraction p as a function of t as

$$p(t) = \frac{a(t)}{a(t) + b(t)}$$
$$= \frac{a_0 e^{\mu t}}{a_0 e^{\mu t} + b_0 e^{\lambda t}}$$

Again, we are not quite done. The initial conditions a_0 and b_0 cannot be measured; only the initial fraction $p(0) = p_0$ can be measured. However, we do know that

$$p_0 = \frac{a_0}{a_0 + b_0}$$

Dividing the numerator and denominator of our expression for $p(t)$ by $a_0 + b_0$, we get

$$p(t) = \frac{\dfrac{a_0}{a_0 + b_0} e^{\mu t}}{\dfrac{a_0}{a_0 + b_0} e^{\mu t} + \dfrac{b_0}{a_0 + b_0} e^{\lambda t}}$$
$$= \frac{p_0 e^{\mu t}}{p_0 e^{\mu t} + (1 - p_0) e^{\lambda t}} \quad (4.7)$$

Does this solution check? First, we need to see whether $p(0)$ matches the initial condition p_0:

$$p(0) = \frac{p_0 e^{\mu \cdot 0}}{p_0 e^{\mu \cdot 0} + (1 - p_0) e^{\lambda \cdot 0}}$$
$$= \frac{p_0}{p_0 + (1 - p_0)}$$
$$= \frac{p_0}{1} = p_0$$

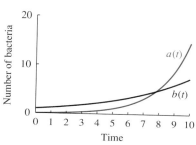

Figure 4.14
Exponential growth of two bacterial populations: $a_0 = 0.1$, $b_0 = 1.0$, $\mu = 0.5$, $\lambda = 0.2$

Second, we take the derivative to make sure that $p(t)$ really solves the differential equation. We get

$$\frac{dp}{dt} = \frac{[p_0 e^{\mu t} + (1-p_0)e^{\lambda t}]\mu p_0 e^{\mu t} - p_0 e^{\mu t}[\mu p_0 e^{\mu t} + \lambda(1-p_0)e^{\lambda t}]}{[p_0 e^{\mu t} + (1-p_0)e^{\lambda t}]^2}$$

$$= \frac{p_0(1-p_0)(\mu-\lambda)e^{\lambda t}e^{\mu t}}{[p_0 e^{\mu t} + (1-p_0)e^{\lambda t}]^2}$$

$$= (\mu - \lambda)\left[\frac{p_0 e^{\mu t}}{p_0 e^{\mu t} + (1-p_0)e^{\lambda t}}\right]\left[\frac{(1-p_0)e^{\lambda t}}{p_0 e^{\mu t} + (1-p_0)e^{\lambda t}}\right]$$

$$= (\mu - \lambda)p(t)[1 - p(t)]$$

We used the fact that

$$1 - p(t) = \frac{[1-p(0)]e^{\lambda t}}{p(0)e^{\mu t} + [1-p(0)]e^{\lambda t}}$$

in the last step.

We can use the methods of reasoning about functions to study this solution. Suppose we use the parameter values $\mu = 0.5$ and $\lambda = 0.2$ and the initial condition

$$p_0 = \frac{a_0}{a_0 + b_0} = \frac{0.1}{0.1 + 1.0} = 0.091$$

(as in Figure 4.14). The solution is

$$p(t) = \frac{0.091 e^{0.5t}}{0.091 e^{0.5t} + (1 - 0.091)e^{0.2t}}$$

$$= \frac{0.091 e^{0.5t}}{0.091 e^{0.5t} + 0.909 e^{0.2t}}$$

Because $0.5 > 0.2$, the leading behavior of the denominator for large values of t is the term $0.091 e^{0.5t}$ (Section 3.7). Therefore, for large t

$$p(t) \approx \frac{0.091 e^{0.5t}}{0.091 e^{0.5t}} = 1$$

The fraction of mutants does approach 1 as t approaches infinity (Figure 4.15).

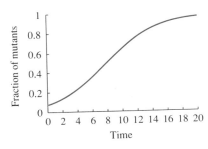

Figure 4.15
Solution for the fraction of mutant bacteria

SUMMARY We have introduced differential equations describing three processes, cooling, diffusion, and selection, and found that the first two processes are governed by the same equation. The equation for selection, derived by returning to the underlying model, is a nonlinear equation, but it can be solved by combining the solutions for each separate population.

4.2 EXERCISES

1. Suppose $\alpha = 0.2$/min and $A = 10°C$ for Newton's law of cooling. Use Euler's method to find the following quantities.

    a. The temperature 2 min after it was $-20°C$.
    b. The temperature 2 min after it was $40°C$.

c. The temperature 1 min after it was 40°C.
d. The temperature 1 min after part **c**.
e. How do the results of parts **a**, **b**, and **d** compare?

2. We will find later (Section 5.3) that the solution for Newton's law of cooling with initial condition $H(0)$ is

$$H(t) = A + [H(0) - A]e^{-\alpha t}$$

Set $\alpha = 0.2/\text{min}$, $A = 10°C$, and $H(0) = 40$ as in Exercise 1.
   a. Write the solution with these parameter values.
   b. Check the solution.
   c. Find the exact temperatures at $t = 1$ and $t = 2$ and compare with your results from Exercise 1.
   d. Sketch the graph of your solution. What happens as t approaches infinity?

3. Use Euler's method to match the results in Figure 4.12.

4. Using the solution for Newton's law of cooling, find the solution expressing the concentration of chemical inside a cell as a function of time in the following examples. Use this expression to find the concentration after 10 s, 20 s, and 60 s. Sketch your solutions for the first minute.
   a. $\beta = 0.01/\text{s}$, $C(0) = 5.0\ \mu\text{mol/cm}^3$, and $\Gamma = 2.0\ \mu\text{mol/cm}^3$.
   b. $\beta = 0.1/\text{s}$, $C(0) = 5.0\ \mu\text{mol/cm}^3$, and $\Gamma = 2.0\ \mu\text{mol/cm}^3$.
   c. $\beta = 0.1/\text{s}$, $C(0) = 3.0\ \mu\text{mol/cm}^3$, and $\Gamma = 5.0\ \mu\text{mol/cm}^3$.

5. When deriving the movement of chemical, we assumed that chemical moved as easily into the cell as out of it. Suppose instead that no chemical can return to the cell.
   a. Draw a version of Figure 4.13 illustrating this situation.
   b. Write the associated differential equation.
   c. Find the solution of this equation.
   d. What does this result mean?

6. Recall that the solution of the updating function

$$b_{t+1} = rb_t$$

is

$$b_t = r^t b_0$$

(Equation 1.29).
   a. For what values of b_0 and r does this solution match $b(t) = 1.0 \times 10^6 e^{2t}$ for all values of t?
   b. For what values of b_0 and r does this solution match $b(t) = 100e^{-3t}$ for all values of t?
   c. For what values of λ do solutions of the differential equation

$$\frac{db}{dt} = \lambda b$$

   grow?
   d. For what values of r do solutions of the discrete-time dynamical system grow?

e. What is the relation between r and λ?

7. Suppose $\mu = 2.0$, $\lambda = 1.0$, and $p(0) = 0.1$ in the equation $\frac{dp}{dt} = (\mu - \lambda)p(1 - p)$. Use Euler's method to estimate the following quantities.
   a. The proportion after 2 min using a time step of $\Delta t = 2.0$.
   b. The proportion after 2 min using a time step of $\Delta t = 0.5$.
   c. Using the formula for the exact solution of the selection equation (Equation 4.7), find the proportion after 2 min. Graph the solution for the first 3 min, and include the estimates from Euler's method.

8. The rate of change in the differential equation

$$\frac{dp}{dt} = (\mu - \lambda)p(1 - p)$$

is a measure of the strength of selection.
   a. Graph the rate of change as a function of p with $\mu = 2.0$ and $\lambda = 1.0$.
   b. Graph the rate of change as a function of p with $\mu = 0.2$ and $\lambda = 0.1$.
   c. Graph the rate of change as a function of p with $\mu = 1.0$ and $\lambda = 2.0$.
   d. For what value of p is the rate of change fastest?

9. Compute $\Delta p = p_{t+1} - p_t$ produced by the discrete-time dynamical system describing selection:

$$p_{t+1} = \frac{sp_t}{sp_t + r(1 - p_t)}$$

After a bit of algebra, your answer should look a bit more like the differential equation of $\frac{dp}{dt} = (\mu - \lambda)p(1 - p)$.

10. **COMPUTER:** Apply Euler's method to solve Newton's law of cooling function assuming that the ambient temperature oscillates with period T according to

$$A(t) = 20.0 + \cos\left(\frac{2\pi t}{T}\right)$$

The differential equation is

$$\frac{dH}{dt} = \alpha[A(t) - H]$$

which, with $A(0) = 20.0$, has a solution for $H(t)$ of

$$20.0 + \frac{\alpha^2 T^2 \cos\left(\frac{2\pi t}{T}\right) + 2\alpha\pi T \sin\left(\frac{2\pi t}{T}\right) - \alpha^2 T^2 e^{-\alpha t}}{\alpha^2 T^2 + 4\pi^2}$$

   a. Use a computer algebra system to check that this answer works.
   b. Plot $H(t)$ for five periods using values of T ranging from 0.1 to 10.0 when $\alpha = 1.0$. Compare $H(t)$ with $A(t)$. When does the temperature of the object most closely track the ambient temperature?

4.3 Solving Pure-Time Differential Equations

Pure-time differential equations describe situations where we have measured the rate of change of a quantity as a function of time and are interested in finding the quantity itself. We have seen two ways to approach solving pure-time differential equations: the method of guessing and Euler's method. We now formalize the method of guessing with a new tool, called the **antiderivative**, or **indefinite integral**.

Pure-Time Differential Equations and the Antiderivative

The general form of a pure-time differential equation is

$$\frac{dF}{dt} = f(t) \tag{4.8}$$

where $F(t)$ is the unknown state variable and $f(t)$ is the measured rate of change. The rate of change $f(t)$ depends only on the time t and not on the state variable F. We studied two pure-time differential equations in Section 4.1: an equation for volume with constant influx (Equation 4.1) and an equation for chemical product formation (Equation 4.2).

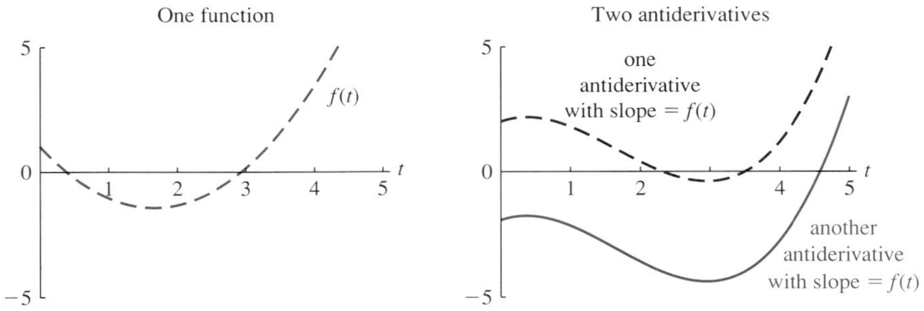

Figure 4.16
The function $f(t)$ and two of its antiderivatives

Pure-time differential equations are among the easiest differential equations to solve—and even they are often impossible. Solving a pure-time differential equation for $F(t)$ requires finding a function that has derivative equal to $f(t)$. Because this process undoes taking the derivative, F is called the **antiderivative** of f; F is also known as the **indefinite integral** of f.

■ **Definition 4.1** The antiderivative, or indefinite integral, of the function f is a function F with derivative equal to f, written

$$F(t) = \int f(t)\,dt$$

■

We say, "$F(t)$ is equal to the integral of f of t dt." The curvy symbol, the **integral sign**, can be thought of as a kind of function that takes one function f as input and returns another function F as output. We call $f(t)$ the **integrand**. The obligatory dt indicates that t is the variable.

Every differentiable function has only one derivative, but the same function has a whole family of antiderivatives. The two functions graphed in Figure 4.16 share the same derivative $f(t)$ and are *both* antiderivatives of f. All antiderivatives must share the same slope, however, and can differ from each other only by a constant. We write

$$\int f(t)\,dt = F(t) + c$$

to indicate we can add any constant c to any particular antiderivative $F(t)$. The constant c is called an **arbitrary constant** because it can take on any value.

It might seem like a nuisance that we must include an arbitrary constant. It is not. Antiderivatives are the way to find the solutions of pure-time differential equations. A differential equation requires an initial condition to get the dynamics started, and the arbitrary constant allows us to find solutions that match the initial condition.

Rules for Antiderivatives

To find the indefinite integrals (antiderivatives) of interesting functions, we begin by using three of the basic rules of differentiation, the power rule (Theorem 2.6), the constant product rule (special case of Theorem 2.9) and the sum rule (Theorem 2.5).

The Power Rule for Integrals

The power rule for derivatives says that

$$\frac{dx^{n+1}}{dx} = (n+1)x^n$$

Dividing both sides by $n + 1$, we get

$$\frac{d}{dx}\left(\frac{x^{n+1}}{n+1}\right) = x^n$$

We have derived the **power rule** for indefinite integrals.

■ **THEOREM 4.1** (The Power Rule for Integrals)

Suppose

$$f(x) = x^n$$

where $n \neq -1$. Then

$$\int x^n\,dx = \frac{x^{n+1}}{n+1} + c$$

We will discuss the case $n = -1$ in Section 4.4.

For example, using the power rule for integrals with $n = 2$, the indefinite integral of x^2 is

$$\int x^2\,dx = \frac{x^{2+1}}{2+1} + c = \frac{x^3}{3} + c$$

Graphically, the slope of the indefinite integral must match that of the original function. In Figure 4.17a, the slope of the indefinite integral $F(x) = x^3/3$ is the original function $f(x) = x^2$. The indefinite integral is always increasing because the original function is always positive, and increases most steeply where $f(x)$

takes on its largest values. With $n = -2$, the indefinite integral of $1/x^2$ is

$$\int \frac{1}{x^2} dx = \int x^{-2} dx = \frac{x^{-2+1}}{-2+1} = -x^{-1} + c$$

The power rule also works with fractional powers. When $n = 0.5$, we have

$$\int x^{0.5} dx = \frac{x^{0.5+1}}{0.5+1} + c = \frac{x^{1.5}}{1.5} + c$$

and when $n = -0.5$,

$$\int x^{-0.5} dx = \frac{x^{-0.5+1}}{-0.5+1} + c = \frac{x^{0.5}}{0.5} + c = 2.0x^{0.5} + c$$

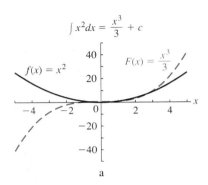

Figure 4.17
Two indefinite integrals found with the power rule

The Constant Product Rule for Integrals

The constant product rule for derivatives states

$$\frac{d}{dx}[af(x)] = a\frac{df}{dx}$$

The derivative of a constant times a function is the constant times the derivative. Antiderivatives work the same way.

■ **THEOREM 4.2** (The Constant Product Rule for Integrals)

Suppose

$$\int f(x) dx = F(x) + c$$

Then

$$\int af(x) dx = a\int f(x) dx = aF(x) + c$$

for any constant c. ■

Notice that we did not multiply the arbitrary constant c by the factor a; we always add the arbitrary constant *last*.

For example, if we multiply the function $f(x) = x^2$ by 5, we multiply the antiderivative by 5, so we have

$$\int 5x^2 dx = 5 \int x^2 dx \qquad \text{pull constant 5 out front}$$

$$= 5\frac{x^3}{3} + c \qquad \text{find integral with power rule, } n = 2$$

Similarly,

$$\int \frac{-3}{x^2} dx = -3 \int x^{-2} dx \qquad \text{pull constant } -3 \text{ outside}$$

$$= \frac{3}{x} + c \qquad \text{find integral with power rule, } n = -2$$

The Sum Rule for Integrals The sum rule for derivatives says that the derivative of the sum is the sum of the derivatives, or

$$\frac{d(f + g)}{dx} = \frac{df}{dx} + \frac{dg}{dx}$$

Again, antiderivatives work the same way.

■ **THEOREM 4.3** (The Sum Rule for Integrals)

Suppose

$$\int f(x) \, dx = F(x) + c$$

$$\int g(x) \, dx = G(x) + c$$

then

$$\int f(x) + g(x) \, dx = \int f(x) \, dx + \int g(x) \, dx = F(x) + G(x) + c \qquad ■$$

Again, there is only a single arbitrary constant c, added at the end.

We have found that

$$\int 5x^2 dx = \frac{5x^3}{3} + c$$

$$\int -3x^{-2} dx = \frac{3}{x} + c$$

Therefore,

$$\int (5x^2 - 3x^{-2}) \, dx = \frac{5x^3}{3} + \frac{3}{x} + c$$

The power rule, constant product rule, and sum rule are summarized in the following table.

The power, constant-product, and sum rules for integrals

Rule	Formula
Power rule	$\int x^n dx = \frac{x^{n+1}}{n+1} + c$ if $n \neq -1$
Constant product rule	$\int af(x) \, dx = a \int f(x) \, dx$
Sum rule	$\int [f(x) + g(x)] \, dx = \int f(x) \, dx + \int g(x) \, dx$

Solving Polynomial Differential Equations

With the power, constant product, and sum rules we can find the indefinite integral of any polynomial and can therefore solve any pure-time differential equation where the rate of change is a polynomial function of time.

The simplest polynomial is a constant. We can solve the differential equation

$$\frac{dV}{dt} = 1.0$$

with the power rule because the rate of change is $1.0 = t^0$. Evaluating the antiderivative, we get

$$\begin{aligned} V(t) &= \int 1.0 \, dt \\ &= \int t^0 \, dt \\ &= \frac{t^{0+1}}{0+1} + c \\ &= t + c \end{aligned}$$

The constant c is determined by the initial conditions. Suppose the volume is 300 at time 0. This gives an equation for the arbitrary constant c,

$$V(0) = 0 + c = 300$$

or $c = 300$. The solution of the pure-time differential equation

$$\frac{dV}{dt} = 1.0$$

with initial condition $V(0) = 300$ is therefore

$$V(t) = t + 300$$

(Figure 4.18).

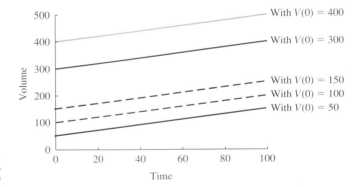

Figure 4.18
Several solutions of $dV/dt = 1.0$

Problems involving finding the indefinite integrals of polynomials arise in the study of the motion of objects with **constant acceleration**. Suppose we are told that a rock is thrown downward from a 100 m tall building with initial downward velocity of 5.0 m/s. The acceleration of gravity is 9.8 m/s^2 (Figure 4.19). How long will it take the rock to hit the ground? How fast will it be going?

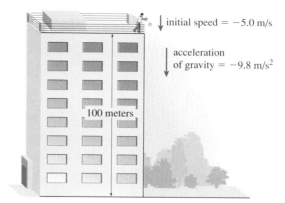

Figure 4.19

A falling rock

Let a represent acceleration, v velocity, and p the position of the rock. The basic differential equations from physics are

$$\frac{dv}{dt} = a \quad \text{acceleration is the rate of change of velocity}$$

$$\frac{dp}{dt} = v \quad \text{velocity is the rate of change of position}$$

The first says that acceleration is the rate of change of velocity and the second that velocity is the rate of change of position. To begin with, we have measured that $a = -9.8$ and that $v(0) = -5.0$. Both values are negative because both are downward. We therefore have enough information to solve the differential equation

$$\frac{dv}{dt} = -9.8$$

with the initial condition $v(0) = -5.0$. As above, the rate of change is constant, so we have

$$v(t) = -9.8t + c$$

We find the constant c with the initial condition by solving

$$v(0) = -9.8 \cdot 0 + c = -5.0$$

so $c = -5.0$. Therefore, the equation for the velocity is

$$v(t) = -9.8t - 5.0$$

As time passes, the velocity becomes more and more negative as the rock falls faster and faster (Figure 4.20a).

We now have enough information to solve for the position p; the pure-time differential equation

$$\frac{dp}{dt} = -9.8t - 5.0$$

and the initial condition $p(0) = 100$. We first solve the differential equation with the indefinite integral, finding

336 Chapter 4 Differential Equations, Integrals, and Their Applications

$$p(t) = \int (-9.8t - 5.0)\, dt \qquad p(t) \text{ is the indefinite integral of } v(t)$$

$$= \int -9.8t\, dt - \int 5.0\, dt \qquad \text{the sum rule}$$

$$= -9.8 \int t\, dt - \int 5.0\, dt \qquad \text{the constant product rule}$$

$$= -9.8 \frac{t^2}{2} - 5.0t + c \qquad \text{the power rule}$$

$$= -4.9t^2 - 5.0t + c$$

We use the initial condition to find the constant c by solving

$$p(0) = -4.9 \cdot 0^2 - 5.0 \cdot 0 + c = 100.0$$

so $c = 100.0$. Therefore, the equation for the position is

$$p(t) = -4.9t^2 - 5.0t + 100$$

(Figure 4.20b).

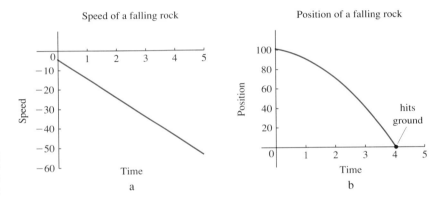

Figure 4.20

The velocity and position of a falling rock

The solution tells when the rock will hit the ground and how fast it will be going. It hits the ground at the time when $p(t) = 0$, the solution of

$$-4.9t^2 - 5.0t + 100 = 0$$

For convenience, we multiply by -1, so

$$4.9t^2 + 5.0t - 100 = 0$$

With the quadratic formula, we get

$$t = \frac{-5.0 \pm \sqrt{5.0^2 - 4 \cdot 4.9 \cdot (-100)}}{2 \cdot 4.9}$$

$$= \frac{-5.0 \pm \sqrt{1985}}{9.8}$$

$$\approx \frac{-5.0 \pm 44.55}{9.8}$$

$$\approx 4.036 \quad \text{or} \quad -5.506$$

The positive solution is the one that makes sense. How fast will the rock be going when it hits the ground? We use the formula for $v(t)$ to find

$$v(4.036) = -9.8 \cdot 4.036 - 5.0 \approx -44.55 \text{ m/s}$$

(Figure 4.21).

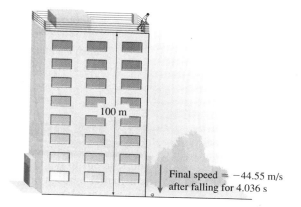

Figure 4.21
A falling rock hits the ground

Polynomial pure-time differential equations also arise in public health policy. During the early years of the AIDS epidemic, workers at the Centers for Disease Control found that the number of new AIDS cases followed the formula

$$\text{rate at which new AIDS cases were reported} \approx 523.8t^2 \quad (4.9)$$

where t is measured in years after the beginning of 1981 (Hyman and Stanley, 1988). Let $A(t)$ denote the total number of people ever infected by this disease. The Centers for Disease Control's formula can be written as a polynomial pure-time differential equation

$$\frac{dA}{dt} = 523.8t^2$$

To solve this equation for the total number of AIDS patients, we require an initial condition. Surveys indicated that about 340 people had been infected at the beginning of 1981 ($t = 0$ in the model), so $A(0) = 340$. We can now solve for $A(t)$, finding

$$A(t) = \int 523.8t^2 \, dt = 523.8 \frac{t^3}{3} = 174.6t^3 + c$$

The constant c is the solution of the equation

$$A(0) = 174.6 \cdot 0^3 + c = 340$$

so $c = 340$. The solution is

$$A(t) = 174.6t^3 + 340$$

(Figure 4.22).

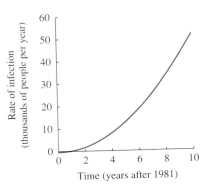
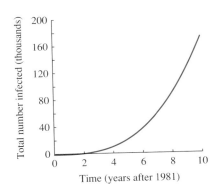

Figure 4.22
The early course of the AIDS epidemic in the United States

SUMMARY

Solving pure-time differential equations requires undoing the derivative. We defined the **antiderivative** of a function f as the function whose derivative is f, and we introduced the notation of the **indefinite integral** to express antiderivatives. Unlike the derivative, the antiderivative includes an **arbitrary constant** to indicate that a whole family of functions share the same slope. From the rules for computing derivatives, we found the **power**, **constant product**, and **sum rules** for integrals. With these rules, we can find the indefinite integral of any polynomial. We applied this method to solve pure-time differential equations that describe the physics of a falling rock and the number of people infected with the AIDS virus.

4.3 EXERCISES

1. Which of the following are pure-time differential equations? Which are autonomous differential equations? Which are neither?
   a. $dV/dt = 2$
   b. $dV/dt = 2Vt$
   c. $dV/dt = 2V$
   d. $dV/dt = 2t$
   e. $dV/dt = 2t^2 + 5$
   f. $dV/dt = 2t^2 + 5V$

2. From the following graphs, sketch one antiderivative of the function.

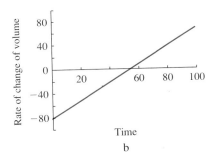
b

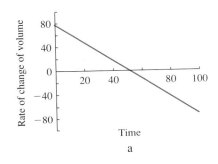
a

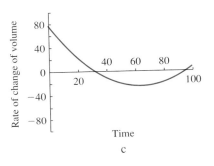
c

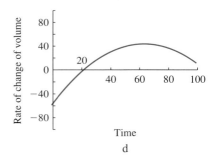

3. Find the antiderivatives of the following functions.
  a. $7x^2$
  b. $72t + 5$
  c. $y^4 + 5y^3$
  d. $10t^9 + 6t^5$
  e. $\dfrac{5}{x^3}$
  f. $3z^{3/7}$
  g. $\dfrac{2}{\sqrt[3]{t}} + 3$

4. Use antiderivatives to solve the following differential equations. Sketch a graph of the rate of change and the solution.
  a. $dV/dt = 2t^2 + 5$ with $V(1) = 19.0$
  b. $df/dx = 5x^3 + 5x$ with $f(0) = -12.0$
  c. $dp/dt = 5t^3 + 5/t^2$ with $p(1) = 12.0$

5. Suppose a cell is taking water into two vacuoles, each of which begins with a volume of 10 μm^3. Let V_1 denote the volume of the first vacuole and V_2 the volume of the second. Suppose that

$$\dfrac{dV_1}{dt} = 2.0t + 5.0$$

$$\dfrac{dV_2}{dt} = 5.0t + 2.0$$

where t is measured in seconds and volume is measured in cubic microns. Define $V = V_1 + V_2$ to be the total vacuole volume of the cell.
  a. Solve for $V_1(t)$ and $V_2(t)$.
  b. Write a differential equation for V. What are the initial conditions?
  c. Show that the solution of the differential equation for V is the sum of the solutions for V_1 and V_2.
  d. Write this fact as an instance of the sum rule for integrals.

6. Pick two power functions $f(x)$ and $g(x)$.
  a. Find the antiderivatives $F(x) = \int f(x)\,dx$ and $G(x) = \int g(x)\,dx$.
  b. Find the function $h(x) = f(x)g(x)$ and show it is a power function.
  c. Find the antiderivative $H(x)$ of $h(x)$.
  d. Show that $H(x) \neq F(x)G(x)$.
  e. Use f, g, and h to show that the following is *not* a correct integral product rule:

$$\int f(x)g(x)\,dx = \left[\int f(x)\,dx\right]\left[\int g(x)\,dx\right]$$

7. Suppose an organism grows in mass according to the differential equation

$$\dfrac{dM}{dt} = \dfrac{\alpha}{\sqrt{t}}$$

where M is measured in grams, t is measured in days, and $\alpha = 2.0$ g/$\sqrt{\text{days}}$.
  a. Check that the units match.
  b. Suppose that $M(0) = 5.0$ g. Find the solution.
  c. Sketch a graph of the rate of change and the solution.
  d. What happens to the mass of the organism? Is your result consistent with the fact that the organism is growing more and more slowly?

8. Suppose a person standing near the edge of a 100 m-tall building tosses a rock straight up with velocity 5.0 m/s, but fails to catch it on the way down, allowing it to hurtle toward the ground with a constant acceleration of 9.8 m/s^2.
  a. Find the velocity and position of the rock as functions of time.
  b. How high will the rock get?
  c. How long will it take to pass the person on the way down? How fast will it be moving?
  d. How long will it take to hit the ground? How fast will it be moving? How fast is this in miles per hour?
  e. Graph the velocity and position as functions of time.

9. Suppose the velocity of an object measured at discrete times is found to be

Time	Velocity
0	1.0
1	3.0
2	6.0
3	10.0
4	15.0

The position at $t = 0$ is $p(0) = 10.0$.
  a. Graph these data.
  b. Using what you know at $t = 0$, use Euler's method (the tangent line approximation) to estimate the position at $t = 1$.
  c. Continue in this way and estimate the position at $t = 5$.

10. Suppose the velocity of an object is

$$v(t) = 0.5t^2 + 1.5t + 1.0$$

and the initial position is $p(0) = 10.0$.
  a. Find the acceleration and the initial velocity.
  b. Find the position as a function of time.
  c. Estimate the position at $t = 5$ with Euler's method and a time step of $\Delta t = 1$, and compare your answer with the answer to Exercise 9.

4.4 Integration of Special Functions and Integration by Substitution

We have seen how to solve any pure-time differential equation with the antiderivative when the rate of change can be expressed as a combination of power functions. There are not many other functions for which we can find the antiderivative, and thus not many pure-time differential equations we can solve. In this section, we learn how to find integrals of exponential, logarithmic, and cosine functions. By using the chain rule for derivatives, we can derive the method of **substitution**, which can be used to evaluate integrals involving parameters.

Integrals of Special Functions

The derivatives of special functions help us to find a few more antiderivatives. Recall that

$$\frac{d}{dx} \ln x = \frac{1}{x}$$

$$\frac{d}{dx} e^x = e^x$$

Because the indefinite integral, the antiderivative, undoes the action of the derivative, we have

$$\int \frac{1}{x} dx = \ln |x| + c \tag{4.10}$$

$$\int e^x dx = e^x + c \tag{4.11}$$

Try not to be confused by the absolute value bars inside the function $\ln |x| + c$ (Figure 4.23). The function $1/x$, unlike the natural logarithm, is well defined for negative values of x. It must be the slope of some function. The graph of $1/x$ for negative x is the negative of the graph for positive x. The antiderivative for negative x must be the mirror image of the antiderivative for positive x. The power rule for integrals (Theorem 4.1) left out $n = -1$; this special integral fills this gap.

Figure 4.23

The integral of $1/x$

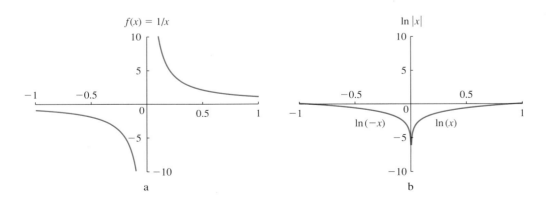

Furthermore, using the derivatives of sine and cosine

$$\frac{d}{dx}\sin(x) = \cos(x)$$

$$\frac{d}{dx}\cos(x) = -\sin(x)$$

we see that

$$\int \cos(x)\,dx = \sin(x) + c \qquad (4.12)$$

$$\int \sin(x)\,dx = -\cos(x) + c \qquad (4.13)$$

Using the constant product and sum rules, we can integrate a few more complicated functions. For example,

$$\int \frac{1}{x} + 2e^x\,dx + 5\sin(x)\,dx = \int \frac{1}{x}\,dx + 2\int e^x\,dx + 5\int \sin(x)\,dx$$
$$= \ln|x| + 2e^x - 5\cos(x) + c$$

according to the sum rule (Theorem 4.3).

These elements form the basic building blocks of integration. The edifice that can be built from these blocks, even with several additional techniques, is rather paltry. Many perfectly reasonable functions cannot be integrated at all. Computer algebra systems know all the sophisticated techniques and save us from having to learn them. Only one, integration by substitution, is of sufficiently general importance to warrant attention in this book.

The Chain Rule and Integration by Substitution

The chain rule (Theorem 2.11) tells how to differentiate the compositions of functions, stating that

$$\frac{d(f \circ g)}{dx} = \frac{df}{dg}\frac{dg}{dx}$$

as expressed in differential notation. Integration by substitution is easier to understand in differential notation, and we will use it here. Because the indefinite integral, the antiderivative, undoes the action of the derivative,

$$\int \frac{df}{dg}\frac{dg}{dx}\,dx = (f \circ g)(x) + c$$

If we were lucky enough to find and recognize an integral of this form, we could use this method to integrate.

Suppose we are asked to find

$$\int 2xe^{x^2}\,dx$$

If we were clever or lucky enough, we might notice that

$$2xe^{x^2} = \frac{dg}{dx}\frac{df}{dg}$$

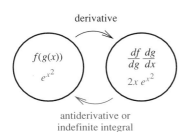

Figure 4.24
The chain rule and the indefinite integral

if $g(x) = x^2$ and $f(g) = e^g$ (Figure 4.24). In other words,

$$\frac{d(f \circ g)}{dx} = \frac{df}{dg}\frac{dg}{dx}$$
$$= e^g \cdot 2x$$
$$= 2xe^{x^2}$$

Therefore, because the indefinite integral undoes the action of the derivative, we have

$$\int 2xe^{x^2} dx = e^{x^2} + c$$

Integration by substitution is a way to recognize these patterns, when they exist, without relying entirely on inspired guessing. There is, however, no guarantee that the technique will work.

■ **Algorithm 4.2** (Integration by substitution)

1. Define a new variable as some function of the old variable.
2. Take the derivative of the new variable with respect to the old variable.
3. Treat the derivative like a fraction and move the dx to the other side.
4. Put everything in the integral in terms of the new variable.
5. Hope you can integrate the expression in terms of the new variable.
6. After integrating, try to put everything back in terms of the old variable.

■

For our example, Algorithm 4.2 proceeds as follows:

1. Define a new variable to be the function of x most tangled up in the expression. In this case, x^2 is inside the exponential, so define y by

$$y = x^2$$

Our hope is to write everything in terms of the new variable y. The e^{x^2} term is now e^y, which we know how to integrate. However, the piece $2x\,dx$ remains.

2. Find the derivative of y with respect to x:

$$\frac{dy}{dx} = 2x$$

3. This step looks weird but is legal. Multiply both sides by dx to find

$$dy = 2x\,dx$$

4. By good luck, dy exactly matches the left-over piece $2x\,dx$. In terms of y, the integral is

$$\int 2xe^{x^2}\,dx = \int e^y\,dy$$

5. By more good luck, this is an expression we know how to integrate, finding

$$\int e^y\,dy = e^y + c$$

6. Put everything back in terms of x (using $y = x^2$), so

$$\int 2xe^{x^2} \, dx = e^{x^2} + c$$

We will follow these steps to evaluate

$$\int \frac{e^{2t}}{(1 + e^{2t})^2} \, dt$$

1. It is far from obvious what to substitute here. The right guess is

$$u = 1 + e^{2t}$$

2. Take the derivative, finding

$$\frac{du}{dt} = 2e^{2t}$$

3. Solve for du:

$$du = 2e^{2t} \, dt$$

4. Put everything in terms of u. Note that $e^{2t} \, dt = du/2$, so

$$\int \frac{e^{2t}}{(1 + e^{2t})^2} \, dt = \int \frac{1}{u^2} \frac{du}{2}$$

5. We can attack the new integral with the constant product and power rules, finding

$$\int \frac{1}{u^2} \frac{du}{2} = \frac{1}{2} \int u^{-2} \, du$$

$$= \frac{1}{2}(-u^{-1}) + c = -\frac{1}{2u} + c$$

6. Put everything back in terms of t, arriving at

$$\int \frac{e^{2t}}{(1 + e^{2t})^2} \, dt = -\frac{1}{2u} + c = -\frac{1}{2(1 + e^{2t})} + c$$

The problem with integration by substitution is that terms must match up exactly. For example, suppose we wanted to evaluate

$$\int 2x^2 e^{x^2} \, dx \quad (4.14)$$

1. Try the substitution $y = x^2$ as before.

2. Then we get

$$\frac{dy}{dx} = 2x$$

3. We have $dy = 2x \, dx$.

4. Putting everything in terms of y is messy:

$$2x^2 \, dx = x \cdot 2x \, dx = x \, dy$$

To get rid of the remaining x, we have to solve for x in terms of y, finding $x = \sqrt{y}$. Plugging this in gives the new integral,

$$\int \sqrt{y} e^y \, dy$$

5. We have no idea how to integrate this.

6. Give up.

In fact, there is no way to compute this indefinite integral in terms of basic functions.

Getting Rid of Excess Constants

Substitution is guaranteed to work in one situation: when the integral would be possible if we could remove some excess constants. For example, the differential equation

$$\frac{dL}{dt} = 6.48 e^{-0.09t} \tag{4.15}$$

can be used to describe the growth of a fish. If we could just get rid of the -0.09 in the exponent there would be no problem finding the integral and solving the differential equation. Substitution can remove the excess constant. Alternatively, suppose we found that the rate of energy loss from an organism is proportional to the temperature above $36°C$, so that

$$\frac{dE}{dt} = -P_d(t) = -0.8 - 0.3 \cos[2\pi(t - 0.583)]$$

where we assumed that temperature follows the daily temperature cycle studied in Section 1.9. We would have no problem integrating and finding the total energy loss if we could make that $2\pi(t - 0.583)$ inside the cosine into a simple t. Again, substitution is guaranteed to work.

Suppose fish size begins at $L(0) = 0.0$ (measured from fertilization) and follows the differential equation

$$\frac{dL}{dt} = 6.48 e^{-0.09t}$$

This equation expresses that fact that fish grow more and more slowly but never stop growing throughout their lives. These match the growth data for walleye in North Caribou Lake, Ontario (Charnov, 1993). To solve it, we need to find the indefinite integral of the rate of change $6.48 e^{-0.09t}$ and then compute the arbitrary constant. To find the integral, we follow Algorithm 4.2.

1. Define a new variable, $z = -0.09t$.

2. We have $\dfrac{dz}{dt} = -0.09$.

3. Then $dz = -0.09\, dt$.

4. To get dt in terms of dz, solve to find

$$dt = -\frac{dz}{0.09}$$

and write

$$\int 6.48 e^{-0.09t}\, dt = \int -6.48 \frac{e^z}{0.09}\, dz = \int -72.0 e^z\, dz$$

5. Integrate, finding

$$\int -72.0 e^z\, dz = -72.0 \int e^z\, dz = -72.0 e^z + c$$

6. Put everything back in terms of t, arriving at

$$\int 6.48 e^{-0.09t}\, dt = -72.0 e^z + c = -72.0 e^{-0.09t} + c \quad (4.16)$$

To finish solving the differential equation, we must find the arbitrary constant. The equation is

$$L(0) = 0.0 = -72.0 e^{-0.09 \cdot 0} + c = -72.0 + c$$

so $c = 72.0$. The solution is

$$L(t) = 72.0 - 72.0 e^{-0.09t} = 72.0(1 - e^{-0.09t}) \quad (4.17)$$

This is called the **Bertalanffy growth equation** (Figure 4.25). The size approaches a limit of 72.0 cm, although the fish never stop growing.

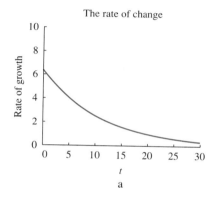

Figure 4.25
Growth of fish

We can use this solution to find the age at which fish mature. Walleye begin to reproduce when they reach about 45 cm in length (but continue to grow after that). How long will it take these fish to mature? We must solve

$$L(t) = 45$$

or

$$72.0(1 - e^{-0.09t}) = 45.0$$

$$1 - e^{-0.09t} = \frac{45.0}{72.0} = 0.625$$

$$e^{-0.09t} = 0.375$$

$$-0.09t = \ln(0.375) = -0.98$$

$$t = \frac{-0.98}{-0.09} = 10.9$$

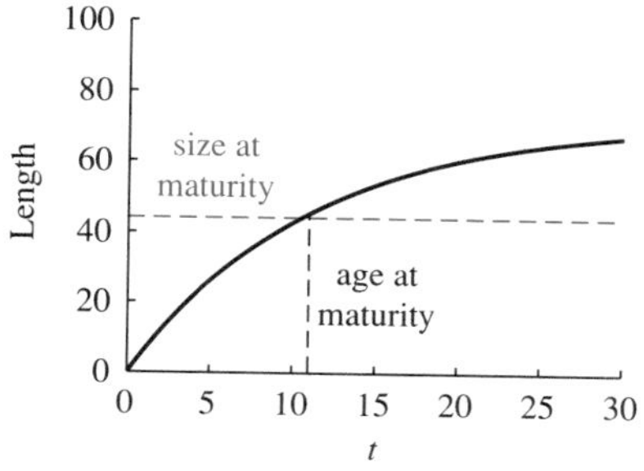

Figure 4.26
Finding the age at maturity of a walleye

These fish will take 10.9 years to mature (Figure 4.26).

Similarly, we can use substitution to find the total energy used in a day from the differential equation

$$\frac{dE}{dt} = -P_d(t) = -0.8 - 0.3 \cos[2\pi(t - 0.583)]$$

with initial condition $E(0) = 0$. To solve the equation, we must find the indefinite integral,

$$\int \{-0.8 - 0.3 \cos[2\pi(t - 0.583)]\} \, dt$$

The hard term to deal with is $\cos[2\pi(t - 0.583)]$. We attack with the method of substitution.

1. Define a new variable $y = 2\pi(t - 0.583)$, the linear function of t inside the cosine.

2. We have $\dfrac{dy}{dt} = 2\pi$.

3. Then $dy = 2\pi \, dt$.

4. To write dt in terms of dy, solve to find

$$dt = \frac{dy}{2\pi}$$

and write

$$\int \cos[2\pi(t - 0.583)] = \int \cos(y) \frac{dy}{2\pi}$$

5. Integrate, finding

$$\int \cos(y) \frac{dy}{2\pi} = \frac{1}{2\pi} \int \cos(y) \, dy = \frac{1}{2\pi} \sin(y) + c$$

6. Put everything back in terms of t, arriving at

$$\int \cos[2\pi(t - 0.583)] = \frac{1}{2\pi} \sin[2\pi(t - 0.583)] + c$$

The solution of the whole equation is

$$E(t) = \int -0.8 - 0.3 \cos[2\pi(t - 0.583)] \, dt$$

$$= -0.8t - \frac{0.3}{2\pi} \sin[2\pi(t - 0.583)] + c$$

$$= -0.8t - 0.048 \sin[2\pi(t - 0.583)] + c$$

To find the initial condition, we must solve

$$E(0) = 0 = -0.8 \cdot 0 - 0.048 \sin[2\pi(0 - 0.583)] + c = -0.024 + c$$

so that $c = 0.024$. The solution is

$$E(t) = -0.8t - 0.048 \sin[2\pi(t - 0.583)] + 0.024$$

(Figure 4.27).

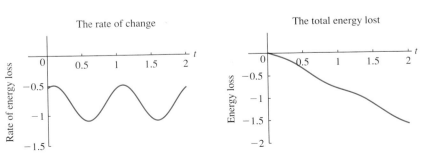

Figure 4.27

Loss of energy due to temperature

4.4 EXERCISES

1. Find the indefinite integrals of the following functions.
   a. $e^x + \dfrac{1}{x} + \sin(x) + \cos(x)$
   b. $3e^x + 2x^3$
   c. $\dfrac{2}{t} + \dfrac{t}{2}$
   d. $\dfrac{3}{z^2} + \dfrac{z^2}{3}$
   e. $2\sin(x) + 3\cos(x)$

2. Find the indefinite integrals of the following functions.
   a. $3e^{x/5}$
   b. $5x^6$
   c. $\left(1 + \dfrac{t}{2}\right)^4$
   d. $(1 + 2t)^{-4}$
   e. $\dfrac{1}{4 + t}$
   f. $\dfrac{1}{1 + 4t}$
   g. $\cos[2\pi(x - 2)]$

3. Use the given substitution to find the indefinite integrals of the following functions.
   a. $\dfrac{e^x}{1 + e^x}$, substitution $y = 1 + e^x$
   b. $2y\sqrt{1 + y^2}$, substitution $z = 1 + y^2$
   c. $\cos(x)e^{\sin(x)}$, substitution $y = \sin(x)$
   d. $\tan(\theta)$ (write it as $\dfrac{\sin(\theta)}{\cos(\theta)}$ and use the substitution $y = \cos(\theta)$)
   e. $e^t(1 + e^t)^4$, substitution $y = 1 + e^t$
   f. $\dfrac{t}{1 + t}$
   (HINT: Write this as $1 - \dfrac{1}{1 + t}$)

4. Suppose that the rate at which a chemical product is formed in a reaction is
$$\dfrac{dP}{dt} = \dfrac{5}{1 + 2.0t}$$
where t is measured in minutes and P in moles.
   a. If there is no product at time 0, use one step of Euler's method to estimate how much there is after 0.1 minute.
   b. Write the solution for all time of this equation.
   c. Compare with the result of part **a** at $t = 0.1$.
   d. Sketch the solution.
   e. What happens to the amount of product as t becomes large?

5. Suppose that the rate at which a chemical product is formed in a reaction is
$$\dfrac{dP}{dt} = 5.0e^{-2.0t}$$
where t is measured in minutes and P in moles.
   a. If there is no product at time 0, use one step of Euler's method to estimate how much there is after 0.1 minute. It should match the rate found in part **a** of Exercise 4.
   b. Write the solution for all time of this equation.
   c. Compare with the result of part **a** at $t = 0.1$.
   d. What happens to the amount of product as t becomes large?
   e. Sketch the solution.
   f. Why do you think the behavior of this solution is so different from that in Exercise 4?

6. The population of lemmings at the top of a cliff is increasing exponentially by the formula
$$L_t(t) = 1000e^{0.2t}$$
where t is measured in years. However, lemmings leap off the cliff at a rate equal to $0.1L_t$ and pile up at the bottom.
   a. Write a pure-time differential equation for the number of lemmings $L_b(t)$ piled up at the bottom of the cliff.
   b. Solve using the initial condition $L_b(0) = 1000$.
   c. Graph the number of lemmings at the top and the number at the bottom of the cliff.
   d. Find the limit of the ratio $\dfrac{L_b(t)}{L_t(t)}$ at t approaches infinity.
   e. Try the same steps with $L_t = 1000e^{0.05t}$. Do you have any idea why the answers seem so different?

7. Walleye in Texas grow much more quickly than those in Ontario (Figure 4.25), following the equation
$$\dfrac{dL}{dt} = 64.3e^{-1.19t}$$
   a. Solve this equation, assuming that $L(0) = 0$.
   b. Find the limit of size as t approaches infinity.
   c. Assume (as is nearly true) that the walleye mature at the same size of 45 cm in length. How old are the Texas walleye when they mature?
   d. Graph the size and compare with Figure 4.25.

8. The Centers for Disease Control usually reports data where t represents the actual year. They give the differential equation describing the number of AIDS cases (Equation 4.9)

as

$$\frac{dA}{dt} = 523.8(t - 1981)^2$$

Remember, here t takes on values larger than 1981. The initial condition is $A(1981) = 340$.
a. Multiply this equation out and solve.
b. Use substitution to solve.
c. Compare your answer with that given in the text.

9. Suppose that temperature follows

$$T(t) = 20.0 + 10.0\cos\left[\frac{2\pi(t - 190.0)}{365}\right]$$

where t is measured in days and temperature is measured in degrees centigrade. Suppose that the growth of an insect follows

$$\frac{dL}{dt} = 0.1T(t)$$

with $L(0) = 0.1$ cm.
a. Sketch a graph of the temperature over the course of a year.
b. Suppose an insect starts growing on January 1. How big will it be after 30 days?
c. Suppose an insect starts growing on June 1 (day 151). How big will it be after 30 days?
d. What is the best day for a bug to hatch?

10. **COMPUTER:** Consider again the fish in Caribou Lake, growing according to

$$\frac{dL}{dt} = \beta e^{-\alpha t}$$

where $\beta = 6.48$ and $\alpha = 0.09$. However, suppose there is some variability among fish in the values of these two parameters.
a. Solve for growth trajectories of five fish with values of β evenly spread from 10% below to 10% above 6.48.
b. Solve for growth trajectories of five fish with values of α evenly spread from 10% below to 10% above 0.09.
c. Which parameter has a greater effect on the size of the fish?

11. **COMPUTER:** Clever genetic engineers design a set of fish that grow according to the following equations.

$$\frac{dL_1}{dt} = 1$$

$$\frac{dL_2}{dt} = \frac{1}{1 + \sqrt{t}}$$

$$\frac{dL_3}{dt} = \frac{1}{1 + t}$$

$$\frac{dL_4}{dt} = \frac{1}{1 + t^2}$$

$$\frac{dL_5}{dt} = \frac{1}{1 + t^3}$$

$$\frac{dL_6}{dt} = e^{-t}$$

$$\frac{dL_7}{dt} = e^{-t^2}$$

Suppose they all start at size 0. Use your computer to sketch the growth of these fish for $0 \leq t \leq 5$. Then zoom in near $t = 0$. Could you tell which fish was which?

4.5 Integrals and Sums

When we cannot guess the answer or use one of the rules of integration to solve a pure-time differential equation, we can use Euler's method to approximate the solution. A graphical analysis of this method shows the link between integrals and sums. Using this link, we define the **definite integral** as the limit of **Riemann sums**. The definite integral represents the total amount of change during some period of time. This interpretation of the integral provides insight into a wide range of applications.

Approximating Integrals with Sums

Consider the pure-time differential equation for the volume, V, of water in a vessel,

$$\frac{dV}{dt} = t^2$$

where t is measured in seconds and V is measured in cm^3. We are asked to find the total quantity of water that entered during the first second. We could solve the differential equation with the indefinite integral to find this quantity. Alternatively, we can take the limit of an approximation in the spirit of Euler's method.

Suppose we had measured the rate at which water was entering the vessel only every 0.2 s. For lack of knowledge of what happens between measurements, we assume that the rate is *constant* between measurements.

There are two ways to use this information to approximate the total amount of water entering during the first second. In one, called the **left-hand estimate**, we pretend that the rate at which water enters between measurements is exactly equal to the value at the *beginning* of the interval (Figure 4.28a). In the other, the **right-hand estimate**, we pretend that the rate at which water enters between measurements is exactly equal to the value at the *end* of the interval (Figure 4.28b).

Time (s)	Rate (cm³/s)
0.0	0.00
0.2	0.04
0.4	0.16
0.6	0.36
0.8	0.64
1.0	1.00

Time interval	Left-hand estimate			Right-hand estimate		
	Rate during interval	Influx during interval	Net influx	Rate during interval	Influx during interval	Net influx
0.0–0.2	0.00	0.000	0.000	0.04	0.008	0.008
0.2–0.4	0.04	0.008	0.008	0.16	0.032	0.040
0.4–0.6	0.16	0.032	0.040	0.36	0.072	0.112
0.6–0.8	0.36	0.072	0.112	0.64	0.128	0.240
0.8–1.0	0.64	0.128	0.240	1.00	0.200	0.440

The estimate replaces the curve with a **step function**, a series of horizontal lines anchored on the actual function. Step functions are easy to deal with because rates are constant during each interval.

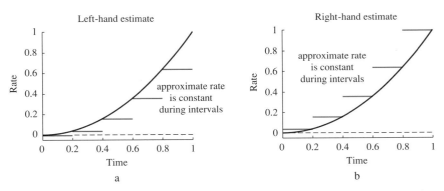

Figure 4.28
Left-hand and right-hand estimates of the integral

The computation of the net, or total, influx proceeds as follows. In the first time period, from $t = 0.0$ to $t = 0.2$, the left-hand estimate approximates the influx with the rate of flow at the beginning of the period, or 0.0. The influx is thus approximately 0.0 during this first interval. In the second period, from 0.2 to 0.4, the left-hand estimate approximates the rate of influx to be 0.04, producing an influx of 0.008 (multiplying the rate by the time). Adding this to the 0.0 from the first interval gives a net influx of 0.008. Continuing in this way, we estimate a total influx of water of 0.24 cm³ during the first second.

The right-hand estimate approximates the rate of influx during the first interval to be 0.04, the rate at the end of the period. The total influx during this period is estimated as 0.008. During the second interval, the right-hand estimate approximates the rate of influx to be 0.16, and the influx by 0.032. Adding this to the 0.008 from the first interval gives a net of 0.040. Continuing in this way, we estimate a total influx of water of 0.44 cm^3 during the first second.

We have computed two estimates of the total, a left-hand estimate of 0.24 and a right-hand estimate of 0.44. The two differ by quite a bit. Figure 4.28 tells us why. The step function associated with the left-hand estimate lies well below the exact measurement, and the step function associated with the right-hand estimate lies well above the exact measurement.

By using more measurements, we expect to get a more accurate answer. We can follow the same steps assuming that we have measured every 0.1 seconds (Figure 4.29).

Time interval	Left-hand estimate			Right-hand estimate		
	Rate during interval	Influx during interval	Net influx	Rate during interval	Influx during interval	Net influx
0.0–0.1	0.00	0.000	0.000	0.01	0.001	0.001
0.1–0.2	0.01	0.001	0.001	0.04	0.004	0.005
0.2–0.3	0.04	0.004	0.005	0.09	0.009	0.014
0.3–0.4	0.09	0.009	0.014	0.16	0.016	0.030
0.4–0.5	0.16	0.016	0.030	0.25	0.025	0.055
0.5–0.6	0.25	0.025	0.055	0.36	0.036	0.091
0.6–0.7	0.36	0.036	0.091	0.49	0.049	0.140
0.7–0.8	0.49	0.049	0.140	0.64	0.064	0.204
0.8–0.9	0.64	0.064	0.204	0.81	0.081	0.285
0.9–1.0	0.81	0.081	0.285	1.00	0.100	0.385

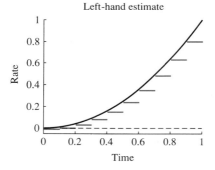
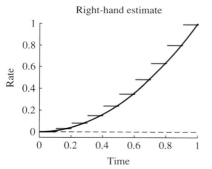

Figure 4.29
Left-hand and right-hand estimates of the integral: $n = 10$

The two estimates are closer to each other, as well as closer to the exact answer.

Approximating Integrals in General

Our example was rather specific, working on a particular function (t^2) and two particular numbers of intervals (five and ten). We will write the estimates for a general number of intervals n, and then we write them for a general function f.

Let n denote the number of intervals to be used. Let t_i be the time at the end of the ith interval. With $n = 5$, we found that $t_1 = 0.2$, $t_2 = 0.4$, and so on. Set t_0 to be the time at the beginning, so $t_0 = 0.0$ in this case. Finally, set Δt to be the length of the intervals, or 0.2 when $n = 5$ (Figure 4.30). We can then write the left-hand and right-hand estimates of volume as sums:

$$I_l = 0.0^2(0.2) + 0.2^2(0.2) + 0.4^2(0.2) + 0.6^2(0.2) + 0.8^2(0.2)$$
$$= t_0^2 \Delta t + t_1^2 \Delta t + t_2^2 \Delta t + t_3^2 \Delta t + t_4^2 \Delta t \quad (4.18)$$

and

$$I_r = 0.2^2(0.2) + 0.4^2(0.2) + 0.6^2(0.2) + 0.8^2(0.2) + 1.0^2(0.2)$$
$$= t_1^2 \Delta t + t_2^2 \Delta t + t_3^2 \Delta t + t_4^2 \Delta t + t_5^2 \Delta t \quad (4.19)$$

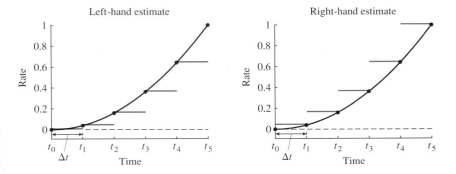

Figure 4.30
Writing the left- and right-hand estimates as sums: $n = 5$

We use **summation notation** to write these more compactly. For example, we write

$$\sum_{i=1}^{3} x_i = x_1 + x_2 + x_3 \quad (4.20)$$

The symbol at the beginning is a capital Greek "sigma," standing for "sum". This expression is read, "The sum from i equal 1 to 3 of x sub i." This means that we add x_i for i taking on values from 1 to 3. The letter i is the **index**. The values x_1, x_2, and x_3 are called **terms**. For example, if $x_1 = 5$, $x_2 = 3$, and $x_3 = 1$, then

$$\sum_{i=1}^{3} x_i = x_1 + x_2 + x_3 = 5 + 3 + 1 = 9$$

and

$$\sum_{i=1}^{3} x_i^2 = x_1^2 + x_2^2 + x_3^2 = 5^2 + 3^2 + 1^2 = 35$$

We can use this notation to rewrite our expressions for I_l and I_r:

$$I_l = \sum_{i=0}^{4} t_i^2 \Delta t$$

$$I_r = \sum_{i=1}^{5} t_i^2 \Delta t$$

These equations, a convenient shorthand for Equations 4.18 and 4.19, are called **Riemann sums**.

With this notation, we can quickly write what happens if we break the interval into ten pieces. The length Δt is 0.1, and $t_1 = 0.1$, $t_2 = 0.2$, up to $t_{10} = 1.0$ (Figure 4.31). We can write our estimates as

$$I_l = \sum_{i=0}^{9} t_i^2 \Delta t$$

$$I_r = \sum_{i=1}^{10} t_i^2 \Delta t$$

The only thing that has changed is the number of terms added. We are adding twice as many terms, but each is smaller because Δt is half as big. Again, these Riemann sums should approximate the total change during this time interval.

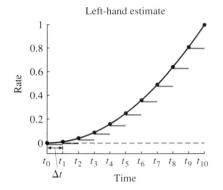
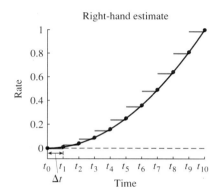

Figure 4.31
Writing the left- and right-hand estimates as sums: $n = 10$

Each term in this sum corresponds to one piece of the step function. The first term is the approximate total amount that entered during the first time interval of length Δt. In the left-hand estimate, the first term is the product of t_0^2, the rate at the beginning of this interval, with the width of the interval Δt.

We are now ready to write the general formula. Instead of 5 or 10, we substitute n into our formulas for I_l and I_r. Breaking 0 to 1 into n intervals produces intervals of length $\Delta t = 1/n$. The times are $t_i = i\Delta t$. For instance, $t_1 = \Delta t$, $t_2 = 2\Delta t$, up to $t_n = 1.0$ (Figure 4.32). We can write

$$I_l = \sum_{i=0}^{n-1} t_i^2 \Delta t$$

$$I_r = \sum_{i=1}^{n} t_i^2 \Delta t$$

for the Riemann sums.

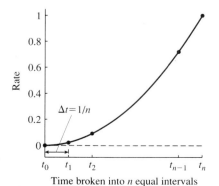

Figure 4.32
Writing the left- and right-hand estimates as sums: general case

In Table 4.1 we show the results of computing these estimates for different values of n. The last column, denoted by I_a, is the average of I_l and I_r and converges rapidly to 0.333333 (Exercise 7). We call this the **averaged estimate**.

Table 4.1 Estimates of the Integral with Different n

n	Δt	I_l	I_r	I_a
5	0.200	0.240000	0.440000	0.340000
10	0.100	0.285000	0.385000	0.335000
20	0.050	0.308750	0.358750	0.333750
30	0.033	0.316852	0.350185	0.333519
40	0.025	0.320938	0.345938	0.333438
50	0.020	0.323400	0.343400	0.333400
100	0.010	0.328350	0.338350	0.333350
500	0.002	0.332334	0.334334	0.333334
1000	0.001	0.332833	0.333833	0.333333

The Definite Integral

We have used sums to approximate the total volume that has been added according to the differential equation

$$\frac{dV}{dt} = t^2$$

during the time interval from $t = 0$ to $t = 1$. The general problem is to find the total change during the time interval from $t = a$ to $t = b$ when the state variable M obeys the pure-time differential equation

$$\frac{dM}{dt} = f(t)$$

The function f appears as the term t_i^2 in each Riemann sum. Sums that approximate the integral of the function f are then

$$I_l = \sum_{i=0}^{n-1} f(t_i) \Delta t \tag{4.21}$$

and

$$I_r = \sum_{i=1}^{n} f(t_i) \Delta t \tag{4.22}$$

The **Riemann integral** or the **definite integral**, is defined as the *limit* of I_r or I_l as n approaches infinity. For example,

$$\int_0^1 t^2 dt = \lim_{n \to \infty} \sum_{i=0}^{n-1} t_i^2 \Delta t \tag{4.23}$$

$$= \lim_{n \to \infty} \sum_{i=1}^{n} t_i^2 \Delta t \tag{4.24}$$

where $\Delta t = 1/n$. The expression

$$\int_0^1 t^2 \, dt$$

is pronounced "the integral from 0 to 1 of $t^2 dt$." The little numbers on the integral, which indicate the interval to be integrated over, are called the **limits of integration**. In this case, we are integrating (finding the total amount that entered) from time 0 to time 1. The dt in the integral can be thought of as the width of an infinitesimally small interval Δt. When limits of integration are present, the expression is called a **definite integral**. Remember that the definite integral is a number and the indefinite integral is a function. In Section 4.6 we will see the connection between them.

In general, we substitute general limits of integration from a to b and a general function f.

■ **Definition 4.2** The Riemann integral of a function f on the interval from a to b is

$$\int_a^b f(t)\,dt = \lim_{n\to\infty} \sum_{i=0}^{n-1} f(t_i)\Delta t$$

$$= \lim_{n\to\infty} \sum_{i=1}^{n} f(t_i)\Delta t$$

where the values $t_0, \ldots, t_n$ break the interval from a to b into n equal pieces of length

$$\Delta t = \frac{b-a}{n}$$ ■

The elements of this definition are illustrated in Figure 4.33.

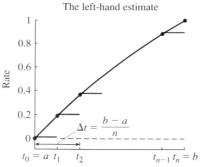
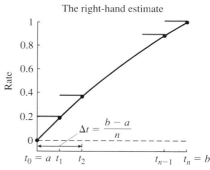

Figure 4.33
The Riemann integral in general

We need to prove a couple of things with regard to this definition. First, we should show that the limits exist. Second, we should show that the limits of the left-hand and right-hand Riemann sums are equal. For our purposes, we will believe these statements. Proofs can be found in standard calculus texts. As long as the function f does not go to infinity, it takes a mathematician with a passion for pathology to cook up a function for which either of these statements is false.

SUMMARY We have approximated integrals with **Riemann sums**, expressing our results in **summation notation**. By making the sums correspond more and more closely to the function, we get more and more accurate estimates. The **definite integral** is defined as the limit of the Riemann sums between particular **limits of integration**.

4.5 EXERCISES

1. Find the value of Δt and the values of $t_0, t_1, \ldots, t_n$ for the following values.
   a. $a = 0, b = 2, n = 5$
   b. $a = 0, b = 2, n = 10$
   c. $a = 2, b = 3, n = 5$
   d. $a = 2, b = 3, n = 100$
   e. $a = -2, b = 2, n = 10$

2. For the given number of intervals, find the left- and right-hand estimates for the definite integrals of the following functions.
   a. $f(t) = 2t$, limits of integration 0 to 1, $n = 5$
   b. $f(t) = 2t$, limits of integration 0 to 2, $n = 5$
   c. $f(t) = t^2$, limits of integration 0 to 2, $n = 5$
   d. $f(t) = 1 + t^3$, limits of integration 0 to 1, $n = 5$

3. Use summation notation to write the Riemann sums from Exercise 2.

4. Assuming that the initial condition is $V(0) = 0$, solve the following pure-time differential equations and find $V(t)$ at the specified time. Compare with the results from Exercise 2.
   a. $\dfrac{dV}{dt} = 2t$; find $V(1)$.
   b. $\dfrac{dV}{dt} = 2t$; find $V(2)$.
   c. $\dfrac{dV}{dt} = t^2$; find $V(2)$.
   d. $\dfrac{dV}{dt} = 1 + t^3$; find $V(1)$.

5. Use Euler's method to estimate the solutions of the following differential equations (Exercise 4) with the following parameters. Suppose that $V(0) = 0$ in each case. Each answer should exactly match one of the estimates in Exercise 2. Can you explain why?
   a. $\dfrac{dV}{dt} = 2t$; estimate $V(1)$ using $\Delta t = 0.2$.
   b. $\dfrac{dV}{dt} = 2t$; find $V(2)$ using $\Delta t = 0.4$.
   c. $\dfrac{dV}{dt} = t^2$; find $V(2)$ using $\Delta t = 0.4$.
   d. $\dfrac{dV}{dt} = t^3$; find $V(1)$ using $\Delta t = 0.2$.

6. Evaluate the following sums.
   a. $\sum_{i=1}^{5} x_i$, for $x_i = 1/i$
   b. $\sum_{i=1}^{5} x_i^2$, for $x_i = 1/i$
   c. $\sum_{j=1}^{6} y_j$, for $y_j = 1/2^j$
   d. $\sum_{k=1}^{7} z_k$, for $z_k = 2^k$

7. Another way to think about the column labeled I_a in Table 4.1 is to think that the value of the function is approximated by the average of the values at the beginning and the end of the time interval. When $f(t) = t^2$, we pretend that the function value is $(t_{i+1}^2 + t_i^2)/2$ during the interval from t_i to t_{i+1}. Use the range of values from 0 to 1.
   a. Draw a graph illustrating this estimate for $n = 5$.
   b. Write an expression for I_a using summation notation.
   c. Using this expression, check the value for $n = 5$ in Table 4.1.
   d. Why do you think this estimate is so much more accurate than I_l and I_r?

8. We can also think of the method of Exercise 7 as the area of a **trapezoid**.
   a. Draw an approximation of the rate of change function which replaces the curve connecting the points $(t_i, f(t_i))$ and $(t_{i+1}, f(t_{i+1}))$ with a secant line.
   b. Notice that the rectangles are now trapezoids. Find the area of each trapezoid.
   c. Try this in general, and show that the area of the trapezoid is exactly equal to the area of the rectangle with height equal to the average.
   d. Which picture looks like a better approximation?

9. One other estimate of the integral, called I_m, can be computed by pretending that the value during the interval from t_i to t_{i+1} is the value of the function at the midpoint, or $f[(t_{i+1} + t_i)/2]$. Do the following for the function $f(t) = t^2$.
   a. Draw a graph illustrating this estimate for $n = 5$. Make sure you see the subtle difference from I_a.
   b. Write an expression for I_m using summation notation.
   c. Using this expression, compute the value for $n = 5$.
   d. Why do you think this estimate is more accurate than I_l and I_r?

10. Suppose the speed of a bee is given in the following table.

Time (s)	Speed (cm/s)
0.0	127.0
1.0	122.0
2.0	118.0
3.0	115.0
4.0	113.0
5.0	112.0
6.0	112.0
7.0	113.0
8.0	116.0
9.0	120.0
10.0	125.0

a. Graph the speed of the bee as a function of time.
b. Using the measurements on even-numbered seconds, find the left-hand and right-hand estimates for the distance the bee moved during the experiment.
c. Using all the measurements, find the left-hand and right-hand estimates for the distance the bee moved during the experiment.
d. Use the method of the last column of Table 4.1 (or Exercise 7) to get two other estimates of the distance moved. Which do you think is the best?

11. **COMPUTER:** We will compare various methods used to estimate the solution of

$$\frac{dV}{dt} = t^2$$

with $V(0) = 0$. We wish to find $V(2)$.
a. Graph the rate of change as a function of t.
b. Use the right-hand estimate with $\Delta t = 0.2, 0.1,$ and 0.02.
c. Use the left-hand estimate with $\Delta t = 0.2, 0.1,$ and 0.02.
d. Find the average of these two estimates.

12. **COMPUTER:** We will compare various methods used to estimate the solution of

$$\frac{dp}{dt} = \ln(1 + \sqrt{t} - t^3)$$

with $p(0) = 0$. We wish to find $p(1)$.
a. Graph the rate of change as a function of t.
b. Use the right-hand estimate with $\Delta t = 0.2, 0.1,$ and 0.02.
c. Use the left-hand estimate with $\Delta t = 0.2, 0.1,$ and 0.02.
d. Try to figure out from your graph why the two estimates are the same.

4.6 Definite and Indefinite Integrals

We now have two types of integral. The *indefinite integral* is a *function* that solves the pure-time differential equation

$$\frac{dF}{dt} = f(t)$$

We write

$$\int f(t)\, dt = F(t) + c$$

to indicate this solution, found by computing or guessing the *antiderivative*. This integral is *indefinite* because it includes an arbitrary constant c, which must be computed from the initial conditions of the differential equation.

The *definite integral* is a *number*, the change in value between the two times represented by the limits of integration. If the rate of change is $f(t)$, then

$$\text{the total change between } a \text{ and } b = \int_a^b f(t)\, dt$$

The definite integral is defined as the limit of *Riemann sums*. We will now learn how to compute the definite integral from the indefinite integral with the **Fundamental Theorem of Calculus**.

The Fundamental Theorem of Calculus: Computing Definite Integrals with Indefinite Integrals

Consider again the pure-time differential equation for volume,

$$\frac{dV}{dt} = t^2$$

The change between times 0 and 1 is given by the definite integral

$$\int_0^1 t^2\, dt$$

We can use Riemann sums to approximate the definite integral, but the procedure requires a great deal of calculation.

The total amount of water that entered between times 0 and 1 must equal the difference between the volumes at times 0 and 1 (Figure 4.34). In particular, if V is a solution of the differential equation, then

$$\int_0^1 t^2\, dt = V(1) - V(0)$$

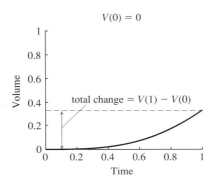
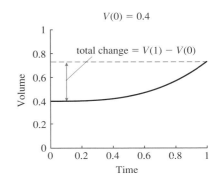

Figure 4.34
Total change

We can use the indefinite integral to compute $V(1)$. Suppose first that $V(0) = 0$. We solve the differential equation with the indefinite integral and the power rule, finding

$$V(t) = \int t^2\, dt = \frac{t^3}{3} + c$$

With the initial condition $V(0) = 0$, $c = 0$, and the solution is

$$V(t) = \frac{t^3}{3}$$

With this formula, we can find the total change during the first minute by subtracting, or

$$\text{total change between times 0 and 1} = V(1) - V(0)$$
$$= \frac{1^3}{3} - \frac{0^3}{3} = \frac{1}{3} = 0.3333$$

This closely matches the estimates we found using Riemann sums (Table 4.1) but took a lot less work.

Suppose instead that $V(0) = 0.4$. Then we get

$$V(t) = \frac{t^3}{3} + c$$

as before, but with the initial condition $V(0) = 0.4$, $c = 0.4$. The solution is

$$V(t) = \frac{t^3}{3} + 0.4$$

We can again find the total change during the first minute by subtracting, or

total change between times 0 and 1 $= V(1) - V(0)$

$$= \left(\frac{1^3}{3} + 0.4\right) - \left(\frac{0^3}{3} + 0.4\right) = 0.3333$$

Although the two solutions do not match, they are *parallel* (Figure 4.34). Both increase by exactly the same amount.

More generally, suppose we denote the volume at time 0 by the unknown value V_0. The solution of the differential equation is still

$$V(t) = \frac{t^3}{3} + c$$

but now the arbitrary constant c must satisfy

$$V(0) = V_0 = \frac{0^3}{3} + c = c$$

or $c = V_0$. The solution is

$$V(t) = \frac{t^3}{3} + V_0$$

and the change in volume is

$$V(1) - V(0) = \left(\frac{1^3}{3} + V_0\right) - \left(\frac{0^3}{3} + V_0\right)$$

$$= \frac{1}{3} = 0.3333$$

The total change does not depend on the initial condition of the pure-time differential equation. The initial condition is required to answer a question like, "Where are you after driving 5 miles due north?" But the *change* in position is clear: You are 5 miles north of where you started.

This statement is the essence of the **Fundamental Theorem of Calculus**. First, we need some new notation. To represent the change in the value of $F(x)$ between $x = a$ and $x = b$, we use the shorthand

$$F(x)|_a^b = F(b) - F(a)$$

Instead of reading this notation in a new way, we say simply "$F(b)$ minus $F(a)$."

We can now state the theorem.

■ **THEOREM 4.4** **(The Fundamental Theorem of Calculus)**

For any reasonable function $f(x)$ and any indefinite integral

$$F(x) = \int f(x)\,dx$$

$$\int_a^b f(x)\,dx = F(b) - F(a) = F(x)|_a^b \qquad ■$$

"Reasonable" includes quite a broad range of functions and includes any function that you can graph (any function that is continuous at all but a finite number of points). This theorem is *fundamental* because it describes the relation between definite and indefinite integrals. Proving this theorem requires showing that taking the limit of Riemann sums corresponds to the very different operation of finding the indefinite integral with the antiderivative. A proof can be found in any standard calculus textbook.

Suppose that the position of a falling rock $p(t)$ follows the differential equation

$$\frac{dp}{dt} = v(t) = -9.8t - 5.0$$

(as in Section 4.3), where t is measured in seconds and p is measured in meters. How far does the rock fall between $t = 1$ and $t = 3$? The total change in position is given by the definite integral of the rate of change of position,

$$\int_1^3 v(t)\,dt$$

According to the Fundamental Theorem of Calculus, we can compute this value by finding *any* indefinite integral of $v(t)$ and subtracting the values at $t = 1$ and $t = 3$. One indefinite integral is

$$\int v(t)\,dt = \int (-9.8t - 5.0)\,dt$$
$$= -9.8\frac{t^2}{2} - 5.0t$$
$$= -4.9t^2 - 5.0t$$

We have left out the arbitrary constant for convenience. Then

total change in position between $t = 1$ and $t = 3$
$$= -4.9t^2 - 5.0t\,\big|_1^3$$
$$= (-4.9 \cdot 3^2 - 5.0 \cdot 3) - (-4.9 \cdot 1^2 - 5.0 \cdot 1)$$
$$= -49.2$$

The rock will have fallen 49.2 m during this time.

The total distance fallen between times 1 and 3 can also be found by solving the differential equation with initial conditions. We can think of the position $p(t)$ as the change in position starting from $t = 1$. It follows the differential equation

$$\frac{dp}{dt} = v(t) = -9.8t - 5.0$$

as before, but with the initial condition $p(1) = 0$ indicating that the distance fallen at $t = 1$ is 0. Solving, we get

$$p(t) = \int v(t)\,dt$$
$$= \int (-9.8t - 5.0)\,dt$$
$$= -9.8\frac{t^2}{2} - 5.0t + c$$
$$= -4.9t^2 - 5.0t + c$$

The constant c satisfies

$$p(1) = 0 = -4.9 \cdot 1^2 - 5.0 \cdot 1 + c$$
$$c = 4.9 \cdot 1^2 + 5.0 \cdot 1 = 9.9.$$

The solution is

$$p(t) = -4.9t^2 - 5.0t + 9.9$$

and the change in position by time $t = 3$ is

$$p(3) = -4.9 \cdot 3^2 - 5.0 \cdot 3 + 9.9 = -49.2$$

(Figure 4.35b). The answers match, but the first method is easier because we did not need to compute the arbitrary constant.

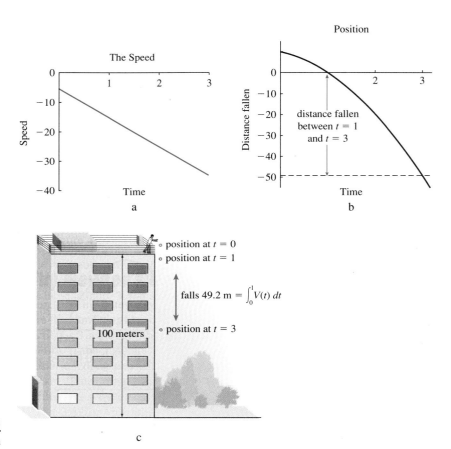

Figure 4.35
Velocity and position

As a final example, suppose the change in length of a fish follows the equation

$$\frac{dL}{dt} = 6.48e^{-0.09t}$$

(Equation 4.15) with t measured as age in years and L measured in centimeters. How much does the fish grow between ages 2 and 5? The total change is

$$\int_2^5 6.48e^{-0.09t}\, dt$$

To evaluate, we find the indefinite integral,

$$\int 6.48e^{-0.09t}\, dt = -72.0e^{-0.09t}$$

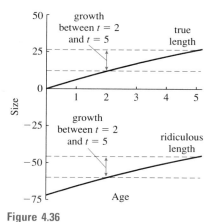

Figure 4.36
The growth of a walleye

(this is the answer we found using substitution in Equation 4.16), but with the arbitrary constant set to 0 for convenience. The change in length is then

$$-72.0e^{-0.09t}\big|_2^5 = -72.0e^{-0.09 \cdot 5}(-)72.0e^{-0.09 \cdot 2} = -45.9 - (-60.1) = 14.2$$

As we can see from this calculation, neither of the component terms (-45.9 and -60.1) makes biological sense. Their difference, however, gives the correct answer (Figure 4.36).

The solution we find using the realistic initial condition $L(0) = 0$ is

$$L(t) = 72.0 - 72.0e^{-0.09t} = 72.0(1 - e^{-0.09t})$$

(Equation 4.17). The change is

$$L(t)\big|_2^5 = L(5) - L(2)$$
$$= 72.0(1 - e^{-0.09 \cdot 5}) - 72.0(1 - e^{-0.09 \cdot 2})$$
$$\approx 26.1 - 11.9 = 14.2$$

The difference, 14.2, is now between the actual lengths at ages 5 and 2 (Figure 4.36).

The Summation Property of Definite Integrals

What happens if the function we want to integrate takes on both positive and negative values? If the function represents the rate at which water enters a vessel, a positive value means that water is entering and a negative value means that water is leaving (Figure 4.37a). Suppose

$$\frac{dV}{dt} = t^2 - t$$

where t is measured in seconds and V is measured in cubic centimeters. Water flows out during the first second when the rate is negative, and in during the next second when the rate is positive (Figure 4.37).

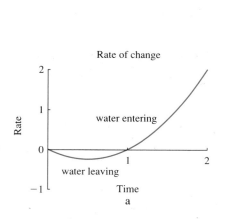
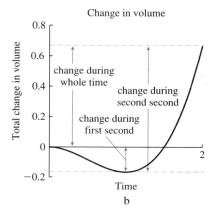

Figure 4.37
Positive and negative rates

To find the total change in volume from time 0 until time 1, we use the indefinite integral and the Fundamental Theorem of Calculus to compute

$$\int_0^1 t^2 - t\, dt = \left(\frac{t^3}{3} - \frac{t^2}{2}\right)\bigg|_0^1$$

$$= \left(\frac{1}{3} - \frac{1}{2}\right) - (0 - 0)$$
$$\approx -0.167 \text{ cm}^3$$

This means that 0.167 cm^3 of water *left* the vessel during the first second. During the next second, the change in volume is

$$\int_1^2 (t^2 - t)\, dt = \left(\frac{t^3}{3} - \frac{t^2}{2}\right)\Big|_1^2$$
$$= \left(\frac{8}{3} - 2\right) - \left(\frac{1}{3} - \frac{1}{2}\right)$$
$$\approx 0.833 \text{ cm}^3$$

This means that 0.833cm^3 entered the vessel during the second second.

What happens during the whole period between $t = 0$ and $t = 2$? Because 0.167 cm^3 left during the first second and 0.833 cm^3 entered during the second second,

(change between $t = 0$ and $t = 2$) = (change between $t = 0$ and $t = 1$)
$\qquad\qquad\qquad\qquad\qquad\quad$ + (change between $t = 1$ and $t = 2$)
$\qquad\qquad\qquad\qquad\qquad\quad \approx -0.167 + 0.833 = 0.666$

Alternatively, evaluating the definite integral from $t = 0$ to $t = 2$, we get

$$\int_0^2 (t^2 - t)\, dt = \left(\frac{t^3}{3} - \frac{t^2}{2}\right)\Big|_0^2$$
$$= \left(\frac{8}{3} - 2\right) - (0 - 0) = 0.667$$

This calculation points out an important property of the definite integral. The total change between times a and b is the change between a and some intermediate time c plus the change between time c and the final time b (Figure 4.38). We give the formal definition of the summation property of the definite integral.

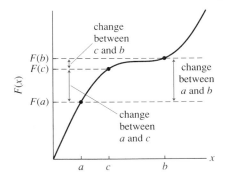

Figure 4.38
The summation property of the definite integral

■ **THEOREM 4.5** **(Summation Property of the Definite Integral)**

$$\int_a^b f(x)\, dx = \int_a^c f(x)\, dx + \int_c^b f(x)\, dx$$

■

Proof: The proof follows from the Fundamental Theorem. Suppose $F(x)$ is any antiderivative of $f(x)$. Then

$$\int_a^b f(x)\,dx = F(b) - F(a)$$

But for any value c it is certainly true that

$$F(b) - F(a) = F(b) - F(c) + F(c) - F(a)$$

because we added and subtracted the same number, $F(c)$. Rearranging, we have

$$F(b) - F(c) + F(c) - F(a) = F(c) - F(a) + F(b) - F(c).$$

Applying the Fundamental Theorem twice, we get

$$F(c) - F(a) = \int_a^c f(x)\,dx$$

$$F(b) - F(c) = \int_c^b f(x)\,dx$$

as we wished to prove. ∎

If we wish to find how far our rock falls between $t = 1$ and $t = 3$, we can add the change of position between $t = 1$ and $t = 2$ to the change of position between $t = 2$ and $t = 3$. Between $t = 1$ and $t = 2$, the change of position is

$$\int_1^2 v(t)\,dt = \int_1^2 (-9.8t - 5.0)\,dt$$
$$= -4.9t^2 - 5.0t\,\big|_1^2$$
$$= (-4.9 \cdot 2^2 - 5.0 \cdot 2) - (-4.9 \cdot 1^2 - 5.0 \cdot 1) = -19.7$$

and between $t = 2$ and $t = 3$, the change of position is

$$\int_2^3 v(t)\,dt = \int_2^3 -9.8t - 5.0\,dt$$
$$= -4.9t^2 - 5.0t\,\big|_2^3$$
$$= (-4.9 \cdot 3^2 - 5.0 \cdot 3) - (-4.9 \cdot 2^2 - 5.0 \cdot 2) = -29.5$$

The total change of position, found earlier to be -49.2 m, is the sum of -19.7 m and -29.5 m (Figure 4.39).

Figure 4.39

The summation property of distance traveled

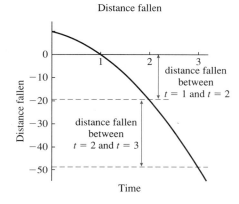
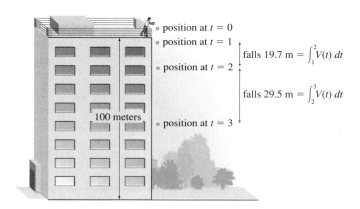

The General Solution of a Pure-Time Differential Equation

We can use the Fundamental Theorem of Calculus to write the solution of a pure-time differential equation with the definite integral. Consider the general pure-time differential equation,

$$\frac{dM}{dt} = f(t)$$

The quantity M has rate of change equal to the measured value $f(t)$. We wish to write a formula for $M(t)$. The idea is the same one we have been using throughout: Where something ends up is where it started plus how much it changed. Suppose we know that the quantity starts with the value $M(t_0)$ at time t_0:

$$M(t) = \text{where it started} + \text{how much it changed}$$
$$= M(t_0) + [M(t) - M(t_0)] \qquad \text{started at } M(t_0), \text{ changed by } M(t) - M(t_0)$$
$$= M(t_0) + \int_{t_0}^{t} f(s)\,ds \qquad \text{apply the Fundamental Theorem to the change}$$

We changed the variable we are using to integrate from t to s to avoid using the same letter for two different things. The variable s is called a **dummy variable** because it disappears when we finish the calculation.

Consider again the differential equation describing the growth of a walleye,

$$\frac{dL}{dt} = 6.48e^{-0.09t}$$

with initial condition $L(0) = 0$. The solution is

$$L(t) = L(0) + \int_{0}^{t} 6.48e^{-0.09s}\,ds \qquad \text{the general formula}$$
$$= 0.0 - 72.0e^{-0.09s}\big|_{0}^{t} \qquad \text{evaluate definite integral by using any indefinite integral}$$
$$= 0.0 - 72.0e^{-0.09t} - (-72.0e^{-0.09 \cdot 0}) \qquad \text{evaluate at the endpoints } t \text{ and } 0$$
$$= 72.0 - 72.0e^{-0.09t} \qquad \text{simplify}$$

For the rock with position $p(t)$ following the differential equation,

$$\frac{dp}{dt} = v(t) = -9.8t - 5.0$$

with $p(0) = 100$, the solution is

$$p(t) = p(0) + \int_{0}^{t} -9.8s - 5.0\,ds$$
$$= 100.0 - -4.9s^2 - 5.0s\big|_{0}^{t}$$
$$= 100.0 - (-4.9t^2 - 5.0t) - (-4.9 \cdot 0^2 - 5.0 \cdot 0)$$
$$= 100 - 4.9t^2 - 5.0t$$

just as we found before.

SUMMARY   The **Fundamental Theorem of Calculus** describes the connection between definite and indefinite integrals; the definite integral is equal to the difference between the values of the indefinite integral at the limits of integration. The Fundamental Theorem simplifies calculations of total change from pure-time differential equations by eliminating the need to solve for the arbitrary constant. From the Fundamental Theorem, we proved the **summation property of definite integrals**, which says that the total change over two time intervals is the sum of the changes in each. Finally, we applied the Fundamental Theorem to write the **general solution of a pure-time differential equation** with the definite integral.

4.6 EXERCISES

1. Compute the following definite integrals and compare with your answers from Exercises 2 and 4 in section 4.5.

   a. $\int_0^1 2t\, dt$
   b. $\int_0^2 2t\, dt$
   c. $\int_0^2 t^2\, dt$
   d. $\int_0^1 t^3\, dt$

2. Compute the following definite integrals.

   a. $\int_1^2 \left[e^x + \dfrac{1}{x} + \sin(x) + \cos(x)\right] dx$
   b. $\int_{-1}^1 (3e^x + 2x^3)\, dx$
   c. $\int_1^4 \left(\dfrac{2}{t} + \dfrac{t}{2}\right) dx$
   d. $\int_1^3 \left(\dfrac{3}{z^2} + \dfrac{z^2}{3}\right) dx$
   e. $\int_0^\pi [2\sin(x) + 3\cos(x)]\, dx$

3. By computing the definite integrals from 1 to 2, 2 to 3, and 1 to 3 of the following functions (from Exercise 1), verify the summation property of definite integrals.

   a. $f(t) = 2t$
   b. $g(t) = t^2$
   c. $h(t) = t^3$

4. Find the change in the state variable between the given times first by solving the differential equation with the given initial conditions and then by using the definite integral.

   a. The change of position of a falling rock between times $t = 1$ and $t = 5$, where its position follows the differential equation
   $$\dfrac{dp}{dt} = -9.8t - 5.0$$
   and initial condition $p(0) = 100$.

   b. The amount a fish grows between ages $t = 1$ and $t = 5$ if it follows the differential equation
   $$\dfrac{dL}{dt} = 6.48e^{-0.09t}$$
   with initial condition $L(0) = 5.0$.

   c. The amount a fish grows between ages $t = 0.5$ and $t = 1.5$ if it follows the differential equation
   $$\dfrac{dL}{dt} = 64.3e^{-1.19t}$$
   with initial condition $L(0) = 5.0$.

   d. The number of new AIDS cases between 1985 and 1987 if the number of AIDS cases follows the differential equation
   $$\dfrac{dA}{dt} = 523.8(t - 1981)^2$$
   with initial condition $A(1981) = 13{,}400$.

   e. The amount of chemical produced between times $t = 5$ and $t = 10$ if the amount P follows the differential equation
   $$\dfrac{dP}{dt} = \dfrac{5}{1 + 2.0t}$$
   with initial condition $P(0.0) = 2.0$, and, where t is measured in minutes and P in moles.

   f. The amount of chemical produced between times $t = 5$ and $t = 10$ if the amount P follows the differential equation
   $$\dfrac{dP}{dt} = 5.0e^{-2.0t}$$
   with initial condition $P(0.0) = 2.0$, and where t is measured in minutes and P in moles.

5. Check the summation property for the solutions of the differential equations in Exercise 4. Compute the value of the solution halfway through the interval (at $t = 3$ in part a). Show that the changes during the first half (from $t = 1$ to $t = 3$ in part a) and the second half of the interval (from $t = 3$ to $t = 5$ in part a) add up to the total change.

6. Consider the differential equation for volume,
   $$\dfrac{dV}{dt} = 2t$$
   with $V(0) = 5.0$.

   a. To find the solution, use the general formula for a solution.
   b. Graph this solution.
   c. Find the change in volume between times $t = 1$ and $t = 2$.
   d. Now suppose $V(0) = 0.0$. Find and graph the solution, and find the change in volume between times $t = 1$ and $t = 2$.

7. Use the general formula to find the solutions of the differential equations in Exercise 4.

366 Chapter 4 Differential Equations, Integrals, and Their Applications

8. A rocket is shot from the ground with upward acceleration of 12.0 m/s². It runs out of fuel after 10 s and begins to fall with an acceleration of −9.8 m/s².
   a. Write and solve differential equations describing the velocity and position of the rocket while it still has fuel.
   b. Find the velocity and height of the rocket when it runs out of fuel.
   c. Write down and solve differential equations describing the velocity and position of the rocket after it has run out of fuel. What is the initial condition for each?
   d. Find the maximum height reached by the rocket. Does it rise more with or without fuel? Why?
   e. Find the velocity when it hits the ground.

9. What is the value of
$$\int_a^a f(x)\,dx$$
for any function $f(x)$? Explain this result both in terms of change and with the Fundamental Theorem of Calculus.

10. **COMPUTER:** Toward the end of the universe, acceleration due to gravity will break down. Suppose that
$$a = -9.8\frac{1}{1+t}$$
where time is measured in seconds after the beginning of the end. An object begins falling from 10.0 m above the ground.
   a. Find the velocity at time t.
   b. Find the position at time t.
   c. Graph acceleration, velocity, and position on the same graph. Which of these measurements are integrals of each other?
   d. When will this object hit the ground?

4.7 Applications of Integrals

In this section, we introduce several remarkable applications of the definite integral. First, we notice that the graph describing the Riemann sum can be interpreted geometrically as a way to approximate the **area under a curve**. In fact, geometric problems of this sort provided the motivation for Archimedes' near discovery of the integral about 2000 years ago, long before Newton introduced the study of differential equations. The idea of chopping quantities into small bits, adding them with Riemann sums, and computing exact answers with the definite integral has many other applications, including finding the **average value of a function** and finding total mass when **density** is known.

Integrals and Areas

Suppose we want to find the area under the curve $f(x) = x^2$ between $x = 0$ and $x = 1$ (Figure 4.40). We can approximate the area with little rectangles, producing a picture identical to those used to find the left-hand Riemann sum (Figure 4.28). In particular, when we break the area into five rectangles as shown in Figure 4.40, their areas match the values found in computing the left-hand estimate.

Rectangle	Base	Height	Area	Total
1	0.2	0.0	0.0	0.0
2	0.2	0.04	0.008	0.008
3	0.2	0.16	0.032	0.040
4	0.2	0.36	0.072	0.112
5	0.2	0.64	0.128	0.240

Because the Riemann sums converge both to the area and to the definite integral, the area under $f(x) = x^2$ from 0 to 1 = $\int_0^1 x^2\,dx$

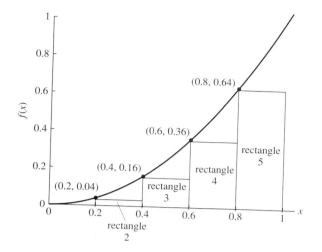

Figure 4.40
The area under $f(x) = x^2$

We have already computed this integral several times, finding the surprisingly simple answer of $1/3$.

In general, we can find the area under the positive curve $f(x)$ from a to b by computing the definite integral between the limits of integration:

$$\text{the area under } f(x) \text{ from } a \text{ to } b = \int_a^b f(x)\,dx \quad (4.25)$$

The summation property of the definite integral (Theorem 4.5) has a convenient geometric interpretation also (Figure 4.41). The area between a and c is the sum of the area between a and b and the area between b and c.

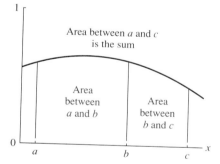

Figure 4.41
Areas and the summation property of definite integrals

Areas are positive, but definite integrals can give negative values. For example, the integral

$$\int_0^2 x^2 - x\,dx$$

does not give the sum of the two shaded areas in Figure 4.42 but *subtracts* the area below the x-axis from the area above the x-axis. Keep this in mind when using integrals to compute areas. To find the total shaded area, it is necessary to integrate the *absolute value* of the function, or

$$\int_0^2 |x^2 - x|\,dx$$

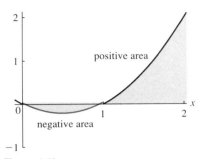

Figure 4.42

Positive and negative area

The only way to evaluate this is to find where the integrand is positive and negative. In this case, the function is negative between 0 and 1 and positive between 1 and 2. Therefore, $|x^2 - x| = x - x^2$ for $0 \le x \le 1$ and $|x^2 - x| = x^2 - x$ for $1 \le x \le 2$:

$$\int_0^2 |x^2 - x|\, dx = \int_0^1 x - x^2\, dx + \int_1^2 x^2 - x\, dx$$

$$= \left(\frac{x^2}{2} - \frac{x^3}{3}\right)\bigg|_0^1 + \left(\frac{x^3}{3} - \frac{x^2}{2}\right)\bigg|_1^2$$

$$= \left(\frac{1}{2} - \frac{1}{3}\right) - (0 - 0) + \left(\frac{8}{3} - 2\right) - \left(\frac{1}{3} - \frac{1}{2}\right) = 1$$

Another quirk of using definite integrals to find areas arises when the limits of integration are in the "wrong" order: when the "lower" limit is a larger number than the "upper" limit. For example,

$$\int_1^0 t^2\, dt = \frac{t^3}{3}\bigg|_1^0 = 0 - \frac{1}{3} = -\frac{1}{3}$$

The answer is negative because the definite integral treats left to right as the positive direction and models areas measured from right to left as being taken away. When computing areas, make sure that the limits of integration are in the right order.

When finding the indefinite integrals of complicated functions, we often use substitution. Substitution works for definite integrals but requires an additional step. Suppose we wish to find the area under the curve $f(t) = (1 + 2t)^2$ from $t = 0$ to $t = 1$ (Figure 4.43a). One way is to multiply out the function and integrate,

$$\int_0^1 (1 + 2t)^2\, dt = \int_0^1 1 + 4t + 4t^2\, dt$$

$$= \left(t + 2t^2 + \frac{4t^3}{3}\right)\bigg|_0^1$$

$$= 1 + 2 + \frac{4}{3} \approx 4.333$$

To use substitution, we use a modified version of Algorithm 4.2:

■ **Algorithm 4.3** (Computing a definite integral with substitution)

1. Define a new variable as some function of the old variable.
2. Take the derivative of the new variable with respect to the old variable.
3. Treat the derivative like a fraction, and move the dx to the other side.
4. Put everything in the integral in terms of the new variable.
5. Change the limits of integration into the new variable.
6. Try to integrate. ■

For our example, we perform the following steps:

1. Set $u = 1 + 2t$.
2. Then $du/dt = 2$.

3. Therefore $du = 2\,dt$, or $dt = du/2$.
4. Put the integrand in terms of u, or

$$(1 + 2t)^2\,dt = u^2 \frac{du}{2}$$

5. The new step is to *change the limits of integration*. The original limits are from $t = 0$ to $t = 1$. When $t = 0$, $u = 1 + 2 \cdot 0 = 1$, and when $t = 1$, $u = 1 + 2 \cdot 1 = 3$. The new integral, after substituting for every t, is

$$\int_1^3 \frac{u^2}{2}\,du$$

6. Work this out to find

$$\int_1^3 \frac{u^2}{2}\,du = \left.\frac{u^3}{6}\right|_1^3 = \frac{27}{6} - \frac{1}{6} = 4.333$$

We do not have to convert back to the old variable at the end because the answer is a number rather than a function. We can summarize these changes in this chart. The graph of the area under this function is shown in Figure 4.43b. The height of the u curve is half that of the t curve, but its width has been doubled, thus preserving the area.

In terms of t	In terms of u
$1 + 2t$	u
$(1 + 2t)^2$	u^2
dt	$\dfrac{du}{2}$
$t = 0$	$u = 1 + 2 \cdot 0 = 1$
$t = 1$	$u = 1 + 2 \cdot 1 = 3$

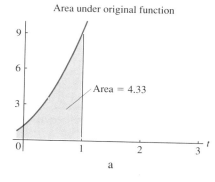
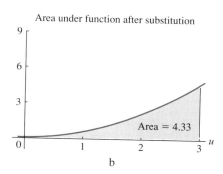

Figure 4.43
Computing an area with substitution

Integrals and Averages

Suppose that water is flowing into a vessel at a rate of $1 - e^{-t}$ cm³/s for the 2 s between $t = 0.0$ and $t = 2.0$. What is the *average* rate at which water enters during this time? The average is the total amount of water that enters divided by the time, or

$$\text{average rate} = \frac{\text{total water entering}}{\text{total time}}$$

The total amount of water that enters is

$$\text{total water entering} = \int_{0.0}^{2.0} 1 - e^{-t}\,dt$$

$$= t + e^{-t}\,\Big|_{0.0}^{2.0}$$

$$= 2.0 + e^{-2.0} - (0.0 + e^{-0.0}) = 1.135$$

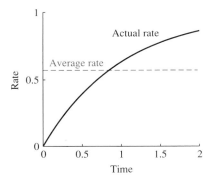

Figure 4.44
The average value of a rate

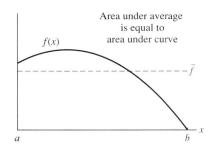

Figure 4.45
The average value in general

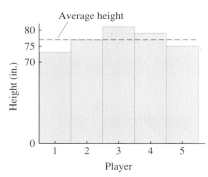

Figure 4.46
The average height of players on a basketball team

The average rate is

$$\text{average rate} = \frac{\text{total water entering}}{\text{total time}} = \frac{1.135}{2.0} = 0.568$$

(Figure 4.44). If water enters at the constant rate of 0.568 cm³/s for 2.0 s, then 1.135 cm³ would enter, equal to the amount of water that entered at the variable rate $1 - e^{-t}$ during those same 2.0 s. Geometrically, the area under the horizontal line at 0.568 is equal to the area under the curve $1 - e^{-t}$.

The general formula for the average value of a function f, often denoted by $\bar{f}$, on the interval from a to b is

$$\text{average value of } f = \frac{1}{b-a}\int_a^b f(x)\,dx \quad \textbf{(4.26)}$$

The area under the horizontal line representing the average is equal to the area under the curve (Figure 4.45).

How does this compare with the usual meaning of "average?" Suppose you wanted to find the average height of players on a starting basketball team. If h_i denotes the height of player i, the average $\bar{h}$ is ordinarily found by adding up the heights and dividing by 5, or

$$\bar{h} = \frac{1}{5}\sum_{i=1}^{5} h_i$$

Figure 4.46 shows a graphical representation of this team. Each player is represented by a bar with height equal to his height. The bars define a function f as follows:

$$f(x) = \begin{cases} h_1 & \text{for } 0 \le x < 1 \\ h_2 & \text{for } 1 \le x < 2 \\ h_3 & \text{for } 2 \le x < 3 \\ h_4 & \text{for } 3 \le x < 4 \\ h_5 & \text{for } 4 \le x < 5 \end{cases}$$

From this graph, we can find the total height of the players as the integral

$$\text{total height} = \int_0^5 f(x)\,dx$$
$$= \int_0^1 h_1\,dx + \int_0^1 h_2\,dx + \int_0^1 h_3\,dx + \int_0^1 h_4\,dx + \int_0^1 h_5\,dx$$
$$= h_1 + h_2 + h_3 + h_4 + h_5$$

To integrate a function that is defined in pieces, integrate each piece separately and add the results. Substituting the definition of the average (Equation 4.26), we have

$$\text{average height} = \frac{\text{total height}}{\text{total number}} = \frac{1}{5}(h_1 + h_2 + h_3 + h_4 + h_5)$$

which matches the usual way to compute the average.

Integrals and Mass

Integration can be used to find the mass of an object when its *density* is known. Here we consider only a one-dimensional case, such as a thin rod. Suppose the density of the bar, measured in grams per centimeter, is $\rho(x)$ at x (Figure 4.47). To

4.7 Applications of Integrals

estimate the mass, we break the bar into n small pieces of length Δx. The mass of the piece between x_i and $x_i + \Delta x$ is approximately $\rho(x_i)\Delta x$, the density at the left end of the piece times the length. Adding all the little pieces, we get

$$\text{mass of bar} \approx \sum_{i=1}^{n} \rho(x_i)\Delta x$$

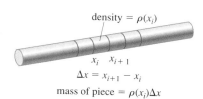

Figure 4.47
The mass of a bar

This has the exact form of a Riemann sum. The limit as Δx approaches 0 and n approaches infinity is equal to the definite integral (Definition 4.2), so

$$\text{mass of bar} = \int_a^b \rho(x)\,dx \qquad (4.27)$$

Consider a 100-cm vertical bar composed of a substance that has settled and become denser near the ground. Let z denote the height above ground, and suppose the density is given by

$$\rho(z) = e^{-0.01z}$$

in grams per centimeter. The mass is

$$\int_0^{100} e^{-0.01z}\,dz = \frac{1}{-0.01}(e^{-0.01z})\Big|_0^{100}$$
$$= -100(e^{-0.01z})\Big|_0^{100}$$
$$= -100(e^{-0.01 \cdot 100}) - [-100(e^{-0.01 \cdot 0})]$$
$$= 100(1 - e^{-1.0}) \approx 63.21$$

Does this result make sense? The density at the bottom of the bar is $\rho(0) = 1.0$ g/cm. If the entire bar had this maximum density, the mass would be 100 g. The density at the top of the bar is $\rho(100) = e^{-1} \approx 0.368$ g/cm, so the mass of the bar would be 36.8 g if the entire bar had this minimum density. Our result lies between these extremes. Furthermore,

$$\text{average density} = \frac{\text{total mass}}{\text{total length}}$$
$$= \frac{\int_0^{100} \rho(x)\,dx}{100}$$
$$\approx \frac{63.2}{100.0}$$
$$= 0.632$$

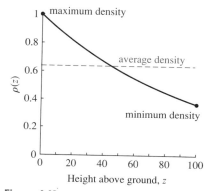

Figure 4.48
The average density of a bar

The result lies between the minimum density of 0.368 and the maximum of 1.0 (Figure 4.48).

The same technique can be used to find totals when density is measured in other units. Suppose the density of otters along the coast of California is

$$f(x) = 3.0 \times 10^{-4} x(1000 - x)$$

in otters per mile, where x is measured in miles from the Mexican border and can take on values between 0 and 1000. The population density takes on a maximum value of 75 otters per mile halfway up the coast at $x = 500$ and a minimum value of 0 at $x = 0$ and $x = 1000$ (Figure 4.49).

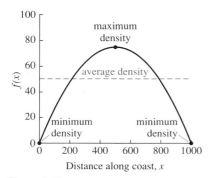

Figure 4.49
The average density of otters

The total number T is the definite integral of the density, or

$$T = \int_0^{1000} 3.0 \times 10^{-4} x(1000 - x)\, dx$$

$$= \int_0^{1000} 0.3x - 3.0 \times 10^{-4} x^2\, dx$$

$$= 0.15 x^2 \big|_0^{1000} - 1.0 \times 10^{-4} x^3 \big|_0^{1000}$$

$$= 1.5 \times 10^5 - 1.0 \times 10^5$$

$$= 0.5 \times 10^5 = 50{,}000 \text{ otters}$$

The average density is

$$\text{average density} = \frac{\text{total number}}{\text{total length}}$$

$$= \frac{50{,}000 \text{ otters}}{1000 \text{ miles}}$$

$$= 50.0 \frac{\text{otters}}{\text{mile}}$$

Once again, this value lies between the maximum density of 75 otters per mile and the minimum density of 0 otters per mile.

SUMMARY

The definite integral can be used to find the **area under a curve**, with special attention paid to negative functions and the order of the limits of integration. The process of integration by substitution, introduced for indefinite integrals, can be extended to definite integrals but requires the additional step of expressing the limits of integration in terms of the new variable. We used the definite integral to calculate **average values of functions** by dividing the integral (the total amount) by the length of the interval. Similarly, we computed masses or total numbers from **densities**. In each case, the underlying idea is that of the Riemann sum: chopping things into small pieces and adding the results.

4.7 EXERCISES

1. Find the areas under the following curves. If you use substitution, draw a graph to compare the original area with that in transformed variables.
   a. Area under $f(x) = 3x^3$ from $x = 0$ to $x = 3$.
   b. Area under $g(x) = e^x$ from $x = 0$ to $x = \ln 2$.
   c. Area under $h(x) = e^{x/2}$ from $x = 0$ to $x = \ln 2$.
   d. Area under $f(t) = (1 + 3t)^3$ from $t = 0$ to $t = 2$.
   e. Area under $G(y) = (3 + 4y)^{-2}$ from $y = 0$ to $y = 2$.
   f. Area under $s(z) = \sin(z)$ from $z = 0$ to $z = \pi$.

2. The definite integral can be used to find the area between two curves.
   a. Sketch the graphs of $f(x) = x^2$ and $g(x) = x^3$ for $0 \le x \le 2$, and shade the area between the curves.
   b. Sketch the graph of $f(x) - g(x)$, and show that the area under the curve matches the area between the curves.
   c. Find the area under the curve in part **b**, remembering to use absolute value.

3. Suppose a math class has four equally weighted tests. A student gets 60 on the first test, 70 on the second, 80 on the third, and 90 on the last. Find this student's average score directly, then find it as an integral of some function. Shade the area computed with this integral.

4. Suppose water is entering a tank at a rate of

$$g(t) = 360t - 39t^2 + t^3$$

where g is measured in liters per hour and t is measured in hours. The rate is 0 at times 0, 15, and 24.
  a. Find the total amount of water entering during the first 15 h, from $t = 0$ to $t = 15$.
  b. Find the average rate at which water entered during this time.
  c. Find the total amount and average rate from $t = 15$ to $t = 24$.
  d. Find the total amount and average rate from $t = 0$ to $t = 24$.
  e. Suppose that energy is produced at a rate of
  $$E(t) = |g(t)|$$
  in joules per hour. Find the total energy generated from $t = 0$ to $t = 24$.
  f. Find the average rate of energy production.

5. A very skinny 2.0-m-long snake has density $\rho(x)$, given by
$$\rho(x) = 1.0 + 2.0 \times 10^{-8} x^2 (300 - x)$$
where ρ is measured in grams per centimeter and x is measured in centimeters from the tip of the tail.
  a. Find the minimum and maximum density of the snake. Where does the maximum occur?
  b. Find the total mass of the snake.
  c. Find the average density of the snake. How does this compare with the minimum and maximum?
  d. Graph the density and average.

6. A piece of E. coli DNA has about 4.7×10^6 nucleotides, and is about 1.6×10^6 nm long. The genetic code consists of four nucleotides, called A, C, G, and T. Suppose that the number of As per thousand increases linearly from 150 at one end of the DNA strand to 300 at the other, the number of Cs per thousand decreases linearly from 350 at one end to 200 at the other, and the number of Gs per thousand increases linearly from 220 at one end to 320 at the other.
  a. Find the formula for the number of As, Cs, and Gs per thousand as a function of distance along the DNA strand.
  b. Find the formula for the number of Ts per thousand as a function of distance along the DNA strand.
  c. Find the total number of As, Cs, Gs, and Ts in the DNA.
  d. Find the mean number of As, Cs, Gs, and Ts in the DNA per thousand.

7. Water is entering one vessel at a rate of t^3 cm^3/s, a second vessel at a rate of $\sqrt{t}$ cm^3/s, and a third vessel at a rate of t cm^3/s.
  a. Find the total amount of water entering the first vessel during the first second, and find the average rate.
  b. Find the total amount of water entering the second vessel during the first second, and find the average rate.
  c. Find the total amount of water entering the third vessel during the first second, and find the average rate.
  d. Compare the average rates with rates at the "average time," halfway through the time period from 0 to 1. For which vessel is the average rate greater than the rate at the average time?
  e. Graph the two flow-rate functions, and mark the flow rate at the average time for each vessel. Can you guess what it is about the shapes of the two graphs that produces the difference in how the average rate compares to the rate at the average time?

8. The otter data presented in the text are an idealization. Suppose instead that 1-mile samples were taken every 50 miles, with results as in the following table.

Position	Number	Position	Number
0	14	500	74
50	27	550	72
100	38	600	69
150	48	650	63
200	56	700	57
250	63	750	48
300	68	800	39
350	72	850	27
400	74	900	15
450	75	950	8

Graph these results. How would you estimate the total number of otters?

9. We have been using little vertically oriented rectangles to compute areas. There is no reason why little horizontal rectangles cannot be used. Here are the steps to find the area under the parabola $y = x^2$ from $x = 0$ to $x = 1$ by using such horizontal rectangles.
  a. Draw a picture with five horizontal rectangles, each of height 0.2, approximately filling the region to the right of the curve.
  b. Calculate an upper and lower estimate of the length of each rectangle based on the length of the upper and lower boundaries.
  c. Add these to find upper and lower estimates of the area.
  d. Think now of a very thin rectangle at height y. How long is the rectangle?
  e. Write down a definite integral expression for the area.
  f. Evaluate the integral, and check that the answer is correct.

10. Consider again the expression for the average value of a function given by Equation 4.26,
$$\text{average value of } f = \frac{1}{b - a} \int_a^b f(x)\, dx$$
Think of the average as a function of b, and write it as $\bar{f}(b)$.
  a. Use the Fundamental Theorem of Calculus and the quotient rule to take the derivative of $\bar{f}(b)$.

b. Rewrite the term involving the integral in terms of $\bar{f}(b)$.
c. Show that this derivative is positive if $f(b) > \bar{f}(b)$.
d. Interpret this result. The idea is that the average is increasing when the functional value exceeds the average.

11. The natural logarithm function is sometimes defined with the definite integral as the function $l(a)$ for which

$$l(a) = \int_1^a \frac{1}{x} dx$$

Using this definition, we can prove the laws of logs (Section 1.6).
a. Show that $l(1) = 0$.
b. Show that $l(6) - l(3) = l(2)$. (Use the summation property of the definite integral to write the difference as an integral, and then use the substitution $y = x/3$.)
c. Find the integral from a to $2a$ by following the same steps (make the substitution $y = x/a$.)
d. What law of logs does this correspond to?
e. Show that $l(a^b) = b \cdot l(a)$. (Try the substitution $y = \sqrt[b]{x}$ in $\int_1^{a^b} 1/x\, dx$.)

12. **COMPUTER:** Suppose the volume of water in a vessel obeys the differential equation

$$\frac{dV}{dt} = f(t) = 1 + 3t + 3t^2$$

with $V(0) = 0$.

a. Graph the functions f and V for $t = 0$ to $t = 2$, and label the curves.
b. Find the volume at time $t = 10$. What definite integral has the same answer? Shade the area on your graph and write the associated integral.
c. Define a function $A(T)$ that gives the average rate of change of volume as a function of time (the total volume added between times $t = 0$ and $t = T$ divided by the elapsed time). Graph this on the same graph as $f(t)$. Label the curves (and write the formula for $A(t)$ as a definite integral). Why is the average rate A greater than the instantaneous rate f?
d. Graph $f(t)$ between $t = 0$ and $t = 10$ and the constant function with rate equal to the average at time $T = 10$. What is the area under the line? Does it match what you found in part **b**? Why should it? Mark the point where the average and instantaneous rates are equal. Use your computer to solve for this point.

13. **COMPUTER:** Find the area between the two curves $f(x) = \cos(x)$ and $g(x) = 0.1x$ for $0 \leq x \leq 10$.
a. Graph the two functions. There should be three separate regions between them.
b. Have your computer find where each region begins and ends.
c. Integrate to find the area of each region.
d. Add the areas.

4.8 Improper Integrals

So far, we have considered only definite integrals of functions that do not blow up to infinity between finite limits of integration. Nice integrals of this sort are called **proper integrals**. **Improper integrals** are of two types: integrals with infinite limits of integration and integrals of functions that blow up somewhere between the limits of integration. We now learn how to compute and apply improper integrals.

Infinite Limits of Integration

"Infinite" measurements cannot crop up in biological experiments. Nonetheless, infinity is a useful mathematical abstraction of "very long" or "very far." Consider again the equation for chemical production with exponentially declining rate (Equation 4.2),

$$\frac{dP}{dt} = e^{-t}$$

in moles per second. The amount of chemical produced between $t = 0$ and $t = T$ is given by the definite integral,

$$\text{production between 0 and } T = \int_0^T e^{-t}\, dt$$

The longer we wait, the more chemical has been produced. Let P_∞ denote the amount that would be produced if the experiment ran forever. We would like to write

$$P_\infty = \int_0^\infty e^{-t}\,dt$$

To be honest, however, we have never defined what this expression means. We *defined* the definite integral to be the limit of Riemann sums. Computing Riemann sums requires breaking the region between the limits of integration into n equally sized regions. But the infinite region from 0 to infinity cannot be broken into n finite regions of equal size. The Riemann sum approach never even gets started.

Instead, we can think of this **improper integral** as the *limit* of **proper integrals** with formula

$$\int_0^\infty e^{-t}\,dt = \lim_{T \to \infty} \int_0^T e^{-t}\,dt$$

This formula captures the spirit of what we want. The proper integral gives the amount of production until time T, and the limit allows us to make T "very large" (Figure 4.50). In this case,

$$\int_0^T e^{-t}\,dt = -e^{-t}\Big|_0^T = -e^{-T} + 1$$

Therefore

$$\int_0^\infty e^{-t}\,dt = \lim_{T \to \infty} 1 - e^{-T}$$
$$= \lim_{T \to \infty} 1 - \lim_{T \to \infty} e^{-T}$$
$$= 1 - 0 = 1.0$$

Exactly 1 mol of chemical would be produced after an infinite amount of time. If we wait a "long time," for example, 10 s, the exact amount is

$$1 - e^{-10.0} \approx 0.99995$$

which is quite close to the limit.

We define formally the improper integral:

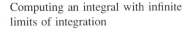

Figure 4.50
Computing an integral with infinite limits of integration

■ **Definition 4.3** The improper integral of the function f from a to ∞ is

$$\int_a^\infty f(t)\,dt = \lim_{T \to \infty} \int_a^T f(t)\,dt \qquad ■$$

Examples of Improper Integrals

Our definition of the improper integral does not guarantee that the limit is finite. Nothing mathematical prevents the definite integral from 0 to T from getting larger and larger as T approaches infinity. Consider a different chemical produced at the diminishing rate

$$\frac{dQ}{dt} = \frac{1}{1+t}$$

in moles per second. This rate decreases to 0 more slowly than the exponential function (Exercise 3). How much chemical would this reaction produce after a long time?

The total chemical ever produced, Q_∞, is computed with the improper integral,

$$Q_\infty = \int_0^\infty \frac{1}{1+t}\, dt$$

$$= \lim_{T\to\infty} \int_0^T \frac{1}{1+t}\, dt$$

To compute this integral, we must use the substitution $u = 1 + t$. Then $dt = du$, and the limits of integration become 1 and $1 + T$ (Exercise 2). Then we have

$$Q_\infty = \lim_{T\to\infty} \int_0^T \frac{1}{1+t}\, dt$$

$$= \lim_{T\to\infty} \int_1^{T+1} \frac{1}{u}\, du$$

$$= \lim_{T\to\infty} \ln u \Big|_1^{T+1}$$

$$= \lim_{T\to\infty} \ln(T+1) - 0$$

$$= \infty$$

If this rule were followed forever, the amount of chemical produced would be infinite (Figure 4.51a). In such a case, we say that the integral **diverges**. Such a result might seem absurd and irrelevant. But this very absurdity provides a valuable negative result; no real system could follow this law indefinitely. Even though the rate gets smaller and smaller, the total production increases without bound.

When the limit exists, we say that the integral **converges**. What laws can be maintained indefinitely without producing an infinite amount of product? First of all, any function that does not decrease to 0 has an infinite improper integral (Figure 4.51b) because the total production gets larger and larger without bound. In terms of area, the region under the curve can be thought of as including a rectangle with positive height and infinite length.

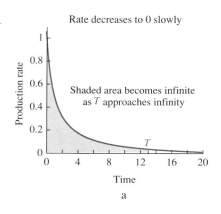
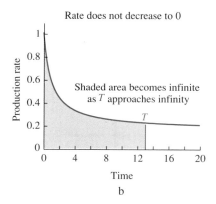

Figure 4.51
Two divergent integrals

For functions that decrease to 0 as their arguments approach infinity, we have learned to integrate only those of the form $1/t^p$ for $p > 0$ and $e^{-\alpha t}$ for $\alpha > 0$. We can use the power rule to compute the integral of $1/t^p$ when $p \neq 1$. (We set the lower limit of integration to 1 to keep the integrand from blowing up at $t = 0$.)

When $p > 1$, we have

$$\int_1^\infty \frac{1}{t^p}\,dt = \lim_{T\to\infty}\int_1^T \frac{1}{t^p}\,dt$$

$$= \lim_{T\to\infty} \left.\frac{t^{1-p}}{1-p}\right|_1^T$$

$$= \lim_{T\to\infty} \frac{T^{1-p}}{1-p} - \frac{1}{1-p}$$

When $p > 1$, the power $1 - p$ is negative. Therefore T^{1-p} approaches 0 as T approaches infinity. The improper integral is

$$\int_1^\infty \frac{1}{t^p}\,dt = \lim_{T\to\infty} \frac{T^{1-p}}{1-p} - \frac{1}{1-p}$$

$$= 0 - \frac{1}{1-p}$$

$$= \frac{1}{p-1}$$

This integral converges. Furthermore, the value of the improper integral (the total area under the curve, or the limiting amount of product produced) becomes smaller as the value of p becomes larger (Figure 4.52).

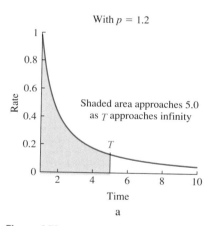
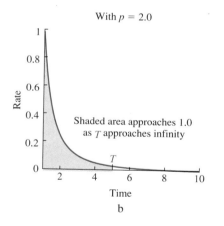
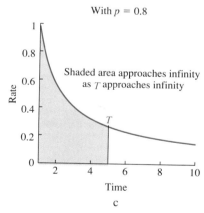

Figure 4.52
Integrals of power functions $1/t^p$

With $p < 1$, the integration is the same. But when we take the limit, we find

$$\int_1^\infty \frac{1}{t^p}\,dt = \lim_{T\to\infty} \frac{T^{1-p}}{1-p} - \frac{1}{1-p}$$

$$= \infty$$

because T is taken to the *positive* power $1 - p$ when $p < 1$. This integral diverges. Recall that $1/t^p$ decreases to 0 faster for larger values of p. If the value of p is sufficiently large (greater than 1), the area under the curve is finite. When p is small (less than 1), the function decreases to 0 more slowly and the area is infinite (Figure 4.52c).

When the rate decreases exponentially with formula $e^{-\alpha t}$, the improper integral is

$$\int_0^\infty e^{-\alpha t}\,dt = \lim_{T\to\infty} \int_0^T e^{-\alpha t}\,dt$$

$$= \lim_{T\to\infty} \int_0^{\alpha T} \frac{e^{-y}}{\alpha}\,dy$$

$$= \lim_{T\to\infty} -\frac{e^{-y}}{\alpha}\Big|_0^{\alpha T}$$

$$= \lim_{T\to\infty} -\frac{e^{-\alpha T}}{\alpha} + \frac{1}{\alpha}$$

$$= \frac{1}{\alpha}$$

where we used the substitution $y = \alpha t$ (Exercise 2). This exponential integral converges for every positive value of α, but the integral takes on a larger value for smaller α, consistent with the fact that $e^{-\alpha t}$ decreases to 0 more slowly when α is small (Figure 4.53). These results on power and exponential functions are summarized in the following table.

Function	Limits of integration	Behavior
$\frac{1}{t^p}$ with $p > 1$	1 to ∞	Converges
$\frac{1}{t^p}$ with $p < 1$	1 to ∞	Diverges
$\frac{1}{t}$	1 to ∞	Diverges
$e^{-\alpha t}$ with $\alpha > 0$	0 to ∞	Converges

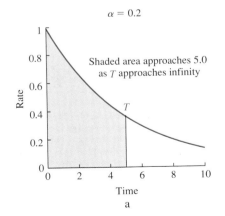
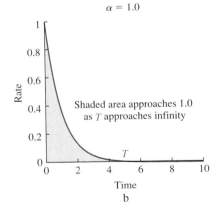

Figure 4.53
Improper integrals with different parameters in the exponent

The **method of leading behavior** (Section 3.7) can be used to extend this table to more complicated functions. The idea is that the behavior of the integral is determined by the leading behavior of the integrand, the piece that decreases to

0 most slowly. Suppose we wish to know whether the integral

$$\int_1^\infty \frac{1}{t} + \frac{1}{e^t} \, dt$$

converges. The leading behavior of the integrand is the largest part, the part that decreases most slowly. Because power functions always grow more slowly than exponential functions, the leading behavior of $f(t) = (1/t) + (1/e^t)$ is

$$f_\infty(t) = \frac{1}{t}$$

The original integral will converge if the integral of the leading behavior converges. In this case,

$$\int_1^\infty f_\infty(t) \, dt = \int_1^\infty \frac{1}{t} \, dt = \infty$$

(Figure 4.54a). Therefore, the original integral also diverges.

How about the similar-looking integral

$$\int_1^\infty \frac{1}{t + e^t} \, dt$$

The leading behavior of the integrand must be found by computing the leading behavior of the denominator $t + e^t$. Because the exponential is larger, we replace $t + e^t$ with e^t. The integral of the leading behavior is

$$\int_1^\infty \frac{1}{e^t} \, dt = \int_1^\infty e^{-t} \, dt$$

which converges (Figure 4.54b). It is impossible to compute exactly the integral

$$\int_1^\infty \frac{1}{t + e^t} \, dt$$

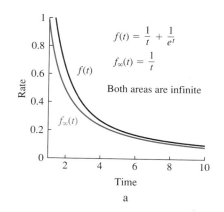
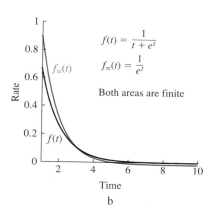

Figure 4.54
Improper integrals computed with the method of leading behavior

Infinite Integrands

Although functions with infinite integrands do not appear frequently in biological problems, it is useful to be familiar with their behavior. Consider trying to find the area under the curve $f(x) = 1/\sqrt{x}$ between $x = 0$ and $x = 1$ (Figure 4.55). This function approaches infinity at $x = 0$.

380 Chapter 4 Differential Equations, Integrals, and Their Applications

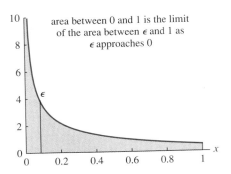

Figure 4.55
Computing an integral with an infinite integrand

Although one can, with some care, define this integral as a limit of right-hand Riemann sums, we instead use the limit to define and compute this type of improper integral.

■ **Definition 4.4** If the function f approaches infinity at $x = 0$ but nowhere else between $x = 0$ and $x = a$, the improper integral of the function f from 0 to a is defined by

$$\int_0^a f(x)\,dx = \lim_{\epsilon \to 0^+} \int_\epsilon^a f(x)\,dx \qquad ■$$

The limit with the 0^+ indicates that ϵ approaches 0 from the right (Section 2.2). With this definition, we have

$$\int_0^1 \frac{1}{\sqrt{x}}\,dx = \lim_{\epsilon \to 0^+} \int_\epsilon^1 \frac{1}{\sqrt{x}}\,dx$$
$$= \lim_{\epsilon \to 0^+} 2\sqrt{x}\,\big|_\epsilon^1$$
$$= \lim_{\epsilon \to 0^+} 2 - 2\sqrt{\epsilon}$$
$$= 2$$

Although the function is ill-behaved, the area is perfectly well defined. In contrast, if $g(x) = 1/x^2$, we have

$$\int_0^1 \frac{1}{x^2}\,dx = \lim_{\epsilon \to 0^+} \int_\epsilon^1 \frac{1}{x^2}\,dx$$
$$= \lim_{\epsilon \to 0^+} -\frac{1}{x}\,\bigg|_\epsilon^1$$
$$= \lim_{\epsilon \to 0^+} -1 + \frac{1}{\epsilon}$$
$$= \infty$$

The rules for improper integrals of functions that approach infinity at 0 are given in the table on the following page. The larger the power in the denominator, the faster the integrand approaches infinity near 0 and the larger the value of the integral.

As before, these rules can be extended with the method of leading behavior. The portion that counts is the part that approaches infinity most quickly. The

Function	Limits of integration	Behavior
$\frac{1}{x^p}$ with $p > 1$	0 to 1	Diverges
$\frac{1}{x^p}$ with $p < 1$	0 to 1	Converges
$\frac{1}{x^p}$ with $p = 1$	0 to 1	Diverges

integrand of the integral

$$\int_0^1 \left(\frac{1}{\sqrt{t}} + \frac{1}{t^2} \right) dt$$

has two components, the larger of which is $1/t^2$. Therefore, the leading behavior is

$$\int_0^1 \frac{1}{t^2} dt = \infty$$

because the power is greater than 1, meaning that the original integral diverges (Figure 4.56a). On the other hand, the leading behavior of the denominator of the integrand of

$$\int_0^1 \frac{1}{\sqrt{t} + t^2} dt$$

must be found by finding the leading behavior of the denominator. Both terms $\sqrt{t}$ and t^2 approach 0, but $\sqrt{t}$ approaches 0 more slowly, is larger, and is the leading behavior of the denominator. The integral of the leading behavior is

$$\int_0^1 \frac{1}{\sqrt{t}} dt$$

which converges because the power is less than 1 (Figure 4.56b).

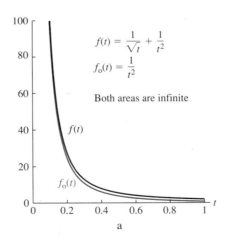
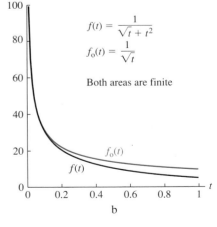

Figure 4.56
Improper integrals with infinite integrands computed with the method of leading behavior

4.8 EXERCISES

1. Evaluate the following improper integrals, or say why they don't converge.

   a. $\int_0^\infty e^{-3t}\, dt$
   b. $\int_0^\infty e^t\, dt$
   c. $\int_1^\infty \frac{1}{\sqrt{x}}\, dx$
   d. $\int_5^\infty \frac{1}{x^2}\, dx$
   e. $\int_0^\infty \frac{1}{(1+3x)^{3/2}}\, dx$

2. Check the following integrals from the text.

   a. $\int_0^\infty \frac{1}{1+t}\, dt$
   b. $\int_0^\infty e^{-\alpha t}\, dt$

3. Remind yourself of the following comparisons. Use L'Hôpital's rule when needed.
   a. Show that e^{-x} approaches 0 faster than $1/x$ as x approaches infinity.
   b. Show that $1/x^2$ approaches 0 faster than $1/x$ as x approaches infinity.
   c. Show that $1/x^2$ approaches infinity faster than $1/x$ as x approaches 0.
   d. Show that $1/\ln(x)$ approaches 0 more slowly than $1/x$ as x approaches infinity.
   e. Do you think that
   $$\int_2^\infty \frac{1}{\ln(x)}\, dx$$
   converges?

4. Write pure-time differential equations to describe the following situations; find out what happens over the long term; and state whether the rule could be followed indefinitely.
   a. The volume of a cell is increasing at a rate of $100/(1+t)^2$ µm³/s, starting from a size of 500 µm³.
   b. The concentration of a toxin in a cell is increasing at a rate of $50e^{-2t}$ µmol/L/s, starting from a concentration of 10 µmol/L. If the cell is poisoned when the concentration exceeds 30 µmol/L, can this cell survive?
   c. A population of bacteria is increasing at a rate of $1000/(2+3t)^{0.75}$ bacteria per hour, starting from a population of 10^6. Can this sort of growth be maintained indefinitely? When would the population reach 2.0×10^6? Would you say that this population is growing quickly?

5. Evaluate the following improper integrals, or say why they don't converge.

   a. $\int_0^1 \frac{1}{x}\, dx$
   b. $\int_0^{0.001} \frac{1}{x^2}\, dx$
   c. $\int_0^{0.001} \frac{1}{\sqrt[3]{x}}\, dx$
   d. $\int_0^\infty \frac{1}{\sqrt[3]{x}}\, dx$

6. State whether the following improper integrals converge.

   a. $\int_0^1 \frac{1}{\sqrt[3]{x}+x^3}\, dx$
   b. $\int_1^\infty \frac{1}{\sqrt[3]{x}+x^3}\, dx$
   c. $\int_0^1 \frac{1}{x^2+e^x}\, dx$
   d. $\int_1^\infty \frac{1}{\sqrt{x}+e^x}\, dx$

7. Some improper integrals can be attacked with tricky substitutions. Try the following.
   a. Compute
   $$\int_0^1 \frac{1}{\sqrt{x}}\, dx$$
   with the substitution $y = 1/x$. Make sure to change the limits of integration.
   b. Compute
   $$\int_0^1 \frac{1}{x^p}\, dx$$
   with the substitution $y = 1/x$. Use this to compare the conditions for convergence of improper integrals of power functions with an infinite limit of integration and those with infinite integrand at $x = 0$.
   c. For what value of p does the substitution $y = 1/x$ leave the integrand unchanged? What is special about integrals involving this power?

8. According to tables of integrals,
   $$\int \ln(x)\, dx = x\ln(x) - x$$

a. Check this indefinite integral.
b. Find $\int_0^1 \ln(x)\, dx$ (use L'Hôpital's rule).
c. Draw a graph illustrating your result.

9. **COMPUTER:** Have your machine find the exact values of the integrals in Exercise 6, and compare with the integrals of the leading behavior.

Supplementary Problems for Chapter 4

EXERCISE 1
The voltage v of a neuron follows the differential equation

$$\frac{dv}{dt} = 1.0 + \frac{1}{1 + 0.02t} - e^{0.01t}$$

over the course of 100 ms, where t is measured in milliseconds and v in millivolts. We start at $v(0) = -70$.
a. Sketch a graph of the rate of change. Indicate on your graph the times when the voltage reaches minima and maxima (you don't need to solve for the numerical values).
b. Sketch a graph of the voltage as a function of time.
c. What is the voltage after 100 ms?

EXERCISE 2
Consider again the differential equation in Exercise 1,

$$\frac{dv}{dt} = 1.0 + \frac{1}{1 + 0.02t} - e^{0.01t}$$

with $v(0) = -70$.
a. Use Euler's method to estimate the voltage after 1 ms, and again 1 ms after that.
b. Estimate the voltage after 2 ms using left-hand and right-hand Riemann sums.
c. Which of your estimates matches Euler's method, and why?

EXERCISE 3
A neuron in your brain sends a charge down an 80-cm-long axon (a long, skinny thing) toward your hand at a speed of 10 m/s. At the time when the charge reaches your elbow, the voltage in the axon is -70 mV except on the 6-cm-long piece between 47 cm and 53 cm from your brain. On this piece, the voltage is given by

$$v(x) = -70.0 + 10.0[9.0 - (x - 50.0)^2]$$

where $v(x)$ is the voltage at a distance of x cm from the brain.
a. How long will it take the information to get to your hand? How long did it take to reach your elbow?
b. Sketch a graph of the voltage along the whole axon.
c. Find the average voltage of the 6-cm piece.
d. Find the average voltage of the whole axon.

EXERCISE 4
The charge in a dead neuron decays according to

$$\frac{dv}{dt} = \frac{1}{\sqrt{1 + 4t}} - \frac{2}{(1 + 4t)^{3/2}}$$

starting from $v(0) = -70$ at $t = 0$.
a. Is the voltage approaching 0 as $t \to \infty$? How do you know that it will eventually reach 0?
b. Write an equation (but don't solve it) for the time when the voltage reaches 0.
c. What is wrong with this model?

EXERCISE 5
Consider the differential equations

$$\frac{db}{dt} = 2b$$

and

$$\frac{dB}{dt} = 1 + 2t$$

a. Which is a pure-time differential equation? Describe circumstances when you might find each of these equations.
b. Suppose $b(0) = B(0) = 1$. Use Euler's method to find estimates for $b(0.1)$ and $B(0.1)$.
c. Use Euler's method again to find estimates for $b(0.2)$ and $B(0.2)$.

EXERCISE 6
Consider the differential equation

$$\frac{dp}{dt} = e^{-4t}$$

where $p(t)$ is product in moles at time t and t is measured in seconds.
a. Explain in words what is going on.
b. Suppose $p(0) = 1$. Find $p(1)$.

EXERCISE 7
Consider the differential equation

$$\frac{dV}{dt} = 4 - t^2$$

where $V(t)$ is volume in liters at time t and t is measured in minutes.
a. Explain in words what is going on.
b. At what time is the volume a maximum?

c. Break the interval from $t = 0$ to $t = 3$ into three parts, and find the left-hand and right-hand estimates of the volume at $t = 3$ (assume $V(0) = 0$).
d. Write the definite integral expressing volume at $t = 3$, and evaluate.

EXERCISE 8
Find the area under the curve $f(x) = 3 + (1 + x/3)^2$ between $x = 0$ and $x = 3$.

EXERCISE 9
The population density of trout in a stream is

$$\rho(x) = |-x^2 + 5x + 50|$$

where ρ is measured in trout per mile and x is measured in miles, and x runs from 0 to 20.
a. Graph $\rho(x)$ and find the minimum and maximum.
b. Find the total number of trout in the stream.
c. Find the average density of trout in the stream.
d. Indicate on your graph how you would find where the actual density is equal to the average density.

EXERCISE 10
The amount of product is described by the differential equation

$$\frac{dp}{dt} = \frac{1}{\sqrt{1 + 3t}}$$

starting at time $t = 0$. Suppose p is measured in moles and t in hours, and that $p(0) = 0$.
a. Find the limiting amount of product.
b. Find the average rate at which product is produced as a function of time, and compute the limit as $t \to \infty$.
c. Find the limit as $t \to 0$ of the average rate at which product is produced.

EXERCISE 11
A student is hooked up to an EEG machine during a test, and her α brain wave power follows

$$A(t) = \frac{50}{2.0 + 0.3t} + 10e^{0.0125t}$$

where t runs from 0 min to 120 min.
a. Convince yourself that brain wave activity has a minimum value some time during the test. Sketch a graph of the function. Find the maximum.
b. Find the total brain wave power during the test.
c. Find the average brain wave power during the test. Sketch the corresponding line on your graph.
d. Estimate the minimum value from your average value.
e. Draw a graph showing how you would estimate the total using the right-hand approximation with $n = 6$. Write the associated sum. Do you think your estimate would be high or low?

EXERCISE 12
Consider the function $G(h)$, which gives the density of nutrients in a plant stem as a function of the height h,

$$G(h) = 5 + 3e^{-2h}$$

where G is measured in moles per meter and h is measured in meters.
a. Find the total amount of nutrient if the stem is 2.0 m tall.
b. Find the average density in the stem.
c. Find the exact and approximate amount between 1.0 m and 1.01 m.

Projects for Chapter 4

PROJECT 1
As noted in the text, Euler's method is not a very good way to solve differential equations numerically. In this project, you will compare Euler's method with the **midpoint** (or second-order Runge-Kutta) **method** and the **implicit Euler method**.

Suppose we know (or estimate) that the solution of the general differential equation

$$\frac{dy}{dt} = f(t, y)$$

is some value $y = y_0$ at time $t = t_0$. Our goal is to estimate the value of $y_1 = y(t_0 + h)$ with *step size* h. We will use y_1, y_2, and so forth to represent these approximate values.

We have seen that Euler's method uses the tangent line to estimate

$$y_1 = y(t_0) + hf(t_0, y_0)$$

This method has the problem that it uses the derivative only at the beginning of the interval between times t_0 and $t_0 + h$. If the derivative changes significantly during this interval, the method can be very inaccurate.

An alternative, called the midpoint method, uses an estimate of the derivative at time $t_0 + (h/2)$ instead. The next step is

$$k = hf(t_0, y_0)$$

$$y_1 = y(t_0) + hf\left(t_0 + \frac{h}{2}, y_0 + \frac{k}{2}\right)$$

where k is the estimated change computed by Euler's method, and we use it to guess the value of y halfway through the interval. This method is more accurate than Euler's method.

Euler's method and the midpoint method are *unstable*, meaning that the approximate solutions they produce can fail to approach a stable equilibrium if the step size h is too large. The **implicit Euler scheme** is stable, for linear equations. The idea is to use the derivative at the end of the interval instead of at the beginning, or

$$y_1 = y(t_0) + hf(t_0 + h, y_1)$$

Because we do not know the value of y_1 used on the right-hand side, we have to solve this equation to find the next value.

Compare these three methods on the four differential equations

$$\frac{db}{dt} = b$$

$$\frac{dx}{dt} = -x$$

$$\frac{dV}{dt} = -e^{-t}$$

$$\frac{dy}{dt} = -e^{-t} + e^{-y}$$

Suppose the state variable in each equation has the initial value 1.0 at $t = 0$.

a. First, find the solution of each at $t = 1$. Try each method (if you can figure out how to get the implicit method to work on the last equation) with values of h ranging from 1.0 to 0.001.

b. The equations for x and V should both approach 0 exponentially as t becomes large. Use large values of h (such as 10 or 100) in each of the three methods. How well do they do for large t?

c. The midpoint method is related to approximating the solution with a quadratic Taylor polynomial (Section 3.8). Develop an extension of Euler's method based on the quadratic approximation, and compare this with the midpoint method.

PROJECT 2

In ancient times, the Greeks were fascinated by such geometric problems as finding the area of a circle. Archimedes, perhaps the greatest mathematician in the ancient world, came up with an idea closely related to the Riemann sum to solve this problem. Using some modern tools, we can use his ideas to compute the value of π.

a. Break a circle of radius 1 into n wedges, each with angle $\theta = (2\pi)/n$. Show that the area of a right triangle inside each wedge is $\sin(\theta)/2$. Make sure to draw a picture.

b. Approximate the area of a circle using $n = 4$. Look up (or remember) the half-angle formula giving $\sin(\theta/2)$ in terms of $\cos(\theta)$ to approximate the area with $n = 8$.

c. Continue in this way to approximate the area with $n = 16$, and so forth.

d. Use the same approach but approximate each wedge with a right triangle that lies outside the circle. Show that the area of the triangle is $\tan(\theta)/2$.

e. Approximate the area of a circle using $n = 8$. (What happens when $n = 4$?) Look up (or remember) the half-angle formula giving $\tan(\theta/2)$ in terms of $\tan(\theta)$ to approximate the area with $n = 8$.

f. Continue in this way to approximate the area with $n = 16$, and so forth.

g. Can you think of a better method Archimedes could have used to compute the value of π?

Bibliography for Chapter 4

Section 4.1

Blanchard, P., R. Devaney, and G. Hall. *Differential Equations*. Brooks/Cole, Pacific Grove, Calif., 1998.

Section 4.3

Hyman, J. M., and E. A. Stanley. Using mathematical models to understand the AIDS epidemic. *Mathematical Biosciences* 90:415–473, 1988.

Section 4.4

Charnov, E. L., *Life History Invariants: Some Explorations of Symmetry in Evolutionary Ecology*. Oxford University Press, New York, 1993.

Projects for Chapter 4

Beckmann, P., *A History of π*. Golem Press, Boulder, Colo., 1982.

Press, W. H., S. A. Teukolsky, W. T. Vettering, and B. P. Flannery. *Numerical Recipes in FORTRAN: The Art of Scientific Computing*. Cambridge University Press, Cambridge, England, 1992.

Chapter 5

Analysis of Differential Equations

We have developed the indefinite and definite integrals as tools to solve pure-time differential equations, and we have linked them with the Fundamental Theorem of Calculus. We now use integrals and several graphical tools to study **autonomous differential equations** in more detail. In particular, we use a method called the **phase-line diagram** to illustrate **equilibria** and **stability**.

Because most biological systems involve several interacting measurements, we then extend our methods to address **systems of autonomous differential equations**, where the rate of change of each measurement depends on the value of the others. Generalization of the phase-line to the **phase-plane** allows study of two-dimensional systems. We conclude with a careful study of a fundamental two-dimensional system of differential equations, called the **FitzHugh-Nagumo equations**, which describes the firing of a neuron.

5.1 Equilibria and Display of Autonomous Differential Equations

Review of Autonomous Differential Equations

We have been developing techniques to solve pure-time differential equations with the general form

$$\frac{dV}{dt} = f(t) \tag{5.1}$$

In these equations the rate of change is a function of time and not of the state variable (volume in this case). Change is imposed from outside the system; for example, the level of a lake is controlled by the weather or by an engineer. In most biological systems, however, the rate of change depends on the current state of the system as well as external factors. The level of a lake that experienced increased evaporation while high would change as a function of the level itself, not as a function of just the weather or the decisions of engineers.

The class of differential equations where the rate of change depends only on the state of the system and not on external circumstances are particularly important. These types of equations are called **autonomous differential equations** (Section 1.1). We have already seen autonomous differential equations that describe population growth, cooling, movement of a chemical, and the invasion of mutant bacteria. For instance, in the model for growth of bacteria,

$$\frac{db}{dt} = \lambda b$$

the rate of change of the population size b depends only on the population size itself and not on the time t.

The general autonomous differential equation for a measurement m is

$$\frac{dm}{dt} = g(m) \tag{5.2}$$

This equation differs from the general pure-time differential equation in that the rate of change $g(m)$ depends on m rather than on t. Autonomous differential equations are slightly harder to solve than pure-time differential equations, but the techniques of solution are different and the solutions are often more interesting.

When the rate of change depends on both the state variable and the time, the equation is called a **nonautonomous differential equation**. The general form is

$$\frac{dm}{dt} = g(m, t) \tag{5.3}$$

These equations are generally much more difficult to solve than either pure-time or autonomous differential equations.

Although the *rate of change* in an autonomous differential equation does not depend explicitly on time, the *solution* does. The size of a growing population changes over time. Suppose that two bacterial cultures with 1.0×10^6 bacteria each are stored in a refrigerator to prevent reproduction. One is warmed to room temperature at 9:00 A.M. and allowed to reproduce. The population b_1 follows the autonomous differential equation

$$\frac{db_1}{dt} = 2b_1$$

If $t = 0$ represents 9:00 A.M., the solution is
$$b_1(t) = 1.0 \times 10^6 \cdot e^{2t}$$
The other culture is warmed and begins reproduction at 10:00 A.M. The population b_2 follows the same differential equation
$$\frac{db_2}{dt} = 2b_2$$
If we let $t = 0$ represent 10:00 A.M., the solution is the same,
$$b_2(t) = 1.0 \times 10^6 \cdot e^{2t}$$
When we graph these apparently identical solutions, however, we must recall that time is measured from different starting points (Figure 5.1).

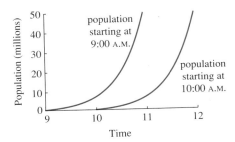

Figure 5.1
Two autonomous bacterial populations started at different times

Equilibria

One of the most important tools for analyzing discrete-time dynamical systems is the **equilibrium**, a state of the system that remains unchanged by the updating function. Solutions starting from an equilibrium do not change over time. For example, solutions of the bacterial competition model (Equation 1.49) approach an equilibrium at $p = 1$ when the mutants are superior to the wild type whether the model is a discrete-time dynamical system (Figure 5.2a) or an autonomous differential equation (Figure 5.2b). In either case, if we started with all mutant bacteria, the population would never change.

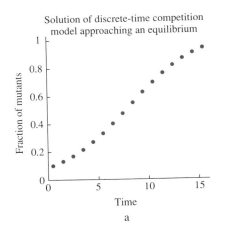
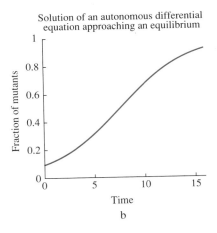

Figure 5.2
Solutions and equilibria: discrete-time dynamical systems and differential equations

5.1 Equilibria and Display of Autonomous Differential Equations

Equilibria are as useful in analyzing autonomous differential equations as in analyzing discrete-time dynamical systems. As before, an equilibrium is a value of the state variable from which nothing happens.

Definition 5.1 A value m^* of the state variable is called an equilibrium of the autonomous differential equation

$$\frac{dm}{dt} = f(m)$$

if

$$f(m^*) = 0$$

At a point where $f(m^*) = 0$, the rate of change is 0. A rate of change of 0 indicates precisely that the value of the measurement remains the same.

The steps for finding equilibria resemble those for finding equilibria of discrete-time dynamical systems.

Algorithm 5.1 (Finding equilibria of an autonomous differential equation)

1. Make sure that the differential equation is autonomous.
2. Write the equation for the equilibria.
3. Factor.
4. Set each factor equal to 0 and solve for the equilibria.
5. Meditate upon the results.

The simplest autonomous differential equation we have considered is the equation for bacterial growth:

$$\frac{db}{dt} = \lambda b$$

What are the equilibria of this differential equation? The rate of change depends only on b and not on the time t, so this equation is autonomous. The algebraic steps are

$$\lambda b^* = 0 \quad \text{the equation for the equilibrium}$$
$$\lambda = 0 \text{ or } b^* = 0 \quad \text{set both factors to 0}$$

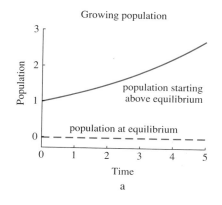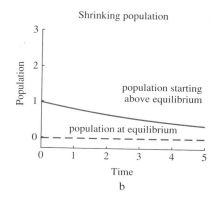

Figure 5.3
Solutions of the bacterial growth model

If $\lambda = 0$, every value of b is an equilibrium because the population does not change. This equilibrium corresponds to the case $r = 1$ of the related discrete-time dynamical system, $b_{t+1} = rb_t$. The other equilibrium at $b^* = 0$ indicates extinction, again matching the results for the discrete-time dynamical system. A solution starting from $b = 0$ remains there forever (Figure 5.3).

For Newton's law of cooling, the differential equation is

$$\frac{dH}{dt} = \alpha(A - H)$$

The rate of change depends only on H and not on the time t, so this equation is autonomous. Following the steps, we find

$\alpha(A - H^*) = 0$	the equation for the equilibrium
$\alpha = 0$ or $(A - H^*) = 0$	set both factors to 0
$\alpha = 0$ or $H^* = A$	solve each term

The solutions are $\alpha = 0$ and $H^* = A$. If $\alpha = 0$, no heat is exchanged with the outside world. Such an object is always at equilibrium. This corresponds to the case $q = 0$ of the lung model (Equation 1.44); a lung that exchanges no air with the outside world is at equilibrium. The other equilibrium is $H^* = A$. The object is in equilibrium when its temperature is equal to the ambient temperature. A solution with initial condition $H(0) = A$ remains there forever. This corresponds to the equilibrium $c^* = \gamma$ for the lung model; a lung that exchanges some air eventually comes to match the external environment.

We follow the same algorithm to find the equilibria of the competition model,

$$\frac{dp}{dt} = (\mu - \lambda)p(1 - p)$$

The equation is autonomous because the rate of change is not a function of t. The equation for the equilibria is

$(\mu - \lambda)p^*(1 - p^*) = 0$	the equation for the equilibrium
$\mu - \lambda = 0$ or $p^* = 0$ or $1 - p^* = 0$	set each factor to 0
$\mu = \lambda$ or $p^* = 0$ or $p^* = 1$	solve each term

If $\mu = \lambda$, the reproductive rates of the two types of bacteria match and the fraction of mutants does not change. The other two equilibria correspond to extinction. If $p^* = 0$, there are no mutants; if $p^* = 1$, there are no wild type bacteria. Solutions starting from either of these points remain there (Figure 5.4). Although the differential equation describing these dynamics looks very different from the discrete-time dynamical system model describing the same process (Equation 1.49), the equilibria match.

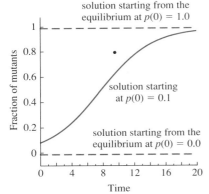

Figure 5.4
Three solutions of the bacterial competition equations

Graphical Display of Autonomous Differential Equations

The graphical technique of cobwebbing helped us sketch solutions of discrete-time dynamical systems with a minimum of algebra. There is a similarly useful graphical way to display autonomous differential equations, called the **phase-line diagram**. The idea is to figure out where the state variable is increasing, decreasing, and unchanged. According to the interpretation of the derivative, the state variable is increasing when the rate of change is positive, decreasing when the rate of change is negative, and unchanged when the rate of change is 0. By figuring out where the rate of change is positive, negative, and 0 we can get a good idea of how solutions behave.

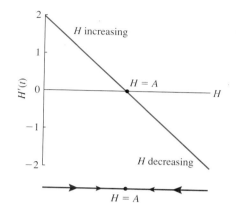

Figure 5.5
The rate of change and phase-line diagram for Newton's law of cooling

The rate of change for Newton's law of cooling is graphed as a function of the state variable H in the top of Figure 5.5. The rate of change is positive when $H < A$, 0 at $H = A$, and negative when $H > A$. The temperature is therefore increasing when $H < A$, constant at the equilibrium $H = A$, and decreasing when $H > A$.

This description can be translated into a one-dimensional drawing of the dynamics, called a *phase-line diagram*, shown at the bottom of Figure 5.5. The line represents the temperature, sometimes referred to as the **phase** of the system. To construct the diagram, draw right-pointing arrows at points where the temperature is increasing, left-pointing arrows at points where temperature is decreasing, and big dots where the temperature is fixed. The directions of the arrows come from the values of the rate of change. When the rate of change is positive, the temperature is increasing and the arrow points to the right.

Solutions follow the arrows. Starting below the equilibrium, the arrows push the temperature up toward the equilibrium. Starting above the equilibrium, the arrows push the temperature down toward the equilibrium. Unlike the solutions of discrete-time dynamical systems, the temperature cannot overshoot the equilibrium.

More information can be encoded on the diagram by making the size of the arrows correspond to the magnitude of the rate of change. In Figure 5.5, the arrows are larger when H is farther from the equilibrium, where the rate of change of temperature is larger. The solutions (Figure 5.6) change rapidly far from the equilibrium A and more slowly near A.

The rate of change of the competition model,

$$\frac{dp}{dt} = (\mu - \lambda)p(1 - p)$$

as a function of the state variable p in the case $\mu > \lambda$ is positive except at the equilibria (Figure 5.7). All the arrows on the phase-line diagram point to the right, pushing solutions toward the equilibrium at $p = 1$. The rate of change takes on

392 Chapter 5 Analysis of Differential Equations

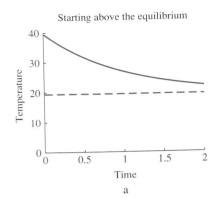
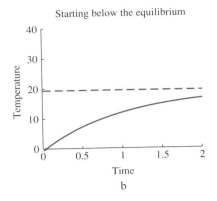

Figure 5.6
Two solutions of Newton's law of cooling

its maximum value at $p = 0.5$, meaning that the largest arrow is at $p = 0.5$ (see Exercise 8). A solution starting near $p = 0$ begins by increasing slowly, increases faster as it passes $p = 0.5$, and then slows down as it approaches $p = 1$ (Figure 5.4). Compared to the complicated algebra required to find a formula for this solution (Section 4.2), this method requires only the ability to graph and interpret a quadratic equation.

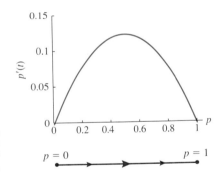

Figure 5.7
The rate of change and phase-line diagram for the competition model

The technique of phase-line diagrams is appropriate only for autonomous differential equations. Were we to try drawing arrows for a nonautonomous differential equation, the arrows would change with time. This is difficult to achieve in a drawing (although some computers can do it).

The phase-line diagram for the competition model (Figure 5.7) has two different kinds of equilibria. The arrows push the solution away from the equilibrium at $p = 0$ and toward the equilibrium at $p = 1$. We propose an informal definition of stable and unstable equilibria similar to that for discrete-time dynamical systems (Definition 1.5).

■ **Definition 5.2** An equilibrium of an autonomous differential equation is *stable* if solutions that begin near the equilibrium approach the equilibrium. An equilibrium of an autonomous differential equation is *unstable* if solutions that begin near the equilibrium move away from the equilibrium. ∎

SUMMARY

The rate of change of an **autonomous differential equation** depends on the state variable, not on the time. The rate of change of a **nonautonomous differential equation** can depend on both the state variable and the time. **Equilibria** of an autonomous differential equation are found by setting the rate of change equal to 0 and solving for the state variable. By graphing the rate of change as a function of the state variable, we can identify values where the state variable is increasing or decreasing. This information can be translated into a **phase-line diagram** on which right-pointing arrows indicate increasing solutions and left-pointing arrows indicate decreasing solutions.

5.1 EXERCISES

1. Identify the following as pure-time, autonomous, or nonautonomous differential equations.
   a. $\dfrac{dF}{dt} = F^2 + t$
   b. $\dfrac{dx}{dt} = \dfrac{x^2}{(x - \lambda)}$
   c. $\dfrac{dy}{dt} = \mu e^{-t}$
   d. $\dfrac{dm}{dt} = \dfrac{e^{\alpha m} m^2}{(m - \lambda)} + \ln(2me^{3\beta t} - 1)$

2. Consider the pure-time differential equation
   $$\dfrac{dp}{dt} = t$$
   a. Use integration to solve this equation starting from the initial condition $p(0) = 1$.
   b. Find $p(1)$.
   c. Solve starting with the initial condition $p(1) = 1$, and find $p(2)$.
   d. Sketch your solutions.
   e. Both parts **b** and **c** give the value of p one time unit after it took on the value 1. Why don't the two answers match? This behavior is typical of pure-time differential equations.

3. Consider the autonomous differential equation
   $$\dfrac{db}{dt} = b$$
   a. Solve this equation starting from the initial condition $b(0) = 1$.
   b. Find $b(1)$.
   c. Solve this equation starting from the initial condition $b(1) = 1$, and find $b(2)$.
   d. Sketch the two solutions.
   e. Both parts **b** and **c** give the value of b one time unit after it took on the value 1. Why do the two answers match? This behavior is typical of autonomous differential equations.

4. Consider the differential equation describing competition:
   $$\dfrac{dp}{dt} = (\mu - \lambda)p(1 - p)$$
   with $\mu < \lambda$.
   a. Draw the phase-line diagram.
   b. Sketch a solution starting from $p = 0.9$.
   c. Sketch a solution starting from $p = 0.1$.

5. Many cell membranes allow chemicals in and out of a cell at different rates.
   a. Sketch a diagram showing a cell that allows some chemical to enter only half as fast as it allows that chemical to leave.
   b. Write a differential equation describing this process (denote the rate out by β and the rate in by $\beta/2$).
   c. Suppose the external concentration is 2.0 mol/L. Find the equilibria.
   d. Draw a phase-line diagram.

6. Suppose a population is growing at constant rate $\lambda = 2.0$ per hour, but that individuals are harvested at a rate of 1000 per hour. The differential equation is
   $$\dfrac{db}{dt} = 2.0b - 1000$$
   a. For what values of the state variable does this equation fail to make sense?
   b. Find the equilibrium.
   c. Draw the phase-line diagram.
   d. Sketch one solution with initial condition below the equilibrium and another with initial condition above the equilibrium.
   e. Explain this result in words.

7. A common model to describe population growth with an autonomous differential equation is the **logistic model**,
   $$\dfrac{dN}{dt} = \lambda N \left(1 - \dfrac{N}{K}\right)$$
   It has the same formula as the discrete logistic updating function (Equation 3.1) but with quite a different behavior. The value K is called the **carrying capacity**. Use the values $\lambda = 2.0$ and $K = 1000$ if you have trouble.

a. Find the equilibria of this model.
b. Draw the phase-line diagram.
c. Sketch one solution with initial condition below the positive equilibrium and another with initial condition above the positive equilibrium.
d. Explain this result in words.

8. Suppose the population size of some species of organism follows the model

$$\frac{dN}{dt} = \frac{3N^2}{2+N^2} - N$$

where N is measured in hundreds.
a. Find the equilibria.
b. Draw the phase-line diagram.
c. Which of the equilibria seem to be stable?
d. Interpret your diagram in biological terms. Why might this population behave as it does at small values?

9. The drag on a falling object is proportional to the square of its speed. The differential equation is

$$\frac{dv}{dt} = a - Dv^2$$

where v is speed, a is acceleration, and D is drag. Suppose that $a = 9.8$ m/s^2 and that $D = 0.0032$ per meter (values for a falling sky-diver).
a. Check that the units in the differential equation are consistent.
b. Find the equilibrium speed. What does it mean?
c. Convert the equilibrium speed into miles per hour.

10. According to Torricelli's law of draining, the rate at which a fluid flows out of a cylinder through a hole at the bottom is proportional to the square root of the depth of the water. Let y represent the depth of water in centimeters. The differential equation is

$$\frac{dy}{dt} = -c\sqrt{y}$$

where $c = 2.0$ cm$^{-1/2}$/s.
a. Show that the units are consistent.
b. Compute the equilibrium, draw the phase-line diagram, and sketch solutions starting from $y = 10.0$ and $y = 1.0$.
c. Write a differential equation describing the depth in a cylinder where water enters at a rate of 4.0 cm/s but continues to drain out as above.
d. Compute the equilibrium, draw the phase-line diagram, and sketch solutions starting from $y = 10.0$ and $y = 1.0$ for the equation in part c.

11. One of the most important differential equations in chemistry uses the **Michaelis-Menten** or **Monod** equation. Suppose S is the concentration of a substrate that is being converted into a product. Then

$$\frac{dS}{dt} = -k_1 \frac{S}{k_2 + S}$$

describes how substrate is used up. Set $k_1 = k_2 = 1$.

a. How does this equation differ from Torricelli's law of draining (Exercise 10)?
b. Find the equilibrium.
c. Write a differential equation describing the amount of substrate if substrate is added at rate R but continues to be converted to product as before.
d. Find the equilibrium.
e. Draw the phase-plane diagram and a representative solution with $R = 0.5$ and $R = 1.5$. Can you explain your results?

12. Small organisms can take in food at rates proportional to their surface area but use energy at rates proportional to their mass. In particular, suppose that

$$\frac{dV}{dt} = a_1 V^{2/3} - a_2 V$$

where V represents the volume in cm^3 and t is time measured in days. The first term says that surface area is proportional to volume to the 2/3 power. The constant a_1 gives the rate at which energy is taken in and has units of centimeters per day; a_2 is rate at which energy is used and has units of per day.
a. Check the units.
b. Compute the equilibrium.
c. What happens to the value of the equilibrium as a_1 becomes smaller? Does this make sense?
d. What happens to the value of the equilibrium as a_2 becomes smaller? Does this make sense?

13. **COMPUTER:** Consider the differential equation

$$\frac{db}{dt} = -b^{-p}$$

for various positive values of the parameter p starting from $b(0) = 1.0$. For which values of p does the solution approach the equilibrium at $b = 0$ most quickly? Plot the solution on a semilog graph. When does the solution approach 0 faster than an exponential function?

14. **COMPUTER:** In Exercise 10 in Section 4.2 we considered the equation

$$\frac{dH}{dt} = \alpha(A(t) - H)$$

where

$$A(t) = 20.0 + \cos\left(\frac{2\pi t}{T}\right)$$

Using either the solution given in that problem or a computer system that can solve the equation, show that the solution always approaches the "equilibrium" at $H = A(t)$. Use the parameter values $\alpha = 1.0$ and values of T ranging from 0.1 to 10.0. For which value of T does the solution get closest to $A(t)$? Can you explain why?

5.2 Stable and Unstable Equilibria

Phase-line diagrams for autonomous differential equations have two types of equilibrium, stable and unstable. Solutions starting near a stable equilibrium move closer to the equilibrium. The equilibrium of Newton's law of cooling (Figure 5.5) and the equilibrium at $p = 1$ of the selection equation (Figure 5.7) seem to be stable. Solutions starting near an unstable equilibrium move farther from the equilibrium. The equilibrium at $p = 0$ of the selection equation (Figure 5.7) seems to be unstable. We now find a way to recognize stable and unstable equilibria.

Recognizing Stable and Unstable Equilibria

Consider again Newton's law of cooling,

$$\frac{dH}{dt} = \alpha(A - H)$$

(Figure 5.8). Our phase-line diagram and our intuition suggest that the equilibrium $H = A$ is stable. Arrows to the left of the equilibrium point right, pushing solutions up toward the equilibrium. Arrows to the right of the equilibrium point left, pushing solutions down toward the equilibrium. The equilibrium must be stable. In other words, the equilibrium is stable if the rate of change is positive when the value of the state variable is less than the equilibrium and negative when the value of the state variable is greater than the equilibrium.

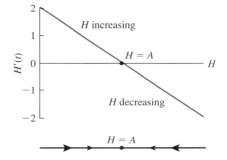

Figure 5.8

The rate of change and phase-line diagram for Newton's law of cooling revisited

When is the rate of change positive just to the left and negative just to the right of an equilibrium? When the rate of change is decreasing. When is a function decreasing? When its derivative is negative. An equilibrium is stable if the derivative of the rate of change *with respect to the state variable* is negative at the equilibrium. The right-hand equilibrium in Figure 5.9 satisfies this criterion.

At an unstable equilibrium, the arrows push solutions away. Solutions starting above the equilibrium increase, and those starting below decrease. In terms of the rate of change, an equilibrium is unstable if the rate of change is negative below the equilibrium and positive above it. When does this occur? When the rate of change is increasing. When is a function increasing? When the derivative is positive. An equilibrium is unstable if the derivative of the rate of change with respect to the state variable is positive at the equilibrium. The left-hand equilibrium in Figure 5.9 satisfies this criterion.

396 Chapter 5 Analysis of Differential Equations

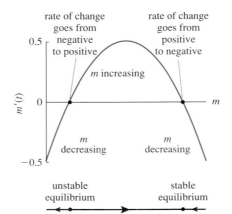

Figure 5.9
Behavior of functions with positive and negative derivatives

For example, the function $\alpha(A - H)$ gives the rate of change of H in Newton's law of cooling. If H is slightly larger than the equilibrium value A, the rate of change is negative. If H is slightly smaller than A, the rate of change is positive. To check stability with our derivative criterion, we compute the derivative of the rate of change with respect to the state variable H,

$$\frac{d}{dH}\alpha(A - H) = -\alpha < 0$$

The rate of change is a decreasing function at the equilibrium, and therefore the equilibrium is stable (Figure 5.8).

Imagine a perverse version of Newton's law of cooling given by

$$\frac{dH}{dt} = \alpha(H - A)$$

with $\alpha > 0$. Objects cooler than A decrease in temperature and objects warmer than A increase in temperature. The rate of change and the phase-line diagram are shown in Figure 5.10. The derivative of the rate of change with respect to H at $H = A$ is

$$\frac{d}{dH}\alpha(H - A) = \alpha > 0$$

The rate of change is an increasing function, and the equilibrium is unstable.

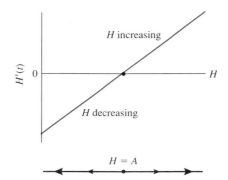

Figure 5.10
A perverse version of Newton's law of cooling

5.2 Stable and Unstable Equilibria

We summarize these results in the following simple and powerful theorem.

■ THEOREM 5.1 (Stability Theorem for Autonomous Differential Equations)

Suppose

$$\frac{dy}{dt} = f(y)$$

is an autonomous differential equation with an equilibrium at y^*. Let $f'(y)$ represent the derivative of f with respect to y. The equilibrium at y^* is stable if

$$f'(y^*) < 0$$

and unstable if

$$f'(y^*) > 0$$

Applications of the Stability Theorem

We now apply the stability theorem for autonomous differential equations to our models of population growth and selection to check whether the phase-line diagrams are correct. The equation for population growth is

$$\frac{db}{dt} = \lambda b$$

If $\lambda \neq 0$, the only equilibrium is $b^* = 0$. The derivative of the rate of change λb with respect to b is

$$\frac{d}{db}\lambda b = \lambda$$

According to the stability theorem for autonomous differential equations, the equilibrium is stable if the derivative is negative. If $\lambda > 0$ the equilibrium is unstable (Figure 5.11a), and if $\lambda < 0$ the equilibrium is stable (Figure 5.11b).

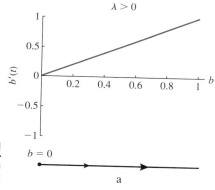
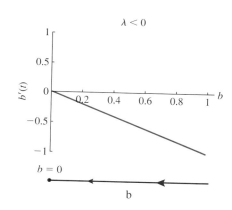

Figure 5.11
Phase-line diagrams for the bacterial growth model

We have found the solution of the bacterial growth model by guessing. With initial condition $b(0) = b_0$, the solution is

$$b(t) = b_0 e^{\lambda t}$$

If $\lambda > 0$, the solution increases exponentially without bound from any positive initial condition, consistent with the instability of the equilibrium in this case

(Figure 5.11a). If $\lambda > 0$, the solution decreases exponentially toward 0, consistent with the stability of the equilibrium in this case (Figure 5.11b).

There are two important differences between using the stability theorem for autonomous differential equations and solving the equation to figure out whether the equilibrium is stable. The stability theorem helps us recognize stable and unstable equilibria with a minimum of work; we solve for the equilibria and compute the derivative of the rate of change with respect to b. Finding what happens near an equilibrium by computing the solution requires more work; we must find the solution and then evaluate its limit as t approaches infinity. We seem to have gotten something for nothing. In mathematics, as in life, this is rarely the case. The other difference between the two methods is that the simpler calculation based on the stability theorem gives less information about the solutions, telling us only what happens to solutions that start *near* the equilibrium.

When finding the exact solution is difficult or impossible, this *local information* about solutions near an equilibrium may be all we can find. Computing the solution of the competition equation,

$$\frac{dp}{dt} = f(p) = (\mu - \lambda)p(1 - p)$$

is possible, but tricky (Section 4.2). Finding the stability of the equilibria with the stability theorem is straightforward. The differential equation has equilibria at $p = 0$ and $p = 1$ (Section 5.1). The derivative of the rate of change $f(p)$ with respect to p is

$$f'(p) = \frac{d}{dp}(\mu - \lambda)p(1 - p) = (\mu - \lambda)(1 - 2p)$$

Then

$$f'(0) = \mu - \lambda$$
$$f'(1) = \lambda - \mu$$

If $\mu > \lambda$, the equilibrium $p = 0$ has a positive derivative and is unstable, and the equilibrium $p = 1$ has a negative derivative and is stable, matching Figure 5.7. These results make biological sense; the mutant bacteria are superior in this case. If $\mu < \lambda$, the mutant is at a disadvantage; the $p = 0$ equilibrium is stable, and the $p = 1$ equilibrium is unstable (Exercise 1) (Figure 5.12).

Figure 5.12

The rate of change and phase-line diagram for the competition model revisited

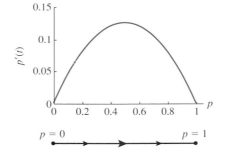

A Model of a Disease

Suppose a disease is circulating in a population. Individuals recover from this disease unharmed but are susceptible to reinfection. How many people will be sick at any given time? Is there any way that such a disease will die out? The factors affecting the dynamics are sketched in Figure 5.13.

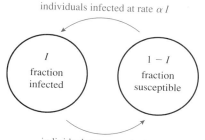

Figure 5.13
Factors involved in disease dynamics

Let I denote the fraction of infected individuals in the population. Each uninfected, or susceptible, individual has a chance of getting infected when she encounters an infectious individual (depending, perhaps, on whether she gets sneezed on). It seems plausible that a susceptible individual will run into infectious individuals at a rate proportional to the number of infectious individuals. Then

per capita rate at which a susceptible individual is infected $= \alpha I$

The parameter α combines the rate at which people are encountered and the probability that an encounter produces an infection.

What fraction of individuals are susceptible? Every individual is either infected or not, so a fraction $1 - I$ are susceptible. The overall rate at which new individuals are infected is the per capita rate times the total number or

rate at which susceptible individuals are infected = per capita rate
$$\times \text{ number of individuals}$$
$$= \alpha I(1 - I)$$

Individuals recover from the disease at a rate proportional to the number of infected individuals,

rate at which infected individuals recover $= \mu I$

Putting the two processes together, we have

$$\frac{dI}{dt} = \alpha I(1 - I) - \mu I \qquad (5.3)$$

Suppose we start with a few infected individuals, a small value of I. Will the disease persist? The equilibria and their stability can provide the answer. To find the equilibria, we use Algorithm 5.1. Equation 5.3 is autonomous because the rate of change depends only on the state variable I and not on time.

$\alpha I^*(1 - I^*) - \mu I^* = 0$ the equation for the equilibrium
$I^*(\alpha(1 - I^*) - \mu) = 0$ factor out I^*
$I^* = 0$ or $\alpha(1 - I^*) - \mu = 0$ set each factor to 0
$I^* = 0$ or $I^* = 1 - \dfrac{\mu}{\alpha}$ solve each term

Do these solutions make sense? There are no sick people if $I = 0$. In the absence of some process bringing in infected people (migration, aliens, etc.), a population without illness will remain so. The other equilibrium depends on the parameters α and μ. There are two cases, depending on which parameter is larger.

Case 1: If $\alpha < \mu$, $1 - (\mu/\alpha) < 0$, which is nonsense. When the recovery rate is high, the only biologically plausible equilibrium is $I = 0$.

Case 2: If $\alpha > \mu$, $1 - (\mu/\alpha) > 0$, which is biologically possible. There are two equilibria if the infection rate is larger than the recovery rate.

To draw the phase-line diagrams for the two cases, we must compute the stability of the equilibria. The derivative of the rate of change with respect to the state variable I is

$$\frac{d}{dI}\alpha I(1 - I) - \mu I = \alpha - 2\alpha I - \mu \qquad (5.4)$$

Case 1: If $\alpha < \mu$, the derivative at $I = 0$ is

$$\alpha - 2\alpha \cdot 0 - \mu = \alpha - \mu < 0$$

The single equilibrium is stable.

Case 2: If $\alpha > \mu$, the derivative at $I = 0$ is

$$\alpha - 2\alpha \cdot 0 - \mu = \alpha - \mu > 0$$

and this equilibrium is unstable. At $I^* = 1 - (\mu/\alpha)$, the derivative of the rate of change is

$$\alpha - 2\alpha \cdot \left(1 - \frac{\mu}{\alpha}\right) - \mu = \mu - \alpha < 0$$

The positive equilibrium is stable.

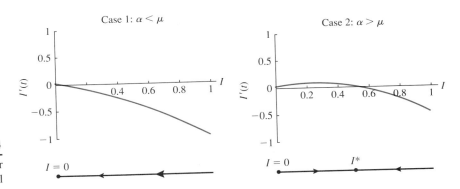

Figure 5.14

Phase-line diagrams for the disease model

The phase-line diagrams for the two cases are shown in Figure 5.14. We can deduce the behavior of solutions from our phase-line diagram. In case 1, with $\alpha < \mu$, all solutions converge to the equilibrium at $I = 0$ (Figure 5.15a). If control measures could be implemented to increase the recovery rate μ or reduce the transmission rate α, the disease could be completely eliminated from a population. Notice that α (transmission) need not be reduced to 0 to eliminate the disease. This is an example of the threshold theorem referred to in Section 1.1. In case 2, with $\mu < \alpha$, the equilibrium I^* is positive and stable. Solutions starting near 0 increase to I^* and the disease remains present in the population (Figure 5.15b). Such a disease is called **endemic**.

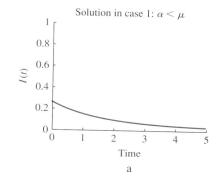
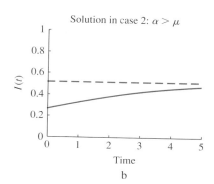

Figure 5.15
Solutions of the disease equation in two cases

a

b

SUMMARY

By carefully examining the phase-line diagram near an equilibrium, we found the **stability theorem for autonomous differential equations**. An equilibrium of an autonomous differential equation is stable if the derivative of the rate of change with respect to the state variable is negative and unstable if this derivative is positive. Calculating stability in this way is easier than solving the equation but does not provide exact information about the behavior of solutions far from the equilibrium. We applied this method to several familiar models and to a new model of a disease, showing that a disease can be eradicated without entirely stopping transmission.

5.2 EXERCISES

1. Consider the differential equation describing competition,
$$\frac{dp}{dt} = (\mu - \lambda)p(1-p)$$
with $\mu < \lambda$ (as in Exercise 4 in Section 5.1).
   a. Sketch the rate of change as a function of p, and show that it is decreasing at $p = 0$ and increasing at $p = 1$.
   b. Apply the stability theorem for autonomous differential equations to show that the equilibrium at $p = 0$ is stable and the equilibrium at $p = 1$ is unstable.

2. Consider the model of chemical concentration varying by diffusion with absorption:
$$\frac{dC}{dt} = (0.05 - C) - \frac{C}{1+C}$$
The first term describes a case of chemical diffusion (Equation 4.6 with $\beta = 1.0$ per minute, and $\Gamma = 0.05$ mol/L). The second term represents absorption.
   a. Describe how the absorption rate depends on the concentration.
   b. Find the equilibrium of this equation.
   c. Find its stability.
   d. Compare the value of this equilibrium with that found for the model without absorption.

3. Analyze Equation 5.3 (page 399) for $\alpha = \mu$.
   a. Find the equilibria.
   b. Draw the phase-line diagram.
   c. Find the stability of the equilibria. (Theorem 5.1 does not work in this case so you must use the phase-line diagram.)

4. Apply the stability theorem for autonomous differential equations to the following equations found in Section 5.1. Show that your results match what you found in your phase-line diagrams, and give a biological interpretation.
   a. The equation in Exercise 5.
   b. The equation in Exercise 6.
   c. The equation in Exercise 7.
   d. The equation in Exercise 8.
   e. The equation in Exercise 9.
   f. The equation in Exercise 10 (without added water).
   g. The equation in Exercise 11 (without added chemical).
   h. The equation in Exercise 12.

5. Try to draw a phase-line diagram with two stable equilibria in a row. Use the Intermediate Value Theorem (Theorem 3.2) to sketch a proof of why this is impossible.

6. Suppose $\mu = 1$ in Equation 5.3. Graph the two equilibria as functions of α for values of α between 0 and 2. Use a solid

line when an equilibrium is stable and a dashed line when an equilibrium is unstable. Notice that stability is exchanged when $\alpha = 1$. Your picture is called a **bifurcation diagram**, and the change that occurs at $\alpha = 1$ is called a **transcritical bifurcation**.

7. **COMPUTER:** Consider the variant of Equation 5.3 given by

$$\frac{dI}{dt} = \alpha I^2(1 - I) - \mu I$$

Suppose $\mu = 1.0$. Use your computer to graph the equilibria as functions of α for values of α between 0 and 5. Use a solid line when an equilibrium is stable and a dashed line when an equilibrium is unstable (do this by hand if necessary). Describe in words what happens at $\alpha = 4$. This picture is another bifurcation diagram, and the change that occurs at $\alpha = 4$ is called a **saddle-node bifurcation**.

8. **COMPUTER:** Consider the equation

$$\frac{dx}{dt} = -x + ax^3$$

for both positive and negative values of x. Use your computer to graph the equilibria as functions of a for values of a between -1 and 1. Use a solid line when an equilibrium is stable and a dashed line when an equilibrium is unstable (do this by hand if necessary). Describe in words what happens at $a = 0$. This picture is yet another bifurcation diagram, and the change that occurs at $a = 0$ is called a **pitchfork bifurcation**.

5.3 Solving Autonomous Differential Equations

The method of phase-line diagrams allows us to sketch solutions with a minimum of algebra. What if want an exact formula for the solution? Guessing sometimes works, as with pure-time differential equations. For a more dependable method, we would like to transform the problem of solving an autonomous differential equation into an integration problem. The technique for doing this, called **separation of variables**, is one of the most powerful techniques in applied mathematics.

Separation of Variables

Consider the autonomous differential equation for bacterial growth,

$$\frac{db}{dt} = \lambda b \tag{5.5}$$

We cannot integrate to solve this as we could with a pure-time differential equation, because integrating the function λb requires knowing the solution $b(t)$.

The trick of separation of variables, as the name implies, is to separate the two variables, the bs and the ts. To separate variables, we divide both sides by b to get all the bs on the left-hand side and multiply by dt to get all the ts on the right-hand side, arriving at

$$\frac{db}{b} = \lambda dt$$

As with substitution, we are allowed to treat dt like the denominator of a fraction. The next step is to compute the indefinite integral of each side. The left-hand side has a db and can be integrated using b as the variable, or

$$\int \frac{1}{b} db = \ln(|b|) + c_1$$

The right-hand side has a dt and can be integrated using t as the variable, or

$$\int \lambda \, dt = \lambda t + c_2$$

The absolute value bars around b (required by the integral of $1/x$ in Equation 4.10) are unnecessary because b represents a population and therefore must be positive.

The idea of separation of variables is that if $(db/b) = \lambda dt$, the integrals must also be equal. Therefore,

$$\ln b + c_1 = \lambda t + c_2$$

Because c_1 and c_2 are arbitrary constants, we can combine the two constants into one by setting $c = c_2 - c_1$,

$$\ln b = \lambda t + c$$

Finally, we solve for b in terms of t. In this case, we can do so by exponentiating both sides, finding

$$b = e^{\lambda t + c}$$

This solution might look strange because the arbitrary constant is inside the exponential function. It is often more convenient to move the constant outside the exponential function. In particular, if we define the new constant

$$K = e^c$$

we find that

$$e^{\lambda t + c} = e^{\lambda t} e^c = K e^{\lambda t}$$

We find the constant K by substituting the initial conditions. If $b(0) = 0.1$, then

$$0.1 = b(0) = K e^{\lambda \cdot 0} = K$$

For example, with $\lambda = 2.0$ and $b(0) = 0.1$ (measured in millions), the solution is

$$b(t) = 0.1 e^{2.0t}$$

(Figure 5.16a). With $\lambda = -1.5$ and $b(0) = 0.9$, the solution is

$$b(t) = 0.9 e^{-1.5t}$$

(Figure 5.16b). Separations of variables helps us find the solution without guessing.

Figure 5.16

Solutions of the bacterial growth equation found with separation of variables

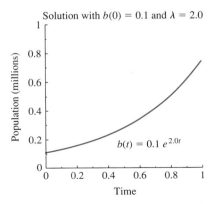
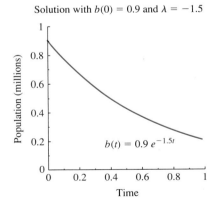

The following are the steps for separation of variables.

Algorithm 5.2 (Separation of variables)

1. Move all instances of the state variable to the left-hand side of the equation and all instances of the time t to the right-hand side.
2. Integrate both sides.
3. Set the two integrals equal to each other.
4. Combine the two arbitrary constants into one.
5. Solve for the state variable in terms of the time t, rewriting the constant in a more convenient form if necessary.
6. Solve for the constant with the initial conditions. ∎

In our example, the first step of separating the state variable b from the time t could have been done differently had we chosen to move λ to the left-hand side:

$$\frac{db}{\lambda b} = dt$$

Following Algorithm 5.2, we find

$$\frac{1}{\lambda}\ln b + c_1 = t + c_2 \quad \text{integrating}$$

$$\frac{1}{\lambda}\ln b = t + c \quad \text{combining constants}$$

$$b = e^{\lambda t}e^{\lambda c} \quad \text{solving for } b$$

This has the same form we found before, but the constant looks different. If we set $K = e^{\lambda c}$, we again find the solution $b(t) = Ke^{\lambda t}$.

Solving Pure-Time Differential Equations with Separation of Variables

The technique of separation of variables is essentially the same method we used to solve pure-time differential equations. For example, consider the pure-time differential equation for volume,

$$\frac{dV}{dt} = t^2$$

with initial condition $V(0) = 1.0 \text{ cm}^3$. We first solved this equation by finding the antiderivative of t^2. The method of separation of variables requires computing the same integral. To begin, we multiply both sides by dt to move all instances of the time t to the right-hand side. All instances of the state variable V are already on the left-hand side because this is a pure-time differential equation. Following Algorithm 5.2, we get

$$dV = t^2 dt \quad \text{separating variables}$$

$$V + c_1 = \frac{t^3}{3} + c_2 \quad \text{integrating}$$

$$V = \frac{t^3}{3} + c \quad \text{combining constants}$$

The last step, solving for V in terms of t, is already done. Furthermore, the constant c appears in its traditional place, added to the antiderivative. We find the constant c with the initial condition,

$$1.0 = V(0) = \frac{0^3}{3} + c = c$$

The solution of the equation is therefore $V = (t^3/3) + 1.0$.

Because the only V in this pure-time differential equation appears as dV, the left-hand side is easy to integrate. The function of t, however, might be difficult or impossible to integrate. In contrast, the only t in an autonomous differential equation appears as dt, making the t integral simple to find. The function of V might be difficult or impossible to integrate. With an autonomous differential equation, there is the additional difficult or impossible step of solving for the state variable (b in the example above). Exercise 8 gives an example of an autonomous differential equation for which this last step is impossible.

Applications of Separation of Variables

We can use separation of variables to solve Newton's law of cooling,

$$\frac{dH}{dt} = \alpha(A - H)$$

Following the algorithm, we find

$$\frac{dH}{A - H} = \alpha \, dt \quad \text{separating variables}$$
$$-\ln|A - H| + c_1 = \alpha t + c_2 \quad \text{integrating}$$
$$-\ln|A - H| = \alpha t + c \quad \text{combining constants}$$
$$|A - H| = e^{-\alpha t - c} \quad \text{solving for } |A - H|$$
$$|A - H| = K e^{-\alpha t} \quad \text{rewriting the constant as } K = e^{-c}$$

To get rid of the absolute value bars, we consider two cases. If $H < A$, $|A - H| = A - H$ so

$$|A - H| = A - H = K e^{-\alpha t}$$

We can now solve for H, finding

$$H(t) = A - K e^{-\alpha t}$$

For example, if $\alpha = 0.1$, $A = 40°C$, and $H(0) = 10°C$, the solution is

$$H(t) = 40 - K e^{-0.1t}$$

We find the arbitrary constant K by substituting $t = 0$,

$$10 = H(0) = 40 - K e^{-0.1 \cdot 0} = 40 - K$$

This has solution $K = 30$, so that

$$H(t) = 40 - 30 e^{-0.1t}$$

(Figure 5.17a). The solution approaches the ambient temperature $40°C$ because

$$\lim_{t \to \infty} 40 - 30 e^{-0.1t} = 40$$

which is consistent with our finding from the phase-line diagram that the equilibrium is stable. The solution gives more information by showing that the solution approaches the equilibrium exponentially.

If $H > A$, $|A - H| = H - A$ and
$$|A - H| = H - A = Ke^{-\alpha t}$$

Next, we solve for H, finding
$$H(t) = A + Ke^{-\alpha t}$$

For example, if $\alpha = 0.1$, $A = 40°C$, and $H(0) = 60°C$, the solution is
$$H(t) = 40 + Ke^{-0.1t}$$

We find the arbitrary constant K by substituting $t = 0$,
$$60 = H(0) = 40 + Ke^{-0.1 \cdot 0} = 40 + K$$

This has solution $K = 20$, so that
$$H(t) = 40 + 20e^{-0.1t}$$

(Figure 5.17b). Again, the solution approaches the ambient temperature $40°C$ because
$$\lim_{t \to \infty} 40 + 20e^{-0.1t} = 40$$

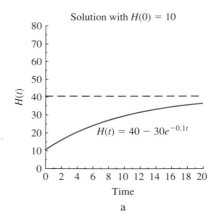
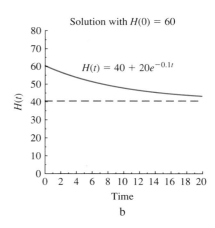

Figure 5.17
Solutions of Newton's law of cooling found with separation of variables

As a more unusual example, suppose that the per capita reproduction of some population with size b is
$$\text{per capita reproduction} = b$$

This means that individuals reproduce faster and faster the more of them there are. We expect that this population will grow very quickly, and we can use separation of variables to compute how quickly. The differential equation for growth is
$$\frac{db}{dt} = \text{per capita reproduction} \cdot b$$
$$= b \cdot b = b^2$$

Separation of variables proceeds as follows:

$$\frac{db}{b^2} = dt \quad \text{separating variables}$$

$$-\frac{1}{b} + c_1 = t + c_2 \quad \text{integrating}$$

$$-\frac{1}{b} = t + c \quad \text{combining constants}$$

$$b = \frac{-1}{t + c} \quad \text{solving for } b$$

The result looks dangerously negative. Proceeding anyway, we suppose that $b(0) = 10$. We solve for c by substituting the initial conditions, finding

$$10 = b(0) = \frac{-1}{c}$$

so that $c = -0.1$. Therefore,

$$b(t) = \frac{-1}{t - 0.1} = \frac{1}{0.1 - t}$$

(Figure 5.18). The population blasts off to infinity at time $t = 0.1$. What happens at $t = 0.11$? The solution does not exist. As far as this differential equation is concerned, the world ceases to exist at $t = 0.1$. Models that include self-enhancing growth tend to have this sort of "chain-reaction" property.

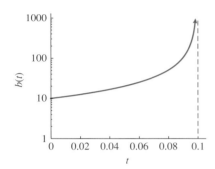

Figure 5.18
Semilog plot of a population with self-enhancing growth

SUMMARY

We developed the method of **separation of variables** to solve autonomous differential equations. The method works by isolating the state variable on the left-hand side of the equation and time on the right-hand side, and then integrating to find a solution. Autonomous differential equations describing population growth and Newton's law of cooling can be solved by this technique. Solving pure-time differential equations with integration is a special case of the separation of variables method.

5.3 EXERCISES

1. Use separation of variables to solve the following autonomous differential equations. Check your answers by differentiating.

   a. $\frac{db}{dt} = 0.01b, \, b(0) = 1000$

   b. $\frac{db}{dt} = -3b, \, b(0) = 1.0 \times 10^6$

   c. $\frac{dN}{dt} = 1 + N, \, N(0) = 1$

   d. $\frac{db}{dt} = 1000 - b, \, b(0) = 500$

   e. $\frac{db}{dt} = 1000 - b, \, b(0) = 1000$

2. Use separation of variables to solve the following pure-time differential equations. Check your answers by differentiating.

   a. $\frac{dP}{dt} = \frac{5}{1 + 2t}$, with $P(0) = 0$

   b. $\frac{dP}{dt} = 5e^{-2t}$, with $P(0) = 0$

   c. $\frac{dL}{dt} = 1000e^{0.2t}$, with $L(0) = 1000$

   d. $\frac{dL}{dt} = 64.3e^{-1.19t}$, with $L(0) = 0$

3. Using the solution derived for Newton's law of cooling, find the solution of the chemical diffusion equation,

$$\frac{dC}{dt} = \beta(\Gamma - C)$$

with the following parameter values and initial conditions. Find the concentration after 10 s. How long would it take for the concentration to get halfway to the equilibrium value?
  a. $\beta = 0.01$ per second, $C(0) = 5.0$ mmol/cm^3, and $\Gamma = 2.0$ mmol/cm^3
  b. $\beta = 0.01$ per second, $C(0) = 1.0$ mmol/cm^3, and $\Gamma = 2.0$ mmol/cm^3
  c. $\beta = 0.1$ per second, $C(0) = 5.0$ mmol/cm^3, and $\Gamma = 2.0$ mmol/cm^3
  d. $\beta = 0.1$ per second, $C(0) = 1.0$ mmol/cm^3, and $\Gamma = 2.0$ mmol/cm^3
  e. Why do the times to get halfway to the equilibrium value match in parts **a** and **b** and parts **c** and **d**?

4. Recall Torricelli's law of draining,

$$\frac{dy}{dt} = -2\sqrt{y}$$

(Exercise 10 in Section 5.1) with the constant set to 2. Suppose the initial condition is $y(0) = 4$.
  a. Find the solution with separation of variables.
  b. Graph your solution.
  c. What really happens at time $t = 2$?
  d. How does this differ from the solution of the equation $\frac{dy}{dt} = -2y$?

5. Find the solution of $\frac{dI}{dt} = \alpha I(1-I) - \mu I$ when $\alpha = \mu = 2.0$ and with initial condition $I(0) = 0.5$. Is your result consistent with what you found in Exercise 3 in Section 5.2? In what way is this solution different from the solution of an equation that decreases exponentially toward 0?

6. Find the solution of the differential equation

$$\frac{db}{dt} = b^p$$

for the following cases. When does it blow up? Sketch a graph.
  a. $p = 2$ (as in the text) and $b(0) = 100$
  b. $p = 2$ and $b(0) = 0.1$
  c. $p = 1.1$ and $b(0) = 100$
  d. $p = 1.1$ and $b(0) = 0.1$

7. Separation of variables can help to solve some nonautonomous differential equations. For example, suppose that the per capita reproduction rate is $\lambda(t)$, a function of time, so that

$$\frac{db}{dt} = \lambda(t)b$$

We can separate this equation into parts depending on only b and t by dividing both sides of the equation by b and multiplying by dt. For the following functions $\lambda(t)$, give an interpretation of the equation and solve the equation with the initial condition $b(0) = 10^6$. Sketch each solution. Check your answer by substituting into the differential equation.

  a. $\lambda(t) = t$
  b. $\lambda(t) = \frac{1}{1+t}$
  c. $\lambda(t) = \cos(t)$
  d. $\lambda(t) = e^{-t}$

8. Consider the autonomous differential equation

$$\frac{dx}{dt} = \frac{x}{1+x}$$

with $x(0) = 1$. This describes a population with per capita reproduction that decreases according to $1/(1+x)$.
  a. Solve the equation with separation of variables.
  b. It is impossible to solve algebraically for x in terms of t. However, you can still find the arbitrary constant. Find it.
  c. Although you cannot find x as a function of t, you can find t as a function of x. Graph this function.
  d. Sketch the solution for x as a function of t.

9. Separation of variables and a trick known as **integration by partial fractions** can be used to solve the competition model of Equation 4.7. For simplicity, we consider the equation

$$\frac{dp}{dt} = p(1-p)$$

  a. Separate variables.
  b. Show that

$$\frac{1}{p(1-p)} = \frac{1}{p} + \frac{1}{1-p}$$

  and rewrite the left-hand side.
  c. Integrate the rewritten left-hand side.
  d. Combine the two natural log terms into one using a law of logs.
  e. Write the equation for the solution with a single constant, c.
  f. Exponentiate both sides, and solve for p.
  g. Using the initial condition $p(0) = 0.01$, find the value of the constant.
  h. Evaluate the limit of your solution as t approaches infinity.

10. **COMPUTER:** We have seen that the differential equation

$$\frac{db}{dt} = b^2$$

blows up to infinity in a finite amount of time. We will try to stop it by multiplying the rate of change by a decreasing function of t in the nonautonomous equation:

$$\frac{db}{dt} = g(t)b^2$$

Try the following three functions for $g(t)$:

$$g_1(t) = e^{-t}$$
$$g_2(t) = \frac{1}{1+t^2}$$

$$g_3(t) = \frac{1}{1+t}$$

a. Which of these functions decreases fastest and should best be able to stop $b(t)$ from blowing up?
b. Have your computer solve the equation for each case with initial conditions ranging from $b(0) = 0.1$ to $b(0) = 5.0$. Which solutions blow up to infinity?
c. Use separation of variables to try to find the solution for each case (have your computer help with the integral of $g_2(t)$). Can you figure out when the solutions should blow up? How well does this match your results from part **b**?

5.4 Two-Dimensional Differential Equations

We now begin the study of problems in two dimensions. The discrete-time dynamical systems and differential equations we have considered have described the dynamics of a single state variable: a population, a concentration, a fraction, and so forth. Most biological systems cannot be described fully without multiple measurements and multiple state variables. The tools for studying these systems, differential equations and discrete-time dynamical systems, remain the same, as does the goal of figuring out what will happen. The methods for getting from the problem to the answer, however, are more complicated. In this section we introduce two important equations for population dynamics and an extension of Newton's law of cooling that takes into account the changing temperature of a room. These are called **systems of autonomous differential equations**, or **coupled autonomous differential equations**. Euler's method can be applied to find approximate solutions of these systems of equations.

Predator-Prey Dynamics

Our basic model of population growth followed a single species that existed undisturbed in isolation. Life is rarely so peaceful. Imagine our bacterial population disrupted by the arrival of a predator, perhaps some sort of amoeba. We expect our bacteria to do worse when more predatory amoebas are around. Conversely, we expect our amoebas to do better when more of their bacterial prey are around. We now translate these intuitions into differential equations.

Denote the population of bacteria at time t by $b(t)$ and the population of amoebas by $p(t)$ to represent predation. We can build our equations by considering the factors affecting the population of each organism. Suppose that the bacteria would grow exponentially in the absence of predators, according to the equation

$$\frac{db}{dt} = \lambda b$$

How might predation affect the bacterial population? One ecologically naive but mathematically convenient approach is to assume that the organisms obey the principle of mass action.

The Principle of Mass Action: Individual bacteria encounter amoebas at a rate proportional to the number of amoebas.

Doubling the number of predators therefore doubles the rate at which bacteria run into predators. If running into an amoeba spells doom, this doubles the rate at which each bacterium risks being eaten.

Expressed as an equation, we have

$$\text{rate at which an individual bacterium is eaten} = \varepsilon p$$

where ε is the constant of proportionality. Therefore, we have

$$\text{per capita growth rate of bacteria} = \lambda - \varepsilon p$$

This equation is illustrated with $\lambda = 1.0$ and $\varepsilon = 0.001$ in Figure 5.19a. Because the growth rate of a population is the product of the per capita growth rate and the population size, we have

$$\frac{db}{dt} = \text{per capita growth rate} \times \text{population size}$$
$$= (\lambda - \varepsilon p)b$$

Suppose that the predators have a negative growth rate of δ in the absence of their prey, so that

$$\frac{dp}{dt} = -\delta p$$

Solutions of this equation converge to 0. These predators must eat to live. How might predation affect the amoeba population? As above, we assume mass action: doubling the number of bacteria doubles the rate at which predators run into bacteria, and thus doubles the rate at which they eat:

$$\text{rate at which an amoeba eats bacteria} = \eta b$$

(η is the Greek letter "eta," used to remind us of eating). If the per capita reproduction is increased by the eating rate, we have

$$\text{per capita growth rate of predators} = -\delta + \eta b$$

These per capita growth rates are illustrated with $\delta = 1.0$ and $\eta = 0.001$ in Figure 5.19b. Multiplying the per capita growth rate of the amoebas by their population size gives

$$\frac{dp}{dt} = \text{per capita growth rate} \times \text{population size}$$
$$= (-\delta + \eta b)p$$

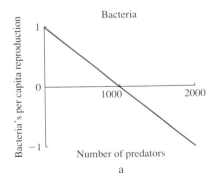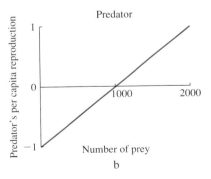

Figure 5.19
Per capita growth rates of prey and predator species

We combine the differential equations into the following **system of autonomous differential equations**:

$$\frac{db}{dt} = (\lambda - \varepsilon p)b$$

$$\frac{dp}{dt} = (-\delta + \eta b)p \tag{5.6}$$

These are also called **coupled autonomous differential equations** because the rate of change of the bacterial population depends on both their own population size and that of the amoebas, and the rate of change of the amoebas depends on both their own population and that of the bacteria. Two separate measurements are required to find the rate of change of either population. These equations are autonomous because neither rate of change depends explicitly on time.

We can use similar reasoning to describe the competitive interaction between two populations. Our basic model of selection is based on the *uncoupled* pair of differential equations:

$$\frac{da}{dt} = \mu a$$

$$\frac{db}{dt} = \lambda b$$

Although there are two measurements, the rate of change of each organism's population does not depend on the population size of the other. This lack of interaction is an idealization. Two populations in a single vessel will probably interact. Suppose that the per capita growth rate of each organism declines as a linear function of the total number $a + b$ according to

$$\text{per capita growth rate of type } a = \mu \left(1 - \frac{a+b}{K_a}\right)$$

$$\text{per capita growth rate of type } b = \lambda \left(1 - \frac{a+b}{K_b}\right)$$

We have written the growth rates in this form to separate the maximum per capita reproduction μ of type a and λ of type b. Reproduction of type a becomes negative when the total population exceeds K_a, and reproduction of type b becomes negative when the total population exceeds K_b. These per capita rates of change are illustrated in Figure 5.20 with $\mu = 2.0$, $\lambda = 2.0$, $K_a = 1000$, and $K_b = 500$.

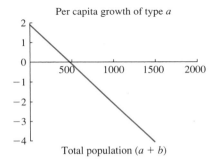
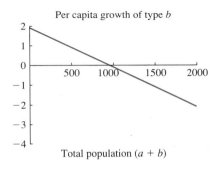

Figure 5.20 Per capita growth rates of competing species

We find the rates of change of the populations by multiplying the per capita growth rates by the population sizes, finding the coupled system of differential equations,

$$\frac{da}{dt} = \mu \left(1 - \frac{a+b}{K_a}\right) a$$

$$\frac{db}{dt} = \lambda \left(1 - \frac{a+b}{K_b}\right) b \tag{5.7}$$

The behavior of each population depends on both its own size and that of the other species. Neither depends explicitly on time, so this system of equations is autonomous.

Newton's Law of Cooling

When discussing both Newton's law of cooling and the model of chemical diffusion across a membrane, we ignored the fact that the ambient temperature or concentration might also change. A small, hot object placed in a large room will have little effect on the room's temperature, but a large, hot object will both cool off itself and heat the room. We require a system of autonomous differential equations to describe both temperatures simultaneously (Figure 5.21). We now derive the coupled differential equations that describe this situation.

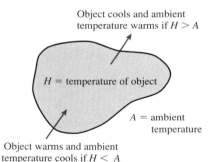

Figure 5.21
Newton's law of cooling revisited

Newton's law of cooling expresses the rate of change of the temperature, H, of an object as a function of the ambient temperature, A, by the equation

$$\frac{dH}{dt} = \alpha(A - H)$$

This equation is valid even if the ambient temperature A itself is changing. If $A < H$, heat is leaving the object and warming the room. The room follows the same law as the object, but we expect that the factor α, which depends on the size, shape, and material of the object, will be different from that of the object, or that

$$\frac{dA}{dt} = \alpha_2(H - A)$$

The rate of change of temperature of each object depends on the temperature of the other. Together, these equations give the system of coupled autonomous differential equations:

$$\frac{dH}{dt} = \alpha(A - H)$$

$$\frac{dA}{dt} = \alpha_2(H - A)$$

What is the relationship between α and α_2? In general, α_2 will be smaller for a larger room. If the "object" is made of the same stuff as the "room," the ratio of α to α_2 is equal to the ratio of the sizes. For example, if the room is three times bigger than the object, we have

$$\alpha_2 = \frac{\alpha}{3}$$

(see Figure 5.22).

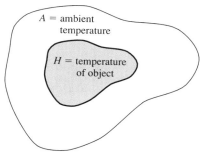

Figure 5.22
Newton's law of cooling applied to objects of different sizes

If ambient space is 3 times larger than object, rate of change of the ambient temperature is 3 times smaller than that of the object

Applying Euler's Method to Systems of Autonomous Differential Equations

How do we find solutions of systems of autonomous differential equations? In general, this is a difficult problem. In Section 5.5, we will extend the method of equilibria and phase-line diagrams to sketch solutions. Perhaps surprisingly, Euler's method for finding approximate solutions with the tangent line approximation works exactly the same way for systems as for single equations. We will apply the method to figure out how solutions behave for the systems describing predator-prey dynamics and competition.

Consider the predator-prey equations (Equation 5.6) with $\lambda = 1$, $\delta = 1$, $\varepsilon = 0.001$, and $\eta = 0.001$,

$$\frac{db}{dt} = (1.0 - 0.001p)b$$

$$\frac{dp}{dt} = (-1.0 + 0.001b)p$$

Suppose the initial condition is $b(0) = 800$ and $p(0) = 200$. Initially, there are 800 prey and 200 predators. What will happen to these populations after 2.0 time units? We break this interval into 10 units of length $\Delta t = 0.2$. After one of these intervals, we can approximate each population with the tangent line by

$\hat{b}(0 + \Delta t)$
$\quad = b(0) + b'(0)\Delta t$ equation for tangent line
$\quad = 800 + (1.0 - 0.001 \cdot 200)$ $b(0) = 800$ (initial condition) and
$\quad\quad \times 800 \cdot \Delta t$ $b'(0) = [1.0 - 0.001p(0)]b(0)$
$\quad = 800 + 640 \cdot \Delta t$ compute that $b'(0) = 640$

After a time $\Delta t = 0.2$, the approximate value of b is

$$\hat{b}(0.2) = 800 + 640 \cdot 0.2 = 928$$

At the same time, the value for p can be found with the tangent line approximation to be

$$\hat{p}(0 + \Delta t)$$
$$= p(0) + p'(0)\Delta t \qquad \text{equation for tangent line}$$
$$= 200 + (-1.0 + 0.001 \cdot 800) \qquad p(0) = 800 \text{ (initial condition) and}$$
$$\times 200 \cdot \Delta t \qquad p'(0) = [-1.0 + 0.001b(0)]p(0)$$
$$= 200 - 40 \cdot \Delta t \qquad \text{compute that } p'(0) = -40$$

After a time $\Delta t = 0.2$, the approximate value of p is

$$\hat{p}(0.2) = 200 - 40 \cdot 0.2 = 192$$

(Figure 5.23).

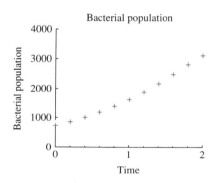
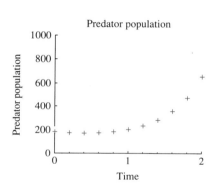

Figure 5.23
Euler's method applied to the predator-prey equations

The next step uses the same idea, but requires the approximate values $\hat{b}(0.2)$ and $\hat{p}(0.2)$ found in the first step.

$$\hat{b}(0.2 + \Delta t)$$
$$= \hat{b}(0.2) + b'(0.2)\Delta t \qquad \text{equation for tangent line}$$
$$= 928 + (1.0 - 0.001 \cdot 192) \qquad b(0.2) = 928 \text{ (initial condition) and}$$
$$\times 928 \cdot \Delta t \qquad b'(0.2) = [1.0 - 0.001\hat{p}(0.2)]\hat{b}(0.2)$$
$$= 928 + 749.8 \cdot \Delta t \qquad \text{compute that } b'(0.2) = 749.8$$

Therefore,

$$\hat{b}(0.4) = 928 + 749.8 \cdot 0.2 = 1078$$

Things seem to be going pretty well for the prey. For the predators, we get

$$\hat{p}(0.2 + \Delta t)$$
$$= \hat{p}(0.2) + p'(0.2)\Delta t \qquad \text{equation for tangent line}$$
$$= 192 + (-1.0 + 0.001 \cdot 928) \qquad p(0.2) = 192 \text{ (initial condition) and}$$
$$\times 192 \cdot \Delta t \qquad p'(0.2) = [-1.0 + 0.001\hat{b}(0.2)]\hat{p}(0.2)$$
$$= 192 - 13.8 \cdot \Delta t \qquad \text{compute that } p'(0.2) \approx -13.8$$

Therefore,
$$\hat{p}(0.4) = 192 - 13.8 \cdot 0.2 = 189.2$$

The predator numbers are fading.

We can continue following these steps to find the approximate solution at time $t = 2.0$. By following along step by step, we would see that eventually the bacterial population begins to decline.

t	$\hat{b}$	$\hat{p}$
0.0	800.0	200.0
0.2	928.0	192.0
0.4	1078.0	189.2
0.6	1252.8	192.2
0.8	1455.2	201.9
1.0	1687.4	220.3
1.2	1950.6	250.6
1.4	2242.9	298.2
1.6	2557.8	372.3
1.8	2878.8	488.3
2.0	3173.5	671.8

Euler's method is an effective way to compute solutions but requires a great deal of numerical calculation. In Section 5.5 we begin to develop methods to predict how solutions will behave using graphical techniques.

SUMMARY

We have introduced three **coupled systems of autonomous differential equations**, pairs of differential equations in which the rate of change of each state variable depends on its own value and on the value of the other state variable. We derived models of a predator and its prey, a competitive analogue of the system studied to describe selection, and a version of Newton's law of cooling that keeps track of the change in room temperature. Each system is **autonomous** because the rates of change depend only on the state variables and not on time. **Euler's method** can be used to compute approximate solutions of these equations using the tangent line approximation.

5.4 EXERCISES

1. Consider the following types of predator-prey interactions. Graph the per capita rates of change, and write the associated system of autonomous differential equations.

   a. The system in Equation 5.6 (page 411) with $\lambda = \delta = 1.0$, $\varepsilon = 0.05$, and $\eta = 0.02$.

   b. Suppose
   $$\text{per capita growth of prey} = 2.0 - 0.01p$$
   $$\text{per capita growth of predators} = 1.0 + 0.01b$$

   How does this differ from Equation 5.6?

   c. Suppose
   $$\text{per capita growth of prey} = 2.0 - 0.0001p^2$$
   $$\text{per capita growth of predators} = -1.0 + 0.01b$$

   d. Suppose
   $$\text{per capita growth of prey} = 2.0 - 0.01p$$
   $$\text{per capita growth of predators} = -1.0 + 0.0001b^2$$

2. Write systems of differential equations to describe the following situations.

a. Two predators that must eat each other to survive.
b. Two competitors: The per capita reproduction of a is decreased when the total population is large, and the total population of b is decreased only when the population of b is large.
c. Two competitors: The per capita reproduction of each is affected only by the population size of the other.

3. Write a system of autonomous differential equations to describe an object placed in a room. The size of the room is 10.0 times that of the object, but the specific heat of the room is 0.2 times that of the object, meaning that a small amount of heat can produce a large change in the temperature of the room.

4. Consider the following special cases of the predator-prey equations (Equation 5.6). Write the differential equations and state what they mean.
   a. $\varepsilon = \eta = 0$ b. $\eta = 0$ c. $\varepsilon = 0$

5. Find the equilibria and draw phase-line diagrams for types a and b of Equation 5.7 supposing that they do not interact (equivalent to setting $a = 0$ in the differential equations for b and $b = 0$ in the differential equations for a).

6. Follow these steps to derive the equations for chemical exchange between two adjacent cells of different size. Suppose that the concentration in the first cell is designated by the variable C_1, the concentration in the second cell is designated by the variable C_2, and the size of the first cell is 2.0 μL, and the size of the second is 5.0 μL.
   a. Write an expression for the total amount of chemical A_1 in the first cell and A_2 in the second.
   b. Suppose that the amount of chemical moving from the first cell to the second cell is β times C_1, and the amount of chemical moving from the second cell to the first cell is β times C_2. Write equations for the rates of change of A_1 and A_2.
   c. Divide by the volumes to find differential equations for C_1 and C_2.
   d. In which cell is the concentration changing faster?

7. The spring equation, or simple harmonic oscillator,
$$\frac{d^2x}{dt^2} = -x$$
which we studied in Section 2.9 can be written as a system of differential equations. Recall that x represents the position of an object at the end of the spring.
   a. Write the velocity v in terms of the derivative of the position x.
   b. Write the acceleration in terms of the derivative of the velocity v.
   c. Write the spring equation as a pair of equations for position and velocity.
   d. Find the equilibrium. What does it mean?
   e. We know that one solution is $x(t) = \cos(t)$. Find $v(t)$ and check that the solution matches your system of equations.

8. Apply Euler's method to the competition equations,
$$\frac{da}{dt} = \mu\left(1 - \frac{a+b}{K_a}\right)a$$
$$\frac{db}{dt} = \lambda\left(1 - \frac{a+b}{K_b}\right)b$$
starting from the following initial conditions. Assume that $\mu = 2.0$, $\lambda = 2.0$, $K_a = 1000$, and $K_b = 500$. Take two steps, with a step length of $\Delta t = 0.1$.
   a. Start from $a = 750$ and $b = 500$.
   b. Start from $a = 250$ and $b = 500$.
   c. Start from $a = 150$ and $b = 200$.

9. Apply Euler's method to Newton's law of cooling,
$$\frac{dH}{dt} = 0.3(A - H)$$
$$\frac{dA}{dt} = 0.1(H - A)$$
starting from the following initial conditions. Take two steps with $\Delta t = 0.1$.
   a. $H = 60$, $A = 20$
   b. $H = 0$, $A = 20$
   c. $H = 20$, $A = 60$

10. An extension of the disease model (Equation 5.3, page 399) is the following:
$$\frac{dI}{dt} = \alpha IS - \mu I$$
$$\frac{dS}{dt} = -\alpha IS$$
where I represents the infected population and S represents the susceptible population. The transfer term is the same as in Equation 5.3. Now, however, individuals who recover become permanently immune rather than susceptible. Compare and contrast this model with the predator-prey model (Equation 5.6).

11. **COMPUTER:** Euler's method for systems of differential equations can be implemented as an **updating system**, a coupled pair of discrete-time dynamical systems. For example, with the predator-prey equations,
$$\frac{db}{dt} = (1.0 - 0.001p)b$$
$$\frac{dp}{dt} = (-1.0 + 0.001b)p$$
Euler's method is
$$\hat{b}(t + \Delta t) = \hat{b}(t) + b'(t)\Delta t$$
$$= \hat{b}(t) + [1.0 - 0.001\hat{p}(t)]\hat{b}(t)\Delta t$$
$$\hat{p}(t + \Delta t) = \hat{p}(t) + p'(t)\Delta t$$
$$= \hat{p}(t) + [-1.0 + 0.001\hat{b}(t)]\hat{p}(t)\Delta t$$
Starting from the initial condition $(b(0), p(0)) = (800, 200)$,

follow this system until it loops around near its initial condition. Use the following values of Δt.
   a. $\Delta t = 1.0$
   b. $\Delta t = 0.2$
   c. $\Delta t = 0.1$
   d. $\Delta t = 0.01$

12. **COMPUTER:** Follow the same steps as in Exercise 11 for the spring equations,

$$\frac{dx}{dt} = v \quad \frac{dv}{dt} = -x$$

which we derived in Exercise 7. How close is your estimated solution to the exact solution? What happens if you keep running for many cycles?

5.5 The Phase Plane

Systems of autonomous differential equations are generally impossible to solve exactly. We have seen how to use Euler's method to find approximate solutions. As with autonomous differential equations and discrete-time dynamical systems, we can deduce a great deal about the behavior of solutions from an appropriate graphical display. For systems of autonomous differential equations, the tool is the **phase-plane diagram**, an extension of the phase-line diagram. Our goal is again to find **equilibria**, points where each of the state variables remain unchanged. Finding these points on the phase plane requires a new tool, the **nullcline**, a graph of the set of points where each state variable separately remains unchanged.

Equilibria and Nullclines: Predator-Prey Equations

A single autonomous differential equation has an equilibrium where the rate of change of the state variable is 0. An **equilibrium** of a two-dimensional system of autonomous differential equations is a point where the rate of change of **each** state variable is 0.

Consider again a predator and its prey described by the system of autonomous differential equations,

$$\frac{db}{dt} = (1.0 - 0.001p)b$$

$$\frac{dp}{dt} = (-1.0 + 0.001b)p$$

(Equation 5.6 with $\lambda = 1$, $\delta = 1$, $\varepsilon = 0.001$, and $\eta = 0.001$). The rates of change of both b and p are equal to 0 when the following hold simultaneously:

$$\frac{db}{dt} = (1.0 - 0.001p)b = 0$$

$$\frac{dp}{dt} = (-1.0 + 0.001b)p = 0$$

Solving **simultaneous equations** can be much harder than solving a single equation. Our technique is to break the problem into pieces and use a graph to combine the results.

How do we graph the solutions of an equation? We are used to graphing functions written in the form $y = f(x)$, placing x on the horizontal axis and y on the vertical axis. Such equations are easy to graph because we have *solved* for y.

To graph the solutions of

$$\frac{db}{dt} = (1.0 - 0.001p)b = 0$$

we must pick one of the variables b or p to place on the vertical axis instead of y. We can pick either one; here we choose p. We therefore set up a graph with the vertical axis labeled p and the horizontal axis labeled b (Figure 5.24). This is a graph of the **phase plane**. Like a phase-line diagram, a phase-plane diagram is a picture of all possible values the state variables can take. For example, the point (1200, 1500) represents the system with 1200 prey and 1500 predators.

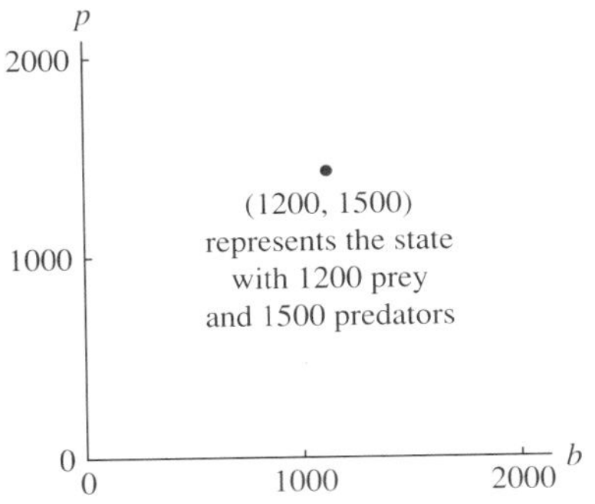

Now we can solve the equation

$$\frac{db}{dt} = (1.0 - 0.001p)b = 0$$

to find all values of b and p where the state variable b does not change. The first step in solving any equation is to factor and set each term equal to 0. This equation, conveniently enough, is already factored as $1.0 - 0.001p$ times b. The two equations to solve are therefore

$$1.0 - 0.001p = 0 \quad \text{and} \quad b = 0$$

Figure 5.24
The phase plane with p on the vertical axis and b on the horizontal axis

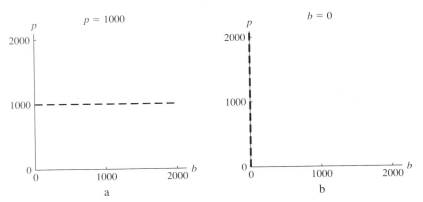

Figure 5.25
The two components of the b-nullcline

Solving means isolating p, the variable represented on the vertical axis. The solution of the first equation is

$$p = 1000$$

In the phase plane, the graph of this function is a horizontal line at $p = 1000$ (Figure 5.25a). The second equation, $b = 0$, does not include the variable p. This means that *any* value of p is a solution when $b = 0$. The graph of the second equation is a vertical line at $b = 0$ (Figure 5.25b).

These solutions represent the set of values where the rate of change of b is 0. This entire set, called the b-nullcline, is graphed on the phase plane in Figure 5.26. This graph might look strange, consisting as it does of two distinct pieces. Although no function could have a graph like this, it is typical of the behavior of nullclines.

We use the same method to find the p-nullcline, the set of points where the rate of change of p is 0. The equation

$$\frac{dp}{dt} = (-1.0 + 0.001b)p = 0$$

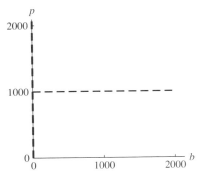

Figure 5.26
The b-nullcline

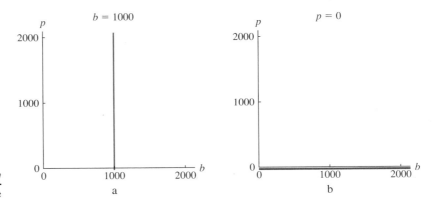

Figure 5.27
The two components of the p-nullcline

has a solution where either of the factors is equal to 0. Because the equation is in factored form with factors $-1.0 + 0.001b$ and p, its solutions are

$$-1.0 + 0.001b = 0 \quad \text{and} \quad p = 0$$

To graph them, we again solve for the vertical variable p. Remember to use the *same* vertical variable for both nullclines. The first factor lacks any occurrence of the vertical variable p to solve for. This indicates that the solution is a vertical line, found by solving for the horizontal variable. The solution of $-1.0 + 0.001b = 0$ is $b = 1000$, so the first factor of the p-nullcline is a vertical line at $b = 1000$ (Figure 5.27a). The second factor is the horizontal line $p = 0$. The entire p-nullcline is the combination of these two factors (Figure 5.28).

The state variable b does not change on the b-nullcline; the state variable p does not change on the p-nullcline. Therefore, neither b nor p changes at any point where the two nullclines intersect. These intersections, therefore, are the equilibria. To find equilibria, plot both nullclines on the same graph, being careful to distinguish which factor belongs to which nullcline (Figure 5.29). There are two intersections of the nullclines, at $(0, 0)$ and $(1000, 1000)$. At the first equilibrium, both populations are extinct. At the second, both populations are positive and the system is balanced.

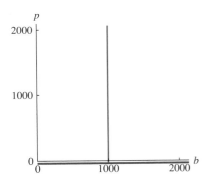

Figure 5.28
The p-nullcline

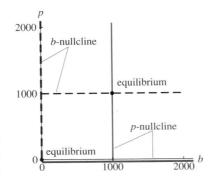

Figure 5.29
The nullclines of the predator-prey system

The points $(0, 1000)$ and $(1000, 0)$ are not equilibria because they do not lie at the intersection of the two nullclines. The point $(0, 1000)$ lies on both pieces of the b-nullcline but not on the p-nullcline. We can check whether a point is an equilibrium by substituting into the system of differential equations. At $b = 0$ and $p = 1000$,

$$\frac{db}{dt} = (1.0 - 0.001p)b = (1.0 - 0.001 \cdot 1000) \cdot 0 = 0$$

$$\frac{dp}{dt} = (-1.0 + 0.001b)p = (-1.0 + 0.001 \cdot 0) \cdot 1000 = -1000$$

Because the rate of change of the predator population p is not equal to 0, this is not an equilibrium. At this point, there are no prey and the predator population declines. Similarly, the point $(1000, 0)$ lies on both pieces of the p-nullcline but not on the b-nullcline. At $b = 1000$ and $p = 0$,

$$\frac{db}{dt} = (1.0 - 0.001p)b = (1.0 - 0.001 \cdot 0) \cdot 1000 = 1000$$

$$\frac{dp}{dt} = (-1.0 + 0.001b)p = (-1.0 + 0.001 \cdot 1000) \cdot 0 = 0$$

Again, the rate of change of one state variable is not 0, and this point is not an equilibrium.

The following algorithm gives the steps to find the nullclines and equilibria of a system of autonomous differential equations:

■ **Algorithm 5.3** (Finding the nullclines and equilibria of a pair of autonomous differential equations)

1. Pick one of the variables to act as the vertical variable in the phase plane.
2. Write the equations for one of the nullclines by setting the rate of change equal to 0.
3. Factor.
4. Solve each factor for the vertical variable. If there is no vertical variable in the factor, solve for the horizontal variable.
5. Graph each factor in the phase plane. If the equation contains no vertical variable, the graph is a vertical line.
6. Do the same for the other nullcline, graphing it in a different color or style.
7. Find the intersections, which are the equilibria of the system. ■

Solutions that begin at an equilibrium remain there. To figure out what other solutions do, we would need to assess the stability of our equilibria. The techniques to do this, extensions of the methods used in one dimension, involve *linear algebra*, which lies beyond the scope of this book. In Section 5.6, however, we will learn to analyze at least some situations by drawing direction arrows on our phase-plane diagram.

Equilibria and Nullclines: Competition Equations

We can use Algorithm 5.3 to find the equilibria of competing types of bacteria that follow the equations

$$\frac{da}{dt} = 2.0\left(1 - \frac{a+b}{1000}\right)a$$

5.5 The Phase Plane

$$\frac{db}{dt} = 2.0\left(1 - \frac{a+b}{500}\right)b$$

(Equation 5.7 with $\mu = 2.0$, $\lambda = 2.0$, $K_a = 1000$, and $K_b = 500$.) We begin by choosing the vertical variable; it doesn't matter which one we choose, so we pick b because it comes later in the alphabet. To find the a-nullcline, we solve

$$\frac{da}{dt} = 2.0\left(1 - \frac{a+b}{1000}\right)a = 0$$

The factors are $2.0[1 - (a+b)/1000]$ and a. Solutions occur where

$$2.0\left(1 - \frac{a+b}{1000}\right) = 0 \quad \text{or} \quad a = 0$$

Solving the first factor for b, the vertical variable, we find

$$2.0\left(1 - \frac{a+b}{1000}\right) = 0$$

$$\left(1 - \frac{a+b}{1000}\right) = 0$$

$$\frac{a+b}{1000} = 1$$

$$a + b = 1000$$

$$b = 1000 - a$$

The second factor has no vertical variable b and is therefore the vertical line at $a = 0$ (Figure 5.30a).

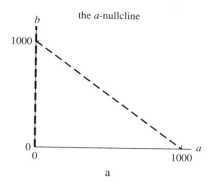

Figure 5.30
The nullclines of the competition system

To find the b-nullcline, we solve

$$\frac{db}{dt} = 2.0\left(1 - \frac{a+b}{500}\right)b$$

The factors are $2.0[1 - (a+b)/500]$ and b. Solutions occur where

$$2.0\left(1 - \frac{a+b}{500}\right) = 0 \quad \text{or} \quad b = 0$$

Solving the first factor for the vertical variable, we find

$$2.0\left(1 - \frac{a+b}{500}\right) = 0$$

$$\left(1 - \frac{a+b}{500}\right) = 0$$

$$\frac{a+b}{500} = 1$$

$$a + b = 500$$

$$b = 500 - a$$

The second factor is a horizontal line at $b = 0$ (Figure 5.30b).

To find the equilibria, we plot both nullclines on the same graph of the phase plane (Figure 5.31). There are three intersections and thus three equilibria: at $(0, 0)$, $(1000, 0)$, and $(0, 500)$. Both types are extinct at $(0, 0)$, type b dominates the population at $(1000, 0)$, and type a dominates the population at $(0, 500)$.

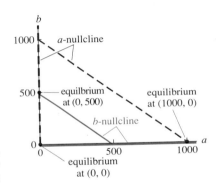

Figure 5.31
The nullclines and equilibria of the competition system

The points $(0, 1000)$ and $(500, 0)$ are not equilibria. At $(0, 1000)$, we have

$$\frac{db}{dt} = 2.0\left(1 - \frac{1000}{500}\right) = -2.0 < 0$$

The point lies on both pieces of the a-nullcline but not on the b-nullcline. Similarly, at $(500, 0)$, we have

$$\frac{da}{dt} = 2.0\left(1 - \frac{500}{1000}\right) = 1.0 > 0$$

This point lies on both branches of the b-nullcline but not on the a-nullcline.

Equilibria and Nullclines: Newton's Law of Cooling

We use the same steps to find the equilibria for Newton's law of cooling:

$$\frac{dH}{dt} = \alpha(A - H)$$

$$\frac{dA}{dt} = \alpha_2(H - A)$$

To begin, we choose the temperature of the object H as the vertical variable. The A-nullcline is the set of points where

$$\frac{dA}{dt} = \alpha_2(H - A) = 0$$

This has only one component and does not need to be factored (unless $\alpha_2 = 0$). Solving for the vertical variable H, we get

$$H = A$$

(Figure 5.32a). The H-nullcline is the set of points where

$$\frac{dH}{dt} = \alpha(A - H) = 0$$

Solving for the vertical variable H, we again find

$$H = A$$

(Figure 5.32b).

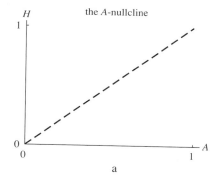
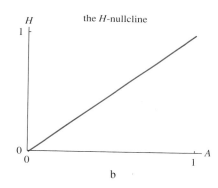

Figure 5.32
The nullclines of Newton's law of cooling

The two nullclines exactly overlap. Because all points where the two nullclines intersect are equilibria, every point on the line $H = A$ is an equilibrium (Figure 5.33). This peculiar result makes sense. When the two temperatures are the same, there is no further change in temperature by either the object or the room.

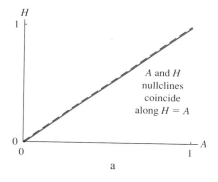

Figure 5.33
The nullclines and equilibria of Newton's law of cooling

SUMMARY

We have introduced the **phase plane**, **nullclines**, and **equilibria** as tools to study autonomous systems of differential equations. Equilibria occur where the rate of change of each state variable is 0. They can be found on the phase plane (the Cartesian plane with axes labeled by the state variables) by graphing the two nullclines. The nullcline associated with a state variable is the set of values where the state variable remains unchanged.

5.5 EXERCISES

1. Graph the nullclines in the phase plane and then find the equilibria of the following.
   a. Predator-prey model (Equation 5.6, page 411) with $\lambda = 1.0$, $\delta = 3.0$, $\varepsilon = 0.002$, $\eta = 0.005$.
   b. Predator-prey model with $\lambda = 1.0$, $\delta = 3.0$, $\varepsilon = 0.005$, $\eta = 0.002$.
   c. Newton's law of cooling with $\alpha = 0.01$ and $\alpha_2 = 0.1$.
   d. Competition model (Equation 5.7, page 412) with $\lambda = 2.0$, $\mu = 1.0$, $K_a = 10^6$, $K_b = 10^7$.
   e. Competition model with $\lambda = 1.0$, $\mu = 2.0$, $K_a = 10^6$, $K_b = 10^7$. If you can, explain why the results exactly match those of the previous part.

2. Redraw the phase-plane diagrams for the problems in Exercise 1 by making the other choice for the vertical variable. Check that you get the same equilibrium.

3. Find the nullclines and equilibria for the models for Exercise 1 **b** through **d** in Section 5.4.

4. Find and graph the nullclines and find the equilibria for the models in Exercise 2 in Section 5.4.

5. The prey in the predator-prey model increase in numbers exponentially in the absence of predators. Consider the model with competition among the prey given by the equation

$$\frac{db}{dt} = (\lambda - \varepsilon p - \alpha b)b$$

Suppose that $\lambda = 1.0$, $\varepsilon = 0.001$, and $\alpha = 0.0002$. The predators follow the equation

$$\frac{dp}{dt} = (-\delta + \eta b)p$$

with $\delta = 1.0$ and $\eta = 0.001$.
   a. Find and graph the nullclines in the phase plane.
   b. Find the equilibria.
   HINT: Consider the predator equation first, solving as in the text. Then substitute these results into the new prey equation.

6. For the model of diffusion we derived in Exercise 6 in Section 5.4, we assumed that the membrane between the vessels is equally permeable in both directions. Suppose instead that the membrane reduces the rate at which chemical enters the first vessel by a factor of 3.

   a. Find the rate at which chemical moves from the smaller to the larger vessel.
   b. Find the rate at which chemical moves from the larger to the smaller vessel.
   c. Find the rate of change of the amount of chemical in each vessel.
   d. Divide by the volumes V_1 and V_2 to find the rate of change of concentration.
   e. Find and graph the nullclines.
   f. What are the equilibria? Do they make sense?

7. In our model of competition, the per capita growth rate of type a and b bacteria are functions of only the total population size. This means that reproduction is reduced just as much by an individual of type a as by an individual of type b. In many systems, type a bacteria interfere more with each other than with type b bacteria and vice versa. In particular, suppose that individuals of type b reduce the per capita growth rate of type a by half as much as individuals of type a, and that individuals of type a reduce the per capita growth rate of type b by half as much as individuals of type b.
   a. Show that the following equations match the assumptions:

$$\frac{da}{dt} = \mu\left(1 - \frac{a + b/2}{K_a}\right)a$$

$$\frac{db}{dt} = \lambda\left(1 - \frac{a/2 + b}{K_b}\right)b$$

   b. Suppose $\lambda = \mu = 1.0$ and $K_a = K_b = 10^4$. Find and graph the nullclines.
   c. Find the equilibria (there should be four).
   d. Try to explain the meaning of the new equilibrium where the two types of bacteria coexist.

8. Consider again the disease model,

$$\frac{dS}{dt} = -\alpha IS$$

$$\frac{dI}{dt} = \alpha IS - \mu I$$

(Exercise 10 in Section 5.4). Find the nullclines and equilibria of this model when $\alpha = 2.0$ and $\mu = 1.0$.

9. There are several ways to include births and deaths in the basic disease model from Exercise 8. First, suppose that all births are into the susceptible group at rate bS, and that there are deaths at rate kI in the infected group and at rate kS in the susceptible group.
   a. What does the birth term mean? Do infected individuals reproduce?
   b. Write the system of differential equations.
   c. Find the nullclines and equilibria of this model if $b > k$ (try $b = 2.0$, $k = 1.0$, $\alpha = 2.0$, and $\mu = 1.0$).
   d. Find the nullclines and equilibria of this model if $b < k$ (try $b = 0.5$, $k = 1.0$, $\alpha = 2.0$, and $\mu = 1.0$).

10. Suppose that everything is the same as in Exercise 9, but that births are into the susceptible class at rate $b(S+I)$.
    a. Explain in biological terms how the two models differ.
    b. Write the system of differential equations.
    c. Find the nullclines and equilibria of this model if $b > k$ (try $b = 2.0$, $k = 1.0$, $\alpha = 2.0$, and $\mu = 1.0$).
    d. Find the nullclines and equilibria of this model if $b < k$ (try $b = 0.5$, $k = 1.0$, $\alpha = 2.0$, and $\mu = 1.0$).

11. **COMPUTER:** One complicated equation for chemical kinetics is the Schnakenberg reaction. Let A and B denote the concentrations of two chemicals, A and B. A is added at constant rate k_1, B is added at constant rate k_4, A breaks down at rate k_2, and B is converted into A by an *autocatalytic reaction*. The equations are

$$\frac{dA}{dt} = k_1 - k_2 A + k_3 A^2 B$$

$$\frac{dB}{dt} = k_4 - k_3 A^2 B$$

The final term, which is somewhat like the term αIS in the epidemic equation in Exercises 8–10, differs in that the rate of the reaction becomes faster the larger the concentration of A. Suppose that $k_2 = k_3 = 1.0$.
   a. Have your computer draw the nullclines and find the equilibria for $k_1 = 0.2$ and $k_4 = 2.0$.
   b. Do the same with $k_1 = -0.2$ and $k_4 = 2.0$.
   c. Try to explain your results.

12. **COMPUTER:** A variant of the Schnakenberg reaction has the chemical A inhibiting its own production. In particular, assume that the transfer term takes the form

$$k_3 \frac{AB}{1+A}$$

Suppose that $k_2 = k_3 = 1.0$.
   a. Write the equations describing this system.
   b. Have your computer draw the nullclines and find the equilibria for $k_1 = 0.2$ and $k_4 = 2.0$.
   c. Do the same with $k_1 = -0.2$ and $k_4 = 2.0$.
   d. Try to explain your results.

5.6 Solutions in the Phase Plane

We have seen how to find nullclines and equilibria in the phase plane. Our real goal is to find *solutions*, descriptions of how the state variables change over time. We begin by graphing in the phase plane the results from Euler's method. To deduce the behavior of solutions without all the calculations necessary for Euler's method, we will add *direction arrows* to the phase-plane diagram. Like the ones that appear in phase-line diagrams, these arrows indicate where solutions are increasing or decreasing and can be used to sketch **phase-plane trajectories**, or solutions in the phase-plane.

Euler's Method in the Phase Plane

We applied Euler's method to the predator-prey equations

$$\frac{db}{dt} = (1.0 - 0.001p)b$$

$$\frac{dp}{dt} = (-1.0 + 0.001b)p$$

from Section 5.4, and we computed the values for 10 steps of length $\Delta t = 0.2$ starting from the initial condition $b(0) = 800$ and $p(0) = 200$. Figure 5.34 shows these values plotted on the phase plane. The value at $t = 0$ is plotted as the point $(800, 200)$ in the phase-plane, the value at $t = 0.2$ is plotted as $(928, 192)$, and so forth. Following the points through time, we see that the prey population increases steadily and the predator population first decreases and then increases.

t	$\hat{b}$	$\hat{p}$
0.0	800.0	200.0
0.2	928.0	192.0
0.4	1078.0	189.2
0.6	1252.8	192.2
0.8	1455.2	201.9
1.0	1687.4	220.3
1.2	1950.6	250.6
1.4	2242.9	298.2
1.6	2557.8	372.3
1.8	2878.8	488.3
2.0	3173.5	671.8

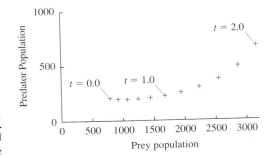

Figure 5.34

Results from Euler's method plotted on the phase-plane

Euler's method does not provide an exact solution, however, and takes a lot of calculation. More precise techniques that correct the errors produced by using the tangent line approximation can be programmed on the computer to generate solutions accurate to any desired level. A solution generated with one such method (called the **Runge-Kutta method**) is plotted in Figure 5.35. The initial conditions in this graph are $b(0) = 800$ and $p(0) = 200$. The solution consists of two curves, one for each of the state variables. At any time t, the values of b and p can be read from the graphs of b and p.

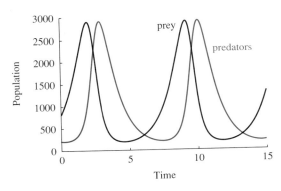

Figure 5.35

Solutions of the predator-prey system

The populations oscillate, with the peak predator population occurring after the peak prey population. When prey are plentiful, the predators increase in number. They eat the prey, eventually decreasing the prey population. This reduces the predator's food supply, which leads to an eventual reduction in the predator population. The prey can then increase, beginning the cycle again.

As with the results from Euler's method, we can graph these solutions in the phase plane as a **phase-plane trajectory** (Figure 5.36). The initial condition $(800, 200)$ is plotted at the point $(800, 200)$. At $t = 2$, the population of prey is 2894 and the population of predators is 1051, so the point $(2894, 1051)$ is plotted in the phase plane. Because time does not appear explicitly as part of a phase-plane trajectory, we have labeled several points with the time. The initial condition (b_0, p_0) is labeled with $t = 0$. We have graphed three things on one graph: the time, the prey population, and the predator population. This is called a **parametric graph** because the time does not appear on either axis.

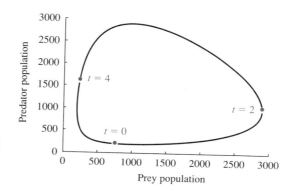

Figure 5.36
Solution of the predator-prey equations on the phase plane

How can we translate between the solutions plotted as functions of time and the phase-plane trajectory? Starting from the solution, you can plot the number of predators against the number of prey at several times (such as $t = 0$, $t = 1$, and so forth) and connect the dots. Starting from the phase-plane trajectory, you can sketch the solutions by tracing along the graph at a constant speed. The horizontal location of your pencil gives the prey population, or the height of the graph of $b(t)$. The vertical location of your pencil gives the predator population, or the height of the graph of $p(t)$. In our example, we see that the prey population begins by increasing, reaches a maximum value of nearly 3000, then decreases to a value of about 200 before beginning to increase again. The predator population decreases slightly below 200 before beginning an increase to nearly 3000. The predators reach their maximum after the prey, and then decrease again.

Direction Arrows: Predator-Prey Equations

The computer solutions (Figures 5.35 and 5.36) give a nearly exact description of the dynamics for our predator-prey example. Our goal, however, is to understand the behavior of the populations without solving the equations. For one-dimensional systems, we sketched solutions from the direction arrows and equilibria on a phase-line diagram. We know how to draw nullclines and find equilibria on a phase-plane diagram. The next step is to figure out how to draw those arrows.

We have redrawn the nullclines and equilibria in the phase plane for the predator prey system (Figure 5.37). The nullclines break the phase plane into four regions, labeled I, II, III, and IV. In each region, we wish to determine whether the populations of predators and prey are increasing or decreasing.

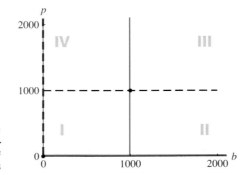

Figure 5.37

Regions of the phase plane for the predator-prey equations

There are three approaches to finding this out:

- Method I: Picking a pair of values (b, p) in the region and substitute into the differential equation.
- Method II: Algebraic manipulation of inequalities.
- Method III: Reasoning about the equations.

Which method is best depends on the equations. We apply all three to the predator-prey phase plane.

In region I, the predator population is below the b-nullcline $p = 1000$ and the prey population is below the p-nullcline $b = 1000$. Method I requires picking a pair of values in this region. One point in this region is $(500, 500)$. We can substitute this into the differential equation to check whether the populations are increasing or decreasing. With these values, we get

$$\frac{db}{dt} = (1.0 - 0.001 \cdot 500) \cdot 500 = 250$$

$$\frac{dp}{dt} = (-1.0 + 0.001 \cdot 500) \cdot 500 = -250$$

The prey population is increasing and the predator population is decreasing, which we indicate by an arrow pointing toward larger values of the prey (to the right) and toward smaller values of the predator (down) (Figure 5.38). This is a direction arrow.

Method II is algebraic. If we are in region I, then we have

$$0 < b < 1000$$
$$0 < p < 1000$$

Then we get

$$\frac{db}{dt} = (1.0 - 0.001p)b > 0$$

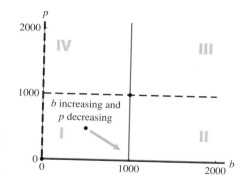

Figure 5.38

Direction arrow in region I of the predator-prey phase plane

because both $1.0 - 0.001p > 0$ and $b > 0$. The prey population increases in this region. Similarly, we get

$$\frac{dp}{dt} = (-1.0 + 0.001b)p < 0$$

because $-1.0 + 0.001b < 0$ and $p > 0$. The predator population decreases. Again, the direction arrow points to the right and down.

Method III uses reasoning about the equations. In region I, the prey and predator populations are both low. This means that the prey are happy (few predators to eat them) and the predators are sad (too few prey to eat). The prey population will increase and the predator population will decrease.

We can use any of the three methods to find the direction arrows in the three remaining regions (Figure 5.39). Using the reasoning method, for region II, we find that the prey population is large and the predator population is small. Both species should be happy and increase, generating a direction arrow that points up and to the right. In region III, both populations are large. This is bad news for the prey and good news for the predators. The direction arrow therefore points toward lower values of prey (to the left) and larger values of predators (up). Finally, in region IV, the prey population is small and the predator population is large. Neither species does well under these circumstances, and the direction arrow points left and down.

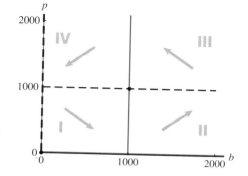

Figure 5.39

Direction arrows for the predator-prey equations

As on a phase-line diagram, solutions on phase-plane trajectories follow the arrows (Figure 5.40). Starting in region I, where predator and prey populations are small, the arrow points down and to the right, pushing the solution into region II. Both populations then increase, following the arrows up and to the right into region

III. The predators then increase while the prey decrease, moving the population into region IV, from which both populations decrease into region I. We cannot tell from this description whether the phase-plane trajectory circles around, spirals toward, or spirals away from the equilibrium.

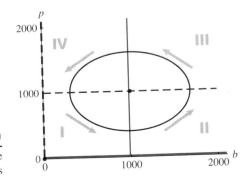

Figure 5.40

Direction arrows and solution for the predator-prey equations

What happens to direction arrows right on the nullclines? For example, what happens to a population with $b = 1000$ and $p < 1000$? This point lies on the p-nullcline, meaning that the population of predators remains unchanged. If an increasing predator population is associated with an upward-pointing arrow and a decreasing predator population is associated with a downward-pointing arrow, an unchanging predator population must be associated with an arrow that points neither up nor down. Such an arrow is horizontal. The population of prey, however, is changing. Because the number of predators is small, the prey population will increase and the arrow will point to the right (Figure 5.41). The other four direction arrows on nullclines can be found in a similar way.

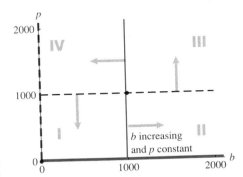

Figure 5.41

Direction arrows on nullclines

Finding the direction arrows on the nullclines is useful for two reasons. First, it provides a useful check on the rest of the direction arrows. Notice that the directions can change only one at a time; when we move from region I to region II, the arrow switches from pointing down and to the right to pointing up and to the right. The only change was the vertical direction, and this change happens right at the nullcline where the arrow is horizontal. Second, these direction arrows can help in sketching more accurate phase-plane trajectories. Because solutions must follow the arrows, solutions must be horizontal when they cross the p-nullcline and vertical when they cross the b-nullcline. The solution sketched in Figure 5.40 satisfies these criteria.

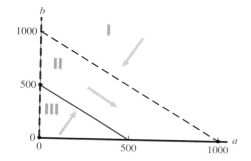

Figure 5.42
Direction arrows for the competition system

Direction Arrows for the Competition Equations and Newton's Law of Cooling

For the competition model, the nullclines break the phase plane into three regions (Figure 5.42). We have again used the parameter values $K_a = 1000$ and $K_b = 500$, and we assume that μ and λ are positive. The a-nullcline lies above the b-nullcline. We can determine the direction arrows by using each of the three methods: substituting values, using algebra, and reasoning. With the first method, we substitute a point in region I, $(1000, 1000)$. Then we get

$$\frac{da}{dt} = \mu \left(1 - \frac{1000 + 1000}{1000}\right) \cdot 1000 = -1000\mu < 0$$

$$\frac{db}{dt} = \lambda \left(1 - \frac{1000 + 1000}{500}\right) \cdot 1000 = -3000\lambda < 0$$

Because both a and b are decreasing, the direction arrow points down and to the left. It is a bit harder to find a point in region II. This region is defined by

$$1000 > a + b > 500$$

so one point is the region is $(400, 400)$. Then

$$\frac{da}{dt} = \mu \left(1 - \frac{400 + 400}{1000}\right) \cdot 400 = 80\mu > 0$$

$$\frac{db}{dt} = \lambda \left(1 - \frac{400 + 400}{500}\right) \cdot 400 = -240\lambda < 0$$

The direction arrow points down and to the right. The point $(100, 100)$ lies in region III, where

$$\frac{da}{dt} = \mu \left(1 - \frac{100 + 100}{1000}\right) \cdot 100 = 80\mu > 0$$

$$\frac{db}{dt} = \lambda \left(1 - \frac{100 + 100}{500}\right) \cdot 100 = 60\lambda > 0$$

The arrow points up and to the right.

Algebraically, in region I, $a + b > 1000 > 500$, so we have

$$1 - \frac{a+b}{1000} < 0$$

$$1 - \frac{a+b}{500} < 0$$

Therefore, we get

$$\frac{da}{dt} = \mu\left(1 - \frac{a+b}{1000}\right)a < 0$$

$$\frac{db}{dt} = \lambda\left(1 - \frac{a+b}{500}\right)b < 0$$

Both populations decrease, and the direction arrow points down and to the left. In region II, $500 < a + b < 1000$ and

$$\frac{da}{dt} = \mu\left(1 - \frac{a+b}{1000}\right)a > 0$$

$$\frac{db}{dt} = \lambda\left(1 - \frac{a+b}{500}\right)b < 0$$

The population of b decreases, and the population of a increases; the direction arrow points down and to the right. In region III, $a + b < 500 < 1000$, so that

$$\frac{da}{dt} = \mu\left(1 - \frac{a+b}{1000}\right)a > 0$$

$$\frac{db}{dt} = \lambda\left(1 - \frac{a+b}{500}\right)b > 0$$

Both populations increase, and the direction arrow points up and to the right.

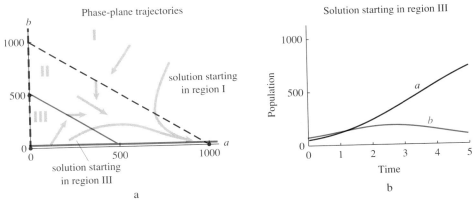

Figure 5.43

Solutions of the competition system

With the reasoning method, think of $K_a = 1000$ and $K_b = 500$ as the largest total populations that types a and b can tolerate. In region I, the total population exceeds both K_a and K_b. Both types suffer from overpopulation and have shrinking populations, generating a direction arrow that points down and to the left. In region II, the total population lies between 500 and 1000. Type b cannot withstand competition that type a can tolerate, producing a direction arrow that points down and to the right. In region III, the total population is less than both K_a

and K_b. Both types can grow, generating a direction arrow pointing up and to the right.

The two phase-plane trajectories plotted in Figure 5.43a follow the arrows and are forced toward the equilibrium at $(1000, 0)$ where type b has become extinct. This figure includes the direction arrows on the nullclines. On the a-nullcline, the population of b is decreasing because the total population is 1000, exceeding the tolerance of type b. The direction arrow therefore points straight down. On the b-nullcline, the population of a is increasing because the total population is 500, below the carrying capacity of type a. The direction arrow points straight to the right. The solution starting in region III is plotted as a function of time in Figure 5.43b. Even though type b is doomed to extinction, it has an initial period of growth. The solution for b reaches a maximum when the trajectory crosses the b-nullcline.

This model differs in several ways from the differential equations

$$\frac{da}{dt} = \mu a$$

$$\frac{db}{dt} = \lambda b$$

(which we reduced to a single equation for the fraction p of type a in Section 4.2). When $\lambda < \mu$, type b grows more slowly and goes extinct. Here, the type with the larger carrying capacity K wins, whether or not it grows faster. In this model, the type better able to withstand competition eventually wins out, even if its growth rate is lower.

The direction arrows are simpler for Newton's law of cooling. The nullclines coincide and break the plane into only two regions (Figure 5.44). In region I, $H > A$, meaning that H decreases and A increases. The direction arrow points down and to the right. In region II, $H < A$, meaning that H increases and A decreases. The direction arrow points up and to the left. Solutions are pushed toward the line of equilibria along the direction arrows.

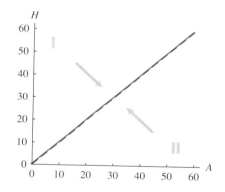

Figure 5.44
Direction arrows for Newton's law of cooling

SUMMARY

We have seen how to plot solutions of two-dimensional differential equations as functions of time and as **phase-plane trajectories**. The nullclines break the phase plane into regions, in each of which we can find a *direction arrow* indicating whether the state variables are increasing or decreasing. For additional information

5.6 Exercises

1. Suppose the following functions are solutions of some differential equation. Graph them as functions of time and as phase-plane trajectories for $0 \leq t \leq 2$. Mark the position at $t = 0$, $t = 1$, and $t = 2$.
   a. $x(t) = t$, $y(t) = 3t$
   b. $a(t) = 2e^{-t}$, $b(t) = e^{-2t}$
   c. $f(t) = 1 + t$, $g(t) = e^{-t}$
   d. $x(t) = t - 2t^2 + t^3$, $y(t) = 4 - t^2$

2. From the following graphs of solutions of differential equations as functions of time, graph the matching phase-plane trajectory.

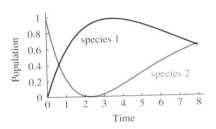

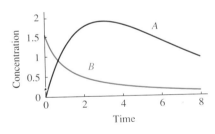

3. From the following graphs of phase-plane trajectories, graph the matching solutions of differential equations as functions of time.

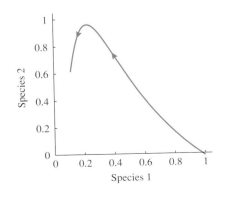

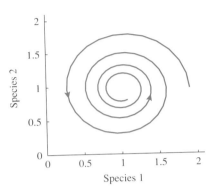

4. On the following phase-plane diagrams, use the direction arrows to sketch phase-plane trajectories starting from two initial conditions.

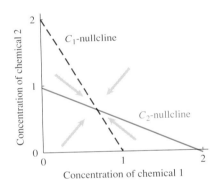

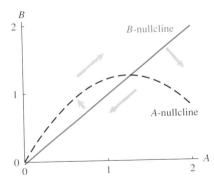

5. Using the information in Exercise 4, draw direction arrows on the nullclines.

6. Draw the nullclines and direction arrows for the spring equation as written in Exercise 7 in Section 5.4. Be sure to in-

clude positive and negative values for the position x and the velocity v.

7. For the problems in Exercise 1 in Section 5.5, draw direction arrows, sketch a phase-plane trajectory, and graph the associated solutions as functions of time.

8. Draw the direction arrows and a sample solution for the following models found in Section 5.5. Explain in words what your solution is doing.
    a. The model presented in Exercise 5.
    b. The model presented in Exercise 6.
    c. The model presented in Exercise 7.
    d. The model presented in Exercise 8.
    e. The model presented in Exercise 9 (with $b > k$).
    f. The model presented in Exercise 10 (with $b < k$).

9. **COMPUTER:** Consider the following differential equations describing diffusion and utilization of a chemical:

$$\frac{dC}{dt} = \alpha(\Gamma - C) - \frac{\delta C}{1 + C}$$

$$\frac{d\Gamma}{dt} = \frac{\alpha}{K}(C - \Gamma) + S$$

The parameters have the following meanings.

Parameter	Meaning	Values to use
α	Diffusion rate	1.0
δ	Use efficiency	4.0 and 1.0
K	Ratio of volumes	2.0
S	Supplementation rate	1.0

   a. Set $\delta = 4$ and the rest of the parameters to their designated values. Plot the nullclines, and find the equilibrium.
   b. Follow the steps in part **a** with $\delta = 1$. Is there an equilibrium? Can you say why not? (No math jargon allowed.) Sketch C and Γ as functions of time.
   c. Try to figure out the critical value of δ where the behavior changes.

10. **COMPUTER:** Many biological systems need to be able to respond to changes in the level of some signal (like a hormone) without responding to the actual level. For example, a cell might have no response to a low level of hormone. If the hormone level rapidly increases, the cell responds. But if the hormone level then remains constant at the higher level, the cell again stops responding. This process is sometimes called *adaptation*.

    One mechanism for this process is summarized in the following model. Internal response is a function of the fraction, p, of cell surface receptors that are bound by the hormone. This fraction increases when the hormone level, H, is high. However, the hormone also dissociates from bound receptors. Assume this happens at a rate A but that this rate is controlled by the cell. One possible set of equations is

$$\frac{dp}{dt} = k_1 H(1 - p) - Ap$$

$$\frac{dA}{dt} = \varepsilon(H - A)$$

Suppose that $k_1 = 0.5$ and that ε is a small value (such as 0.1 or 0.01). The value of H, determined by conditions external to the cell, is not described by a differential equation.
   a. Find the nullclines and equilibria of this model assuming that H is a constant. Does H appear in your final results? Explain why the cell should respond in the same way to any constant level of H.
   b. Use your computer to simulate the response when the level of H jumps quickly from $H = 1$ to $H = 10$. One way to do this is to solve the equations with $H = 10$, using as initial conditions the equilibrium values of p and A when $H = 1$. Draw graphs of p and A in the phase plane and as functions of time. Explain what is happening.
   c. Do part **b** assuming that the level of H drops rapidly from $H = 10$ to $H = 1$.

5.7 The Dynamics of a Neuron

In Section 1.13, we studied a discrete-time dynamical system describing the heart, an important excitable system that responds to a periodic stimulus. To follow more precisely the response to a single stimulus, or to figure out mechanisms for **creating** a periodic stimulus, we need to use differential equations. In this section, we present a simplified (but venerable and valuable) model of a neuron.

A Mathematician's View of a Neuron

Some basic properties of a neuron are illustrated in Figure 5.45. The key measurements are of the concentrations of two ions, sodium and potassium. Like most cells, a resting neuron maintains an excess of potassium and a deficit of sodium. A neuron maintains a negative **resting potential**, meaning that there is an overall excess of negative charge inside the cell. Sodium and potassium ions are both positively charged. Because sodium concentrations are much higher than potassium concentrations, the deficit of these positive charges inside the cell accounts for most of the negative charge to the cell. Cells use a significant amount of energy to maintain this distribution of ions.

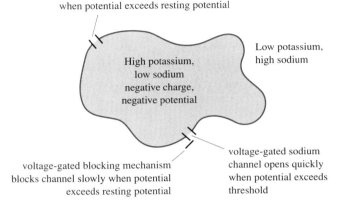

Figure 5.45
The basic elements of neural dynamics

The neuron uses these gradients of sodium, potassium, and charge across the cell membrane to amplify and transmit information. The process depends on a set of **voltage-gated channels** for each ion. Such channels open and close in response to voltage (another name for potential) differences and are closed when the cell is at rest.

When a burst of positive charges enters the cell (and makes the potential of the cell less negative), voltage-gated sodium channels open (see Figure 5.46). Because there is an excess of sodium outside the cell, more sodium ions enter, further increasing the potential of the cell until it becomes positive.

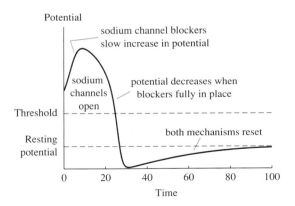

Figure 5.46
An action potential

Two things then occur. One slow mechanism acts to block the voltage-gated sodium channels, and another slow mechanism begins to open voltage-gated potassium channels. Both of these processes act to diminish the build-up of positive charge in the cell. Blocking the sodium channels halts the entry of sodium, and opening potassium channels allows the exit of (positively charged) potassium. Neither process happens as quickly as the opening of the sodium channels, and neither can halt the increase in potential immediately, just as a weak backward force applied to a heavy moving object only gradually slows it down and reverses its direction. When the potential of the cell has again decreased to or below the resting potential, these mechanisms slowly turn off. Soon the cell is ready to begin the cycle again.

How does a neuron use this ability? This sharp peak of electrical excitation, called an **action potential**, can be transmitted precisely to other neurons. Action potentials are part of the language of the brain, like the 0s and 1s used by computers.

Our model is designed to study the structure of the interaction of two processes: a fast process with **positive feedback** (a slight increase in cell potential quickly induces sodium channels to open and increase the potential further) and a slow process with **negative feedback** (the increase in cell potential induces two mechanisms that halt the increase). Because this structure requires only the two mechanisms acting on sodium channels (fast opening and slow blocking), we concentrate on these. Cells without functioning potassium channels produce action potentials that are less precise.

The Mathematics of Sodium Channels

The potential in a cell can be scaled to more convenient mathematical values. Denoting the scaled potential by v, we set

$$\begin{cases} v = 0 & \text{resting potential} \\ v = a & \text{the threshold above which the neuron fires} \\ v = 1 & \text{potential with sodium channels open} \end{cases} \quad (5.8)$$

Small deviations above resting potential are not amplified, and the potential returns to rest. However, if the cell potential is raised above the threshold a, the cell moves toward the higher potential found when the sodium channels are open. A phase-line diagram for cell potential following this description is shown in Figure 5.47.

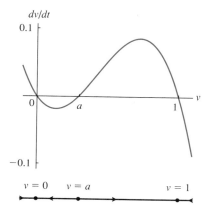

Figure 5.47
The phase-line diagram without the slow mechanism for blocking sodium channels

One convenient equation consistent with this diagram is

$$\frac{dv}{dt} = -v(v - a)(v - 1) = f(v) \quad (5.9)$$

The right-hand side is in factored form, and the values $v = 0$, $v = a$, and $v = 1$ are all equilibria. We can take the derivative of the rate of change to check the stability of the equilibria (Theorem 5.1), finding

$$f'(v) = -(v-a)(v-1) - v(v-1) - v(v-a)$$

Then we have

$$f'(0) = -a < 0$$
$$f'(a) = a(1-a) > 0$$
$$f'(1) = -(1-a) < 0$$

The equilibrium at $v = 0$ is stable, the one at $v = a$ is unstable, and the one at $v = 1$ is stable. The graph of f and the associated phase-line diagram are consistent with our biological assumptions (Figure 5.47). This cubic equation (Equation 5.9) is the simplest equation that describes the basic rules of sodium dynamics, and we use it to illustrate the behavior of the system.

Suppose a cell at resting potential receives an influx of positive charges from another neuron that raises its potential slightly above the threshold a. The solution moves upward, amplifying the signal, the first task of a functioning neuron. However, this neuron gets stuck at the higher equilibrium $v = 1$ (Figure 5.48), and therefore cannot be restimulated.

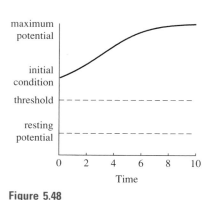

Figure 5.48
Solution without the slow mechanism for blocking sodium channels

The Mathematics of Sodium Channel Blocking

Without the slow mechanism for blocking sodium channels, a neuron could respond to only a single stimulus. How does blocking the sodium channels help the neuron function?

Let w represent the strength of the blocking mechanism. At $v = 0$, this mechanism is turned off, so $w = 0$. As v gets closer to 1, the mechanism becomes stronger, so w takes on larger and larger values. A simple equation that *seems* to model this behavior is

$$\frac{dw}{dt} = \varepsilon v$$

for positive ε. However, if v remains positive (above resting potential) for a long time, w might approach infinity. There must be an upper bound on the strength of the blocking mechanism.

To incorporate the requirement that the blocking mechanism has a maximum possible strength, we use an equation resembling Newton's law of cooling,

$$\frac{dw}{dt} = \varepsilon(v - \gamma w) \tag{5.10}$$

For every fixed value of v, this equation has an equilibrium at

$$w = \frac{v}{\gamma}$$

If $v = 0$, w approaches 0 and the mechanism does not operate. If, however, $v = 1$, w increases to an equilibrium value of $1/\gamma$. This represents the maximum possible strength of the blocking mechanism. A smaller γ produces a larger equilibrium value. If $\gamma = 1.0$ and $v = 1$, the equilibrium is $w = 1.0$. If $\gamma = 10.0$ and $v = 1$, the equilibrium is $w = 0.1$ (Figure 5.49).

The parameter ε does not affect the equilibrium level. Instead, it changes the rate at which the equilibrium is approached. A small value of ε produces a small rate of change and a slow response of w. Because the blocking mechanism is slow, ε has a fairly small value.

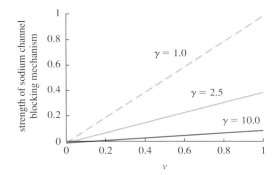

Figure 5.49
Equilibrium strength of blocking mechanism for fixed voltage

The FitzHugh-Nagumo Equations

How does the blocking mechanism modify the rate of change of the potential v? In other words, how do we couple the dynamics of v to the value of w? When w is large, the sodium channels have been blocked, stopping the entry of more sodium ions. The cell then tends to return to its resting potential. For convenience, we assume that the rate at which the potential decreases toward the resting potential is proportional to w. This gives the coupled system of equations

$$\frac{dv}{dt} = -v(v-a)(v-1) - w \quad (5.11)$$

$$\frac{dw}{dt} = \varepsilon(v - \gamma w) \quad (5.12)$$

These are called the **FitzHugh-Nagumo equations**. Can they reproduce the firing behavior of the neuron?

Figure 5.50 shows the nullclines and direction arrows for the FitzHugh-Nagumo equation with $\gamma = 2.5$ and $a = 0.3$, where we have chosen w as the vertical variable. The v-nullcline can be found by solving

$$\frac{dv}{dt} = -v(v-a)(v-1) - w = 0$$

for w, with solution

$$w = -v(v-a)(v-1)$$

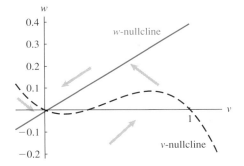

Figure 5.50
The nullclines and direction arrows for the FitzHugh-Nagumo equations

The w-nullcline is found by solving

$$\frac{dw}{dt} = \varepsilon(v - \gamma w)$$

for w, with solution

$$w = \frac{v}{\gamma}$$

In the case shown, the nullclines intersect only at the point $(0, 0)$. At this equilibrium, the cell is at resting potential ($v = 0$) and the sodium channel blocking mechanism is off ($w = 0$). In other words, the only equilibrium describes a cell completely at rest. Although we do not have the techniques to prove it (linear algebra is needed), this equilibrium is stable, meaning that solutions that start nearby return to the equilibrium. This hardly seems to be the recipe for useful dynamics.

We can draw direction arrows by reasoning about the equations. The strength of the blocking mechanism, w, is subtracted from the rate of change of v, so large values of w (above the v-nullcline) correspond to decreasing v and arrows that point to the left. Small values of w (below the v-nullcline) correspond to increasing v and arrows that point to the right. Similarly, $\varepsilon\gamma w$ is subtracted from the rate of change of w, so large values of w (above the w-nullcline) correspond to decreasing w and arrows that point down. Small values of w (below the w-nullcline) correspond to increasing w and arrows that point up (Figure 5.50).

What happens when the potential of the cell is raised above the threshold a? Mathematically, this corresponds to initial conditions $(v_0, 0)$ with $v_0 > a$. When we solve the equation for sodium channels opening alone (Equation 5.9), we find that the potential moves up to the equilibrium at $v = 1$ and remains there, leaving the cell paralyzed and useless.

The results of the same experiment with the slow mechanism for blocking sodium channels are quite different (Figure 5.51). Initially, the potential of the cell increases (the phase-plane trajectory moves to the right) because of the rapid opening of the sodium channels (Figure 5.51). Slowly, however, the blocking mechanism kicks in (the trajectory moves toward larger values of w). When the trajectory crosses the v-nullcline, the potential begins to decrease (the trajectory moves to the left), and when it crosses the w-nullcline, the blocking mechanism begins to turn off (the trajectory heads down). The trajectory undershoots $v = 0$ before the sodium channel blocking mechanism returns to 0. This system has the basic properties of a neuron, the ability quickly to amplify a signal and return to a state of readiness for the next signal. In addition, the mathematical analysis predicts an unexpected (but real) phenomenon, the undershoot of the potential.

Figure 5.51

Results of perturbing above resting potential

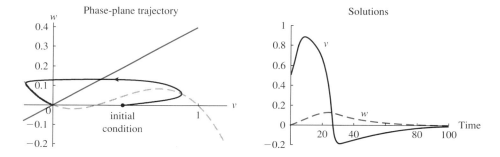

This undershoot results from the slowness of the mechanism for blocking the sodium channels.

If the perturbation of the cell is insufficient to open the sodium channels ($v_0 < a$), the cell does not amplify the signal. In this way, the cell acts as a filter that can ignore small stimuli (Exercise 6).

Weak Channel Blocking Mechanism

The larger the value of γ, the smaller the response of the sodium channel blocking mechanism. A cell without this mechanism (Equation 5.9) can respond only to a single stimulus. What happens if this mechanism is weak?

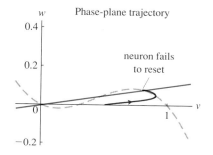
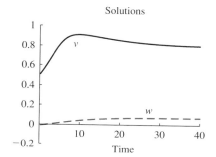

Figure 5.52
Dynamics with weak sodium channel blocking response

When the value of γ is large, the w-nullcline swings down to intersect the v-nullclines in three places (Figure 5.52). Although we cannot prove it without linear algebra, the central equilibrium is unstable and the outer two are stable. If we stimulate the cell by raising the potential of the cell above the threshold a, the potential rises further as the sodium channels open. However, the blocking mechanism cannot overcome the opening, and the trajectory is trapped at the upper equilibrium (Figure 5.52).

If the w-nullcline is low enough to cross the v-nullcline, the neuron does not fire. Further increases in γ change where the potential gets stuck but not the *qualitative* behavior of the cell. In effect, a cell with a weak blocking mechanism acts like a cell with no blocking mechanism at all.

The Effects of Constant Applied Current

We have imagined a resting cell receiving a single pulse of positively charged ions. One interesting experiment alters this situation by giving the cell a constant input of positive ions. Can the cell convert a constant input into a usable output?

The modification of the FitzHugh-Nagumo equations is not complicated. The current describes the rate at which ions enter the cell and is therefore proportional to the derivative of the potential. If the current applied is I_a, we add I_a to the rate of change of potential, finding

$$\frac{dv}{dt} = -v(v-a)(v-1) - w + I_a \qquad (5.13)$$

Because we are not studying actual numerical values, we have set the constant of proportionality in front of I_a to 1, assuming that 1 unit of current raises the potential by 1 unit in 1 unit of time. The constant is positive because current increases the potential.

The v-nullcline is

$$w = -v(v-a)(v-1) + I_a$$

and the w-nullcline is unchanged (Figure 5.53a). As in the original case (Figure 5.50), the nullclines intersect at only a single point. In this case, however, the equilibrium is unstable (we would again need linear algebra to prove it). A phase-plane trajectory and the corresponding values of v and w as functions of time are shown in Figure 5.53.

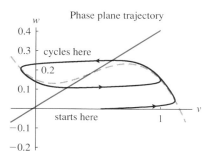
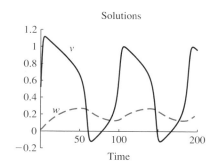

Figure 5.53
Dynamics with a constant applied current

The neuron shows a periodic "bursting" behavior. Real neurons produce periodic spikes when subjected to a constant current. This periodic output, which translates a current, an analog input, into periodic bursts, a digital output, can be used by the body as a pacemaker for a periodic process.

A cautionary note is in order: The equations we have been studying are a simplified version of the *four-dimensional* Hodgkin-Huxley equations (a system of four coupled autonomous differential equations). The value of studying this simplified model lies in its ability to mimic qualitatively more complex and accurate models while remaining easy to understand with phase-plane techniques.

SUMMARY

We used some basic facts about sodium channels to derive equations describing **action potentials**, the firing of neurons. In particular, we have seen how one mechanism that opens **voltage-gated sodium channels** can amplify an incoming stimulus and how a slower voltage-gated mechanism can block those channels. We combined these two processes into the **FitzHugh-Nagumo equations** and used phase-plane analysis to study these equations in several circumstances. The basic equations display excitability, the ability to amplify a signal temporarily and reset. Reducing the strength of the sodium channel blocking mechanism can make the cell unable to respond to more than a single stimulus. An excitable cell responds to a constant stimulus by producing a periodic sequence of action potentials.

5.7 EXERCISES

1. Give an equation like Equation 5.9 to describe a system with two thresholds. The system has a stable equilibrium at a resting potential at 0 but will be pushed to a higher positive equilibrium if the potential is raised above a particular positive threshold, and to a negative equilibrium if the potential is dropped below a particular negative threshold. Draw a phase-line diagram, and check that your equations match it.

2. Draw phase-line diagrams for Equation 5.9 for the following cases.
   a. $a = 0.01$. How might this neuron malfunction?
   b. $a = 0.5$.
   c. $a = 0.99$. Why might this neuron work poorly?

3. Use the method of separation of variables to solve

$$\frac{dw}{dt} = \varepsilon(v - \gamma w)$$

(Equation 5.10). Assume that v is constant. Show how the dynamics are slowed when ε is small, but that ε does not affect the equilibrium.

4. Assuming that w is constant (perhaps the sodium channel blocking mechanism is jammed in a particular state) and $a = 0.3$, figure out the dynamics of

$$\frac{dv}{dt} = -v(v - a)(v - 1) - w$$

(Equation 5.11), which can be thought of as a one-dimensional differential equation. Try it with $w = 0.01$, $w = 0.05$, and $w = 0.1$.

5. Draw direction arrows on the nullclines for the FitzHugh-Nagumo equations.

6. Sketch the phase-plane trajectory and solutions of the FitzHugh-Nagumo equations when the initial stimulus is less than the threshold.

7. Sketch the phase-plane diagram and solution for the FitzHugh-Nagumo equations for the following values of ε. (Think of changing ε as changing the direction arrows: The arrows are nearly horizontal when ε is small because w, the vertical variable, changes slowly.)
   a. ε very small
   b. ε rather large

8. With positive applied current there could be three intersections of the nullclines (as in Figure 5.52). Draw a phase-plane diagram illustrating this scenario, and take a guess at the dynamics. Try to make sense of the results biologically.

9. There could also be a single intersection with positive applied current, but on the rightmost decreasing part of the v-nullcline. Draw such a phase-plane diagram. Assuming that this equilibrium is stable, sketch the dynamics. Why might this cell also be thought of as excitable?

10. What happens to the phase-plane diagram and the cell if the applied current is negative? Can the cell lose its ability to respond if the applied current is negative and large?

11. Suppose that a higher applied current I_a in Equation 5.13 produces **faster** bursting. How might the body use this ability to translate signal strength into response speed?

12. **COMPUTER:** Use a computer to study a cell that is forced by an external current I_a that oscillates. Try different periods and amplitudes of the oscillation. What happens? Do you see any strange behaviors?

13. **COMPUTER:** The behaviors observed in Exercise 12 can occur when an object that naturally oscillates at one frequency is **forced** at a different frequency. For example, a spring that follows the equation

$$\frac{d^2x}{dt^2} = -x$$

naturally oscillates with period 2π. Forcing can be added with the modified equation

$$\frac{d^2x}{dt^2} = -x + A\cos\left(\frac{2\pi t}{T}\right)$$

Study this equation with the following values of T, trying a range of values of A from 0.1 to 10.0.
   a. $T = 2\pi$
   b. $T = \frac{\pi}{2}$
   c. $T = 4\pi$
   d. $T = 3.0$
   e. $T = 4.0$
   f. $T = 3.14$

Supplementary Problems for Chapter 5

EXERCISE 1
Consider the differential equation

$$\frac{dC}{dt} = 3(\Gamma - C) + 1$$

where C is the concentration of some chemical in a cell, measured in moles per liter, and Γ is a constant.
   a. What kind of differential equation is this? Explain the terms in the equation.
   b. Draw the phase-line diagram.
   c. Verify the stability of the equilibrium by using the derivative.
   d. Sketch solutions as functions of time starting from the initial conditions $C(0) = 0$ and $C(0) = \Gamma + 1$.

EXERCISE 2
Consider again the differential equation

$$\frac{dC}{dt} = 3(\Gamma - C) + 1$$

where Γ is a constant.
   a. Solve the equation for $C(0) = 0$.
   b. Check your answer.
   c. Find $C(0.4)$.
   d. After what time will the solution be within 5% of its limit?

EXERCISE 3
Consider the system of equations

$$\frac{d\Gamma}{dt} = (C - \Gamma) - \frac{\Gamma^2}{3}$$

$$\frac{dC}{dt} = 3(\Gamma - C) + 1$$

where C is the internal concentration of a chemical and Γ is the external concentration.

a. Describe in words the two processes affecting concentration in the external environment. How big is the external environment relative to the cell?
b. Draw a phase-plane diagram replete with nullclines, equilibria, and direction arrows.
HINT: It is easier to put C on the vertical axis.

EXERCISE 4
Describe conditions when you might observe population growth described by the following.

a. A one-dimensional autonomous differential equation.
b. A one-dimensional nonautonomous differential equation.
c. A one-dimensional pure-time differential equation.
d. A two-dimensional autonomous differential equation.

EXERCISE 5
Consider the differential equation

$$\frac{dV}{dt} = 12 - t^2$$

where $V(t)$ is volume in liters at time t and t is measured in seconds.

a. What kind of differential equation is this?
b. Graph the rate of change, and use the graph to sketch a graph of the solution.
c. At what time does V take on its maximum?
d. Suppose $V(0) = 0$. Use Euler's method to estimate $V(0.1)$.
e. Suppose $V(0) = 0$. Find the time T when $V(t)$ is 0 again.
f. What is the average volume between 0 and T?

EXERCISE 6
Consider the differential equation

$$\frac{dx}{dt} = 3x(x - 1)^2$$

a. Draw the phase-line diagram of this equation.
b. Find the stability of the equilibria using the derivative.
c. Sketch trajectories of x as a function of time for initial conditions $x(0) = -0.5$, $x(0) = 0.5$, and $x(0) = 1.5$.

EXERCISE 7
Suppose the per capita reproduction rate of a population of bacteria is given by

$$\text{per capita reproduction} = \frac{1}{\sqrt{b}}$$

where $b(t)$ is the population size at time t and t is measured in hours.

a. Find the differential equation describing this population.
b. Solve the equation and check your answer.
c. What is the population after 2 h if the population starts at $b(0) = 10{,}000$?
d. Does this population grow faster or slower than one growing exponentially? Why?

EXERCISE 8
Differential equations to describe an epidemic are sometimes given as

$$\frac{dS}{dt} = \beta(S + I) - cSI$$

$$\frac{dI}{dt} = cSI - \delta I$$

where S measures the number of susceptible people and I the number of infected people.

a. Compare these equations with the predator-prey equations. What is different in these equations? What biological process does each term on the right-hand sides describe?
b. Sketch the nullclines and find the equilibria if $\beta = 1$ and $c = \delta = 2$. (Draw only the parts where S and I are positive.)
c. Sketch direction arrows on your phase-plane diagram.

EXERCISE 9
Consider the following differential equation that describes the concentration of sodium ions in one of your neurons after you consume a bag of Doritos at time $t = 0$ seconds:

$$\frac{dN}{dt} = 2 - 10t$$

Suppose $N(0) = 50$ mmol/cm^3.

a. What kind of differential equation is this?
b. Sketch a graph of the rate of change as a function of time.
c. Sketch a graph of the concentration as a function of time.
d. Use Euler's method to estimate $N(0.1)$.
e. Find $N(1)$ exactly.
f. At what times is $N(t) = 50$?

EXERCISE 10
Consider the following differential equation that describes the concentration of sodium ions in a cell:

$$\frac{dN}{dt} = 2(N - 50) - (N - 50)^2$$

Assume this equation works only for $45 \leq N \leq 55$.

a. What kind of differential equation is this? What does each term mean?
b. Sketch a graph of the rate of change as a function of concentration.
c. Draw the phase-line diagram.
d. Sketch solutions starting from $N(0) = 48$, $N(0) = 51$, and $N(0) = 54$.
e. Suppose $N(0) = 51$. Estimate $N(0.1)$.
f. What method would you use to solve this equation?

EXERCISE 11
Two types of bacteria, a and b, are living in a culture. Suppose

$$\frac{da}{dt} = 2a\left(1 - \frac{a}{500} - \frac{b}{200}\right)$$

$$\frac{db}{dt} = 3b\left(1 + \frac{a}{1000} - \frac{b}{100}\right)$$

a. What kind of differential equation is this? Explain the terms.
b. Draw the phase-plane diagram, including nullclines, equilibria, and direction arrows.

EXERCISE 12
Suppose the scaled potential v and level of potassium channel opening w in a neuron were described by

$$\frac{dv}{dt} = v(1-v) - w$$

$$\frac{dw}{dt} = v - 2w$$

a. Draw a phase-line diagram for v assuming that w is fixed at 0.
b. Draw the phase-plane diagram for the full system, including equilibria, nullclines, and direction arrows.

EXERCISE 13
The density of sugar in a hummingbird's 20-mm-long tongue is

$$s(x) = \frac{1.2}{1.0 + 0.2x}$$

where x is measured in millimeters from the end of the tongue and s is measured in moles per meter.

a. Find the total amount of sugar in the hummingbird's tongue.
b. Find the average density of sugar in the tongue.
c. Compare the average with the minimum and maximum densities. Does your answer make sense?

EXERCISE 14
The length L of a microtubule is found to follow the differential equation

$$\frac{dL}{dt} = -L(2-L)(1-L)$$

where L is measured in microns.

a. Draw the phase-line diagram.
b. Check the stability of the equilibria using the derivative.
c. Sketch trajectories starting from $L(0) = 0.5$, $L(0) = 1.5$, and $L(0) = 2.5$.

EXERCISE 15
A different lab finds that

$$\frac{dL}{dt} = -2.0L + 5.4LE$$

$$\frac{dE}{dt} = 3.5 - LE$$

where E is the level of some component of the microtubules and L is the length of the microtubule.

a. Explain the terms in these equations.
b. Draw the phase-plane diagram, including nullclines, equilibria, and direction arrows.

EXERCISE 16
Consider the differential equation that describes a population of mathematically sophisticated bacteria,

$$\frac{db}{dt} = b \ln\left(\frac{2b+1}{2+b}\right)$$

a. What kind of differential equation is this?
b. Draw the derivative as a function of the state variable, and draw the phase-line diagram.
c. Sketch trajectories starting from $b(0) = 0.8$ and $b(0) = 1.2$.
d. Check the stability of the equilibrium at $b = 0$ by taking the derivative of the rate of change.
e. Use Euler's method to estimate $b(0.01)$ if $b(0) = 0.5$.

Projects for Chapter 5

PROJECT 1
Exercise 10 (Section 5.6) presents one possible model of adaptation. This project studies a simpler alternative model (proposed by H. G. Othmer). Because this model is so simple, we can compare the results of phase-plane analysis with actual solutions of the equations.
Consider the equations

$$\frac{dp}{dt} = k(H - A - p)$$

$$\frac{dA}{dt} = \varepsilon(H - A)$$

where p represents the response of the cell and H the external condition driving the response. The variable A describes some internal state of the cell. First, suppose that H is constant.

a. Find the nullclines and equilibria and draw the phase plane, including direction arrows.
b. The equation for A is the same as Newton's law of cooling. Write down the solution for A with an arbitrary initial condition.
c. Plug this solution into the equation for p. There is a clever way to solve the resulting nonautonomous equation. Make up a new variable $q(t) = e^{kt} p(t)$. With a bit of manipulation, you can write a pure-time differential

equation for q. Solve this equation and find the solution for $p(t)$.

d. Plot this solution as a phase-plane trajectory for several different initial conditions. Are the results consistent with the phase plane?

Using these results, we can study what happens when H changes. Suppose first that the cell is at equilibrium with $H = 1$, and then H jumps from 1 up to 10. Set $k = 1$ and $\varepsilon = 0.1$.

e. How long will it take before A has roughly reached its new equilibrium value?

f. What is the value of p at this time? How long will it take p to nearly reach its equilibrium value again?

g. Suppose now that $H = 1$ for a time T, jumps to $H = 10$ for a time T, jumps back to 1, and so forth. Experiment with different values of T. How large must T be before the cell produces a healthy response to each change? What happens when T is much smaller than this value? What might the cell do to be able to respond more quickly?

PROJECT 2

Consider an interaction between two mutually inhibiting proteins with concentrations x and y, given by the differential equations

$$\frac{dx}{dt} = f(y) - x$$

$$\frac{dy}{dt} = g(x) - y$$

Both $f(y)$ and $g(x)$ are decreasing functions. (Based on the research of J. Cherry.)

a. Explain each of the terms in these equations.

b. Try to imagine a biological situation they might describe.

c. Sketch the nullclines (remember that both f and g are decreasing).

d. Show that equilibria occur where $f[g(x)] = x$. What discrete-time dynamical system shares the equilibria of the system of differential equations?

Next try the following steps for functions of three different forms:

Case 1: $f(y) = \dfrac{1}{1+\alpha y}, g(x) = \dfrac{1}{1+\alpha x}$

Case 2: $f(y) = e^{-\alpha y}, g(x) = e^{-\alpha x}$

Case 3: $f(y) = \dfrac{1}{1+\alpha y^2}, g(x) = \dfrac{1}{1+\alpha x^2}$

Experiment with different values of α.

a. Find the function $f[g(x)]$ and graph it on a cobwebbing diagram. Does the equilibrium look stable?

b. Draw the nullclines and direction arrows for the differential equation. Is the equilibrium stable?

c. Why does the stability you "found" in **a** match that in **b**?

d. Try to figure whether it is possible to have three equilibria (it is possible in Cases 2 and 3).

Why might it be important for a biological system to have three equilibria? How could it operate as a switch?

Bibliography for Chapter 5

Section 5.1

Anderson, R. M., and R. M. May. *Infectious Diseases of Humans.* Oxford University Press, Oxford, England, 1992.

Blanchard, P., R. Devaney, and G. Hall. *Differential Equations.* Brooks/Cole, Pacific Grove, Calif., 1998.

Section 5.2

Edelstein-Keshet, L. *Mathematical Models in Biology.* Random House, New York, 1988.

Murray, J. D. *Mathematical Biology.* Springer-Verlag, New York, 1993.

Section 5.6

Rinzel, J. Electrical excitability of cells, theory and experiment: Review of the Hodgkin-Huxley foundation and an update. *Bulletin of Mathematical Biology* 52:5–23, 1990.

Chapter 6

Probability Theory and Descriptive Statistics

The dynamical models studied in Chapters 1 through 5 of this course, discrete-time dynamical systems and differential equations, are **deterministic**. The state at one time exactly **determines** the state at every future time. When population size is accurately described by a discrete-time dynamical system, the population size at any future time can be computed exactly by repeatedly applying the updating function starting from a known initial condition. When population size is accurately described by a differential equation, the population size at any future time can be computed exactly by integrating the differential equation starting from a known initial condition.

These accurate models and exact solutions are a mathematical idealization. If we track several experimental populations, each experiencing "identical" conditions, the results will not be identical. Various more or less "random" events can affect the outcome. Lower temperatures might decrease reproduction. An infection might diminish one population. All the offspring in one experiment might be males. Describing these chance events and deducing their consequences is the realm of **probability theory**.

Conversely, when faced by data that seem to include these "noisy" factors, our goal of figuring out what happened is more difficult. How do we distinguish the signal from the noise? This is the realm of **statistics**, the art of describing and reasoning about real data. The final three chapters are motivated by a single principle:

Correctly understanding and applying statistics requires understanding the underlying model.

So far, we have learned how to write and analyze models of important biological situations. This has given insight into the implications of different sets of assumptions. Our goal now is to use modeling to begin the analysis of data.

These final three chapters are concerned with developing the language and techniques of **probability theory** and **descriptive statistics**. Probability theory provides a precise mathematical description of *chance*, and descriptive statistics (such as the average) give ways to summarize results.

6.1 Introduction to Probabilistic Models

We introduce the basic terminology of **probability theory** and the distinction and relation between probability and **statistics**. Using examples from population growth, we illustrate three types of stochastic model and the sorts of questions probabilists and statisticians might ask about them.

Probability and Statistics

Describing chance events and deducing their consequences is the realm of **probability theory**. To avoid the ambiguous colloquial connotations of the words "random" and "chance," we use the word **stochastic** to describe unpredictable effects. A stochastic model describes a biological process that includes the "chance" events. The goals of a stochastic model are the same as the goals of a deterministic model: to produce an accurate description of a biological process and to deduce its consequences.

Conceptually, using probability theory to analyze a stochastic model is similar to using discrete-time dynamical systems and differential equations to analyze a deterministic model. First, we must come up with a sufficiently detailed description of a biological system to write a model. Then, we use appropriate mathematical tools to figure out what the model does. In the probabilistic case, however, *describing* the results is more difficult because a single number will not suffice.

Because of this difficulty with description, probabilistic systems are inextricably linked with **statistics**. Suppose we ran an experiment 1000 times, perhaps allowing a bacterial population to grow for 24 h. We would get 1000 different, though related, numbers. Listing them all would convey little information. It would be better to state one or two numbers, perhaps the average and minimum, to summarize the data in a meaningful and appropriate way. The choice and computation of these numbers is the realm of **descriptive statistics**, and the numbers chosen are called *statistics*. More precisely,

■ **Definition 6.1** A **statistic** is one number that summarizes many numbers. ■

Even if we had computer-like minds and could contemplate enormous data sets in a single thought, we need statistics to *compare* data sets. Two sets of 1000 measurements might have no exact values in common but could be essentially identical when considered in their entirety. This similarity can be revealed by comparing appropriate statistics. The average, for example, might be the same for each data set.

We often wish to do more with our results than just summarize or describe them. Instead, we seek to use them to make **deductions** or **inferences** about the underlying biological process. If the world were deterministic, this task would be relatively easy. In a stochastic world, this task is more difficult, requiring the techniques of **inferential statistics**. These techniques use statistical descriptions of results to make inferences about what happened. For example, we might wish to infer a numerical estimate (such as the per capita reproduction) or the answer to a biological question (for example, is one population growing more rapidly than another?). The relations among probability theory, descriptive statistics, and inferential statistics are illustrated in Figure 6.1.

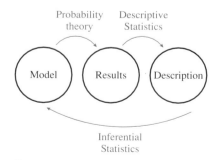

Figure 6.1
The relations among probability theory, descriptive statistics, and inferential statistics

In Sections 6.1–6.3, we develop stochastic models of the three fundamental processes studied in this book: population growth, diffusion, and selection. For each model, we pose questions that we answer later using probabilistic and statistical methods.

Stochastic Population Growth

Recall the discrete-time dynamical system describing population growth,

$$b_{t+1} = rb_t$$

(Equation 1.28). The new population can be computed as the old population multiplied by the per capita reproduction r. In a real-world system, the per capita reproduction varies stochastically. We indicate this by rewriting the equation:

$$b_{t+1} = r(t)b_t \tag{6.1}$$

The dependence of the per capita reproduction r on the time t indicates that the reproduction varies over time, perhaps because of such stochastic factors as temperature, availability of food, and predation. We do not expect, however, that it varies in any predictable way.

Figure 6.2 illustrates three related situations. In Figure 6.2a, the per capita reproduction is the fixed constant $r = 1.1$. The population size increases by 10% each year, and the population b_t after t years can be found with the exact formula

$$b_t = 1.1^t b_0$$

(Equation 1.29). If $b_0 = 1$, $b_{50} = 117.4$. Furthermore, the logarithm of population size follows a straight line with equation

$$\ln(b_t) = \ln(1.1^t b_0) = \ln(b_0) + t \ln(1.1) \approx 0.0953t$$

if $b_0 = 1$.

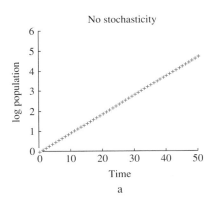
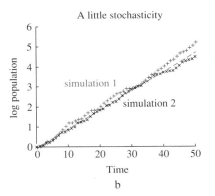
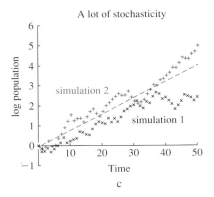

Figure 6.2
Stochastic population growth

Suppose instead that the per capita reproduction r only *averages* 1.1. If it is chosen from the range 1.0 to 1.2, then there are some "bad" years when the population does not grow at all ($r = 1.0$) and some "good" years when the population grows by 20% ($r = 1.2$). The results are similar to those without stochasticity, but the population wiggles around the line (Figure 6.2b). If we run the same experiment twice, we get different results because different stochastic factors affect the output. The per capita reproduction could be even more variable, ranging from 0.7 (population shrinks by 30%) to 1.5 (population grows by 50%). Again, the population growth roughly follows the straight line, but it jumps around even more (Figure 6.2c).

As probabilists, we want to ask and answer questions about these results. In the deterministic case (Figure 6.2a), we can ask what the population is after 50 yr or how long it would take the population to reach exactly 100 (Exercise 1). In the stochastic cases, we can ask about the average population after 50 yr, or the *probability* that the population is greater than 125 after 50 yr (which is impossible in the deterministic case). Alternatively, we can ask how long it would take, on average, for the population to reach 100. As we can see in Figure 6.2c, the population declines for the first few years. What is the probability that the population is less than the original population after 10 yr?

As statisticians, we are faced with different problems. Instead of a model, we have only the data in Figure 6.2. Our goal is to describe and answer questions about what happened. For example, we might want to know the average per capita reproduction over this time. This is straightforward with deterministic data (Figure 6.2a). We can compute the slope by dividing the change in log population by the change in time. With the two points $[0, \ln(b_0)] = [1, \ln(1)] = (1, 0)$ and $[50, \ln(b_{50})] = [50, \ln(117.4)] = (50, 4.765)$, we find the slope,

$$\text{slope} = \frac{4.765 - 0}{50 - 0} = 0.0953$$

This is the natural log of the per capita reproduction, $e^{0.0953} = 1.1$.

In the stochastic cases, however, our task is not so simple. How do we go about drawing a line that seems to "fit" the data? Is the slope of that line a good estimate of the average logarithm of the per capita reproduction?

Alternatively, a scientist might want to answer specific questions about the data. From looking at the two populations in Figure 6.2c, it might seem that one is growing faster than the other. Can we use the data to tell whether it "really" is? More generally, the data could be used to ask whether the simple growth model (Equation 6.1) is an adequate description. Does the per capita reproduction become smaller for larger populations? These sorts of questions lie at the heart of the scientific process: using data to test hypotheses about biological mechanisms.

Populations can grow by reproduction, immigration, or both. In real world situations, both reproduction and immigration are stochastic. The simplest model of deterministic immigration into a population of size N_t at time t is

$$N_{t+1} = N_t + 1$$

meaning that exactly one immigrant arrives per day. This population grows linearly with slope 1 (Figure 6.3a). Suppose instead that these animals travel in pairs, and that one pair, or two individuals, arrives each day with probability 0.5. This population, denoted S_t at time t, obeys the rule

$$S_{t+1} = \begin{cases} S_t + 2 & \text{with probability 0.5} \\ S_t & \text{with probability 0.5} \end{cases} \quad (6.2)$$

This can be thought of as a coin tossing experiment, where "heads" means that a pair of immigrants arrives and "tails" means that no pair arrives. Two simulations are shown in Figure 6.3b. Although we expect 100 immigrants after 100 days, there are 96 in one simulation and 106 in the other because of the vagaries of chance.

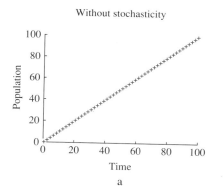
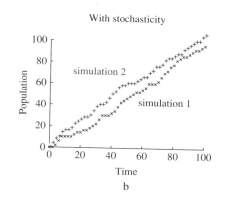

Figure 6.3
Stochastic immigration

As probabilists, our goal is to describe what happens to this population after a given number of days. We might want to know how likely it is that exactly 86 immigrants arrived during the first 100 days, or that more than 120 immigrants arrived in this same time. These questions would be silly in the deterministic case; we know that exactly 100 immigrants arrived. A useful tool for describing the results of many stochastic simulations is the **histogram** (Figure 6.4). The height of each bar indicates how many of 1000 simulations ended up exactly at the population shown on the x-axis. For example, 24 populations ended up with exactly 86 immigrants.

As statisticians, our task is again different. Suppose we have observed one or more populations over a period of time, as shown in Figure 6.3b. We might then try to estimate the average rate at which immigrants arrive. Again, this is simple in the deterministic case; the line with slope 1 indicates that exactly 1 immigrant arrived per year. The data in Figure 6.3b seem roughly to follow a line, but which line?

Alternatively, we might wish to ask particular scientific questions about the data. Is the immigration rate higher in one population than the other? Is it more likely that an immigrant arrives in the second population in years when an immigrant arrives in the first? Is the immigration rate becoming higher over time?

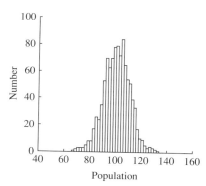

Figure 6.4
A summary of the results of many simulations of the immigration model

Markov Chains

In our stochastic equations for population growth by reproduction (Equation 6.1) and immigration (Equation 6.2), population growth does not depend on population size. The per capita reproduction $r(t)$ depends on only external factors and does not become smaller when the population is large. Pairs of immigrants are no more likely to join the population described by Equation 6.2 when the population is small than when it is large. As in realistic deterministic models of population growth, however, the growth of a population often depends on the size of a population. One tool for describing and analyzing this situation in the presence of stochastic effects is called a **Markov chain**.

■ **Definition 6.2**   A discrete-time Markov chain is a stochastic dynamical system in which the probability of arriving in a particular state at a particular time depends on the state at the previous time.

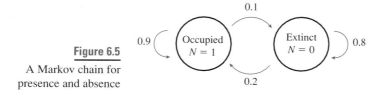

Figure 6.5
A Markov chain for presence and absence

For example, the probability that the population is 100 this year depends on the population last year.

A simple and widely applicable Markov chain follows a population on an island, noting each year only whether the population is extant or extinct. This description requires two values: 0 (to represent absence) and 1 (to represent presence). The dynamics depend on the probabilities of transitions between these two states. Suppose that an empty island is settled in a given year with probability 0.2 and remains empty otherwise, and that the population on an occupied island goes extinct with probability 0.1 in a given year and remains occupied otherwise (Figure 6.5). Expressed as equations, we have

$$N_{t+1} = 1 \begin{cases} \text{with probability } 0.2 \text{ if } N_t = 0 \\ \text{with probability } 0.9 \text{ if } N_t = 1 \end{cases} \quad (6.3)$$

$$N_{t+1} = 0 \begin{cases} \text{with probability } 0.8 \text{ if } N_t = 0 \\ \text{with probability } 0.1 \text{ if } N_t = 1 \end{cases} \quad (6.4)$$

We will soon develop a more convenient probabilistic language for writing and describing this model. An island following this model might start out occupied, become unoccupied for nine years, be occupied for five years, and so forth (Figure 6.6).

Presented with this model, a probabilist might ask how often or for how long the island will be occupied. Presented with the data in Figure 6.6, a statistician might attempt to estimate the probability that an occupied island becomes empty. Scientific questions about the data might include whether occupied islands are more likely to become unoccupied after they have been occupied for a while, or whether conditions on the island are getting better or worse over time.

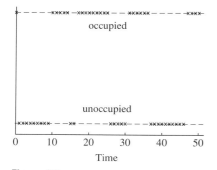

Figure 6.6
An island switching between occupied and unoccupied according to a Markov chain

SUMMARY

We compared and contrasted **deterministic models**, in which the future is exactly determined, with **stochastic models**, in which "chance" events make the future unpredictable. It is the realm of **probability theory** to write and analyze mathematically models that describe these situations. Description and comparison of large quantities of data require **statistics**, single numbers that summarize important properties of the data. **Inferential statistics** are used to figure out what sort of process might have generated that data. We examined three generalizations of deterministic models of population growth: a model in which per capita reproduction changes stochastically over time, a model in which immigration is a random event, and a **Markov chain** describing how an island might alternate stochastically between being occupied and being empty.

6.1 EXERCISES

1. a. How long will it take a population reproducing with $r = 1.1$ to reach 100 starting from a population of 1.0?
    b. Estimate from Figure 6.2b how long it took each of the two populations to reach 100 (remember that the logarithm of the population size has been plotted).
    c. Estimate from Figure 6.2c how long it took each of the two populations to reach 100.

2. Suppose the growth of two populations is given in the following table.

Generation	Population 1	Population 2
0	640	640
1	960	943
2	1440	1349
3	2160	2084
4	3240	3210
5	4860	4840

    a. Graph each population.
    b. Find the per capita reproduction for each population in each generation.
    c. Which population acts deterministically? What updating function describes it?
    d. How would you describe the behavior of the other population?

3. Suppose a population grows with per capita reproduction $r(t)$, which alternates between 0.6 and 1.5. Is the average per capita reproduction higher than 1? Do you think this population will grow or shrink?

4. Think of two biological factors that are neglected in the stochastic model $b_{t+1} = r(t)b_t$ (Equation 6.1).

5. Suppose the growth of two populations by immigration is given in the following table.

Generation	Population 1	Population 2
0	64	64
1	96	100
2	128	142
3	160	164
4	192	200
5	224	220

    a. Graph each population.
    b. Find the number of immigrants received by each population in each generation.
    c. Which population acts deterministically? What updating function describes it?
    d. How would you describe the behavior of the other population?

6. Suppose a population receives two immigrants in a day with probability 0.5 and loses one emigrant with probability 0.5. What is the average number of immigrants each day? If the population started at 60, what population would you expect after 40 days?

7. Think of three factors neglected in the stochastic model of immigration (Equation 6.2).

8. Describe what would happen to a population following the Markov chain for occupation and extinction (Equations 6.3 and 6.4) for the following cases.
    a. The probability of an empty island being occupied and the probability of an occupied island becoming empty are both 1.
    b. The probability of an empty island being occupied is 0 and the probability of an occupied island becoming empty is less than 1.

9. Suppose two islands are described in the following table.

Year	Island 1	Island 2
0	occupied	occupied
1	extinct	occupied
2	occupied	extinct
3	extinct	occupied
4	occupied	occupied
5	occupied	extinct
6	occupied	occupied
7	extinct	occupied
8	occupied	extinct

    a. Illustrate what is happening on each island with a graph.
    b. Does one island have a pattern? Can it be described deterministically?
    c. How would you describe the behavior of the other island?
    d. What would you expect to happen on each island in year 15?

10. Think of two factors neglected in the Markov chain model of presence and absence on an island (Equations 6.3 and 6.4).

11. **COMPUTER:** Your computer should have several types of **random number generators**. One type chooses a random number between an upper and a lower limit. Consider a population growing according to the rule

    $$b_{t+1} = rb_t$$

    where r is a random number chosen from the range 0.5 to 1.5. Start two simulations with $b_0 = 100$, and run them for

50 generations. Try this a few times until you get a plot that looks interesting. Why are the two solutions so different? If someone showed you these data without telling you they were generated on a computer, how would you describe and interpret the results?

12. **COMPUTER:** Consider a population growing due to immigration:

$$N_{t+1} = \begin{cases} N_t + 1 & \text{with probability } 0.5 \\ N_t & \text{with probability } 0.5 \end{cases}$$

We can define an updating function,

$$g(N) = N + p$$

where p is a random number chosen by the computer to take on the value 0 with probability 0.5 and 1 with probability 0.5. Generate two 50-generation solutions starting from populations of 0. If someone showed you these data, how would you describe and interpret the results?

13. **COMPUTER:** Consider a population on an island described by $M = 1$ if the island is occupied and $M = 0$ if the island is unoccupied. Suppose this population follows the rule,

$$M_{t+1} = 1 \begin{cases} \text{with probability } 0.3 \text{ if } M_t = 0 \\ \text{with probability } 0.9 \text{ if } M_t = 1 \end{cases}$$

$$M_{t+1} = 0 \begin{cases} \text{with probability } 0.7 \text{ if } M_t = 0 \\ \text{with probability } 0.1 \text{ if } M_t = 1 \end{cases}$$

There is a way to program the updating function with a random number by

$$h(M) = (1 - M)q_{0.3} + Mq_{0.9}$$

where $q_{0.3}$ is a random number that is equal to 1 with probability 0.3 and 0 with probability 0.7, and $q_{0.9}$ is a random number that is equal to 1 with probability 0.9 and 0 with probability 0.1. If $M = 0$, then $h(M) = q_{0.3}$ and M switches to 1 with probability 0.3, and if $M = 1$, then $h(M) = q_{0.9}$ and M remains at 1 with probability 0.9. Generate a solution for 100 generations of this population starting from the occupied state. When does the population first go extinct? For how long? What is the final state of the population? Is the island occupied more often than unoccupied? Does this make sense?

6.2 Stochastic Models of Diffusion

We have developed two types of deterministic model to describe the movement or diffusion of a chemical: a discrete-time dynamical system (the lung model of Section 1.10) and an autonomous differential equation (diffusion across a membrane in Section 4.2). These models are appropriate only for large numbers of molecules because the movement of individual molecules is governed by chance. We develop three stochastic models in this section: a discrete-time dynamical system for the probability that a molecule remains inside a cell after a certain amount of time, a differential equation describing the same process, and a Markov chain to model a molecule that can reenter the cell.

Stochastic Diffusion: Discrete-Time Model

At a macroscopic level, molecules come in huge numbers like Avogadro's number, 6.02×10^{23}. Within a single cell, however, important molecules can come in much smaller numbers. When there are very few molecules, chance effects can be important. For example, if only ten molecules of some enzyme exist in a cell, it is quite possible that all of them will wander simultaneously into one half of the cell. It is virtually impossible that all of the many millions of water molecules in the cell would do the same.

To study enzymes or toxins that can have important effects on cell function even in small numbers, we must use stochastic models of diffusion. These models take into account the fact that molecules bounce around unpredictably. Stochastic models are also appropriate for those plentiful molecules that take on functional importance on the rare occasions when they bind with particular proteins.

6.2 Stochastic Models of Diffusion

Suppose a single molecule of some toxin is drifting around in a cell. It leaves during any 1-min interval with probability 0.1 and cannot reenter once it has left (Figure 6.7).

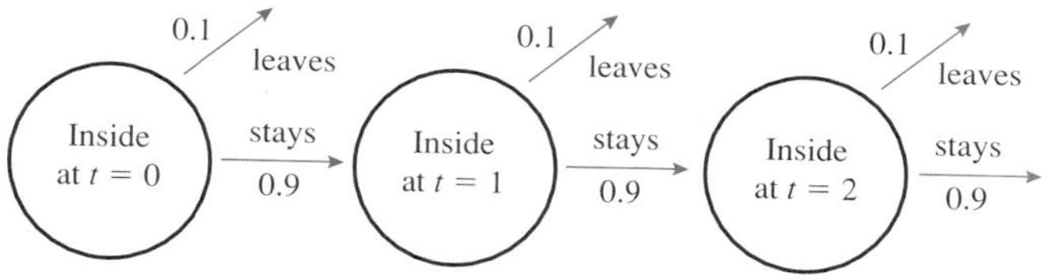

Figure 6.7 A stochastic, toxic molecule

The fates of several such molecules are given in the following table. The first molecule left at time 10, the second at time 2, and so forth. Molecules 5 and 9 remained inside at time 10. But what is the *probability* that a molecule is still in the cell after 10 minutes?

	Molecule									
Time	1	2	3	4	5	6	7	8	9	10
1	in	in	in	in	in	in	in	in	in	in
2	in	out	in	in	in	in	out	in	in	in
3	in	out	in	in	in	in	out	in	in	in
4	in	out	out	in	in	in	out	in	in	in
5	in	out	out	in	in	in	out	in	in	in
6	in	out	out	out	in	in	out	out	in	in
7	in	out	out	out	in	out	out	out	in	in
8	in	out	out	out	in	out	out	out	in	out
9	in	out	out	out	in	out	out	out	in	out
10	out	out	out	out	in	out	out	out	in	out

The first step in writing a model is to define the **state variable**, the quantity we want to follow. Let p_t be the probability that the toxic molecule is still in the cell after t minutes. We can derive a discrete-time dynamical system describing this new kind of state variable. The molecule begins inside the cell, so $p_0 = 1$ (a probability of 1 corresponds to certainty). After 1 min, the molecule has left with probability 0.1 and remains with probability 0.9, so $p_1 = 0.9$. After the second minute, the molecule is inside only if it was inside at time 1 (probability 0.9) and remained there during the next minute (probability 0.9). Therefore,

$$p_2 = 0.9 \cdot 0.9 = 0.81$$

In general, we have

$$p_{t+1} = 0.9 p_t \qquad (6.5)$$

The probability that the molecule is inside at the beginning of 1 min is 90% of the probability that it was inside at the beginning of the previous minute.

This updating function has the same form as the equation $b_{t+1} = rb_t$, which describes a bacterial population (Equation 1.28) with $r = 0.9$. With the initial condition $p_0 = 1$, the solution is

$$p_t = 0.9^t \qquad (6.6)$$

The probability that a molecule is inside the cell decreases exponentially. The more time passes, the more cetain we are that the molecule has left.

What does this probability mean? If a large number of molecules follow this process, a *fraction* of approximately p_t will be inside the cell after t minutes. For example, $p_2 = 0.81$, so about 81 of 100 molecules will remain inside the cell after 2 min. After 10 min, $p_{10} = 0.9^{10} = 0.349$, so we expect about 35 molecules of 100 to remain. These predictions are more accurate the larger the number of molecules. Starting with 100 molecules, we predict that

$$\text{number inside of } 100 = 100 p_t = 100 \cdot 0.9^t$$

Figure 6.8a compares a computer simulation of 100 toxic molecules with the predicted number. Although the two curves match pretty well, the actual number of molecules inside is not exactly equal to the predicted number. Starting with 10^4 molecules, we predict that

$$\text{number inside of } 10^4 = 10^4 p_t = 10^4 \cdot 0.9^t$$

Figure 6.8b compares a computer simulation of 10^4 molecules with the predicted number. The prediction is too close to distinguish on the graph, but it is still not exact.

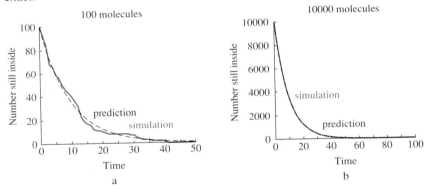

Figure 6.8
The fate of many toxic molecules

Our result does not tell us what a particular molecule will do because the behavior of a particular molecule depends on stochastic factors. Each molecule in the table behaves quite differently from the others. Molecule 2 left after the first minute; molecules 5 and 9 did not leave during the first 10 min. When probabilities are very small, however, certain possibilities can be ruled out. After 100 min, the probability of finding a particular molecule inside is $p_{100} = 0.9^{100} \approx 2.656 \times 10^{-5}$. Only about 26 of 1 million molecules will remain. If toxic effects are not observed in 100 min, we can be pretty sure they will never occur.

Stochastic Diffusion: Differential Equation

We can describe the same process more precisely with a differential equation. Nothing constrains molecules to leave a cell at either the beginning or end of a minute. Actual departure times might be as in the table at the top of the next page. For example, molecule 1 left at time 5.85509 and remained outside thereafter. How can we describe this process mathematically?

Define the state variable $P(t)$ as the probability that the molecule remains inside the cell at time t. To write a differential equation, we must compute the probability that it is inside a short time, Δt, later. The basic principle is that a

6.2 Stochastic Models of Diffusion 457

Molecule	Time	Molecule	Time
1	5.85509	6	4.85506
2	2.71001	7	11.1952
3	4.85506	8	12.0252
4	10.52519	9	2.99502
5	3.78004	10	35.3734

molecule is inside at $t + \Delta t$ if it was inside at time t and did not leave in the next Δt, or

probability inside at $t + \Delta t$ = (probability inside at t)
$\times$ (probability did not leave in time Δt)

What is the probability the molecule did not leave in time Δt? It is certainly true that

probability did not leave in time Δt = 1 − (probability did leave in time Δt)

because it either does or does not leave. We make the assumption that

probability did leave in time $\Delta t = 0.1 \Delta t$

The factor 0.1 is the *rate* at which an individual molecule leaves. The units of this probabilistic rate are 1 per time, and we find totals with the formula

total = rate $\times$ time

The total probability that something happened in the short time Δt is the rate 0.1 times the time Δt.

Putting these pieces together, we get

probability inside at $t + \Delta t$ = (probability inside at t)
$\times$ (probability did not leave in time Δt)
$$P(t + \Delta t) = P(t)(1 - 0.1\Delta t)$$

This model would be exact if we let Δt approach 0. Taking the limit, we get

$$\lim_{\Delta t \to 0} P(t + \Delta t) = \lim_{\Delta t \to 0} P(t)(1 - 0.1\Delta t) \quad \text{take limit}$$
$$P(t) = P(t) \lim_{\Delta t \to 0} (1 - 0.1\Delta t) \quad \text{assume that } P(t) \text{ is continuous}$$
$$P(t) = P(t) \cdot 1.0 \quad \text{substitute } \Delta t = 0 \text{ into } 1 - 0.1\Delta t$$

This result is true but completely useless.

Instead, we use some algebra to rewrite the equation in the form of the *derivative of P*:

$$P(t + \Delta t) = P(t)(1 - 0.1\Delta t) \quad \text{original equation}$$
$$P(t + \Delta t) = P(t) - 0.1P(t)\Delta t \quad \text{multiply out right-hand side}$$
$$P(t + \Delta t) - P(t) = -0.1P(t)\Delta t \quad \text{subtract } P(t) \text{ from both sides}$$
$$\frac{P(t + \Delta t) - P(t)}{\Delta t} = -0.1P(t) \quad \text{divide both sides by } \Delta t$$

The left-hand side matches the definition of the derivative. If we now take the limit as $\Delta t \to 0$, we get

$$\lim_{\Delta t \to 0} \frac{P(t + \Delta t) - P(t)}{\Delta t} = \lim_{\Delta t \to 0} -0.1 P(t)$$

$$\frac{dP}{dt} = -0.1 P(t) \tag{6.7}$$

because we have eliminated Δt from the right-hand side.

This differential equation matches Equation 4.4, which describes bacterial death, with the exponential solution

$$P(t) = P(0) e^{-0.1t}$$

The initial condition is $P(0) = 1$ because we are certain that the molecule begins inside the cell, so the solution is

$$P(t) = e^{-0.1t} \tag{6.8}$$

At time $t = 10$, the probability a molecule remains inside is $P(10) = e^{-1} \approx 0.368$. In terms of fractions, this means that about 37/100 molecules would remain inside. After 20 min, $P(20) = 0.135$, meaning that only 13 or 14 of 100 molecules would remain. The prediction for 100 molecules is

$$\text{number inside} = 100 e^{-0.1t}$$

The behavior of 100 simulated molecules is compared with the predicted number in Figure 6.9. The probability that the molecule is inside the cell decays exponentially. Individual molecules, however, do not leak out gradually like the probability, but depart sharply at particular instants.

What is the difference between the discrete-time model and the differential equation? In the first, we check the position of the molecule only at discrete times, or once per minute. In the second, we check the position of the molecule continuously, and record exactly when it leaves (compare the two tables just given).

Although it might seem that we have solved the problem completely, there are many other relevant questions. For example, we might want to compute the probability that exactly 10 of 100 molecules have left after 20 min. This requires the **Poisson distribution**, which we study in detail in Section 7.7.

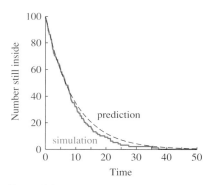

Figure 6.9

The fates of 100 toxic molecules revisited

Stochastic Diffusion: Markov Chain Model

Suppose now that our molecule can reenter the cell. Possible data might appear as in the following table.

	Time									
Molecule	1	2	3	4	5	6	7	8	9	10
1	in	in	in	in	in	in	in	in	in	out
2	in	out	out	out	out	out	out	out	out	out
3	in	in	in	in	in	out	out	out	out	out
4	in	in	out	out	out	out	out	out	out	out
5	in	in	in	in	in	in	in	in	in	in
6	in	in	in	in	in	in	in	in	out	out
7	in	in	out	out	out	out	out	out	out	out
8	in	in	in	in	out	out	out	out	out	out
9	in	in	in	in	in	in	in	in	in	in
10	out	out	out	out	out	out	out	out	out	out

We want to know how likely we are to find such a molecule inside. Let $C_t = 1$ indicate that the molecule is inside at time t and $C_t = 0$ indicate that the molecule is outside at time t. Make the following assumptions: If the molecule is inside, it leaves during the next minute with probability 0.2 and otherwise remains inside; if it is outside, it enters during the next minute with probability 0.1 (Figure 6.10). This molecule, perhaps a toxin, is more likely to leave when inside than to enter when outside. This could be a consequence of filtering by the cell membrane or of a larger volume outside than inside the cell.

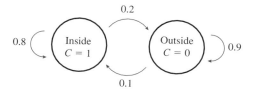

Figure 6.10
A Markov chain model of a molecule

Our model can be written mathematically as

$$C_{t+1} = 1 \begin{cases} \text{with probability } 0.1 \text{ if } C_t = 0 \\ \text{with probability } 0.8 \text{ if } C_t = 1 \end{cases}$$

$$C_{t+1} = 0 \begin{cases} \text{with probability } 0.9 \text{ if } C_t = 0 \\ \text{with probability } 0.2 \text{ if } C_t = 1. \end{cases}$$

The molecule follows a rule much like the one the occupied or empty island follows (Equations 6.3 and 6.4). What is the long-term probability of finding the molecule inside the cell?

Let p_t be the probability that the molecule is inside at time t. The probability that it is outside is then $1 - p_t$. There are two ways the molecule can be inside at time $t + 1$: It was inside at time t and remained inside; or it was outside at time t and entered. If it was inside at time t (probability p_t), it remains with probability 0.8. If it was outside at time t (probability $1 - p_t$), it enters with probability 0.1. The updated probability is

(probability inside at $t + 1$) = (probability inside at t)
$\times$ (probability did not leave)
+ (probability outside at t)
$\times$ (probability entered)

$$p_{t+1} = 0.8 p_t + 0.1(1 - p_t) \tag{6.9}$$

This updating function has the same form as the updating function for the lung (Equation 1.44). We can apply the method of cobwebbing (Section 1.10) to figure out what this updating function does (Figure 6.11). This equation has a single stable equilibrium, and the value of p_t converges to the equilibrium. The equilibrium p^* satisfies the equation

$$p^* = 0.8 p^* + 0.1(1 - p^*)$$

This has the solution $p^* = 1/3$.

What does this tell us? The probability of finding a particular molecule inside the cell after a long time is $1/3$. If many molecules were observed, about $1/3$ of them would be inside the cell after a long time. The fates of 100 computer-

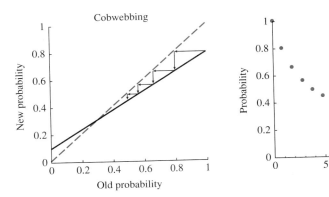

Figure 6.11
Cobwebbing and solution for the probability updating function

simulated molecules that began inside the cell are shown in Figure 6.12. The fraction inside approaches 1/3, but there is much variation around this average. Without this variation, the model matches our model of deterministic chemical exchange (Exercise 9).

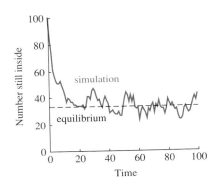

Figure 6.12
The fates of 100 molecules

SUMMARY We have introduced and analyzed three models describing the behavior of a molecule diffusing out of a cell. First, we assumed that the molecule has a fixed probability of leaving during each minute. In this case, the probability that the molecule is inside follows a simple discrete-time dynamical system and decreases exponentially. Next, we assumed that the molecule left the cell at a fixed rate. In this case, the probability that the molecule is inside follows a simple differential equation and again decreases exponentially. Finally, we assumed that the molecule could reenter the cell and found a Markov chain describing the probability that a molecule is inside the cell. In this case, the probability approaches an intermediate equilibrium.

6.2 EXERCISES

1. Suppose a herd of lemmings is standing at the top of a cliff. Each jumps off with probability 0.2 each hour.
   a. What is the probability that a particular lemming remains on top of the cliff after 3 h?
   b. If 5000 lemmings are standing around on top of the cliff, about how many remain after 3 h?
   c. About how many hours would it take for about 90% of the lemmings to jump?

2. A molecule has a 5.0% chance of binding to an enzyme each second and remains permanently attached thereafter. If the molecule starts out unbound, find the following.
   a. The probability it is not bound after 1 s.
   b. The probability it is not bound after 2 s.
   c. The probability it is not bound after t s.
   d. How long would you wait before you felt confident that the molecule had bound?

3. Refer to Exercise 1. Suppose there are only two lemmings at the top of the cliff.
   a. What is the probability that both are still there at time t?
   b. What is the probability that both have jumped at time t?
   c. What is the probability that exactly one has jumped at time t?

4. Suppose that each lemming in Exercise 1 jumps off the cliff at a rate of 0.2 per hour. This means that it jumps with probability $0.2\Delta t$ during a short time Δt.
   a. What is the approximate probability a given lemming jumps during 1 min?
   b. What is the approximate probability a given lemming jumps during 1 s?
   c. Write a differential equation describing the probability that the lemming is still on top of the cliff.
   d. Solve the equation to find the exact probability that a lemming remains on the cliff after 1 h.
   e. About how many of 5000 lemmings would still be on top of the cliff after 10 h?

5. A certain type of light bulb blows out at a rate of 0.001 per hour.
   a. Write a differential equation to describe the probability that the light bulb is still good.
   b. Solve the differential equation, and find the probability that the light bulb is still good after 500 h.
   c. Use this information to estimate how long the light bulb will last "on average."

6. Suppose the probability that a toxic molecule remains inside a cell obeys the differential equation

   $$\frac{dP}{dt} = -\lambda P(t)$$

   Suppose that $\lambda = 0.1$ and that there are 1000 toxic molecules inside at $t = 0$.
   a. Find the approximate number inside at time t.
   b. The rate at which damage is done to the cell is proportional to the number of toxic molecules inside the cell. Write and solve a differential equation describing the damage done.
   c. How much damage will have been done when all the toxic molecules have left?
   d. Do the same if $\lambda = 0.5$ and compare with the result of part b. Does the result make sense?

7. Consider the population model

   $$N_{t+1} = 1 \begin{cases} \text{with probability 0.2 if } N_t = 0 \\ \text{with probability 0.9 if } N_t = 1 \end{cases}$$

   $$N_{t+1} = 0 \begin{cases} \text{with probability 0.8 if } N_t = 0 \\ \text{with probability 0.1 if } N_t = 1 \end{cases}$$

   (Equations 6.3 and 6.4).
   a. Find an updating function (like Equation 6.9) for the probability that the population is equal to 1.
   b. Find the equilibrium.
   c. Of 100 islands that follow these dynamics, about how many would be occupied after a long time?

8. Suppose that bound molecules in Exercise 2 have a 2.0% chance of unbinding from the enzyme each second. Use the Markov chain approach to find the fraction of molecules that are bound in the long run.

9. Consider the following model of chemical exchange. Each minute, 20% of the chemical in container 1 enters container 2 and 10% of the chemical in container 2 returns to container 1.
   a. Draw a picture of this system.
   b. Let c_t represent the amount in container 1 and d_t the amount in container 2 at time t. Write a discrete-time dynamical system for the amount of chemical in each container.
   c. Define p_t to be the fraction of chemical in container 1, and write a discrete-time dynamical system for p.
   d. Why is the equation the same as Equation 6.9? What is different about the interpretation?

10. **COMPUTER:** Consider a molecule that leaves a cell each minute with probability 0.1. If $M_t = 1$ when the molecule is inside at time t and $M_t = 0$ if the molecule is outside, then

    $$M_{t+1} = q_{0.9}M_t$$

    where $q_{0.9}$ takes on the value 1 with probability 0.9 and 0 with probability 0.1, as in Exercise 13 in Section 6.1. Define an updating function F, and replicate the computer experiment 10 times. Count how many times the molecule is inside at $t = 5$.

    The probability m_t that the molecule is inside the cell at time t has updating function

    $$f(m) = 0.9m$$

    Plot the solution of this deterministic discrete-time dynamical system with $m_0 = 1$. Compare the mathematically expected fraction and the fraction you counted in the stochastic version. What could you do to make the fraction closer to the probability?

11. **COMPUTER:** We can simulate a Markovian molecule that has a 20% chance of jumping back into the cell with the updating function

    $$G(M) = Mq_{0.9} + (1 - M)q_{0.2}$$

The probability m that the molecule is inside follows the related updating function,

$$g(m) = 0.9m + 0.2(1 - m)$$

Plot a solution of the probability equation and a simulation of a single molecule starting from $M = m = 1$ (use enough steps to see what is going on). Solve for the equilibrium probability. Why doesn't the simulation seem to approach an equilibrium? What do the two curves have to do with each other?

6.3 Stochastic Models of Genetics

Genetic systems are unavoidably probabilistic. Models of genetics (along with models of gambling) provided much of the motivation for the development of probability theory. The results of breeding experiments and the attempt to deduce the underlying genetics provided a primary impetus behind the development of modern statistics. Here we introduce three basic models of genetics. The first deduces some of the consequences of inbreeding, the second the consequences of outcrossing, and the third the consequences of "blending" inheritance.

The Genetics of Inbreeding

One of the consequences of inbreeding (breeding with close relatives) is loss of genetic variability. The simplest form of inbreeding is **selfing** (or self-pollination) in plants. Rather than wait for pollen to arrive from other plants in the wind or on the bodies of bees, some plants use their own pollen. Selfing makes finding a mate easy. Many important crop plants use this dependable system of fertilization.

Suppose a plant is **diploid**, meaning that it has two copies of each gene (one from the ovule, or egg, and one from the pollen). Plants with two different **alleles** (variants) of this gene are called **heterozygous**; those with two copies of the same allele are called **homozygous**. What is the probability that an offspring of a heterozygous plant is heterozygous?

Let **a** and **A** denote the two different alleles. A heterozygous plant has genotype **Aa** (plants of type **aA** are the same). There are four possible ways to combine genes in the offspring: an offspring could get **a** from the ovule and **a** from the pollen, **a** from the ovule and **A** from the pollen, **A** from the ovule and **a** from the pollen, or **A** from the ovule and **A** from the pollen (Figure 6.13). In the absence of other factors, each of these possibilities is equally likely and has probability 0.25 (Figure 6.14). Two of these four possibilities produce indistinguishable heterozygous offspring, so that the probability an offspring is heterozygous is 0.5.

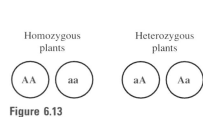

Figure 6.13
The four possible genotypes

We can set up a discrete-time dynamical system to compute the probability that a plant will be heterozygous after t generations of selfing. Let h_t represent the probability a plant is heterozygous in generation t. Numbering the parent plant as generation 0, we have assumed that $h_0 = 1$. Half the offspring are heterozygous after one generation, so $h_1 = 0.5$. What is the updating function? In the absence of mutation, a homozygous parent in generation t will produce only homozygous offspring. If the parent is heterozygous, its offspring will be heterozygous with probability 0.5 (Figure 6.15). The updating function is

$$h_{t+1} = 0.5 h_t \tag{6.10}$$

Half the offspring of heterozygous plants are heterozygous. This updating function again has the form of the bacterial growth equation, and has solution

$$h_t = 0.5^t$$

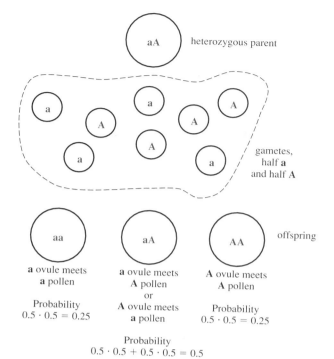

Figure 6.14
The probabilities of the four genotypes

After ten generations, the probability that a plant is heterozygous is

$$h_{10} = 0.5^{10} = 0.001$$

Only about 1 of 1000 progeny of this plant would remain heterozygous after ten generations.

An additional implication of this model is that about half the plants will eventually end up with the **aa** genotype and the other half with the **AA** genotype. Although fewer and fewer individuals carry the two alleles as a heterozygote, the two alleles are retained in the whole population.

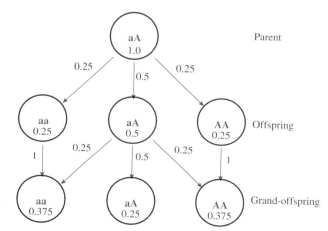

Figure 6.15
Selfing and heterozygosity

The Dynamics of Height

Even with modern techniques, it remains difficult to identify the genetic basis of such complex traits as plant height. Based on a general knowledge of genetics, we can propose various hypotheses about the underlying genetics and use probability theory to deduce the resulting dynamics. As statisticians, we can compare data with the results of probabilistic models to check whether the original hypotheses make sense.

The basic model of underlying genetics dates back to Gregor Mendel in the 19th century. Consider a population of plants that *cannot* self-pollinate. Assume first that height is determined by a single gene. Plants with at least one copy of the **B** allele are tall, and those with two copies of the **b** allele are short (the **B** allele is said to be **dominant**) (Figure 6.16). A breeding experiment is diagrammed in Figure 6.17. A short plant with genotype **bb** is crossed with a tall **BB** parent. All the offspring are of type **Bb** and are tall. These offspring are then crossed with each other.

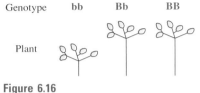

Figure 6.16
The action of a dominant gene

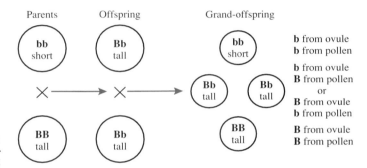

Figure 6.17
An experiment in outcrossing

What fraction of the next generation will be short? Short plants must have received a **b** allele from each parent. The probability of getting the **b** allele from the ovule is 0.5, equal to that of getting the **b** allele from the pollen. The probability of getting both is the product $0.5 \cdot 0.5 = 0.25$. As before, this probability describes the fraction of short plants in an infinitely large number of offspring. With 100 offspring, we might find 28 short plants and 72 tall plants.

Suppose instead that the effects of the genes are **additive**, meaning that a plant with genotype **bb** is short, a plant with genotype **BB** is tall, and a plant with genotype **Bb** is intermediate (Figure 6.18). Computing as above, we see that 25% of second-generation plants will have genotype **bb**, and 25% will have genotype **BB**, leaving the remaining 50% with genotype **Bb**. The heights of 100 simulated plants are shown in Figure 6.19; 21 are short, 48 medium, and 31 tall. These fractions do not match the probabilities exactly because we measured only 100 plants.

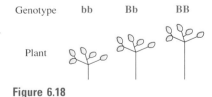

Figure 6.18
The action of an additive gene

This figure illustrates two widely observed properties: there is a single most common observation, and this most common observation is in the middle. These facts become more clear when height is determined by many genes. Suppose a plant has ten genes that affect height. Let b_i be the "short" allele of gene i and B_i be the "tall" allele. Each tall allele makes the plant 1 cm taller. A plant with no tall alleles has a baseline height of 40 cm. A plant with two copies of the tall allele of

6.3 Stochastic Models of Genetics

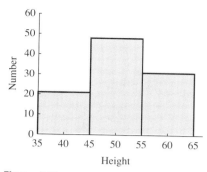

Figure 6.19
Heights of 100 plants produced by a single additive gene

each gene will have a height of 60 cm. In general,

$$\text{height of plant} = 40 + \text{total number of } \mathbf{B} \text{ alleles} \qquad (6.11)$$

(Figure 6.20a). We can use these plants to conduct the same experiment shown in Figure 6.17, crossing a very short plant with a very tall plant for two generations. After the first generation, all plants have a height of exactly 50 cm. What happens in the next generation? The result of a simulation of 200 plants is shown in Figure 6.20b. The distribution of heights is approximately bell-shaped and is an instance of the **binomial distribution**, which we study in detail in Section 7.4.

Although the sizes of these second-generation plants cluster around the average of 50 cm, there is still significant variability. If these plants continued to breed among themselves, little further variability would be lost. This variability provides the material used by breeders.

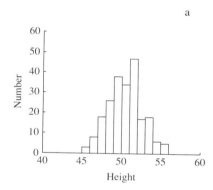

Figure 6.20
Heights produced by ten additive genes

The Dynamics of Blending Inheritance

Consider an alternative situation where offspring height is *exactly* intermediate between the heights of their parents, called **blending inheritance**. If two parents with heights 50 were crossed, all offspring would have height 50. Suppose we raised a population of plants, half with height 40 and half with height 60, and crossed them. The offspring would all have height 50. Unlike the cases we have studied hitherto, however, the grand-offspring produced by crossing the offspring would also have height 50, and so forth (Figure 6.21). All variability disappears permanently.

When Charles Darwin was developing the theory of evolution by natural selection, little was known about the mechanisms of inheritance. It was thought that offspring traits resulted from a process like blending some "tall substance" from the mother with some "short substance" from the father. When a mathematician demonstrated these results, Darwin was concerned that his theory would not work because natural selection, for example, plant breeding, depends on the persistence of extreme types. Darwin modified later editions of his book, *The Origin of Species* (1859), to include the now discredited idea of inheritance of acquired

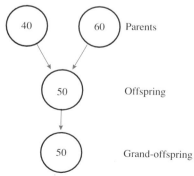

Figure 6.21
The action of blending inheritance

SUMMARY

We have examined the dynamic consequences of three different genetic systems: inbreeding, outcrossing, and blending. With a single gene, we saw first that selfing reduces **heterozygosity** by 50% in each generation. The distribution of offspring heights determined by a single gene was examined for plants that do not inbreed for both **dominant** and **additive** genes, and we compared these results with what happens when height is determined by many genes. Population variability in height is not lost in this situation. With **blending inheritance**, wherein offspring have height exactly equal to the average of the heights of their parents, all plants converge to the same height.

6.3 EXERCISES

1. Suppose that heterozygous plants are tall and that homozygous plants are short.
   a. What fraction of the offspring of a tall plant will be tall if it self-pollinates?
   b. What fraction of the offspring of a tall plant will be tall if it crosses with a short plant?
   c. What fraction of the offspring of a short plant will be tall if it crosses with another short plant?

2. Suppose we allow a plant and its offspring to self-pollinate for many generations; and suppose the original parent is heterozygous at locus 1 with alleles **a** and **A** and is also heterozygous at locus 2 with alleles **b** and **B**.
   a. What fraction of the first-generation offspring will have the completely heterozygous genotype **AaBb**?
   b. What fraction will be heterozygous at at least one locus after a long time?
   c. How many different genotypes at the two loci will remain after many generations?
   d. What fraction of offspring will have the genotype **AABB** after a long time?
   e. How many different genotypes would remain after a long time if there were ten different loci?

3. Compute the following values from Figure 6.15 (page 463).
   a. The fraction of grand-offspring (second generation) with genotype **AA**.
   b. The fraction of third-generation offspring with genotype **AA**.
   c. The fraction of offspring with genotype **AA** in generation t.

4. Geneticists often want to change one allele in an outcrossing organism while keeping the rest of the genome the same. For example, they might want to take a specially designed stock of flies and alter the eye color from red to white. Suppose that the white-eye allele is dominant. One procedure used is to take some white-eyed fly and cross it with the red-eyed stock. The white-eyed offspring are then crossed with the red-eyed stock, and their white-eyed offspring are crossed with the red-eyed stock, and so forth.
   a. What is the genotype at the eye color locus in the first, second, and subsequent generations?
   b. Suppose the special red-eyed stock is homozygous for the desirable allele **A** at some other locus, but the white-eyed fly is homozygous for the inferior **a** allele at that locus. What fraction of flies will have the **a** allele after t generations?
   c. How many back-crosses would be necessary to purge 99.9999% of the inferior genes from the white-eyed fly?

5. Suppose plants with genotype **bb** have height 52 cm, those with genotype **Bb** have height 54 cm, and those with genotype **BB** have height 55 cm. A **bb** plant is crossed with a **BB** plant.
   a. What are the heights of the offspring?
   b. These offspring are crossed with each other. For the resulting offspring, give the heights and their probabilities.
   c. The **bb** parent is crossed with one of the offspring from part a. For the resulting offspring, give the heights and their probabilities.

6. One force that can alter the ratios of heterozygotes produced by a selfing heterozygote is **meiotic drive**. This means that one allele, say **A**, pushes its way into more than half the gametes (ovules or pollen).
   a. Suppose meiotic drive affects the pollen only and that 80% of the pollen grains from a heterozygote carry the **A** allele. What fraction of offspring from a selfing heterozygote will be heterozygous?
   b. Suppose meiotic drive affects both pollen and ovules and that 80% of the pollen grains and ovules from a heterozy-

gote carry the **A** allele. What fraction of offspring from a selfing heterozygote will be heterozygous?

c. Suppose meiotic drive affects both pollen and ovules but that 80% of the pollen grains carry the **A** allele and 80% of the ovules carry the **a** allele. What fraction of offspring from a selfing heterozygote will be heterozygous?

d. How many generations would it take before the probability of a descendent of a plant described in part **c** would have less than a 0.01 chance of being a heterozygote? Compare this with the number of generations required in the absence of meiotic drive.

7. Suppose height is governed by two genes and follows the rule that each **B** allele increases height by 1 cm (as in Equation 6.11). A short plant with two copies of the short allele at each locus (genotype $b_1b_1b_2b_2$) is crossed with a tall plant with two copies of the tall allele at each locus (genotype $B_1B_1B_2B_2$).
   a. For the offspring of these plants, find the heights and their probabilities.
   b. Suppose two of these offspring are crossed. For the resulting offspring, find the heights and their probabilities.
   c. Use a coin (heads for **B** and tails for **b**) to create ten offspring from this cross. How many were of each height?

8. Consider the same situation as in Exercise 7, but suppose that the **B** genes are dominant. That is, a plant gets a 2-cm boost in height from the first locus if it has at least one allele of type B_1, and a 2-cm boost in height from the second locus if it has at least one allele of type B_2.
   a. Find the height of the offspring of these plants.
   b. For the next generation of plants, find the heights and their probabilities.
   c. Use a coin (heads for **B** and tails for **b**) to create ten offspring from this cross. How many were of each height?

9. Consider the following case of blending inheritance. A population of plants starts out with individuals of heights 40, 50, and 60 cm. Assume plants cannot breed with plants of the same height.
   a. What are the possible heights of offspring in the first generation?
   b. What are the possible heights of offspring in the second generation?
   c. What are the possible heights of offspring after n generations?

10. Heterozygosity in inbreeding organisms can be restored by mutation. Suppose that mutations always create brand new alleles. Suppose that each parental allele has a probability 0.01 of mutating.
    a. Suppose a plant is homozygous **AA**. Without mutation, all offspring will be **AA**. What might happen with mutation?
    b. What is the probability that the allele that came from the pollen is type **A**?
    c. What is the probability that the allele that came from the ovule is type **A**?
    d. The probability that both alleles in the offspring are type **A** is the *product* of the probabilities in parts **b** and **c** (we will derive this in Section 6.6). What is the probability that the offspring of a homozygous parent is homozygous?
    e. What is the probability that the offspring of a homozygous parent is heterozygous?

11. Continuing with Exercise 10, we will find the probability that an offspring of a heterozygous selfing parent is heterozygous.
    a. Suppose a plant is heterozygous **Aa**. Without mutation, all offspring will be **AA**, **Aa**, or **aa**. What might happen with mutation?
    b. What is the probability that the allele that came from the pollen is type **A**?
    c. What is the probability that the allele that came from the ovule is type **A**?
    d. Multiply these to find the probability that the offspring is **AA**.
    e. Find the probability that the offspring is **aa**.
    f. What is the probability that the offspring of a heterozygous parent is homozygous?
    g. What is the probability that the offspring of a heterozygous parent is heterozygous?

12. Using the results of Exercises 10 and 11, find the updating function for the probability that an offspring is a heterozygote. What is the equilibrium fraction of heterozygotes? Does this result make sense?

13. **COMPUTER:** We can simulate the process in Exercises 10 and 11. Using the function p from Exercise 13, the stochastic updating function is
$$G(H) = q_{0.50995}H + q_{0.0199}(1-H)$$
where $H = 1$ represents a heterozygote and $H = 0$, a homozygote. Let h represent the probability that an offspring is a heterozygote. The updating function for the probability is
$$g(h) = 0.50995h + 0.0199(1-h)$$
Plot the results of these two systems for 500 steps, starting from the initial condition 1 for each system. Why do the results of the simulation look so odd? How would you go about estimating and computing the fraction of heterozygous plants in a large population?

14. **COMPUTER:** Plants with many genes affecting a trait can be simulated by adding the effects of each gene. For example, suppose that each of 20 genes adds 1 to the height with probability 0.5 and 0 with probability 0.5. The total height is the sum of 20 such numbers. Find a way to create 100 such plants. Which height is most common? What is the largest plant? What is the smallest plant?

6.4 Probability Theory

In presenting models of populations, diffusion, and genetics, we have been rather vague about the mathematical underpinnings of stochastic events and their probabilities. We now begin the study of **probability theory** to make precise these concepts and help us to analyze models. The basic language of probability theory comes from **set theory** but is rephrased to describe **events**, or possible occurrences. Probability theory requires assigning numbers, probabilities, to events in mathematically consistent and biologically useful ways.

Sample Spaces, Experiments, and Events

Probability theory exists in the abstract world of mathematics. We imagine idealized experiments that have a specific set of possible outcomes, called the **sample space**. The value taken by $r(t)$ in the growth model (Equation 6.1) can be thought of as the result of an experiment. In a given year, the result is a particular positive number equal to the per capita reproduction. The sample space is the set of all positive numbers (Figure 6.22). The random component of the immigration model (Equation 6.2) is whether or not a pair of immigrants arrive. The sample space consists of the two events, "pair" or "no pair."

Probability theory will help us to compute the probabilities of the different outcomes of an experiment. First, we need some technical language to describe outcomes. Any particular outcome is known as a **simple event**. For example, $r(t) = 1.2542$ is a simple event in the bacterial growth model (Figure 6.22), and "pair" is a simple event in the immigration model (Equation 6.2). An **event** is a set of simple events or, equivalently, a subset of the sample space. Events for Equation 6.1 include $r(t) > 1$ and $2.7 \leq r(t) \leq 3.6$. Events are sometimes referred to with a single letter; the letter A in Figure 6.22 refers to the event $2.7 \leq r(t) \leq 3.6$.

Suppose 100 molecules begin inside a cell and each leaves with probability 0.1 each minute for ten minutes (Equation 6.5). Many different measurements could describe the results of this experiment. A scientist might count the number of molecules left in the cell, denoting it by N. The sample space is the set of integers from 0 to 100. The simple events are $N = 0, N = 1, \ldots, N = 100$, the specific outcomes of the experiment. Events include $40 \leq N \leq 60$, $N < 9$ and "N is prime." Each event is a set of possible specific outcomes.

Alternatively, the scientist could measure exactly when each molecule left the cell. The sample space in this case is much larger, all sets of 100 positive integers. A simple event is one precise listing of when the various molecules left, such as "molecule 1 left at $t = 47$, molecule 2 left at $t = 13$, etc." Events include "all the molecules left before time 20," "half the molecules left between times 3 and 7," and "the prime numbered molecules left at prime times." Each of these events could occur in many different ways.

One can think of the sample space as a dartboard and an experiment as throwing a dart (Figure 6.23). A simple event corresponds to a single point on the dartboard. Events correspond to regions of the board. One simple event is hitting exactly in the middle of the top half of the dartboard. Events include hitting the bullseye (hitting any of the points in this region), scoring 20, and hitting the top half of the board.

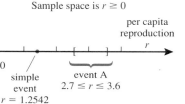

Figure 6.22

Sample space, simple event, and event for per capita reproduction

Figure 6.23

A dartboard

Set Theory

A probability model assigns a probability to each event. Doing this requires knowledge of what the events are (the experiment and its possible outcomes) and how "likely" they are to occur. For a dartboard, one probability model is that the chance of hitting a region is proportional to the size of that region. This means that the dart thrower has terrible aim and is just as likely to hit near the edge as near the bullseye. A quite different probability model would be associated with a skillful British thrower who gets mainly bullseyes.

To explain how one assigns probabilities to all possible events (there can be quite a few) we need to review some **set theory**. Sets can be visually portrayed on **Venn diagrams**, which are essentially abstracted drawings of dartboards. Figure 6.24 illustrates the three basic operations of set theory: intersection, union, and complementation. The region inside the rectangle represents the whole sample space, all the possible things that can happen. The circular regions represent events, particular things that can happen, to which we would like to assign probabilities.

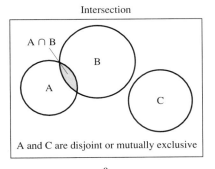
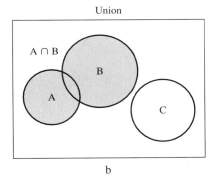
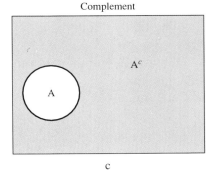

Figure 6.24

Venn diagrams for the basic operations of set theory

The **intersection** of two sets is the region contained in both sets (this region may be the empty, or null, set), corresponding to the English word "and." It means that both events occurred. In Figure 6.24a, the hatched region indicates where both events A *and* B occurred. This is written A ∩ B and read "A intersection B." On a dartboard, if event A is "I hit the bullseye," and event B is "I hit the top half of the dartboard," the intersection is "I hit the top half of the bullseye." Notice that A ∩ B = B ∩ A. If the intersection of two sets is the null set (as with sets A and C in Figure 6.24) the sets are said to be **disjoint**, and their associated events are said to be **mutually exclusive**. Mutually exclusive events cannot occur simultaneously. On a dartboard, hitting the bullseye and scoring a 3 are mutually exclusive events. Any pair of distinct simple events are mutually exclusive because it is impossible for one dart to hit two different points.

The **union** of two sets is the region contained in either set, corresponding to the English word "or." It means that one or the other event occurred (Figure 6.24b) and includes the possibility that both occurred. This is written A ∪ B and read "A union B." On a dartboard, if the events are "I hit the bullseye," and "I hit in the top half of the dartboard," the union is hitting either the bullseye or the top half of the board.

The final operation of set theory is taking the **complement**. The complement of a set is everything outside the set (Figure 6.24c). This corresponds to the English word "not." If the event is "I hit the bullseye," the complement is "I did not hit the bullseye." We denote the complement of the set A by A^c (it is sometimes denoted A').

Assigning Probabilities to Events

A **probability model** assigns probabilities to all events. To make sense, these assigned probabilities must satisfy several requirements or axioms. We write the probability of event A by Pr(A), said "the probability of A."

The four requirements for a probability model:

1. If S is the sample space, $\Pr(S) = 1$ (Figure 6.25a).

2. $0 \leq \Pr(A) \leq 1$ for any event A (Figure 6.25b).

3. If A and B are mutually exclusive events, $\Pr(A \cup B) = \Pr(A) + \Pr(B)$ (Figure 6.25c).

4. If A^c is the complement of A, $\Pr(A^c) = 1 - \Pr(A)$ (Figure 6.25d).

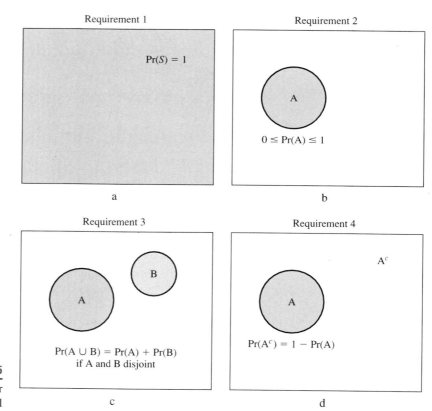

Figure 6.25
The four requirements for a probability model

What do these requirements mean? A probability of 1 means certainty. Requirement 1 says that if you do the experiment, something must happen: Every dart that hits the dartboard must hit the dartboard somewhere. Requirement 2 says that the probability of any event lies between 0 (never happens) and 1 (always happens). Requirement 3 is the most important mathematically, abstracting the intuitive idea of probability. The probability of hitting one of two nonintersecting regions on a dartboard is the sum of the probabilities of hitting either region. If you have a 0.1 probability of hitting the bullseye and a 0.2 probability of scoring a 3, the probability of hitting the bullseye *or* getting a 3 is $0.1 + 0.2 = 0.3$. Requirement 4 is a special case of requirement 3, because a set and its complement do not intersect and have a union equal to the whole space S. If the probability of hitting the bullseye is 0.1, the probability of missing it, the complement, is $1 - 0.1 = 0.9$.

A valuable way to think about probabilities on a Venn diagram is as *areas*. The area of the whole universe (the sample space S) is 1. The area of any subset, such as the event A, must be between 0 and 1 (requirement 2). The total area of any two subsets that do not overlap is the sum of their areas (requirement 3). The area outside a subset can be found by subtracting the area of the subset from the total area of 1 (requirement 4).

How do we assign probabilities to all events in ways consistent with the four requirements? There are two mathematically distinct cases: a finite number of simple events and an infinite number. In the model of immigration, there are only two simple events ("pair" and "no pair"), and the model of per capita reproduction for one generation has an infinite number of possible outcomes (r could take on any positive value). We address the finite case here, saving the more difficult infinite case for Section 6.7.

When the number of possible outcomes (simple events) is finite, we can construct consistent assignments of probabilities directly. Denote the simple events by $E_1, \ldots, E_n$, and associate each with a probability satisfying requirement 2, or $0 \leq \Pr(E_i) \leq 1$ for each i (illustrated with $n = 6$ in Figure 6.26). To guarantee that the rest of the requirements are met, we need only make sure that the sum of the probabilities of all simple events is equal to 1, or that

$$\sum_{i=1}^{n} \Pr(E_i) = 1 \tag{6.12}$$

This combines requirements 1 and 3. Because simple events are mutually exclusive, the probability of their union is equal to the sum of their probabilities. But the union of all the simple events is the whole sample space S, which has probability 1 (requirement 1).

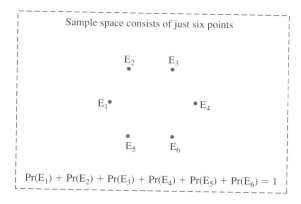

Figure 6.26
A finite set of simple events

How can we use this procedure to assign probabilities in the immigration model? We could suppose that the probability that a pair of immigrants arrives is 0.5. Because the complement of this event is "no pair," the probability of the complement must be $1 - 0.5 = 0.5$. More generally, if we pick any probability p

of there being a pair of immigrants (with $0 \leq p \leq 1$), the probability of "no pair" must be $1 - p$. These are *all* the possible consistent assignments of probabilities to the events in this case.

Models with two events are particularly simple. Consider three molecules leaving a cell. We measure the number, N, left in the cell after 10 min. The simple events are $N = 0, N = 1, N = 2$, and $N = 3$ (Figure 6.27). One assignment of probabilities is $\Pr(N = 0) = 0.4$, $\Pr(N = 1) = 0.3$, $\Pr(N = 2) = 0.2$, and $\Pr(N = 3) = 0.1$. Because these add to 1, the assignment is mathematically consistent.

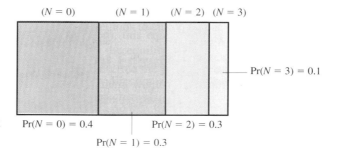

Figure 6.27

A Venn diagram for number of molecules left in a cell

With this assignment we can find the probability of any event by writing it as the union of mutually exclusive simple events. The event "N is odd" is the union of the simple events $N = 1$ and $N = 3$. We can use requirement 3 to compute

$$\Pr(N \text{ is odd}) = \Pr(N = 1) + \Pr(N = 3) = 0.3 + 0.1 = 0.4$$

The probability of the event "$N \neq 1$" can be computed in two ways. First, this event is the complement of the event $N = 1$, and can be found with requirement 4 as

$$\Pr(N \neq 1) = 1 - \Pr(N = 1) = 1 - 0.3 = 0.7$$

Alternatively, "$N \neq 1$" is the union of the disjoint events $N = 0$, $N = 2$, and $N = 3$, so

$$\Pr(N \neq 1) = \Pr(N = 0) + \Pr(N = 2) + \Pr(N = 3) = 0.4 + 0.2 + 0.1 = 0.7$$

There are many consistent ways to assign probabilities to the simple events. Each assignment represents a different model of the underlying process. We can determine which is right from the biological mechanisms driving the underlying model. It is *mathematically* consistent to set the probability of a heterozygous offspring produced by selfing from a heterozygous parent to be 0.31416, but this is not *biologically* consistent with our understanding of genetics. The mathematical requirements define the universe of *possible* assignments. The biological mechanisms restrict our attention to *reasonable* assignments. Much of the work in the rest of the book will be to derive these biologically reasonable assignments from probability models.

Using our new terminology, we can restate the connection between probability and statistics. Suppose we are given the results of an experiment and want to understand the process that produced those results. First, we use our understanding of biology to develop a stochastic model of the system. Next, we use techniques from probability theory to assign probabilities to every possible model outcome. Finally, as statisticians, we compare the measured results with the mathematical expectation. If the results match, the data are consistent with the model and we can estimate parameters and make predictions about other experiments. If the results do not match, we must reject or revise our model.

SUMMARY This section introduced and defined the basic language of probability theory. Probabilistic models are mathematical experiments with possible outcomes known as **simple events**. The set of all possible simple events is called the **sample space**, subsets of which are called **events**. We portrayed the sample space on a **Venn diagram**, illustrating the **intersection** of events A and B (both events happened), the **union** of events A and B (one or both events happened), and the **complement** of A (event A did not happen). Sets with null intersection are called **disjoint**, and their associated events are called **mutually exclusive**. We presented the consistency requirements for assigning probabilities to different events and showed how to make a mathematically consistent assignment when the number of simple events is finite.

6.4 EXERCISES

1. Give the sample spaces associated with the following experiments. Say how many simple events there are, and list them if there are fewer than ten.
   a. We cross two plants with genotype **bB** and check the genotype of one offspring.
   b. We cross two plants with genotype **bB** and check the genotypes of two offspring.
   c. A molecule jumps in and out of a cell. We record whether the molecule is inside or outside the cell at times 2, 5, and 10.
   d. We count how many of 16 plants are taller than 50 cm.
   e. We measure the heights of 16 plants.

2. We start with 100 molecules in a cell and count the number, N, left after 10 min. Give five simple events that are included in the following events.
   a. $N > 90$.
   b. N is odd.
   c. $30 \leq N \leq 32$ or $68 \leq N \leq 70$.

3. We follow four individually labeled molecules and record the minute, t_i, when molecule i leaves the cell. Give three simple events that are included in the following events.
   a. All molecules left before minute 20.
   b. Exactly half the molecules left between minutes 3 and 7 (inclusive).
   c. All prime numbered molecules left at prime times.

4. Draw Venn diagrams with sets A, B, and C satisfying the following requirements.
   a. A and B disjoint, B and C disjoint, A and C not disjoint.
   b. A and B disjoint, B and C not disjoint, A and C not disjoint.
   c. No two sets disjoint, but $A \cap B \cap C$ empty.
   d. No two sets disjoint, and $A \cap B \cap C$ nonempty.

5. Find the union and intersection of the following events.
   a. Events **a** and **b** in Exercise 2
   b. Events **a** and **c** in Exercise 2
   c. Events **b** and **c** in Exercise 2
   d. Events **a** and **b** in Exercise 3

6. Give two assignments, different from those in the text, of probabilities when counting the number of molecules inside a cell starting from an initial number of three. Compute $Pr(N$ is odd$)$ and $Pr(N \neq 1)$ for each case.

7. Give two mathematically consistent ways of assigning probabilities to the results of the following experiments. Try to make one of your assignments biologically reasonable.
   a. Exercise 1a
   b. Exercise 1b
   c. Exercise 1c

8. For the following examples, show that $Pr(A \cup B) \neq Pr(A) + Pr(B)$. Let event A be "$N$ is odd" and event B be "$N \geq 2$," as in Exercise 6.

a. Assignment of probabilities given in Figure 6.27.
b. First assignment used in Exercise 6.
c. Second assignment used in Exercise 6.

9. Draw a Venn diagram and convince yourself that

$$Pr(A \cup B) = Pr(A) + Pr(B) - Pr(A \cap B)$$

Test the formula on the examples in Exercise 8.

6.5 Conditional Probability

Probability theory can be thought of as the study of scientific problems with limited information. For example, we sometimes wish to make deductions about a process that we cannot measure directly. **Conditional probability** is the tool we need to compute the probability of one event conditional on if another occurred. We will use conditional probability to break up the sample space using the **law of total probability** and to reason about complicated systems using **Bayes' theorem**.

Conditional Probability

Suppose plant height is governed by a single dominant allele **B**. Two plants of genotype **Bb** (created by crossing tall and short inbred plants) are crossed. We are given a tall offspring from this cross. What is the probability that its genotype is **BB**? (It could be either **Bb** or **BB**). It might seem tempting to guess that these genotypes are equally probable, meaning that the probability of **BB** is 1/2. As is so often the case, temptation must be resisted. We must use **conditional probability** to figure out the probability correctly.

We can work out these probabilities by thinking of a large number of offspring, perhaps 100. If everything behaved exactly according to theory, there would be 25 of type **bb**, 50 of type **Bb**, and 25 of type **BB**. Knowing that our plant is tall eliminates the 25 short **bb** plants. Our plant is one of the 75 tall offspring, 1/3 of which are of type **BB**. This is the probability of genotype **BB conditional** on the plant being tall. This probability exceeds the probability 1/4 of genotype **BB** in the absence of information about height.

To express this more compactly, we need some new notation. Let G represent the event "genotype **BB**" and T represent the event "tall." We know that Pr(G) = 1/4 and Pr(T) = 3/4. To denote the probability of B conditional on T, we write

$$Pr(G \mid T) = \text{the probability of G conditional on T} \qquad (6.13)$$

The expression Pr(G | T) means the probability of G given that T is true.

The mathematical definition of conditional probability looks rather different.

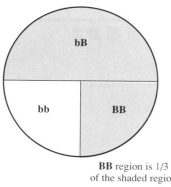

Figure 6.28
A Venn diagram for genetics

■ **Definition 6.3** The probability of event A conditional on event B is

$$Pr(A \mid B) = \frac{Pr(A \cap B)}{Pr(B)}$$

if Pr(B) ≠ 0. ■

In our example, Pr(G ∩ T) = 1/4 (the probability that an offspring has genotype **BB** and is tall) and Pr(T) = 3/4, so

$$\Pr(G \mid T) = \frac{\Pr(G \cap T)}{\Pr(T)} = \frac{1/4}{3/4} = 1/3$$

as we found above.

Why does this rule work? The probability of G conditional on T can be thought of as the probability of G after we restrict our attention to cases in which T is true. Using the area interpretation of probabilities, the probability of G when restricted to the smaller universe T is the fraction of area of T [Pr(T)] that is also included in G [Pr(G ∩ T)].

Using the diagram of Figure 6.28, we can find the probability that a plant is homozygous (event H) conditional on it being tall (event T):

$$\Pr(H \mid T) = \frac{\Pr(H \cap T)}{\Pr(T)}$$

The region H ∩ T is precisely the event G, plants with genotype **BB**. Therefore,

$$\Pr(H \mid T) = \frac{\Pr(G)}{\Pr(T)} = \frac{1/4}{3/4} = 1/3$$

The conditional probability is the same as we found before because the only way that a homozygous plant could be tall is by having genotype **BB**.

The definition of conditional probability (Definition 6.3) can be rewritten in an alternative, multiplicative form as

$$\Pr(A \cap B) = \Pr(A \mid B) \Pr(B) \tag{6.14}$$

The left-hand side means that both A and B occurred. This happens when B occurs [probability Pr(B)] and A occurs **conditional on B** [probability Pr(A | B)]. The probability that both A and B occurred is the product.

Two special cases can help clarify the meaning of conditional probability (Figure 6.29). First, consider two disjoint events B and C. B ∩ C is empty, so Pr(B ∩ C) = 0 and

$$\Pr(B \mid C) = \frac{\Pr(B \cap C)}{\Pr(C)} = 0$$

This means that once we know that C has occurred, we can be sure that B did **not** occur. The event "the dart scored less than 10" is disjoint from the event "the dart hit the bullseye." Once you know that your shot hit the bullseye, you can be sure that your dart did not score less than ten.

$$\Pr(C \mid B) = \frac{\Pr(C \cap B)}{\Pr(B)} = 0$$

When your dart scores less than ten points, you can be sure you did not hit the bullseye. Knowledge about one event provides complete information about the other.

Second, consider two events A and C where A is a subset of C (Figure 6.29). Therefore, A ∩ C = A and Pr(A ∩ C) = Pr(A). From the definition of conditional probability, we have

$$\Pr(C \mid A) = \frac{\Pr(C \cap A)}{\Pr(A)} = \frac{\Pr(A)}{\Pr(A)} = 1$$

[as long as Pr(A) ≠ 0]. Knowing that A occurred guarantees that C occurred. Furthermore,

$$\Pr(A \mid C) = \frac{\Pr(A \cap C)}{\Pr(C)} = \frac{\Pr(A)}{\Pr(C)}$$

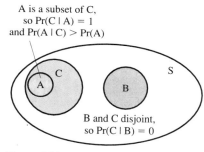

Figure 6.29
Conditional probability: special cases

[as long as Pr(C) ≠ 0]. If Pr(C) < 1, knowing that C occurred makes it more probable, although not certain, that A occurred. Information about one event has increased our confidence about predicting the other.

For example, consider the events C, "the dart scored more than 15 points with that shot," and A, "the dart hit the bullseye." The event A is a subset of the event C. Knowing that the shot was a bullseye guarantees that the shot scored more than 15 points. Knowing that a shot scored more than 15 points makes it more probable, although not certain, that the shot was a bullseye.

In biological terms, knowing that a plant is heterozygous (for the case in Figure 6.28) guarantees that the plant is tall. Knowing that the plant is tall increases the probability that it is heterozygous.

The Law of Total Probability

Conditional probability allows us to break events into their component parts, which we can then reassemble with **the law of total probability**. Let $E_1, E_2, \ldots, E_n$ form a set of mutually exclusive (no pair intersect) and collectively exhaustive (the union is the full set) events (Figure 6.30). The sets E_i break the whole sample space into disjoint pieces, so the sum of their probabilities must be 1, or

$$\sum_{i=1}^{n} \Pr(E_i) = 1 \qquad (6.15)$$

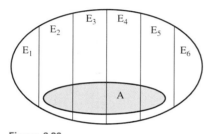

Figure 6.30
$E_1, \ldots, E_6$ form a mutually exclusive and collectively exhaustive set of events

For example, the three events $E_1 = \{r(t) < 1.0\}$, $E_2 = \{1.0 \leq r(t) < 2.0\}$, $E_3 = \{r(t) \geq 2.0\}$ break possible growth rates $r(t)$ into three mutually exclusive and collectively exhaustive sets. Exactly one of these events must occur.

We use this decomposition of the sample space to decompose other events and their probabilities.

■ **THEOREM 6.1** (The Law of Total Probability)

Suppose that $E_1, E_2, \ldots, E_n$ form a set of mutually exclusive and collectively exhaustive events. Then for any event A,

$$\Pr(A) = \sum_{i=1}^{n} \Pr(A \mid E_i) \Pr(E_i) \qquad ■$$

Proof: This theorem follows from the requirements on probabilities. The set A can be written as a union of the disjoint sets $A \cap E_i$, so that by requirement 3 we have

$$\Pr(A) = \sum_{i=1}^{n} \Pr(A \cap E_i) \qquad ■$$

But $\Pr(A \cap E_i) = \Pr(A \mid E_i) \Pr(E_i)$, from which the theorem follows.

Many useful instances require breaking the sample space into only two disjoint sets, E and E^c. For example, suppose 1% of females and 5% of males are color blind. What is the probability that a person chosen at random is color blind? Let C represent the event "color blind," F the event "female," and F^c the event "male." If the number of males and females is equal, $\Pr(F) = \Pr(F^c) = 0.5$. By the law of total probability,

$$\Pr(C) = \Pr(C \mid F) \Pr(F) + \Pr(C \mid F^c) \Pr(F^c)$$
$$= (0.01)(0.5) + (0.05)(0.5) = 0.03$$

Bayes' Theorem and the Rare Disease Example

If we know $\Pr(A)$, $\Pr(B)$, and $\Pr(A \mid B)$, we can find the probability of B conditional on A using **Bayes' theorem**, which forms the basis for **Bayesian statistics** and **Bayesian decision theory**.

■ **THEOREM 6.2** (Bayes' Theorem)

For any events A and B where $\Pr(A) \neq 0$,

$$\Pr(B \mid A) = \frac{\Pr(A \mid B)\Pr(B)}{\Pr(A)}$$

Proof: From the multiplicative formula (Definition 6.3), we have

$$\Pr(B \mid A) = \frac{\Pr(A \cap B)}{\Pr(A)}$$

but $\Pr(A \cap B) = \Pr(A \mid B)\Pr(B)$ (Equation 6.14). Substituting this into Equation 6.16 proves the theorem.

No discussion of conditional probability and Bayes' theorem is complete without the rare disease example. Suppose a disease infects only 1% of people. A diagnostic test always picks up the disease but also generates 5% false positives. A patient walks into a doctor's office having found she tested positive. What is the probability that she has the disease? This is a conditional probability problem; how likely is it that she has the disease *conditional* on her testing positive? We solve this problem first by working out the component probabilities explicitly and then using Bayes' theorem.

Before applying the formal rules of conditional probability, we can do a more intuitive calculation. Out of 1000 people, 10 would have the disease. All 10 of these would test positive. Of the 990 who did not have the disease, 5% (49 or 50) would test positive. Out of the 59 or 60 individuals testing positive, then, only 10 have the illness, or about 17%.

This calculation can be formalized with the laws of conditional probability (Figure 6.31). Let D denote the event of an individual having the disease, N the event of not having the disease, and P be the event of a positive result on the test. We can translate our assumptions as follows:

$$\Pr(D) = 0.01$$
$$\Pr(N) = 0.99$$
$$\Pr(P \mid D) = 1.00$$
$$\Pr(P \mid N) = 0.05$$

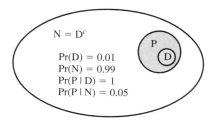

Figure 6.31
The probabilities in the rare disease model

We are interested in computing $\Pr(D \mid P)$.

The definition of conditional probability (Definition 6.3) says that

$$\Pr(D \mid P) = \frac{\Pr(D \cap P)}{\Pr(P)}$$

We need to find both $\Pr(D \cap P)$ and $\Pr(P)$. How do we compute $\Pr(D \cap P)$? Because the event P is certain when D is true, D is a subset of P, $D \cap P = D$, and

$$\Pr(D \cap P) = \Pr(D) = 0.01$$

Next, we need to find Pr(P). For this, we use the law of total probability (Theorem 6.1). Breaking the sample space into the mutually exclusive and collectively exhaustive events N and D, we get

$$\Pr(P) = \Pr(P \mid D)\Pr(D) + \Pr(P \mid N)\Pr(N)$$
$$= (1.0)(0.01) + (0.05)(0.99) = 0.0595$$

This says that 5.95% of the people test positive for either reason (corresponding to the 59 or 60 of 1000 found above). Therefore,

$$\Pr(D \mid P) = \frac{0.01}{0.0595} = 0.168$$

Alternatively, we can use Bayes' theorem, finding

$$\Pr(D \mid P) = \frac{\Pr(P \mid D)\Pr(D)}{\Pr(P)}$$
$$= \frac{1.0 \cdot 0.01}{0.0595} = 0.168$$

Although this calculation looks much shorter, we still had to compute Pr(P) with the law of total probability.

Even though this diagnostic test is fairly good, an individual with a positive result is still rather unlikely to have the disease. For this reason, many screenings for disease do not test the entire population but focus on risk groups where the disease is more common. Within a risk group, the test can be more useful. If $\Pr(D) = 0.2$ in this group, $\Pr(D \mid P) = 0.80$ (Exercise 7).

SUMMARY

Understanding probabilistic systems requires understanding relations between different events. The key tool is **conditional probability**, the probability that one event occurs *conditional* on some second event having occurred, found as the probability that both occurred divided by the probability that the second event occurred. By breaking the sample space into components, probabilities can be computed using the **law of total probability**. The order of the conditional probabilities can be reversed with **Bayes' theorem**. We applied these rules to find the probability of a patient having a rare disease conditional on a positive test result.

6.5 EXERCISES

1. Give a set of three mutually exclusive and collectively exhaustive sets for each of the sample spaces in Exercise 1.

2. An ecologist is looking for the effects of eagle predation on jackrabbits. She sees an eagle with probability 0.2 during an hour of observation, a jackrabbit with probability 0.5, and both with probability 0.05.
   a. Draw a Venn diagram to illustrate this situation.
   b. Find the probability that she saw a jackrabbit conditional on her seeing an eagle. How might you interpret this result? Compare with the overall probability of seeing a jackrabbit.
   c. Find the probability that she saw an eagle conditional on her seeing a jackrabbit. How might you interpret this result?

3. Suppose another ecologist repeats the experiment described in Exercise 2 and finds the same results except that the probability of seeing both is 0.15. Redo Exercise 2 with the new data.

4. A lab is attempting to stain many cells. Young cells stain properly 90% of the time, and old cells stain properly 70% of the time.

a. If 30% of the cells are young, what is the probability that a cell stains properly?
b. If 70% of the cells are young, what is the probability that a cell stains properly?
c. What happens when all the cells are young? Show this is a special case of the law of total probability.

5. Further study of the cell-staining problem (Exercise 4) reveals the following: New cells stain properly with probability 0.95, 1-day-old cells stain properly with probability 0.9, 2-day-old cells stain properly with probability 0.8, and 3-day-old cells stain properly with probability 0.5. Suppose

$$\Pr(\text{cell is 0 days old}) = 0.4$$
$$\Pr(\text{cell is 1 day old}) = 0.3$$
$$\Pr(\text{cell is 2 days old}) = 0.2$$
$$\Pr(\text{cell is 3 days old}) = 0.1$$

a. Find the probability that a cell stains properly.
b. The lab finds a way to eliminate the oldest (more than 3 days old) cells from its stock. What is the probability of proper staining? Write this as a conditional probability.
c. The lab finds a way to eliminate all cells more than 2 days old from its stock. What is the probability of proper staining?
d. The lab finds a way to eliminate all cells more than 1 day old from its stock. What is the probability of proper staining?

6. Find the following. Say whether the stain is a good indicator of the age of the cell.
a. For the cells in Exercise 4a, what is the probability that a cell that stains properly is young?
b. For the cells in Exercise 4b, what is the probability that a cell that stains properly is young?
c. For the cells in Exercise 5, what is the probability that a cell that stains properly is less than 1 day old?

7. For the rare disease example in this section, find the following.
a. Find $\Pr(D \mid P^c)$, the probability that a person who did not test positive has the disease.
b. Suppose $\Pr(D) = 0.2$. Find and interpret $\Pr(D \mid P)$.
c. Suppose $\Pr(P \mid D) = 0.95$ and $\Pr(D) = 0.2$. Find and interpret $\Pr(D \mid P)$ and $\Pr(D \mid P^c)$.

8. Suppose there are three different alleles, **A**, **B**, and **C**. The allele **A** is dominant, so individuals with genotype **AB** or **AC** are indistinguishable from individuals with genotype **AA**. An **AB** individual is crossed with an **AC** individual. Find the probability that an offspring that looks like an **AA** individual has genotype **AB** for the following cases.
a. The situation described above.
b. It is later realized that the parent thought to be of type **AB** was really the offspring of an **AA** and **AB** cross with unknown genotype.

c. It is found that this organism self-fertilizes 10% of the time (so that 80% of the offspring come from the cross, 10% from **AB** selfing, and 10% from **AC** selfing).

9. Four balls are placed in a jar, two red, one blue, and one yellow. Two are removed "at random."
a. You are told that the first ball removed was red. What is the probability that the second ball removed is red?
b. You are told that at least one of the two removed is red. What is the probability that both are?
c. Why are the results different?
d. The first ball is replaced (but remembered) before the second ball is drawn. Find the answers to parts **a** and **b** for this situation.

10. A well known expert on statistics has asserted that 97.6% of statistics are false. What is the probability that this statement is true?

11. **COMPUTER**: Use the command q_p (Exercise 13 in Section 6.1) that returns 1 with probability p and 0 with probability $1 - p$ to simulate the rare disease example.
a. Simulate 100 people who have the disease with probability 0.05, and count the number with the disease.
b. For each of the remaining people, assume that the probability of a false positive is 0.1. Simulate the remaining people, and count the number of positives.
c. What fraction of positive tests identify people who are sick? How does this compare with the mathematical expectation?
d. Try this experiment assuming that the probability that each person has the disease is 0.4.

12. **COMPUTER**: Suppose cells fall into three categories: those that are dead, those that are alive but do not stain properly, and those that are alive and do stain properly. Let D_t denote the number of cells that are dead, N_t the number that are alive but do not stain properly, and S_t those that are alive and do stain properly. Each day, there are two possible transitions:

- Cells that are alive die with probability 0.9.
- Cells that stain properly cease to stain properly with probability 0.8.

a. Start with $S_0 = 100$, $D_0 = 0$, and $N_0 = 0$. Use your computer to simulate the numbers for the next generation.
b. Follow these cells until there are no more cells that stain properly. How long did it take?
c. At each time, what is the fraction of cells that stain properly? What is the fraction of *living* cells that stain properly? Estimate the probability that a cell stains properly and the probability it stains properly conditional on being alive.

6.6 Independence and Markov Chains

Conditional probability can be used to describe whether or not two events are related. If knowing about one gives no additional information about the other, we say that the events are **independent**. Markov chains, probabilistic models that evolve over time, describe events that are related over time and can be conveniently and powerfully formulated in terms of conditional probability.

Independence

Suppose two parents, each with genotype **Bb**, are crossed. It is generally thought that the allele contributed by the mother is *independent* of the allele contributed by the father. Scientifically, this means that knowing that the mother provided allele **b** tells us nothing new about the allele provided by the father. In other words, knowledge of one allele in the offspring gives no new information about the other.

How do we describe independence mathematically? Let B_m designate the event that the allele from the mother is **B**, b_m be the event that the allele from the mother is **b**, and B_f and b_f describe the allele from the father in the same way (Figure 6.32). Suppose we learn that the mother provided allele **b**, so that event b_m occurred. What is the probability that the father provided allele **b** also? Biologically, this probability is the same as what it was without the additional information. In the mathematical language of conditional probability,

$$\Pr(b_f \mid b_m) = \Pr(b_f) = 0.5$$

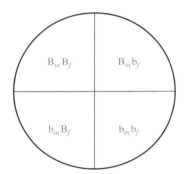

Figure 6.32
Genetic probabilities: independent case

The father, at least under ordinary circumstances, provides the **b** allele with probability 0.5 *independent* of what the mother does.

The fact that knowing one event tells nothing about the other means that the two are unrelated. We formalize the idea in the definition of independence.

■ **Definition 6.4**   Event A is **independent** of event B if

$$\Pr(A \mid B) = \Pr(A)$$

This compact definition matches our finding in the genetic case. The additional information (event B in the definition or b_m in the example) does not change our calculation of the probability of the other event. On a dartboard, suppose B is "I hit the right side of the dartboard," and A is "I hit the bullseye." For most players, these events are independent. If someone tells you that he hit the right side of the board, you are unlikely to change your guess of whether he hit the bullseye.

Independence has some useful properties. First, the relation is **reciprocal**, meaning that if A is independent of B, then B is independent of A. Suppose we know that A is independent of B, but we are told that event A occurred. Then

$$\Pr(B \mid A) = \frac{\Pr(A \mid B)\Pr(B)}{\Pr(A)} \qquad \text{Bayes' theorem}$$

$$= \frac{\Pr(A)\Pr(B)}{\Pr(A)} \qquad \text{A is independent of B}$$

$$= \Pr(B) \qquad \text{divide out factors of } \Pr(A)$$

Because the relation is reciprocal, we can simply say that A and B are independent.

In the genetic example, we assumed that the allele contributed by the father is independent of the allele contributed by the mother. The reciprocal property guarantees that the converse is also true: The allele contributed by the mother is independent of the allele contributed by the father. In terms of conditional probability,

$$\Pr(b_m \mid b_f) = \Pr(b_m) = 0.5$$

If information about a first event (the allele contributed by the mother) gives information about a second (the allele contributed by the father), then the converse must also be true; information about the second event (the allele contributed by the father) gives information about the first (the allele contributed by the mother). In the dart example, knowing that the dart hit the bullseye tells nothing about whether it hit the right or the left side of the board.

The Multiplication Rule for Independent Events

A second extremely useful property of independent events follows from the multiplication rule for conditional probabilities:

$$\Pr(A \cap B) = \Pr(A \mid B) \Pr(B)$$

(Equation 6.14). When events A and B are independent, $\Pr(A \mid B) = \Pr(A)$. This gives the following theorem.

■ **THEOREM 6.3** (Multiplication Rule for Independent Events)

Suppose A and B are any two independent events. Then

$$\Pr(A \cap B) = \Pr(A) \Pr(B)$$

■

In words, the probability that two independent events both occur is equal to the product of the probabilities for each.

This is the rule we have used implicitly to derive the probabilities of genotypes **bb** and **BB** (Section 6.3). When meiosis is fair, each allele has a 0.5 chance of appearing in the offspring. Therefore, the probability of a **b** from the mother is 0.5, as is the probability of a **b** from the father. An offspring has genotype **bb** if it gets a **b** from *each* parent. If the alleles from the parents arrive *independently*, then

$$\Pr(\mathbf{bb}) = \Pr(\mathbf{b} \text{ from mother}) \times \Pr(\mathbf{b} \text{ from father})$$
$$= 0.5 \cdot 0.5 = 0.25$$

It is important to realize that we had to make *three* assumptions in this calculation: fair meiosis in the mother, fair meiosis in the father, and independence. Even if the first two assumptions hold, the last might fail. For example, it is possible that both parents contribute the **B** allele with probability 0.5 but do not contribute them independently. Figure 6.33 shows a case in which the probability that the mother contributes the **B** allele is much greater when the father contributes the **B** allele than when he contributes the **b** allele.

The multiplication rule works only when the events are independent. For the example shown in Figure 6.33,

$$\Pr(B_f \mid B_m) = \Pr(b_f \mid b_m) = 0.8$$

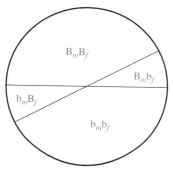

Allele contributed by father tends to match the allele contributed by mother

Figure 6.33

Genetic probabilities: dependent case

Therefore,
$$\Pr(B_f \cap B_m) = \Pr(B_f \mid B_m) \Pr(B_m) = 0.8 \cdot 0.5 = 0.4$$

Because the alleles contributed by the two parents tend to match, the offspring is more likely to be homozygous.

The multiplication rule can be used to define the idea of **mutual independence**. We think of events $A_1, \ldots, A_n$ as **mutually independent** if knowledge of all but one gives no information about the remaining one.

■ **Definition 6.5** The set of events $A_1, \ldots, A_n$ are mutually independent if for any group of k events numbered $A_{i_1}, A_{i_2}, \ldots, A_{i_k}$,

$$\Pr(A_{i_1} \cap A_{i_2} \cap \cdots \cap A_{i_k}) = \Pr(A_{i_1}) \cdot \Pr(A_{i_2}) \cdots \Pr(A_{i_k})$$ ■

For example, four events A, B, C, and D are mutually independent if

$$\Pr(A \cap B \cap C) = \Pr(A) \Pr(B) \Pr(C)$$
$$\Pr(B \cap C \cap D) = \Pr(B) \Pr(C) \Pr(D)$$
$$\Pr(A \cap B) = \Pr(A) \Pr(B)$$
$$\Pr(B \cap D) = \Pr(B) \Pr(D)$$

and so forth. When we know or believe that a set of events are mutually independent, probabilities can be computed by multiplication.

As an example of the need to establish independence before applying the multiplication rule for many events, consider the following situation. A class consists of ten students, each of whom skips class with probability 0.2. What is the probability no students come to class on a particular day? Suppose first that students behave independently; each makes the decision about whether or not to come without regard for what her classmates do. We find the probability that nobody comes to class by multiplying the probabilities for each student (Definition 6.5),

$$\Pr(\text{nobody came to class}) = 0.2^{10} \approx 0.0000001$$

or one day in 10 million.

If students do pay attention to each other's behavior, the results can be quite different. First, suppose two of the students detest each other. Each immediately goes to class when he hears that the other has decided not to. This *response* means that these two events are not independent and that the multiplication rule will not work. At least one of these surly students is sure to attend, meaning that the probability that nobody comes is 0. Alternatively, suppose that the students have a leader who pursuades the other students to do exactly as she does. Again, the multiplication rule fails. In this case, the professor would find herself teaching to an empty room on one day out of five.

Whether events are independent is generally a matter of scientific judgment. Do we really know that alleles behave independently? Are we sure that hitting the right side of the bullseye is exactly as likely as hitting the left side? Testing these apparently obvious assumptions can provide novel insights into the way things work. We often build the assumption of independence explicitly into a mathematical model. If the assumption is later shown to be false or the results of the model do not match those of the experiment, the model must be modified.

Markov Chains and Conditional Probability

Much philosophical energy has been expended regarding the distinction between related events and cause and effect. We will avoid this issue and think of cause and effect as a lack of temporal independence; if one event occurred it is more or less likely that some other event will occur later. Our study of discrete-time dynamical systems and autonomous differential equations was essentially the study of cause and effect in deterministic systems. We defined a Markov chain (Definition 6.2) as a stochastic dynamical system in which the probability of arriving in a particular state at a particular time depends on the state at the previous time. Markov chains are one way to formalize cause and effect for stochastic systems.

We have considered two Markov chains, one for presence and absence (Section 6.1) and one for molecular position (Section 6.2). For comparison, we will also consider a sequence of independent coin tosses and the behavior of a shy rabbit jumping in and out of its hole. In each case, the system has two possible states ("in" and "out," or "heads" and "tails"). Does the probability that the molecule is inside the cell during one minute depend on whether it was inside during the previous minute? We would generally expect so. Does the probability of tossing heads depend on whether the previous toss was tails? We would generally expect not.

Recall the model for molecular position (Figure 6.34a). Let I_t denote the event that the molecule was inside during minute t and O_t denote the event that the molecule was outside during minute t. We can translate the assumptions into conditional probability notation:

$$\Pr(I_{t+1} \mid I_t) = 0.8$$
$$\Pr(I_{t+1} \mid O_t) = 0.1$$
$$\Pr(O_{t+1} \mid I_t) = 0.2$$
$$\Pr(O_{t+1} \mid O_t) = 0.9$$

Is the position at time $t + 1$ independent of the position at time t? If it were, it would be true that

$$\Pr(I_{t+1}) = \Pr(I_{t+1} \mid I_t) = \Pr(I_{t+1} \mid O_t)$$

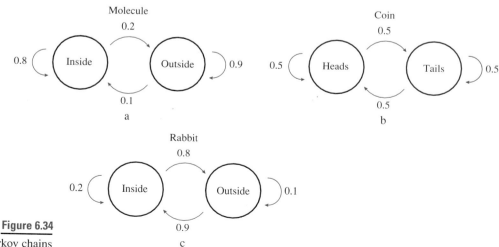

Figure 6.34
Three Markov chains

In this case, however,

$$\Pr(I_{t+1} \mid I_t) = 0.8 \neq \Pr(I_{t+1} \mid O_t) = 0.1$$

Because it is much more likely that the molecule remains inside than that it enters, the positions at different times are not independent.

How about a fair coin? Let H_t denote the event "heads" and T_t denote the event "tails" on toss t. We expect

$$\Pr(H_{t+1} \mid H_t) = 0.5$$
$$\Pr(H_{t+1} \mid T_t) = 0.5$$
$$\Pr(T_{t+1} \mid H_t) = 0.5$$
$$\Pr(T_{t+1} \mid T_t) = 0.5$$

(Figure 6.34b). The probability of heads on toss $t + 1$ is the same whether toss t was heads or tails. These conditional probabilities encode the assumption of independence.

Finally, consider a rabbit jumping in and out of its burrow, with position denoted by the same notation as for the molecule (Figure 6.34c). The rabbit, however, is very nervous and tends to switch positions. In fact,

$$\Pr(I_{t+1} \mid I_t) = 0.2$$
$$\Pr(I_{t+1} \mid O_t) = 0.9$$
$$\Pr(O_{t+1} \mid I_t) = 0.8$$
$$\Pr(O_{t+1} \mid O_t) = 0.1$$

The behavior during one minute is not independent of the behavior in the previous minute because, for instance, the rabbit is much more likely to be inside after it was outside (probability 0.9) than after it was inside (probability 0.1).

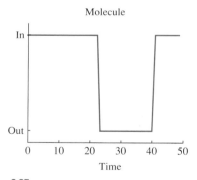
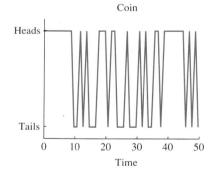
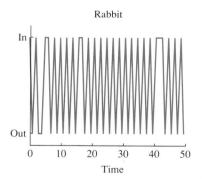

Figure 6.35
Results from three Markov chains

What are the consequences of dependence? Figure 6.35 compares computer simulations of molecular position, independent coin tosses, and the nervous rabbit. The molecule maintains position for long times with infrequent switches, the rabbit jumps back and forth with no long periods in the same position, while the coin switches between heads and tails "randomly" (in colloquial usage, we tend to confound "random" and "independent").

SUMMARY

Two events are **independent** if the probability of one conditional on the other is equal to the unconditional probability, meaning that information about one event conveys nothing about the other. The probability that two independent events happen simultaneously can be found with the **multiplication rule**. The multiplication rule generalizes to sets of **mutually independent** events. Markov chains provide an example of temporal dependence, where the state at one time depends probabilistically on the state at the previous time.

6.6 EXERCISES

1. Suppose the rabbit and eagle in Exercise 2 in Section 6.5 behave independently. Find the probability that the ecologist sees both a rabbit and an eagle during a particular hour of observation, and draw a Venn diagram of this situation.

2. Someone invents a cut-rate test for a rare disease. This test gives a positive result with probability 0.5 whether or not the patient has the disease. Suppose 1% of people actually have the disease. Find the probability of having the disease conditional on a positive test in two ways.
   a. Work it out directly, as in Section 6.5.
   b. Use independence.

3. Show that the multiplication rule (Theorem 6.3) does not work in the following cases.
   a. When events are disjoint.
   b. When one event is a subset of the other.

4. A coin is flipped 100 times and observed to come up heads 90 times. Can this happen if tosses are independent? Can you think of a way this could happen if the coin is fair?

5. Suppose a molecule is transferred among three cells according to a Markov chain. Write down conditional probabilities to describe the following situations. (It can help to draw a picture.)
   a. The position of the molecule in one minute is independent of the position in the previous minute.
   b. The molecule rarely leaves a cell. When it does so it enters each of the other cells with equal probability.
   c. Imagine the three cells arranged in a ring. The molecule rarely leaves a cell, and when it does so it always moves to the right.
   d. Imagine the three cells arranged in a line. The molecule rarely leaves a cell. If it is at the end, it moves to the middle. If it is in the middle, it enters either of the end cells with equal probability.

6. A creative geneticist discovers a variety of bizarre systems when crossing parents with genotype **Bb**. Find the probability that an offspring is **BB**, **Bb**, or **bb** in each case.
   a. Each parent contributes the **B** allele independently, but the father provides it with probability 0.8 and the mother with probability 0.5.
   b. Each parent contributes the **B** allele independently, but both provide it with probability 0.8.
   c. The father provides the **B** allele with probability 0.8. The mother provides the **B** allele with probability 0.4 if the father provides **B** and with probability 0.8 if the father provides **b**.

7. A species of bird comes in three colors, red, blue, and green; 20% are red, 30% are blue, and 50% are green.
   a. Suppose females prefer red to blue and blue to green, but mate with the first male they meet. What is the probability a female mates with a red bird? A blue bird? A green bird?
   b. Females prefer red to blue and blue to green, but pick the best of the first two males they meet. What is the probability a female mates with a green bird? What did you have to assume?

8. A small class has only three students. Each student comes to class with probability 0.9. Find the probability that all the students come to class and the probability that no students come to class in the following circumstances.
   a. The students act independently.
   b. Student 2 comes to class with probability 1.0 if student 1 does. Student 3 ignores them.
   c. Student 2 comes to class with probability 8/9 if student 1 does. Student 3 ignores them.
   d. Student 3 comes to class with probability 1.0 if both the others come. Students 1 and 2 ignore each other.

9. Using the method presented in Equation 6.9, find the long-term probability that the rabbit in Figure 6.34 is in its burrow. What is the probability that it is in for 2 minutes straight?

10. A popular probability problem refers to a once popular game show called "Let's Make a Deal." In this game, the host (Monty Hall) hands out large prizes to contestants for no reason at all. In one situation, Monty would show the contestant three doors, named door 1, door 2, and door 3. One would hide a new car, one $500 worth of false eyelashes, and the other a goat (deemed worthless by the purveyors of the show). The contestant picks door 1. But instead of showing her the prize, Monty opens door 3 to reveal the goat.
   a. Should the contestant switch her guess to door 2?
   b. If she uses the right strategy, what is her probability of getting the new car?
   c. It is later revealed that Monty does not always show what is behind one of the other doors, but does so only when the contestant guessed right in the first place (the so-called "Machiavellian Monty"). Knowing this, should

the contestant switch if Monty shows another door? Should she switch if he does not?

d. If she uses the right strategy in this case, what is her probability of getting the new car?

e. Monty becomes more devious yet and shows another door with probability 1/3 if the first guess was right and 2/3 if it was wrong. What is the best strategy now?

11. A man fancies himself a meteorologist. He knows that it rains on 40% of mornings and on 40% of afternoons, and that the probability that it rains in the afternoon if it rains in the morning is 75%. He therefore decides to bring an umbrella on 75% of the mornings when it rains.

a. What is the conditional probability that it rains in the afternoon if it does not rain in the morning?

b. If his guess in the morning is independent of what happens in the afternoon, what is the probability that he will be caught in the rain in the afternoon if he neglects to bring his umbrella on a rainy morning?

c. On dry mornings, our man decides to bring an umbrella with probability equal to the answer of part a. How often does he bring an umbrella?

d. Under the assumptions of part c, what is the best he can do? Can he always avoid getting caught in the rain in the afternoon?

e. Under the assumptions of part c, what is the worst he can do? How often can he get caught in the rain in the afternoon?

12. **COMPUTER**: Figure out a way (using the function q defined in Exercise 13 in Section 6.1) to simulate 100 offspring from each of the three mechanisms in Exercise 6. How close are your results to the mathematically expected results?

13. **COMPUTER**: The updating function for the position of a molecule is given by

$$h(x) = q_{0.7}x + q_{0.3}(1 - x)$$

where $x = 1$ represents inside, $x = 0$ outside, and q is defined in Exercise 13 in Section 6.1. Run this system for 50 steps. Based on your data, estimate $\Pr(x_{t+1} = 1)$, $\Pr(x_{t+1} = 1 \mid x_t = 1)$, and $\Pr(x_{t+1} = 1 \mid x_t = 0)$. Compare these results with what you would expect based on the updating function.

6.7 Displaying Probabilities

In order to understand fully the assignment of probabilities in a model, we need an efficient and readable way to display them. For small numbers of simple events, we have used **pie charts**. In general, however, the best and most widely used method is the **histogram**, or bar graph, which plots the probability of each event as the height of the bar. A related graph is the **cumulative distribution**, which adds probabilities in order before graphing them. A more subtle problem arises when specifying and characterizing the probabilities of simple events when the sample space is infinite. By a limiting process, we develop two mathematical and graphical tools, the **probability density function (p.d.f.)** and the **cumulative distribution function (c.d.f.)**

Probability and Cumulative Distributions

Suppose we cross two heterozygous plants of intermediate height, each with one short gene and one tall gene. If a single offspring were produced, there are three simple events: the offspring could be short, tall, or intermediate. Under the assumptions that the alleles each have a fair chance of 0.5 and come from the two parents independently, the offspring will be short with probability 0.25, tall with probability 0.25, and intermediate with probability 0.5. The biological assumptions guide us to an assignment of probabilities to all of the simple events. These probabilities add to 1, so our biologically consistent assignment is also mathematically consistent.

These probabilities can be displayed as a **probability distribution**, or **histogram** (Figure 6.36). The vertical axis is the probability. The height of the bar above the event that the height of the offspring is 40 is 0.25, the probability.

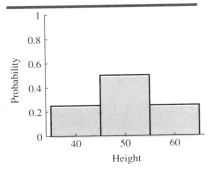

Figure 6.36
A histogram with three simple events

Histograms are convenient because we can see immediately which event is most probable (height is 50). Furthermore, the fact that the distribution is **symmetric** (would look the same in a mirror) means that a plant is just as likely to be short as to be tall.

Consider now ten toxic molecules diffusing out of a cell. Suppose that the underlying mathematical model is that each molecule has a 10% chance of leaving the cell each minute. We learn later how to compute the probability that N molecules remain after different lengths of time (Section 7.6). Results after 1, 4, and 8 min are displayed in Figure 6.37 and the following table.

Number left	Probability at $t = 1$	Probability at $t = 4$	Probability at $t = 8$
0	0.000	0.000	0.004
1	0.000	0.000	0.027
2	0.000	0.004	0.092
3	0.000	0.019	0.186
4	0.000	0.064	0.246
5	0.001	0.147	0.223
6	0.011	0.234	0.141
7	0.057	0.255	0.061
8	0.194	0.183	0.017
9	0.387	0.077	0.003
10	0.349	0.015	0.000

In the figure, the simple events are arrayed along the horizontal axis, and above each rises a bar with height equal to the probability of that simple event.

With a bit of experience, a lot of information can be read from a histogram. The tallest bar indicates the most probable simple event. The distributions at times $t = 4$ (Figure 6.37b) and $t = 8$ (Figure 6.37c) look almost symmetric around the most probable events of $N = 7$ and $N = 4$, respectively. However, there is slightly more weight (taller bars) to the left of the peak at $t = 4$ and slightly more weight to the right of the peak at $t = 8$.

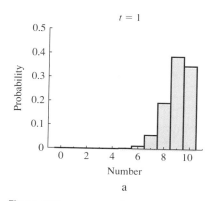
a

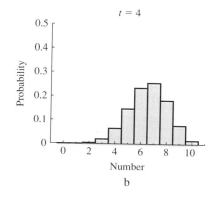
b

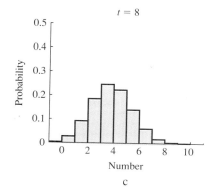
c

Figure 6.37
Diffusion of ten molecules

When the simple events have a natural numerical order, the **cumulative distribution** provides another useful way to plot results. Instead of plotting the probability that N molecules remain, we plot the probability that N or fewer molecules remain (Figure 6.38). For example, the probability that five or fewer molecules remain after 8 min is

$$\Pr(N \le 5) = \Pr(N = 0) + \Pr(N = 1) + \Pr(N = 2)$$
$$+ \Pr(N = 3) + \Pr(N = 4) + \Pr(N = 5)$$
$$= 0.004 + 0.027 + 0.092 + 0.186 + 0.246 + 0.223 = 0.778$$

The complete results are compiled in the following table.

Number left	Probability at $t = 1$	Probability at $t = 4$	Probability at $t = 8$
0	0.000	0.000	0.004
1	0.000	0.000	0.031
2	0.000	0.004	0.123
3	0.000	0.024	0.309
4	0.000	0.088	0.555
5	0.002	0.235	0.778
6	0.013	0.470	0.919
7	0.070	0.725	0.980
8	0.264	0.908	0.997
9	0.651	0.985	1.000
10	1.000	1.000	1.000

The probabilities must increase to 1.0 because there are always ten or fewer molecules remaining. Graphically, the heights of the bars in a cumulative distribution plot always increase to 1. The fact that the bars increase more quickly when t is larger means that there are *fewer* molecules left at the later times.

Figure 6.38
Diffusion of 10 molecules: cumulative distributions

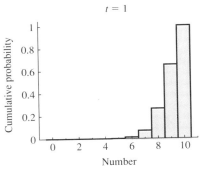
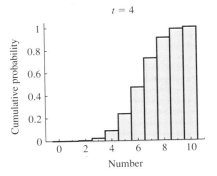
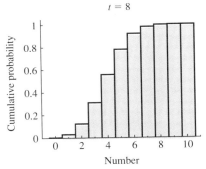

The Probability Density Function

So far, we have computed and displayed probabilities only when the number of simple events is finite. We now address the case when the simple events form a continuum. As with the derivative and the integral, the key is to think of the continuum as the limit of smaller and smaller units.

To understand the process, we consider again a single molecule diffusing out of a cell and think of measuring the time of departure with a more and more precise clock. With a rough measuring device, we might measure to the nearest 10 s. With a more precise device, we might measure to the nearest millisecond. But as mathematicians we can imagine measuring exactly, to an infinite number of decimal places.

The simple events are "the molecule left exactly at time t" for any positive value of t. To build probabilities of other events, such as "the molecule left after time 1.374," we must find the probabilities of the simple events and add them. What is the probability that the molecule left after exactly $t = \pi$ seconds? Strictly speaking, this probability is 0. A difference even in the millionth decimal place would mean a completely different simple event. But this logic holds for any time t. The probability that it left exactly at **any** particular time is 0, which might sound rather odd. How can we add 0s and get a positive number? (As a preview, this is one way to think of *integration*.)

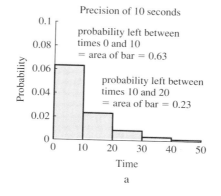
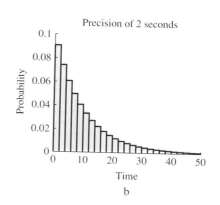

Figure 6.39

Probability the molecule left the cell during a particular interval, measured with two devices

The way to break out of this apparent paradox is to think back to our less precise measuring device. If our device has precision of only 10 s, we cannot distinguish measurements in the intervals from 0 to 10, 10 to 20, and so forth. A true value of 3.14159 cannot be told from a true value of 2.71828, for example. We can break the possibilities from 0 to 50 into five blocks of width 10. The results (which we derive in Section 7.6) are summarized in the following table.

Time range	Probability	Height of bar
0–10	0.632	0.063
10–20	0.233	0.023
20–30	0.086	0.009
30–40	0.031	0.003
40–50	0.012	0.001

We draw a new type of histogram, where the *area* (rather than the height) of a bar represents the probability that the result lies in that interval (Figure 6.39a). For example, the probability that the molecule left between times 0 and 10 is 0.63, so we draw a bar with height 0.063 and width 10. Using a more precise device with accuracy 2.0 s, we break the possibilities into intervals from 0 to 2, 2 to 4, and so forth, finding the results in the following table. We again draw bars with area equal to the probability (Figure 6.39b). For example, the probability that the molecule left between times 2 and 4 is 0.148, so the bar has height half that, or 0.074.

490 Chapter 6 Probability Theory and Descriptive Statistics

Time range	Probability	Height of bar	Time range	Probability	Height of bar
0–2	0.181	0.091	26–28	0.013	0.007
2–4	0.148	0.074	28–30	0.011	0.006
4–6	0.122	0.061	30–32	0.009	0.005
6–8	0.099	0.050	32–34	0.007	0.004
8–10	0.081	0.041	34–36	0.006	0.003
10–12	0.067	0.033	36–38	0.005	0.002
12–14	0.055	0.027	38–40	0.004	0.002
14–16	0.045	0.022	40–42	0.003	0.002
16–18	0.037	0.018	42–44	0.003	0.001
18–20	0.030	0.015	44–46	0.002	0.001
20–22	0.025	0.012	46–48	0.002	0.001
22–24	0.020	0.010	48–50	0.001	0.001
24–26	0.016	0.008			

With an accuracy of 0.5 s (Figure 6.40a), the histogram looks a lot like the curve sketched in Figure 6.40b. What does this curve mean? Suppose we wanted to find the probability that the molecule left between times 0 and 10. With a precision of 10 (Figure 6.39a), this is the area under the first bar. With a precision of 2 (Figure 6.39a), this is the sum of the areas under the first five bars. These areas are approximately the area under the curve. With a mathematically exact device, the probability is equal to the area under the curve between 0 and 10. If we knew the formula for the curve, we could find probabilities by integrating.

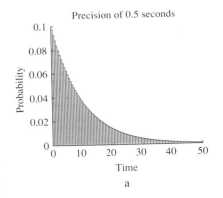
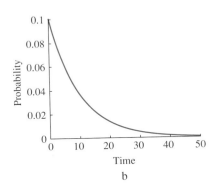

Figure 6.40
Probability density function for a diffusing molecule

a

b

Such a curve is called a **probability density function (p.d.f.)**. What curves can act as a p.d.f.? The requirements are given in the following definition.

■ **Definition 6.6** Suppose simple events can be indexed by the real number x. Let a be the smallest possible simple event and b the largest. A function $f(x)$ can be a p.d.f. if

1. $f(x) \geq 0$ for all x with $a \leq x \leq b$.
2. $\int_a^b f(x)\,dx = 1$.

The first requirement says that probabilities must be positive, and the second that they must add to 1. The definition does *not* require the value of the function to be less than 1, as in a probability distribution.

How do we use a p.d.f.? The probability is the area, and the area is the integral. Therefore, if X denotes the outcome of the experiment and f is the p.d.f, then

$$\Pr(a \leq X \leq b) = \int_a^b f(x)\,dx \tag{6.16}$$

A probability density acts like an ordinary density (such as grams per centimeter). Even though there is no mass exactly at a particular point, the total mass can be found by integrating.

The exact p.d.f. for the time T when a diffusing toxic molecule leaves the cell is

$$f(t) = 0.1e^{-0.1t}$$

when the rate of departure is 0.1. The maximum value of the p.d.f. occurs at the time when the molecule is most likely to leave. In this case, the highest value is at $t = 0$, meaning that the molecule is most likely to leave near time 0, a rather counterintuitive result.

To find the probability that the molecule left between time 0 and time 10, integrate the p.d.f. from $t = 0$ to $t = 10$:

$$\begin{aligned}\Pr(0 \leq T \leq 10) &= \int_0^{10} 0.1e^{-0.1t}\,dt \\ &= -e^{-0.1t}\Big|_0^{10} \\ &= -e^{-1.0} + e^{-0.0} \approx 0.632\end{aligned}$$

If we ran this experiment 1000 times, about 623 molecules would leave before time 10 (Figure 6.41a). Similarly, the probability a molecule leaves between time 5 and time 15 is

$$\begin{aligned}\Pr(5 \leq T \leq 15) &= \int_5^{15} 0.1e^{-0.1t}\,dt \\ &= -e^{-0.1t}\Big|_5^{15} \\ &= -e^{-1.5} + e^{-0.5} \approx 0.383\end{aligned}$$

(Figure 6.41b).

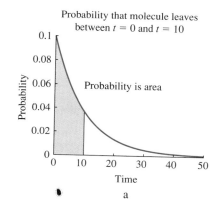

Figure 6.41
Finding probabilities as areas under a p.d.f.

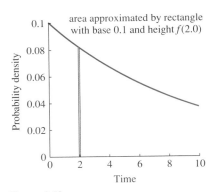

Figure 6.42
Approximating area with a rectangle

To understand the definition better, it helps to think of finding the probability that the molecule leaves between two times that are close together, such as $t = 2.0$ and $t = 2.01$. The exact probability is

$$\Pr(2.0 \leq T \leq 2.1) = \int_{2.0}^{2.1} 0.1 e^{-0.1t}\, dt$$
$$= -e^{-0.1t}\big|_{2.0}^{2.1}$$
$$= -e^{-0.21} + e^{-0.2} \approx 0.00815$$

Because the interval is narrow, we can approximate this region as a rectangle with base 0.1 and height $f(2.0)$

$$\Pr(2.0 \leq T \leq 2.1) \approx \text{base} \times \text{height}$$
$$= 0.1 f(2.0)$$
$$= 0.1 \cdot 0.1 e^{-0.2} = 0.00819$$

(Figure 6.42). The probability that the molecule leaves the cell near time 2.0 is *proportional* to the height of the p.d.f. at that point.

Three other examples of probability density functions are shown in Figure 6.43. Figure 6.43a shows the p.d.f. for a departing molecule with $\lambda = 10$. The vertical range rises above 1. This is mathematically consistent because the height is not *equal* to the probability, but only *proportional* to it. The large value of the p.d.f. at $t = 0$ indicates that the molecule is very likely to leave near time $t = 0$, consistent with its high rate of departure. Figure 6.43b shows a hypothetical p.d.f. with a finite domain. For values outside this domain, the probability of events is 0.

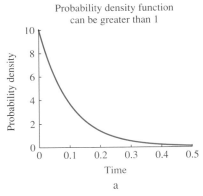
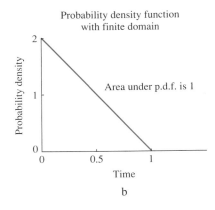
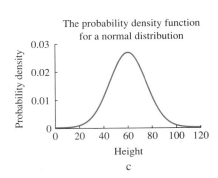

Figure 6.43
Three probability density functions

The bell curve, or normal density function, is shown in Figure 6.43c. This density function, which we study in great detail later, characterizes many measurements and is the most important shape in statistics. This function takes on its maximum where the height is equal to 60, which is right in the middle of this symmetric picture. Strictly speaking, the domain of this distribution includes all values from $-\infty$ to ∞. However, because the value of the p.d.f. becomes very small, the probabilities of absurd results (such as negative height) are negligible.

The Cumulative Distribution Function

What is the probability that our original molecule with p.d.f.

$$f(t) = 0.1 e^{-0.1t}$$

left before some given time t? To find the probability it left before time $t = 10$, we integrate:

$$\text{Pr}(\text{molecule left before time } 10) = \int_0^{10} f(t)\, dt$$

The probability it left before a general time t is

$$\text{Pr}(\text{molecule left before time } t) = \int_0^{t} f(s)\, ds$$

(We changed t to s inside the integral to avoid using the same letter to mean two different things.)

Thought of as a function of t, this probability defines the **cumulative distribution function (c.d.f.)**. Formally, we have the following definition.

■ **Definition 6.7** Suppose $f(x)$ is a p.d.f. with smallest simple event a (which could be $-\infty$) and largest simple event b (which could be ∞). Then the function $F(x)$ defined by

$$F(x) = \int_a^x f(y)\, dy$$

is the cumulative distribution function with domain $a \leq x \leq b$. ■

In general, we use a capital letter, such as F, to denote the c.d.f. associated with a p.d.f. denoted by the related lowercase letter, such as f. The cumulative distribution functions for the p.d.f.s in Figure 6.43 are shown in Figure 6.44. Like a cumulative distribution, the cumulative distribution function must increase to 1. The curve is steepest where the p.d.f. takes on its largest value.

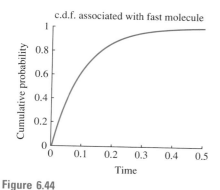
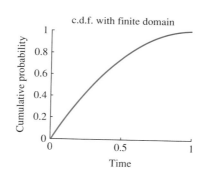
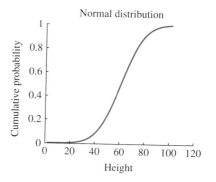

Figure 6.44
Three cumulative distribution functions

In the example with p.d.f $f(t) = 0.1e^{-0.1t}$, the c.d.f. is

$$\begin{aligned} F(t) &= \int_0^t f(s)\, ds \\ &= \int_0^t 0.1 e^{-0.1s}\, ds \\ &= -e^{-0.1s}\big|_0^t \\ &= 1 - e^{-0.1t} \end{aligned}$$

The probability that the molecule left at or before time t is given by $F(t)$. The probability that the molecule left before time 10 is

$$\text{Pr(molecule left before time 10)} = F(10.0) = 1 - e^{-0.2} \approx 0.632$$

(Figure 6.45). We can check that the cumulative distribution increases to a value of 1, meaning that the molecule must eventually leave, by taking the limit

$$\lim_{t \to \infty} F(t) = \lim_{t \to \infty} 1 - e^{-0.1t} = 1$$

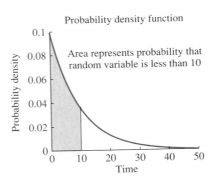 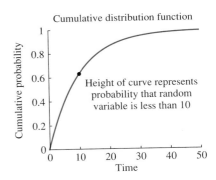

Figure 6.45

The probability distribution function and cumulative distribution function for a molecule

The Fundamental Theorem of Calculus (Theorem 4.4) makes working with the c.d.f. convenient. First, because the c.d.f. is the integral of the p.d.f., the p.d.f. must be the derivative of the c.d.f., or

$$\frac{d}{dt}(1 - e^{-0.1t}) = 0.1 e^{-0.1t}$$

Second, we can use the c.d.f. to find probabilities by subtracting. The Fundamental Theorem says that

$$\int_a^b f(x)\,dx = F(b) - F(a)$$

Therefore, the probability that the departure time lies between $t = 5$ and $t = 15$ is

$$F(15) - F(5) = (1 - e^{-0.1 \cdot 15}) - (1 - e^{-0.1 \cdot 5}) \approx 0.776 - 0.393 = 0.383$$

as we found before (Figure 6.46).

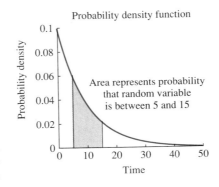 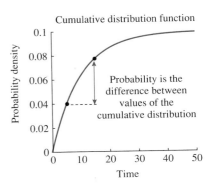

Figure 6.46

Computing probabilities with the cumulative distribution function

6.7 EXERCISES

1. Draw histograms to describe the following probabilities.

Number of mutants	Experiment a	Experiment b	Experiment c
0	0.1	0.6	0.3
1	0.2	0.3	0.2
2	0.3	0.1	0.2
3	0.3	0.0	0.2
4	0.1	0.0	0.1

2. On the histograms, find the most and least likely simple events. Which are symmetric?

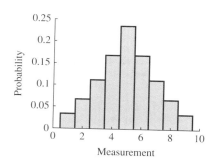

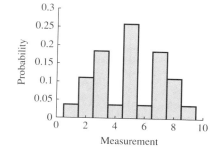

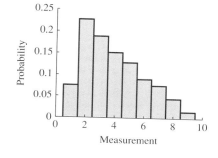

3. Using the histograms shown for Exercise 2, estimate the probabilities of the following compound events.
  a. The measurement is less than or equal to 4.
  b. The measurement is equal to 7.
  c. The measurement is exactly 1 unit away from the peak (either above or below)

4. Sketch the cumulative distributions associated with the histograms of Exercise 2.

5. On the graphs of the p.d.f.s, find the most and least likely simple events. Which of the pictures are symmetric? Estimate the probability that the measurement is less than 0.2, and the probability that the measurement is greater than 0.8.

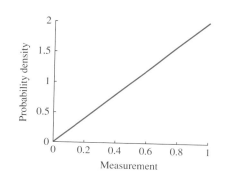

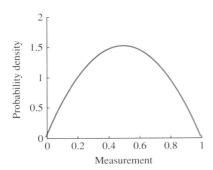

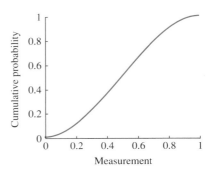

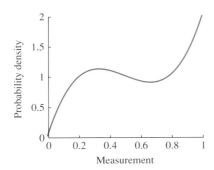

7. Suppose the p.d.f. for a measurement X is given by $g(x) = 0.5e^{-0.5x}$ for positive x with associated c.d.f. $G(x) = 1 - e^{-0.5x}$.
   a. Plot the p.d.f. and c.d.f.
   b. Check that the p.d.f. is really the derivative of the c.d.f.
   c. Find $\Pr(X \leq 1)$. Indicate this on both graphs.
   d. Write $\Pr(X \geq 1)$ in terms of both the p.d.f. (as an integral) and the c.d.f. Compute the value.
   e. Compute $\Pr(1 \leq X \leq 3)$.
   f. $\Pr(1 \leq X \leq 1.01)$. Show it is approximately $g(1) \cdot 0.01$.

8. Draw the following.
   a. A symmetric histogram with the shortest bar in the middle.
   b. Sketch the cumulative distribution associated with the histogram of part **a**.
   c. A symmetric p.d.f. with the minimum value in the middle.
   d. Sketch the c.d.f. associated with the p.d.f. of part **c**.

6. The three pictures show cumulative distribution functions. Match them with the p.d.f.s shown in Exercise 5.

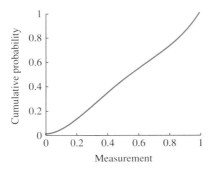

9. Find and graph the cumulative distribution functions associated with the following p.d.f.s. Show that the c.d.f.s increase to 1. Find the probability that the measurement is less than 0.3, and indicate this on your graph.
   a. $f(x) = 1$ for $0 \leq x \leq 1$
   b. $f(x) = 2 - 2x$ for $0 \leq x \leq 1$
   c. $f(x) = 6x(1-x)$ for $0 \leq x \leq 1$

10. **COMPUTER:** Suppose that the p.d.f. for the time a molecule leaves a cell is equal to

$$f(x) = 2.5e^{-2.5x}$$

for $x \geq 0$. Use integration to compute the c.d.f., $F(x)$. Plot f and F on one graph for $0 \leq x \leq 2$. Compute the probability that a number chosen according to this p.d.f. lies between 1 and 1.5, and mark the associated area on your graph of f. Compute the probability that the number is less than 0.6, and indicate this on your graphs of f and F. At which value of x does F cross 0.5? What does this mean?

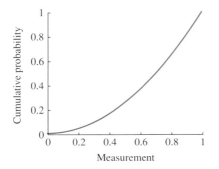

11. **COMPUTER:** If a molecule leaves a cell at rate 0.5 per second, the probability that it leaves in any short time interval of length Δt is approximately $0.5\Delta t$.
   a. Suppose we choose $\Delta t = 1.0$. Write an updating function to describe the probability it is inside at time $t + \Delta t$ in terms of the probability it is inside at time t.
   b. Plot a solution assuming that it begins inside with probability 1.
   c. Do part **a** with $\Delta t = 0.1$. Try to plot the solution in a way to match the results of part **b**.
   d. Do part **a** with $\Delta t = 0.01$.
   e. Graph the p.d.f. $f(x) = 2.0e^{-0.5t}$. How does it compare with the results in **b**, **c**, and **d**?

6.8 Random Variables

Many measurements can be made from a single experiment. Rather than setting up a whole new sample space and set of simple events for each measurement, we exploit the common probabilistic structure by using **random variables**. Here we introduce the two main types of random variables, discrete and continuous. The probabilities associated with different values of random variables are displayed in the same way as simple events, with probability distributions and probability density functions. We conclude with the mathematical version of the **average** of a measurement, the **expectation** of a random variable.

Types of Random Variables

Throwing a dart at a dartboard is a sort of experiment and can be thought of as a metaphor for any experiment. There is a continuum of simple events, one for each point on the board. Often enough, we do not care *exactly* where the dart hit but only what it scored (Figure 6.47). The score is an example of a **random variable**. Because we can figure out the score from the exact position where the dart hit, the score is a **function** of the exact location. This function has domain equal to the sample space (all the points on the board) and range equal to the set of possible scores. We give the following formal definition.

■ **Definition 6.8** A **random variable** is a function from the sample space to some subset of the real numbers. When the number of values of the function is finite, it is called a **discrete random variable.** When the number of values of the function is infinite, it is called a **continuous random variable**. ■

In more scientific terms, a random variable is a measurement that depends on the result of an experiment. When the measurement can take on only a finite number of values, the associated random variable is a **discrete random variable**. When the measurement can take on an infinite range of values, the associated random variable is a **continuous random variable**.

Random variables are displayed in the same way as probabilities. Discrete random variables are shown with histograms and cumulative distributions. Continuous random variables are plotted with probability density functions and cumulative distribution functions.

There are many random variables associated with a given experiment. A simplified scoring system might give a player 1 for hitting the board and 0 for hitting anything else. A random variable taking on only the values 0 and 1 is called a **Bernoulli random variable**.

Figure 6.47
The dart score is a random variable

Definition 6.9 A Bernoulli random variable is a discrete random variable that takes on the value 1 with probability p and the value 0 with probability $1 - p$.

If the probability that a dart hits the board and scores 1 is 0.8, the random variable has the histogram shown in Figure 6.48a. As before, the height of the bar gives the probability of a particular event, here the event of scoring 1. In a simpler game, where a player scores 1 for flipping heads and 0 for flipping tails, the histogram consists of two bars of equal height (Figure 6.48b).

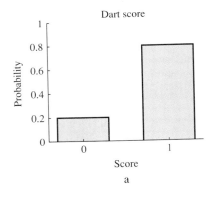
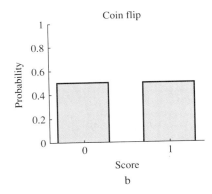

Figure 6.48
Histograms for Bernoulli random variables

The dart score is an intermediate case. The number of possible scores is finite, but greater than 2. We usually denote random variables by capital letters, such as S for score. To indicate that a dart scored 20, we write $S = 20$. In our probabilistic terminology, $S = 20$ represents the event "the dart scored 20." This is not a simple event because there are many different ways it could happen.

To plot the histogram for the discrete random variable S representing the score, we array the different possible scores along the horizontal axis and indicate their probabilities with the bars (Figure 6.49a). The scores have been resolved into the 44 possible scores from a dart throw, each with its own probability. The probability that $S = 0$ is 0.2 and the probability that $S = 20$ is 0.06. Figure 6.49b shows the associated cumulative distribution, the probability that the score is less than or equal to a particular value.

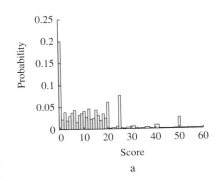
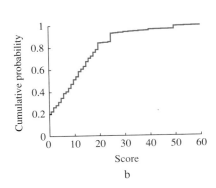

Figure 6.49
Histogram and cumulative distribution for dart score

Many random variables take on a continuum of values and are called **continuous random variables**. Assume that the dartboard has a radius of 4 in. The random variable that measures how far the dart hits from the center of the board can take on any value between 0 and 4. Denoting this random variable by R, we indicate the event "the dart hit 2.5 in. from the center" by $R = 2.5$.

Like the exact time that a molecule left a cell, the probability that R takes on any particular value is 0. To find actual probabilities, we must integrate a **probability density function**. For the random variable R measuring the distance of the dart from the exact center of the board, the p.d.f. $g(r)$ is the function satisfying

$$\Pr(a \leq R \leq b) = \int_a^b g(r)\,dr$$

A possible p.d.f. for the dart is

$$g(r) = \frac{1}{8}(4 - r)$$

(Figure 6.50a). This is a mathematically consistent p.d.f. because it takes on only positive values, and because

$$\int_0^4 \frac{1}{8}(4 - r)\,dr = \frac{r}{2} - \frac{r^2}{16}\Big|_0^4$$

$$= \frac{4}{2} - \frac{4^2}{16} = 1$$

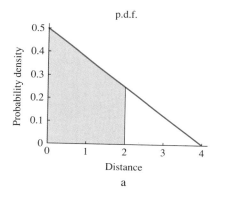
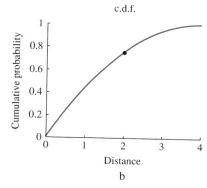

Figure 6.50
The p.d.f. and c.d.f. for the first dart player

As before, $g(r)$ is proportional to the probability that R take on a value near r. Because the p.d.f. is decreasing, this player is more likely to hit near the center of the board ($R = 0$) than near the edge ($R = 4$). We integrate to find the probability that she hits within 2 in. of the center,

$$\Pr(0 \leq R \leq 2) = \int_0^2 \frac{1}{8}(4 - r)\,dr$$

$$= \frac{r}{2} - \frac{r^2}{16}\Big|_0^2$$

$$= \frac{2}{2} - \frac{2^2}{16} = 0.75$$

The cumulative distribution function $G(r)$ gives the probability that $R \leq r$. The c.d.f. associated with the dart is

$$\begin{aligned} G(r) &= \Pr(R \leq r) \\ &= \int_0^r g(s)\,ds \\ &= \int_0^r \frac{1}{8}(4-s)\,ds \\ &= \frac{r}{2} - \frac{r^2}{16} \end{aligned}$$

(Figure 6.50b). Using the c.d.f., we could find the probability that the dart hits within 2 in. of the center by

$$G(2) = \frac{2}{2} - \frac{2^2}{16} = 0.75$$

matching the answer found earlier.

A different player might have a different probability density function. One possibility is

$$h(r) = \frac{r}{8}$$

(Figure 6.51a). This player is more likely to hit far from the center than near the center. In fact, the probability that he hits within 2 in. of the center is only

$$\int_0^2 \frac{r}{8}\,dr = \left.\frac{r^2}{16}\right|_0^2 = \frac{2^2}{16} = 0.25$$

The c.d.f. for this player is

$$\begin{aligned} H(r) &= \int_0^r h(s)\,ds \\ &= \int_0^r \frac{s}{8}\,ds \\ &= \left.\frac{s^2}{16}\right|_0^r \\ &= \frac{r^2}{16} \end{aligned}$$

(Figure 6.51b).

Figure 6.51
The p.d.f. and c.d.f. for the second dart player

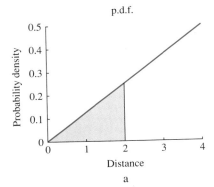
p.d.f.
a
Distance

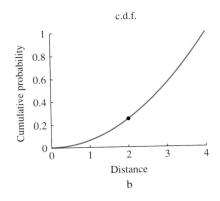
c.d.f.
b
Distance

6.8 Random Variables

Expectation of a Discrete Random Variable

Expectation is the mathematician's word for average. If ten students show up for class 75% of the time and six show up 25% of the time (Figure 6.52), the **average** number who show up is

$$\text{average number of students in class} = 10 \cdot 0.75 + 6 \cdot 0.25 = 9$$

The average (also known as the **mean** or **arithmetic mean**) number of students is nine, even though the exact number never takes on this value. We multiplied each value of the measurement (the number of students) by the probability of that measurement and added the resulting products.

The definition of the **expectation** of a random variable formalizes this approach. Let $v_1, \ldots, v_n$ denote the n possible values of a random variable V. Suppose that $V = v_i$ with probability p_i (Figure 6.53). The expectation of V is often denoted by $E(V)$ or $\overline{V}$, and is defined as follows.

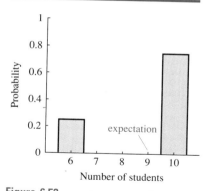

Figure 6.52
A simple discrete random variable

■ **Definition 6.10** The **expectation** $E(V)$ or $\overline{V}$ of a discrete random variable V taking on the value v_i with probability p_i is

$$E(V) = \overline{V} = \sum_{i=1}^{n} v_i p_i$$ ■

To find the expectation, multiply the values by the probabilities and add the resulting products. In the class, $v_1 = 10$, $v_2 = 6$, $p_1 = 0.75$, and $p_2 = 0.25$. The expectation is

$$\text{expected number of students in class} = v_1 p_1 + v_2 p_2$$
$$= 10 \cdot 0.75 + 6 \cdot 0.25 = 9.0$$

Consider again the number of molecules left in a cell at three different times:

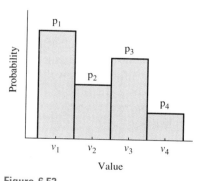

Figure 6.53
The general notation for a discrete random variable

Number left	Probability at $t = 1$	Probability at $t = 4$	Probability at $t = 8$
0	0.000	0.000	0.004
1	0.000	0.000	0.027
2	0.000	0.004	0.092
3	0.000	0.019	0.186
4	0.000	0.064	0.246
5	0.001	0.147	0.223
6	0.011	0.234	0.141
7	0.057	0.255	0.061
8	0.194	0.183	0.017
9	0.387	0.077	0.003
10	0.349	0.015	0.000

First of all, we can think of the measurements at times 1, 4, and 8 as three random variables, which we denote by $N_1, N_4,$ and N_8, respectively. We can use our formula to find the expectation of each. At $t = 1$, for example,

$$E(N_1) = \overline{N}_1 = 0 \cdot 0.0 + 1 \cdot 0.0 + 2 \cdot 0.0 + 3 \cdot 0.0 + 4 \cdot 0.0 + 5 \cdot 0.001$$
$$+ 6 \cdot 0.011 + 7 \cdot 0.057 + 8 \cdot 0.194 + 9 \cdot 0.387 + 10 \cdot 0.349$$
$$= 9.0$$

On average, there are exactly nine molecules in the cell after 1 min. Similarly, we can use the formula for expectation to compute

$$E(N_4) = \overline{N}_4 = 6.56$$
$$E(N_8) = \overline{N}_8 = 4.30$$

The expected number of molecules remaining in the cell declines over time. The expectation is always near the center of the distribution (Figure 6.54) but not quite in the middle of the tallest bar.

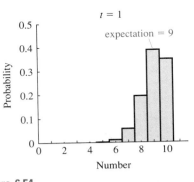 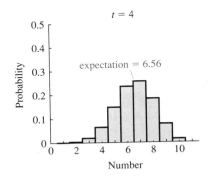 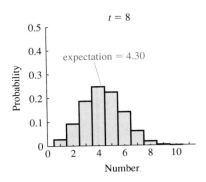

Figure 6.54

Probability distributions and their expectations

The expectation is probably the most widely used **statistic**. Recall that a statistic is one number that summarizes many numbers (Definition 6.1). The expectation summarizes the probability distribution in one way. The information given by this statistic is not complete; we cannot figure out the values and probabilities from the expectation. Knowing that nine students attend class on average does not tell us whether exactly nine ever do, or with what probability.

Expectation of a Continuous Random Variable

The idea of summing the products of the values and probabilities to find the expectation does not quite work for continuous random variables because each simple event has probability 0. By returning to the Riemann sum approach, we will find the expectation of a continuous random variable using an integral.

Figure 6.55

Approximating a continuous random variable

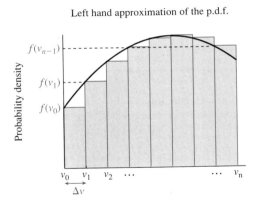 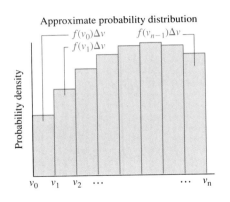

By breaking the set of all possible values of the random variable into pieces (Figure 6.55), we can approximate a continuous random variable with a discrete random variable. As in the left-hand Riemann approximation to a function (Section 4.5), we approximate the values of the continuous random variable with a discrete random variable that takes on the values $v_0, \ldots, v_{n-1}$ with probabilities $f(v_0)\Delta v, \ldots, f(v_{n-1})\Delta v$. In scientific terms, this corresponds to being unable to distinguish the values in the interval v_i to $v_i + \Delta v$ from v_i itself. The expectation of the approximate discrete random variable is

$$\sum_{i=0}^{n-1} v_i f(v_i) \Delta v$$

This has precisely the form of a Riemann sum. Taking the limit as $n \to \infty$ gives a definite integral, leading us to the definition of the expectation of a continuous random variable.

Definition 6.11 The **expectation** $E(V)$ or $\overline{V}$ of a continuous random variable V taking on values between a and b with p.d.f. $f(v)$ is

$$E(V) = \overline{V} = \int_a^b v f(v) \, dv$$

Again, we can think of the expectation as the "sum" of the value times the probability, where here the sum is an integral and the probability is the probability density function.

Consider the p.d.f. that describes the distance R that the dart hits from the center of the board,

$$g(r) = \frac{1}{8}(4 - r)$$

defined for $0 \leq r \leq 4$ (Figure 6.56a). To find the expectation $\overline{R}$, we integrate:

$$\overline{R} = \int_0^4 r g(r) \, dr$$
$$= \int_0^4 r \frac{1}{8}(4 - r) \, dr$$
$$= \int_0^4 \frac{1}{8}(4r - r^2) \, dr$$
$$= \frac{r^2}{4} - \frac{r^3}{24} \Big|_0^4 = \frac{4}{3}$$

The mean distance is less than halfway out because the p.d.f. has extra weight near 0.

For the second player, the p.d.f. is

$$h(r) = \frac{r}{8}$$

with expectation

$$\overline{R} = \int_0^4 r h(r) \, dr = \int_0^4 r \frac{r}{8} \, dr = \int_0^4 \frac{r^2}{8} \, dr = \frac{r^3}{24} \Big|_0^4 = \frac{8}{3}$$

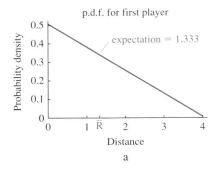
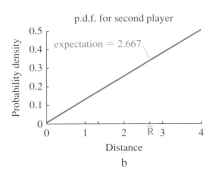

Figure 6.56
The p.d.f. and expectations for two dart players

The mean distance is more than halfway out because the p.d.f. has extra weight near 4 (Figure 6.56b).

SUMMARY

When more than one measurement is made from an experiment, the measurements can be thought of as **random variables**, functions from the sample space to the real numbers. Important examples include **Bernoulli random variables**, which take on the two values 0 and 1, **discrete random variables**, which take on a finite number of values, and **continuous random variables**, which take on a continuum of values. Random variables can be plotted with the same tools as probabilities: histograms and cumulative distributions in the discrete case and probability density functions and cumulative distribution functions in the continuous case. As scientists, most of the probabilities we think about are associated with some random variable taking on a particular value. The **expectation** of a random variable, defined to match the intuitive idea of average, is found by multiplying the value by the probability and summing (in the discrete case) or integrating (in the continuous case).

6.8 EXERCISES

1. Think about one or more molecules leaving a cell.
   a. Make up two Bernoulli random variables for different measurements associated with this process, and give plausible probabilities to the two values. Explain why you chose the probabilities you did.
   b. Make up two discrete random variables (not Bernoulli random variables) for different measurements, and draw a plausible histogram for each case. Explain why you chose the probabilities you did.
   c. Make up two continuous random variables for different measurements, and sketch a plausible p.d.f. for each. Explain why you drew the p.d.f. as you did.

2. In addition to the two dart players with p.d.f.s $g(r) = \frac{1}{8}(4-r)$ and $h(r) = \frac{r}{8}$, suppose another has p.d.f. $j(r) = 0.25$. Sketch each p.d.f., and indicate the following on your graphs.
   a. Find the probability that each player hits within 1 in. of the center.
   b. Find the probability that each player hits between 1 and 3 in. from the center.
   c. Find the probability that each player hits more than 3 in. from the center.

3. Recall Exercise 4 in Section 6.5. Imagine that older cells, instead of not staining as often, do not stain as well. In particular, they get a score of 7 on the stainometer, whereas young cells score 9.
   a. If 30% of the cells are young, what is the expected score of a cell on the stainometer?
   b. If 70% of the cells are young, what is the expected score of a cell on the stainometer?
   c. What happens when all the cells are young?

4. Further study of the cell staining problem (Exercise 3) reveals the following: New cells score 9.5 on the stainometer, 1-day-old cells score 9.0 on the stainometer, 2-day-old cells score 8.0 on the stainometer, and 3-day-old cells score 5.0

on the stainometer. Suppose

$$Pr(\text{cell is 0 days old}) = 0.4$$
$$Pr(\text{cell is 1 day old}) = 0.3$$
$$Pr(\text{cell is 2 days old}) = 0.2$$
$$Pr(\text{cell is 3 days old}) = 0.1$$

a. Find the expected score on the stainometer.
b. The lab finds a way to eliminate the oldest cells (more than 3 days old) from its stock. What is the expected score?
c. The lab finds a way to eliminate all cells more than 2 days old from its stock. What is the expected score?
d. The lab finds a way to eliminate all cells more than 1 day old from its stock. What is the expected score?

5. Suppose immigrants arrive into three populations with the following probabilities.

	Population a		Population b		Population c
Number	Probability	Number	Probability	Number	Probability
−1	0.4	−1	0.4	−10	0.4
0	0.2	0	0.2	0	0.2
1	0.3	1	0.3	1	0.3
2	0.1	100	0.1	2	0.1

Find the expected number arriving in each population. About how many immigrants would arrive in ten years? Which of the populations will grow? Does the expectation seem close to the "middle" of the distribution?

6. Approximate the p.d.f.

$$g(r) = \frac{1}{8}(4 - r)$$

by breaking the interval from 0 to 4 into four pieces.
a. Draw the histogram for the approximate discrete random variable.
b. Find the probability in each interval.
c. Find the approximate expectation, and compare with the actual expectation.
d. Add the approximate probabilities. Does anything seem wrong?

7. Approximate the p.d.f.

$$h(r) = \frac{r}{8}$$

by breaking the interval from 0 to 4 into eight pieces.
a. Draw the histogram for the approximate discrete random variable.
b. Find the probability in each interval.
c. Find the approximate expectation, and compare with the actual expectation.
d. Add the approximate probabilities. Does anything seem wrong?

8. Find expectations of random variables with the given probability density functions.
a. $f(x) = 1$ for $0 \leq x \leq 1$
b. $f(x) = e^{-x}$ for $0 \leq x \leq \infty$
   HINT: The antiderivative of the function xe^{-x} is $-xe^{-x} - e^{-x}$.
c. $f(x) = 5e^{-5x}$ for $x \geq 0$
   HINT: Use substitution to get the integral into the same form as in part **b**.
d. $f(x) = 6x(1 - x)$ for $0 \leq x \leq 1$. Could you have guessed the answer by looking at the graph?

9. Check that the following could be probability density functions and compute their expectations. Does anything seem odd about them?

a. $f(x) = \dfrac{1}{2\sqrt{x}}$ for $0 < x \leq 1$

b. $g(t) = \dfrac{1}{x^2}$ for $1 \leq x < \infty$

c. $h(y) = \dfrac{1}{2\sqrt{1-y}}$ for $0 \leq y < 1$

10. **COMPUTER:** Your computer should have a way to roll a random die (giving results 1 through 6 each with equal probability). Roll such a die 5, 10, 20, 50, and 100 times, and find the average score. How close is each to the expectation?

11. **COMPUTER:** Consider again the population growing by stochastic immigration as in Exercise 12 in Section 6.1, where an immigrant arrives each year with probability 0.5.
a. Generate two 50-generation solutions starting from populations of 0.
b. What is the expected population as a function of time?
c. Does the solution get closer to or farther from the expectation?
d. Find the average number of immigrants that have arrived by each time (if 13 immigrants arrived in the first 25 time steps, the average is $13/25 = 0.52$ per year).
e. Does the average get closer or farther from the expectation?

12. **COMPUTER:** Your computer should have a way to choose a random number between 0 and 1 (using the uniform p.d.f.). Pick 5, 10, 20, 50, and 100 such numbers and find the average. How close is each to the expectation?

6.9 Descriptive Statistics

In Section 6.8, we computed one important statistic, the expectation, that gives one idea of the "middle" of a probability distribution for a discrete random variable or of a probability density function for a continuous random variable. Many other statistics are used to summarize probability distributions and random variables. Like the expectation, they give an idea of what to "expect" from an experiment, but they have different applications. The **median** lies right in the middle of the probabilities, the **mode** is the most probable single value, and the **geometric mean** is the appropriate average for random variables that are *multiplied* rather than added.

The Median

The expectation might seem the most natural measure of the central value of a random variable. The expectation, however, can be rather far from the "middle" of a distribution (Exercise 5 in Section 6.8). A rare extreme value can pull the expectation far from the bulk of the values. Two alternative statistics, the **median** and the **mode**, capture different intuitive ideas of the middle.

The median is defined to be right in the middle of the probabilities. Picking a value less than the median is exactly as likely as picking a value greater than the median. Consider the probability distribution of dart scores (Figure 6.57a). The median is the score that gets beaten exactly half the time. The best way to find the median is to plot the cumulative distribution and find where it crosses 0.5 (Figure 6.57b). In this case, the median is near 10. It is just as likely to get a score less than 10 as it is to get a score greater than 10.

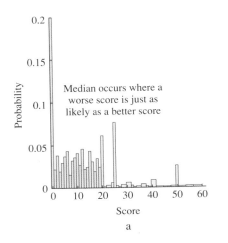 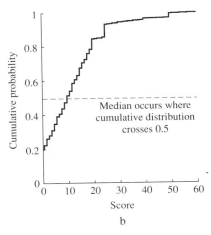

Figure 6.57
The median of the dart scores

When is the median a good measure of the middle? Unlike the expectation, the median is not sensitive to a few large values. In the figure, the probability of getting a bullseye is 0.025. If the score for a bullseye were increased to 6.02×10^{23}, the expectation would increase to 1.5×10^{22}. The median would remain the same. When we are interested in a "typical" score, the median can be a better measure of the middle.

6.9 Descriptive Statistics

■ Algorithm 6.1 (Finding the median of a discrete random variable)

1. Sort the values of the random variable in order.
2. Compute the cumulative distribution.
3. Find the point where the cumulative distribution crosses 0.5. If it hits exactly 0.5 at some value v_i, this is the median. If it hits between values v_i and v_{i+1}, the median lies between v_i and v_{i+1}. ■

The median is generally harder to compute than the mean. To find the median of the scores, we had to sort them into order (which was easy in this case), find the cumulative distribution, and find when it is equal to 0.5. The expectation can be computed by multiplying and adding, without sorting or solving anything.

Recall again the probability distribution for the number of molecules out of ten left in a cell after 8 min, given in the following table.

Number left	Probability at $t = 8$	Cumulative probability
0	0.004	0.004
1	0.027	0.031
2	0.092	0.123
3	0.186	0.309
4	0.246	0.555
5	0.223	0.778
6	0.141	0.919
7	0.061	0.980
8	0.017	0.997
9	0.003	1.000
10	0.000	1.000

From the cumulative distribution, we can see that the probabilities cross 0.5 between 3 and 4 (Figure 6.58). This is the median number of molecules.

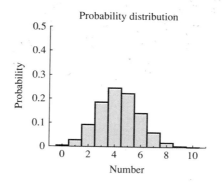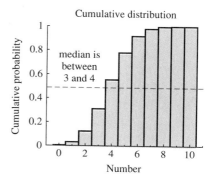

Figure 6.58
The median number of molecules

We use the same idea to compute the median of a continuous random variable.

■ Algorithm 6.2 (Finding the median of a continuous random variable)

1. Compute the cumulative distribution.
2. Solve for the point where the cumulative distribution crosses 0.5. ■

The median, sometimes denoted $\tilde{X}$, is the solution of the equation

$$\Pr(X \leq \tilde{X}) = F(\tilde{X}) = 0.5$$

(Figure 6.59).

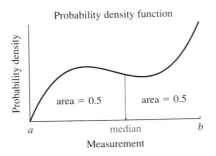
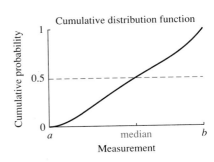

Figure 6.59
The median of a continuous random variable

For example, consider again the probability density function for a dart given by

$$g(r) = \frac{1}{8}(4 - r)$$

for $0 \leq r \leq 4$. We have found that the cumulative distribution function is

$$G(r) = \frac{r}{2} - \frac{r^2}{16}$$

To find the median, we must solve

$$G(r) = \frac{r}{2} - \frac{r^2}{16} = 0.5$$

For this, we use the quadratic formula,

$$\frac{r}{2} - \frac{r^2}{16} = 0.5 \quad \text{original equation}$$

$$\frac{r^2}{16} - \frac{r}{2} + 0.5 = 0 \quad \text{move everything to left-hand side and multiply by } -1$$

$$r^2 - 8r + 8 = 0 \quad \text{multiply by 16}$$

Solving, we get

$$r = \frac{8 \pm \sqrt{8^2 - 4 \cdot 8}}{2}$$

$$= \frac{8 \pm \sqrt{32}}{2}$$

$$= 4 \pm 2\sqrt{2}$$

$$\approx 6.82 \text{ or } 1.18$$

The only reasonable solution is at 1.18, the median. Half the darts will hit within 1.18 in. of the center, and the other half will be farther away (Figure 6.60). The median is close to, but not exactly equal to, the mean, which was 1.33 in this case.

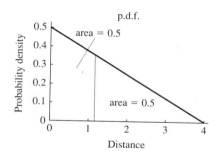
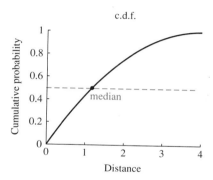

Figure 6.60
Median for the first dart player

For the second dart player, the p.d.f. is

$$h(r) = \frac{r}{8}$$

for $0 \leq r \leq 4$. The c.d.f. is

$$H(r) = \frac{r^2}{16}$$

and the median is the solution of

$$H(r) = \frac{r^2}{16} = 0.5$$

In this case, $r = \sqrt{8} \approx 2.83$ (Figure 6.61). Again the median is close to, but not equal to, the mean of 2.67.

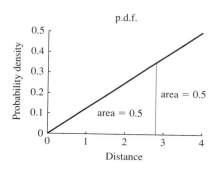
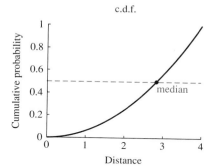

Figure 6.61
Median for the second dart player

Like most concepts in probability theory, the median can be interpreted in terms of gambling. Because the median score is 10, a single dart is just as likely to score more than 10 as less than 10. Suppose two people had to agree on a number x for the following bet: if the score is less than x, the first person wins, if greater, the second person wins. What value produces a fair bet? Two gamblers trained in probability theory would choose the median.

The Mode

For a random variable with many values, the **mode**, defined as the most common measurement, is a useful measure of the middle. The mode of the dart scores is 0, the single most likely score (the tallest bar in Figure 6.62a). If you threw only one dart, your best guess of the result would be a score of 0. In gambling terms, if you had to bet on any exact score for the dart, you would be wise to bet on 0. Sometimes, we say that the *modal score* is 0. The modal number of molecules left at $t = 8$ is 4 (Figure 6.62b). One of the nice things about the mode is how easy it is to compute.

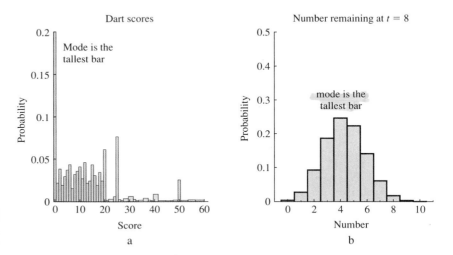

Figure 6.62
The modes of dart score and molecule number

The **mode** of a continuous random variable occurs where the p.d.f. takes on its maximum. Consider the p.d.f. $f(x) = 6x(1 - x)$ (Figure 6.63b). To find the mode, we find where the p.d.f. takes on its maximum value. In this case,

$$f'(x) = 6(1 - 2x) = 0$$

at $x = 0.5$. Because $f(0) = f(1) = 0$, the p.d.f. takes on its maximum value at 0.5. For the density function,

$$f(t) = 0.1e^{-0.1t}$$

which describes the time T when a molecule leaves a cell (Figure 6.40b), the mode is 0. The mode of the dart distances in Figure 6.60 is 0, and the mode of the dart distances in Figure 6.61 is 4. In cases like this, the mode does not give a great deal of information.

■ **Algorithm 6.3** (Finding the mode of a random variable)

1. For a discrete random variable, find the largest value in the probability distribution.

2. For a continuous random variable, find the maximum of the probability density function. ■

The mode, the most easily computed measure of the middle, requires us to find only the maximum.

The mean, median, and mode are equal for symmetric distributions and symmetric p.d.fs that have their maximum value in the center (Figure 6.63). In this situation, which comes up quite often, we do not have to worry about which statistic is most appropriate.

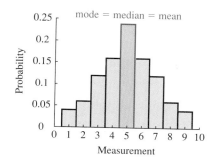
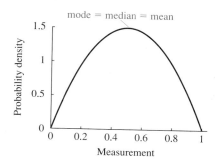

Figure 6.63
Symmetric distributions

The Geometric Mean

The expectation is computed with addition. Traditionally, mathematical objects computed with addition are called **arithmetic**, hence the name "arithmetic mean." This mean is appropriate for use with additive biological processes, such as the arrival of immigrants. Some biological processes, such as population growth through reproduction, are multiplicative. For example, suppose a population increases by 50% (per capita reproduction of 1.5) with probability 0.6 and decreases by 50% (per capita reproduction of 0.5) with probability 0.4. Will this population grow?

The *expectation* of the per capita reproduction is

$$1.5 \cdot 0.6 + 0.5 \cdot 0.4 = 1.1$$

Because this "average" growth is greater than 1, we might guess that the population grows.

This reasoning is inappropriate because growth is a multiplicative process. Let R_t be a random variable representing the per capita growth in year t. Suppose the population starts at a value of N_0. The population after 1 yr is $R_1 N_0$, after 2 yr is $R_2 R_1 N_0$, and after t years is

$$N_t = R_t R_{t-1} \cdots R_2 R_1 N_0 \qquad (6.17)$$

(Figure 6.64). Do we expect N_t to be larger or smaller than N_0?

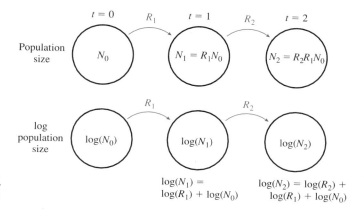

Figure 6.64
A stochastic reproduction model

The trick is to convert the multiplication into addition with the logarithm. Taking logarithms of both sides of Equation 6.17, we get

$$\ln(N_t) = \ln(R_t) + \ln(R_{t-1}) + \cdots + \ln(R_2) + \ln(R_1) + \ln(N_0)$$

Every year, we *add* the *logarithm* of the per capita reproduction to the logarithm of the population. If the expectation of the terms we are adding is positive, the sum will increase; if the expectation is negative, the sum will decrease. The key quantity is therefore the expectation of the logarithm of the per capita reproduction.

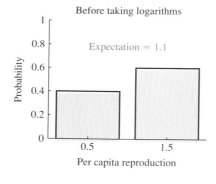
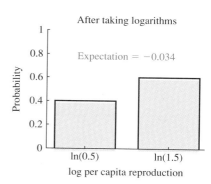

Figure 6.65
Expectations before and after taking the logarithm

What is the expectation of the logarithm of the per capita reproduction? We use the same formula for the expectation: Multiply the values times the probabilities and add them. In this case, the values are $\ln(1.5) = 0.405$ with probability 0.6 and $\ln(0.5) = -0.693$ with probability 0.4, so we have

$$E[\ln(R)] = \ln(1.5) \cdot 0.6 + \ln(0.5) \cdot 0.4 = -0.034$$

This is negative, meaning that, on average, the logarithm of the population size becomes smaller the more terms we add (Figure 6.65).

What happens to the population size itself? The logarithm of the population size decreases by about -0.034 per generation; therefore,

$$\ln(N_{t+1}) \approx \ln(N_t) - 0.034$$

The population sizes themselves can be found by exponentiating, so we have

$$N_{t+1} \approx e^{-0.034} N_t \approx 0.967 N_t$$

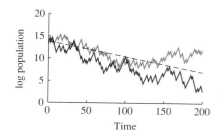

Figure 6.66
Two simulations of a stochastic population

This population, on average, decreases by more than 3% per generation. A simulation of two such populations is shown in Figure 6.66. Even though the populations jump up and down quite a bit, both decrease from an initial size of 1 million down to essentially 0 in 200 generations.

This average per capita reproduction, called the geometric mean, is defined as follows.

■ **Definition 6.12** Suppose R is a random variable that takes on only positive values. If R is a discrete random variable that takes on the values r_i with probability p_i, then

$$\text{the geometric mean of } R = e^{E[\ln(R)]}$$
$$= e^{\sum \ln(r_i) p_i}$$

If R is a continuous random variable with p.d.f. $f(r)$, minimum value $a \geq 0$, and maximum value b, then

$$\text{the geometric mean of } R = e^{E(\ln(R))}$$
$$= e^{\int_a^b \ln(r) f(r) \, dr}$$

The geometric mean can be computed only for random variables that take on strictly positive values because we cannot take the logarithm of a negative number.

■ **Algorithm 6.4** (Finding the geometric mean)

1. Take the logarithms of the measurements.
2. Find the expectation of the logarithm.
3. Exponentiate the result.

■

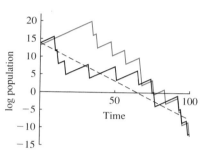

Figure 6.67
Two simulations of a stochastic population with occasional large decreases

If the geometric mean is less than 1, the population shrinks; if the geometric mean is greater than 1, the population grows. The key to computing the geometric mean correctly is remembering *not to change the probabilities*. Think of $\ln(R)$ as a new random variable, a new way to measure per capita reproduction. The underlying probabilities remain the same; only the values change (Figure 6.65).

Unlike the arithmetic mean, which emphasizes large values, the geometric mean emphasizes small values. If per capita reproduction near 0 is possible, it can eliminate the effects of many good years. For example, if a population grows by 22% (per capita reproduction of 1.22) in 90% of years and decreases by 98% (per capita reproduction of 0.02) in 10% of years, the arithmetic mean of the per capita reproduction is $1.22 \cdot 0.90 + 0.02 \cdot 0.1 = 1.10$. The geometric mean, however, is

$$e^{\ln(1.22)0.9 + \ln(0.02)0.1} = 0.809$$

The rare bad years mean that this population will, on average, decline very rapidly (Figure 6.67).

The geometric mean of a continuous random variable is also found by exponentiating the expectation of the logarithm of the random variable. Suppose the per capita reproduction R has p.d.f. $g(r) = 1.0$ for $0.5 \leq r \leq 1.5$. The per capita reproduction is equally likely to take on any value between 0.5 and 1.5. The arithmetic mean of this symmetric distribution of the per capita reproduction is

$$\int_{0.5}^{1.5} rg(r)\,dr = \int_{0.5}^{1.5} r\,dr = \left.\frac{r^2}{2}\right|_{0.5}^{1.5} = 1.0$$

The geometric mean is found by computing the expectation of the logarithm,

$$\begin{aligned} E[\ln(R)] &= \int_{0.5}^{1.5} \ln(r)g(r)\,dr \\ &= \int_{0.5}^{1.5} \ln(r)\,dr \\ &= [r\ln(r) - r]|_{0.5}^{1.5} \\ &= -0.045 \end{aligned}$$

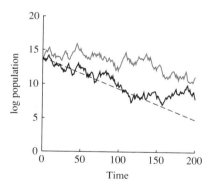

Figure 6.68
Two simulations of a stochastic population with per capita growth rates between 0.5 and 1.5

Therefore, the geometric mean is $e^{-0.045} \approx 0.955 < 1.0$. This population will decline (Figure 6.68). Once again, the small values pull the geometric mean below the arithmetic mean.

There is a general inequality relating the arithmetic and geometric means of random variables.

■ **THEOREM 6.4** (Arithmetic–Geometric Inequality)

The geometric mean is less than or equal to the arithmetic mean.

■

Chapter 6 Probability Theory and Descriptive Statistics

When the geometric mean is appropriate, the arithmetic mean always overestimates the value. We have seen this inequality at work in each of our examples.

SUMMARY We introduced three statistics to describe the central value of a random variable: the median, the mode, and the geometric mean. The value of a random variable is just as likely to be less than the **median** as it is to be greater. The **mode** is the single most likely value of the random variable. The mean, median, and mode are all equal for a **symmetric distribution**, that has its mode in the center. The **geometric mean**, defined as the exponential of the expectation of the logs, is the appropriate mean to use for multiplicative processes.

6.9 EXERCISES

1. (From *How to Lie with Statistics*) Suppose that incomes in a company have the following probabilities.

Income	Probability
20,000	0.48
30,000	0.04
35,000	0.16
50,000	0.12
57,000	0.04
100,000	0.08
150,000	0.04
450,000	0.04

   a. Find the mean, median, and mode.
   b. Which statistic is most informative about the distribution of salaries?
   c. If the highest-paid person got a raise to $4,500,000, what would happen to the mean, median, and mode?

2. Find the mean, median, and mode of the random variables in the table. Which statistic would you use to guess the population 10 yr into the future? Which would you use to guess the population next year?

Value	Population 1 Probability	Population 2 Probability	Population 3 Probability
10	0.05	0.20	0.01
5	0.15	0.25	0.02
−3	0.15	0.20	0.03
4	0.20	0.15	0.90
15	0.25	0.10	0.03
−10	0.20	0.10	0.01

3. Find the median, mean, and mode of random variables with the following p.d.f.s.

   a. $f(x) = 0.5$ for $0 \leq x \leq 2$
   b. $g(x) = 2x$ for $0 \leq x \leq 1$
   c. $h(x) = 4x$ for $0 \leq x \leq 0.5$ and $h(x) = 4 - 4x$ for $0.5 \leq x \leq 1$
   d. $f(t) = 0.1e^{-0.1t}$ for $t \geq 0$

4. Find the arithmetic and geometric means of the random variables R for per capita reproduction for the following cases. Check that the arithmetic–geometric inequality holds in each case. Which describes a growing population?

   a. $R = 4$ with probability 0.5, $R = 0.25$ with probability 0.5
   b. $R = 4$ with probability 0.25, $R = 0.25$ with probability 0.75
   c. $R = 4$ with probability 0.75, $R = 0.25$ with probability 0.25
   d. $R = 4$ with probability 0.25, $R = 0.25$ with probability 0.25, $R = 1$ with probability 0.5

5. Find the geometric mean of the following p.d.fs. Verify that the geometric mean is less than the arithmetic mean.

   a. $g(x) = 5.0$ for $1.0 \leq x \leq 1.2$ (the values used to generate Figure 6.2b). Use the fact that
   $$\int \ln(x)\,dx = x\ln(x) - x$$
   b. $g(x) = 1.25$ for $0.7 \leq x \leq 1.5$ (the values used to generate Figure 6.2c)
   c. $g(x) = x$ for $0.5 \leq x \leq 1.5$. Use the fact that
   $$\int x\ln(x)\,dx = \frac{x^2}{2}\ln(x) - \frac{x^2}{4}$$

6. Suppose populations start at 1 (thought of as thousands). Estimate the log population size and the actual population size after 50 generations for the following cases.
   a. Exercise 4a
   b. Exercise 4b
   c. Exercise 4c
   d. Exercise 4d
   e. Exercise 5a. Compare with Figure 6.2b.

f. Exercise **5b**. Compare with Figure 6.2c.
g. Exercise **5c**

7. A store has two managers, one who believes that high profits come from lowering prices and getting more customers, and another who believes that high profits come from raising prices and making more profit per customer. In week 1, the "low price" manager cuts prices by 50%. In week 2, the "high price" manager raises prices by 50%, and so forth. Suppose an item started out at $100.
   a. Find the price of the item after 1, 2, 3, and 4 weeks.
   b. Find a formula for the price of the item after t weeks (break into the two cases t even and t odd).
   c. Which of the managers wins?
   d. What does this have to do with the geometric mean? In what way is this different from a population that grows by 50% with probability 0.5 and shrinks by 50% with probability 0.5?

8. The term geometric mean is based on the following argument. Suppose a random variable R takes on each of the values r_1 and r_2 with probability 0.5.
   a. Use the laws of logs to show that the geometric mean is $\sqrt{r_1 r_2}$.
   b. Show that a square with sides of length equal to the geometric mean has the same area as a rectangle with sides of lengths r_1 and r_2. Draw the associated picture.
   c. Fix $r_1 = 1$. Find the value of r_2 that maximizes the ratio of the geometric mean to the arithmetic mean.
   d. Try to prove this special case of the arithmetic–geometric inequality.

9. **COMPUTER:** Suppose that the p.d.f. for the time a molecule leaves a cell is $f(x) = 6.25xe^{-2.5x}$ for $x \geq 0$. Use your computer to find the expectation and the median.

Graph the p.d.f. and c.d.f. on one graph (pick a reasonable range for x), and indicate the mean and median on the appropriate curves.

10. **COMPUTER:** The p.d.f. for a random variable taking on values between 0.8 and 1.1 with equal probability is $f(x) = 3.3333$ for $0.8 \leq x \leq 1.1$. Find the geometric mean r of this random variable by exponentiating the integral of the value $[\ln(x)]$ times the probability (3.3333) over all values of x. Would a population with this random variable as its per capita reproduction grow in the long run? Define two updating functions, a deterministic g for a population that has per capita reproduction of exactly r, and a stochastic G. Compare the dynamics of the two populations for 100 steps starting from an initial condition of 100. How similar do they look?

11. **COMPUTER:** Suppose that the per capita reproduction of a population is a random variable with uniform (flat) p.d.f. on the interval from 0.8 to y, where y is an unknown value. To guarantee that the integral is equal to 1, the height of the p.d.f. must be $1/(y - 0.8)$ rather than 3.333. Why is this?
   a. Compute a function $H(y)$ that gives the geometric mean as a function of y.
   b. Solve for the value $y = y_{\max}$ for which the geometric mean is 1.0.
   c. Define a random variable R to produce random numbers in the interval from 0.8 to $y_{\max}$ and an associated updating function G to describe a population with per capita reproduction equal to R.
   d. Generate two trajectories of 100 generations using the updating function G starting from $N = 100$.
   e. Where do you expect the trajectories to end up? How close are they?

6.10 Descriptive Statistics for Spread

A measure of the center can be thought of as a best guess of the value of a random variable. Our *confidence* in that guess depends on how spread out the probability distribution or probability density function is. Knowing that a score has a mean of 12.5 (or a median of 10) does not tell us whether likely scores range from 5 to 15 or from 0 to 60. We introduce the most widely used statistics for describing spread: **range**, **percentiles**, the **variance**, and the **standard deviation**. As always, different statistics are appropriate in different circumstances.

Range and Percentiles

The simplest measure of spread is the **range**, a description of all the values taken on by a random variable, usually summarized by the lowest and highest values. The range of dart scores is from 0 to 60 (Figure 6.69a). The range of the number of molecules remaining out of 10 is from 0 to 10 (Figure 6.69b). Many continuous random variables have a mathematically possible range from 0 (or negative infinity) to infinity. In these cases, the range gives no useful information. If random variables are thought of as functions, the range coincides with the *range of the function*. Perhaps confusingly, however, the range of the random variable is the *domain* of the probability density function.

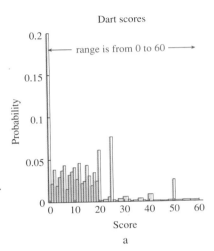
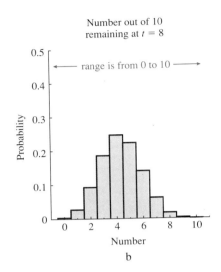
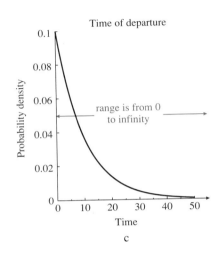

Figure 6.69
The ranges of three random variables

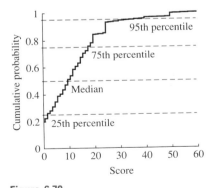

Figure 6.70
Finding percentiles from a cumulative distribution

Percentiles generalize the median. Exactly 50% of the values of the random variable lie below the median. We define the *p*th percentile to be the value of the random variable that exceeds exactly *p*% of the values. The median is the 50th percentile. Like the median, percentiles must be computed from the cumulative distribution.

Consider again the cumulative distribution for dart scores (Figure 6.70). We found the median by solving for where the cumulative distribution crossed the horizontal line at 0.5. We find the 95th percentile by finding where the cumulative distribution crosses the horizontal line at 0.95, in this case at a score of 34.

The lower and upper **quartiles**, defined as the 25th and 75th percentiles, are often used to describe random variables. The 25th percentile of the dart scores is 2, and the 75th percentile is 18. The quartiles and the median divide the values of the random variable into four equally probable sets. One quarter of dart throws score less than 2, another quarter between 2 and 10, another quarter between 10 and 18, and a quarter more than 18. Quartiles can also be thought of as dividing the data into two equal pieces, the central half between the quartiles (between 2 and 18) and the outlying half outside the quartiles (less than 2 or greater than 18).

Percentiles of continuous random variables are computed by finding where the cumulative distribution function crosses the appropriate horizontal line. The time T when a diffusing molecule leaves a cell is described by the p.d.f.

$$f(t) = 0.1e^{-0.1t}$$

and the c.d.f.

$$F(t) = 1 - e^{-0.1t}$$

(Figure 6.71). We found the median value of 6.93 by solving the equation
$$F(t) = 0.5$$
We find the percentiles in the same way. To find the 95th percentile, solve
$$F(t) = 1 - e^{-0.1t} = 0.95$$
$$e^{-0.1t} = 0.05$$
$$t = -10\ln(0.05) \approx 29.96$$
Only 5% of measurements exceed 29.96. The 5th percentile is
$$F(t) = 1 - e^{-0.1t} = 0.05$$
$$e^{-0.1t} = 0.95$$
$$t = -10\ln(0.95) \approx 0.51$$
Only 5% of measurements are less than 0.51. On the graph of probability density function, the area above the 95th percentile is 0.05, equal to the area below the 5th percentile.

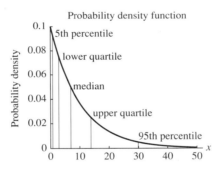
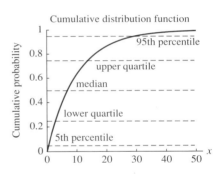

Figure 6.71
Percentiles of the probability density function for a diffusing molecule

Using the same procedure, we find that the quartiles are 2.88 and 13.86. These quartiles, along with the median, break the random variable into four equally likely intervals (each with area 0.25). The results of an experiment are precisely as likely to lie between 0 and 2.88 as between 6.93 and 13.86.

Mean Absolute Deviation

Percentiles generalize the median. Another group of statistics generalize the mean, measuring the "average distance" from the mean. Their calculation proceeds in two steps: Define a new random variable measuring distance from the mean, and compute its expectation.

For example, consider again the number of molecules left in a cell after 8 min. In addition to tabulating the values and the probabilities, we have also tabulated how far the values are from the mean (see the table at the top of the next page). As a measure of spread, we would like to know how far *on average* the values are from the mean.

Distance is generally computed with the absolute value. If there are 0 molecules left in the cell, this is 4.3 below the mean. The absolute value of the difference is 4.3. If there are six molecules left, this is 1.7 above the mean, and the

Number left	Mean	Difference from mean	Absolute difference from mean	Probability
0	4.3	−4.3	4.3	0.004
1	4.3	−3.3	3.3	0.027
2	4.3	−2.3	2.3	0.092
3	4.3	−1.3	1.3	0.186
4	4.3	−0.3	0.3	0.246
5	4.3	0.7	0.7	0.223
6	4.3	1.7	1.7	0.141
7	4.3	2.7	2.7	0.061
8	4.3	3.7	3.7	0.017
9	4.3	4.7	4.7	0.003
10	4.3	5.7	5.7	0.000

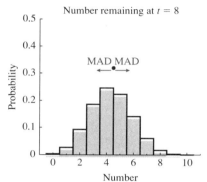

Figure 6.72
Mean absolute deviation of molecule number

absolute value of the difference is 1.7. We find the expectation of these distances in the usual way, by taking the values times the probabilities and adding them,

$$\text{mean distance from mean} = 4.3 \cdot 0.004 + 3.3 \cdot 0.027 + 2.3 \cdot 0.092$$
$$+ 1.3 \cdot 0.186 + 0.3 \cdot 0.246 + 0.7 \cdot 0.223$$
$$+ 1.7 \cdot 0.141 + 2.7 \cdot 0.061 + 3.7 \cdot 0.017$$
$$+ 4.7 \cdot 0.003 + 5.7 \cdot 0.000 \approx 1.27$$

This means that the measurement is, on average, 1.27 away from the mean. This gives a rough idea of how "spread out" the distribution is (Figure 6.72).

This measure is often called the **mean absolute deviation**, or **MAD**. The formal definition of the mean absolute deviation is as follows.

Definition 6.13 Suppose that the expectation of the random variable X is $\overline{X}$. Consider the new random variable $Y = |X - \overline{X}|$. Then

$$\text{mean absolute deviation} = E(Y) = E(|X - \overline{X}|)$$

More particularly, if X is a discrete random variable that takes on the value x_i with probability p_i, then

$$\text{MAD} = \sum_{i=1}^{n} |x_i - \overline{X}| p_i$$

If X is a continuous random variable with p.d.f. $f(x)$ and range from a to b, then

$$\text{MAD} = \int_a^b |x - \overline{X}| f(x)\,dx \qquad \blacksquare$$

Suppose we wish to find the mean absolute deviation of the continuous random variable R with p.d.f.

$$g(r) = \frac{1}{8}(4 - r)$$

for $0 \leq r \leq 4$ (representing the distance a dart hits from the center of the board). The mean of this random variable is $4/3$ in. The mean absolute deviation is found by integrating:

$$\text{MAD} = \int_0^4 \left| r - \frac{4}{3} \right| g(r)\, dr$$

$$= \int_0^4 \left| r - \frac{4}{3} \right| \frac{1}{8}(4-r)\, dr$$

$$= \int_0^{4/3} \left(\frac{4}{3} - r \right) \frac{1}{8}(4-r)\, dr + \int_{4/3}^4 \left(r - \frac{4}{3} \right) \frac{1}{8}(4-r)\, dr$$

$$= \left(\frac{1}{24}r^3 - \frac{1}{3}r^2 + \frac{2}{3}r \right) \bigg|_0^{4/3} - \left(\frac{1}{24}r^3 - \frac{1}{3}r^2 + \frac{2}{3}r \right) \bigg|_{4/3}^4$$

$$\approx 0.79$$

Solving this integral requires a lot of algebra, which is one reason why the mean absolute deviation is rarely used.

Although the variance, which we introduce next, has much nicer mathematical properties than the mean absolute deviation, the mean absolute deviation is a perfectly reasonable measure of spread.

Variance, Standard Deviation, and Coefficient of Variation

A more widely-used statistic, based on the same sort of reasoning, is the **variance**. When we measured the distance of our various measurements from the mean, we used the absolute value. An alternative way to make sure that all the distances are positive is to *square* them. The results are shown in the following table.

Number left	Mean	Difference from mean	Squared difference from mean	Probability
0	4.3	−4.3	18.49	0.004
1	4.3	−3.3	10.89	0.027
2	4.3	−2.3	5.29	0.092
3	4.3	−1.3	1.69	0.186
4	4.3	−0.3	0.09	0.246
5	4.3	0.7	0.49	0.223
6	4.3	1.7	2.89	0.141
7	4.3	2.7	7.29	0.061
8	4.3	3.7	13.69	0.017
9	4.3	4.7	22.09	0.003
10	4.3	5.7	32.49	0.000

To find the mean squared deviation from the mean, we take these new values, multiply by the probabilities, and add them. In this case, we have

mean squared distance from mean
$$= 18.49 \cdot 0.004 + 10.89 \cdot 0.027 + 5.29 \cdot 0.092$$
$$+ 1.69 \cdot 0.186 + 0.09 \cdot 0.246 + 0.49 \cdot 0.223$$
$$+ 2.89 \cdot 0.141 + 7.29 \cdot 0.061 + 13.69 \cdot 0.017$$
$$+ 22.09 \cdot 0.003 + 32.49 \cdot 0.000 = 2.45$$

This new statistic, called the **variance**, is formally defined as follows.

Definition 6.14 The variance of a random variable X, denoted by σ^2 or Var(X), is equal to

$$\sigma^2 = E[(X - \overline{X})^2]$$

More particularly, if X is a discrete random variable that takes on the value x_i with probability p_i, then

$$\sigma^2 = \sum_{i=1}^{n}(x_i - \overline{X})^2 p_i$$

If X is a continuous random variable with p.d.f. $f(x)$ and range from a to b, then

$$\sigma^2 = \int_a^b (x - \overline{X})^2 f(x)\, dx$$

In the continuous case, consider once again the random variable with p.d.f.

$$g(r) = \frac{1}{8}(4 - r)$$

$0 \le r \le 4$. Using the mean of $4/3$, we find that the variance is

$$\sigma^2 = \int_0^4 \left(r - \frac{4}{3}\right)^2 g(r)\, dr$$

$$= \int_0^4 \left(r - \frac{4}{3}\right)^2 \frac{1}{8}(4 - r)\, dr$$

$$= -\frac{1}{32}r^4 + \frac{5}{18}r^3 - \frac{7}{9}r^2 + \frac{8}{9}r \bigg|_0^4$$

$$= \frac{8}{9} = 0.889$$

This still requires some messy algebra, but at least it is not necessary to split the integral into two pieces.

An important advantage of the variance over the mean absolute deviation is the existence of the following computational formula.

THEOREM 6.5 (Computational Formula for the Variance)

If X is a random variable with mean $\overline{X}$, the variance can be written

$$\sigma^2 = E(X^2) - E(X)^2 = E(X^2) - \overline{X}^2$$

More particularly, if X is a discrete random variable that takes on the value x_i with probability p_i, then

$$\sigma^2 = \sum_{i=1}^{n} x_i^2 p_i - \overline{X}^2$$

If X is a continuous random variable with p.d.f. $f(x)$ and range from a to b, then

$$\sigma^2 = \int_a^b x^2 f(x)\, dx - \overline{X}^2$$

The proof of this theorem is outlined in Exercise 8. In words, the variance is the expectation of the square minus the square of the expectation.

Why are these formulas easier to use? Consider the discrete random variable for molecule number. Instead of finding how far each possible measurement is from the mean, we have only to find the square of each measurement, as in the following table.

Number left	Probability	Square of number left
0	0.004	0
1	0.027	1
2	0.092	4
3	0.186	9
4	0.246	16
5	0.223	25
6	0.141	36
7	0.061	49
8	0.017	64
9	0.003	81
10	0.000	100

The variance is the mean of the squares minus the square of the mean. We find the mean of the squares as usual, the sum of the values times the probabilities, or

$$\text{mean of squared number} = 0 \cdot 0.004 + 1 \cdot 0.027 + 4 \cdot 0.092 + 9 \cdot 0.186$$
$$+ 16 \cdot 0.246 + 25 \cdot 0.223 + 36 \cdot 0.141$$
$$+ 49 \cdot 0.061 + 64 \cdot 0.017 + 81 \cdot 0.003$$
$$+ 100 \cdot 0.000$$
$$= 20.94$$

The variance is then

$$\text{mean of squared number} - \text{square of mean} = 20.94 - 4.3^2 = 2.45$$

as we found before.

For the random variable R with p.d.f.

$$g(r) = \frac{1}{8}(4 - r)$$

for $0 \le r \le 4$, the computational formula for the variance is

$$\sigma^2 = \int_0^4 r^2 \frac{1}{8}(4 - r) \, dr - \left(\frac{4}{3}\right)^2$$
$$= \int_0^4 \frac{r^2}{2} - \frac{r^3}{8} \, dr - \left(\frac{4^2}{3}\right)$$
$$= \left. \frac{r^3}{6} - \frac{r^4}{32} \right|_0^4 - \left(\frac{4^2}{3}\right)$$
$$= \frac{8}{9} \approx 0.889$$

again matching our earlier result.

An important variance to know and remember is the variance of a *Bernoulli random variable*, a random variable B that takes on the value 1 with probability p

and 0 with probability $1 - p$. The expectation of B is

$$E(B) = 0 \cdot (1 - p) + 1 \cdot p = p$$

The variance is

$$Var(B) = 0^2 \cdot (1 - p) + 1^2 \cdot p - p^2 = p - p^2 = p(1 - p)$$

What does the variance mean? A first step in interpreting a new quantity is to figure out its units. The variance of the number of molecules has units of molecules^2, because we squared the difference from the mean. Similarly, the variance of dart distances is inches^2. These quantities are rather hard to interpret. To get back to units of molecules or inches, we can take the square root. The square root of the variance is the **standard deviation**, denoted σ.

■ **Definition 6.15** The standard deviation σ is the square root of the variance. ■

The standard deviation of molecule number is $\sqrt{2.45} = 1.56$, which is close to, but not equal to, the mean absolute deviation of 1.27. The standard deviation of dart distances is $\sqrt{0.889} = 0.94$. What does the standard deviation tell us? Like the mean absolute deviation, it is a measure of how spread out the distribution is. Roughly speaking, most dart distances lie within two standard deviations of the mean, or between $1.33 - 2 \cdot 0.94 = -0.55$ and $1.33 + 2 \cdot 0.94 = 3.21$ (Figure 6.73). In this case, the standard deviation does not tell us very much.

The standard deviation is most useful for describing random variables with a bell-shaped distribution or probability density function (Figure 6.74). In this case, the standard deviation is the basis of two rules of thumb.

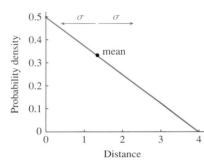

Figure 6.73
Standard deviation of dart distances

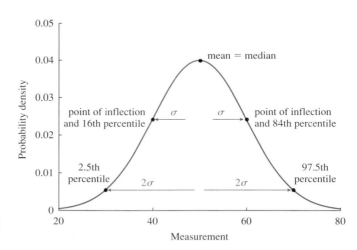

Figure 6.74
A bell-shaped distribution

Rules of Thumb for Bell-Shaped Distributions:

1. The points of inflection are one standard deviation from the mean.

2. Approximately 95% of the probability lies within two standard deviations of the mean.

Recall that the points of inflection are where the curvature changes from concave up (curved upward) to concave down (curved downward) (Section 2.4).

In terms of percentiles, the point of inflection one standard deviation below the mean is the 16th percentile and the point of inflection one standard deviation above the mean is the 84th percentile. The second rule says that the 2.5th percentile is approximately two standard deviations below the mean and the 97.5th percentile is approximately two standard deviations above the mean. If the mean is designated by μ (the convention for distributions of this shape), these rules are summarized in the following table.

Point	Shape	Percentile
$\mu - 2\sigma$		2.5th
$\mu - \sigma$	point of inflection	16th
μ	maximum	50th
$\mu + \sigma$	point of inflection	84th
$\mu + 2\sigma$		97.5th

These rules are illustrated in Figure 6.74. The mean is equal to the mode, and looks to be about 50. The points of inflection are at 40 and 60, about 10 away from the mean. According to the first rule of thumb, we estimate that the standard deviation is about 10. According to the second rule of thumb, the 2.5th percentile will be about 30 and the 97.5th percentile about 70. Therefore, 95% of measurements lie between 30 and 70.

These rules work *only for symmetric bell-shaped distributions*. Because the variance and standard deviation weight outlying points heavily, these statistics are very sensitive to values far from the mean (Exercise 3).

Sometimes we wish to use a number with no units to compare different distributions. One example is the **coefficient of variation**, or **CV**.

■ **Definition 6.16** Suppose X is a random variable that takes on only positive values. The coefficient of variation, or CV, is the ratio of the standard deviation to the mean, or

$$\text{CV} = \frac{\sigma}{\overline{X}}$$ ■

Because the standard deviation has the same units as the measurement itself, the coefficient of variation has no units and is a dimensionless measure of spread. For the random variable shown in Figure 6.74, the coefficient of variation is

$$\text{CV} = \frac{\sigma}{\overline{X}} = \frac{10.0}{50.0} = 0.2$$

For the molecules, the coefficient of variation is

$$\text{CV} = \frac{1.56}{4.3} = 0.36$$

The larger the coefficient of variation, the more spread out the distribution. The coefficient of variation can be thought of as a measure of uncertainty. If the coefficient of variation is close to or greater than 1, values of the random variable have a large range and are likely to be far from the mean.

SUMMARY

We have introduced several statistics that describe the spread of a random variable. The **range** gives the minimum and maximum values. The **percentiles** generalize the median to give the values where a random variable exceeds a given fraction of the distribution. The lower and upper **quartiles** indicate the values lying above

25% and 75% of the distribution, respectively. Statistics of spread that generalize the mean include the **mean absolute deviation**, the **variance** (the mean squared deviation from the mean), the **standard deviation** (the square root of the variance), and the **coefficient of variation** (the standard deviation divided by the mean).

6.10 EXERCISES

1. For the three random variables described in Exercise 2 in Section 6.9, find the range, the quartiles, the MAD, the variance, the standard deviation, and the coefficient of variation. Where appropriate, mark these values on plots of the probability distribution and cumulative distribution.

2. For the four continuous random variables given in Exercise 3 in Section 6.9, find the range, the 5th and 95th percentiles, the quartiles, the MAD, the variance, the standard deviation, and the coefficient of variation. Where appropriate, mark these values on graphs of the p.d.f. and c.d.f.

3. Find the MAD, the variance, and the coefficient of variation for the salary distributions in Exercise 1 in Section 6.9 (first with the high salary of $450,000 and then with the high salary of $4,500,000). Which statistics seem to be most sensitive to large values? What are the units of each statistic?

4. Consider a Bernoulli random variable that takes on the value 1 with probability p and 0 with probability $1 - p$.
   a. Find the MAD, the variance, the standard deviation, and the coefficient of variation.
   b. What is the maximum possible value of the coefficient of variation (as a function of p)? Does this make sense?

5. Try the same calculations as in Exercise 4 for the following.
   a. A random variable that takes on the value 10 with probability p and 0 with probability $1 - p$. Which part of your answer remains the same?
   b. A random variable that takes on the value 5 with probability p and -5 with probability $1 - p$.

6. Estimate the standard deviation, coefficient of variation, 2.5th percentile, and 97.5th percentile from the following figures.

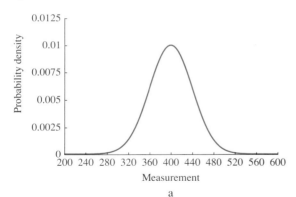

a

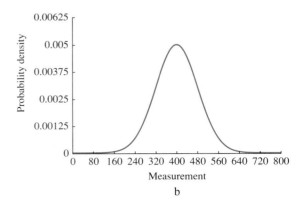

b

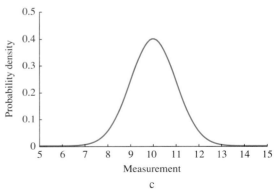

c

7. Consider the bell-shaped p.d.f. in Figure 6.74 (page 522).
   a. Draw a p.d.f. with the same standard deviation of 10, but with a mean of 500. Calculate the coefficient of variation.
   b. Draw a p.d.f. with the same standard deviation of 10, but with a mean of 5. Calculate the coefficient of variation.
   c. Draw a p.d.f. with mean of 50 and coefficient of variation of 0.4.

8. The following outlines the proof of the computational formula for the variance (Theorem 6.5).
   a. Multiply out the squared term into three terms.
   b. Break the sum into three sums.
   c. Factor constants out of the sums. Remember that $\overline{X}$ is a constant.
   d. Recognize certain sums to be equal to the mean. Write the sums in terms of the mean and group together like terms.

9. Suppose a population follows the rule

$$N_{t+1} = R_t N_t$$

where R_t is a random variable that takes on the value 1.5 with probability 0.6 and 0.5 with probability 0.4. Suppose $N_0 = 1$.
  a. Find the variance of the random variable N_1.
  b. Find the variance of the random variable $\ln(N_1)$.

10. A general inequality that gives information about any random variable X is **Chebyshev's inequality**. Suppose X has mean μ and standard deviation σ. Then
$$\Pr(|X - \mu| \geq k\sigma) \leq \frac{1}{k^2}$$
for any value of k.
  a. What does this mean for $k = 1$? Does this tell us anything?
  b. What does this mean for $k = 2$? How much of the probability must lie within two standard deviations of the mean?
  c. How much of the probability must lie within three standard deviations of the mean?
  d. Compare the result of part b with the second rule of thumb for bell-shaped distributions. Which gives more precise information?

11. **COMPUTER:** Consider again the p.d.f. $f(x) = 6.25xe^{-2.5x}$ from Exercise 9 in Section 6.9. Use your computer to find the variance, the standard deviation, the coefficient of variation, the MAD, and the 5th, 25th, 75th, and 95th percentiles.

12. **COMPUTER:** The bell-shaped, or normal, p.d.f. has the unlikely looking equation
$$f(x) = \frac{1}{\sqrt{2\pi}\sigma} e^{-(x-\mu)^2/(2\sigma^2)}$$
where μ is the mean and σ the standard deviation. Set $\mu = 50$ and $\sigma = 20$.
  a. Plot this function between some reasonable limits.
  b. Use integration to compute the mean and standard deviation. Remember that the limits of integration are from $-\infty$ to ∞.
  c. Find the coefficient of variation.
  d. We can check the rules of thumb regarding the shape of the curve and the standard deviation. Find the first derivative and solve for the maximum. Does it match the mean?
  e. Find the second derivative and solve for the points of inflection. Do they match the rules of thumb?

13. **COMPUTER:** Find and graph the cumulative distribution F associated with f from Exercise 12. Compute the percentiles associated with points one and two standard deviations above and below the mean. Mark these on your graph. How well do they match the rules of thumb? Find the 5th and 95th percentiles and the lower and upper quartiles.

Supplementary Problems for Chapter 6

EXERCISE 1
An intriguing species of worm spends its time wandering around in search of food. It is on a plate, 2% of which is covered with food. Worms occupy 10% of 1.0 mm² regions containing food. Worms occupy 4% of 1.0 mm² regions without food.
  a. Draw a diagram illustrating the events and probabilities.
  b. Find the probability that a 1 mm² region containing a worm also contains food.
  c. Are the worms and food independent? What does this mean?
  d. If the plate is 1000 cm², estimate the total number of worms on and off the food.

EXERCISE 2
An experiment involves placing a single worm on a plate. After 3 days, there are no worms with probability 0.1, one worm with probability 0.35, two worms with probability 0.15, three worms with probability 0.3, and four worms with probability 0.1.
  a. Sketch the probability distribution.
  b. Sketch the cumulative distribution.
  c. Find the expected number of worms.
  d. What does this mean?

EXERCISE 3
Baby worms grow from a length of 0.5 mm to a length of 1.0 mm. After a day of growth, the measurement L representing length has p.d.f.
$$f(l) = 1.5(l + l^2)$$
for $0.5 \leq l \leq 1.0$.
  a. Graph the p.d.f. and check that it is consistent.
  b. Find the probability that a worm is less than 0.75 mm long.
  c. Find the approximate probability that a worm is between 0.75 and 0.76 mm long.
  d. How many worms out of 1000 would you expect to be longer than 0.75 mm?

EXERCISE 4
During a 10-min interval, adult worms switch from eating to egg-laying with probability 0.1 and from egg-laying to eating with probability 0.15.
  a. Draw a diagram illustrating this process.
  b. Derive an updating equation for the probability the worm is eating.

c. Find the long-term fraction of time spent eating.
d. If egg-laying takes place at a rate of 1.5 per 10 min, about how many eggs would a worm lay during its 15-day life (worms never sleep)?

EXERCISE 5
Two codons of DNA are being compared (each consists of three nucleotides). The experiment is to check whether the two match. For example, if the first is AAG and the second ATG, the result would be "match, no match, match."
  a. List all the simple events.
  b. What simple events are included in the event "they match at the second nucleotide."
  c. Find the union and intersection of the events "they match at the second nucleotide" and "they do not match at the first nucleotide."
  d. Suppose the probability of a change is 0.1 at the first site, 0.2 at the second, and 0.4 at the third. What is the probability of no change at any site if the events are independent?

EXERCISE 6
A bacterium switches between moving and stopped. During a given second, it switches from moving to stopped with probability 0.2 and from stopped to moving with probability 0.1. The bacterium moves at 10 μm/s.
  a. How would you simulate this system?
  b. What is an updating function for the probability the bacterium is moving?
  c. Find the average speed of the bacterium after a long time.

EXERCISE 7
Suppose the per capita reproduction of a population takes on the three values 0.5, 1.0, and 1.5 with the following probabilities:

$$\Pr(\text{per capita reproduction} = 0.5) = 0.3$$
$$\Pr(\text{per capita reproduction} = 1.0) = 0.3$$
$$\Pr(\text{per capita reproduction} = 1.5) = 0.4$$

  a. Draw the histogram.
  b. Find the expectation.

EXERCISE 8
Suppose a measurement X has p.d.f.

$$f(x) = 0.5 + x$$

for $0 \leq x \leq 1$.
  a. Find $\Pr(0.3 \leq X \leq 0.7)$. Sketch the region corresponding to this probability on a graph of the p.d.f.
  b. Find and graph the c.d.f.
  c. Find the expectation.

EXERCISE 9
A laboratory is testing for a rare mutant bacteria, which appears in 1% of cells. A test finds the mutant with probability 0.9 and gives a false positive with probability 0.02.
  a. Find the probability of a positive test.
  b. Find the probability that a cell that tests positive has the mutation.

EXERCISE 10
The per capita reproduction R of a population is a function of the annual mean temperature T, with formula

$$R = 1.5 - \frac{(T-50)^2}{100}$$

Suppose the annual mean temperature T has the probability distribution

$$\Pr(T = 40) = 1/3, \Pr(T = 50) = 1/3, \Pr(T = 60) = 1/3$$

  a. Draw histograms for the temperature and the per capita reproduction.
  b. Find the expectations of T and R.

EXERCISE 11
Suppose a measurement Z has p.d.f.

$$f(z) = \frac{1}{z}$$

for $1 \leq z \leq e$.
  a. Sketch this p.d.f., and show it satisfies the necessary requirements.
  b. Find $\Pr(1.2 \leq Z \leq 2.4)$, and indicate it on your graph.
  c. Find the expectation, and mark it on your graph.

EXERCISE 12
After studying probability theory in the third grade, a boy fancies himself a meteorologist. He advises his family to cancel a picnic with probability 0.8 if it is cloudy, because that is the probability of rain if it is cloudy. He advises his family to cancel a picnic with probability 0.2 if it is sunny, because that is the probability of rain if it is sunny. Thirty percent of days are cloudy.
  a. Find the probability that a picnic is cancelled.
  b. Find the probability that it rains on the picnic if it is cloudy.
  c. Find the probability that it rains on the picnic.

EXERCISE 13
Birds switch between being infested with parasites and being parasite-free. A bird with parasites has a 10% probability of getting rid of them over the winter, and a bird without parasites has a 5% chance of getting infested.
  a. Use a diagram or mathematical notation to describe this process.
  b. Write an updating function for the probability a bird is infested.
  c. Find the equilibrium fraction of infested birds.
  d. Suppose only 80% of birds survive the winter (independent of their being infested) and are replaced by parasite-free baby birds. Find the equilibrium fraction of infested birds.

EXERCISE 14
Careful observation reveals that your probability of receiving a phone call during any 20-min interval while you are awake

is 0.1. You spend 20 min per day in the shower (which is 2% of your waking hours), during which time the phone rings with probability 0.2.
  a. Are the events of showering and receiving phone calls independent? Prove it.
  b. Find the probability of receiving a phone call during a 20-min period when you are not in the shower.
  c. Find the probability that you are in the shower when the phone rings.

EXERCISE 15
A certain basketball player makes a shot with probability 0.3 after making a shot and with probability 0.6 after missing. She complains to the officials about having been fouled with probability 0.2 after making a shot and with probability 0.7 after missing.
  a. What is her long-term probability of making a shot?
  b. What fraction of times does she complain about a foul?
  c. You turn on the TV just in time to see her complaining to the officials. What is the probability she hit her shot?

EXERCISE 16
Researchers in the Department of Human Genetics are rumored to have discovered a gene for mathematical ability. They have identified three alleles (variants) of this gene, creatively named **A, B**, and **C**. Let G represent the trait "good at math." Testing has shown that $\Pr(G \mid A) = 0.2$, $\Pr(G \mid B) = 0.5$, $\Pr(G \mid C) = 0.1$. Furthermore, $\Pr(A) = 0.6$ and $\Pr(B) = 0.3$.
  a. Draw a picture illustrating this situation.
  b. Find Pr(C) and Pr(G) and Pr(G and C).
  c. Find $\Pr(A \mid G)$ and $\Pr(A \mid G^c)$, where G^c means "not good at math." Do your answers match? Why or why not?
  d. If someone walked into your office and said "I have allele C and should not take math," how would you respond?

EXERCISE 17
Researchers in the Department of Horticulture are rumored to be developing data on the last frost. They measure the time t continuously in months after April 1 (for example, $t = 0.5$ corresponds to the exact middle of April, or midnight on the 15th). Let T be a random variable representing the time of the last frost. They have found that

$$\Pr(T \le t) = 2t - t^2$$

for $0 \le t \le 1$.
  a. Check that this cumulative distribution function makes sense. When is the last possible day for the last frost?
  b. Find the median date of the last frost.
  c. Find the probability that plants will get frosted if planted at midnight on April 20.
  d. Find the probability density function of T.
  e. Find the mean date of the last frost. Compare with the median and explain why one is greater than the other.
  f. (For gardeners only.) When would you plant, and why?

EXERCISE 18
Researchers in the Department of Ecology and Evolution are rumored to have discovered a population of tropical birds whose numbers jump between 5 and 10 and take on no other values. Let H represent the event that the population is "high" (10), and L represent the event that it is "low" (5). Suppose that $\Pr(H \mid L) = 0.3$ and $\Pr(L \mid H) = 0.2$.
  a. Give a complete description of the dynamics of this population.
  b. Write an updating function for the probability that the population is high.
  c. Find the long-term probability.
  d. The population can double, halve, or stay the same. Find the probability of each of these events after a long time, and compute the geometric mean growth rate. Why does your answer make sense?

Projects for Chapter 6

PROJECT 1
We have used the geometric mean to study the growth of populations that follow the updating function

$$N_{t+1} = RN_t$$

where R is some random variable. This analysis was possible because the updating function is *linear*, meaning that nothing complicated happened to the population N_t. Consider the following stochastic variant of the Ricker equation (Section 3.2),

$$N_{t+1} = RN_t e^{-N_t/K}$$

where the per capita reproduction R and the carrying capacity K might both be random variables.

  a. Suppose that $K = 100$ is a fixed value. Choose different random variables for R (with positive values only, of course) and simulate to see when the population survives and when it goes extinct. Does the geometric mean still identify those populations that will survive? Why or why not?
  b. Suppose that $R = 1.1$ is a fixed value. Choose different random variables for K (again with positive values only) and simulate to see when the population survives and when it goes extinct. Does it ever go extinct? What is the average population? Let $\overline{K}$ be the expectation of K. Compare the behavior of your stochastic population

with one following the updating function

$$N_{t+1} = RN_t e^{-N_t/K}$$

Is the average of the stochastic population higher or lower than the equilibrium of this deterministic population?

c. Try cases where both R and K are random variables. When the values of R and K are independent, do we need to know anything about K to figure out whether the population will go extinct?

d. Can you find a case where the geometric mean of R is greater than 1 but the population goes extinct? Why or why not?

PROJECT 2

Research the following for at least ten diseases.

- Fraction of people with the disease.
- Fraction of people tested.
- Fraction of false positives.
- Fraction of false negatives.
- Cost of the test.
- Availability of alternative tests.
- Treatment options and effectiveness.

Our analysis predicts that for rare diseases, the fraction of people tested should be small, particularly when the probability of a false positive is high. Is this what you find? How might the last three factors affect a doctor's decisions?

Bibliography for Chapter 6

Section 6.1

Huff, D., *How to Lie with Statistics*. Norton, New York, 1982.

Section 6.3

Falconer, D. S., and T. F. C. Mackay. *Introduction to Quantitative Genetics*. Longman, Essex, England, 1996.

Hartl, D. L., and A. G. Clark. *Principles of Population Genetics*. Sinauer, Sunderland, Mass., 1989.

Section 6.9

Zar, J. H., *Biostatistical Analysis*. Prentice-Hall, Englewood Cliffs, N.J., 1984.

Chapter 7

Probability Models

The language of probability theory provides a set of raw materials. To succeed in building useful structures, these raw materials must first be assembled into a set of bricks, particular tools that are used over and over again. The probability models introduced in this chapter provide such a set of bricks. In probability theory, as in all of applied mathematics, the key is to describe the biological system and translate the description into mathematical form.

Each model describes a different way to combine simple measurements, such as counts (with the **binomial** and **Poisson distributions**), waiting times (with the **geometric** and **exponential distributions**), and sums (with the **normal distribution**). These basic distributions, like polynomial, exponential, and trigonometric functions, describe a remarkable range of applications and form the basis for the most important methods in statistics.

7.1 Joint Distributions

We have described the behavior of a single measurement with one random variable and its associated probability distribution or probability density function. The chief tool for studying the behavior of *two* measurements simultaneously is the **joint distribution**, which gives the probability of each possible *pair* of values for the random variables. From the joint distribution, we derive the **marginal distributions**, which give the probabilities for each random variable separately, and the **conditional distributions**, which give the probabilities for one random variable when the other takes on a particular value.

Joint Distributions

Two dart players, each of whom scores according to the same probability distribution (perhaps that in Figure 6.49) face off in a high-stakes match at their local pub. Do we have enough information to guess what will happen? Suppose one player always throws first. The results depend on how the second player responds to good and bad throws by the first. If the second player ignores what the first does, the throws will be **independent**; the probability that the second player scores a bull's-eye is the same whether the first got a bull's-eye or missed the board. Knowing what the first player did gives no information about the second. Most players, however, are incapable of ignoring the success or failure of their opponent. Some might do better under pressure and be more likely to hit the bull's-eye after their opponent did than after their opponent missed. Other players tighten up after a good shot and are less likely to hit the bull's-eye after a bull's-eye than after a miss. Each of these types of second player might have the same histogram. But no sports commentator worth her salary would think these players equivalent.

How can we summarize this information mathematically and distinguish these different types of player? Consider a simplified dart game in which hitting the board scores 1 and missing the board scores 0. Suppose that each player scores 1 with probability 0.3. What is the probability that both players score 1? The answer depends on whether the player who throws second responds to the success of the first. Let S_1 denote the event "the first player scored," S_2 the event "the second player scored," M_1 the event "the first player missed" and M_2 the event "the second player missed."

If the second player is oblivious to the first, the events will be independent and

$$\Pr(S_1 \text{ and } S_2) = \Pr(S_1)\Pr(S_2) = 0.3 \cdot 0.3 = 0.09$$
$$\Pr(S_1 \text{ and } M_2) = \Pr(S_1)\Pr(M_2) = 0.3 \cdot 0.7 = 0.21$$
$$\Pr(M_1 \text{ and } S_2) = \Pr(M_1)\Pr(S_2) = 0.7 \cdot 0.3 = 0.21$$
$$\Pr(M_1 \text{ and } M_2) = \Pr(M_1)\Pr(M_2) = 0.7 \cdot 0.7 = 0.49$$

where we have used the multiplication rule for the probabilities of independent events (Theorem 6.3). We summarize these probabilities in the following table:

	S_1	M_1
S_2	0.09	0.21
M_2	0.21	0.49

(7.1)

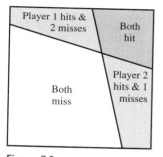

Figure 7.1
Venn diagram for independent dart players

A Venn diagram of this situation is shown in Figure 7.1.

If the second player becomes more competitive and throws better after the first scores, the probability of S_2 *conditional* on S_1 will be greater than 0.3. Assume that this conditional probability is 0.8, or $\Pr(S_2 \mid S_1) = 0.8$. Applying the multiplication rule for conditional probabilities (Equation 6.15), we get

$$\Pr(S_1 \text{ and } S_2) = \Pr(S_2 \mid S_1)\Pr(S_1) = 0.3 \cdot 0.8 = 0.24$$

The probability that both score is higher. How do we find the rest of the probabilities? First, we know that the probability that the first player scores is 0.3. This is the combination of two mutually exclusive events: either both players scored (probability 0.24) or the first scored and the second missed (probability unknown). In equations, we have

$$\Pr(S_1) = \Pr(S_1 \text{ and } S_2) + \Pr(S_1 \text{ and } M_2)$$
$$0.3 = 0.24 + \Pr(S_1 \text{ and } M_2)$$

Solving, we get

$$\Pr(S_1 \text{ and } M_2) = 0.06$$

We can use the same reasoning to compute that

$$\Pr(M_1 \text{ and } S_2) = 0.06$$

To find the probability that both players missed, we use the fact that all the probabilities must add to 1:

$$1 = \Pr(S_1 \text{ and } S_2) + \Pr(S_1 \text{ and } M_2) + \Pr(M_1 \text{ and } S_2)$$
$$1 = 0.24 + 0.06 + 0.06 + \Pr(M_1 \text{ and } M_2)$$

so that

$$\Pr(M_1 \text{ and } M_2) = 0.64$$

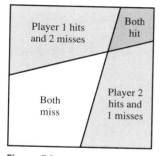

Figure 7.2
Venn diagram for competitive dart players

Our table is

	S_1	M_1
S_2	0.24	0.06
M_2	0.06	0.64

(7.2)

It is more likely that both players score or that both miss than in the independent case, but less likely that exactly one scores. This situation is shown as a Venn diagram in Figure 7.2.

If the second player is intimidated when the first scores, the probability of S_2 conditional on S_1 will be diminished, perhaps to 0.1. In this case,

$$\Pr(S_1 \text{ and } S_2) = \Pr(S_1)\Pr(S_2 \mid S_1) = 0.3 \cdot 0.1 = 0.03$$

Using the law of total probability, we write the rest of the table.

	S_1	M_1
S_2	0.03	0.27
M_2	0.27	0.43

(7.3)

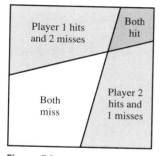

Figure 7.3
Venn diagram for intimidated dart players

(Exercise 1). The probability that both players score or both miss is less than in the independent case. This situation is shown as a Venn diagram in Figure 7.3.

Each of these tables gives the probability of each possible pair of events and the information they give constitutes the **joint probability distribution**. Like an ordinary probability distribution, the joint probability distribution associates a probability with every possible measured result. The set of these probabilities must add to 1. In a joint distribution, however, results consist of two measurements rather than one.

■ **Definition 7.1** Suppose X and Y are discrete random variables taking on the values $x_1, \ldots, x_n$ and $y_1, \ldots, y_m$, respectively. The joint distribution consists of the probabilities p_{ij} of each composite event $X = x_i$ and $Y = y_j$, or

$$p_{ij} = \Pr(X = x_i \text{ and } Y = y_j)$$ ■

A table of this general joint distribution is as follows.

	$Y = y_1$	$Y = y_2$	...	$Y = y_j$	...	$Y = y_m$
$X = x_1$	p_{11}	p_{12}	...	p_{1j}	...	p_{1m}
$X = x_2$	p_{21}	p_{22}	...	p_{2j}	...	p_{2m}
...	...	...	...	...	...	...
$X = x_i$	p_{i1}	p_{i2}	...	p_{ij}	...	p_{im}
...	...	...	...	...	...	...
$X = x_n$	p_{n1}	p_{n2}	...	p_{nj}	...	p_{nm}

All the probabilities in a joint distribution must add to 1, or

$$\sum_{i=1}^{n} \sum_{j=1}^{m} p_{ij} = 1$$

because the joint probability distribution describes a mutually exclusive and collectively exhaustive set of events.

The joint distribution for more than two random variables is based on the same idea. The joint distribution for three discrete random variables X, Y, and Z is the set of probabilities

$$p_{ijk} = \Pr(X = x_i \text{ and } Y = y_j \text{ and } Z = z_k) \quad (7.4)$$

Again, we assign a probability to every possible combination of measurements. This joint distribution gives the probability that the three measurements simultaneously take a particular set of values.

For continuous random variables, one can define a **joint probability density function** analogous to the ordinary probability density function (Definition 6.6). The statement and application of this definition requires the **double integral**, a generalization of the integral to functions of two variables. Here we concentrate on the discrete case, which is the most important for statistical applications.

Marginal Probability Distributions

Suppose we know the joint distribution of two measurements. How can we figure out the probability distribution for each measurement singly? To compute the probability that the first player scored from the joint distribution, we return to the argument behind the law of total probability (Theorem 6.1). The event "the first player scored" consists of two disjoint events: "The first player scored and the second player scored" and "the first player scored and the second player did not." The probabilities of mutually exclusive events add (requirement 3 for probabilities in Section 6.4), so we have

$$\Pr(S_1) = \Pr(S_1 \text{ and } S_2) + \Pr(S_1 \text{ and } M_2)$$

We see that $\Pr(S_1) = 0.24 + 0.06 = 0.3$, which is consistent with our assumptions. We used this reasoning already to fill in the table of probabilities.

The probabilities of measurements taken singly can be computed conveniently from the tables. $\Pr(S_1)$ is the sum of the values in the first column, the probabilities associated with events involving a score by the first player. The probability that the second player scores is the sum of the values in the first row, and so forth. In our table showing a joint distribution, we can display this with arrows as follows:

	S_1	M_1	
S_2	0.24	0.06	→ $\Pr(S_2) = 0.3$
M_2	0.06	0.64	→ $\Pr(M_2) = 0.7$
	↓	↓	
	$\Pr(S_1) = 0.3$	$\Pr(M_1) = 0.7$	

In general, we have

$$\Pr(X = x_i) = \sum_{j=1}^{m} \Pr(X = x_i \text{ and } Y = y_j)$$

added over all possible values y_j. We call this the marginal probability distribution, defined as follows.

■ **Definition 7.2** Suppose discrete random variables X and Y have the joint distribution

$$p_{ij} = \Pr(X = x_i \text{ and } Y = y_j)$$

where X takes on the values $x_1, x_2, \ldots, x_n$ and Y takes on the values $y_1, y_2, \ldots, y_m$. The **marginal probability distributions** $p_{X,i}$ for the random variable X and $p_{Y,j}$ for the random variable Y are

$$p_{X,i} = \sum_{j=1}^{m} p_{ij}$$

$$p_{Y,j} = \sum_{i=1}^{n} p_{ij}$$

■

The marginal distributions have a convenient relation to the joint distribution when the random variables are independent. The multiplicative rule for independent events (Theorem 6.3) says that

$$\Pr(X = x_i \text{ and } Y = y_j) = \Pr(X = x_i) \cdot \Pr(Y = y_j)$$

so

$$p_{ij} = p_{X,i} p_{Y,j}$$

In fact, this is often used as the **definition** of independent random variables.

534   Chapter 7   Probability Models

■ **Definition 7.3**   Two discrete random variables X and Y are independent if the joint probability distribution is equal to the product of the marginal distributions.   ■

Joint Distributions and Conditional Distributions

A more complex example is shown in Table 7.1. Four types of bird are infected by two types of parasites, lice and mites. Because these parasites are rather large, no bird has more than two of either one. Data are collected on the joint probabilities of parasite numbers for each type of bird. For example, the probability that a bird of the first type has one mite and two lice is 0.09, and the probability that it has two mites and one louse is 0.06.

Table 7.1   Lice and mites on four types of bird

Bird 1

Mites	Lice 0	1	2
0	0.20	0.15	0.15
1	0.12	0.09	0.09
2	0.08	0.06	0.06

Bird 2

Mites	Lice 0	1	2
0	0.21	0.13	0.16
1	0.13	0.07	0.10
2	0.06	0.10	0.04

Bird 3

Mites	Lice 0	1	2
0	0.33	0.08	0.09
1	0.05	0.13	0.12
2	0.02	0.09	0.09

Bird 4

Mites	Lice 0	1	2
0	0.06	0.21	0.23
1	0.21	0.03	0.06
2	0.13	0.06	0.01

To find the marginal distribution describing the number of lice, add up all the numbers in each column. For example, the probability that the first bird has one louse is the sum $0.15 + 0.09 + 0.06 = 0.3$, as summarized in the following table.

	$L = 0$	$L = 1$	$L = 2$	
$M = 0$	0.20	0.15	0.15	→ $\Pr(M = 0) = 0.5$
$M = 1$	0.12	0.09	0.09	→ $\Pr(M = 1) = 0.3$
$M = 2$	0.08	0.06	0.06	→ $\Pr(M = 2) = 0.2$
	↓	↓	↓	
	$\Pr(L = 0) = 0.4$	$\Pr(L = 1) = 0.3$	$\Pr(L = 2) = 0.3$	

Continuing in this way, we find that the four types of birds match; the probability of no lice is 0.4, of one louse is 0.3 and two lice is 0.3 (Figure 7.4a). Similarly, the marginal distribution of the number of mites can be found by adding the rows. The probability that the first type of bird has one mite is $0.12 + 0.9 + 0.9 = 0.3$. Again, the marginal distributions have been chosen to match. Each type of bird has 0 mites with probability 0.5, one mite with probability 0.3 and two mites with probability 0.2 (Figure 7.4b).

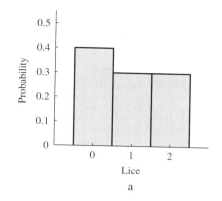
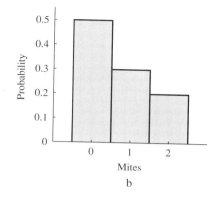

Figure 7.4
Marginal distributions of lice and mites

For the first type of bird, the product of these marginal distributions matches the probabilities. For example, the probability of one mite and two lice is 0.06, equal to the product of the marginal probability of one mite (0.3) with the marginal probability of two lice (0.2). The two parasites are distributed independently in this type of bird. A bird with no mites is just as likely to have two lice as is a bird with two mites.

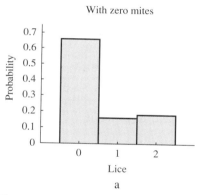
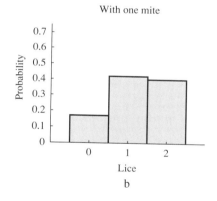
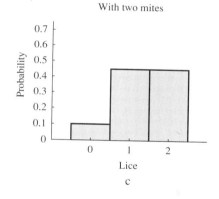

Figure 7.5
Conditional probabilities of mites on the third type of bird

When analyzing joint distributions, a new tool called the **conditional distribution** provides a clear way to depict the relation between these two parasites.

■ **Definition 7.4** Suppose random variables X and Y have joint distribution p_{ij} and marginal distributions $p_{X,i}$ and $p_{Y,j}$. The **conditional distribution** of X conditional on $Y = y_j$ is

$$\Pr(X = x_i \mid Y = y_j) = \frac{\Pr(X = x_i \text{ and } Y = y_j)}{\Pr(Y = y_j)}$$

$$= \frac{p_{ij}}{p_{Y,j}}$$

Similarly, the **conditional distribution** of Y conditional on $X = x_i$ is

$$\Pr(Y = y_j \mid X = x_i) = \frac{\Pr(Y = y_j \text{ and } X = x_i)}{\Pr(X = x_i)}$$

$$= \frac{p_{ij}}{p_{X,i}}$$

■

What does the conditional distribution mean? Instead of ignoring the number of mites and finding the marginal probability distribution for the number of lice, we fix the number of mites to compute the conditional distribution. What is the probability distribution describing the number of lice on birds with no mites? For the third type of bird, the probability of 0 lice conditional on 0 mites is

$$\Pr(L = 0 \mid M = 0) = \frac{\Pr(L = 0 \text{ and } M = 0)}{\Pr(M = 0)} = \frac{0.33}{0.5} = 0.66$$

Similarly,

$$\Pr(L = 1 \mid M = 0) = \frac{\Pr(L = 1 \text{ and } M = 0)}{\Pr(M = 0)} = \frac{0.08}{0.5} = 0.16$$

$$\Pr(L = 2 \mid M = 0) = \frac{\Pr(L = 2 \text{ and } M = 0)}{\Pr(M = 0)} = \frac{0.09}{0.5} = 0.18$$

(Figure 7.5a). A bird without mites is more likely to have no lice. Perhaps these birds are healthier or live in better locations. Perhaps mites and lice depend on each other. The conditional distributions when the number of mites is fixed at $M = 1$ is

$$\Pr(L = 0 \mid M = 1) = \frac{\Pr(L = 0 \text{ and } M = 1)}{\Pr(M = 1)} = \frac{0.05}{0.3} = 0.17$$

$$\Pr(L = 1 \mid M = 1) = \frac{\Pr(L = 1 \text{ and } M = 1)}{\Pr(M = 1)} = \frac{0.13}{0.3} = 0.43$$

$$\Pr(L = 2 \mid M = 1) = \frac{\Pr(L = 2 \text{ and } M = 1)}{\Pr(M = 1)} = \frac{0.12}{0.3} = 0.4$$

(Figure 7.5b). A bird with one mite is less likely to have zero lice. When the number of mites is fixed at $M = 2$, the conditional distribution is

$$\Pr(L = 0 \mid M = 2) = \frac{\Pr(L = 0 \text{ and } M = 2)}{\Pr(M = 2)} = \frac{0.02}{0.2} = 0.1$$

$$\Pr(L = 1 \mid M = 2) = \frac{\Pr(L = 1 \text{ and } M = 2)}{\Pr(M = 2)} = \frac{0.09}{0.2} = 0.45$$

$$\Pr(L = 2 \mid M = 2) = \frac{\Pr(L = 2 \text{ and } M = 2)}{\Pr(M = 2)} = \frac{0.09}{0.2} = 0.45$$

(Figure 7.5c). These birds are similar to the birds with one mite.

Like any other probability distribution, the probabilities in a conditional distribution must add to 1. This makes comparing these distributions easy. A good way to understand how a joint distribution deviates from independence is to compare the conditional distributions with the marginal distribution. For example, the conditional distributions in Figure 7.5 all differ from the marginal distribution for lice numbers (Figure 7.4a). As noted, a bird with zero mites is much less likely to have zero lice than a bird with one or more mites.

This sort of reasoning is particularly useful because of the following theorem about independent random variables.

■ **THEOREM 7.1** Suppose X and Y are discrete random variables. Then their conditional distributions are equal to the marginal distributions if and only if X and Y are independent.

Proof: If X and Y are independent, then

$$p_{ij} = p_{X,i} p_{Y,j}$$

(Definition 7.3). Therefore,

$$\Pr(X = x_i \mid Y = y_j) = \frac{p_{ij}}{p_{Y,j}} = \frac{p_{X,i} p_{Y,j}}{p_{Y,j}} = p_{X,i}$$

By the same reasoning,

$$\Pr(Y = y_j \mid X = x_i) = p_{Y,j}$$

so the conditional distributions are both equal to the marginal distributions.

Conversely, suppose that the conditional distributions are equal to the marginal distributions. Then

$$\Pr(X = x_i \mid Y = y_j) = \frac{\Pr(X = x_i \text{ and } Y = y_j)}{\Pr(Y = y_j)} = \Pr(X = x_i)$$

Therefore,

$$\Pr(X = x_i \text{ and } Y = y_j) = \Pr(Y = y_j) \Pr(X = x_i)$$

and the random variables are independent (Definition 7.3). ∎

If we find that the conditional distributions of one measurement are the same no matter what the value of the other measurement, then the two measurements are independent. This is consistent with our interpretation of independence as a lack of information. Knowing about one measurement gives no new information about another independent measurement.

SUMMARY

To describe two discrete random variables simultaneously we use the **joint distribution**, the set of probabilities describing all possible pairs of measurements. Summing over all possible values of one discrete random variable gives the **marginal probability distribution** for the other. Two random variables are independent when their joint distribution is equal to the product of their marginal distributions. By fixing the value of one random variable, we can find the **conditional distributions**. The conditional distributions are the same when the two random variables are independent. Differences between the conditional distributions are useful for reasoning about the relation between two measurements.

7.1 EXERCISES

1. Verify the probabilities in Equation 7.3 with the following steps.
   a. Find $\Pr(S_1 \text{ and } M_2)$ with the law of total probability and the fact that $\Pr(S_1) = 0.3$.
   b. Use the fact that $\Pr(S_2) = 0.3$ and the law of total probability to find $\Pr(M_1 \text{ and } S_2)$.
   c. Use the fact that probabilities must add to 1 to find $\Pr(M_1 \text{ and } M_2)$.

2. Recall the ecologist observing eagles and rabbits in Exercise 2 in Section 6.5, who found an eagle with probability 0.2 during an hour of observation, a jackrabbit with probability 0.5, and both with probability 0.05.

   a. What are the two random variables and their possible values?
   b. Draw a table of the probabilities (use the law of total probability to fill in squares not given explicitly).
   c. Compute and graph the marginal probability distributions.
   d. Compute and graph the conditional probability distributions.

3. Recall the rare disease model described in Section 6.5. Translate the information into a joint distribution.

4. Careful observation indicates that bus A arrives with probability 0.1 in any given minute, and that bus B arrives each

minute with the same probability. However, 9 times out of 10, bus A arrives before bus B. How is this possible? What if bus A was "jackrabbit" and bus B was "eagle"?

5. Recall the situation in Exercise 5 in Section 6.5. New cells stain properly with probability 0.95, one-day-old cells stain properly with probability 0.9, two-day-old cells stain properly with probability 0.8, three-day-old cells stain properly with probability 0.5. Suppose

$$\Pr(\text{cell is 0 days old}) = 0.4$$
$$\Pr(\text{cell is 1 day old}) = 0.3$$
$$\Pr(\text{cell is 2 days old}) = 0.2$$
$$\Pr(\text{cell is 3 days old}) = 0.1$$

   a. What are the two random variables and their possible values?
   b. Draw a table of the probabilities (use the law of total probability to fill in squares not given explicitly).
   c. Compute the marginal probability distributions.
   d. Compute and graph the conditional probability distributions for cell age. Compare with the marginal distribution.

6. Suppose one baseball player gets a hit 25% of the time and another gets a hit 35% of the time.
   a. Suppose they hit independently of one another. Find the joint probability distribution.
   b. Make up a joint distribution where the players tend to hit together more often than in the independent case (as in Equation 7.2). Make sure your joint distribution is consistent. Find the conditional probabilities.
   c. Do part **b** for players who tend to hit separately (as in Equation 7.3).

7. Suppose two dart players A and B throw alternately, and that

   Pr(A makes a good shot | B just made a good shot) = 0.8
   Pr(A makes a good shot | B just made a poor shot) = 0.2
   Pr(B makes a good shot | A just made a good shot) = 0.8
   Pr(B makes a good shot | A just made a poor shot) = 0.2

   a. Write this as a Markov chain.
   b. What is different from the cases studied in the text?
   c. What is the probability that A follows one good shot with another?
   d. What is the probability that A follows a bad shot with a good shot?
   e. What is the long-term probability that A makes a good shot?
   f. If you saw just two shots, one by A and one by B, what is the joint distribution of what you saw?

8. Find the conditional distributions for the number of lice on birds with no, one, and two mites for birds of types 1, 2, and 4. By comparing these distributions with each other and with the marginal distribution, try to think of a mechanism that could produce these results.

9. Draw the conditional distribution for the number of mites on birds with no, one, and two lice for birds of types 1, 2, 3, and 4. Compare your results with those in Figure 7.5 and Exercise 8.

10. Many matings are observed in a species of bird. Both female and male birds can take on three colors: red, blue, and green.

		Male	
Female	R	B	G
R	0.125	0.195	0.180
B	0.225	0.027	0.048
G	0.090	0.102	0.008

Find the marginal distributions for both sexes and the conditional distributions of male color for red, blue, and green females, respectively. What might be going on with these birds?

11. Prove that the probabilities in a conditional distribution add to 1.

12. **COMPUTER:** Suppose two die are rolled, but the second die must be rerolled until its score is less than or equal to that on the first. For example, if the first die rolls a 3, then the second must be rolled again and again until its value is 3 or less. The total score is the sum.
    a. Roll 100 computer die with these rules, and record your results.
    b. Try to figure out the mathematical joint distribution for this process.

7.2 Covariance and Correlation

The joint distribution or set of conditional distributions completely describes the simultaneous behavior of two random variables. We have seen situations in which one random variable is large when the other is large (the dart players in Equation 7.2

and the birds of type 3 in Table 7.1) and others in which one random variable is large when the other is small (the dart players in Equation 7.3 and the birds of type 4 in Table 7.1). These different relationships can be summarized by two statistics, the **covariance** and a scaled version called the **correlation**. A *positive covariance* indicates that two random variables tend to be large or small simultaneously; a *negative covariance* indicates that one tends to be large when the other is small.

Covariance

Intuitively, two measurements are "correlated" if they take on high values or low values simultaneously. More precisely, X and Y are positively correlated if X tends to exceed its mean $\overline{X}$ when Y exceeds its mean $\overline{Y}$. We quantify this relation first with the **covariance**.

■ **Definition 7.5** Suppose X and Y are two random variables with expectations $\overline{X}$ and $\overline{Y}$, respectively. The **covariance**, denoted by $\text{Cov}(X, Y)$, of X and Y is

$$\text{Cov}(X, Y) = E[(X - \overline{X})(Y - \overline{Y})]$$

More particularly, if X and Y are discrete random variables where X takes on the values $x_1, \ldots, x_n$, Y takes on the values $y_1, \ldots, y_m$, and the joint distribution is p_{ij}, then

$$\text{Cov}(X, Y) = \sum_{j=1}^{m} \sum_{i=1}^{n} (x_i - \overline{X})(y_j - \overline{Y}) p_{ij}$$

■

As for every other expectation, we take the value (here the product of the deviations from the means) times the probability (the joint probability p_{ij}) and add the products. The values involved are listed in the following table.

X	Y	Difference of X from mean	Difference of Y from mean	Product of differences	Joint probability
x_i	y_j	$x_i - \overline{X}$	$y_j - \overline{Y}$	$(x_i - \overline{X})(y_j - \overline{Y})$	p_{ij}

X	Y	Product of differences
$x_i > \overline{X}$	$y_j > \overline{Y}$	positive
$x_i < \overline{X}$	$y_j < \overline{Y}$	positive
$x_i > \overline{X}$	$y_j < \overline{Y}$	negative
$x_i < \overline{X}$	$y_j > \overline{Y}$	negative

The product of the differences is positive in two cases: if both X and Y are larger than their means, and if both X and Y are smaller than their means. The product of the differences is negative if exactly one of X and Y is larger than its mean. If for many terms the product of the differences is positive, the covariance is positive.

As with the variance, there is a useful computational formula for the covariance.

■ **THEOREM 7.2** (Computational Formula for the Covariance)

Suppose X and Y are two random variables with expectations $\overline{X}$ and $\overline{Y}$, respectively. The covariance can be written

$$\text{Cov}(X, Y) = E(XY) - \overline{XY}$$

More particularly, if X and Y are discrete random variables where X takes on the values $x_1, \ldots, x_n$, Y takes on the values $y_1, \ldots, y_m$, and the joint distribution is p_{ij},

then
$$\text{Cov}(X,Y) = \sum_{j=1}^{m}\sum_{i=1}^{n} x_i y_j p_{ij} - \overline{XY}$$

The proof is almost identical to the proof for the computation formula for the variance (Theorem 6.5). In words, the covariance is the expectation of the product minus the product of the expectations.

Consider the dart players with the joint distribution shown in the table. Recall that a score is worth 1 and a miss is worth 0. The covariance can be computed using either the definition or the computational formula. Let P_1 and P_2 be random variables giving the scores of players 1 and 2, respectively. Recall that both have mean 0.3. With the definition, we can tabulate the data as shown in the following table.

	S_1	M_1
S_2	0.09	0.21
M_2	0.21	0.49

P_1	P_2	Difference of P_1 from mean	Difference of P_2 from mean	Product of differences	Joint probability
1	1	1 − 0.3 = 0.7	1 − 0.3 = 0.7	0.49	0.09
0	1	0 − 0.3 = −0.3	1 − 0.3 = 0.7	−0.21	0.21
1	0	1 − 0.3 = 0.7	0 − 0.3 = −0.3	−0.21	0.21
0	0	0 − 0.3 = −0.3	0 − 0.3 = −0.3	0.09	0.49

The covariance is
$$\text{Cov}(P_1, P_2) = 0.09 \cdot 0.49 - 0.21 \cdot 0.21 - 0.21 \cdot 0.21 + 0.49 \cdot 0.09 = 0.0$$

The covariance is 0, as we might expect with *independent* random variables. With the computational formula, we have the following table.

P_1	P_2	Product	Joint probability
1	1	1	0.09
0	1	0	0.21
1	0	0	0.21
0	0	0	0.49

The covariance is
$$\text{Cov}(P_1, P_2) = 1 \cdot 0.09 + 0 \cdot 0.21 + 0 \cdot 0.21 + 0 \cdot 0.49 - 0.09 = 0.0$$

The results match, but the calculation is easier.

When the second player throws better when the first player scores, we expect the covariance to be positive. The joint distribution is

	S_1	M_1
S_2	0.24	0.06
M_2	0.06	0.64

Using the computational formula, our table is

P_1	P_2	Product	Joint probability
1	1	1	0.24
0	1	0	0.06
1	0	0	0.06
0	0	0	0.64

and the covariance is

$$\text{Cov}(P_1, P_2) = 1 \cdot 0.24 + 0 \cdot 0.06 + 0 \cdot 0.06 + 0 \cdot 0.64 - 0.09 = 0.15$$

This positive value indicates that these two players tend to do well together.

When the second player throws better when the first player misses, we expect the covariance to be negative. The joint distribution is

	S_1	M_1
S_2	0.03	0.27
M_2	0.27	0.43

Using the computational formula, our table is

P_1	P_2	Product	Joint probability
1	1	1	0.03
0	1	0	0.27
1	0	0	0.27
0	0	0	0.43

and the covariance is

$$\text{Cov}(P_1, P_2) = 1 \cdot 0.03 + 0 \cdot 0.27 + 0 \cdot 0.27 + 0 \cdot 0.43 - 0.09 = -0.06$$

Covariances of joint distributions with more than two values are computed in the same way. Consider again bird 3 from Table 7.1:

	Bird 3		
	Lice		
Mites	0	1	2
0	0.33	0.08	0.09
1	0.05	0.13	0.12
2	0.02	0.09	0.09

Using the conditional distributions, we observed that birds that have no mites also tend to have no lice. We expect, therefore, that the covariance should be positive. To get started, we compute the expected number of mites $\overline{M}$ and lice $\overline{L}$ from the marginal distribution,

$$\overline{M} = 0 \cdot 0.5 + 1 \cdot 0.3 + 2 \cdot 0.2 = 0.7$$
$$\overline{L} = 0 \cdot 0.4 + 1 \cdot 0.3 + 2 \cdot 0.3 = 0.9.$$

Using the computational formula, we can retabulate the values as follows.

M	L	Product	Joint probability
0	0	0	0.33
0	1	0	0.08
0	2	0	0.09
1	0	0	0.05
1	1	1	0.13
1	2	2	0.12
2	0	0	0.02
2	1	2	0.09
2	2	4	0.09

The covariance is therefore

$$\text{Cov}(M,L) = 0 \cdot 0.33 + 0 \cdot 0.08 + 0 \cdot 0.09 + 0 \cdot 0.05 + 1 \cdot 0.13$$
$$+ 2 \cdot 0.12 + 0 \cdot 0.02 + 2 \cdot 0.09 + 4 \cdot 0.09 - 0.7 \cdot 0.9$$
$$= 0.28$$

This new statistic, the covariance, successfully captures the positive relation between these two measurements.

Correlation

Although the sign of the covariance tells us whether two random variables have a positive or negative relation, its magnitude and units are difficult to interpret. The covariance we just computed is 0.28 mite-lice. To get rid of the units and make the values easier to use, we transform the covariance into the **correlation**, denoted by ρ ("rho"), by dividing by the product of the standard deviations.

■ **Definition 7.6** Suppose X and Y are two random variables with standard deviations σ_X and σ_Y, respectively (computed from the marginal distributions). The correlation $\rho_{X,Y}$ of X and Y is

$$\rho_{X,Y} = \frac{\text{Cov}(X,Y)}{\sigma_X \sigma_Y}$$ ■

The units cancel, giving a pure number mathematically guaranteed to lie between -1 and 1 (for a proof, see a more advanced text on probability theory).

A correlation of 1 means that Y is large exactly when X is large. A positive correlation gives a measure of how strong is the tendency of one measurement to be large when the other is. Roughly speaking, if the correlation is 0.7, knowing that the value of X is large makes us 70% certain that Y is also large. If the correlation is -0.7, knowing that the value of X is large makes us 70% certain that Y is small. The magnitude of the correlation is a measure of information, and its sign a measure of how the two random variables covary. A correlation of 0 might seem to mean that knowing the value of X gives no information about Y, but we will see shortly that this interpretation can be misleading.

In the dart score example, P_1 and P_2 are Bernoulli random variables with $p = 0.3$. Therefore

$$\text{Var}(P_1) = \text{Var}(P_2) = p(1-p) = (0.3)(0.7) = 0.21$$

(Equation 6.10). The standard deviations are therefore

$$\sigma_{P_1} = \sigma_{P_2} = \sqrt{0.21} \approx 0.458$$

For independent random variables, the covariance is 0, as is the correlation:

$$\rho_{P_1,P_2} = \frac{0.0}{0.458 \cdot 0.458} = 0.0$$

When the two players tend to hit together, the covariance has the positive value 0.15. This is hard to interpret, but if we normalize this into the correlation, we get

$$\rho_{P_1,P_2} = \frac{0.15}{0.458 \cdot 0.458} = 0.762$$

The correlation is quite high, indicating a strong relationship. When the two players tend to hit separately, the covariance takes on the negative value of -0.06. The correlation is

$$\rho_{P_1,P_2} = \frac{-0.06}{0.458 \cdot 0.458} = -0.286$$

Because computing the correlation takes many steps, it helps to organize them into an algorithm.

■ **Algorithm 7.1** **(Finding the correlation of two discrete random variables)**

1. Find the marginal distribution of each variable.
2. Compute the expectation of each variable from the marginal distribution.
3. Compute the variance of each variable from the marginal distribution, and take the square root to find the standard deviation.
4. Compute the covariance with the computational formula.
5. Divide the covariance by the product of the standard deviations to find the correlation. ■

We have already found the expectation and covariance for the numbers of mites and lice on birds. To find the correlation, we must compute the variance of each marginal distribution. For the mites, we use the computational formula (Theorem 6.5) to find

$$\sigma_M^2 = 0^2 \cdot 0.5 + 1^2 \cdot 0.3 + 2^2 \cdot 0.2 - 0.7^2 = 0.61$$
$$\sigma_M = \sqrt{0.61} \approx 0.781$$

For the lice, we find

$$\sigma_L^2 = 0^2 \cdot 0.4 + 1^2 \cdot 0.3 + 2^2 \cdot 0.3 - 0.9^2 = 0.69$$
$$\sigma_L = \sqrt{0.69} \approx 0.831$$

The correlation for birds of type 3 is therefore

$$\rho_{M,L} = \frac{\text{Cov}(M,L)}{\sigma_M \sigma_L}$$
$$= \frac{0.28}{0.781 \cdot 0.831} \approx 0.43$$

Knowing the number of lice on a bird gives us significant information about the number of mites.

Perfect Correlation

What are the covariance and correlation of a random variable with itself? From Definition 7.5, we have

$$\text{Cov}(X, X) = E[(X - \overline{X})(X - \overline{X})] = E[(X - \overline{X})^2] = \text{Var}(X) = \sigma_X^2$$

because it matches the definition of the variance (Definition 6.14). The correlation is

$$\rho_{X,X} = \frac{\text{Cov}(X, X)}{\sigma_X \cdot \sigma_X} = \frac{\sigma_X^2}{\sigma_X^2} = 1$$

A random variable is perfectly correlated with itself. If we think of correlation as a measure of how well the value of one random variable predicts another, it is not surprising that X is a perfect predictor of itself.

Are there situations in which two *different* random variables X and Y are perfectly correlated ($\rho_{X,Y} = 1$)? Knowledge of X provides perfect knowledge of Y when Y is a *function* of X. For example, suppose V measures the volume and M the mass of some rocks. If the density of rocks is exactly 5.5 g/cm³, when $V = 1.0$ cm³ we know that $M = 5.5$ g. The distribution of rock sizes could be described equally well by either measurement. We expect that this situation guarantees a perfect correlation.

Suppose that rocks have volume 1.0 cm³ with probability 0.5, volume 8.0 cm³ with probability 0.3, and volume 27.0 cm³ with probability 0.2. The joint distribution is

	$V = 1.0$	$V = 8.0$	$V = 27.0$
$M = 5.5$	0.5	0.0	0.0
$M = 44.0$	0.0	0.3	0.0
$M = 148.5$	0.0	0.0	0.2

We follow the steps in Algorithm 7.1 to compute the correlation. The marginal distribution for V is

V	Probability
1.0	0.5
8.0	0.3
27.0	0.2

Therefore, we have

$$E(V) = 1.0 \cdot 0.5 + 8.0 \cdot 0.3 + 27.0 \cdot 0.2 = 8.3$$
$$\text{Var}(V) = 1.0^2 \cdot 0.5 + 8.0^2 \cdot 0.3 + 27.0^2 \cdot 0.2 - 8.3^2 = 96.61$$
$$\sigma_M \approx 9.83$$

The marginal distribution for M is

M	Probability
5.5	0.5
44.0	0.3
148.5	0.2

Therefore, we have

$$E(M) = 5.5 \cdot 0.5 + 44.0 \cdot 0.3 + 148.5 \cdot 0.2 = 45.65$$
$$\text{Var}(M) = 5.5^2 \cdot 0.5 + 44.0^2 \cdot 0.3 + 148.5^2 \cdot 0.2 - 45.65^2 = 2922.45$$
$$\sigma_M \approx 54.06$$

Next, we find the covariance (simplifying our work by ignoring terms with probability 0) as follows:

$$\text{Cov}(V, M) = E(VM) - E(V)E(M)$$
$$= 1.0 \cdot 5.5 \cdot 0.5 + 8.0 \cdot 44.0 \cdot 0.2 + 27.0 \cdot 148.5 \cdot 0.2 - 8.3 \cdot 45.65$$
$$\approx 531.4$$

The correlation is

$$\rho_{V,M} = \frac{\text{Cov}(V, M)}{\sigma_V \sigma_M} \approx \frac{531.4}{9.83 \cdot 54.06} = 1.0$$

The correlation between these two measurements is indeed perfect.

It is interesting, however, that even when measurement Y is a function of measurement X, we find perfect correlation only when Y is a **linear function** of X. The results are summarized in the following theorem.

■ **THEOREM 7.3** Let X and Y be two random variables.

- If $Y = aX + b$ for $a > 0$, $\rho_{X,Y} = 1$.
- If $Y = aX + b$ for $a < 0$, $\rho_{X,Y} = -1$.
- If $Y = aX + b$ for $a = 0$, $\rho_{X,Y}$ is undefined.

■

We defer the proof until Section 7.3, when we learn how to compute expectations and variances of combinations of random variables. For the example just discussed, $M = 5.5V$, which corresponds to $a = 5.5$ and $b = 0$ in Theorem 7.3. The case with $a = 0$ occurs when the random variable Y takes on only the single value b. Both $\text{Cov}(X, Y)$ and σ_Y are 0 in this case, leading to legal trouble in computing the correlation (Exercise 10).

Perfect negative correlation of -1 occurs when the random variables behave in exactly opposite ways. Suppose a vain bird spends all its time either sleeping (a time S) or preening (a time P). If the bird sleeps for 10 h, it must preen for 14 h. One set of probabilities is

	Sleeping		
Preening	14	15	16
10	0.0	0.0	0.4
9	0.0	0.3	0.0
8	0.3	0.0	0.0

(7.5)

In general, $P = 24 - S$, fitting Theorem 7.3 with $a = -1$ and $b = 24$. These random variables are perfectly negatively correlated because P is small whenever S is large and vice versa.

However, the correlation of two random variables will not be perfect if the second variable is a **nonlinear function** of the first. Suppose that rocks are perfect cubes with sides of length L. They come in three sizes, $L = 1.0$, $L = 2.0$, and

$L = 3.0$. The volume V is the function $V = L^3$ of the side length and can take on values $V = 1.0$, $V = 8.0$, and $V = 27.0$. The joint probability distribution is

	$L = 1.0$	$L = 2.0$	$L = 3.0$	
$V = 1.0$	0.5	0.0	0.0	(7.6)
$V = 8.0$	0.0	0.3	0.0	
$V = 27.0$	0.0	0.0	0.2	

It seems reasonable to expect that side length and volume will be perfectly correlated because we can compute the exact value of V when we know the exact value of L.

To check this intuition, we compute the correlation. The mean and standard deviation of the marginal distributions of L and V are $\bar{L} = 1.7$, $\bar{V} = 8.3$, $\sigma_L = 0.843$, and $\sigma_V = 9.83$ (Exercise 4). Furthermore, $\text{Cov}(L, V) = 7.39$. The correlation is

$$\rho_{L,V} = \frac{\text{Cov}(L, V)}{\sigma_L \sigma_V} = \frac{7.39}{9.72 \cdot 0.843} = 0.90$$

Even though the volume is a function of the length, the correlation is not perfect because the function $V = L^3$ is nonlinear. The correlation measures the strength of the linear relation between two random variables. In Section 8.7, we will apply these concepts to data using the statistical techniques of linear and nonlinear regression.

SUMMARY

We have seen how to compute the **covariance** and the **correlation**, two statistics that describe the relation between two random variables. A positive covariance indicates that the two random variables tend to be both large or both small. The correlation is a normalized version of the covariance, guaranteed to take on values between -1 and 1. Values near -1 indicate a strong negative relation, values near 0 indicate no relation, and values near 1 indicate a strong positive relation. The correlation measures the strength of the *linear relation* between two random variables.

7.2 EXERCISES

1. Compute the covariances and correlations for birds 1, 2, and 4 in Table 7.1.

2. Show that the correlation between time sleeping S and time preening P is really -1 for the vain bird that spends all its time either sleeping or preening (Equation 7.5).

3. Compute the covariance and correlation from the joint distributions defined in the following exercises.
   a. Exercise 2 in Section 7.1
   b. Exercise 5 in Section 7.1

4. Check the calculation of the statistics describing the joint distribution in Equation 7.6. Draw the conditional distributions for V. How do they show that V is a function of L?

5. Suppose the following are measurements of the temperature T and insect size S.

	$T = 10$	$T = 20$	$T = 30$
$S = 5$	0.2	0.1	0.0
$S = 20$	0.1	0.2	0.1
$S = 80$	0.0	0.1	0.2

   a. Find the correlation of S with T.
   b. Find the correlation of $\ln(S)$ with T.

6. Consider the following data for cell age A and the number of toxic molecules N inside the cell.

	$A=0$	$A=1$	$A=2$	$A=3$	$A=4$
$N=0$	0.00	0.008	0.026	0.048	0.070
$N=1$	0.00	0.064	0.092	0.100	0.097
$N=2$	0.20	0.128	0.082	0.052	0.033

Find the correlation of A with N.

7. Suppose the random variable X takes on the values 0, 1, and 2 with probabilities 0.2, 0.3, and 0.5, respectively. Suppose $Y = 3X + 1$.
   a. Find the joint distribution of X and Y.
   b. Compute the covariance explicitly.
   c. Compute the correlation explicitly.

8. Consider the random variable X in Exercise 7.
   a. Suppose $Y = X^2$. Compute the joint distribution, covariance, and correlation.
   b. Suppose $Y = (X-1)^2$. Compute the joint distribution, covariance, and correlation.

9. Consider the random variable X defined in Exercise 7. Suppose $Y = (X-1)^2$ and a new random variable Z is defined by

$$Z = -\frac{3}{2}X^2 + \frac{7}{2}X$$

   a. Find the covariance between X and Z.
   b. Find the covariance between Y and Z. Although Y and Z are positively correlated with X, they are negatively correlated with each other.

10. Let X and Y be two random variables where $Y = aX + b$ as in Theorem 7.3, but suppose that $a = 0$. Try to find the correlation.

7.3 Sums and Products of Random Variables

We often want to combine measurements by adding or multiplying random variables that might or might not be independent. Many scientific measurements can be thought of best as combinations of simpler measurements in this way. Mathematically, understanding these combined measurements is greatly simplified by three powerful theorems. The first states that the expectation of the sum is the sum of the expectations. The second states that the expectation of the product is the product of the expectations when the two random variables are independent. The third states that the variance of a sum is the sum of the variances, again only when the two random variables are independent. These mathematical theorems form the basis of a great deal of statistical reasoning.

Expectation of a Sum

You throw a dart two times. Each time, the score is drawn from the same distribution (such as that in Figure 6.49) with expectation 12.55. What is the expectation of the sum of the two scores? The answer, whether or not the throws are independent, is twice the expectation of each score, or 25.1. This is formalized in the following theorem.

■ **THEOREM 7.4** If X and Y are any two random variables with finite expectation, then

$$E(X + Y) = E(X) + E(Y)$$

Proof: Suppose X takes on values $x_1, \ldots, x_n$ and Y takes on values $y_1, \ldots, y_m$. Suppose they have marginal distributions $p_{X,i}$ and $p_{Y,j}$, respectively, and joint distribution p_{ij}. Then

$$E(X+Y) = \sum_{i=0}^{n}\sum_{j=0}^{m}(x_i+y_j)p_{ij} \qquad \text{definition of expectation}$$

$$= \sum_{i=0}^{n}\sum_{j=0}^{m}x_i p_{ij} + \sum_{i=0}^{n}\sum_{j=0}^{m}y_j p_{ij} \qquad \text{breaking up the sum}$$

$$= \sum_{i=0}^{n}x_i \sum_{j=0}^{m}p_{ij} + \sum_{j=0}^{m}y_j \sum_{i=0}^{n}p_{ij} \qquad \text{switching sums and removing constants}$$

$$= \sum_{i=0}^{n}x_i p_{X,i} + \sum_{j=0}^{m}y_j p_{Y,j} \qquad \text{definition of marginal distributions}$$

$$= E(X) + E(Y) \qquad \text{definition of expectation}$$

The proof for continuous random variables is similar but requires manipulating double integrals. ∎

Consider the lice and mites presented for two of the birds in Table 7.1. We found that the two measurements were independent for bird 1 and positively correlated for bird 3:

	Bird 1				Bird 3		
		Lice				Lice	
Mites	0	1	2	Mites	0	1	2
0	0.20	0.15	0.15	0	0.33	0.08	0.09
1	0.12	0.09	0.09	1	0.05	0.13	0.12
2	0.08	0.06	0.06	2	0.02	0.09	0.09

Suppose we are interested in the total number of parasites found by adding the number of mites and number of lice. How can we find the expectation of this combined random variable? The straightforward method is to use the joint distribution and compute the probability of 0, 1, 2, 3, or 4 total parasites. There is only one way for a bird to have no parasites, so for bird 1, we have

$$\Pr(0 \text{ parasites}) = \Pr(0 \text{ lice and } 0 \text{ mites}) = 0.20$$

There are two ways to have one parasite (one louse and no mites, or one mite and no lice). Therefore for bird 1, we have

$$\Pr(1 \text{ parasite}) = \Pr(0 \text{ lice and } 1 \text{ mite}) + \Pr(1 \text{ louse and } 0 \text{ mites})$$
$$= 0.12 + 0.15 = 0.27$$

Continuing in this way, we find

$$\Pr(2 \text{ parasites}) = 0.08 + 0.09 + 0.15 = 0.32$$
$$\Pr(3 \text{ parasites}) = 0.06 + 0.09 = 0.15$$
$$\Pr(4 \text{ parasites}) = 0.06$$

We have just found the probability distribution for the total number of parasites in bird 1, a random variable we denote by P_1 (Figure 7.6a). This new measurement is the sum of L and M, or $P_1 = L + M$. Remember to think of this as adding *measurements*, not as adding *probability distributions*. Armed with its probability distribution, the expectation is

$$E(P_1) = 0 \cdot 0.20 + 1 \cdot 0.27 + 2 \cdot 0.32 + 3 \cdot 0.15 + 4 \cdot 0.06 = 1.6$$

Similarly, the probability distribution for the total number of parasites P_3 on birds of type 3 is

$$\Pr(0 \text{ parasites}) = 0.33$$
$$\Pr(1 \text{ parasite }) = 0.05 + 0.08 = 0.13$$
$$\Pr(2 \text{ parasites}) = 0.02 + 0.13 + 0.09 = 0.24$$
$$\Pr(3 \text{ parasites}) = 0.09 + 0.12 = 0.21$$
$$\Pr(4 \text{ parasites}) = 0.09$$

(Figure 7.6b). This distribution looks quite different from that of P_1. However, the expected total number

$$E(P_3) = 0 \cdot 0.33 + 1 \cdot 0.13 + 2 \cdot 0.24 + 3 \cdot 0.21 + 4 \cdot 0.09 = 1.6$$

is exactly the same.

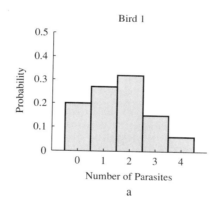
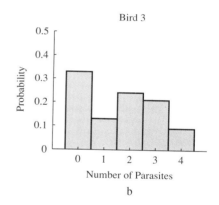

Figure 7.6 The total number of parasites on two types of bird

Theorem 7.4 shows that the expected total *must* be the same for each type of bird and gives a simpler way to compute them. We can compute the expectation of the sum as the sum of the expectations. Using the fact that $E(L) = 0.9$ and $E(M) = 0.7$ (Figure 7.7), we get

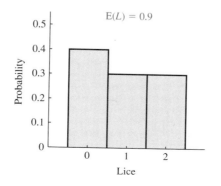
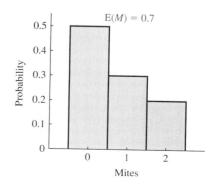

Figure 7.7 Marginal distributions of lice and mites revisited

$$E(P) = E(L + M) = E(L) + E(M) = 0.9 + 0.7 = 1.6$$

no matter what the joint distribution is. The same calculation works for birds 2 and 4 even though we have not found the probability distributions for P_2 and P_4. This is a remarkable, and convenient, achievement.

Another useful property of the expectation is that multiplying a random variable by a constant multiplies the expectation by the same constant.

THEOREM 7.5 If X is a random variable with finite expectation and a is a constant, then

$$E(aX) = aE(X)$$

Proof: Suppose X takes on values $x_1, \ldots, x_n$ with probability distribution p_i. Then

$$E(aX) = \sum_{i=0}^{n} a x_i p_i \qquad \text{definition of expectation}$$

$$= a \sum_{i=0}^{n} x_i p_i \qquad \text{move constant outside sum}$$

$$= a E(X) \qquad \text{definition of expectation}$$

The proof for continuous random variables is similar but removes the constant a from an integral at the second step. ∎

For example, suppose each louse weighs exactly 0.05 g. What is the mean weight of lice on a bird? The mean number is 0.9, so we expect that the mean weight is $0.05 \cdot 0.9 = 0.045$. This is precisely what is guaranteed by Theorem 7.5. The new random variable W_L representing the total weight of lice is $W_L = 0.05L$. Therefore, we have

$$E(W_L) = E(0.05L) = 0.05 E(L) = 0.05 \cdot 0.9 = 0.045$$

We can combine Theorems 7.4 and 7.5 into a single general theorem that says that the expectation of a new random variable constructed by multiplying simple random variables by constants and adding can be found by multiplying the expectations of the simple random variables by constants and adding.

THEOREM 7.6 If $X_1, X_2, \ldots, X_n$ are random variables with finite expectation and $a_1, a_2, \ldots, a_n$ are constants, then

$$E\left(\sum_{i=1}^{n} a_i X_i\right) = \sum_{i=1}^{n} a_i E(X_i)$$

Proof:

$$E\left(\sum_{i=1}^{n} a_i X_i\right) = \sum_{i=1}^{n} E(a_i X_i) \qquad \text{Theorem 7.4}$$

$$= \sum_{i=1}^{n} a_i E(X_i) \qquad \text{Theorem 7.5} \qquad ∎$$

For example, suppose that each louse weighs 0.05 g and that each mite weighs 0.02 g. The total weight of parasites on the bird is

$$W = 0.05L + 0.02M$$

The mean weight of parasites on a bird is

$$E(W) = E(0.05L + 0.02M)$$
$$= 0.05E(L) + 0.02E(M)$$
$$= 0.05 \cdot 0.9 + 0.02 \cdot 0.7 = 0.059$$

This expectation can be computed by finding the probability distribution of the random variable W (Exercise 2), but the "straightforward" method requires much more work.

Expectation of a Product

If the expectation of a sum is the sum of the expectations, is the expectation of a product the product of the expectations? As a test, consider a species of bird that can raise either one or two chicks. In some nests, the chicks are smaller (mass 40 g); in others, the chicks are larger (mass 50 g). Let the random variable S denote the mass of each chick, and let C be the number of chicks. The total weight of chicks in the nest is

| | Weight of each chick ||
Number of chicks	$S = 40$	$S = 50$
$C = 1$	40	50
$C = 2$	80	100

Suppose that the joint probability distribution is

	$S = 40$	$S = 50$
$C = 1$	0.42	0.28
$C = 2$	0.18	0.12

This means that 42% of nests have one small chick, 18% have two small chicks, and so forth.

The marginal distributions are

$$\Pr(S = 40) = 0.6$$
$$\Pr(S = 50) = 0.4$$

and

$$\Pr(C = 1) = 0.7$$
$$\Pr(C = 2) = 0.3$$

Therefore,

$$E(S) = 40 \cdot 0.6 + 50 \cdot 0.4 = 44$$
$$E(C) = 1 \cdot 0.7 + 2 \cdot 0.3 = 1.3$$

The total weight T of chicks in the nest is the **product** of the weight of each chick S and the number of chicks C. Can we compute the expectation of the product as the product of the expectations? Is the following true?

$$E(T) = E(S)E(C) = 44 \cdot 1.3 = 57.2$$

From our tables, the probability distribution for T is

$$\Pr(T = 40) = 0.42$$
$$\Pr(T = 50) = 0.28$$
$$\Pr(T = 80) = 0.18$$
$$\Pr(T = 100) = 0.12$$

(Figure 7.8a). Therefore, we find

$$E(T) = 40 \cdot 0.42 + 50 \cdot 0.28 + 80 \cdot 0.18 + 100 \cdot 0.12 = 57.2$$

They match.

Suppose we try an alternative probability distribution:

	$S = 40$	$S = 50$
$C = 1$	0.6	0.1
$C = 2$	0.0	0.3

The marginal distributions for S and C are the same, but the probability distribution for the product T is

$$\Pr(T = 40) = 0.6$$
$$\Pr(T = 50) = 0.1$$
$$\Pr(T = 80) = 0.0$$
$$\Pr(T = 100) = 0.3$$

(Figure 7.8b). The expectation is now

$$E(T) = 40 \cdot 0.6 + 50 \cdot 0.1 + 80 \cdot 0.0 + 100 \cdot 0.3 = 59.0$$

which is slightly higher.

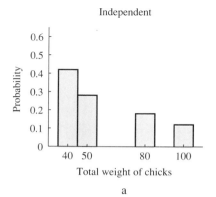
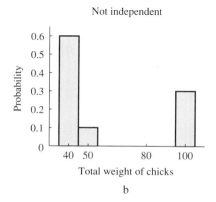

Figure 7.8
The total weight of chicks in the nest

What is the difference between these two distributions? A little examination reveals that the first is independent, because the joint probability distribution is equal to the product of the marginal distributions. The expectation of the product is equal to the product of the expectations **only** when the random variables are independent.

■ **THEOREM 7.7** If X and Y are independent random variables with finite expectation, then

$$E(XY) = E(X)E(Y)$$

Proof: Suppose X takes on values $x_1, \ldots, x_n$ and Y takes on values $y_1, \ldots, y_m$, with marginal distributions $p_{X,i}$ and $p_{Y,j}$, respectively, and joint distribution p_{ij}. Then

$$\begin{aligned}
E(XY) &= \sum_{i=0}^{n} \sum_{j=0}^{m} x_i y_j p_{ij} && \text{definition of expectation} \\
&= \sum_{i=0}^{n} \sum_{j=0}^{m} x_i y_j p_{X,i} p_{Y,j} && \text{independence} \\
&= \sum_{i=0}^{n} x_i p_{X,i} \sum_{j=0}^{m} y_j p_{Y,j} && \text{breaking up sums} \\
&= E(X)E(Y) && \text{definition of expectation}
\end{aligned}$$

The proof for continuous random variables is similar but requires manipulating double integrals. ■

Although useful in itself, Theorem 7.7 is more often used to prove the following theorem that verifies our intuition about the covariance.

■ **THEOREM 7.8** If X and Y are independent random variables, then

$$\text{Cov}(X, Y) = 0$$

Proof:

$$\begin{aligned}
\text{Cov}(X, Y) &= E(XY) - E(X)E(Y) && \text{computation formula for covariance (Theorem 7.2)} \\
&= E(X)E(Y) - E(X)E(Y) && \text{Theorem 7.7} \\
&= 0
\end{aligned}$$

■

Variance of a Sum

Expectations add. Is this true of any other statistic? It is quite difficult to concoct random variables for which the median, mode, or mean absolute deviation add. The reason the variance is the second most widely used statistic is the fact that the variance of the sum of *independent* random variables is the sum of the variances.

■ **THEOREM 7.9** If X and Y are two independent random variables with finite variance,

$$\text{Var}(X + Y) = \text{Var}(X) + \text{Var}(Y)$$

Proof: Suppose X and Y are two independent random variables. Then

$$\begin{aligned}
\text{Var}(X+Y) &= E[(X+Y)^2] - [E(X+Y)]^2 && \text{computational formula for variance}\\
&= E(X^2 + 2XY + Y^2) - E(X)^2 && \text{expand the squares}\\
&\quad - 2E(X)E(Y) - E(Y)^2\\
&= E(X^2) + 2E(XY) + E(Y^2) - E(X)^2 && \text{Theorem 7.6}\\
&\quad - 2E(X)E(Y) - E(Y)^2\\
&= E(X^2) - E(X)^2 + 2E(XY) && \text{rearrange the terms}\\
&\quad - 2E(X)E(Y) + E(Y^2) - E(Y)^2\\
&= \text{Var}(X) + \text{Var}(Y) + 2\,\text{Cov}(X,Y) && \text{recognize the pieces}\\
&= \text{Var}(X) + \text{Var}(Y) && \text{Theorem 7.8} \quad \blacksquare
\end{aligned}$$

To check this theorem, we find the variance of the total number of parasites P_1 on bird 1 and P_3 on bird 3, using the probability distributions found earlier. When finding the correlation, we computed that

$$\text{Var}(M) = 0.61$$
$$\text{Var}(L) = 0.69$$

Because M and L are independent for birds of type 1, we expect that

$$\begin{aligned}
\text{Var}(P_1) &= \text{Var}(M) + \text{Var}(L)\\
&= 0.61 + 0.69 = 1.3
\end{aligned}$$

Using the probability distribution found above, we get

$$\text{Var}(P_1) = 0^2 \cdot 0.20 + 1^2 \cdot 0.27 + 2^2 \cdot 0.32 + 3^2 \cdot 0.15 + 4^2 \cdot 0.06 - 1.6^2 = 1.3$$

In contrast, for birds of type 3, for whom the distributions are not independent, we get

$$\text{Var}(P_3) = 0^2 \cdot 0.33 + 1^2 \cdot 0.13 + 2^2 \cdot 0.24 + 3^2 \cdot 0.21 + 4^2 \cdot 0.09 - 1.6^2 = 1.86$$

Why is the variance of parasite numbers on birds of type 3 higher? There are two ways of thinking about this. The graph of the probability distribution for P_3 is more spread out, with shorter bars in the middle (Figure 7.6 on page 549). Alternatively, we can look to the proof of Theorem 7.9. Before the last step, we found the **general addition formula for variances**:

$$\text{Var}(X+Y) = \text{Var}(X) + \text{Var}(Y) + 2\,\text{Cov}(X,Y) \qquad (7.7)$$

We computed that $\text{Cov}(X,Y) = 0.28$ for bird 3, so

$$\begin{aligned}
\text{Var}(P_3) &= \text{Var}(M) + \text{Var}(L) + 2\,\text{Cov}(M,L)\\
&= 0.61 + 0.69 + 2 \cdot 0.28 = 1.86
\end{aligned}$$

The positive covariance spreads out the distribution and increases the variance because there are more birds with no or four parasites.

Multiplying a random variable by a constant multiplies the variance by the *square* of that constant.

■ **THEOREM 7.10** If X is a random variable with finite variance and a is a constant, then

$$\text{Var}(aX) = a^2\,\text{Var}(X)$$

∎

Proof:

$$\begin{aligned}\text{Var}(aX) &= \text{E}(a^2X^2) - \text{E}(aX)^2 &&\text{computational formula for variance}\\&= a^2\text{E}(X^2) - [a\text{E}(X)]^2 &&\text{Theorem 7.5}\\&= a^2\text{E}(X^2) - a^2\text{E}(X)^2 &&\text{algebra}\\&= a^2\,\text{Var}(X) &&\text{recognize variance}\end{aligned}$$
∎

Combining Theorem 7.9 with Theorem 7.10 gives us the following theorem.

■ **THEOREM 7.11** If $X_1, X_2, \ldots, X_n$ are independent random variables with finite variance and $a_1, a_2, \ldots, a_n$ are any constants, then

$$\text{Var}\left(\sum_{i=1}^{n} a_i X_i\right) = \sum_{i=1}^{n} a_i^2 \,\text{Var}(X_i)$$
∎

Proof:

$$\begin{aligned}\text{Var}\left(\sum_{i=1}^{n} a_i X_i\right) &= \sum_{i=1}^{n} \text{Var}(a_i X_i) &&\text{Theorem 7.9}\\&= \sum_{i=1}^{n} a_i^2 \,\text{Var}(X_i) &&\text{Theorem 7.10}\end{aligned}$$
∎

For example, we defined the random variable W,

$$W = 0.05L + 0.02M$$

to represent the total weight of parasites. For bird 1, L and M are independent. Therefore,

$$\begin{aligned}\text{Var}(W) &= 0.05^2\,\text{Var}(L) + 0.02^2\,\text{Var}(M)\\&= 0.05^2 \cdot 0.69 + 0.02^2 \cdot 0.61 = 0.00197\end{aligned}$$

The only tricky thing about this theorem is that negative constants are added, not subtracted. Suppose we are interested in the excess of L over M, or the random variable

$$D = L - M$$

From Theorem 7.6, we find

$$\text{E}(D) = \text{E}(L) - \text{E}(M) = 0.9 - 0.7 = 0.2$$

meaning that these birds have an average of 0.2 more lice than mites. To find the variance of the difference without finding the probability distribution, however, we must assume that the random variables are independent (as on bird 1) and be careful about the negative sign. For bird 1,

$$\text{Var}(D) = \text{Var}(L - M) = \text{Var}(L) + (-1)^2\,\text{Var}(M) = 0.69 + 0.61 = 1.3$$

SUMMARY

We have seen that the expectation of a sum of random variables is equal to the sum of the expectations, whether or not the random variables are independent. Only if the random variables are independent, however, is the expectation of the product equal to the product of the expectations and the variance of the sum equal to the sum of the variances. Multiplying a random variable by a constant multiplies the expectation by the constant and multiplies the variance by the *square* of the constant. These theorems provide the foundation for building the important probability distributions we study next.

7.3 EXERCISES

1. Consider again the birds suffering from mites and lice (Table 7.1).
   a. Find the probability distribution of $P = L + M$ for birds 2 and 4.
   b. Compute the expectations directly.
   c. Show how you could have found the expectation using Theorem 7.4.
   d. Compute the variances directly.
   e. Show how you could have found the variance with the general addition formula for variances (Equation 7.7).

2. Find the probability distribution of $W = 0.05L + 0.02M$ for birds 1 and 3. Check that the expectation is 0.059 in both cases (from Table 7.1)

3. Find the expected number of each type of immigrant from the following table. Find the expected total number of immigrants from all species combined.

Number of immigrants	Species 1	Species 2	Species 3
0	0.3	0.6	0.1
1	0.6	0.2	0.3
2	0.1	0.2	0.6

4. Suppose species 1 in Exercise 3 has mass 10 kg, species 2 has mass 5 kg and species 3 has mass 15 kg.
   a. Find the expected mass of the immigrants of each species that arrive.
   b. Find the expected total mass of all immigrants.

5. Ignore the third immigrant species in Exercise 3, and suppose that species 1 and 2 arrive independently.
   a. Give the joint probability distribution for species 1 and 2.
   b. Find the probability of each possible number of immigrants.
   c. Find the expected number of immigrants of each of these two species and compare with the sum of the expected numbers.
   d. Find the variance in the total number of immigrants and compare with the sum of the variances.

6. Consider the situation described in Exercise 5 but suppose that the two species do not arrive independently.
   a. Find a set of probabilities for outcomes, consistent with the probabilities in Exercise 3, that you think will have higher variance than the independent case. Compute the expectation and variance.
   b. Redo part a but shoot for lower variance.

7. The first two immigrant species in Exercise 3 are appealing to eco-tourists, and each additional individual brings in $1000. The third species is repellent to eco-tourists and reduces revenue by $500.
   a. Find the expected revenue from each species separately and for all species together.
   b. Find the variance in revenue due to each species separately.
   c. Find the variance in the total revenue if the species immigrate independently.
   d. Interpret these results.

8. Consider the following table giving the probability that three types of birds consume two different species of caterpillars that come in two different sizes. The first has food quality (in kcal/cm³) of 1.0 and the second has food quality of 2.0. The two volumes are 0.5 and 1.5 cm³.

Quality	Volume eaten by bird 1		Volume eaten by bird 2		Volume eaten by bird 3	
	V = 0.5	1.5	0.5	1.5	0.5	1.5
1.0	0.35	0.25	0.24	0.36	0.10	0.50
2.0	0.05	0.35	0.16	0.24	0.30	0.10

   a. Which of the joint distributions is independent?
   b. Find the mean quality and volume of caterpillars.
   c. Find the expected total calories per caterpillar for each bird.

9. Consider the lice and mites on birds 1 and 3 (Table 7.1).
   a. Find the probability distribution of $D = L - M$ for birds 1 and 3.
   b. Compute the expectation and variance directly.
   c. Show how you could have found the expectation using Theorem 7.4.
   d. Use some algebra and try to find the variance with an equation like Equation 7.7 (the general computational formula for variance). Does this result make sense?

10. Let $Y = X + X$. Find the variance of Y in two ways.
    a. Notice that $Y = 2X$.

b. Use Equation 7.7 and the covariance of X with itself.

c. Suppose X_1 and X_2 are independent random variables each having the same distribution as X. Why is $\text{Var}(X_1 + X_2) < \text{Var}(Y)$?

11. Prove that the geometric mean of the product of any two random variables is equal to the product of the geometric means.

12. **COMPUTER:** Consider Markov chains where

$$\Pr(I_{t+1} \mid I_t) = p$$
$$\Pr(I_{t+1} \mid O_t) = 1 - p$$
$$\Pr(O_{t+1} \mid I_t) = 1 - p$$
$$\Pr(O_{t+1} \mid O_t) = p$$

for various values of p. In each case, the long-term probability that a molecule is inside a cell is 0.5. Suppose you get 1 point when the molecule is inside the cell and 0 points when it is outside.

a. Add the random variables I_1 through I_{100} starting from $I_0 = 1$ when $p = 0.01$, $p = 0.1$, $p = 0.25$, $p = 0.5$, $p = 0.75$, $p = 0.9$, and $p = 0.99$.

b. What should the value be? Which value is closest?

c. Try again with the initial value I_0 set to 1 with probability 0.5 and to 0 with probability 0.5. Does this change your results?

13. **COMPUTER:** Recall the dice rolling rules in Exercise 12 in Section 7.1. Redo the experiment, and find the average value for each die. Is the sum of the averages equal to the average of the sums?

7.4 The Binomial Distribution

We have developed tools to display and combine probability distributions and probability density functions. We now begin the study of special distributions generated by particular biological models. These distributions arise repeatedly in remarkably diverse contexts. By understanding these distributions and the underlying models, we have a baseline against which to compare and reason about data. It is a good idea to start a table of these special distributions, including for each the underlying model, the range, the formula, the mean, and the variance. In this section we analyze the **binomial distribution**, the distribution constructed from sums of independent Bernoulli random variables.

The Binomial Distribution Defined

Suppose that each year a population receives an immigrant with probability p and no immigrant with probability $1 - p$. The arrivals of immigrants are independent; knowing that an immigrant arrived one year says nothing about whether an immigrant will arrive in future years. What is the probability that exactly two immigrants arrive in 3 yr? What is the probability that exactly six immigrants arrive in 10 yr? We can answer these questions with the **binomial distribution**.

Let the random variable N represent the number of immigrants that arrived. After n years, this random variable could take on any value between 0 and n. To understand this process completely, we must compute the probability that $N = k$ for every value of $0 \leq k \leq n$. We define the binomial probability distribution b as

$$b(k; n, p) = \Pr(N = k) \tag{7.8}$$

This gives the probability of exactly k immigrants after n years, where p is the probability that an immigrant arrives in any particular year. The probability that six immigrants arrive in 10 yr if the probability of an immigrant in each year is 0.7 can be written

$$\Pr(6 \text{ immigrants in 10 yr}) = b(6; 10, 0.7)$$

The binomial distribution can describe the probability of flipping exactly 50 heads in 100 tosses of a fair coin. The probability per try is 0.5, so this probability can be written with the binomial distribution as

$$\Pr(50 \text{ heads in } 100 \text{ flips}) = b(50; 100, 0.5)$$

At the moment, we have no formula for this quantity.

The total number of immigrants in n years is the *sum* of the number of immigrants in each year. The number of immigrants in each year is a Bernoulli random variable, a random variable that takes on only the values 0 and 1 (Definition 6.9). Suppose the number of immigrants over 10 yr is

	\multicolumn{10}{c}{Year}									
	1	2	3	4	5	6	7	8	9	10
Immigrants	0	1	1	0	1	0	1	1	1	0
Total	0	1	2	2	3	3	4	5	6	6

The total number after 10 yr is $0 + 1 + 1 + 0 + 1 + 0 + 1 + 1 + 1 + 0 = 6$. The compound random variable N is a *sum* of simpler random variables.

The number of heads, or immigrants (often called "successes") has a binomial distribution only if the conditions in the following definition are satisfied.

■ **Definition 7.7** A random variable N has a binomial distribution with parameters n and p if it is the sum of n independent Bernoulli random variables, each of which has a probability p of being equal to 1. The range of N is the set of integers from 0 to n, and the probability that $N = k$ is indicated by $b(k; n, p)$. ■

This definition gives the two key conditions for a random variable to have a binomial distribution:

- The random variable is the sum of n *independent* Bernoulli random variables.
- Each Bernoulli random variable takes on the value 1 with equal probability p.

"All" that remains to be done is to compute $b(k; n, p)$. Before attacking this problem, we can use the theorem guaranteeing that the expectation of the sum is the sum of the expectations (Theorem 7.6) to find $E(N)$, the mean number of immigrants after n years. Let I_i represent the number of immigrants in year i, so that

$$N = \sum_{i=1}^{n} I_i$$

(Figure 7.9). The expectation of N is the sum of the expectations of the I_i. First, recall that the expectation of a Bernoulli random variable is

$$E(I_i) = 0 \cdot (1 - p) + 1 \cdot p = p$$

Therefore, we have

$$E(N) = \sum_{i=1}^{n} E(I_i) = \sum_{i=1}^{n} p = np \quad (7.9)$$

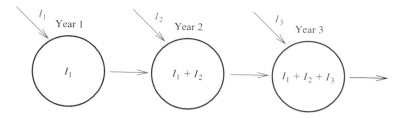

Figure 7.9
A binomial random variable as a sum of Bernoulli random variables

The mean number of successes is the number of trials times the probability of success per trial. If the probability of an immigrant is 0.7 in each year, the mean is 7 immigrants in 10 yr. If the probability of a head is 0.5 on each flip, the mean is 50 heads in 100 flips.

Similarly, because the Bernoulli random variables are independent (first condition), we can find the variance by adding (Theorem 7.9). The variance of each Bernoulli random variable is

$$\text{Var}(I_i) = 1^2 \cdot p + 0^2 \cdot (1-p) - p^2 = p - p^2 = p(1-p) \quad (7.10)$$

(Equation 6.10). The variance of N is therefore

$$\text{Var}(N) = \sum_{i=1}^{n} \text{Var}(I_i) = \sum_{i=1}^{n} p(1-p) = np(1-p) \quad (7.11)$$

For example, when immigrants arrive with probability $p = 0.7$ for $n = 10$ yr, the variance is

$$\text{Var}(N) = 10 \cdot 0.7 \cdot 0.3 = 2.1$$

When we flip for heads with probability $p = 0.5$ for $n = 10$ flips, the variance is

$$\text{Var}(H) = 100 \cdot 0.5 \cdot 0.5 = 25$$

Using the rule of thumb that most of the distribution is within two standard deviations of the mean (Section 6.10), we find that the number of heads will usually be between $50 - 2\sqrt{25} = 40$ and $50 + 2\sqrt{25} = 60$.

Computing the Binomial: Special Cases

We begin finding the binomial probabilities $b(k; n, p)$ with small values of n. In these cases, we can list all the possibilities. After 1 yr:

Year 1	Total number of immigrants k	Probability	$b(k; n, p)$
0	0	$1-p$	$b(0; 1, p) = 1-p$
1	1	p	$b(1; 1, p) = p$

Either no immigrant arrived ($k = 0$ with probability $1 - p$), or one did ($k = 1$ with probability p). Therefore $b(0; 1, p) = 1 - p$ and $b(1; 1, p) = p$. This table merely repeats our assumptions.

After 2 yr, $n = 2$, there are four possibilities:

Year 1	Year 2	Total number of immigrants k	Probability	$b(k; n, p)$
0	0	0	$(1-p)^2$	$b(0; 2, p) = (1-p)^2$
1	0	1	$p(1-p)$	
0	1	1	$(1-p)p$	$b(1; 2, p) = 2p(1-p)$
1	1	2	p^2	$b(2; 2, p) = p^2$

The probabilities in the fourth column were computed using the assumption of independence and the multiplicative law for independent probabilities. For example, the second line multiplies the probability of an immigrant in the first year, p, by the probability of no immigrant in the second year, $1 - p$.

The difficult part arises in computing the last column, the probability distribution itself. There is one way to receive no immigrants (none in either year) and one way to receive two (one in each year). But there are two ways to receive one immigrant, or have $k = 1$. The probabilities of the second and third lines must be added to find $b(1; 2, p) = 2p(1 - p)$. Addition of probabilities is justified because the events in these two lines are mutually exclusive. This is the same logic we used to calculate the probability distribution for the total number of parasites on a bird in Section 7.3.

How do we use this distribution? Suppose first that $p = 0.5$, like a fair coin. The number of heads after two flips, denoted by the random variable H, has the probability distribution

$$\Pr(H = 0) = b(0; 2, 0.5) = (1 - 0.5)^2 = 0.25$$
$$\Pr(H = 1) = b(1; 2, 0.5) = 2(0.5)(1 - 0.5) = 0.5$$
$$\Pr(H = 2) = b(2; 2, 0.5) = (1 - 0.5)^2 = 0.25$$

(Figure 7.10a). These probabilities match those of getting the three different types of offspring from a cross of two heterozygous parents (Section 6.3). $H = 0$ corresponds to inheriting a **b** allele from each parent, $H = 1$ corresponds to inheriting a **b** allele from one parent and a **B** allele from the other, and $H = 2$ corresponds to inheriting a **B** allele from each parent. The probabilities are not all equal because there are two ways to inherit different alleles from the parents, just as there are two ways to get exactly one immigrant in 2 yr.

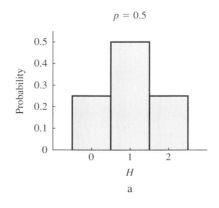
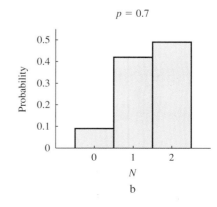

Figure 7.10 Two binomial distributions with $n = 2$

What is the distribution of the number of immigrants after two years? Recall that we assumed that an immigrant arrives each year with probability $p = 0.7$. Then

$$\Pr(N = 0) = b(0; 2, 0.7) = (1 - 0.7)^2 = 0.09$$
$$\Pr(N = 1) = b(1; 2, 0.7) = 2(0.7)(1 - 0.7) = 0.42$$
$$\Pr(N = 2) = b(2; 2, 0.7) = (1 - 0.7)^2 = 0.49$$

(Figure 7.10b). The probability of two "successes" is much higher than with the fair coin.

We follow the same method to work out the probabilities after three years.

Year 1	Year 2	Year 3	Total number of immigrants k	Probability	$b(k; n, p)$
0	0	0	0	$(1-p)^3$	$b(0; 3, p) = (1-p)^3$
1	0	0	1	$p(1-p)^2$	
0	1	0	1	$(1-p)p(1-p)$	
0	0	1	1	$(1-p)^2 p$	$b(1; 3, p) = 3p(1-p)^2$
1	1	0	2	$p^2(1-p)$	
1	0	1	2	$p(1-p)p$	
0	1	1	2	$(1-p)p^2$	$b(2; 3, p) = 3p^2(1-p)$
1	1	1	3	p^3	$b(3; 3, p) = p^3$

Each probability was again computed with the multiplication rule for independent events. For example, in the second row, the probability that there is an immigrant in the first year, p, is multiplied by the probability of no immigrant in the second year, $1 - p$, and the probability that there is no immigrant in the third year, another $1 - p$. Because there are three different ways to have exactly one immigrant in 3 yr, the probability of $k = 1$ is $3p(1-p)^2$.

We can use this table to find the probability of flipping 0, 1, 2, or 3 heads in three tosses of a fair coin:

$$\Pr(H = 0) = b(0; 3, 0.5) = (1 - 0.5)^3 = 0.125$$
$$\Pr(H = 1) = b(1; 3, 0.5) = 3(0.5)(1 - 0.5)^2 = 0.375$$
$$\Pr(H = 2) = b(2; 3, 0.5) = 3 \cdot 0.5^2(1 - 0.5) = 0.375$$
$$\Pr(H = 3) = b(3; 3, 0.5) = (1 - 0.5)^3 = 0.125$$

(Figure 7.11a). After 3 yr of immigrants arriving with $p = 0.7$, we have

$$\Pr(N = 0) = b(0; 3, 0.7) = (1 - 0.7)^3 = 0.027$$
$$\Pr(N = 1) = b(1; 3, 0.7) = 3(0.7)(1 - 0.7)^2 = 0.189$$
$$\Pr(N = 2) = b(2; 3, 0.7) = 3 \cdot 0.7^2(1 - 0.7) = 0.441$$
$$\Pr(N = 3) = b(3; 3, 0.7) = (1 - 0.7)^3 = 0.343$$

(Figure 7.11b).

We can use these probability distributions to check our calculation of the expectation. With $p = 0.5$ and $n = 3$, the expectation is

$$E(H) = 0 \cdot 0.125 + 1 \cdot 0.375 + 2 \cdot 0.375 + 3 \cdot 0.125 = 1.5$$

This matches the formula (Equation 7.9) that states that

$$E(H) = np = 3 \cdot 0.5 = 1.5$$

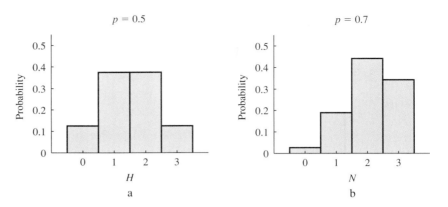

Figure 7.11
Two binomial distributions with $n = 3$

Similarly, the variance can be found explicitly as follows:

$$\text{Var}(H) = 0^2 \cdot 0.125 + 1^2 \cdot 0.375 + 2^2 \cdot 0.375 + 3^2 \cdot 0.125 - 1.5^2 = 0.75$$

Again, this matches the formula (Equation 7.11) that states that

$$E(H) = np(1 - p) = 3 \cdot 0.5(1 - 0.5) = 0.75$$

Binomial Distribution: the General Case

Each value of $b(k; n, p)$ we have found is a product of two terms: a constant and a polynomial in p. We can figure out the general formula by working out each term separately, beginning with the polynomial. Each of the three ways of getting one immigrant in 3 yr includes 1 yr with an immigrant and 2 yr with none, although in different orders. In each order, the single immigrant arrives in a particular year with probability p, and no immigrant arrives in each of the other years with probability $1 - p$. Multiplying these three terms gives the polynomial term $p(1 - p)^2$.

How does the probability of three immigrants in 5 yr depend on the probability p? The following table lists two ways this could happen.

Year 1	Year 2	Year 3	Year 4	Year 5	Probability
1	0	1	1	0	$p^3(1-p)^2$
0	1	1	0	1	$p^3(1-p)^2$

The three immigrants correspond to three factors of p, or p^3. The 2 yr without immigrants correspond to two factors of $1 - p$, or $(1 - p)^2$.

In general, to get k immigrants in n years, there must be k successes (a factor of p^k) and $n - k$ failures [a factor of $(1 - p)^{n-k}$]. Therefore,

$$\text{polynomial term of } b(k; n, p) = p^k(1 - p)^{n-k} \quad (7.12)$$

Finding the constant term requires some combinatoric reasoning, also known as *advanced counting*. The factor of 2 in

$$b(1; 2, p) = 2p(1 - p)$$

arises from the fact that there are two ways that one immigrant could arrive in 2 yr. The factor of 3 in

$$b(1; 3, p) = 3p(1 - p)^2$$

7.4 The Binomial Distribution 563

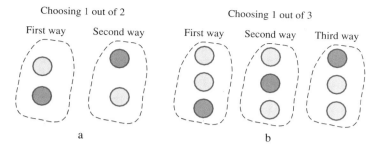

Figure 7.12
Choosing k out of n

Figure 7.13
Choosing two things out of four

Figure 7.14
Correcting for counting the same possibility more than once

arises from the fact that there are three ways that one immigrant could arrive in 3 yr.

Our more general problem is counting the number of ways exactly k immigrants can arrive in n years. If we think of the years with immigrants as "chosen," how many ways are there to choose k years out of n? There are two ways to choose 1 yr out of 2 (Figure 7.12a) and three ways to choose 1 yr out of 3 (Figure 7.12b).

To begin, we name the unknown quantity:

$$\text{the number of ways } k \text{ immigrants can arrive in } n \text{ years} = \binom{n}{k} = \text{"}n \text{ choose } k\text{"} \quad \textbf{(7.13)}$$

This object is called "n choose k" because it represents the number of ways to choose k things from a set of n. We have found that

$$\binom{2}{1} = 2$$

and

$$\binom{3}{1} = 3$$

How do we go about computing these in general? The trick is thinking of doing the choosing sequentially. Consider computing $\binom{4}{2}$, the number of ways of choosing two things out of four. There are four ways to choose the first element and only three ways to choose the second, because the second must be different from the first (Figure 7.13). We might guess that $\binom{4}{2}$ is given by $4 \cdot 3 = 12$ because we multiply the number of ways to choose the first by the number of ways to choose the second. But we will see that this plausible argument is **not** correct.

We can check by listing all possibilities:

$$(1, 2), (1, 3), (1, 4), (2, 1), (2, 3), (2, 4), (3, 1), (3, 2), (3, 4), (4, 1), (4, 2), (4, 3)$$

We found 12 possibilities, but each is listed twice. We could get the pair (2, 3) in two ways, either by picking element 2 then element 3, or by picking element 3 and then element 2 (Figure 7.14). The correct value is

$$\binom{4}{2} = 6$$

Following these steps, we find that in general,

$$\binom{n}{k} = \frac{n(n-1)\cdots(n-k+2)(n-k+1)}{\text{number of times each set was counted}}$$

The numerator is the product of the number of possible first choices (n) multiplied by the number of second choices ($n-1$) down to the number of kth choices ($n-k$).

The denominator is the number of different ways to order a set of k things. There are two ways to order two things, $(1, 2)$ and $(2, 1)$. This is why we counted each of our sets of size 2 twice. There are six ways to order three things,

$$(1, 2, 3), (1, 3, 2), (2, 1, 3), (2, 3, 1), (3, 1, 2), (3, 2, 1)$$

We can use the same sort of reasoning to figure out the pattern. There are three possible first choices, leaving two second choices and one third choice. We multiply these to find the total number.

In general, with k elements, there are k possible first choices, $k-1$ second choices, and so on. Multiplying these, we find the following.

■ **Definition 7.8** The number of ways to order k things is called k factorial, with formula

$$k! = k(k-1) \cdots 2 \cdot 1$$

For convenience, we define $0! = 1$. ■

k	$k!$
0	1
1	1
2	2
3	6
4	24
5	120
6	720
7	5040
8	40,320
9	362,880
10	3,628,800

The values of $k!$ become large very quickly.

Using this notation, we can write n choose k as

$$\binom{n}{k} = \frac{n(n-1) \cdots (n-k+2)(n-k+1)}{k!}$$

To write this more simply, note that

$$n(n-1) \cdots (n-k+2)(n-k+1) = \frac{n(n-1) \cdots 2 \cdot 1}{(n-k)(n-k-1) \cdots 2 \cdot 1}$$

$$= \frac{n!}{(n-k)!}$$

We have finally found the number in front of the polynomial.

■ **THEOREM 7.12** The number of ways to choose k things out of n has the formula

$$\binom{n}{k} = \frac{n!}{(n-k)! \, k!}$$

■

With $n = 4$ and $k = 2$, we get

$$\binom{4}{2} = \frac{4!}{(4-2)! \, 2!} = \frac{4 \cdot 3 \cdot 2 \cdot 1}{(2 \cdot 1)(2 \cdot 1)} = \frac{24}{2 \cdot 2} = 6$$

Larger values are much harder to compute. For instance,

$$\binom{10}{4} = \frac{10!}{6! \, 4!} = \frac{3628800}{720 \cdot 24} = 210$$

There are 210 different ways to choose four things out of ten, or 210 ways that exactly four immigrants arrive in 10 yr.

Combining our computation of the constant term n choose k with the polynomial term (Equation 7.12), we have proven the following:

$$b(k; n, p) = \begin{cases} \binom{n}{k} p^k (1-p)^{n-k} = \frac{n!}{(n-k)! \, k!} p^k (1-p)^{n-k} & \text{if } 0 \leq k \leq n \\ 0 & \text{otherwise} \end{cases}$$

(7.14)

The probabilities of 0 correspond to impossible events, such as 11 immigrants in 10 yr or −3 heads in 50 coin flips.

Using this formula, we can answer some questions about immigration and coins. What is the probability of exactly six heads in ten flips of a coin? With the binomial distribution, we get

$$b(6; 10, 0.5) = 210(0.5)^6(1 - 0.5)^4 = 0.205$$

The entire probability distribution with $n = 10$ and $p = 0.5$ is shown in Figure 7.15a. The mode of the distribution is at $H = 5$, near the mean, and the distribution is roughly bell-shaped.

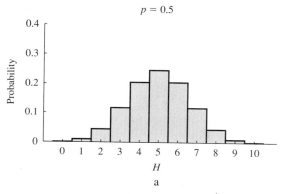
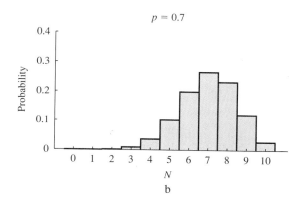

Figure 7.15
Binomial distribution: $n = 10$

The probability of exactly six immigrants in ten years with $p = 0.7$ is

$$b(6; 10, 0.7) = 210(0.7)^6(1 - 0.7)^4 = 0.200$$

The entire distribution has a very different shape, being shifted up toward the mean and mode of 7.

Computing the probability of 50 heads in 100 coin flips requires a computer because the factorials are enormous. With a computer, we find that

$$b(50, 100, 0.5) \approx 0.0796$$

A fair coin will produce *exactly* 50 heads (the expected number) less than 8% of the time.

SUMMARY

The number of successes (for example, immigrants or "heads") in n independent trials of a Bernoulli random variable is described by the **binomial distribution**. The coefficient $b(k; n, p)$ gives the probability of exactly k successes in n trials when the probability of success in a single trial is p. The expectation and variance of the binomial distribution can be computed using the theorems about the expectation and variance of sums of random variables. With some combinatoric reasoning, we found a formula for $b(k; n, p)$. Along the way, we defined and computed "n choose k," the number of different ways to pick k things out of n in terms of the **factorial**, the number of ways to order k things.

7.4 EXERCISES

1. Give three experiments not mentioned in the text that might have outcomes described by the binomial distribution. What would you need to assume?

2. Using the formula for the binomial probability distribution, check the formulas for the mean and variance of the binomial (Equations 7.9 and 7.11) with $p = 0.7$ for $n = 2$ and $n = 3$.

3. Under the conditions for the binomial distribution, make a table of all possibilities for $n = 4$. Check that your values of $b(k; 4, p)$ match the formula for the binomial (Equation 7.14).

4. Find the number of ways the following can be ordered.
   a. The items in a four-course meal.
   b. A five-card poker hand.
   c. Six occupants of offices along a hall.

5. Compute the following probabilities.
   a. Each of five patients has been prescribed a different medication, but the prescriptions were accidentally shuffled. What is the probability that each patient gets the correct medication?
   b. What is the probability that any given patient gets the right medicine?
   c. Why can't the value in part **b** be raised to the fifth power to match the probability in part **a**?
   d. What is the probability that the second patient gets the right medicine conditional on the first getting the right medicine?

6. A group of identical quintuplets named Aaron, Bill, Carl, Dave, and Ed enjoy confusing the teachers at their school. List the number of different possibilities for each case, and then count them. Make sure your counts match the appropriate value of n choose k.
   a. Only one goes to school.
   b. Two go to school.
   c. Three go to school.
   d. Four go to school.
   e. Five go to school.

7. Suppose a heterozygous plant self-pollinates and produces five offspring. Find the probability distribution for the number of heterozygous offspring. Find the expectation and variance. Sketch the distribution.

8. Refer to Exercise 7. Suppose that one allele in a heterozygous plant is dominant and produces tall plants.
   a. Find the probability distribution for the number of tall offspring.
   b. Find the expectation and variance of the number of tall offspring.
   c. Find the probability distribution for the number of short offspring.
   d. Find the expectation and variance of the number of short offspring.
   e. Sketch the probability distributions. Why are they mirror images?

9. Compute the coefficient of variation of the binomial distribution for the following.
   a. $p = 0.5$, $n = 100$
   b. $p = 0.5$, $n = 25$. Why is the coefficient of variation larger than in part **a**?
   c. $p = 0.9$, $n = 25$
   d. $p = 0.1$, $n = 25$. Why is the coefficient of variation larger than in part **c**?

10. (For the mathematically inclined.)
    a. Show that
    $$\binom{n}{k} = \binom{n}{n-k}$$
    Explain why this must be true.
    b. Figure out why the following induction should hold, and show that it does.
    $$\binom{n}{k} = \binom{n-1}{k} + \binom{n-1}{k-1}$$
    c. The values of n choose k are also called binomial coefficients. Expand $(x + 1)^3$ and $(x + 1)^4$ and explain why the coefficients of the different powers of x are the binomial coefficients. Explain the connection with part **b**.

11. **COMPUTER:** There is a remarkable approximation for $n!$ called the **Stirling approximation** (derived by methods related to the method of leading behavior),
    $$n! \approx \sqrt{\frac{2\pi}{n}} \left(\frac{n}{e}\right)^n$$
    a. Compare the values for n ranging from 1 to 10.
    b. Plot the values of $n!$ and the Stirling approximation for n ranging from 1 to 100 on a semilog graph. How close are they?
    c. Stirling's approximation can be made more accurate by multiplying it by $1 + (1/12n)$. How much better is it for $n = 10$? For $n = 100$?

12. **COMPUTER:** The following values of n and p all give an expectation of 10. Use your computer to plot the binomial distribution for each. Describe how they are different.
    a. $n = 10$, $p = 1$
    b. $n = 15$, $p = 2/3$
    c. $n = 20$, $p = 0.5$
    d. $n = 30$, $p = 1/3$
    e. $n = 50$, $p = 0.2$
    f. $n = 100$, $p = 0.1$
    g. $n = 1000$, $p = 0.01$

13. **COMPUTER:** If the probability that a team wins any particular game is 0.6, then the probability that it wins a 5-game series is $b(3, 5, 0.6) + b(4, 5, 0.6) + b(5, 5, 0.6)$. (Think about it—who would win if teams kept playing even after one team had won three games?) Figure out how to write this in terms of the cumulative distribution B, and then how to write it in general for a series of n games with probability q of winning any particular game (assume n is odd). On a single graph, plot the probabilities that a team wins a series of length 1, 5, and 101 as functions of q. Why is each curve increasing? Explain the shape of the curve with $n = 1$. Explain the shape of the curve with $n = 101$. What is the value of each curve at $q = 0.5$ and why?

7.5 Applications of the Binomial Distribution

Many processes in biology and throughout the sciences fit (or nearly fit) the conditions for the binomial distribution. Using the mean, variance, and formula for the binomial probability distribution, we explore a few of these many applications.

Application to Genetics and Calculation of the Mode

Suppose that plants with genotypes **BB** and **Bb** are tall and plants with genotype **bb** are short (**B** is a dominant allele). We cross two heterozygous plants with genotype **Bb**. According to our basic probability models (sections 6.3 and 6.6), an offspring is short with probability $1/4$. This does not mean that exactly one out of four offspring will be short. Suppose that a plant produces 8 offspring. What is the probability that exactly two are short?

Let S be the random variable giving the number of short offspring; S will follow the binomial distribution if it satisfies the assumptions that each offspring must have the same probability of being short, and the offspring must be independent. The second assumption fails if some of the offspring are genetically identical, for example, identical twins. If we believe that these assumptions hold, S follows the binomial distribution with $n = 8$ and $p = 0.25$. The probability that two offspring are short is

$$\Pr(S = 2) = b(2; 8, 0.25)$$
$$= \binom{8}{2} 0.25^2 0.75^6$$
$$= \frac{8 \cdot 7 \cdot 6 \cdot 5 \cdot 4 \cdot 3 \cdot 2 \cdot 1}{6 \cdot 5 \cdot 4 \cdot 3 \cdot 2 \cdot 1 \cdot 2 \cdot 1} 0.25^2 0.75^6$$
$$= \frac{8 \cdot 7}{2 \cdot 1} 0.25^2 0.75^6 \approx 0.311$$

The whole probability distribution is shown in Figure 7.16a. The probability that four offspring are short is

$$\Pr(S = 4) = b(4; 8, 0.25)$$

$$= \binom{8}{4} 0.25^4 0.75^4$$

$$= \frac{8 \cdot 7 \cdot 6 \cdot 5 \cdot 4 \cdot 3 \cdot 2 \cdot 1}{4 \cdot 3 \cdot 2 \cdot 1 \cdot 4 \cdot 3 \cdot 2 \cdot 1} 0.25^4 0.75^4$$

$$= \frac{8 \cdot 7 \cdot 6 \cdot 5}{4 \cdot 3 \cdot 2 \cdot 1} 0.25^4 0.75^4 \approx 0.086$$

The deceptive result that half the offspring are short would occur fully 1/4 as often as the "expected" result. It might be difficult to guess the underlying genetics from eight offspring.

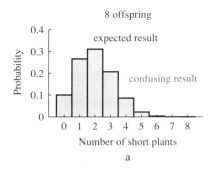
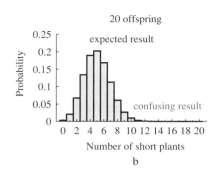

Figure 7.16
Probability distribution for plant height

Similar results are shown for 20 offspring in Figure 7.16b. Although the distribution is more tightly piled up around the mean of 5, there is still a great deal of variation. The deceptive result of 10 out of 20 short plants occurs with probability

$$b(10; 20, 0.25) \approx 0.01$$

which was calculated with the aid of a computer. There is still a 1% chance of being thoroughly misled. The probability that 3/4 of the offspring are short (the expected result if the **b** allele were dominant) is the tiny value

$$b(15; 20, 0.25) \approx 3.42 \times 10^{-6}$$

With 20 plants, it is virtually impossible that our plants would so thoroughly deceive us by chance.

The coefficient of variation of the number of short plants N out of n offspring is

$$\text{CV} = \frac{\sigma_N}{\overline{N}}$$

$$= \frac{\sqrt{0.25 \cdot 0.75 n}}{0.25 n}$$

$$= \sqrt{\frac{3}{n}}$$

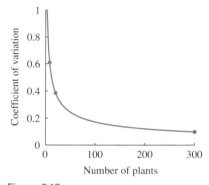

Figure 7.17
Coefficient of variation as a function of number of offspring

(Figure 7.17). With $n = 8$, the coefficient of variation is 0.612, indicating that results are highly unpredictable. With $n = 20$, the coefficient of variation is 0.387, which is smaller but still relatively large. The coefficient of variation does not reach the relatively small value of 0.1 until $n = 300$.

In Figure 7.16, the mode occurs right at the expectation. In fact, the mode is always within 1 of the expectation.

■ **THEOREM 7.13** The mode of the binomial probability distribution with parameters n and p is the smallest integer greater than $np - 1 + p$. ■

Proof: Because the probabilities increase up to some point and then begin decreasing, the mode is where the probabilities start decreasing. This occurs at the smallest value of k for which $b(k + 1; n, p) < b(k; n, p)$. The trick is to find a formula for $b(k + 1; n, p)$ in terms of $b(k; n, p)$ as follows:

$$b(k + 1; n, p) = \frac{n!}{(k + 1)!\,(n - k - 1)!} p^{k+1}(1 - p)^{n-k-1}$$

$$= \left(\frac{n - k}{k + 1}\right)\left(\frac{p}{1 - p}\right) b(k; n, p)$$

Therefore, $b(k + 1; n, p) < b(k; n, p)$ if

$$\left(\frac{n - k}{k + 1}\right)\left(\frac{p}{1 - p}\right) < 1$$

Solving for k, we get

$$k > np - 1 + p$$

Therefore, the mode is the smallest value of k for which $k > np - 1 + p$. ■

To compute the mode, find $np - 1 + p$ and *round up* to the nearest integer. With $n = 8$ and $p = 0.25$, we get

$$np - 1 + p = 8 \cdot 0.25 - 1 + 0.25 = 1.25$$

Rounding up to the nearest integer, we find that the mode is $k = 2$. With $n = 20$ and $p = 0.25$, we get

$$np - 1 + p = 20 \cdot 0.25 - 1 + 0.25 = 4.25$$

Rounding up to the nearest integer, we find that the mode is $k = 5$. In both of these cases, the mode is exactly equal to the mean. If $n = 10$ and $p = 0.25$, the mean is 2.5. The mode is the first integer above

$$np - 1 + p = 10 \cdot 0.25 - 1 + 0.25 = 1.75$$

or 2.0. The mode cannot be *equal* to the mean in this case because the random variable cannot take on the value 2.5.

Application to Markov Chains: Definition of Cumulative Distribution

The binomial probability distribution can be applied to the Markov chain model of a molecule hopping in and out of a cell (sections 6.2 and 6.6). We found that after a long time a given molecule is inside the cell with probability $1/3$. The number out of 50 molecules inside fulfills the conditions for the binomial model if each molecule has the same chance of being inside and behaves independently of the others. Let the random variable M denote the number of molecules in the cell (Figure 7.18). Then

$$E(M) = np = 50 \cdot \frac{1}{3} = 16.67$$

The probability that $M = 17$ is

$$\Pr(M = 17) = b\left(17; 50, \frac{1}{3}\right) = 0.1178$$

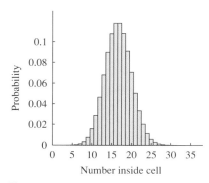

Figure 7.18

Number of molecules inside a cell

Suppose we replicate this experiment 20 times, and count how many times there are exactly 17 molecules inside the cell. As long as the experimental conditions and each experiment is independent, the number R matching the mean will have a binomial distribution with $n = 20$ and $p = 0.1178$ (Figure 7.19).

What is the probability that less than half the expectation, 8 or fewer molecules are inside a cell in a given experiment? This is the *sum* of the probabilities for each value less than or equal to 8, or

$$\Pr(M \leq 8) = \sum_{i=0}^{8} b\left(i; 50, \frac{1}{3}\right)$$

The sum can evaluated term by term on a computer, for a total of 0.0051. Only about 1 in 200 experiments would give a result this extreme. This calculation would be easy if we had a formula for the **cumulative distribution**.

■ **Definition 7.9** The cumulative distribution for the binomial distribution, written $B(k; n, p)$, is equal to

$$B(k; n, p) = \sum_{i=0}^{k} b(i; n, p)$$
■

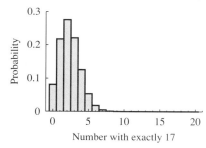

Figure 7.19

Number of experiments out of 20 with exactly 17 molecules inside a cell

Unfortunately, there is no simple formula for the cumulative distribution associated with the binomial probability distribution. Computers or tables must be used to compute its value. There is therefore no convenient way to find the median or other percentiles for a binomial distribution.

The number of molecules inside a cell would be less than half the mean in only 1 in 200 experiments. How likely is it that more than twice the mean number are inside? We wish to compute $\Pr(M \geq 34)$, because the mean is 16.7. To convert into a form more like the cumulative distribution, we write

$$\Pr(M \geq 34) = 1 - \Pr(M \leq 33)$$

In terms of the cumulative distribution, then, we have

$$\Pr(M \geq 34) = 1 - \Pr(M \leq 33)$$
$$= 1 - B\left(33; 50, \frac{1}{3}\right)$$
$$= 1.0 - 0.999999422 \approx 5.78 \times 10^{-7}$$

The probability that the molecules will distribute themselves this differently from the expectation (and the mode) is extremely small.

These applications give an indication of how the binomial distribution is used in statistics. If we observe an outcome that is astronomically unlikely for a given model (such as 15 out of 20 offspring being short, or 34 or more molecules being inside the cell), it is a good bet that something is wrong with our model.

Applications to Diffusion

Suppose 100 molecules begin inside the cell. Each has a 10% chance of leaving each minute and cannot return. What is the probability that exactly 90 molecules remain after 1 min or that fewer than 10 remain after 20 min? What is the mean number inside as a function of t? How spread out is the distribution at different times?

Identical molecules diffusing out of a cell satisfy the conditions for the binomial model as long as they behave independently. Both properties, that the molecules are identical and that they are independent, encode different biological assumptions. If the molecules had different configurations when the experiment began, the assumption that molecules are identical would not be valid. Similarly, if the molecules interacted, the assumption of independence would not be valid.

If the assumptions are valid, we can use the binomial distribution to answer our questions. We find the probability p_t that a given molecule is inside at time t. This probability follows the updating function

$$p_{t+1} = 0.9 p_t$$

with solution

$$p_t = 0.9^t$$

(Equation 6.6). Let N_t be the random variable measuring the number of molecules inside the cell at time t. Then N_t has a binomial distribution with $n = 100$ and $p = p_t$, so we have

$$\Pr(N_t = k) = b(k; 100, p_t)$$

The expected number of molecules in the cell at time t is

$$E(N_t) = n p_t = 100(0.9^t)$$

(Equation 7.9). Starting with 100 molecules, a mean of 90 are inside after the first minute, 81 after the second, and so on.

The probability that exactly 90 molecules remain after 1 min is

$$\Pr(N_1 = 90) = b(90; 100, 0.9)$$
$$= \binom{100}{90}(0.9)^{90}(0.1)^{10}$$
$$\approx 0.1318$$

(Figure 7.20a). The measured number will match the mean number only about 13% of the time. If we ran this experiment 100 times, we would observe exactly 90 molecules about 13 times (Exercise 4). The average number of molecules, however, would be close to 90.

The probability that 10 or fewer molecules remain after 20 min must be found using the cumulative distribution as follows:

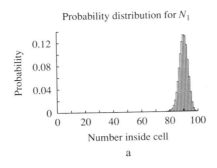
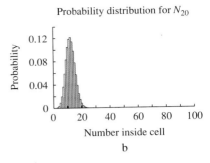

Figure 7.20

Probability distribution for 100 molecules at two times

$$\Pr(N_{20} \leq 10) = B(10; 100, p_{20})$$

$$= \sum_{i=0}^{10} b(i; 100, 0.1216)$$

$$= \sum_{i=0}^{10} \binom{100}{i} 0.1216^i (1 - 0.1216)^{100-i}$$

$$\approx 0.3165$$

Again, calculations of this sort are impossible without a computer.

How predictable is this process? We can estimate this with the coefficient of variation. The variance at time t is

$$\text{Var}(N_t) = nPt(1 - p_t) = 100(0.9)^t(1 - 0.9^t)$$

(Figure 7.21). When does the variance take on its maximum? We can compute this by taking the derivative of $v(t) = \text{Var}(N_t)$ and setting it equal to 0 (Exercise 7). The maximum is at around $t = 6.5$. The coefficient of variation is

$$\text{CV}(N_t) = \frac{\sqrt{\text{Var}(N_t)}}{E(N_t)}$$

$$= \frac{\sqrt{100(0.9)^t(1 - 0.9^t)}}{100(0.9)^t}$$

(shown multiplied by 100 in Figure 7.21). Although the variance increases and then decreases, the coefficient of variation steadily increases. As fewer and fewer molecules remain in the cell, the *relative uncertainty* about the number of molecules increases. The fewer molecules that remain, the less predictable the process.

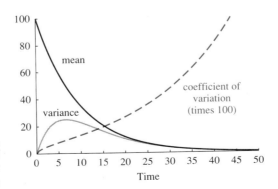

Figure 7.21

Mean, variance, and coefficient of variation of molecule number as functions of time

SUMMARY

We have used the binomial distribution to study problems in genetics and diffusion, computing the probability that a particular number of offspring are short or that a particular number of molecules remain inside a cell. The *mode* of the binomial distribution is always within 1 of the mean and can be found as the smallest integer above $np - 1 + p$. The binomial distribution has **cumulative distribution** $B(k; n, p)$, but there is no convenient formula for this quantity. For a process describing the number of molecules remaining in a cell, the expected number of molecules in a cell decreases, the variance first increases and then decreases, and the coefficient of variation increases steadily.

7.5 EXERCISES

1. For an additive gene, we expect 1/4 of the offspring of a short–tall cross to be short, 1/2 to be intermediate, and 1/4 to be tall.
   a. Find the probability that exactly 3 out of 12 offspring are tall.
   b. Find the probability that exactly 6 out of 12 offspring are intermediate.
   c. Find the probability that between 2 and 4 offspring are short.
   d. Find the probability that between 2 and 4 offspring are intermediate.

2. Consider the islands that have a 0.2 chance of switching from empty to occupied, and a 0.1 chance of switching from occupied to empty (Equation 6.3 and 6.5 and Exercise 7 in Section 6.2). We found that the equilibrium fraction occupied is 2/3.
   a. Find the probability that exactly seven out of ten islands are occupied.
   b. Find the probability that all ten are occupied.
   c. Find the mean of the distribution.
   d. Find the mode of the distribution.

3. Suppose that the islands of Exercise 2 are all occupied if the weather is good (probability 2/3) and empty if the weather is bad (probability 1/3).
   a. Find the probability distribution.
   b. Find the mean number occupied.
   c. Identify the mode.
   d. Why does this not look like a binomial distribution?

4. Suppose that the probability that exactly 90 molecules remain inside a cell after 1 min is 0.1318.
   a. Find the probability that exactly 13 out of 100 experiments find exactly 90 molecules inside after 1 min.
   b. Find the standard deviation of the number of experiments out of 100 that find exactly 90 molecules inside after 1 min.

5. Starting with five molecules inside a cell, each leaving with probability 0.8 per minute, compute and graph the probability distribution at the following times.
   a. 1 min
   b. 2 min
   c. 5 min
   d. 10 min

6. For the five molecules of Exercise 5, find probabilities of the following events.
   a. Fewer than four remain after 1 min.
   b. Fewer than three remain after 5 min.
   c. An odd number remain after 1 min.
   d. An odd number remain after 5 min.
   e. More than two remain after 5 min.
   f. More than two remain after 10 min.

7. Find the time t when the variance is maximized if
   $$\text{Var}(N_t) = 100 \cdot 0.9^t (1 - 0.9^t)$$
   What is the fraction of molecules inside the cell at this time?

8. Forty molecules begin inside a cell. Each leaves independently with probability 0.8 each minute.
   a. Find the expected number remaining inside the cell as a function of time.
   b. Find the mode at $t = 1$, $t = 2$, and $t = 3$. Indicate whether the mode is equal to, greater than, or less than the mean.
   c. Compute the variance. At what time is it a maximum?
   d. Find the coefficient of variation of the number remaining inside as a function of time.

9. Figure 6.3 (page 451) was generated by adding 2 with probability 0.5 and 0 with probability 0.5 for 100 generations. The results in the figure show final populations of 106 and 96.
   a. What is the expected number after 100 generations?
   b. What is the variance after 100 generations?
   c. Write the probability of exactly 106 in terms of the binomial distribution.

10. Unbeknownst to the experimenter, a cell contains two types of molecule; one is inside with probability p_1 and the other is inside with probability p_2. Suppose there are two of each type of molecule.

a. Suppose $p_1 = 0$ and $p_2 = 1$. Find and graph the probability distribution for the total number of molecules inside the cell.
b. Find the expectation and the variance.
c. Suppose $p_1 = 0.25$ and $p_2 = 75$. Find and graph the probability distribution for the total number inside.
d. Find the expectation and the variance (this can be written as the the sum of two binomial random variables).
e. Compare with the results if $p_1 = p_2 = 0.5$.

11. COMPUTER: Simulate groups of 8 plants, each of which has a 0.25 chance of being tall. How long does it take until half the plants are tall? If you can automate the process, try this with groups of 20 plants.

12. COMPUTER: The probability p_t that a molecule is inside a cell after t time steps is

$$p_t = 0.9^t$$

Suppose 100 molecules start out in a cell. On a single graph with reasonable axes, plot the probabilities that 50 molecules are inside at time t and that 10 are inside at time t as functions of t. Explain the shape of the curves. Is it more likely there are exactly 50 or exactly 10 molecules at time 12? What is a more likely number of molecules at this time? Compute the probability that exactly 10 molecules are inside at time 20, and indicate this point on your graph. Do the same for the probability that exactly 50 molecules are inside at time 8. Is the area under each curve equal to 1? Why or why not?

13. COMPUTER: Consider the situation as in Exercise 10, where two types of molecule leave a cell with different probabilities during each minute. The first leaves with probability q_1; the second leaves with probability q_2. Suppose we begin with n_1 of the first type, n_2 of the second type, and a total of $n = n_1 + n_2$. If we do not know the difference, the average probability is

$$q = q_1 \frac{n_1}{n} + q_2 \frac{n_2}{n}$$

Simulate the number of molecules of each type that remain inside after 5 min, and compare with the number out of n that would remain if they left with the average probability q in each of the following cases. How important is it to distinguish between types of molecule?
a. $n_1 = n_2 = 50$, $q_1 = 0.1$, $q_2 = 0.9$
b. $n_1 = 90$, $n_2 = 10$, $q_1 = 0.1$, $q_2 = 0.9$
c. $n_1 = 10$, $n_2 = 90$, $q_1 = 0.1$, $q_2 = 0.9$
d. $n_1 = n_2 = 50$, $q_1 = 0.4$, $q_2 = 0.6$
e. $n_1 = 90$, $n_2 = 10$, $q_1 = 0.4$, $q_2 = 0.6$
f. $n_1 = 10$, $n_2 = 90$, $q_1 = 0.4$, $q_2 = 0.6$

7.6 Waiting Times: Geometric and Exponential Distributions

The binomial distribution gives the probability that a particular *number* of molecules remain inside a cell at a given time. The *time* that a molecule leaves the cell is a different random variable, described by the **geometric distribution** in the discrete case and the **exponential probability density function** in the continuous case. We derive the formulas and properties of these two important distributions.

The Geometric Distribution

Suppose the position of a molecule is checked once per minute. The molecule leaves a cell with probability q each minute. What is the probability that a molecule leaves at time t? Computing this value is related to finding the probability p_t that the molecule is still inside the cell (Equation 6.6). Recall that p_t follows the updating function

$$p_{t+1} = (1-q)p_t$$

with solution

$$p_t = (1-q)^t$$

7.6 Waiting Times: Geometric and Exponential Distributions

Let T be the random variable representing the time at which the molecule leaves, and set

$$g_t = \Pr(T = t)$$

to be the probability that the molecule left during minute t (Figure 7.22). To leave at time $T = t$, the molecule must have remained inside for the first $t - 1$ minutes and then left during the next minute, or

$$\begin{aligned}
g_t &= \Pr(\text{molecule left at time } t) && \text{definition of } g_t \\
&= \Pr(\text{inside at time } t - 1) && \text{break up with definition} \\
&\quad \times \Pr(\text{left between } t - 1 \text{ and } t && \text{of conditional probability} \\
&\quad\quad \text{conditional on in at } t - 1) \\
&= p_{t-1} \cdot q && \text{probability inside at } t - 1 \text{ is } p_{t-1} \\
&= (1 - q)^{t-1} q && p_{t-1} = (1 - q)^{t-1} \text{ (Equation 6.6)}
\end{aligned}$$

The random variable T follows a **geometric distribution** with parameter q. The range of the random variable T is all positive integers (not including 0). When using the geometric distribution, remember to count starting from $t = 1$.

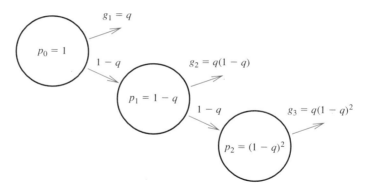

Figure 7.22
The process underlying the geometric distribution

Histograms of the geometric distribution for two values of q are shown in Figure 7.23. Because T can take on any positive integer value, the histograms show only the beginning of the distribution. Larger values of T are so improbable that the bars do not appear on the graph.

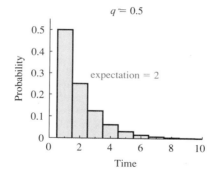
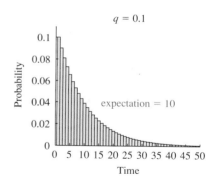

Figure 7.23
The geometric probability distribution: $q = 0.5$ and $q = 0.1$

The cumulative distribution G_t, the probability that the molecule left at or before time t, can be found from p_t, the probability that the molecule remains inside at time t. The molecule left at or before time t precisely if it is not inside, or

$$\begin{aligned} G_t &= \Pr(\text{left at or before time } t) \\ &= 1 - \Pr(\text{still in at time } t) \\ &= 1 - p_t \\ &= 1 - (1-q)^t \end{aligned}$$

Alternatively, using the definition of the cumulative probability distribution, we have

$$\begin{aligned} G_t &= \sum_{i=1}^{t} g_t \\ &= \sum_{i=1}^{t} q(1-q)^{i-1} \\ &= 1 - (1-q)^t \end{aligned}$$

where the last step can be derived algebraically as the sum of a **geometric series** (Exercise 9). The cumulative distribution increases to 1, as it must, because

$$\lim_{t \to \infty} G_t = \lim_{t \to \infty} 1 - (1-q)^t = 1$$

(unless $q = 0$, in which case the molecule never leaves).

The mode of the geometric distribution is $T = 1$ even when q is small. The **expectation** of T is

$$E(T) = \frac{1}{q}$$

This computation requires a difficult infinite sum or a clever trick (Exercise 11). If the value of q is small, meaning that the molecule leaves with low probability each minute, the mean time until departure is large. If the value of q is near 1, meaning that the molecule leaves with high probability each minute, the mean is near 1.

The median $\tilde{T}$ occurs where the cumulative distribution crosses 0.5, or

$$\begin{aligned} G_t &= 0.5 \\ 1 - (1-q)^{\tilde{T}} &= 0.5 \\ (1-q)^{\tilde{T}} &= 0.5 \\ \tilde{T} \ln(1-q) &= \ln(0.5) \\ \tilde{T} &= \frac{\ln(0.5)}{\ln(1-q)} \end{aligned}$$

(Figure 7.24). When $q = 0.5$, the median is 1.0, because the wait is exactly 1 with probability 0.5. When $q = 0.1$, the median is 6.57 (strictly speaking, the median lies between 6 and 7), which is quite a bit smaller than the mean. In fact,

$$G_{E(T)} = G_{10} = 1 - 0.9^{10} = 0.65$$

so that almost 2/3 of observations are less than the mean. This is a case where only 1/3 of the measurements are "above average."

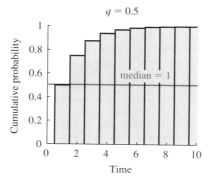
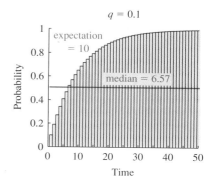

Figure 7.24
Cumulative geometric distribution: $q = 0.5$ and $q = 0.1$

The variance of a geometric random variable T is

$$\text{Var}(T) = \sigma_T^2 = \frac{1-q}{q^2}$$

Again, the computation requires an infinite sum or a clever trick (Exercise 12). The standard deviation is

$$\sigma_T = \frac{\sqrt{1-q}}{q}$$

and the coefficient of variation is

$$\text{CV} = \frac{\sigma_T}{\text{E}(T)} = \frac{\sqrt{1-q}/q}{1/q} = \sqrt{1-q}$$

The coefficient of variation is small only when q is near 1 and all the molecules leave almost immediately.

When $q = 0.5$, we have

$$\text{Var}(T) = \frac{1-0.5}{0.5^2} = 2.0$$

$$\sigma_T = \sqrt{2.0} = 1.414$$

$$\text{CV} = \frac{\sigma_T}{\text{E}(T)} = 0.707.$$

When $q = 0.1$, we have

$$\text{Var}(T) = \frac{1-0.1}{0.1^2} = 90$$

$$\sigma_T = \sqrt{90} = 9.49$$

$$\text{CV} = \frac{\sigma_T}{\text{E}(T)} = 0.949$$

The large value of the coefficient of variation indicates that waiting is highly unpredictable.

We summarize these results in the following theorem.

■ THEOREM 7.14 A random variable T follows the geometric distribution when it measures the time until the first success, where the probability of a success on each step is independent and equal to q on each time step. The probability distribution g_t and the cumulative distribution G_t are

$$g_t = q(1-q)^{t-1}$$
$$G_t = 1 - (1-q)^t$$

The mean $E(T)$, median $\tilde{T}$, and variance σ_T^2 are

$$E(T) = \frac{1}{q}$$

$$\tilde{T} = \frac{\ln(0.5)}{\ln(1-q)}$$

$$\sigma_T^2 = \frac{1-q}{q^2}$$

∎

The Exponential Distribution

The exponential distribution is the continuous-time version of the geometric distribution. This distribution describes a random variable measuring the *exact* time when a molecule leaves a cell. Finding the distribution requires writing and solving a differential equation. This derivation closely follows the derivation in Section 6.2.

Suppose a molecule leaves a cell at rate λ, meaning that the probability it leaves in the short time Δt is $\lambda \Delta t$. Let $P(t)$ be the probability that the molecule is inside at time t. To write a differential equation, we must compute the probability that it is inside a short time Δt later. The basic principle is that a molecule is inside at $t + \Delta t$ if was inside at time t and did not leave in the next Δt, or

(probability inside at $t + \Delta t$) = (probability inside at t)
$\times$(probability did not leave in time Δt)

We have assumed that

probability did leave in time $\Delta t \cong \lambda \Delta t$

Therefore,

$$P(t + \Delta t) \cong P(t)(1 - \lambda \Delta t)$$

Because this equation is exact as $\Delta t \to 0$, we wish to take the limit. Doing so directly does not work (see Section 6.2), so first we rearrange it as follows:

$$\frac{P(t + \Delta t) - P(t)}{\Delta t} = -\lambda P(t)$$

When we take the limit, the left-hand side is the derivative and

$$\lim_{\Delta t \to 0} \frac{P(t + \Delta t) - P(t)}{\Delta t} = \frac{dP}{dt} = -\lambda P(t)$$

This differential equation has the exponential solution

$$P(t) = e^{-\lambda t}$$

with the initial condition $P(0) = 1$.

$P(t)$ is the probability that the molecule has not left by time t. We are trying to find the probability density function and cumulative distribution function describing the exact time T when it leaves. The c.d.f. $F(t)$ is

$$F(t) = \text{Pr}(\text{left by time } t)$$
$$= 1 - \text{Pr}(\text{inside at time } t)$$
$$= 1 - P(t)$$
$$= 1 - e^{-\lambda t}$$

The probability density function is the *derivative* of the cumulative distribution function (Section 6.7), so we have

$$f(t) = F'(t) = \lambda e^{-\lambda t}$$

We summarize these results in the following theorem.

■ **THEOREM 7.15** Suppose an event happens at a constant rate λ. The random variable T measuring the time until the first event has probability density function $f(t)$ and cumulative distribution function $F(t)$ with formulas

$$f(t) = \lambda e^{-\lambda t}$$
$$F(t) = 1 - e^{-\lambda t}$$
■

At the large rate $\lambda = 10.0$, the probability density function begins at a high value and diminishes rapidly (Figure 7.25a), and the cumulative distribution function rapidly increases to 1 (Figure 7.25b). $P(t)$, the probability that the molecule is still inside the cell as a function of t, is called the **survivorship function** (Figure 7.25c).

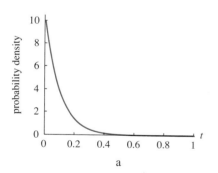
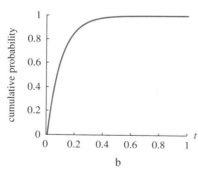
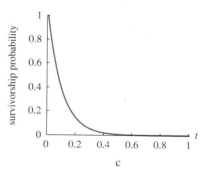

Figure 7.25
Probability density function, cumulative distribution function, and survivorship function with $\lambda = 10$

With the smaller rate $\lambda = 0.5$, the probability density function begins lower and diminishes more slowly (Figure 7.26a), the cumulative distribution function rises more slowly (Figure 7.26b), and the survivorship function declines more slowly (Figure 7.26c).

Using the probability density function, we can find all the basic statistics that describe this distribution.

■ **THEOREM 7.16** Suppose the random variable T follows an exponential distribution with parameter λ. Then

$$E(T) = \frac{1}{\lambda}$$
$$\text{Var}(T) = \frac{1}{\lambda^2}$$
$$\tilde{T} = \frac{\ln(2.0)}{\lambda}$$
■

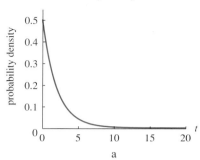
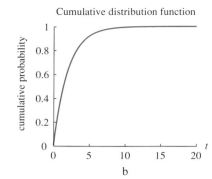
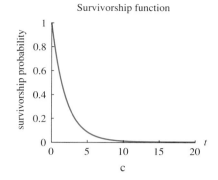

Figure 7.26
Probability density function, cumulative distribution function, and survivorship function with $\lambda = 0.5$

Proof: The expectation of T is

$$E(T) = \int_0^\infty t f(t) \, dt$$

$$= \int_0^\infty t \lambda e^{-\lambda t} \, dt$$

$$= -\frac{\lambda t e^{-\lambda t} + e^{-\lambda t}}{\lambda} \Big|_0^\infty$$

$$= \frac{1}{\lambda}$$

where the integral was evaluated using a computer or table of integrals. We also use a computer or table of integrals to evaluate the variance of T:

$$\mathrm{Var}(T) = \int_0^\infty t^2 f(t) \, dt - E(T)^2$$

$$= -\frac{\lambda^2 t^2 e^{-\lambda t} + 2\lambda t e^{-\lambda t} + 2 e^{-\lambda t}}{\lambda^2} \Big|_0^\infty - \frac{1}{\lambda^2}$$

$$= \frac{1}{\lambda^2}$$

The median is found by solving

$$F(t) = 1 - e^{-\lambda t} = 0.5 \qquad \text{definition of median}$$
$$e^{-\lambda t} = 0.5 \qquad \text{isolate } t$$
$$-\lambda t = \ln(0.5) \qquad \text{take natural log of both sides}$$
$$t = -\frac{\ln(0.5)}{\lambda} \qquad \text{solve for } t$$
$$t = \frac{\ln(2.0)}{\lambda} \qquad \text{use Law 3 of logs to write } \ln(2.0) = -\ln(0.5) \blacksquare$$

The standard deviation is

$$\sigma = \sqrt{\mathrm{Var}(T)} = \frac{1}{\lambda}$$

so we have

$$\mathrm{CV} = \frac{\sigma_T}{E(T)} = 1 \qquad (7.15)$$

for any value of the rate parameter λ. A random variable with the exponential p.d.f. is highly unpredictable, with many results far from the mean.

With the high rate $\lambda = 10$, we get

$$E(T) = 0.1$$
$$\text{Var}(T) = 0.01$$
$$\tilde{T} \approx 0.0693$$

The mean is small because the process is rapid. With the smaller rate $\lambda = 0.5$, we get

$$E(T) = 2.0$$
$$\text{Var}(T) = 4.0$$
$$\tilde{T} \approx 1.386$$

The mean and median are larger. Because the standard deviation is equal to the mean, all we can say with confidence is that the molecule will leave the cell before too long. If the rate is $\lambda = 0.5$ and we wait for 12 time units (6 times larger than the mean), the value of the c.d.f. is

$$F(12) = 1 - e^{-0.5 \cdot 12.0} \approx 0.9975$$

The molecule has almost certainly left.

For some animals, the probability that an organism dies during a short interval of time is independent of age. Exponential survivorship occurs when death results, not from aging, but from a random event that could happen at any time. Because the survivorship function is exponential, such an animal is said to have **exponential survivorship**.

The Memoryless Property

Both the geometric and exponential distributions are **memoryless**; knowledge of the past gives no information about the future. Suppose that the mean time for a molecule to leave a cell is 0.4 min. Observing that the molecule has remained in the cell for 2 min does not change our expectation; it still leaves after a mean of 0.4 *additional* minutes. Similarly, if the average lifespan of an insect with exponential survivorship is 2 yr and we encounter a 2-yr-old insect, the mean *remaining* lifespan is still 2 yr.

These properties can be demonstrated mathematically with conditional probability. For the geometric distribution, the probability of dying t time steps after surviving τ time steps is

$$\begin{aligned}
\Pr(T = \tau + t \mid T > \tau) & \\
= \frac{\Pr(T = \tau + t \text{ and } T > \tau)}{\Pr(T > \tau)} & \quad \text{definition of conditional probability} \\
= \frac{\Pr(T = \tau + t)}{\Pr(T > \tau)} & \quad \text{alive at } t + \tau \text{ implies alive at } \tau \\
= \frac{q(1-q)^{\tau+t-1}}{(1-q)^{\tau}} & \quad \text{formulas for } g_{t+\tau} \text{ and } p_\tau \\
= q(1-q)^{t-1} & \quad \text{simplify} \\
= \Pr(T = t) & \quad \text{recognize formula for } g_t
\end{aligned}$$

The probability of surviving t time units after surviving τ is indeed equal to the probability of surviving t time units in the first place. A similar calculation works in the continuous case.

Suppose a scientist returns from a long lunch to find that somebody started an unknown number of molecules inside a cell at an unknown time. If there are 100 molecules inside when he returns and the molecules follow a geometric or exponential distribution, the probability distribution describing the times when those molecules leave is the same no matter when the experiment started or with how many molecules. The experiment has no way to *remember* when it began.

SUMMARY The time when a molecule leaves a cell is described by the **geometric distribution** in discrete time and by the **exponential distribution** in continuous time. We computed the probabilities, means, and variances associated with these distributions. Each has a large variance, with the standard deviation being equal to the mean for the exponential distribution. Both distributions have the **memoryless** property, meaning that the waiting time is independent of the previous wait.

7.6 EXERCISES

1. Find the probabilities of the following events. Sketch the associated distribution and find its expectation and variance.
   a. A molecule leaves a cell with probability 0.3 each second. Find the probability it leaves during the third second.
   b. A light bulb blows out with probability 0.01 each day. Find the probability that it blows out on the 50th day.
   c. A child is a girl with probability 0.5. Find the probability that the first girl is the sixth child.
   d. Of some items, 10% are defective. Find the probability that the first defective item found is the fifth one inspected.
   e. Newly hatched fish have an 80% chance of getting eaten each day. Find the probability that a fish survives a week.

2. Consider again the inspection process in Exercise 1d.
   a. Use the binomial distribution to find the expected number of defective items after n items have been inspected.
   b. What is the expected number of defective items when $n = E(T)$?
   c. What is the probability that exactly one defective item is found when $n = E(T)$?

3. The items to be inspected were sorted in the factory in order from best to worst. Suppose the first item is defective with probability 0.1, the second with probability 0.2, and so forth.
   a. What is the probability that the second item inspected is the first defective one?
   b. What is the probability that the third item inspected is the first defective one?
   c. Find the whole probability distribution.

4. Find the probabilities of the following events. Find the associated p.d.f., c.d.f., and survivorship function, and give the expectation and variance.
   a. A molecule leaves a cell at rate $\lambda = 0.3$ per second. What is the probability it has left by the end of the third second? By the end of the first millisecond? Compare your last answer with the definition of rate.
   b. A light bulb blows out at a rate of $\lambda = 0.001$ per hour. What is the probability that it blows out in less than 500 hr? In less than 1 hr? Compare your last answer with the definition of rate.
   c. Phone calls arrive at a rate of $\lambda = 0.2$ per hour. What is the probability that there are no calls in 10 hr? What is the probability that a call arrives during the 45 s you spend in the bathroom?
   d. Raindrops hit a leaf at a rate of 7.3 per minute. What is the probability that the first one hits in less than 0.5 min?

5. A population of 100 bacteria are dying independently at rate $\lambda = 2.0$.
   a. Find the probability that a given bacterium is alive at time t.
   b. What distribution describes the population at time t?
   c. Find the expectation and variance of the number of live bacteria as functions of time.
   d. When is the variance at a maximum?

6. Apply the memoryless property to the following.
   a. The molecule in Exercise 1a is still in the cell after 10 s. What is the probability that it leaves during the 13th second?

b. The light bulb in Exercise 4b is still working after 1000 h. What is the probability that it blows out before hour 1500?

c. A couple has five boys. What is the probability that they still have not had a girl by their eighth child?

7. Suppose that some species of insect dies faster as it gets older according to

$$\text{rate of death} = t$$

where t is the age in years.
a. Write a differential equation describing the probability the insect is still alive.
b. Solve the equation with separation of variables (Section 5.3 and Exercise 7a in that section).
c. Graph the survivorship function.
d. Find the c.d.f. and p.d.f.
e. Find the probability the insect survives to age 1 yr.
f. Find the probability it survives to age 2 yr conditional on surviving to age 1 yr. Explain why this shows that this insect does not have the memoryless property.

8. You are trapped behind an annoyingly slow driver (a.s.d.) in a long no-passing zone. A second a.s.d. merges in front of the first, leaving you twice as trapped. To calm yourself, you attempt to guess which driver will exit first.
a. If the length of time that cars remain on the freeway follows an exponential distribution, what is the probability that the new a.s.d. exits first? Under what conditions might this occur?
b. Under what assumptions would you expect the new a.s.d. to exit first?
c. Under what assumptions would you expect the new a.s.d. to exit second?

9. Verify that the cumulative distribution for the geometric distribution is $G_t = 1 - (1-q)^t$.
a. Show that

$$\sum_{i=0}^{n} x^i = \frac{1 - x^{n+1}}{1 - x}$$

for any x by multiplying both sides by $1 - x$.
b. Find $\sum_{i=1}^{t} q(1-q)^{i-1}$ (use $x = 1 - q$).

10. We can find the exponential p.d.f. as the limit of the geometric distribution. Think of dividing time into small units of length Δt.
a. If the probability that a molecule leaves a cell during one minute is q, approximate the probability that it leaves during time Δt.
b. How many steps of length Δt does it take to reach time t?
c. Find the probability that the molecule leaves the cell between times t and $t + \Delta t$.
d. Take the limit as $\Delta t \to 0$.
e. Find the p.d.f.

11. The expectation of the geometric distribution can be found using the memoryless property.
a. A molecule either leaves a cell immediately or does not. Find the probability that it leaves immediately and the time at which it leaves.
b. Find the probability that the molecule does not leave immediately. Use the memoryless property to show that the expected time to leave in this case is $1 + E(T)$.
c. Multiply the values by the probabilities in parts **a** and **b**, and add them.
d. Set the result equal to $E(T)$ and solve.

12. The variance can be found with a clever trick. Define the random variable R, which is 1 with probability q and $1 + T$ with probability $1 - q$.
a. Why does R have the same probability distribution as T?
b. Show that $E(R^2) = q + (1-q)E[(1+T)^2]$.
c. Show that $E(T^2) = q + (1-q)E[(1+T)^2]$.
d. Solve for $E(T^2)$ and use the computational formula to find $\text{Var}(T)$.

13. **COMPUTER:** The probability p_t that a molecule is inside a cell after t time steps is

$$p_t = 0.9^t$$

a. Simulate 100 such molecules. How many are inside at time 3?
b. Of the ones that are inside at time 3, how many are still inside at time 6?
c. What does this have to do with the memoryless property of the geometric distribution?

14. **COMPUTER:** Use your computer to find the mean and variance of the exponential distribution with parameter λ. What fraction of measurements are greater than the mean? What fraction of measurements are greater than the mean plus one standard deviation? The mean plus two standard deviations? Why do these results seem so different from the rules of thumb in Section 6.10?

15. **COMPUTER:** There is a generalization of the geometric distribution called, confusingly enough, the **negative binomial distribution**. Under the same assumptions as the geometric, the random variable T_r is the number of failures that precede the rth success. The case $r = 1$ is the geometric distribution. In general, if the probability of a success on each trial is q, then

$$\Pr(T_r = t) = \binom{t + r - 1}{r - 1} q^r (1-q)^t$$

Set $q = 0.1$.
a. Plot the distribution with $r = 1$. Find the expectation.
b. Plot the distribution with $r = 2$. Find the expectation. Does this result make sense?
c. How can you think of the random variable T_2 as the sum of two other random variables?
d. Plot the distribution with $r = 10$. Find the expectation. Does this result make sense? Why is the shape of the distribution so different from those in **a** and **b**?

7.7 The Poisson Distribution

The binomial and geometric distributions describe two random variables associated with the same process. The binomial distribution describes the number of successes in a given number of trials, and the geometric distribution describes the time until the first success. The exponential density function extends the geometric distribution to describe waiting for events that occur at a continuous *rate*, according to a **Poisson process**. But we have yet to find the distribution of the number of such events that occur in a fixed time. We now study the **Poisson distribution**, the probability distribution for the number of events that are generated by the **Poisson process**. The following chart relates these four fundamental probability distributions.

	Time	
Measurement	Discrete	Continuous
count	binomial	Poisson
waiting time	geometric	exponential

The Poisson Process

Consider a cell floating around in a medium containing many molecules that bind to receptors on the cell at rate λ. Our random variable is N, the number of molecules binding to the cell during some interval of time. The probability that one molecule binds during a short time Δt is $\lambda \Delta t$. During this short time, then, the probability that none binds is $1 - \lambda \Delta t$. Our goal is to figure out the probability that 0, 1, 2, and so on, bind during a longer period of time t.

A simulation of this process with $\lambda = 1.0$ produced the following results:

Event	Time	Event	Time
1	0.219032	6	4.84409
2	0.929085	7	5.54173
3	1.50534	8	6.45588
4	3.79879	9	9.47364
5	4.66062	10	10.6027

(Figure 7.27). Each spike in Figure 7.27a indicates exactly when a molecule bound. We summarize the results by asking how many bound during each of the 12 one second intervals: two during the first second (between times $t = 0.0$ and $t = 1.0$), one during the next second (between $t = 1.0$ and $t = 2.0$), none during the third (between times $t = 2.0$ and $t = 3.0$), and so on. A total of four seconds had zero events, six seconds had one event, and two seconds had two events. Figure 7.27b shows a histogram summarizing these results.

A second simulation with $\lambda = 4.5$ produced the results in Figure 7.28a. There are many more events, producing much larger numbers in the histogram (Figure 7.28b). Our goal is to figure out the mathematically exact probability that exactly k events occurred during an interval of length t.

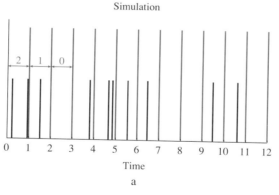
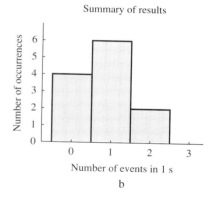

Figure 7.27

The Poisson process with $\lambda = 1.0$

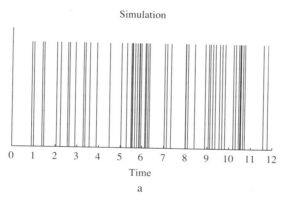
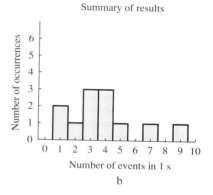

Figure 7.28

The Poisson process with $\lambda = 4.5$

Let N be the random variable counting the number of events in an interval of length t. To begin, we use the exponential distribution to find the probability that $N = 0$. Because events (or departures) occur at rate λ, the probability of 0 events is given by the exponential survivorship function, or

$$\Pr(N = 0) = e^{-\lambda t}$$

With $\lambda = 1.0$ and $t = 1.0$, the probability of 0 events is

$$\Pr(N = 0) = e^{-1.0 \cdot 1.0} = 0.368$$

This matches the simulated result where 4 of 12 seconds experienced 0 events (Figure 7.27b). With $\lambda = 4.5$ and $t = 1$, the probability of 0 events is

$$\Pr(N = 0) = e^{-4.5 \cdot 1.0} = 0.011$$

It is quite unlikely that nothing happens in 1 s, consistent with the simulation (Figure 7.28).

The Poisson distribution fills in the rest of the probabilities, giving $\Pr(N = k)$ for every value of k. We denote $\Pr(N = k)$ by $p(k; \lambda t)$.

■ THEOREM 7.17

Suppose events occur at a constant rate λ. The probability that exactly k events occur in time t is

$$\Pr(N = k) = p(k; \lambda t) = \frac{e^{-\lambda t}(\lambda t)^k}{k!}$$

The range of the random variable N is all nonnegative integers, $0, 1, 2, \ldots$. ■

A hint of one way to derive these probabilities is given in Exercise 11.

The Poisson distribution gives the probability that k molecules bind during time t when the underlying rate is λ. With $k = 0$,

$$\Pr(N = 0) = p(0; \lambda t)$$
$$= \frac{e^{-\lambda t}(\lambda t)^0}{0!}$$
$$= e^{-\lambda t}$$

(recall that $0! = 1$), matching the earlier formula.

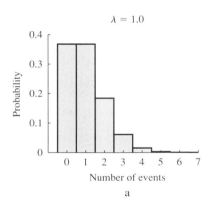
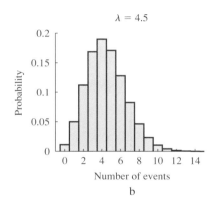

Figure 7.29

The Poisson distribution

Because the Poisson distribution depends on λ and t only through their *product* λt, we often write $\Lambda = \lambda t$ and rewrite the formula for the Poisson distribution.
Simplified formula for the Poisson distribution:

$$\Pr(N = k) = p(k; \Lambda) = \frac{e^{-\Lambda}(\Lambda)^k}{k!}$$

The Poisson distribution describes a random variable under the following conditions.
Conditions for the Poisson distribution:

- The random variable counts events that occur at rate λ, meaning that the probability of one event in a short time Δt is $\lambda \Delta t$.
- Events are independent.

When events obey these rules, they are said to follow a **Poisson process**. These conditions are similar to those for the binomial distribution, counting independent events. In this case, however, events occur continuously, and there is no upper limit on the number of events possible.

Suppose $\lambda = 1.0$ per minute (as in Figure 7.27). We can use the formula for the Poisson distribution to graph the probability distribution after 1 min. The parameter $\Lambda = \lambda t = 1.0 \cdot 1.0 = 1.0$.

$$\Pr(N = 0) = p(0; 1) = \frac{e^{-1.0}(1.0^0)}{0!} \approx 0.3679$$

$$\Pr(N = 1) = p(1; 1) = \frac{e^{-1.0}(1.0^1)}{1!} \approx 0.3679$$

$$\Pr(N = 2) = p(2; 1) = \frac{e^{-1.0}(1.0^2)}{2!} \approx 0.1839$$

$$\Pr(N = 3) = p(3; 1) = \frac{e^{-1.0}(1.0^3)}{3!} \approx 0.0613$$

and so on. The probability distribution, plotted in Figure 7.29a, matches the simulated results in Figure 7.27 reasonably well.

With $\lambda = 4.5$, many more events occur per minute (Figure 7.28). In this case, $\Lambda = \lambda t = 4.5 \cdot 1.0 = 4.5$ and

$$\Pr(N = 0) = p(0; 4.5) = \frac{e^{-4.5}(4.5^0)}{0!} \approx 0.01111$$

$$\Pr(N = 1) = p(1; 4.5) = \frac{e^{-4.5}(4.5^1)}{1!} \approx 0.04999$$

$$\Pr(N = 2) = p(2; 4.5) = \frac{e^{-4.5}(4.5^2)}{2!} \approx 0.11248$$

$$\Pr(N = 3) = p(3; 4.5) = \frac{e^{-4.5}(4.5^3)}{3!} \approx 0.16872$$

The whole probability distribution is plotted in Figure 7.29b.

What if we run the slower process with $\lambda = 1.0$ for a longer time, such as 4.5 min? Because the only value used in computing the probability distribution is the product $\Lambda = \lambda t = 1.0 \cdot 4.5 = 4.5$, the results of a fast process after a short time exactly match those of a slow process after a longer time.

Although they are a bit tricky to derive, the formulas for the mean and variance of the Poisson distribution are easy to remember.

THEOREM 7.18 Suppose the random variable N has a Poisson distribution with parameter Λ. Then

$$E(N) = \Lambda$$
$$\text{Var}(N) = \Lambda$$

Why are these formulas so simple? Because λ is a rate, the expectation Λ is the rate times the time. If water is entering a vessel at rate λ for time t, the total amount is exactly λt. The difference between the Poisson process and constant flux is the *variance*. When the flow rate is 1.0 L/min, exactly 4.5 L enter in 4.5 min. When molecules enter at an average rate of 1.0 per minute, the *expected* number of molecules entering in 4.5 min is 4.5, but with a variance of 4.5 molecules^2 and a standard deviation of $\sqrt{4.5} \approx 2.12$ molecules.

Be careful to distinguish these results from the exponential distribution, which has *standard deviation* equal to the mean. With the Poisson distribution, the coefficient of variation gets *smaller* as the mean gets larger. By sampling more events, we get a more consistent count. The exponential distribution is always highly variable because it concerns counting only a *single* event.

The mode of the Poisson distribution, like that of the binomial distribution, is always close to the mean.

■ **THEOREM 7.19** Suppose the random variable N has a Poisson distribution with parameter Λ. The mode of N is the largest integer less than Λ. ■

Proof: The mode can be found with an iterative formula like that for the binomial distribution (Theorem 7.13). We find

$$p(k+1; \Lambda) = \frac{e^{-\Lambda}(\Lambda)^{k+1}}{(k+1)!}$$

$$= p(k; \Lambda) \frac{\Lambda}{k+1}$$

The mode is the smallest value of k for which $p(k+1; \Lambda) < p(k; \Lambda)$. This occurs when $k + 1 > \Lambda$. Therefore, the mode is the largest integer less than Λ. ■

For example, with $\Lambda = 4.5$, the largest integer less than 4.5 is 4, matching the mode in Figure 7.29b. When $\Lambda = 1$, the mode is 1 (which ties with 0).

The Poisson Distribution in Space

The Poisson distribution has a remarkable array of applications. One of the most useful is the distribution of objects or events in space rather than time. The parallel with the temporal case is strongest in one dimension. Suppose two bacteria clones have been accumulating mutations since they last had a common ancestor, and that an average of 1.3 mutations have become fixed per million nucleotides. We have a piece of DNA 4.7 million nucleotides long from each organism (Figure 7.30). In how many sites will the two differ?

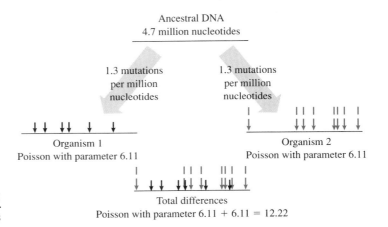

Figure 7.30
The evolution of two organisms

The picture of the mutations, which occur over the *length* of the DNA, looks exactly like the simulation of events occurring in *time* (Figure 7.27). What assumptions must we make to use the Poisson distribution? The rate at which mutations occur along the piece of DNA must be constant, and mutations must occur independently. Both these assumptions have been the source of important debates among geneticists. Assuming that they are true, the number of mutations M_1 in the organism 1 has a Poisson distribution with rate $\lambda = 1.3$ per million and "t" equal to 4.7 million. Therefore, $\Lambda = 1.3 \cdot 4.7 = 6.11$ mutations. The same argument holds for organism 2, so M_2 has a Poisson distribution with $\Lambda = 6.11$.

However, both M_1 and M_2 measure the number of differences from the common ancestor, which is extinct and inaccessible. What is the distribution of the number of **differences** between the two organisms? If we superimpose the two pieces of DNA, the number of differences is the **sum** of the number of mutations in each (except for the unlikely event that two mutations occurred exactly in the same spot). In mathematical notation, the total number of differences D is

$$D = M_1 + M_2$$

The following theorem tells us everything we need to know about D.

■ **THEOREM 7.20** Suppose X and Y are independent Poisson-distributed random variables with parameters Λ_X and Λ_Y. Then the sum $Z = X + Y$ has a Poisson distribution with parameter $\Lambda_X + \Lambda_Y$. ■

Therefore, D has a Poisson distribution with mean $6.11 + 6.11 = 12.22$ and variance 12.22. Our theorems about expectation and variance of the sum of independent random variables (Theorems 7.4 and 7.9) guarantee that these must be the mean and variance of D,

$$E(D) = E(M_1) + E(M_2) = 6.11 + 6.11 = 12.22$$
$$\text{Var}(D) = \text{Var}(M_1) + \text{Var}(M_2) = 6.11 + 6.11 = 12.22$$

Theorem 7.20 tells us more—the entire probability distribution of D. By computing the probability distribution for D, we can compare our result (14 mutations) with the expectation (about 12 mutations). Figure 7.31 plots the distribution, from which we can see that the probability of 14 mutations (0.094) is very close to the probability of the mode (12 mutations with probability 0.114).

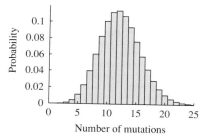

Figure 7.31

The distribution of the number of mutations separating two organisms

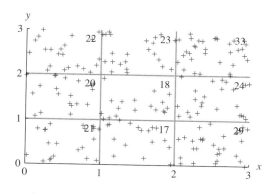

Figure 7.32

Scatter of seeds in two dimensions

The Poisson distribution also applies to processes in two or more dimensions. Suppose seeds fall independently in a region at a rate of 0.0023 seeds per square centimeter. A simulation of this process in a 9 m² region is shown in Figure 7.32, with the number of seeds in each block indicated. The number of seeds S per square meter follows a Poisson distribution. In this case, $\lambda = 0.0023$ seeds per

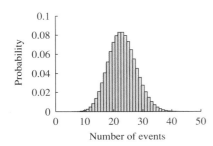

Figure 7.33
The Poisson distribution with $\Lambda = 23$ and the binomial distribution with $p = 0.0023$ and $n = 10{,}000$.

square centimeter and "t" is 10,000 cm^2. Therefore

$$\Lambda = 0.0023 \cdot 10000 = 23.0$$

the mean number of seeds per square meter. The probability of k seeds in a given square meter is $p(k; 23.0)$ (Figure 7.33). The distribution is less spread out and has more of a "bell" shape than with small Λ (Figure 7.29). The probability that S is within 5 of the expectation is

$$\Pr(18 \leq M \leq 28) = \sum_{k=18}^{28} p(k; 23.0) \approx 0.750$$

where we evaluated the various values $p(k; 23.0)$ on the computer (as with the binomial distribution, there is no convenient formula for the cumulative distribution). About 75% of square meters surveyed should have within five seeds of the expectation, consistent with the six out of nine in the simulation.

The Poisson Distribution and the Binomial Distribution

The seed example highlights the link between the Poisson distribution and the binomial distribution. The number of seeds in 1 m^2 is a count of the number in each of 10,000 cm^2, each with the tiny probability 0.0023 of containing a seed. If we ignore the probability that some square centimeter has two seeds, the counts follow a binomial distribution with $n = 10{,}000$ and $p = 0.0023$. The probability distribution for this binomial random variable is indistinguishable from Figure 7.33.

In general, we have the following rule.

Rule for Approximating the Binomial Distribution with the Poisson Distribution If p is small (less than about 0.01) and n is large (greater than about 100), then

$$b(k; n, p) \approx p(k; np) \qquad (7.16)$$

The Poisson distribution can be thought of as the limit of infinitely many trials with infinitesimally unlikely Bernoulli random variables.

Let B be a random variable following a binomial distribution with parameters n and p, and let P be a random variable following a Poisson distribution with parameter $\Lambda = np$. Then

$$E(B) = E(P) = np$$
$$\text{Var}(B) = np(1-p)$$
$$\text{Var}(P) = np$$

The expectations match, and the variances are very close when p is small. How do B and P differ? The random variable P can take on any nonnegative integer value, whereas B can take on only values from 0 to n. However, the probability that $P > n$ (10,000 in the seed example) is astronomically small.

SUMMARY

If events occur independently at rate λ, they are said to follow a **Poisson process**. A random variable that counts the number of events that occur in time t has a **Poisson distribution** with parameter $\Lambda = \lambda t$. The mean and variance are each equal to the product Λ of the rate and the time. The Poisson distribution applies equally well to events that occur independently in space. Independent random variables with Poisson distributions have the convenient property that their sum also has a Poisson distribution. Finally, a binomial distribution with small p and large n can be approximated by a Poisson distribution with $\Lambda = np$.

7.7 EXERCISES

1. Find the average number of hits per second in figures 7.27 and 7.28. Compare with the results from the Poisson distribution.

2. Find the probabilities of the following events. Sketch the associated distribution and find its expectation, variance, and mode.
    a. Molecules leave a cell at rate $\lambda = 0.3$ molecules per second. What is the probability that exactly two have left by the end of the third second?
    b. Phone calls arrive at a rate of $\lambda = 0.2$ per hour. What is the probability that there are exactly five calls in 9 hr?
    c. Cosmic rays hit an organism at a rate of 1.2 per day. What is the probability of being hit ten times in a week?

3. Find the probabilities of the following events.
    a. Molecules leave a cell at rate $\lambda = 0.3$ molecules per second. What is the probability that four or fewer have left by the end of the third second?
    b. Phone calls arrive at a rate of $\lambda = 0.2$ per hour. What is the probability that there are more than five calls in 9 hr?
    c. Cosmic rays hit an organism at a rate of 1.2 per day. What is the probability of being hit between five and ten times in a week?

4. Molecules leave a cell at rate $\lambda = 0.3$ molecules per second. Let $N(t)$ be the random variable measuring the number that have left as a function of t. Consider times t between 0 and 10.
    a. Compute and graph $E[N(t)]$ as a function of time.
    b. Compute and graph $CV[N(t)]$ as a function of time.
    c. Compute and graph $\Pr[N(t) = 1]$ as a function of time. Find the maximum. Why does this graph increase and then decrease?
    d. Compute and graph $\Pr[N(t) = 2]$ as a function of time. Find the maximum. Why is the maximum greater than in part c?

5. Cosmic rays hit the surface of an organism at a rate of 1.2 per square centimeter per week. The surface area is 60 cm².
    a. Find the expectation and the standard deviation of the number to hit the organism in a week.
    b. Find the probability that exactly 70 hit the organism in a week.
    c. Find the expectation and the standard deviation of the number to hit the organism in a day.
    d. Find the probability that exactly 10 hit the organism in a day.

6. Consider a binomial model of phone calls. Assume you get one call in a given hour with probability 0.2 and no calls with probability 0.8.
    a. Find the probability of exactly two calls in 10 hr.
    b. Write the Poisson approximation to this model.
    c. Find the probability of exactly two calls in 10 hr based on the Poisson approximation. How close is your answer?

7. Suppose 100,000 gnats are flying around in a room. Each leaves independently with probability 0.0067 during a test of a new insect repellent.
    a. Describe the random variable with a binomial distribution giving the probability that exactly k have left.
    b. Find the Poisson distribution approximating the probability that exactly k have left.
    c. Find the expected number to leave.
    d. Find the variance of the binomial random variable and the Poisson approximation.

8. A geneticist studies mutational differences between pieces of DNA from a species of bacteria. The first piece is 2.6 million nucleotides long, and mutations occur at rate 0.56 per million nucleotides. The second piece is 1.3 million nucleotides long, and mutations occur at rate 3.11 per million nucleotides.
    a. Find the distribution giving the number of mutations in the first type of DNA.
    b. Find the distribution giving the number of mutations in the second type of DNA.
    c. Find the distribution giving the number of mutations in both types of DNA.

9. Suppose another study of mutations reveals that mutations accumulate at a rate of 1.3 per million nucleotides during the first year of a study and at a rate of 2.2 per million nucleotides during the second year. The DNA is 4.7 million nucleotides long.
    a. Describe the distribution of mutations produced during the first year.
    b. Describe the distribution of mutations produced during the second year.
    c. Describe the distribution of mutations produced during the 2 yr.
    d. How many mutations do you think there would be if the mutation rate μ followed the function $\mu(t) = 1.3 + 0.9t$? (Try breaking up the interval with the left-hand approximation.)

10. Suppose an organism would live forever if it weren't for predators, which attack at a rate λ per year. Fortunately, only a fraction q of attacks are successful. The following steps will help us compute how long the organism will live.
    a. What is the probability the organism has not been attacked at time t?
    b. What is the probability that the organism has been attacked once but survived?
    c. What is the probability that the organism has been attacked twice but survived?
    d. Write a sum giving the probability that the organism is alive at time t.
    e. Write the first few probabilities in a Poisson distribution with parameter $(1 - q)\lambda$. What do they add to?
    f. Use this last fact to determine the probability that the organism is alive at time t. What is the average lifetime?
    g. Based on this result, could you have guessed the answer to **f** without doing any calculations?

11. We can derive a differential equation for the probability that exactly 1 molecule has left a cell as a function of time. Let $P_i(t)$ be the probability that exactly i molecules left by time t.
    a. Write $P_1(t + \Delta t)$ in terms of $P_0(t)$ and $P_1(t)$ (think of the two ways there could have been one molecule at time $t + \Delta t$).
    b. Rearrange terms so your formula looks like the derivative of $P_1(t)$, and take the limit as $\Delta t \to 0$.
    c. Recalling that $P_0(t) = e^{-\lambda t}$, check that $P_1(t) = \lambda t e^{-\lambda t}$ is a solution of your equation. There are fancy ways to solve the equation itself, which can be extended to find all the probabilities $P_i(t)$.

12. **COMPUTER:** Suppose T is an exponentially distributed random variable with parameter $\lambda = 0.3$ that describes the waiting time between events. If the nth event occurs at time t, then the $n + 1$st event occurs at time $t + T$. The updating function describing this process is
    $$g(t) = t + T$$
    a. Starting from an initial condition of 0, simulate this updating function for 20 steps. How many events occurred by time 20?
    b. Mark the two longest waits and their durations.
    c. What is the expected waiting time between events? Sketch the line that should lie close to the graph of your solution.
    d. The number of events at time 20 should be described by a Poisson distribution with parameter $\Lambda = 20\lambda$. Plot the histogram of this distribution.
    e. Using the number of events k that occurred by time 20 in part **a**, compute the probability of exactly k, fewer than k, and more than k events. Indicate the areas associated with each of these events on your histogram.
    f. Do you have reason to think that your result from part **a** is peculiar?

13. **COMPUTER:** In Exercise 12, we simulated a Poisson process as a series of exponentially distributed waiting times. An alternative method uses the the relation between the Poisson and binomial distributions. Suppose we want to simulate a process with $\lambda = 1.5$ per minute for 10 min.
    a. Break each minute into n intervals for some large value of n, for example, 100. What value of p will produce an average of 1.5 successes per minute?
    b. Choose a series of $10n$ independent Bernoulli random variables that equal 1 with probability p. Each value of 1 corresponds to an event in the Poisson process.
    c. Produce a graph showing when the events occur.
    d. Why does this method fail to reproduce exactly the Poisson process?

14. **COMPUTER:** We can use the the relation between the Poisson and binomial distributions to simulate processes in which the value of λ changes over time. Suppose that $\lambda = 0.5t$ per minute for t ranging from 0 to 10 minutes.
    a. Break each minute into n intervals for some large value of n, for example, 100. What value of $p(n)$ should you use during each interval n?
    b. Choose a series of $10n$ independent Bernoulli random variables that equal 1 with probability $p(n)$. Each value of 1 corresponds to an event in the Poisson process.
    c. Produce a graph showing when the events occur.
    d. Describe in words what is going on in the underlying process. Does your graph match your description?
    e. For an extra challenge, try to figure out how you would simulate this process with waiting times.

7.8 The Normal Distribution

We have studied two distributions associated with counting: the binomial and the Poisson. Counting is the simplest form of adding, one for each event that occurred. A more general sum adds random variables that can take on values other than 0 and 1, such as dart scores. The exact distribution resulting from a more general sum is often impossible to compute. Under a wide range of assumptions, however, the **normal distribution** provides a remarkably accurate approximation of the exact distribution. Here we sketch the theory behind this approximation.

The Normal Distribution: An Example

The development of statistics was largely inspired by problems in breeding. Here is one sort of problem breeders face. A type of plant has ten different genes affecting height, with an equally common short and tall allele at each locus. Plants with two copies of the short allele gain no height from that locus, those with one copy of the short allele and one copy of the tall allele gain 1.0 cm, and those with two copies of the tall allele gain 2.5 cm. The complicating fact is that two copies of the tall allele give more than double the height change from one copy (contrast with Equation 6.12). Suppose the total height gain is the sum of the height gained from each locus. What is the distribution of heights of these plants?

We computed the binomial distribution for a similar case by enumerating all the possibilities. Enumeration is virtually impossible in this case because there are many different ways a plant can reach the same height (Exercise 1).

Let H_i be a random variable measuring the height gain from locus i. Because the short and tall alleles are equally common, we have

$$\Pr(H_i = 0.0) = 0.25$$
$$\Pr(H_i = 1.0) = 0.5$$
$$\Pr(H_i = 2.5) = 0.25$$

for each i (Figure 7.34b). Therefore

$$E(H_i) = 0.0 \cdot 0.25 + 1.0 \cdot 0.5 + 2.5 \cdot 0.25 = 1.125$$

and

$$\text{Var}(H_i) = 0.0^2(0.25) + 1.0^2(0.5) + 2.5^2(0.25) - 1.125^2 = 0.797$$

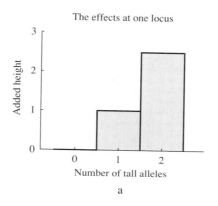
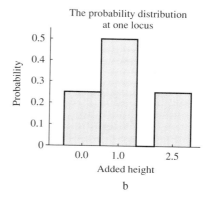

Figure 7.34
The effects of alleles at one locus

Let H represent the total added height. Because heights have been assumed to add, we have

$$H = \sum_{i=1}^{10} H_i$$

The expectation of the sum is the sum of the expectations (Theorem 7.6), so

$$E(H) = \sum_{i=1}^{10} E(H_i) = 11.25$$

Similarly, because we assume that the different loci are independent (the alleles at one locus are independent of those at every other), the variance of the sum is the sum of the variances (Theorem 7.9), and we have

$$\text{Var}(H) = \sum_{i=1}^{10} \text{Var}(H_i) = 7.97$$

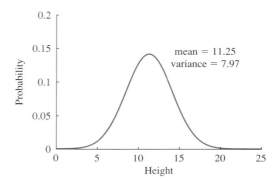

Figure 7.35
What the mean and variance tell us

We have used the underlying structure of the random variable H to find its mean and variance, which give some information about the whole distribution (Figure 7.35). What if we want to find, for example, the probability that the height of the plant is increased by less than 6 cm, by between 10 and 15 cm, or by more than 19 cm? The mean and variance alone cannot help us.

There is a powerful and deep theorem that does. The **Central Limit Theorem** says that the sum of independent random variables with the same distribution can be approximated by a **normal distribution**.

■ **Definition 7.10** A continuous random variable X has a **normal distribution** with mean μ and variance σ^2 if it has probability density function

$$f(x; \mu, \sigma^2) = \frac{1}{\sqrt{2\pi}\sigma} e^{-(x-\mu)^2/2\sigma^2}$$

The range of this random variable is from $-\infty$ to ∞. We write

$$X \sim N(\mu, \sigma^2)$$

■

The normal distribution has a symmetric bell-shaped probability density function that takes on its maximum value at $x = \mu$ and approaches 0 at both ∞ and $-\infty$ (Figure 7.36). The points of inflection are at $x = \mu + \sigma$ and $x = \mu - \sigma$, one standard deviation above and below the mean, which is the first rule of thumb discussed in Section 6.10 (Exercise 8).

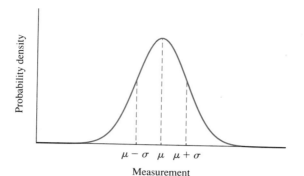

Figure 7.36
The normal distribution

The normal distribution is completely described by two numbers: the mean and the variance. What happens if we graph a normal distribution $X \sim N(11.25, 7.97)$ with mean $\mu = 11.25$ and variance $\sigma^2 = 7.97$ equal to the mean and variance of H? Figure 7.37 shows the normal probability density function superimposed on data plotted for 200 simulated plants.

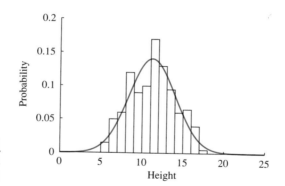

Figure 7.37
The normal distribution compared with histogram for 200 plants

Because this normal probability density function approximates the true distribution, we can use it to estimate probabilities. In particular, we have

$$\Pr(H \leq 6) \approx \Pr(X \leq 6) = \int_{-\infty}^{6} f(x; 11.25, 7.97) \, dx$$

$$\Pr(10 \leq H \leq 15) \approx \Pr(10 \leq X \leq 15) = \int_{10}^{15} f(x; 11.25, 7.97) \, dx$$

$$\Pr(H \geq 19) \approx \Pr(X \geq 19) = \int_{19}^{\infty} f(x; 11.25, 7.97) \, dx$$

Some of the limits of integration might seem absurd. Plant heights cannot be negative or greater than 25 in our model, but the limits of integration extend to $-\infty$ and ∞. The normal approximation is just that, an *approximation*. It is a good approximation because the values of the p.d.f. are extremely small in the absurd region. For example, $f(0; 11.25, 7.97) = 5.0 \times 10^{-5}$ and $f(25; 11.25, 7.97) = 1.0 \times 10^{-6}$.

It is impossible to write a formula for the integral of the normal distribution, and we must use a table or computer to help with the calculation of probabilities. In Section 7.9, we find a convenient way to make these calculations.

The Central Limit Theorem for Sums

Why does the normal distribution approximate H so well? We now present a formal statement of the theorem behind the method. One bit of shorthand is often used in this context. Random variables with the same probability distribution or probability density function are said to be **identically distributed**. The random variables H_i discussed above, or different trials of an experiment, are identically distributed. Such sets of random variables are both **independent and identically distributed**, which is abbreviated **i.i.d.**

■ **THEOREM 7.21** (The Central Limit Theorem for Sums)

Suppose $X_1, \ldots, X_i, \ldots$, are a set of i.i.d. random variables, each with finite expectation μ and finite variance σ^2. Let

$$S_n = \sum_{i=1}^{n} X_i$$

be a random variable representing the sum of the first n variables. Then the probability density function for S_n is approximately $N(n\mu, n\sigma^2)$ for sufficiently large n. ■

The proof of this theorem requires technical tools beyond the scope of this book. However, as with the calculation of $E(H)$ and $Var(H)$, we know that

$$E(S_n) = \sum_{i=1}^{n} E(X_i) \quad \text{expectation of sum is sum of expectations}$$

$$= \sum_{i=1}^{n} \mu \quad E(X_i) = \mu \text{ for all } i$$

$$= n\mu \quad \text{adding } n \text{ } \mu\text{s gives } n\mu$$

and

$$Var(S_n) = \sum_{i=1}^{n} Var(X_i) \quad \begin{array}{l}\text{expectation of variance is sum of variances}\\ \text{(if independent random variables)}\end{array}$$

$$= \sum_{i=1}^{n} \sigma^2 \quad Var(X_i) = \sigma^2 \text{ for all } i$$

$$= n\sigma^2 \quad \text{adding } n \text{ } \sigma^2\text{s gives } n\sigma^2$$

A few hypotheses of the theorem might look a bit odd. We have not met any random variables with infinite expectation or infinite variance (although the issue came up in Section 7.3). Such distributions do occasionally arise in practice (Exercises 9 and 10) but do not cause much trouble because the attempt to compute the approximate normal distribution fails. More important are the assumptions that the random variables be independent and identically distributed. Generalizations of the Central Limit Theorem deal with random variables that are not quite independent (in a suitably precise sense) or that have distributions that are not quite the same. For scientific applications, because the theorem gives an approximate result in the first place, as long as there is good reason to think that the random variables in question are close to satisfying the i.i.d. requirement, application of the Central Limit Theorem is pretty safe.

Why the normal distribution? A hint is given by the **additive property** of the normal distribution. If X and Y are independent random variables with

$$X \sim N(\mu_1, \sigma_1^2)$$
$$Y \sim N(\mu_2, \sigma_2^2)$$

then

$$X + Y \sim N(\mu_1 + \mu_2, \sigma_1^2 + \sigma_2^2)$$

(Figure 7.38). The mean must be $\mu_1 + \mu_2$ because the expectation of the sum is the sum of the expectations (Theorem 7.4). Because the random variables are independent, the variance must be $\sigma_1^2 + \sigma_2^2$ (Theorem 7.9). The normal distribution is special because the sum of independent normally distributed random variables has the same shape, a normal distribution. (We have seen that this special property also holds for the Poisson distribution in Theorem 7.20.)

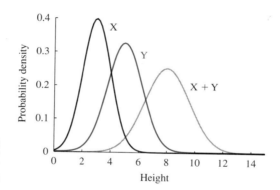

Figure 7.38
The sum of two independent normally distributed random variables

The Central Limit Theorem states that the sum is approximately normal for "sufficiently large n." How large is large?

Rule of Thumb for Applying the Central Limit Theorem:
It is safe to use the Central Limit Theorem when $n \geq 30$.

For example, suppose T_i has an exponential distribution. The random variable T_i is the waiting time between event $i - 1$ and event i. Therefore,

$$S_n = \sum_{i=1}^{n} T_i$$

is the waiting time until event n (Figure 7.39). The probability density function for S_n is called a **Gamma distribution** (computed in Exercise 7), and is compared with

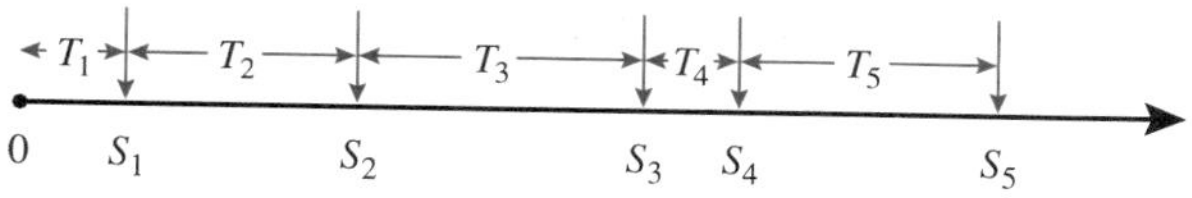

Figure 7.39
The sum of many waiting times

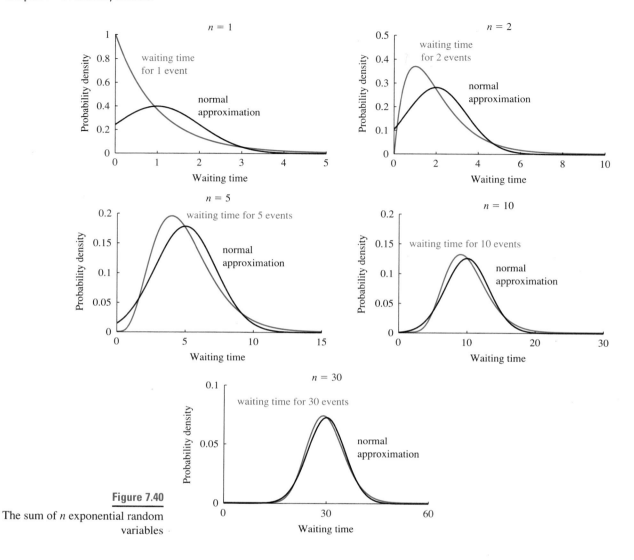

Figure 7.40
The sum of n exponential random variables

the normal approximation for five different values of n in Figure 7.40. Although the probability density function for the waiting time T_1 has an exponential density and looks nothing like a normal distribution, the shape becomes more "normal" as we add more independent terms.

The Central Limit Theorem for Averages

An important statistical application of the normal distribution is a consequence of the following rephrasing of the Central Limit Theorem for Sums (Theorem 7.21).

■ **THEOREM 7.22** **(The Central Limit Theorem for Averages)**

Suppose $X_1, \ldots, X_i, \ldots,$ are a set of i.i.d. random variables, thought of as independent samples from the same distribution, each with finite expectation μ and finite variance σ^2. Let

$$A_n = \frac{1}{n}\sum_{i=1}^{n} X_i$$

be a random variable equal to the average of the first n variables. Then the probability density function for A_n is approximately $N(\mu, \sigma^2/n)$ for sufficiently large n. ∎

Proof: This theorem follows from Theorem 7.21. By definition,

$$A_n = \frac{S_n}{n}$$

Dividing a random variable by a constant divides the expectation of that random variable by the constant and the variance by the square of the constant. Therefore, we have

$$E(A_n) = \frac{E(S_n)}{n} = \mu$$

$$\text{Var}(A_n) = \frac{\text{Var}(S_n)}{n^2} = \frac{n\sigma^2}{n^2} = \frac{\sigma^2}{n}$$ ∎

How do we use the Central Limit Theorem for averages? Suppose we measure the waiting times for ten events, and we repeat the experiment five times.

Molecule	Experiment 1	Experiment 2	Experiment 3	Experiment 4	Experiment 5
1	4.046	0.437	1.737	0.380	2.603
2	0.849	1.498	0.341	1.202	1.709
3	0.356	0.909	1.455	2.683	2.414
4	1.120	0.976	0.211	0.070	0.476
5	0.467	0.427	0.094	0.298	0.837
6	0.194	0.365	0.316	3.132	0.479
7	0.851	3.026	0.043	0.441	1.374
8	2.230	0.727	0.302	0.533	0.543
9	0.870	0.532	0.595	1.725	0.481
10	0.641	0.295	0.133	0.458	0.351
Average	1.162	0.919	0.523	1.092	1.127

If we know that the waiting times follow an exponential distribution with $\lambda = 1.0$, what can we say about the average? Let T_i be a random variable indicating the waiting time for event i. Each of these random variables satisfies

$$E(T_i) = \frac{1}{\lambda} = 1.0$$

$$\text{Var}(T_i) = \frac{1}{\lambda^2} = 1.0$$

The average is the sum of these values divided by 10, or

$$A = \frac{1}{10}\sum_{1} 10T_i$$

According to the Central Limit Theorem for Averages, the new random variable A will have approximately a normal distribution with mean μ equal to the mean of each of T_i, and variance equal to the variance divided by 10. In mathematical terms, we have

$$A \sim N(1.0, 0.1)$$

According to our rules of thumb (Section 6.10), the averages should have a mean at 1.0 and most should lie within two standard deviations of the mean. In this case,

$$\sigma_A = \sqrt{0.1} = 0.316,$$

meaning that most results should be between $1.0 - 2 \cdot 0.316 = 0.368$ and $1.0 + 2 \cdot 0.316 = 1.632$ (Figure 7.41). This accords quite well with our results, even though the number of terms averaged is less than the 30 recommended by the rule of thumb.

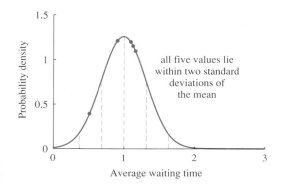

Figure 7.41
Applying the Central Limit Theorem for averages

The Central Limit Theorem for Averages is fundamental to estimation techniques in statistics. If we try to estimate the expectation of a distribution by taking independent samples X_i and computing their average, as we do when estimating a value from a series of experimental replicates, we end up with estimates of the true expectation which have a normal distribution centered at the true expectation *even if we do not know the shape of the underlying distribution.*

SUMMARY

We have approximated the sum of **independent, identically distributed (i.i.d.)** random variables with a **normally distributed** random variable by matching the mean and variance. This approximation is justified by the **Central Limit Theorem for Sums**. This theorem can be reinterpreted as the **Central Limit Theorem for Averages**, showing that the averages of independent samples from one distribution are approximately normally distributed with mean equal to the true mean and variance decreasing with the size of the sample.

7.8 EXERCISES

1. On the basis of the probabilities in Figure 7.34b, find directly the probability that a plant gains 5 cm in height from 10 genes [Pr($H = 5$)].
   a. Find the two ways to add up 0s, 1s, and 2.5s to get 5.
   b. Find the number of different orders each way from **a** could occur.
   c. Find the probability associated with each way.
   d. Add them to find the total probability.

2. Consider the plants in Exercise 1.
   a. Find the normal distribution approximating the added height.
   b. Find the value of the normal p.d.f. at 5.0.
   c. Use the result of part b to approximate the probability that the height is in the interval between 4.5 and 5.5. Compare with the result of Exercise 1.

3. Suppose that X and Y are independent with $X \sim N(5.0, 16.0)$ and $Y \sim N(10.0, 9.0)$.
   a. Find and sketch the p.d.f. of $3X$.
   b. Find and sketch the p.d.f. of $X + Y$.
   c. Find and sketch the p.d.f. of the sum of nine independent samples from X.
   d. Find and sketch the p.d.f. of the mean of nine independent samples from Y.

4. Dart scores (Figure 6.49) have mean 12.55 and variance 154.7.
   a. Find the normal approximation of the sum of 5 dart scores chosen independently from the distribution.
   b. Find the normal approximation of the average of 5 dart scores chosen independently from the distribution. What is the largest value you would expect?
   c. Find the normal approximation of the average of 10 dart scores chosen independently from the distribution. What is the largest value you would expect?
   d. Find the normal approximation of the average of 100 dart scores chosen independently from the distribution. What is the largest value you would expect?

5. Scientists develop a sophisticated new model of human IQ that includes three independent factors: genes of large effect, genes of small effect, and environmental effects. All genes are assumed to be dominant. There are 10 smart genes of large effect, each of which adds 2.5 IQ points. There are 20 smart genes of small effect, each of which adds 0.6 IQ points. There are 30 environmental factors, with favorable ones adding 0.9 IQ points. People start out at 80, the baseline IQ to which all genetic and environmental factors are added. Finally, suppose the probability of receiving each smart gene is 0.75 and each favorable environmental effect is 0.5.
   a. Find the normal approximation for IQ based only on genes of large effect.
   b. Find the normal approximation for IQ based only on genes of small effect.
   c. Find the normal approximation for IQ based only on environmental effects.
   d. Find the normal approximation for IQ with both genetic and environmental effects.
   e. What is the maximum possible IQ according to this model?

6. Recall the growth of populations in Exercise 5 in Section 6.9. Let R_1 be a random variable giving the per capita reproduction in one population with p.d.f. $g(x) = 5.0$ for $1.0 \le x \le 1.2$ (the values used to generate Figure 6.2b).

Let R_2 be a random variable giving the per capita reproduction in another population with p.d.f. $g(x) = 1.25$ for $0.7 \le x \le 1.5$ (the values used to generate Figure 6.2c). We found that $E[\ln(R_1)] = 0.0939$ and $E[\ln(R_2)] = 0.0723$.
   a. Use the fact that
   $$\int \ln(x)^2 dx = x\ln(x)^2 - 2x\ln(x) + 2x$$
   to find $\text{Var}[\ln(R_1)]$ and $\text{Var}[\ln(R_2)]$.
   b. Let $P_1(t)$ be the size of population 1 after t generations. Assume that $P_1(0) = 1$. Find the normal approximation for $\ln[P_1(t)]$.
   c. Do part b for $P_2(t)$.

7. Suppose $T_1, T_2, T_3, \ldots$ are i.i.d. exponential random variables with $\lambda = 1.0$, and that
$$S_n = \sum_{i=1}^{n} T_i$$
(Figure 7.39).
   a. Use the law of total probability to show that
   $$\Pr(S_2 = t) = \int_{s=0}^{t} \Pr(T_1 = s)\Pr(T_2 = t - s)\, ds$$
   b. Solve the integral to find the p.d.f. for S_2.
   c. Use the law of total probability as in a to find the p.d.f. for S_3.
   d. Can you guess the pattern? Why does the answer look so much like the Poisson distribution?

8. Show the following.
   a. The normal p.d.f. takes on its maximum at $x = \mu$.
   b. The normal p.d.f. approaches 0 as $x \to \infty$ and as $x \to -\infty$.
   c. The normal p.d.f. has points of inflection at $x = \mu + \sigma$ and $x = \mu - \sigma$.

9. Consider the following game (called the Petersburg game or Petersburg paradox). The first player flips a coin. If the first head is on the first toss, she pays the second player $1. If the first head is on the second toss, she pays the second player $2. In general, if the first head is on the nth toss, she pays the second player $$2^{n-1}$.
   a. Find the expected payoff to the second player. (Use the geometric distribution to find the probability of the first head being on the nth toss, and then add the probabilities times the payoffs.)
   b. In principle, a person should be willing to pay any amount of money to enter the game as the second player. How much would you be willing to pay for this opportunity?
   c. Why are you unwilling to pay an arbitrarily large amount of money?

10. Show that $g(x) = (3/2)x^{-2.5}$ defined for $1 \le x \le \infty$ is a p.d.f. (it integrates to 1), and that a random variable with this p.d.f. has a finite mean and an infinite variance.

11. **COMPUTER**: Two values often used to describe distributions are the **skewness** and the **kurtosis**. Suppose a random variable has p.d.f. $f(x)$ for $-\infty < x < \infty$ and mean μ. The skewness k_3, which describes symmetry, has a formula like the variance,

$$k_3 = E[(X - \mu)^3] = \int_{-\infty}^{\infty} (x - \mu)^3 f(x)\, dx$$

Instead of taking the difference from the mean and squaring it, we take the difference from the mean and *cube* it. This is sometimes called the **third moment around the mean**. Symmetric distributions have a skewness of 0.

The kurtosis k_4 is based on the **fourth moment around the mean**. However, it includes a correction factor involving the variance σ^2,

$$k_4 = E[(X - \mu)^4] - 3\sigma^4 = \int_{-\infty}^{\infty} (x - \mu)^4 f(x)\, dx - 3\sigma^4$$

The correction was added to make the kurtosis of a normal distribution equal to 0.

  a. Find the skewness and kurtosis of a normal distribution with mean 0 and variance 1.
  b. Find the skewness and kurtosis of a normal distribution with mean 1 and variance 2.
  c. Find the skewness and kurtosis of an exponential distribution with mean 1.0 (remember that the random variable in this case must be positive). The skewness is positive because the distribution is stretched out to the right.
  d. Find the skewness and kurtosis of a uniform distribution with p.d.f. $f(x) = 1$ for $0 \le x \le 1$. The kurtosis is positive because the distribution has "broad shoulders."

12. **COMPUTER**: Program your computer to choose random numbers from an exponential distribution with mean 1.
  a. Pick 100 such numbers, and put them into 15 bins (the numbers between 0 and 0.2, the number between 0.2 and 0.4, etc.). Plot a histogram of how many numbers are in each bin.
  b. Pick 100 pairs of random numbers, and take the average of each pair. Put these values into bins and plot a histogram.
  c. Pick 100 sets of 10 random numbers, and take the average of each set. Put these values into bins and plot a histogram.
  d. Pick 100 sets of 30 random numbers, and take the average of each set. Put these values into bins and plot a histogram. Do the results look more and more like a normal distribution?

7.9 Applying the Normal Approximation

Here we put the Central Limit Theorem and the normal distribution to work to estimate some actual probabilities. Because the normal distribution is impossible to integrate, we first convert questions about normally distributed random variables into questions about the **standard normal distribution**, amenable to evaluation by computer or table. We then find the normal distribution approximating a given binomial or Poisson random variable, using the **continuity correction** to adjust for the fact that the normal distribution is continuous whereas the binomial and Poisson distributions are discrete.

The Standard Normal Distribution

The **standard normal distribution** is a normal distribution with mean 0 and variance 1, or $N(0, 1)$. The p.d.f. is

$$f(x; 0, 1) = \frac{1}{\sqrt{2\pi}} e^{-x^2/2} \tag{7.17}$$

found by substituting $\mu = 0$ and $\sigma^2 = 1$ into the definition of the normal p.d.f. (Definition 7.10). The standard normal p.d.f. is plotted in Figure 7.42.

The standard normal distribution is useful because tables and computers are set up to compute probabilities based on this distribution. Let Z be a random variable with a standard normal distribution. To compute the probability that $-1 \le Z \le 2$, we could either integrate the normal probability density function or

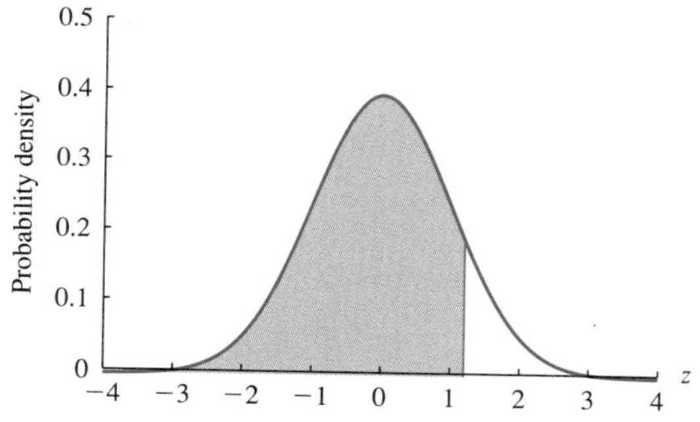

Figure 7.42
The standard normal curve: shaded area is equal to $\Phi(z)$

z	$\Phi(z)$
-4.0	0.00003
-3.0	0.00135
-2.0	0.02275
-1.0	0.15866
0.0	0.50000
1.0	0.84134
2.0	0.97725
3.0	0.99865
4.0	0.99997

use the cumulative distribution function. Although a formula for the c.d.f. of the standard normal distribution cannot be written, the values are widely available in tables and on computers. This cumulative distribution function, denoted by $\Phi(z)$, is defined by

$$\Phi(z) = \Pr(Z \le z) = \int_{-\infty}^{z} \frac{1}{\sqrt{2\pi}} e^{-x^2/2} dx \qquad (7.18)$$

The area associated with this definite integral is shown in Figure 7.42. A few values of this function are given in the table (with more detail in the inside back cover). The probability that Z is less than -1.0 is

$$\Pr(Z \le -1.0) = \Phi(-1.0) \approx 0.1587$$

and the probability that Z is less than 2.0 is

$$\Pr(Z \le 2.0) = \Phi(2.0) \approx 0.9772$$

Relatively few values are less than one standard deviation below the mean of the standard normal distribution, and almost all values are less than two standard deviations above the mean.

We find the probability that Z lies between two values by subtracting:

$$\Pr(z_1 \le Z \le z_2) = \int_{z_1}^{z_2} \frac{1}{\sqrt{2\pi}} e^{-x^2/2} dx = \Phi(z_2) - \Phi(z_1) \qquad (7.19)$$

from the Fundamental Theorem of Calculus (Section 6.7). The associated area is illustrated in Figure 7.43. For example, the probability that a measurement on the standard normal distribution lies between -1.0 and 2.0 is

$$\Pr(-1.0 \le Z \le 2.0) = \int_{-1.0}^{2.0} \frac{1}{\sqrt{2\pi}} e^{-x^2/2} dx$$
$$= \Phi(2.0) - \Phi(-1.0)$$
$$= 0.9772 - 0.1597 = 0.8175$$

Most random variables with normal distributions have neither mean 0 nor variance 1. How can we transform a problem about a random variable X with mean μ and variance σ^2 into a problem about the standard normal distribution?

■ **THEOREM 7.23** Suppose $X \sim N(\mu, \sigma^2)$. Then the random variable

$$Z = \frac{X - \mu}{\sigma}$$

has a standard normal distribution. ■

Figure 7.43
The standard normal curve: shaded area is $\Phi(z_2) - \Phi(z_1)$

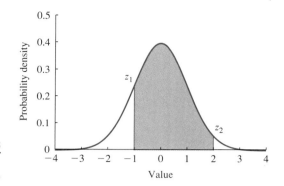

Proof: We know that

$$\Pr(X \le x) = \int_{-\infty}^{x} \frac{1}{\sqrt{2\pi}\sigma} e^{-(y-\mu)^2/2\sigma^2} dy$$

By the definition of Z, we have

$$\Pr(X \le x) = \Pr\left(\frac{X-\mu}{\sigma} \le \frac{x-\mu}{\sigma}\right) = \Pr\left(Z \le \frac{x-\mu}{\sigma}\right)$$

We need to show that

$$\Pr\left(Z \le \frac{x-\mu}{\sigma}\right) = \Phi\left(\frac{x-\mu}{\sigma}\right)$$

We apply the change of variables,

$$z = \frac{x-\mu}{\sigma}$$

Then $dx = \sigma dz$, and the limits of integration become $z = -\infty$ to $z = (x-\mu)/\sigma$. Therefore, we have

$$\Pr(X \le x) = \int_{-\infty}^{x} \frac{1}{\sqrt{2\pi}\sigma} e^{-(y-\mu)^2/2\sigma^2} dy$$

$$= \int_{-\infty}^{(x-\mu)/\sigma} \frac{1}{\sqrt{2\pi}\sigma} e^{-z^2/2} \sigma \, dz$$

$$= \int_{-\infty}^{(x-\mu)/\sigma} \frac{1}{\sqrt{2\pi}} e^{-z^2/2} dz$$

$$= \Phi\left(\frac{x-\mu}{\sigma}\right) \qquad \blacksquare$$

Suppose $X \sim N(\mu, \sigma^2)$ and we wish to compute $\Pr(a \le X \le b)$. We use the following algorithm.

■ **Algorithm 7.2** (Converting into a problem about the standard normal distribution)

Suppose that $X \sim N(\mu, \sigma^2)$. Then

$$\Pr(X \le b) = \Phi\left(\frac{b-\mu}{\sigma}\right)$$

7.9 Applying the Normal Approximation

$$\Pr(a \leq X) = 1 - \Phi\left(\frac{a-\mu}{\sigma}\right)$$

$$\Pr(a \leq X \leq b) = \Phi\left(\frac{b-\mu}{\sigma}\right) - \Phi\left(\frac{a-\mu}{\sigma}\right)$$

∎

(See Figure 67.3.) Be careful to divide by the standard deviation σ, not by the variance σ^2. This algorithm is based on Theorem 7.23 because

$$\Pr(a \leq X \leq b)$$

$$= \Pr\left(\frac{a-\mu}{\sigma} \leq \frac{X-\mu}{\sigma} \leq \frac{b-\mu}{\sigma}\right) \quad \text{subtract mean and divide by standard deviation}$$

$$= \Pr\left(\frac{a-\mu}{\sigma} \leq Z \leq \frac{b-\mu}{\sigma}\right) \quad \text{the transformed } X \text{ has a standard normal distribution}$$

$$= \Phi\left(\frac{b-\mu}{\sigma}\right) - \Phi\left(\frac{a-\mu}{\sigma}\right) \quad \text{the definition of } \Phi$$

The values $(a-\mu)/\sigma$ and $(b-\mu)/\sigma$, sometimes called **z-scores**, should be thought of as the difference from the mean measured in units of standard deviations. A value 2.5 standard deviations above the mean is equally unlikely for any normal distribution.

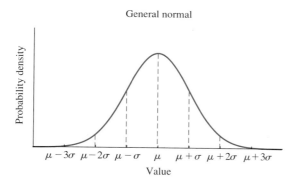
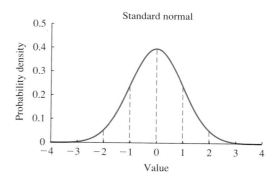

Figure 7.44
A general normal distribution and the standard normal distribution

We found the added height H of plants to be approximately normally distributed with mean 11.25 and variance 7.97 (Figure 7.37). The standard deviation σ is $\sqrt{7.97} \approx 2.823$. The probability that a plant gains less than 6 cm in height is

$$\Pr(H \leq 6) = \Pr\left(\frac{H-11.25}{2.823} \leq \frac{6-11.25}{2.823}\right)$$

$$= \Pr\left(Z \leq \frac{6-11.25}{2.823}\right)$$

$$= \Pr(Z \leq -1.86) = \Phi(-1.86) \approx 0.031$$

(Figure 7.45). Only about 3% of plants will gain less than 6 cm from the ten genes. Using Algorithm 7.2, we find that

$$\Pr(H \leq 6) = \Phi\left(\frac{6-11.25}{2.823}\right) = \Phi(-1.86) = 0.031$$

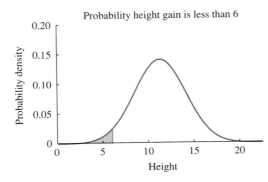
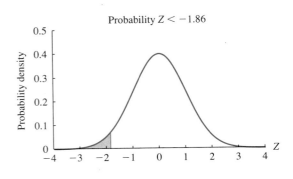

Figure 7.45
Computing the probability that height is less than 6.0 cm

The probability that a plant gains between 10 and 15 cm is

$$\Pr(10 \leq H \leq 15) = \Pr\left(\frac{10 - 11.25}{2.823} \leq \frac{H - 11.25}{2.823} \leq \frac{15 - 11.25}{2.823}\right)$$

$$= \Pr\left(\frac{10 - 11.25}{2.823} \leq Z \leq \frac{15 - 11.25}{2.823}\right)$$

$$= \Pr(-0.443 \leq Z \leq 1.328)$$

$$= \Phi(1.328) - \Phi(-0.443) = 0.908 - 0.329 = 0.579$$

(Figure 7.46). Using the shorter method of Algorithm 7.2, we get

$$\Pr(10 \leq H \leq 15) = \Phi\left(\frac{15 - 11.25}{2.823}\right) - \Phi\left(\frac{10 - 11.25}{2.823}\right)$$

$$= \Phi(1.328) - \Phi(-0.443)$$

$$= 0.908 - 0.329 = 0.579$$

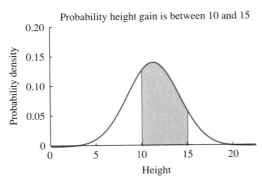
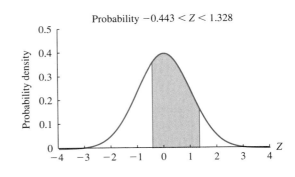

Figure 7.46
Computing the probability that height is between 10.0 and 15.0 cm

Finally, the probability that a plant gains more than 19 cm is

$$\Pr(H > 19) = 1 - \Pr(H \leq 19)$$

$$= 1 - \Pr\left(\frac{H - 11.25}{2.823} \leq \frac{19 - 11.25}{2.823}\right)$$

$$= 1 - \Pr\left(Z \leq \frac{19 - 11.25}{2.823}\right)$$

$$= 1 - \Pr(Z \leq 2.745) = 1 - \Phi(2.745) = 0.003$$

Only 0.3% of plants gain more than 19 cm (Figure 7.47). Using Algorithm 7.2, we get

$$\Pr(H > 19) = 1 - \Pr(H \leq 19) = 1 - \Phi(2.745) = 0.003$$

Figure 7.47
Computing the probability that height is greater than 19 cm

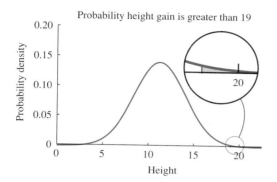
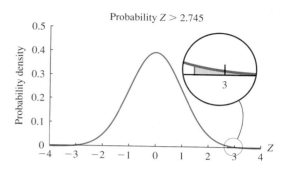

Normal Approximation of the Binomial Distribution

Suppose a heterozygous plant self-pollinates and produces 20 offspring. What is the probability that between 8 and 12 offspring are heterozygous? The number follows a binomial distribution with $n = 20$ and $p = 0.5$. Finding the probability directly requires computing a lot of factorials. Because the number of heterozygous offspring is a *sum*, we can use the Central Limit Theorem for Sums and the normal approximation. Each individual offspring can be described by a Bernoulli random variable with $p = 0.5$:

$$I_i = \begin{cases} 1 & \text{if heterozygous (probability 0.5)} \\ 0 & \text{if homozygous (probability 0.5)} \end{cases}$$

The total number of heterozygotes H is the sum

$$H = \sum_{i=1}^{20} I_i$$

and is therefore a binomial random variable with

$$E(H) = np = 10.0$$
$$\text{Var}(H) = np(1-p) = 5.0$$

As long as the I_i are *independent* (which will be the case if the offspring are not twins), we can apply the Central Limit Theorem for Sums. The probability distribution for H approximately matches that for a normally distributed random variable X with the same mean and variance, or

$$X \sim N(10.0, 5.0)$$

(Figure 7.48).

Although H and X appear to be quite similar, H is a discrete random variable and X is a continuous random variable. We can use this fact to find an estimate better than

$$\Pr(8 \leq H \leq 12) \approx \Pr(8 \leq X \leq 12)$$

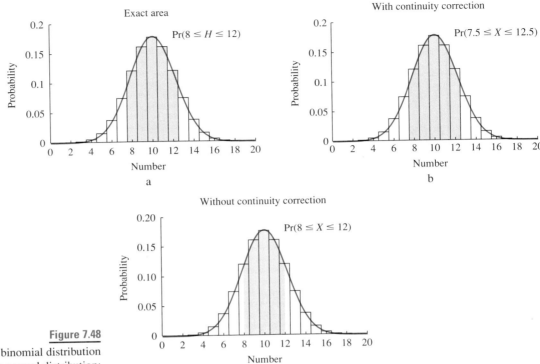

Figure 7.48
Approximating a binomial distribution with a normal distribution: the continuity correction

How can we correct for this difference? The probability that between 8 and 12 offspring are heterozygous is the sum of the areas of the bars centered at 8 through 12 (Figure 7.48a). These bars run from 7.5 to 12.5. To find the area under the normal distribution that approximates the area under the bars, we integrate from 7.5 to 12.5. The extra 0.5s are called the **continuity correction**. With this correction, we get

$$\Pr(8 \leq H \leq 12) \approx \Pr(7.5 \leq X \leq 12.5)$$
$$= \Phi\left(\frac{7.5 - 10}{\sqrt{5}}\right) - \Phi\left(\frac{12.5 - 10}{\sqrt{5}}\right)$$
$$= \Phi(1.118) - \Phi(-1.118)$$
$$= 0.8682 - 0.1318 = 0.7364$$

(Figure 7.48b). The exact answer, found on the computer by adding the binomial probabilities, is 0.7368. The approximation is extremely close. Without the continuity correction, we have

$$\Pr(8 \leq H \leq 12) \approx \Pr(8 \leq X \leq 12)$$
$$= \Phi\left(\frac{8 - 10}{\sqrt{5}}\right) - \Phi\left(\frac{12 - 10}{\sqrt{5}}\right)$$
$$= \Phi(0.894) - \Phi(-0.894)$$
$$= 0.8144 - 0.1855 = 0.7289$$

which is a bit low (Figure 7.48c).

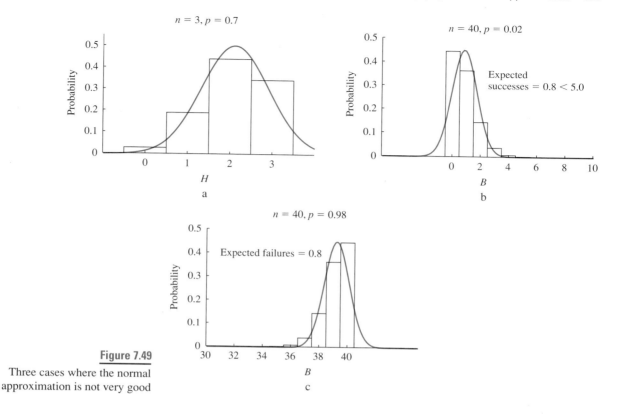

Figure 7.49
Three cases where the normal approximation is not very good

Like the Central Limit Theorem, the normal approximation to the binomial distribution works only when the number of terms is sufficiently large. For example, the binomial distribution with $n = 3$ and $p = 0.7$ does not have much of a bell shape (Figure 7.49a). The normal approximation is not accurate. Furthermore, if p is near 0 (Figure 7.49b) or 1 (Figure 7.49c), we are keeping track of rare events (successes in the first case and failures in the second). Once again, we are not counting enough events for the Central Limit Theorem to apply. To ensure an expectation of at least five successes *and* five failures, we use the following condition.

Condition for Applying the Normal Approximation to the Binomial Distribution:
Apply the approximation distribution to a binomial distribution with parameters n and p only when $np \geq 5$ and $n(1 - p) \geq 5$.

■ **Algorithm 7.3** (Applying the normal approximation to the binomial distribution)

Suppose B has a binomial distribution with parameters n and p satisfying $np \geq 5$ and $n(1 - p) \geq 5$.

1. Define X to be a normally distributed random variable with

$$\mu = E(B) = np$$
$$\sigma^2 = \text{Var}(B) = np(1 - p)$$

2. Then for integers k_1 and k_2,

$$\Pr(B \leq k_2) \approx \Pr(X \leq k_2 + 0.5)$$
$$\Pr(k_1 \leq B) \approx \Pr(k_1 - 0.5 \leq X)$$
$$\Pr(k_1 \leq B \leq k_2) \approx \Pr(k_1 - 0.5 \leq X \leq k_2 + 0.5)$$

To approximate the probability of exactly eight heterozygous offspring, use the algorithm with $k_1 = 8$ and $k_2 = 8$:

$$\Pr(H = 8) \approx \Pr(7.5 \leq X \leq 8.5)$$
$$= \Phi\left(\frac{7.5 - 10}{\sqrt{5}}\right) - \Phi\left(\frac{8.5 - 10}{\sqrt{5}}\right)$$
$$= \Phi(-0.671) - \Phi(-1.118)$$
$$= 0.2511 - 0.1317 = 0.1194$$

The exact answer is $b(8; 12, 0.5) = 0.1201$. Similarly, to find the probability of 15 or more:

$$\Pr(H \geq 15) \approx \Pr(X \geq 14.5)$$
$$= 1 - \Phi\left(\frac{14.5 - 10}{\sqrt{5}}\right)$$
$$= 1 - \Phi(2.012)$$
$$= 1 - 0.9779 = 0.0221$$

Normal Approximation of the Poisson Distribution

Suppose seeds land in a region according to a Poisson process with a mean number of $\Lambda = 23$ per square meter (as in Figure 7.32). What is the probability that between 18 and 28 seeds fall in a given square meter? Finding the exact probability with the Poisson distribution requires a lot of calculation. Again, we can simplify this calculation with the normal approximation to the Poisson distribution.

The Poisson distribution with $\Lambda = 1.0$ (Figure 7.50a) is not at all bell-shaped, becomes reasonably so when $\Lambda = 4.5$ (Figure 7.50b), and quite so when $\Lambda = 23$

Figure 7.50
Poisson distribution and normal approximation for small values of Λ

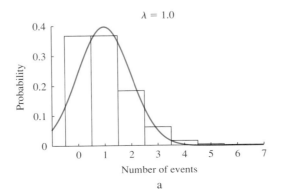

a

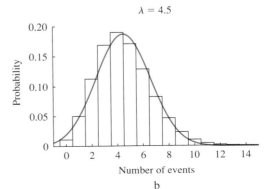

b

(Figure 7.33). To be safe, use the normal approximation to the Poisson only when $\Lambda \geq 5$, or

Condition for Applying the Normal Approximation to the Poisson Distribution:
Apply the normal approximation to a Poisson distribution with parameter Λ only when $\Lambda \geq 5$.

■ **Algorithm 7.4** (Applying the normal approximation to the Poisson distribution)

Suppose P has a Poisson distribution with parameters $\Lambda \geq 5$.

1. Define X to be a normally distributed random variable with
$$\mu = E(P) = \Lambda$$
$$\sigma^2 = \text{Var}(P) = \Lambda$$

2. Then, for integers k_1 and k_2,
$$\Pr(P \leq k_2) \approx \Pr(X \leq k_2 + 0.5)$$
$$\Pr(k_1 \leq P) \approx \Pr(k_1 - 0.5 \leq X)$$
$$\Pr(k_1 \leq P \leq k_2) \approx \Pr(k_1 - 0.5 \leq X \leq k_2 + 0.5)$$

The idea of the algorithm is precisely that for approximating the binomial distribution (Algorithm 7.3): Find a normal random variable X with matching mean and variance, and compute probabilities using the **continuity correction**.

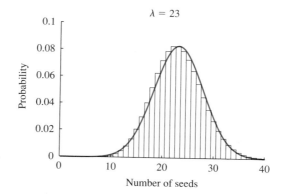

Figure 7.51
The Poisson distribution and the normal distribution

To estimate the probability that there are between 18 and 28 seeds in a given square meter, we first find the normal approximation,
$$X \sim N(23, 23)$$
by matching the mean and variance (Figure 7.51). To find the probability, we use the continuity correction and integrate the normal p.d.f. from 17.5 to 28.5. Let S be a random variable representing the number of seeds. Then
$$\Pr(18 \leq S \leq 28) \approx \Pr(17.5 \leq X \leq 28.5)$$
$$= \Phi\left(\frac{17.5 - 23}{\sqrt{23}}\right) - \Phi\left(\frac{28.5 - 23}{\sqrt{23}}\right)$$

$$= \Phi(1.147) - \Phi(-1.147)$$
$$= 0.8743 - 0.1257 = 0.7486$$

The exact answer is 0.7498.

SUMMARY We transformed a general normally distributed random variable into a **standard normal distribution** by subtracting the mean and dividing by the standard deviation. Because the probability density function for the standard normal distribution is impossible to integrate, we use the cumulative distribution function Φ, which is widely available in tables and computers. To approximate discrete random variables described by the binomial and Poisson distributions, we find a normal distribution with matching mean and variance, and compute probabilities without forgetting the **continuity correction**, extending the limits of integration by 0.5 on each side.

7.9 Exercises

1. Using a table or a program of the cumulative distribution function for the standard normal distribution, find the following probabilities.
   a. The masses of a type of insect are normally distributed, with a mean of 0.38 g and a standard deviation of 0.09 g. What is the probability that a given insect has mass less than 0.40 g?
   b. Scores on a test are normally distributed, with mean 70 and standard deviation 10. What is the probability that a student scores more than 85?
   c. Measurement errors are normally distributed, with a mean of 0 and a standard deviation of 0.01 mm. Find the probability that a given measurement is within 0.015 mm of the true value.
   d. The number of insects captured in a trap on different nights is normally distributed with mean 2950 and standard deviation 550. What is the probability of capturing between 2500 and 3500 insects?

2. Use the normal approximation to the binomial distribution with and without the continuity correction to find the following probabilities. How much difference does the continuity correction make? Exact answers are given in parentheses.
   a. Forty-three percent of trees are infested by a certain insect. What is the chance of randomly choosing 40 trees, fewer than 10 of which are infested? (0.0144).
   b. Thirty percent of the cells in a small organism are not functioning. What is the probability that an organ consisting of 250 cells is functioning if it requires 170 cells to work? (0.777).
   c. In a certain species of wasp, 75% of individuals are female. Find the probability that between 70 and 80 are female in a sample of 100. (0.7967).
   d. Ten percent of people are known to carry a certain gene. In a sample of 200, what is the probability that between 5 and 15 carry the gene? (0.1431).

3. Use the normal approximation to the Poisson distribution to find the following probabilities. Exact answers are given in parentheses.
   a. Insects are caught in a trap at a rate of 0.21 per minute. What is the probability of catching between 10 and 15 in an hour? (0.6040).
   b. Molecules enter a cell at a rate of 0.045 per second. What is the probability that between 150 and 200 enter during an hour? (0.8352).
   c. Seeds have fallen into a region with an average density of 20 seeds per square meter. What is the probability that a particular square meter contains fewer than 15 seeds? (0.1565).
   d. Mutations occur along a piece of DNA at a rate of 0.023 per thousand codons. What is the probability of 30 or more mutations in a piece of DNA 1 million codons long? (0.0915).

4. The normal approximation cannot be used in the following cases. Why not? Find another way to compute the probabilities and compare with the normal approximation.
   a. A fair coin is flipped eight times. Find the probability of no more than one head.
   b. Five percent of people are infected with a disease. Find the probability of choosing 50 people and finding none who are infected.
   c. Refer to Exercise 3a. Find the probability of catching more than one insect in a minute.
   d. Refer to Exercise 3d. Find the probability of no mutation in a piece of DNA 10,000 codons long.

5. Suppose that dart scores are independent and follow the distribution in Figure 6.49. Using the results of Exercise 4 in Section 7.5, find the following.
   a. The probability that the average of 5 scores is between 10 and 15.

b. The probability that the average of 10 scores is between 10 and 15.
c. The probability that the average of 100 scores is between 10 and 15.
d. Would you expect these probabilities to be larger or smaller if dart scores were positively correlated (large scores tending to follow large scores and vice versa)? Why?
e. What if the scores were negatively correlated?

6. Suppose cosmic rays hit an organism at a rate of 1.2 per day. Use the normal approximation to find the following.
   a. An organism is killed by more than 10 hits in a week. What is the probability it survives?
   b. What is the probability it gets exactly 10 hits in a week?
   c. An organism is killed by more than 40 hits in a 30-day month. What is the probability it survives?
   d. What is the probability it gets exactly 40 hits in a month?

7. Consider again the situation in Exercise 5 in Section 7.8.
   a. What is the probability of an IQ above 100 with just the genes of large effect?
   b. What is the probability of an IQ above 100 with just the genes of small effect?
   c. What is the probability of an IQ above 100 with just environmental factors?
   d. What is the probability of an IQ above 100 with both genetics and environemental factors?

8. In Exercise 6 in Section 7.8, we found the normal approximation for the log of population size in two populations. Use it to find the following.
   a. The probability that population 1 is less than 1.0 at time 10.
   b. The probability that population 2 is less than 1.0 at time 10.
   c. What does this say about the two populations?
   d. The probability that population 1 is greater than 10.0 at time 20.
   e. The probability that population 2 is greater than 10.0 at time 20.
   f. What does this say about the two populations? Is it consistent with the results in parts a and b?

9. If
$$X \sim N(\mu, \sigma^2) \quad \text{and} \quad Z = \frac{X - \mu}{\sigma}$$
show, without writing an integral, that $E(Z) = 0$ and $\text{Var}(Z) = 1$.

10. **COMPUTER**: Generate 100 numbers from a binomial distribution with $n = 20$ and $p = 0.3$.
    a. Plot the theoretical histogram for this binomial distribution, and overlay it with the normal approximation.
    b. Plot a histogram of your results.
    c. Overlay the historgram with a graph of the normal distribution with matching mean and variance. Did you have to modify the formula to make the areas match?
    d. Compare the fraction of times your random number was equal to 4 with the fraction based on the binomial distribution. Shade the region on your graph of the theoretical histogram from part **a**, and compare with the area on the normal curve with the continuity correction.
    e. Follow the steps in **d** for the number of times your random number was between 7 and 10, inclusive.

11. **COMPUTER**: Recall the growth of populations in Exercise 6 in Section 7.8. Let R_1 be a random variable giving the per capita reproduction in one population with p.d.f. $g(x) = 5.0$ for $1.0 \leq x \leq 1.2$, and let R_2 be a random variable giving the per capita reproduction in a population with p.d.f. $g(x) = 1.25$ for $0.7 \leq x \leq 1.5$. We found that $E[\ln(R_1)] = 0.0939$ and $E[\ln(R_2)] = 0.0723$.
    a. Simulate each population for 50 steps, starting from an initial condition of 100.
    b. How do the sizes compare with the expectation?
    c. Simulate each population 100 times, and record the resulting values. Place the values into 15 bins, and plot a histogram of the results. Do the results seem to have a normal distribution?
    d. Now take the natural log of each final population size, and plot a histogram. Why do these results have a more nearly normal distribution? The actual population sizes are said to have a **lognormal distribution**.

Supplementary Problems for Chapter 7

EXERCISE 1
A certain student always arrives at office hours the day before the test clutching the supplementary problems and asks about each with probability 0.6.
   a. Under what conditions is the binomial distribution appropriate?
   b. Under these conditions, what is the probability that the first problem the student asks about is Problem 4?
   c. What is the probability that this student asks about exactly four of the first six?
   d. What are the expectation and variance of the number asked about if there are eight supplementary problems?

EXERCISE 2
A second student frequently appears after the first has left. This student also asks about 60% of the problems, but he asks about

a particular problem with probability 0.8 if the first student did not.

    a. Is the behavior of these students independent? How could you tell?
    b. Give the entire joint distribution. What is the probability that neither student asks about Problem 5?
    c. Draw the marginal distribution for the second student.
    d. Draw the conditional distributions for the second student.
    e. Find the covariance if asking is worth -1 and not asking is worth 2.

EXERCISE 3
During a drizzle, raindrops hit a given square meter at a rate of 12.5 per minute.

    a. What is the expected number of raindrops after 30 s?
    b. What is the probability that exactly five raindrops hit in 30 s?
    c. What is the probability that no raindrops hit in 10 s?
    d. If the drizzle lasts 1 h, would you be surprised to find a square meter with exactly 700 wet spots?

EXERCISE 4
During a particularly wet spring, a rainy day follows a rainy day with probability 0.8, and a rainy day follows a dry day with probability 0.5.

    a. Write an updating function to describe this model.
    b. Draw the distributions that describe the weather on one day conditional on the weather of the previous day.
    c. Find the fraction of rainy days after a long time.
    d. Draw the joint distribution of weather on consecutive days after a long time.

EXERCISE 5
A band is composed of 4 guitarists and a drummer. The drummer knows only how to keep a steady beat, and each guitarist knows only the A and D chords. Each time the drummer hits a beat, each guitarist plays an A chord with probability 0.75 and a D chord with probability 0.25.

    a. Under what conditions is the binomial distribution appropriate for describing the number of A chords played on a given beat?
    b. What are the expectation and variance of the number of guitarists playing an A chord on a given beat?
    c. If the binomial distribution is appropriate, find the probability that exactly two guitarists hit the A chord.
    d. Find the probability that all four guitarists play the same chord.
    e. Find the probability that all four guitarists play the same chord on exactly two out of three consecutive beats.

EXERCISE 6
People attending a concert by the band of Exercise 5 are observed to be leaving at a rate of three per minute.

    a. Under what conditions is the Poisson distribution appropriate for describing the number of people who left during a given minute?
    b. Under these conditions, find the probability that nobody leaves during a period of length T, and sketch the function.
    c. Find the expected number and the variance of people leaving during a given minute.
    d. Find the probability that exactly three people leave during a given minute.
    e. Find the probability that exactly three people leave during a 2-min interval.

EXERCISE 7
Refer to Exercise 6. The remaining throngs show wild enthusiasm after a song with probability 0.3, mild enthusiasm with probability 0.2, and hurl insults with probability 0.5. The band reduces the volume by 20% in response to wild enthusiasm, by 10% in response to mild enthusiasm, and increases volume by 20% in response to insults. Suppose the volume begins at 100.

    a. Find the expected volume after one song.
    b. Would you expect the band to be louder or quieter at the end of a long concert?

EXERCISE 8
After a particularly demoralizing performance, two guitarists decide to start working together. Recall that they know only two chords, A and D.

    a. If they began by playing chords independently, give the joint distribution.
    b. If the second player plays an A with probability 90% when the first player does, give the joint distribution.
    c. If playing an A is worth 1 point and playing a D is worth 0 points, find the correlation.
    d. Draw the marginal and conditional distributions.

EXERCISE 9
An aspiring worm biologist decides to use a random procedure to stock his lab. He chooses among three catalogues, Baxter, Fisher, and Sigma. He orders from Baxter with probability 0.2, from Fisher with probability 0.7, and from Sigma with probability 0.1. Products are delivered with probability 0.9 from Baxter, 0.95 from Fisher, and 0.8 from Sigma.

    a. Find the probability that a product is delivered.
    b. Find the probability that a product that was not delivered was ordered from Fisher.
    c. Find the joint distribution.
    d. What strategy would maximize the probability of delivery?

EXERCISE 10
A population switches between phases of growth (G) and decline (D) in subsequent years according to $\Pr(G \mid D) = 0.15$ and $\Pr(D \mid G) = 0.05$. The per capita growth rate is 1.2 during growth and 0.7 during decline.

    a. If the population grows in the first year, find the expected size after 2 yr.
    b. If the population grows in the first year, what is the expected time until the first decline? Sketch the distribution.

c. Will this population grow or decline in the long run?
d. Find the joint distribution and covariance between years after a long time.

EXERCISE 11
A staining technique has a 20% chance of success. Let S_n be the random variable representing the number of successful stainings in n attempts.
a. What are the conditions for S_n to have a binomial distribution? What are the parameters?
b. Find $E(S_{10})$ and $\Pr(S_{10} = 2)$.

EXERCISE 12
Under the same conditions as in Exercise 11, let T be a random variable representing the time of the first successful attempt. For example, $T = 2$ means that the first try was a failure and the second a success.
a. Graph the probability distribution of T.
b. Find the expectation and variance of T.

EXERCISE 13
The density of bacterial colonies on a circular dish of radius 3 cm is 0.12 colonies per square centimeter. Let B be a random variable describing the number of colonies on the dish.
a. Find the approximate probability of there being a colony in 1 mm².
b. Find the expectation and variance of B.
c. Find the probability of there being no colonies on the dish.

EXERCISE 14
The woodpecker *Picoides villosus* enjoys eating both the wood-boring tipulid (*Ctenophora vittata*) and the hickory bark beetle (*Scolytus quadrispinosus*). During any given minute, the woodpecker eats one tipulid with probability 0.5 and none with probability 0.5. Independently, during each minute, it eats one beetle with probability 0.25 and none with probability 0.75.
a. Sketch the distribution that describes the time when the first beetle is eaten.
b. Find the expected time, $\bar{B}$, when the first beetle is eaten.
c. Find the probability that no tipulids had been eaten at or before $\bar{B}$, and the expected number of tipulids eaten by time $\bar{B}$.

EXERCISE 15
A measurement X has normal distribution with mean 10 and variance 9, and an independent measurement Y has normal distribution with mean 20 and variance 4.
a. Find $\Pr(X \leq 11)$.
b. Find $E(X + Y)$ and $\text{Var}(X + Y)$.
c. Find $\Pr(X + Y \leq 13)$.

EXERCISE 16
Bubbles are produced by a chemical reaction at a rate of 1.8 per minute. Two assiduous students observe two bubbles during the first minute, one during the second minute, and fail to pay attention during the third minute because they are studying math. They then knuckle down and pay attention for the next 30 min.

a. What is the probability that they didn't miss anything during that third minute? What is the probability that they missed two or more bubbles?
b. What are the expectation and variance of the number of bubbles seen during the 30 min they pay attention?
c. Write the formula (but don't evaluate it) for the probability of exactly 50 bubbles in 30 min.
d. Two other students pay attention for only 15 min and multiply their results by 2. What is the expectation and variance for them? Explain the difference.

EXERCISE 17
Because students have proven unable to pay close attention to chemical reactions, the professor designs a new version of the experiment, in which bubbles do not pop. Students check for bubbles once per minute, record whether there are any (but don't count them), and pop all the ones that are there. The probability of at least one bubble during a minute is 0.83.
a. What is the probability that the first bubble is seen on the fifth check?
b. Find the expected number of students in a class of 30 who would see no bubble during the first 4 min. What is the variance? What does it mean?
c. What is the probability that a particular team of students will record bubbles in exactly 25 minutes out of 30?
d. Estimate the probability of 25 or more minutes with bubbles during 30 minutes (no factorials allowed). Sketch the probability distribution, and indicate the probability you estimated.

EXERCISE 18
Students begin measuring two numbers each minute, the number of bubbles B and the temperature T. B can take on only two values, 0 and 1, and T can take on the values 20 and 30. The joint distribution is

	$B = 0$	$B = 1$
$T = 20$	0.1	0.1
$T = 30$	0.07	0.73

a. Find the marginal distributions.
b. Find and sketch the conditional distribution of B when $T = 20$ and when $T = 30$.
c. Find the covariance, and say what you would do to make this a more useful statistic.

EXERCISE 19
Two mathematicians propose models for the growth of an elk population. The first says that the number increases by B, where B is a random variable with a Poisson distribution and mean 5. The second says that the population increases by exactly 4% each year. Suppose the actual population begins at 100 and is followed for 10 yr, and is found to be

Year t	Number of elk	Year t	Number of elk
0	100	6	134
1	103	7	153
2	104	8	155
3	115	9	151
4	111	10	156
5	118		

a. Sketch data that could result from each of the models.
b. What is the expected number of elk and the variance after 10 yr according to the models?
c. Which model predicts that the population will grow faster in the long run?
d. What are the strengths and weaknesses of the two models in making sense of the data?
e. If you had to construct a model of this population, what factors would you like to include?

Projects for Chapter 7

PROJECT 1

An important class of stochastic processes that describe many types of populations are called **birth-death processes**. In particular, suppose that a population has size $N(t)$ at some time t. The probability that a new individual is born during the short interval of time between t and $t + \Delta t$ is assumed to be $\lambda N(t)\Delta t$. In other words, births occur at a *rate* proportional to the population size. Similarly, the probability that an individual dies during the short interval of time between t and $t + \Delta t$ is assumed to be $\mu N(t)\Delta t$. Deaths also occur at a *rate* proportional to the population size.

We begin by assuming that $\mu = 0$ (no deaths). This model is called a **birth process**.

a. What is the probability that $N(t + \Delta t) = N(t) + 1$?
b. What is the probability that $N(t + \Delta t) = N(t)$?
c. Let $E(t)$ be the expected number of individuals at time t. What is $E(t + \Delta t)$? (Use the fact that the expectation of the sum is the sum of the expectations.) Write and solve a differential equation for $E(t)$. Use the initial condition $E(0) = N_0$.
d. Let $V(t)$ be the variance at time t. What is $V(t + \Delta t)$? (Use the fact that the variance of the sum of independent random variables is the sum of the variances.) Write and solve a differential equation for $V(t)$. Use the initial condition $V(0) = 0$.
e. Figure out a way to simulate this birth process (use $\lambda = 0.1$).
f. Starting from an initial condition of $N(0) = 30$, compare the values of three simulations of $N(t)$ with the values of $E(t)$ for up to $t = 20$.
g. Using the calculation of $V(t)$, draw curves two standard deviations above and below $E(t)$. Do your simulations always remain within these bounds?

Now assume that $\mu \neq 0$.

h. What is the probability that $N(t + \Delta t) = N(t) + 1$?
i. What is the probability that $N(t + \Delta t) = N(t) - 1$?
j. What is the probability that $N(t + \Delta t) = N(t)$?
k. Write and solve a differential equation for the expected population size $E(t)$ with initial condition $E(0) = N_0$. When do you expect this population to grow?
l. Write and solve a differential equation for the variance $V(t)$ with initial condition $V(0) = 0$. Does the death process increase or decrease the variance?
m. Figure out a way to simulate this birth-death process (use $\lambda = 0.1$ and $\mu = 0.08$).
n. Starting from an initial condition of $N(0) = 30$, compare the values of three simulations of $N(t)$ with the values of $E(t)$ for up to $t = 20$. Discuss your results.
o. Starting from an initial condition of $N(0) = 1$, compare the values of three simulations of $N(t)$ with the values of $E(t)$ for up to $t = 20$. Discuss your results. How does the actual population differ from $E(t)$? Draw curves two standard deviations above and below $E(t)$. Do your simulation results remain within these bounds?
p. Try the same exercise as in steps **m** through **o** with $\lambda = \mu = 0.1$. Do you expect this population to go extinct? Can you guess about how long it will take?

PROJECT 2

One of the classic experiments in genetics was designed to determine whether mutations occur in response to selection or whether they occur randomly before selection. Following an inspirational moment watching slot machines, Salvador Luria realized that random mutations would produce bacterial cultures with either very few or very many mutants, while mutations produced in response to selection would produce a more even distribution of results. This project leads you through the reasoning behind this classic work.

The bacteria in the experiment were subjected to selection by phage, a type of virus that kills any bacterium that is not resistant. The bacteria from a culture were placed on a plate imbued with high concentrations of phage. Any resistant bacterium would survive and grow, producing a colony that could be seen and counted after a day or two. Luria and Delbrück used this count to figure out the exact number of resistant bacteria out of the hundreds of millions in the culture.

The basic population dynamics are assumed to obey the following rules:

- A culture with $N(t)$ individuals at time t reproduces at per capita rate r.

- The population begins with N_0 individuals at time $t = 0$.
- The experiment runs until time $t = T$.

To begin, use these assumptions to find the number of bacteria at time T.

We can now analyze what would happen if mutations occurred in response to selection. If the $N(T)$ bacteria at the end of the experiment had never seen phage, all would be susceptible to attack and none would survive. However, if they had a chance to catch a whiff of phage and mutate before attack, they could survive and produce colonies.

   a. If the chance a bacterium smelled phage and quickly mutated was v and if each bacterium responded independently, what is the distribution of the number of surviving colonies?
   b. What is the expected number of surviving colonies?
   c. What is the variance?
   d. What is the probability of 0 surviving colonies?

The results are quite different if any mutation making bacteria resistant must have already occurred in culture. Any mutant that arose early in the growth phase would produce a large number of progeny, all of which would be resistant. Thus, a few experiments would generate huge numbers of resistent colonies, while most would generate rather few. Denote the number of mutants at time t by $M(t)$ and suppose that bacteria mutate from being susceptible to resistant at a per capita rate μ.

   e. Explain why the expected number of mutants $\overline{M}(t)$ follows the differential equation
   $$\frac{d\overline{M}}{dt} = r[\mu N(t) + \overline{M}]$$
   f. Show that
   $$\overline{M}(t) = N(t)(1 - e^{-\mu r t})$$
   is a solution [remember to use the formula you found for $M(t)$].
   g. Suppose that μrT is very small (which is generally the case). Use the tangent line approximation to show that
   $$\overline{M}(T) \approx \mu r T N(t)$$
   h. Compare this with your result when mutations occur is response to selection. Could you use these results to distinguish the two cases?
   i. Let $P(t)$ denote the probability of 0 mutants at time t. Write and solve a differential equation for $P(t)$. Could this value be used to distinguish the two cases?

The most important difference between the two cases is that the second has higher variance. The following steps will help you compute $V(t)$, the variance of $M(t)$.

   j. What is the probability that a mutation occurs between times t and $t + \Delta t$?
   k. If a mutation occurs at time t, how many offspring will it have at time T?
   l. Multiply the value times the probability and integrate to check your result for the expected number of mutants.
   m. What is the variance of the number produced between t and $t + \Delta t$? What is this approximately equal to if Δt is small?
   n. The total number of mutants is the sum of independent events occurring during all time intervals from t to $t + \Delta t$. The variance of $M(t)$ is therefore the sum (or the integral) of the variances over each interval. Compute $V(T)$.
   o. What is the ratio of $V(t)$ to $\overline{M}(T)$? Will it be large or small when $N(t)$ is large? How does this ratio compare to the case when bacteria respond to selection? Could you use this to distinguish the two cases?
   p. What is the coefficient of variation of $M(t)$? How does this compare to the case when bacteria respond to selection?

Reference Luria, S. E., and M. Delbrück. Mutations of bacteria from virus sensitivity to virus resistance. *Genetics*. 28:491–511, 1943.

PROJECT 3

(Based on "Departure time versus departure rate: how to forage optimally when you are stupid." by F. R. Adler and M. Kotar.) The Marginal Value Theorem (Section 3.3) gives the optimal time for a forager to leave a patch. However, some organisms might be unable to leave at the right moment, perhaps because they are too stupid to remember how long they have been there or have no way to assess the current resource level. Worse yet, some animals might leave a patch simply because they had become lost and could not find the way back. How might such a cognitively challenged organism forage efficiently?

Suppose that the travel time T between patches is equal to 1.0 and that the amount of food gathered by time t is $F(t) = 1 - e^{-\alpha t}$. The parameter α describes how quickly the resource is depleted by the organism.

   a. Find the optimal moment for this organism to leave the patch for different values of α (the equation cannot be solved algebraically).
   b. Suppose that the organism gets lost at rate λ. What distribution describes the time when this organism leaves?
   c. What is the average amount of food found during this time?
   d. What is the average time spent in a patch?
   e. Write a formula for the average amount of food during the total time for a visit (time in patch plus travel time).
   f. Solve for the value of λ that maximizes this (you should be able to do this without picking a particular value of α).
   g. Compare the average time spent by this stupid organism with the optimal time computed in **a** for small and large values of α. Can you make sense of the result?
   h. Think of another model of departure times (such as a normal distibution) and repeat the analysis.

Chapter 8

Introduction to Statistical Reasoning

Probability theory and the basic models of probability are part of **mathematical modeling**. Beginning with a careful description of a biological process, we use mathematical tools to compute and summarize its dynamical implications. When the biological process depends on a count, the binomial or Poisson distribution might describe the results. When the biological process depends on a waiting time, the geometric or exponential distribution would be appropriate.

As statisticians, we begin with a set of observations and try to infer the underlying process. This branch of science is called **inferential statistics**. All such reasoning begins with a model. To use data to test possible mechanisms, we formalize the biological assumptions in a probabilistic model, use mathematical methods to compute and describe the results, and then **compare** the mathematical results with the actual measurements. The following is the guiding principle behind using inferential statistics:

The key to understanding and applying statistics correctly is understanding the underlying probabilistic model.

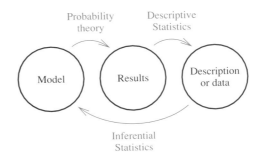

8.1 Statistics: Estimating Parameters

The set of measurements used to make statistical inferences is called a **sample**, thought of as part of a **population**. The population is the set of all possible measurements from which the sample is drawn (Figure 8.1). A common problem in statistics is to estimate characteristics of a population from a sample. The population can be a real entity (all the people in Utah) or a mathematical abstraction (all possible results of flipping a coin 1 million times).

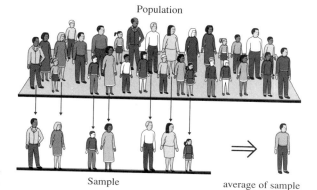

Figure 8.1
A sample taken from a larger population

Statistics (numbers computed from measurements of a sample) are used to estimate the parameters (numerical characteristics) of a population. For example, we might estimate the average height of people in Utah by choosing a sample of 100 people from the phone book and computing the average of their heights. This is an example of **parameter estimation**. Here we use the binomial distribution to help us estimate the fraction p of a population with a particular property, and we use the exponential distributions to estimate the rate λ at which events occur. The technique we use is called **maximum likelihood**.

Estimating the Binomial Proportion

Of 100 people tested for a particular gene, 20 test positive. What proportion p of the total population has the gene? The only reasonable guess seems to be $p = 0.2$. We have estimated a property of a population from a statistic describing a sample.

To work with this estimate, we must know something about the testing and sampling procedures. First, we must know how dependable the test is. Does it always give the right answer? Estimating results from an undependable test can be quite risky. Second, we would like to assume that the individuals chosen for the sample are independent and have the same value of p, thus satisfying the assumptions of the binomial distribution.

When might our sample fail to satisfy the assumptions? If the gene is most common in people with Japanese ancestry and our sample came from an area with many such people, we would tend to overestimate the true proportion. This sample does not accurately reflect the population as a whole. A similar problem arises if we choose many related individuals. In either case, the sample is not composed

Chapter 8 Introduction to Statistical Reasoning

of independent individuals. If one brother has the gene it is much more likely that another does. Choosing a good sample is difficult (techniques are covered in texts on **statistical design**). In general, samples should be chosen "randomly," based on criteria unrelated to any factor influencing the measurement.

Suppose we trust our test and our sample. Our guess that $p = 0.2$ is called an **estimator**.

■ **Definition 8.1** An estimator is a value of some parameter describing a population computed from a sample. ■

Estimated values are generally denoted with a "hat." For example, the estimator $\hat{p}$ is defined by

$$\hat{p} = \frac{\text{number of people with gene}}{\text{number of people tested}}$$

Sample	Number with gene	$\hat{p}$
1	14	0.14
2	20	0.20
3	17	0.17
4	18	0.18
5	16	0.16
6	15	0.15
7	22	0.22
8	14	0.14
9	20	0.20
10	16	0.16

The estimator $\hat{p}$ is a guess of the true proportion p in the whole population. It would be rather naive to believe that p is *exactly* equal to $\hat{p}$. If the total population consists of 1,782,260 people, there would have to be 356,452 people with the gene for p to be equal to $\hat{p}$. If there were 355,987 people with the gene, p would be 0.1997, close enough to the estimator for any practical purpose. Part of the process of finding an estimator is to derive a range around $\hat{p}$ that we can be *confident* contains the true p, as we will study in Section 8.2.

One way to evaluate whether an estimator is good is to check whether it gives the right answer *on average*. Suppose that the true proportion were $p = 0.18$. If we drew ten samples of 100, we might get the results shown in the table. The value of the estimator is different for each sample. The average of the estimators is 0.172, not exactly equal to the true value, but closer than many of the individual estimators.

■ **Definition 8.2** An estimator $\hat{p}$ is **unbiased** if

$$E(\hat{p}) = p$$

where the expectation is taken over samples from the population. An estimator is **biased** if

$$E(\hat{p}) \neq p$$ ■

Is the estimator $\hat{p}$ unbiased? The number of people G with the gene follows a binomial distribution with $n = 100$ and $p = 0.18$. The data in the table are ten such values. Our estimator is

$$\hat{p} = \frac{G}{100}$$

Therefore,

$$E(\hat{p}) = \frac{E(G)}{100} \quad \text{divide out constant, by Theorem 7.6}$$

$$= \frac{0.18 \cdot 100}{100} \quad E(G) = np \text{ because } G \text{ has a binomial distribution}$$

$$= 0.18 = p \quad \text{the expectation of the estimator is the true value}$$

Thus $\hat{p}$ does gives the right answer on average, and the estimator is indeed unbiased.

Maximum Likelihood

We found our estimate $\hat{p}$ of the true proportion using "common sense." The method of **maximum likelihood** uses the uncommon sense derived from understanding the underlying probabilistic model. Unlike common sense, this method works as well on arcane counterintuitive models as it does on the stuff of everyday life.

The method of maximum likelihood is based on a series of thought experiments. We ask how *likely* it is we would have gotten the data we did, 20 out of 100, with different possible values of the unknown parameter p. For example, suppose we guessed that the true proportion in the population is $p = 0.18$. Is this plausible? We quantify this by computing the probability of finding 20 out of 100 with the gene. If $p = 0.18$, the number G with the gene has a binomial distribution with $n = 100$ and $p = 0.18$, so

$$\Pr(20 \text{ out of } 100 \text{ with } p = 0.18) = b(20; 100, 0.18) = 0.0870$$

(Figure 8.2). This does not seem very likely. Does this mean that the proportion $p = 0.18$ is not plausible?

The method of maximum likelihood addresses the question by defining and analyzing the **likelihood function**.

■ **Definition 8.3** Suppose a population is described by an unknown parameter p, and that data have been collected from a particular sample. Then

$$L(p) = \Pr(\text{data if parameter is equal to } p)$$ ■

For example, with a data set consisting of 20 out of 100 testing positive for the gene, we have

$$L(0.18) = \Pr(20 \text{ out of } 100 \text{ with } p = 0.18) = b(20; 100, 0.18) = 0.0870$$

Several values of the likelihood function for this data set are given in the following table.

Guess of p	$L(p)$ = probability of 20 out of 100
0.10	$b(20; 100, 0.10) = 0.0012$
0.12	$b(20; 100, 0.12) = 0.0074$
0.14	$b(20; 100, 0.14) = 0.0258$
0.16	$b(20; 100, 0.16) = 0.0567$
0.18	$b(20; 100, 0.18) = 0.0870$
0.20	$b(20; 100, 0.20) = 0.0993$
0.22	$b(20; 100, 0.22) = 0.0881$
0.24	$b(20; 100, 0.24) = 0.0629$
0.26	$b(20; 100, 0.26) = 0.0369$
0.28	$b(20; 100, 0.28) = 0.0182$
0.30	$b(20; 100, 0.30) = 0.0078$

If p happened to be 0.1, we would see 20 out of 100 with the tiny probability $L(0.1) = 0.0012$. If p were 0.2, we would measure 20 out of 100 with likelihood $L(0.2) = 0.0993$. Oddly enough, even this probability is rather small. Does this mean that the common sense estimator of the proportion, $p = 0.2$, is not plausible either?

The method of maximum likelihood gives one way to interpret such results.

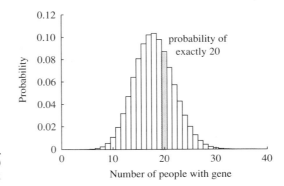

Figure 8.2
The likelihood of 20 out of 100 if $p = 0.18$

■ **Definition 8.4** Suppose $L(p)$ is a likelihood function. The maximum likelihood estimator of the parameter p is the point $\hat{p}$ where $L(p)$ takes on its maximum value. ■

For the example, we have

$$L(p) = b(20; 100, p) = \binom{100}{20} p^{20}(1-p)^{80}$$

(Figure 8.3). The maximum likelihood estimator of p is the point where $L(p)$ takes on its maximum value. Taking the derivative with respect to p, we get

$$\frac{dL}{dp} = \binom{100}{20} \frac{d}{dp} p^{20}(1-p)^{80}$$

$$= \binom{100}{20} \left[20p^{19}(1-p)^{80} - 80p^{20}(1-p)^{79} \right]$$

by the product rule (Theorem 2.9). Setting the result equal to 0 and factoring, we find

$$0 = \binom{100}{20} \left[20p^{19}(1-p)^{80} - 80p^{20}(1-p)^{79} \right]$$

$$= \binom{100}{20} p^{19}(1-p)^{79} \left[20(1-p) - 80p \right].$$

Because this is in factored form, critical points occur where one of the factors is 0, or

$$\binom{100}{20} p^{19}(1-p)^{79} [20(1-p) - 80p] = 0$$

$$p^{19} = 0 \quad \text{or} \quad (1-p)^{79} = 0 \quad \text{or} \quad 20(1-p) - 80p = 0$$

$$p = 0 \quad \text{or} \quad p = 1 \quad \text{or} \quad p = 0.2$$

To find the maximum, we compute the likelihood at the critical points:

$$L(0) = 0$$
$$L(1) = 0$$
$$L(0.2) = 0.0993$$

The maximum likelihood estimator is therefore $p = 0.2$, exactly matching the common sense estimator. The likelihood of 0 at $p = 0$ and $p = 1$ means that the data (20 out of 100) are *impossible* in these two cases (Figure 8.3).

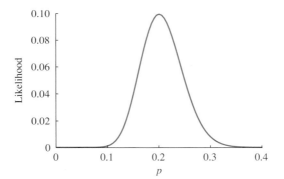

Figure 8.3
The likelihood of 20 out of 100 as a function of p

These calculations do not change the fact that the likelihood of the data with $p = 0.2$ is less than 10%. How can we put our trust in this unlikely result? The question results from thinking about the technique backward. The data may or may not be unlikely, but they are exactly what we observed. No particular result may be likely for an experiment with many possible outcomes. For instance, suppose that an animal is homozygous with the genotype **BB** at each of 100 loci (Figure 8.4). The mother is known to be heterozygous with the genotype **Bb** at every locus. There are two possible fathers, one with the genotype **BB** at every locus, and one, like the mother, heterozygous with the genotype **Bb** at every locus. Which is more likely to be the real father? With father 1, the offspring has genotype **BB** at a particular locus with probability 0.5, and a probability of **BB** at *every* locus of

$$
\begin{aligned}
L(\text{father 1}) &\\
&= \Pr(\textbf{BB} \text{ at every locus}) \\
&= \Pr(\textbf{BB} \text{ at locus 1}) \times \Pr(\textbf{BB} \text{ at locus 2}) \times \cdots \times \Pr(\textbf{BB} \text{ at locus 100}) \\
&= 0.5^{100} \approx 7.89 \times 10^{-31}
\end{aligned}
$$

which is astronomically small. With father 2, the offspring would have genotype **BB** at a particular locus with probability 0.25, and a probability of **BB** at every locus of

$$
\begin{aligned}
L(\text{father 2}) &\\
&= \Pr(\textbf{BB} \text{ at every locus}) \\
&= \Pr(\textbf{BB} \text{ at locus 1}) \times \Pr(\textbf{BB} \text{ at locus 2}) \times \cdots \times \Pr(\textbf{BB} \text{ at locus 100}) \\
&= 0.25^{100} \approx 6.22 \times 10^{-61}
\end{aligned}
$$

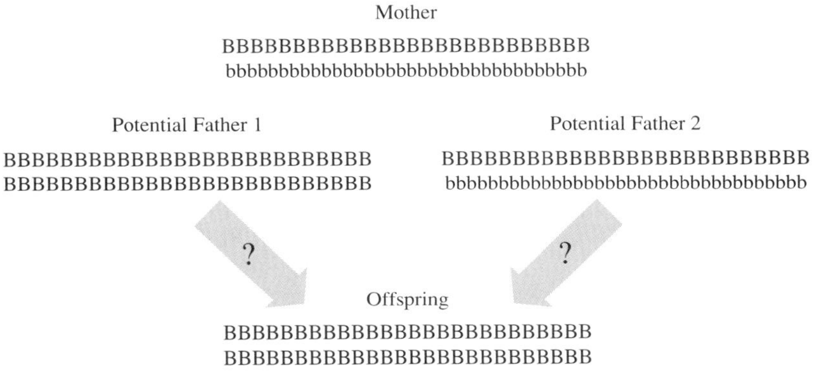

Mother
BBBBBBBBBBBBBBBBBBBBBBBBB
bbbbbbbbbbbbbbbbbbbbbbbbbbbbbb

Potential Father 1
BBBBBBBBBBBBBBBBBBBBBBBBBBB
BBBBBBBBBBBBBBBBBBBBBBBBBBB

Potential Father 2
BBBBBBBBBBBBBBBBBBBBBBBBBBB
bbbbbbbbbbbbbbbbbbbbbbbbbbbbbb

?   ?

Offspring
BBBBBBBBBBBBBBBBBBBBBBBBBBB
BBBBBBBBBBBBBBBBBBBBBBBBBBB

Figure 8.4
The likelihood of two fathers and their unlikely offspring

which is astronomically smaller. It is virtually impossible that either father would produce such an offspring. But *given that the offspring exists*, the first father is much more likely to have produced it. It is far more reasonable to deduce that the first male fathered the offspring than to deduce that the offspring does not exist.

In our example, the value of the likelihood function is particularly small when $p < 0.1$ or $p > 0.35$. If the true proportion p were smaller than 0.1, we could be quite surprised to see 20 out of 100. The method of maximum likelihood finds the single value of p *most likely* to have generated the actual observations. Because $L(0.18)$ is not much lower than $L(0.2)$, that guess is almost as consistent with the data. We later develop a rule of thumb to indicate when we can exclude a particular guess (like $p = 0.1$) with confidence (although never with certainty).

Estimating a Rate

Suppose we observe ten molecules leaving a cell and record the waiting time for each.

Molecule	Waiting time	Estimated rate = reciprocal of waiting time
1	$t_1 = 0.311$	3.208
2	$t_2 = 0.791$	1.263
3	$t_3 = 1.196$	0.836
4	$t_4 = 0.539$	1.854
5	$t_5 = 0.575$	1.738
6	$t_6 = 0.908$	1.100
7	$t_7 = 0.088$	11.315
8	$t_8 = 1.764$	0.566
9	$t_9 = 0.619$	1.614
10	$t_{10} = 0.038$	25.949
Average	0.683	4.945

If we suspect that the results follow an exponential distribution, we would like to use the data to estimate the parameter λ, the rate at which molecules leave.

One approach, indicated in the third column of the table, is to estimate the rate for each molecule as the reciprocal of the waiting time, because the expectation of the waiting time with parameter λ is $1/\lambda$ (Theorem 7.16). For example, the first molecule left after a wait of 0.311, which is the mean time if the true rate λ were

$$\lambda = \frac{1}{0.311} = 3.215$$

The average of these estimated rates is 4.945. Alternatively, we could average the waiting times to find the average wait (0.683 in this case) and estimate λ as the reciprocal (1.464). These estimates are quite different. Is either one right?

With an understanding of the underlying exponential model, we can use the method of maximum likelihood. The first step is to find the likelihood function, $L(\lambda)$, giving the probability of the observed data for different values of λ. With a continuous random variable, any particular measurement is impossible. Our concern is to find the value of λ for which the data are "least impossible."

The probability that the waiting time is near a particular value t is

$$f(t) = \lambda e^{-\lambda t}$$

For example, the probability density at $t_1 = 0.311$ is

$$f(t_1) = \lambda e^{-0.311\lambda}$$

What is the probability of measuring *exactly* the data in the table? Assuming that all the measurements are independent, the probability of the whole set is the product of the probabilities, so

$$\begin{aligned}L(\lambda) &= f(t_1) \cdot f(t_2) \cdots \cdot f(t_{10}) && \text{probability of data is product of probabilities} \\ &= \lambda e^{-\lambda t_1} \cdot \lambda e^{-\lambda t_2} \cdots \cdot \lambda e^{-\lambda t_{10}} && \text{substitute probability density function} \\ &= \lambda^{10} e^{-\lambda t_1} \cdot e^{-\lambda t_2} \cdots \cdot e^{-\lambda t_{10}} && \text{pull } \lambda^{10} \text{ out front} \\ &= \lambda^{10} e^{-\lambda \sum_{i=1}^{10} t_i} && \text{use Law 1 of exponents to change product into sum} \\ &= \lambda^{10} e^{-6.833\lambda} && \text{use the fact that the sum of the } t_i\text{s is 6.833}\end{aligned}$$

(Figure 8.5).

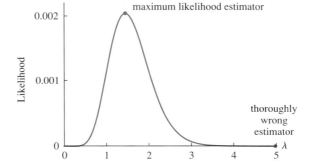

Figure 8.5
The likelihood function for waiting times

To compute the maximum likelihood estimator we differentiate the likelihood function. Let $\overline{T}$ denote the average of the waiting times (here 0.6833). We can rewrite the likelihood function as

$$L(\lambda) = \lambda^{10} e^{-10\lambda \overline{T}}$$

because $\sum t_i = 10\overline{T}$ by the definition of the average. Taking the derivative, we get

$$\begin{aligned}L'(\lambda) &= 10\lambda^9 e^{-10\lambda \overline{T}} - 10\overline{T}\lambda^{10} e^{-10\lambda \overline{T}} \\ &= 10\lambda^9 e^{-10\lambda \overline{T}}(1 - \lambda \overline{T})\end{aligned}$$

This is 0 at $\lambda = 0$ and at $\lambda = 1/\overline{T}$. At these points,

$$L(1.464) = 2.05 \times 10^{-3}$$
$$L(0) = 0$$

so the maximum is 1.464. The maximum likelihood estimator of the rate turns out to be the reciprocal of the average waiting time. The likelihood derived by averaging the rates is $L(4.945) = 1.895 \times 10^{-8}$, far smaller than the maximum.

626   Chapter 8   Introduction to Statistical Reasoning

SUMMARY We have begun the study of **inferential statistics**, the methods used to infer properties of whole **populations** based on the properties of **samples**. Choosing a sample appropriately is the domain of **statistical design**. Assuming the sample was appropriately chosen, we estimated the proportion of a population with a particular trait first with common sense and then with the method of **maximum likelihood**. The **likelihood function** expresses the probability that the observed data would have occurred for different possible parameter values. We applied the method of maximum likelihood to estimate the rate of an exponential process.

8.1   EXERCISES

1. Suppose we want to calculate the proportion of days that the temperature rises above 20°C. Evaluate the following sampling schemes.
   a. Sample 100 consecutive days beginning on January 1.
   b. Sample 100 consecutive days beginning on June 1.
   c. Sample the temperature on March 15 for 100 yr.
   d. What might be a good method if only 100 days could be sampled?

2. A clever pollster decides to find the average income of people by calling random individuals on their cellular phones. Suppose, however, that only 20% of people have cellular phones, and that the distribution of their incomes is $N(40{,}000, 10{,}000)$. The distribution for people without cellular phones is $N(20{,}000, 5000)$.
   a. What would be the results of sampling 20 people with cellular phones?
   b. What would be the results of sampling 20 people without cellular phones?
   c. What would be the results of sampling 20 people at random?

3. Find the likelihood of the following.
   a. Flipping 2 out of 4 heads with a fair coin.
   b. Rolling 2 out of 4 sixes with a fair die. Why is your answer smaller?
   c. Flipping 2 out of 12 heads with a fair coin.
   d. Rolling 2 out of 12 sixes with a fair die. Why is your answer larger?
   e. Flipping 2 out of 4 heads with a fair coin and rolling 2 out of 12 sixes with a fair die.

4. Two individuals are tested for a particular gene, and one has it.
   a. Find and graph the likelihood function for the proportion of individuals in the whole population with this gene.
   b. What is the maximum likelihood?
   c. Why is the likelihood equal to 0 at $p = 0$ and $p = 1$?

5. Three individuals are tested for a particular gene, and all three have it.
   a. Find and graph the likelihood function for the proportion of individuals in the whole population with this gene.

   b. What is the maximum likelihood?
   c. Why is the likelihood equal to 1 at $p = 1$? Does this mean that the true proportion is really $p = 1$?

6. Of 100 individuals, 30 are found to be infected with a disease. Estimate the proportion of infected women and infected men in the following circumstances. Assuming that the population is composed of 50% women, estimate the infected proportion in the whole population.
   a. The sample consists of 20 out of 50 infected women and 10 out of 50 infected men.
   b. The sample consists of 20 out of 40 infected women and 10 out of 60 infected men.
   c. The sample consists of 20 out of 20 infected women and 10 out of 80 infected men.

7. Use the method of maximum likelihood to estimate the rate λ from the following exponential data.

Molecule	Waiting time
1	1.565
2	0.888
3	0.874
4	5.156
5	0.018
6	0.048
7	1.496
8	0.422
9	0.721
10	1.119

   a. Find the likelihood function.
   b. Find the maximum likelihood.
   c. Does it seem very likely that the true rate is 1.0?

8. A couple has children in hope of having a girl. They have seven boys before having a girl. The wife accuses the husband of not having as many female as male sperm.

a. Find the likelihood function for the fraction q of female sperm (use the geometric distribution).
b. Find the maximum likelihood, and compare with the likelihood of $q = 0.5$.
c. Another couple has four boys, then one girl, then three more boys. Find the maximum likelihood estimator of q in this case. Why is it the same?

9. Suppose 14 mutations are counted in 1 million bases.
    a. Write the likelihood function for the rate of mutations per million bases.
    b. Find the maximum likelihood. Does it make sense?

10. Write down the equations that would express the fact that the following estimators are unbiased.
    a. The estimator of q in Exercise 8.
    b. The estimator of Λ in Exercise 9. How can you tell that this estimator is unbiased?

11. **COMPUTER:** Simulate the following experiments.
    a. People have a 0.4 chance of having black hair. Simulate 50 groups of five people. How many have exactly two out five with black hair?
    b. Suppose that two out of five people are found to have black hair. Simulate 50 groups of 5 people with different unknown probabilities p that a person has black hair. Use values of p ranging from 0 to 1.0. Which value produces the most groups that match the data exactly?
    c. Find and plot the likelihood function describing part **b**. Where is the maximum? What does it correspond to in your simulation?
    d. Explain how the simulations of parts **a** and **b** differ. For each, say what is known and what is unknown.

12. **COMPUTER:** Cells are placed for 1 min in an environment where they are hit by X rays, some of which are damaging. Cells not hit by the damaging rays are healthy, those hit exactly once are damaged, and those hit more than once are dead. By measuring the states of a number of cells, we wish to infer the rate at which cells are hit by X rays. Let x denote the unknown parameter of the Poisson distribution. Use the formula for the Poisson distribution to compute the probabilities p_0 of no hits, p_1 of one hit, and p_m of more than one hit in 1 min as *functions* of x.
    a. Suppose the true value of x is 3.0. Plot the histogram.
    b. Simulate 50 cells, and count how many you have of each type. (To keep things interesting, keep sampling until you get at least one cell of each type.)
    c. Compare the results of your simulation with the idealized histogram.
    d. Now pretend that x is unknown. We can use the method of maximum likelihood to analyze our data. Find the likelihood function L of these data (it is the product of the likelihoods for each of the 50 cells) as a function of the unknown parameter y, and let S be the support, the natural log of L. Plot $S(y)$ over a reasonable range.
    e. Find the maximum of the support and mark it on your graph.
    f. Find the support of the hypotheses $x = X_{\max}$, $x = 2$, $x = 4$, and the "truth," $x = 3$, and indicate each on your graph.

8.2 Confidence Limits

The method of maximum likelihood gives a way to estimate parameter values from data. We know, however, that our estimator is not *exactly* right. We want a **range** of values we can be confident contains the true value. The idea is formulated with **confidence limits**, values of the parameter we can be confident lie below and above our estimates. We compute confidence limits exactly in the binomial case and then estimate them with computer simulation and likelihood.

Exact Confidence Limits

When we find that 20 out of 100 people carry a gene, our best guess of the population proportion p is $\hat{p} = 0.2$. How can we find an **interval** that contains the population proportion 95% of the time? We attack this problem in this section with the exact probabilities, and we apply the normal approximation in Section 8.3.

A **confidence interval** is described by two numbers, p_l and p_h, between which the true value should lie ($p_l < \hat{p} < p_h$) (Figure 8.6). The lower confidence limit is a value $p_l < \hat{p}$ that we are confident is less than the true p. To find it, we perform a thought experiment much as in the derivation of the likelihood function. Imagine

628 Chapter 8 Introduction to Statistical Reasoning

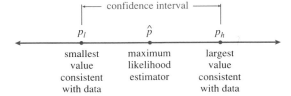

Figure 8.6
Confidence limits

that the true p were smaller than $\hat{p}$, perhaps 0.15. It is *possible* that our sample would include 20 people with the gene, but not likely. How do we quantify this?

To be 95% sure that our interval captures the true value, there should be only a 2.5% chance that our lower confidence limit is too large (saving 2.5% for the upper confidence limit). Computing p_l has one major difference from finding the likelihood. Instead of finding the probability of *exactly* 20 out of 100 with a particular guess of p, we find the probability of *20 or more out of 100*. Why this change? Particular results are usually highly unlikely, while a *more extreme* result is not. We test possible values of p_l, finding one for which the probability of getting 20 or more is only 0.025.

Suppose the true p had been 0.15. The probability of getting 20 or more out of 100 is

$$\Pr(20 \text{ or more out of } 100 \text{ with } p = 0.15) = \sum_{k=20}^{100} b(k; 100, 0.15) = 0.1065$$

(Figure 8.7a). This value had to be found on a computer. A true value of $p = 0.15$ would produce a result at least as extreme as our actual measurement 10.65% of

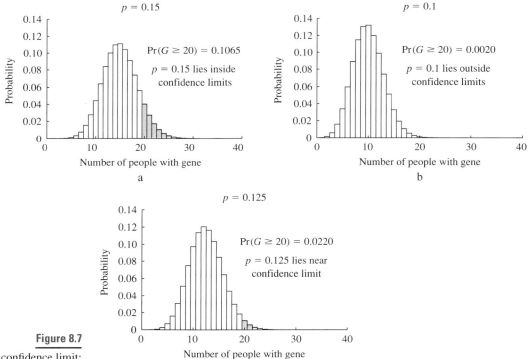

Figure 8.7
Finding the lower confidence limit: three thought experiments

the time. If the true value were $p = 0.1$, then

$$\Pr(20 \text{ or more out of } 100 \text{ with } p = 0.1) = \sum_{k=20}^{100} b(k; 100, 0.1) = 0.0020$$

(Figure 8.7b). Only two out of 1000 random samples would be this extreme. We can be rather confident (although not certain) that a true value of p this small did not produce our observation. Finally, if we guess $p = 0.125$, then

$$\Pr(20 \text{ or more out of } 100 \text{ with } p = 0.125) = \sum_{k=20}^{100} b(k; 100, 0.125) = 0.0220$$

(Figure 8.7c). Approximately 2.2% of samples would exceed our measured value. The precise definition of the lower confidence limit is given here.

■ **Definition 8.5** The lower confidence limit is the largest value of the parameter that produces a result as large as or larger than the measured value only 2.5% of the time. In formal terms, p_l solves the equation

$$\Pr(\text{value greater than or equal to measured data if } p = p_l) = 0.025 \quad ■$$

For our example, the lower confidence limit p_l is the solution of

$$\sum_{k=20}^{100} b(k; 100, p_l) = 0.025$$

The graph of the probability of 20 or more as a function of p_l is shown in Figure 8.8. With a computer, we can solve for $p_l = 0.127$. If p were 0.127, we would get 20 or more out of 100 in exactly 2.5% of experiments. If we excluded this value from our confidence limits, there is a reasonably small chance of throwing out a correct value.

We find the upper confidence limit in the same way, as a value p_h we can be confident (with 0.025 probability of error) *exceeds* the true p.

■ **Definition 8.6** The upper confidence limit is the smallest value of the parameter that produces a result as small as or smaller than the measured value only 2.5% of the time. In formal terms, p_h solves the equation

$$\Pr(\text{value less than or equal to measured data if } p = p_h) = 0.025 \quad ■$$

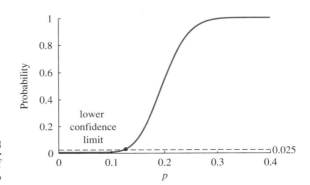

Figure 8.8
The probability of 20 or more out of 100 as a function of p

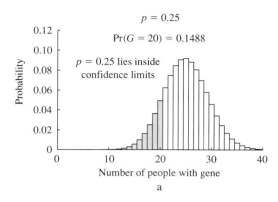
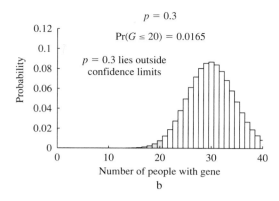

Figure 8.9

Finding the upper confidence limit: two thought experiments

We test values p_h in quest of one for which the probability of getting *20 or fewer* out of 100 is less than 0.025. For instance, with $p_h = 0.25$,

$$\Pr(20 \text{ or fewer out of } 100 \text{ with } p = 0.25) = \sum_{k=0}^{20} b(k; 100, 0.25) = 0.1488$$

(Figure 8.9a). A true value of $p = 0.25$ would produce a result as extreme as our observations nearly one time out of six, which is not particularly unlikely. If $p = 0.3$,

$$\Pr(20 \text{ or fewer out of } 100 \text{ with } p = 0.3) = \sum_{k=0}^{20} b(k; 100, 0.3) = 0.0165$$

(Figure 8.9b). This value lies *outside* our confidence limits because a true p of 0.3 would generate 20 or fewer positives out of 100 *less than* 2.5% of the time. The exact p_h solves

$$\sum_{k=0}^{20} b(k; 100, p_h) = 0.025$$

with solution $p_h = 0.292$ (Figure 8.10). If the true value of p were greater than p_h, it would be unlikely to have generated the observed data.

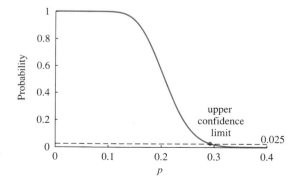

Figure 8.10

The probability of 20 or fewer out of 100 as a function of p

The interval between p_l and p_h is called the **95% confidence interval**. If we followed our procedure on many data sets, we would capture the true population proportion 95% of the time. Is it correct to say that the true p lies between p_l and p_h with probability 0.95? Strictly speaking, no. Either the true p lies between these limits or it does not. Given our information, however, we can think of the confidence limits as being quite likely to contain the true proportion.

The equations for p_l and p_h are impossible to solve unless n is quite small (Exercise 1). When n is large enough for the conditions in Section 7.9 to apply, we will use the normal approximation to the binomial distribution (Section 8.3). With small data sets, however, the binomial probability distribution *must* be used. Suppose we want to estimate the proportion of bears infested with a parasite. At great hazard, ten bears are captured, of which three are infested. We find the 95% confidence interval around the estimate $\hat{p} = 0.3$ by solving for p_l and p_h (Definitions 8.5 and 8.6):

$$\sum_{k=3}^{10} b(k; 10, p_l) = 0.025$$

$$\sum_{k=0}^{3} b(k; 10, p_h) = 0.025$$

With a computer, we find $p_l = 0.0667$ and $p_h = 0.6524$. The 95% confidence interval is quite wide, the price paid for a small sample size. Based on our data, it is quite possible that as few as 7% or as many as 65% of bears are actually infested. More dangers would have to be braved to improve this estimate.

There is nothing magical about the traditional 95%. In this case, we say that the degree of confidence is 0.95. Let α be 1 minus the degree of confidence, or $\alpha = 0.05$. The value 0.025 is half the acceptable uncertainty.

■ **Definition 8.7** The lower confidence limit with degree of confidence $1 - \alpha$ is the largest value of the parameter that produces a result as large as or larger than the measured value only a fraction $\alpha/2$ of the time. In formal terms, p_l solves the equation

$$\text{Pr}(\text{value greater than or equal to measured data if } p = p_l) = \frac{\alpha}{2}.$$

The upper confidence limit with degree of confidence $1 - \alpha$ is the smallest value of the parameter that produces a result as small as or smaller than the measured value only a fraction $\alpha/2$ of the time. In formal terms, p_h solves the equation

$$\text{Pr}(\text{value less than or equal to measured data if } p = p_h) = \frac{\alpha}{2} \quad ■$$

To find confidence limits with degree of confidence $1 - \alpha$ when we count 20 out of 100, solve

$$\sum_{k=20}^{100} b(k; 100, p_l) = \frac{\alpha}{2}$$

$$\sum_{k=0}^{20} b(k; 100, p_h) = \frac{\alpha}{2}$$

If we want 99% confidence in our genetic example, $\alpha = 0.01$ and

$$\sum_{k=20}^{100} b(k; 100, p_l) = \frac{0.01}{2} = 0.005$$

$$\sum_{k=0}^{20} b(k; 100, p_h) = \frac{0.01}{2} = 0.005$$

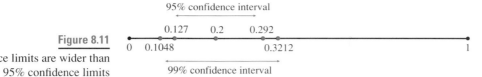

Figure 8.11
99% confidence limits are wider than 95% confidence limits

Again using a computer, we find $p_l = 0.1048$ and $p_h = 0.3212$. More confidence requires a wider interval (Figure 8.11). The appropriate level of uncertainty depends on the application and the tradeoff between greater uncertainty and wider confidence limits.

Monte Carlo Method

Even with small values of n, the equations for p_l and p_h are virtually impossible to solve. For models more complex than the binomial distribution, even writing the equations can be difficult. Fast computers have made an alternative approach, computer simulation, feasible. Because this computer experiment includes random inputs (as in gambling), the technique is called the **Monte Carlo method**. The logic behind confidence limits, however, remains the same.

Our actual experiment involved picking 100 individuals, each with an unknown probability p of having a gene. The computer mimics this, but with different guesses of the value of p. In particular, we pick a test value of p, choose 100 Bernoulli random variables each with probability p, and count the number of "successes." To get a clear idea of the implications of the test value p, we run the computer experiment many times. Our goal is to establish whether p is *consistent* with the observed data.

Suppose first we wish to find p_l, the lower confidence limit or the smallest value of p consistent with the actual results (Definition 8.5). If a particular test value often produces 20 or more out of 100, it is perfectly plausible that it might have generated our experimental results. There is no reason to think that it is too small. If we rarely see 20 or more out of 100, that value is unlikely to have produced the data. Similarly, the upper confidence limit can be found by checking test values larger than $\hat{p}$ and finding the one that infrequently produces 20 or fewer out of 100.

■ **Algorithm 8.1** (Monte Carlo method for finding confidence limits)

1. Compute the maximum likelihood estimator of the parameter.

2. To find the lower confidence limit, simulate the experiment many times on the computer with values of the parameter smaller than the maximum likelihood estimator, and find the fraction of times results are greater than or equal to the measured result.

3. With degree of confidence $1 - \alpha$, the lower confidence limit is the point where this fraction is equal to $\alpha/2$.

4. To find the upper confidence limit, simulate the experiment many times on the computer with a value of the parameter larger than the maximum

likelihood estimator, and find the fraction of times results are less than or equal to the measured result.

5. The upper confidence limit is the point where this fraction is equal to $\alpha/2$. ∎

The results of 1000 repetitions of the experiment of choosing 100 Bernoulli random variables with probability p are given in Table 8.1. For example, with $p = 0.1$, only 1 of 1000 simulations produced more than 20 successes. Because this fraction, 0.001, is less than 0.025, this value is less than the 95% confidence limit. The 95% confidence limits require that no more than 0.025, or 25 of 1000, experiments produce 20 or more by chance. This occurs between $p = 0.12$ and $p = 0.14$, bracketing the exact result of 0.127 (Figure 8.12a). With $p = 0.3$, 22 of 1000 simulations produced 20 or fewer successes. This is quite close to 0.025, meaning that the upper confidence limit is near 0.3, close to the exact result of 0.292 (Figure 8.12b). We would have to test more values of p and run more simulations to get more precise estimates.

In many cases, it is impossible to compute confidence limits mathematically. But any experiment that can be precisely described can be simulated on a computer. The Monte Carlo method works for these difficult models.

Table 8.1 Monte Carlo Simulation Results for Finding Confidence Limits

p	Simulations producing 20 or more	Fraction	Simulations producing 20 or fewer	Fraction
0.10	1	0.001	999	0.999
0.12	8	0.008	997	0.997
0.14	80	0.080	954	0.954
0.16	168	0.168	893	0.893
0.18	312	0.312	780	0.780
0.20	561	0.561	553	0.553
0.22	722	0.722	368	0.368
0.24	852	0.852	219	0.219
0.26	927	0.927	107	0.107
0.28	974	0.974	46	0.046
0.30	990	0.990	22	0.022
0.32	995	0.995	6	0.006
0.34	999	0.999	1	0.001

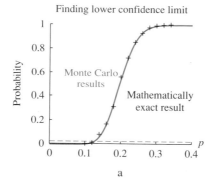
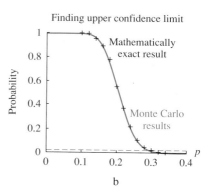

Figure 8.12 Using the Monte Carlo method to estimate confidence limits

Likelihood, Support, and Confidence Limits

Our intuition about likelihood suggests that values of p with likelihood not too far below the maximum should be part of our confidence interval. For example, $p = 0.18$ has likelihood 0.0870, not far below the maximum of 0.0993. On the other hand, $p = 0.1$ has a far smaller likelihood of 0.0012. There is a powerful rule of thumb that uses likelihood to estimate the 95% confidence interval. (Loyal adherents of the likelihood method do not use confidence intervals and recommend the method of this section as an alternative.)

As we have seen, values of the likelihood function can be tiny. The best way to compare them is with ratios. For example,

$$\frac{L(0.18)}{L(0.02)} = \frac{0.0870}{0.0993} = 0.876$$

That fact that this ratio is near 1 indicates the likelihood that $p = 0.18$ is close to the maximum.

A more convenient way to compare likelihoods is to take the natural logarithm:

■ **Definition 8.8** The **support** (or **log likelihood**) for a particular value of an unknown parameter p is

$$S(p) = \ln[L(p)]$$ ■

The numerical value of the support can be thought of as the support that the data provide for a particular hypothesis (a value of the unknown parameter p in this case). For example,

$$S(0.18) = \ln[L(0.18)] = \ln(0.0870) = -2.441$$
$$S(0.2) = \ln[L(0.2)] = \ln(0.0993) = -2.310$$

Values of the support are always negative because the likelihood is always less than 1. The support takes on its maximum precisely where the likelihood does (Exercise 10).

We use the difference between the values of the support as a comparison of two different values of the unknown parameter p.

Rule for Estimating Confidence Limits with the Support:
Suppose $\hat{p}$ is the maximum likelihood estimate of the unknown parameter p. The approximate upper and lower confidence limits (see Figure 8.13) are solutions of

$$S(p) = S(\hat{p}) - 2$$

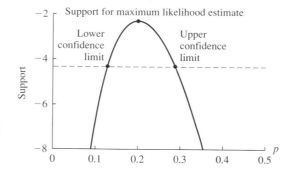

Figure 8.13
Estimating confidence limits with the support

For our example, the support $S(p)$ for the likelihood of exactly 20 out of 100 with a true parameter p is

$$S(p) = \ln[L(p)]$$
$$= \ln\left[\binom{100}{20}p^{20}(1-p)^{80}\right]$$
$$= \ln\binom{100}{20} + 20\ln(p) + 80\ln(1-p)$$

(Figure 8.14). We used Laws 1 and 2 of logs to write the log of a product as the sum of the logs and to move the powers 20 and 80 to the front of the equation.

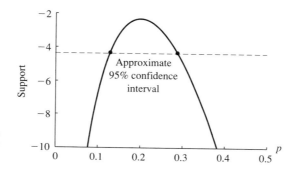

Figure 8.14
Support given 20 out of 100 as a function of p

The approximate upper and lower confidence limits are solutions of

$$S(p) = S(0.2) - 2$$

In formal terms, we have

$$\left[-\ln\binom{100}{20} + 20\ln(p) + 80\ln(1-p)\right] = \ln\binom{100}{20} + 20\ln(0.2)$$
$$+ 80\ln(1-0.2) - 2$$
$$20\ln(p) + 80\ln(1-p) = 20\ln(0.2) + 80\ln(1-0.2) - 2$$

The constants cancel, saving much computation. These equations, although still impossible to solve, are much simpler than equations for the exact confidence limits. By experimentation or using Newton's method (Exercise 9) we find two solutions, $p = 0.128$ and $p = 0.287$ (Figure 8.14). These values are quite close to the exact confidence limits of $p_l = 0.127$ and $p_h = 0.292$.

SUMMARY

We have developed the idea of **confidence limits**. True values less than the lower confidence limit are unlikely to have produced results exceeding what we measured, and true values greater than the upper confidence limit are unlikely to have produced results less than our measurement. When the equations for the confidence limits are hopelessly intractable, we can conduct computer experiments using the **Monte Carlo method**. Finally, we have seen that the log of the likelihood, the **support**, provides a convenient way to estimate confidence limits.

8.2 EXERCISES

1. Consider the tiny data set of Exercise 4 in Section 8.1, where one of two people is found to carry a gene.
   a. Find the exact 95% confidence limits.
   b. Find the exact 99% confidence limits.
   c. How would you use the Monte Carlo method to estimate the 99% confidence limits?

2. Use the method of support to estimate 95% confidence limits for Exercise 1, and compare your results.

3. Consider the data set of Exercise 5 in Section 8.1, where three of three people are found to carry a gene.
   a. Find the exact 95% confidence limits.
   b. Find the exact 99% confidence limits.
   c. Why is the upper confidence limit strange?

4. Use the method of support to estimate 95% confidence limits for Exercise 3, and compare your results.

5. A coin is flipped and gives five heads in six flips.
   a. Flip a fair coin until you get five or more heads out of six flips. How long did it take?
   b. How long would you expect it to take? (Use the geometric distribution.)

6. A person places an advertisement in the newspaper to sell his car and settles down to await calls, which he expects will arrive according to a Poisson process. The first call arrives in 20 min.
   a. Find the maximum likelihood estimator of the rate.
   b. Find the 95% confidence limits.
   c. How many calls might he expect to miss if he went out for a 2-hr hike?

7. 14 mutations are counted in 1 million base pairs (as in Exercise 9 in Section 8.1).
   a. Write the equations for the 95% confidence limits, and solve them if you have a computer handy.
   b. How would you go about finding the confidence limits with the Monte Carlo method?
   c. Use the method of support to estimate the confidence limits.

8. Find 95% confidence limits around the maximum likelihood estimate of q in Exercise 8 in Section 8.1. How do you interpret these results?

9. Suppose that 20 of 100 individuals are measured with a particular gene. Use experimentation and Newton's method to find where the support for the parameter p is 2 less than the support for $\hat{p} = 0.2$.

10. Prove that the support takes on its maximum precisely where the likelihood does.

11. **COMPUTER:** Do a simulation to reproduce the results in Table 8.1. Why are your results different from the table in the text? Do you estimate the same confidence limits? Do more simulations to get a more accurate answer. How close can you get to the exact answer?

12. **COMPUTER:** Consider again the data in Exercise 12 in Section 8.1. Use the method of support to find approximate 95% confidence limits around your estimated value. Do they include the true value $x = 3.0$?

13. **COMPUTER:** Use the Monte Carlo method to estimate the 95% confidence limits for Exercise 7. How close is this method to the others?

8.3 Estimating the Mean

For many quantitative measurements, the mean is a useful statistic for describing a population. This is particularly true when measurements follow a roughly normal distribution. After introducing three methods for estimating the mean, we discuss confidence limits around the **sample mean**, the average of the measured values. When the variance is unknown, we can estimate it as the **sample variance** and use that value to estimate confidence limits.

Estimating the Mean

There are many ways to estimate the population mean from a sample. Some commonly used methods are

$$\overline{X} = \text{the sample mean} \tag{8.1}$$

8.3 Estimating the Mean

$$\tilde{X} = \text{the sample median} \quad (8.2)$$
$$\overline{X}_{tr(10)} = \text{the trimmed mean} \quad (8.3)$$

The **sample mean** $\overline{X}$ is the average of the sample, the "common sense" estimate of the population mean. In many realistic circumstances, this estimator is undependable because it can be changed drastically by a few extreme values. Consider the following measurements.

Measurement	Value	Measurement	Value
x_1	43.2	x_{11}	58.3
x_2	51.7	x_{12}	76.1
x_3	81.2	x_{13}	54.8
x_4	67.9	x_{14}	37.2
x_5	48.5	x_{15}	71.4
x_6	39.2	x_{16}	52.1
x_7	361.1	x_{17}	47.5
x_8	62.2	x_{18}	65.7
x_9	39.9	x_{19}	40.3
x_{10}	50.8	x_{20}	48.3

(See Figure 8.15.) The values seem to be concentrated around 50, except for the extreme measurement x_7. The average of the whole data set is 69.96, larger than all but four measurements. Throwing out the large measurement x_7 reduces the mean to 54.6, much closer to the center of the distribution.

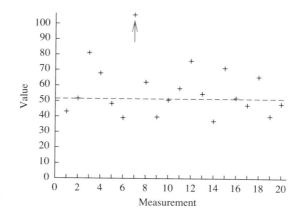

Figure 8.15
A data set and the sample median

There are several ways to deal with this problem. One popular method is to throw out the extreme data, claiming to have had a bad day. Others are the **sample median** and the **trimmed mean**. The sample median $\tilde{X}$ is the middle of the values measured:

$$\text{sample median} = \begin{cases} \text{middle measurement if } n \text{ is odd} \\ \text{average of two middle measurements if } n \text{ is even} \end{cases}$$

If we sort the values in increasing order, x_2 and x_{16} are 10th and 11th. The sample median is the average, or

$$\text{sample median} = \text{average of two middle measurements}$$
$$= \frac{51.7 + 52.1}{2} = 51.9$$

Because the median uses only relative size, it is insensitive to large values. If x_7 had been recorded wrong and was really 61.1, the median would not be affected.

The **trimmed mean** $\overline{X}_{tr(10)}$ lops off the largest and smallest 10% of the measurements (x_6, x_{14}, x_3, and x_7) and averages the rest. In this case,

$$\overline{X}_{tr(10)} = \frac{\text{sum of middle 16 measurements}}{16} = 55.0$$

The subscript indicates the **percentage** of the sample discarded on each end. Like the median, the trimmed mean ignores large values.

All of these statistics perform badly at finding the mean of numbers from a distribution that looks nothing like the normal distribution, such as the exponential distribution. Before trying to take the mean of data like this, it is a good idea to ask why the mean is worth estimating in the first place. It often makes more sense to estimate parameters associated with an explicit underlying model (for example, the rate constant for the exponential). When the underlying model resembles the normal distribution, the mean is, as it were, meaningful. With other models, other statistics can be more appropriate for answering scientific questions.

Confidence Limits for the Sample Mean

Suppose our sample data seem to be clustered around an average value that we wish to estimate. If none of the data are suspect, the sample mean is a good estimator of the population mean. In particular, we use the estimator

$$\hat{\mu} = \frac{\sum_{i=1}^{n} x_i}{n}$$

where $x_1, x_2, \ldots, x_n$ are the n measured values. How can we go about computing confidence limits around this estimator?

The confidence limit depends on the true, unknown, underlying distribution and on the size of our sample. Suppose the true distribution has mean μ and standard deviation σ. If we made the measurements many times, measurement i of the sample can be thought of as a random variable X_i. The sample mean is a random variable defined by

$$\overline{X} = \frac{\sum_{i=1}^{n} X_i}{n}$$

By the Central Limit Theorem for Means (Theorem 7.22), the sample mean $\overline{X}$ of a sample of size n has approximate distribution

$$\overline{X} \sim N\left(\mu, \frac{\sigma^2}{n}\right)$$

If we drew many samples of size n from this population, the sample means would have a normal distribution with mean μ and variance σ^2/n. The narrower this distribution, the more accurately we can estimate the true mean from the sample mean. There are two ways that the variance of $\overline{X}$ can be made smaller: by decreasing the variance of each measurement X_i (perhaps through more accurate measurement), and increasing n (by making more measurements) (Figure 8.16).

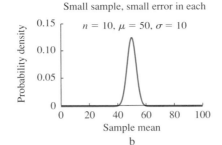
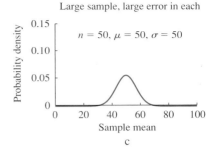

Figure 8.16
Distributions of the sample mean: effects of smaller σ and smaller n

If we know the standard deviation σ of the population (we will estimate it in Section 8.4), we can derive confidence limits around our estimator $\hat{\mu}$. The lower confidence limit is a value μ_l unlikely to produce a sample mean larger than what we measured, $\hat{\mu}$. If the true mean were μ_l, the sample mean X_l would be a random variable with

$$X_l \sim N\left(\mu_l, \frac{\sigma^2}{n}\right)$$

Distributions with different possible values of μ_l are shown in Figure 8.17. The lower end of the 95% confidence interval is the value μ_l that solves the equation

$$\Pr(X_l \geq \hat{\mu}) = 0.025$$

We choose μ_l small enough to virtually guarantee we could not have measured a value of $\hat{\mu}$ as large as it is, making us confident that the true μ is no smaller than μ_l. If μ_l is too large, it is quite probable that the sample mean would exceed $\hat{\mu}$ (Figure 8.17a).

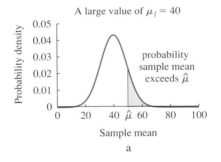
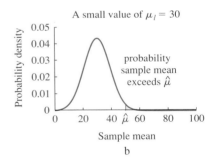

Figure 8.17
Probability of mean greater than $\hat{\mu}$ with different μ_l

How do we solve this equation for μ_l? Because it involves a normal distribution, we transform it into a question about the standard normal distribution (Algorithm 7.2), subtracting the mean and dividing by the standard deviation.

$$\Pr(X_l \geq \hat{\mu}) = \Pr\left(\frac{X_l - \mu_l}{\sigma/\sqrt{n}} \geq \frac{\hat{\mu} - \mu_l}{\sigma/\sqrt{n}}\right)$$

$$= \Pr\left(Z \geq \frac{\hat{\mu} - \mu_l}{\sigma/\sqrt{n}}\right)$$

$$= 0.025$$

We are trying to solve for the value of

$$\frac{\hat{\mu} - \mu_l}{\sigma/\sqrt{n}}$$

that lies below all but 2.5% of the standard normal, or where

$$\Phi\left(\frac{\hat{\mu} - \mu_l}{\sigma/\sqrt{n}}\right) = 0.975$$

(Figure 8.18).

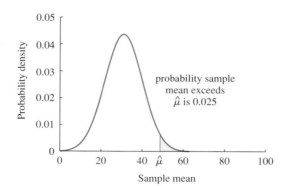

Figure 8.18
The lower confidence limit

From tables of the c.d.f., we find that $\Phi(1.96) = 0.975$. Therefore, the lower confidence limit μ_l satisfies

$$\frac{\hat{\mu} - \mu_l}{\sigma/\sqrt{n}} = 1.96$$

Solving for μ_l, we get

$$\mu_l = \hat{\mu} - 1.96\frac{\sigma}{\sqrt{n}}$$

The lower confidence limit is closer to the sample mean $\hat{\mu}$ when σ is smaller or n is larger.

The upper confidence limit μ_h is found in a similar way. Let X_h be a normally distributed random variable for the sample mean of a sample of size n from a population with mean μ_h and variance σ^2. Then

$$X_h \sim N\left(\mu_h, \frac{\sigma^2}{n}\right)$$

The upper confidence limit μ_h is the value for which

$$\Pr(X_h \leq \hat{\mu}) = 0.025$$

(Figure 8.19). Transforming into a question about the standard normal distribution as above, we get

$$\Pr(X_h \leq \hat{\mu}) = \Pr\left(\frac{X_h - \mu_h}{\sigma/\sqrt{n}} \leq \frac{\hat{\mu} - \mu_h}{\sigma/\sqrt{n}}\right)$$

$$= \Pr\left(Z \le \frac{\hat{\mu} - \mu_h}{\sigma/\sqrt{n}}\right)$$
$$= 0.025$$

The equation

$$\Phi\left(\frac{\hat{\mu} - \mu_h}{\sigma/\sqrt{n}}\right) = 0.025$$

can be solved using the fact that $\Phi(-1.96) = 0.025$. The upper confidence limit μ_h satisfies

$$\frac{\hat{\mu} - \mu_h}{\sigma/\sqrt{n}} = -1.96$$

and

$$\mu_h = \hat{\mu} + 1.96 \frac{\sigma}{\sqrt{n}}$$

The upper and lower confidence limits are centered around $\hat{\mu}$, a consequence of the symmetry of the normal distribution.

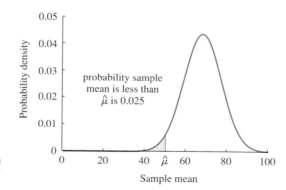

Figure 8.19
The upper confidence limit

We summarize these results in the following theorem.

■ **THEOREM 8.1** If $\hat{\mu}$ is the sample mean of n measurements from a normal distribution with unknown mean μ and known variance σ^2, then the 95% lower and upper confidence limits are

$$\mu_l = \hat{\mu} - 1.96 \frac{\sigma}{\sqrt{n}}$$

$$\mu_h = \hat{\mu} + 1.96 \frac{\sigma}{\sqrt{n}}$$ ■

We can choose a different confidence level if we wish. To find the 95% confidence limits, we had to find where $\Phi(z)$ is equal to 0.025 and 0.975. To find the 99% confidence limits, we solve for where $\Phi(z)$ is equal to 0.005. This occurs at 2.576, so

$$\mu_l = \hat{\mu} - 2.576 \frac{\sigma}{\sqrt{n}}$$

$$\mu_h = \hat{\mu} + 2.576 \frac{\sigma}{\sqrt{n}}$$

Wider limits are required for more confidence.

The confidence interval depends on the sample size n, the sample mean $\hat{\mu}$, the standard deviation σ, and the level of confidence. Confidence limits are different for each experiment, because they are arranged around the measured sample mean rather than the unknown true mean. We know only that the true mean μ lies between the 95% confidence limits μ_l and μ_h with probability 0.95.

Suppose the standard deviation of a plant height distribution is known to be 3.2 cm. If we measure 100 plants and find a sample mean of 40.2 cm, the 95% confidence limits are

$$\mu_l = 40.2 + 1.96 \frac{3.2}{\sqrt{100}} = 40.2 - 0.63 = 39.57$$

$$\mu_h = 40.2 - 1.96 \frac{3.2}{\sqrt{100}} = 40.2 + 0.63 = 40.83$$

(Figure 8.20a). We can be fairly confident that the true mean is somewhere between these limits. The 99% confidence limits are

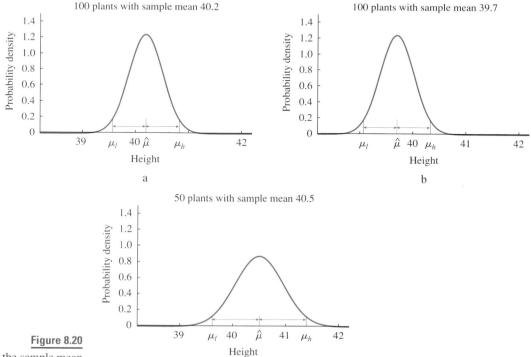

Figure 8.20
Confidence limits for the sample mean with known variance: three replicates

$$\mu_l = 40.2 + 2.576 \frac{3.2}{\sqrt{100}} = 40.2 - 0.82 = 39.38$$

$$\mu_h = 40.2 - 2.576 \frac{3.2}{\sqrt{100}} = 40.2 + 0.82 = 41.02$$

We can be even more confident that the true mean lies between these values.

If we repeat the experiment, each time our confidence limits will be different. If the sample mean were 39.7 with 100 plants, then

$$\mu_l = 39.7 + 1.96 \frac{3.2}{\sqrt{100}} = 39.7 - 0.63 = 39.07$$

$$\mu_h = 39.7 - 1.96 \frac{3.2}{\sqrt{100}} = 39.7 + 0.63 = 40.33$$

(Figure 8.20b). If the sample mean were 40.5 with 50 plants, then

$$\mu_l = 40.5 + 1.96 \frac{3.2}{\sqrt{50}} = 40.5 - 0.89 = 39.61$$

$$\mu_h = 40.5 - 1.96 \frac{3.2}{\sqrt{50}} = 40.5 + 0.89 = 41.39$$

(Figure 8.20c).

Sample Variance and Standard Error

If we do not know the true mean μ, it seems rather unlikely we would have access to the true variance σ. Like the mean, σ can be estimated from the data. Recall that the variance is the mean squared deviation from the mean. If we know μ, the logical estimator is

$$\hat{V} = \frac{\sum_{i=1}^{n}(x_i - \mu)^2}{n}$$

where $x_1, \ldots, x_n$ are again the measured values.

If we do not know μ, we would like to use the sample mean in place of μ. Replacing μ by $\hat{\mu}$ produces an estimate of the variance that tends to be small (the estimator is **biased**). To correct this bias, we use the following definition:

■ **Definition 8.9** The sample variance s^2 when the mean is unknown is

$$s^2 = \frac{\sum_{i=1}^{n}(x_i - \hat{\mu})^2}{n - 1}$$

The $n - 1$ in the denominator corrects for the fact that we used the estimated mean $\hat{\mu}$ rather than the true mean μ (Exercise 11).

As for the variance, there is a simpler computational formula for the sample variance s^2.

■ **THEOREM 8.2** (Computational Formula for the Sample Variance)

If x_i are n independent measurements, the sample variance can be written

$$s^2 = \frac{\sum_{i=1}^{n}x_i^2 - n\hat{\mu}^2}{n - 1}$$

The proof uses the same algebra as the computational formula for the variance (Theorem 6.5).

The **sample standard deviation** is s, the square root of the sample variance. Even though s^2 is an unbiased estimator of σ^2, s is *not* an unbiased estimator of σ. Consider the data in the following table.

Measurement	Value
x_1	43.2
x_2	51.7
x_3	81.2
x_4	67.9
x_5	48.5

The sample mean is

$$\hat{\mu} = \frac{43.2 + 51.7 + 81.2 + 67.9 + 48.5}{5} = 58.5$$

The sample variance is

$$s^2 = \frac{43.2^2 + 51.7^2 + 81.2^2 + 67.9^2 + 48.5^2}{4} = 4523.8$$

and the sample standard deviation is

$$s = 67.26$$

Recall that this is an estimate of the standard deviation of each measurement in the sample. The sample variance is even more sensitive to extreme values than the mean. Be careful about computing the sample variance when measurements include extreme values.

We would like to use s in place of σ in our calculation of confidence limits. When is it justified to do so? Two factors are important. When the underlying distribution is not normal, we need to invoke the Central Limit Theorem for Means, which requires a reasonably large sample. Furthermore, there is additional uncertainty due to the estimated, rather than exact, sample variance. As a rule of thumb, *substitute s for σ only if $n \geq 30$*. Otherwise, an alternative distribution, called **t distribution**, must be used.

■ **THEOREM 8.3** Suppose $\hat{\mu}$ is the sample mean and s the sample standard deviation of $n \geq 30$ measurements from a normal distribution with unknown mean and variance. The 95% lower and upper confidence limits are

$$\mu_l = \hat{\mu} - 1.96 \frac{s}{\sqrt{n}}$$

$$\mu_h = \hat{\mu} + 1.96 \frac{s}{\sqrt{n}}$$

■

The quantity $s/\sqrt{n}$, an estimate of the standard deviation of the sample mean, is called the **standard error of the mean**, or the **standard error**. The upper and lower confidence limits are approximately two standard errors above and below the sample mean.

The binomial distribution is an important special case. Suppose we wish to use the normal approximation to compute confidence limits around the estimator

$\hat{p}$, the fraction of individuals in a sample with a particular property. How do we estimate the variance? For the underlying probability model, if we know the value of p we know that the variance is $p(1-p)$. We can use this idea to estimate the variance from data and apply the normal approximation.

■ **Algorithm 8.2** (Finding confidence limits with the normal approximation: binomial case)

Suppose we measure m successes out of n trials, where $m \geq 5$ and $n - m \geq 5$ (to guarantee that the normal approximation is appropriate).

1. Estimate the probability of success as
$$\hat{p} = \frac{m}{n}$$

2. Estimate the sample variance as
$$s^2 = \hat{p}(1-\hat{p})$$

3. Using Theorem 8.3, estimate the confidence limits by
$$p_l = \hat{p} - 1.96 \frac{s}{\sqrt{n}}$$
$$p_h = \hat{p} + 1.96 \frac{s}{\sqrt{n}}$$

■

If samples include fewer than 30 measurements (or fail to satisfy the condition in Section 7.9), the exact method should be used. (Were we to calculate s^2 from Definition 8.9, we would see that Algorithm 8.2 uses n rather than $n-1$ in the denominator (Exercise 8)).

Consider again the data consisting of 20 out of 100 carrying a particular gene. Then

1. $\hat{p} = 0.2$,
2. $s^2 = 0.2 \cdot 0.8 = 0.16$.
3. The confidence limits are

$$p_l = 0.2 - 1.96 \frac{\sqrt{0.16}}{10} = 0.122$$

$$p_h = 0.2 + 1.96 \frac{\sqrt{0.16}}{10} = 0.278$$

These answers compare well with the exact results, $p_l = 0.127$ and $p_h = 0.292$, and the answers found using the method of support, $p_l = 0.128$ and $p_h = 0.287$.

SUMMARY

We have studied the problem of estimating the mean and computing confidence limits for the mean of a normal population. Estimates of the mean include the **sample mean**, the **sample median**, and the **trimmed mean**. The statistical properties of the sample mean can be analyzed with the Central Limit Theorem for Means when the variance is known. When the variance is unknown, the sample mean can be estimated with the **sample variance** if sample size is at least 30. The 95% confidence limits are approximately two **standard errors** above and below the sample mean. The normal approximation to the binomial distribution can be used to estimate confidence limits in the same way.

8.3 EXERCISES

1. Consider the following weights, height, and yields of 20 plants.

Plant	Weight	Height	Yield	Plant	Weight	Height	Yield
1	7.3	2.1	0.045	11	43.8	6.7	0.276
2	10.8	3.3	0.132	12	44.9	7.6	0.353
3	12.1	3.5	0.187	13	45.8	5.6	0.462
4	18.6	4.2	0.129	14	46.4	8.0	0.561
5	20.1	5.0	0.201	15	52.7	7.8	0.435
6	32.3	6.1	0.184	16	60.3	9.8	0.598
7	38.4	5.5	0.231	17	98.9	11.2	0.723
8	40.2	7.2	0.298	18	178.3	9.2	0.668
9	42.1	7.4	0.287	19	213.1	13.5	1.022
10	43.8	6.9	0.310	20	298.8	17.3	1.745

Find the following for each of W, H, and Y.
   a. The sample mean.
   b. The sample median.
   c. The trimmed means $\overline{X}_{tr(5)}$, $\overline{X}_{tr(10)}$, and $\overline{X}_{tr(20)}$.
   d. The sample variance and sample standard deviation.

2. Find the normal approximation to the following.
   a. The average of 30 numbers chosen from the exponential p.d.f. $g(x) = 2e^{-2x}$.
   b. The average of 40 numbers chosen from a distribution $N(25.0, 16.0)$.
   c. The average of 30 numbers chosen from the p.d.f. $f(x) = 1$ for $0 \le x \le 1$ (Exercise 8a in Section 6.8).

3. Find 95% confidence intervals for the following, assuming that the standard deviations are known to match those in the given exercises. Does the confidence interval include the true mean?
   a. A sample mean of 0.4 is found in Exercise 2a.
   b. A sample mean of 0.6 is found in Exercise 2a.
   c. A sample mean of 23.4 is found in Exercise 2b.
   d. A sample mean of 0.7 is found in Exercise 2c.

4. Find 98% confidence intervals for the cases in Exercise 3.

5. Use the normal approximation to find 95% confidence limits around the estimated proportion $\hat{p}$ for the following.
   a. A coin is flipped 100 times and comes out heads 44 times.
   b. Of 1000 people polled, 320 favor the use of mathematics in biology.
   c. Thirty cells are treated with a detergent; the membranes of six dissolve.

6. In one set of 1 million DNA base pairs, 14 mutations are found. Use the normal approximation to the Poisson distribution to estimate 95% confidence limits around the true mean number of mutations per million base pairs, and compare with Exercise 7 in Section 8.2.

7. Thirty different sets of 1 million base pairs are measured, and the average number of mutations per million is found to be 13.5.
   a. Estimate the standard deviation.
   b. Find the standard error.
   c. Find the 99% confidence limits.
   d. Why are these narrower than in Exercise 6?

8. Use the computational formula for the sample variance to estimate s^2 from a set of measurements consisting of 20 successes in 100 attempts (thought of as 20 values of 1 and 80 values of 0). Compare your result with $\hat{p}(1 - \hat{p})$ used in Algorithm 8.2.

9. Consider measuring n plants with a known standard deviation of 3.2 cm (as in the text). How many plants would have to be measured to achieve the following?
   a. Ninety-five percent confidence limits 2.0 cm wide
   b. Ninety-five percent confidence limits 0.5 cm wide
   c. Ninety-nine percent confidence limits 2.0 cm wide
   d. Ninety-nine percent confidence limits 0.5 cm wide

10. One measurement of 10.0 is made from a normal distribution with known variance $\sigma^2 = 1$.
    a. Write the likelihood function for the unknown parameter μ.
    b. Write the support function.
    c. Find the maximum likelihood estimator of μ.
    d. Use the method of support to estimate the 95% confidence interval.
    e. Compare with the usual method in the text.

11. We show why using a denominator of n in the equation for the sample variance produces a biased estimate. Suppose a population consists of half 0s and half 1s.
    a. Use the variance of a Bernoulli distribution with $p = 0.5$ to find the exact variance of each measurement.
    b. Find all possible samples of size two and their associated probabilities.
    c. Find the mean squared deviation from the mean for each sample and average them to find the expected sample variance.
    d. Compare with the true answer, and show that using a denominator of $n - 1$ rather than n would give the right answer.
    e. Try to explain the bias.

12. **COMPUTER:** Use the Monte Carlo method to estimate the confidence limits for Exercise 3a and b. How close are your results to the exact answer?

8.4 Hypothesis Testing

Estimators and confidence limits help us figure out what happened in an experiment. Much of science, however, is built upon stating and testing a **hypothesis**. An experimental result may be consistent with a **null hypothesis**, which states that nothing interesting happened, in which case we *accept* the null hypothesis. If the results are inconsistent, we *reject* the null hypothesis. The **significance level**, the probability of rejecting a true null hypothesis, and the **power**, the probability of rejecting a false null hypothesis, quantify the effectiveness of a statistical test. We present two approaches to **hypothesis testing**, one based on the same logic as confidence limits, and one based on the method of support.

Hypothesis Testing: An Example

Extensive sampling of an entire population has shown that exactly 13% of people have a particular gene. We select 100 individuals from this population who have been diagnosed with a particular disease and find that 20 of them have the gene. It seems that more than 13% of individuals with the disease have the gene. Would we be justified in reporting that the gene is connected to the disease?

Hypothesis testing begins by setting up a **null hypothesis**, referred to as H_0, and checking whether the data are *consistent* with it. The null hypothesis is a mathematical way of saying that nothing interesting happened. In our example, the null hypothesis is that diseased individuals have the gene with the *same probability* as everybody else. Let q be the true unknown proportion of diseased individuals in the population who have the gene. The null hypothesis is $q = 0.13$.

We compare this null hypothesis with an **alternative hypothesis**, referred to as H_a. There are two possible alternative hypotheses. If we suspect that the gene is positively related to the disease, the alternative hypothesis is $q > 0.13$. If we simply suspect a relationship of some sort, the alternative hypothesis is $q \neq 0.13$. Strictly speaking, a scientist should state which of these alternatives will be tested before examining the data. The null hypothesis and alternative hypothesis must be mutually exclusive.

A **statistical test** is a way to *accept* or *reject* the null hypothesis. Rejecting the null hypothesis means that we are confident that the null hypothesis is not true. Accepting the null hypothesis means that we cannot prove it wrong. Acceptance does *not* mean that the null hypothesis is correct, just that we do not know enough to reject it. There are four possible results.

There are two correct results: accepting a true null hypothesis and rejecting a false null hypothesis. We may also make errors of two types. We make a **type I error** if we reject a true null hypothesis. If individuals with the disease were really the same as everybody else ($q = 0.13$) but we decided otherwise based on our sample and statistical test, we would commit a type I error. The probability of committing a type I error, called the **significance level**, is denoted by α. We make a **type II error** if we accept a false null hypothesis. If the proportion q is really $q = 0.18$ and we fail to distinguish this from the hypothesized value of 0.13, we have committed a type II error.

The probability of a type I error, denoted by α, depends on our statistical test. The probability of a type II error, denoted by β, depends also on the true value of the parameter. If we fail to capture the fact that $q = 0.5$ is different from

	H_0 accepted	H_0 rejected
H_0 true	correct	type I error
H_0 false	type II error	correct

$q = 0.13$, we feel humble. If we fail to capture the fact that $q = 0.131$ is different from $q = 0.13$, we feel fine. The probability of rejecting a false null hypothesis as a function of the true value of the parameter q is called the **power** of a test.

■ **Definition 8.10** The power of a statistical test is the probability that it correctly rejects a false null hypothesis:

$$\text{power of test} = 1 - \text{probability of a type II error} = 1 - \beta$$

Ideally, we could design tests that make neither type I nor type II errors. Because chance plays many tricks, this is impossible. There is a tradeoff between type I and type II errors. A very powerful test that always rejects a false null hypothesis is also more likely to reject a true null hypothesis. A weaker test that never rejects a true null hypothesis will often fail to reject a false null hypothesis. Think of a statistical test as a sort of detector (Figure 8.21). A detector that is not very powerful will detect only genuine signals (no type I errors) but will miss weaker signals (many type II errors). A very powerful detector will not miss any real signals (no type II errors) but will misidentify noise as genuine alien communication (many type I errors).

Small detector captures few signals but all are from aliens

Powerful detector captures many signals, many not from aliens

Few type I errors
Many type II errors

Few type II errors
Many type I errors

a   b

Figure 8.21
Detecting aliens: type I and type II errors

How do we design a test? Consider the null hypothesis $q = 0.13$ and the alternative $q > 0.13$. One method is based on **p-values**.

■ **Definition 8.11** The p-value is the probability of observing a result at least as extreme as the measured result if the null hypothesis is true.

The probability of finding 20 or more people with the gene if $q = 0.13$ is

$$p = \sum_{k=20}^{100} b(k; 100, 0.13) = 0.0319$$

(Figure 8.22a). Because this region includes only one side, this is called a **one-tailed test**. The probability of a result at least as extreme as what we observed is 3.2% if the null hypothesis is true. If we reject the null hypothesis, we have a 3.2% chance of making a type I error, the probability that our extreme result occurred by chance. If we decided to treat diseased individuals with an expensive

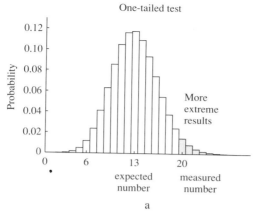
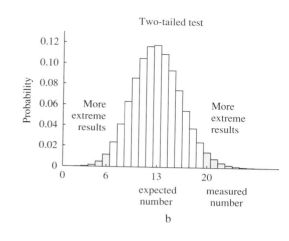

Figure 8.22

Probability of result more extreme than 20 if $q = 0.13$

new gene therapy, there is a 3.2% chance we would be throwing our money away. The chance of making a type I error is called the **significance** of the result. A p-value less than 0.05 is generally considered to be *significant*.

If we had found 23 rather than 20 people out of 100, our p-value would be

$$p = \sum_{k=23}^{100} b(k; 100, 0.13) = 0.00424$$

When the chance of a type I error is less than 0.01, the result is considered to be **highly significant**. Twenty-three out of 100 people is convincing evidence that the true q is greater than 0.13. When the chance of a type I error is less than 0.001, the result is considered to be **very highly significant**. One would be very unlucky indeed to get such an unlikely result by chance.

p-value	Significance
$p > 0.1$	not significant
$0.1 > p > 0.05$	trend toward significance
$0.05 > p > 0.01$	significant
$0.01 > p > 0.001$	highly significant
$p < 0.001$	very highly significant

Figure 8.23

Probability of result more extreme than 23 if $q = 0.13$

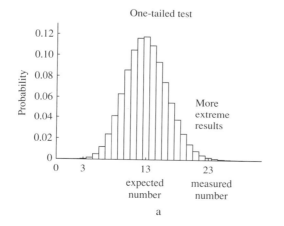
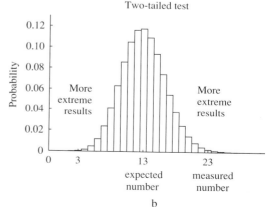

What happens if our alternative hypothesis is $q \neq 0.13$? The identification of the set of results more extreme than our observation is not so simple. In this case, "more extreme" means further from the null expectation (13 out of 100) on either side. A result of 6 or fewer is considered as extreme as a result of 20 or more because both lie at least 7 away from 13 (the shaded region in Figure 8.22b). This is called a **two-tailed test**. The p-value is then

$$p = \sum_{k=0}^{6} b(k; 100, 0.13) + \sum_{k=20}^{100} b(k; 100, 0.13) = 0.0512$$

The result is no longer significant, although it is close.

Testing hypotheses with p-values is closely related to the calculation of confidence intervals. If we counted 20 people out of 100 with the gene, the 95% confidence interval runs from 0.127 to 0.292. The two-tailed test asks precisely whether the null hypothesis lies between these confidence limits. The null hypothesis $q = 0.13$ lies within these limits and thus is not rejected.

Why is the significance level different for the one-tailed test and two-tailed test? Both use the same data, the same null hypothesis, and the same underlying model. The two-tailed test, however, uses a less informative alternative hypothesis. It is appropriate to use the one-tailed test only if

1. you have reason to believe that the results will fall on a particular side of the null hypothesis, and

2. you make a public announcement of this expectation *before* looking at the data.

If laboratory evidence indicates that a particular gene is related to a disease, then a public prediction that the fraction of diseased individuals with the gene will be greater than 0.13 makes the one-tailed test appropriate.

When the sums required to compute p-values cannot be evaluated, the Monte Carlo method can be used. To test whether data is consistent with the null hypothesis, assume the null hypothesis and simulate many times on a computer. The results of 100 simulations with $q = 0.13$ are shown in Table 8.2. In this case, only 3 out of 100 results equal or exceed the measured result of 20 out of 100. If we are using a one-tailed test, this indicates that the significance of the result is 0.03, quite close to the exact significance level of 0.0319. According to this Monte Carlo simulation, the result is indeed significant. Were the Monte Carlo result to exceed the measured result less than 1% of the time, the result would be highly significant.

Table 8.2 Testing a Hypothesis with the Monte Carlo Method

Number out of 100	Number of times out of 100 replicates	Number out of 100	Number of times out of 100 replicates
6	1	15	6
7	5	16	6
8	7	17	10
9	7	18	2
10	9	19	3
11	10	20	0
12	9	21	2
13	10	22	1
14	12		

Power

How many more than 13 people out of 100 would we have to find before we could reject the null hypothesis at the 0.05 significance level? Suppose that only 19 people out of 100 were found with the gene. The significance level with a one-tailed test is

$$p = \sum_{k=19}^{100} b(k; 100, 0.13) = 0.056$$

This result is not considered strong enough to reject the null hypothesis. If our null hypothesis is $q = 0.13$, we reject it only if we find 20 or more people with the gene in a sample of 100. Think of a test as a buzzer that goes off when the data are extreme enough to reject the null hypothesis. The buzzer goes off if 20 or more people are found with the gene.

How powerful is this test? Suppose the "true" q for diseased individuals was 0.15, only a bit larger than 0.13. How likely are we to miss this (commit a type II error) with a one-tailed test? The buzzer fails to go off if we find 19 or fewer people with the gene. This event has probability β given by

$$\beta = \sum_{k=0}^{19} b(k; 100, 0.15) = 0.893$$

(Figure 8.24a). We fail to reject the null hypothesis (by failing to count 20 or more people with the gene) 89% of the time when $q = 0.15$. Our power to find the difference is only 11%. When we compare this with the significance level of about 3%, we see that we are nearly as likely to make a type I error as we are to reject the null hypothesis correctly. A sample of 100 is too small to effectively distinguish 0.13 from 0.15.

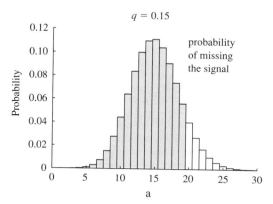
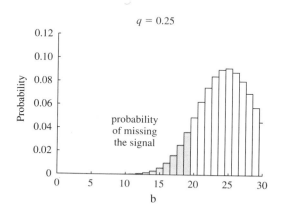

Figure 8.24

Power with two different values of q

If the true q is 0.25, the buzzer fails to go off with probability

$$\beta = \sum_{k=0}^{19} b(k; 100, 0.25) = 0.099$$

We have less than a 10% chance of missing and a 90% chance of recognizing this much stronger signal (Figure 8.24b).

The power is different from the significance level in that it cannot be calculated without knowing the true value of the parameter. Because the true value is unknown, the best we can do is to graph the power as a function of the true parameter value (Figure 8.25). The farther the true parameter is from 0.13, the better we can distinguish it from 0.13. Computing the power function gives an idea of whether a test can recognize signals of a particular strength.

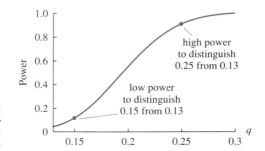

Figure 8.25

Power as a function of the true parameter value

Likelihood and the Method of Support

The computation of p-values includes unobserved data. In our example, we did not observe 21, 22, or more people with the gene. Strictly speaking, there is no logical justification for including these values in our calculation of p-values. It has been said of p-values that "a hypothesis that may be true may be rejected because it has not predicted observable results that have not occurred." Furthermore, in some cases it is not easy to decide what results should be considered "more extreme" than the observation. Is 6 out of 100 really exactly as extreme as 20 out of 100?

One can avoid these problems with the method of support. We compare the null hypothesis with the alternative by comparing the support for each. If the support for the alternative hypothesis exceeds that for the null by more than 2, we place the null hypothesis under doubt (devotees of likelihood do not believe in rejecting hypotheses).

■ **Algorithm 8.3** (Comparing hypotheses with the method of support)

1. Find the likelihood and support functions.

2. Compute the support under the null hypothesis.

3. Compute the maximum of the support under the alternative hypothesis.

4. If the maximum support under the alternative hypothesis exceeds the support under the null hypothesis by more than 2, view the null hypothesis with doubt. ■

For the data set consisting of 20 out of 100 diseased individuals with a gene, the likelihood and support functions are

$$L(p) = b(20; 100, p)$$
$$S(p) = \ln[b(20; 100, p)]$$

The support for the null hypothesis is

$$S(0.13) = \ln[b(20; 100, 0.13)] = -4.215$$

The support of the alternative hypothesis is the *maximum* over all possible values in the alternative. Both alternatives $q > 0.13$ and $q \neq 0.13$ include the maximum likelihood estimator $q = 0.2$, for which

$$S(0.2) = -2.310$$

(Figure 8.26). The difference is

$$S(0.2) - S(0.13) = -2.310 - (-4.215) = 1.905$$

which is slightly less than 2. The null hypothesis might well explain the data. This is consistent with the finding, by the two-tailed test, that the result is not quite significant.

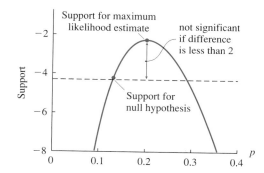

Figure 8.26
Hypothesis testing with the method of support

SUMMARY

To test whether an observation is consistent with a **null hypothesis** or an **alternative hypothesis** we have introduced **p-values**, the probability of measuring a result at least as extreme as the actual measurement when the null hypothesis is true. The test is called a **one-tailed test** when extreme results are included on one side, and a **two-tailed test** if extreme results are included on both sides. The **significance level** is the probability of a **type I error**, the rejection of a true null hypothesis. Accepting a false null hypothesis is called a **type II error**. The efficiency with which a test rejects a false null hypothesis, called the **power** of a test, depends on the true value of the parameter. The method of *support* provides an alternative way to compare hypotheses.

8.4 EXERCISES

1. Certain screening tests for cancer work by examining many cells under a microscope and looking for abnormalities. Discuss how setting the threshold for cell abnormality can affect the number of type I and type II errors. What factors would go into deciding where to set the threshold?

2. A supposedly fair coin is flipped 10 times and comes up heads 9 times. You have reason to suspect however, that the coin produces an excess of heads (and have told this to a friend).
   a. State a null hypothesis and an alternative.
   b. Compute the p-value.
   c. What is the cutoff number of heads above which you reject the null hypothesis at the $\alpha = 0.05$ significance level?
   d. Find the power of your test if the true probability of a head is 0.6.
   e. Find the power of your test if the true probability of a head is 0.9. What does this mean?

3. Phone calls used to arrive at an average rate of 3.5 per hour, but after placing a car ad, you receive 7 in 1 h.

a. State null and alternative hypotheses.
b. Use the Poisson distribution to compute the probability of this event if the null hypothesis is true.
c. Compute the probability of an event at least this extreme if the null hypothesis is true.
d. Is this result significant?

4. Consider again the data of Exercise 3.
   a. Find the cutoff value for a test with $\alpha = 0.05$. Find the power with $\lambda = 4.0$, $\lambda = 7.0$, and $\lambda = 10.0$.
   b. Find the cutoff value for a test with $\alpha = 0.01$. Find the power with $\lambda = 4.0$, $\lambda = 7.0$, and $\lambda = 10.0$.
   c. Why does a higher significance level reduce the power?

5. The survival time for a light bulb is known to be exponentially distributed with a mean of 1000 h. After you apply a special bulb invigorating lotion, one bulb lasts 4000 h.
   a. State null and alternative hypotheses.
   b. At what level can you reject the null hypothesis? Is it significant?
   c. What is the shortest survival time over 1000 h for which you might claim a significant result (at the 0.05 level)?

6. Suppose you adopt the cutoff from Exercise 5b.
   a. What is the power of the test if the true mean is 1500?
   b. What is the power of the test if the true mean is 3500?
   c. Find and graph the power as a function of the true mean.
   d. Solve for when the power is equal to 0.95. Interpret your answer.

7. Consider again a treated light bulb as in Exercise 5. A sceptical friend asserts, "That bulb is going to blow out in no time." The bulb then lasts only 40 h.
   a. State the friend's null and alternative hypotheses.
   b. Find the significance level.
   c. Does it seem appropriate to reject the null hypothesis? (Think about the mode of the exponential distribution.)
   d. What would the friend say if the bulb lasted 4000 h?

8. It is known that 10% of light bulbs are defective. You are testing light bulbs produced by a new manufacturing process and find that the first defective light bulb is the 25th.
   a. State null and alternative hypotheses.
   b. At what level can you reject the null hypothesis?
   c. What is the smallest number of light bulbs greater than 10 for which you can reject at the 0.05 level?

9. Find the difference in support of the following hypotheses.
   a. You flip a coin ten times and get nine heads. The null hypothesis is that $p = 0.5$ with alternative $p > 0.5$.
   b. You receive 7 phone calls in an hour. The null hypothesis is that $\lambda = 3.5$ per hour with alternative that $\lambda > 3.5$ per hour.
   c. You wait 4000 h for an exponentially distributed event to occur. The null hypothesis is that the mean wait is 1000 h with alternative that the mean wait is greater than 1000 h.
   d. You wait 40 h for an exponentially distributed event to occur. The null hypothesis is that the mean wait is 1000 h with alternative that the mean wait is less than 1000 h.
   e. The first defective light bulb is the 25th. The null hypothesis follows a geometric distribution with parameter $q = 0.1$.

10. **COMPUTER:** Repeat the calculations in Table 8.2. Is the hypothesis $q = 0.13$ rejected? Try the experiment until you find that the result is not significant. What does this mean? How many experiments would it take, on average, to produce this result?

11. **COMPUTER:** Generate nine independent random numbers from a normal distribution with mean 10 and variance 9, and find their average $\hat{\mu}$. If your value of $\hat{\mu}$ is within 0.5 of 10.0, try again to guarantee nice pictures and interesting results (real scientists are not allowed to do this sort of thing, of course). The average of your nine measurements comes from a normal distribution with mean 10 and some standard error. Find the standard error and the 95% confidence interval around $\hat{\mu}$, calling the lower limit μ_l and the upper limit μ_h.

    Simulate the following four experiments 100 times: (1) average nine values from a normal distribution with true mean 10.0; (2) average nine values from a normal distribution with true mean μ_l; (3) average nine values from a normal distribution with true mean μ_h; (4) average nine values from a normal distribution with true mean $\hat{\mu}$. Count how many values in each experiment lie below μ_l, between μ_l and 10.0, between 10.0 and μ_h, and above μ_h.

    Define functions f_1, f_2, f_3, and f_4 to be the p.d.f.s describing the distribution of elements in the four experiments. Print a graph of each function, and list below it the number of elements from the appropriate experiment lying in the various intervals. Indicate which ones have values predicted by the theory of confidence intervals and what those values should be. Are you bothered by the fact that more than 5% of the results of experiment 1 lie outside your confidence interval? Why not?

8.5 Hypothesis Testing: Normal Theory

The methods of Section 8.4, p-values and support, apply to distributions generated by any underlying model. Because of the Central Limit Theorem, the normal distribution is often the appropriate underlying model for biological systems. Here we apply the methods of hypothesis testing when the underlying distributions are normal or approximately so, compute the power of the tests, and compare the results with the method of support.

Computing p-Values with the Normal Approximation

Suppose we try out a new fertilizer on plants that previously had a mean height of 39.0 cm. The fertilized plants have a mean height of 40.2 cm. Are we convinced that the fertilizer works?

Our null hypothesis is that the mean height of our fertilized plants is still 39.0 and that the increase was due to chance. If we had announced that the plants would be taller, the alternative hypothesis would be that the fertilized plants have mean height greater than 39.0, leading to a one-tailed test. If we had reason to fear that the fertilizer might inhibit growth, our alternative hypothesis would be that the height of the fertilized plants is different from 39.0, leading to a two-tailed test.

Let μ represent the true mean of the fertilized plants. The null hypothesis is $\mu = 39.0$. When we believe that the fertilizer fertilizes, the alternative hypothesis is $\mu > 39.0$. When we fear that the fertilizer might not fertilize, the alternative hypothesis is $\mu \neq 39.0$. Suppose that plant heights have a normal distribution with known variance σ^2. If we measure n plants, the sample mean has a normal distribution with

$$\text{sample mean} \sim N\left(\mu, \frac{\sigma^2}{n}\right)$$

(Central Limit Theorem for Means).

Suppose we measure 25 plants and know that the standard deviation of fertilized plants is 3.2 cm (so that the variance is 10.24 cm^2). To test the one-tailed alternative hypothesis, we compute the distribution X of sample means under the null hypothesis to find the probability of measuring a sample mean of 40.2 or more with a sample of size 25 from a normal distribution with mean 39.0 and variance 10.24. From the Central Limit Theorem, we have

$$X \sim N\left(39.0, \frac{10.24}{25}\right) = N(39.0, 0.4096)$$

Using Algorithm 7.2 to find the probability with the standard normal distribution, we get

$$\Pr(X \geq 40.2) = \Pr\left(\frac{X - 39.0}{\sqrt{0.4096}} \geq \frac{40.2.0 - 39.0}{\sqrt{0.4096}}\right)$$
$$= \Pr(Z \geq 1.875)$$
$$= 1 - \Phi(1.875) = 0.030$$

(Figure 8.27a). With the one-tailed test, the result is significant at the 0.03 level. There is, however, a 3.0% chance of accepting a fertilizer that does not work.

If we have reason to suspect that fertilized plants may be shorter, we use the alternative hypothesis $\mu \neq 39.0$. Remember that one must decide on the appropriate test before looking at the data. Every data set looks odd in one way or another, and snooping around to find these oddities generally leads to mistakes. If the sample mean of fertilized plants had turned out to be 37.6, we cannot pretend

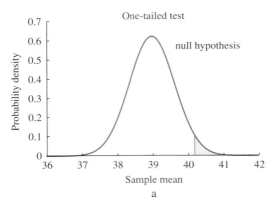
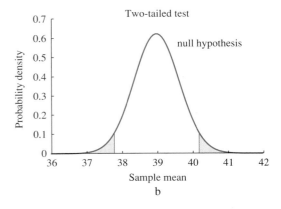

Figure 8.27
One-tailed and two-tailed hypothesis tests

we expected a decrease unless we made a public announcement to that effect before examining the data.

With a two-tailed test, the set of results more extreme than our observation includes all results more than 1.2 away from 39.0, greater than 40.2 and less than 37.8 (Figure 8.27b).

$$\Pr(X \geq 40.2 \text{ or } X \leq 37.8) = \Pr\left(\frac{X - 39.0}{\sqrt{0.4096}} \geq \frac{40.2.0 - 39.0}{\sqrt{0.4096}}\right)$$
$$+ \Pr\left(\frac{X - 39.0}{\sqrt{0.4096}} \leq \frac{37.8.0 - 39.0}{\sqrt{0.4096}}\right)$$
$$= \Pr(Z \geq 1.875) + \Pr(Z \leq -1.875)$$
$$= 1 - \Phi(1.875) + \Phi(-1.875)$$
$$= 0.061$$

With this alternative hypothesis, the result is not significant. The loss of significance is a consequence of our less informative alternative hypothesis. In general, the significance level of the two-tailed test is double that of the one-tailed test when the distribution is symmetric (Figure 8.27).

If we have a sufficiently large sample ($n \geq 30$), we use the sample variance s instead of σ and follow the same steps. Suppose we do not know that the true variance σ^2 is 10.24. If we measured fewer than 30 plants, we could estimate the sample variance in the usual way, but we would have to consult a statistics text for a different test called the **t test**. We decide instead to measure 25 more plants, for a sample of size $n = 50$. Suppose that the sample variance s^2 turns out to be 12.45. The sample mean of 50 plants has variance

$$\frac{s^2}{n} = \frac{12.45}{50} = 0.249$$

so that

$$\text{standard error} = \sqrt{0.249} = 0.499$$

Applying the one-tailed test, we get

$$\Pr(X \geq 40.2) = \Pr\left(\frac{X - 39.0}{0.499} \geq \frac{40.2.0 - 39.0}{0.499}\right)$$

$$= \Pr(Z \geq 2.405)$$
$$= 1 - \Phi(2.405) = 0.008.$$

The result is highly significant, the benefit of using a larger sample size (Figure 8.28). Although the difference in the means is the same and the variance is slightly larger, the larger sample size makes our result more significant.

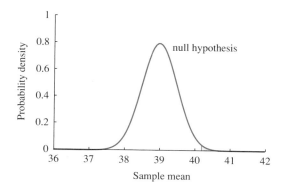

Figure 8.28
One-tailed test with a larger sample size

We can summarize these methods in the following algorithm.

■ **Algorithm 8.4** **(Hypothesis testing with the normal distribution)**

Suppose a set of n measurements have sample mean $\hat{\mu}$, and the null hypothesis is that the true mean is μ_0. If the variance of the underlying distribution is the known value σ^2, then the distribution X of sample means under the null hypothesis is

$$X \sim N\left(\mu_0, \frac{\sigma^2}{n}\right)$$

If the variance of the underlying distribution is unknown but $n \geq 30$, the underlying variance can be estimated as the sample variance s^2. The distribution X of sample means under the null hypothesis is

$$X \sim N\left(\mu_0, \frac{s^2}{n}\right)$$

1. If $\hat{\mu} > \mu_0$, the two-tailed test computes

$$\text{p-value} = \Pr(X \geq \hat{\mu}) + \Pr[X \leq \mu_0 - (\hat{\mu} - \mu_0)]$$
$$= \Pr(X \geq \hat{\mu}) + \Pr(X \leq 2\mu_0 - \hat{\mu})$$

The one-tailed test computes

$$\text{p-value} = \Pr(X \geq \hat{\mu})$$

2. If $\hat{\mu} < \mu_0$, the two-tailed test computes

$$\text{p-value} = \Pr(X \leq \hat{\mu}) + \Pr[X \geq \mu_0 + (\mu_0 - \hat{\mu})]$$
$$= \Pr(X \leq \hat{\mu}) + \Pr(X \geq 2\mu_0 - \hat{\mu})$$

The one-tailed test computes

$$\text{p-value} = \Pr(X \leq \hat{\mu})$$

■

The same methods apply to the binomial distribution. Suppose, as in Section 8.4, that we identify 20 out of 100 people with a gene, and compare with the null hypothesis that 13% have the gene. If the null hypothesis is true, then the mean number G has a binomial distribution with $n = 100$ and $p = 0.13$. Then

$$E(G) = 13.0 \text{ and } Var(G) = 11.31$$

The normal approximation is therefore

$$X \sim N(13, 11.31)$$

The p-value for a one-tailed test is

$$Pr(G \geq 20) = Pr(X \geq 19.5) \quad \text{continuity correction}$$

$$= 1 - \Phi\left(\frac{19.5 - 13.0}{\sqrt{11.31}}\right) \quad \text{use Algorithm 7.2}$$

$$= 0.027 \quad \text{compute value with computer or table}$$

The result is significant, and quite close to the significance level of 0.0319 computed with the exact method (Figure 8.29).

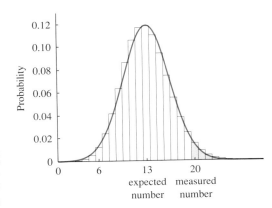

Figure 8.29
One-tailed test with the normal approximation to the binomial distribution

The Power of Normal Tests

Consider again unfertilized plants with a mean height of 39.0 and known standard deviation of 3.2. How much larger than 39.0 would the sample mean have to be for us to reject the null hypothesis at the 0.05 level with a sample size of 25? Let X be a random variable giving the sample mean under the null hypothesis, having mean 39.0 and standard error $3.2/5.0 = 0.64$. We are looking for the value $\overline{X}$ such that

$$Pr(X > \overline{X}) = 0.05$$

We transform the data into a standard normal distribution and find that this occurs when

8.5 Hypothesis Testing: Normal Theory

$$0.05 = \Pr\left(\frac{X - 39.0}{0.64} > \frac{\overline{X} - 39.0}{0.64}\right)$$

$$= \Pr\left(Z > \frac{\overline{X} - 39.0}{0.64}\right)$$

We are looking for the value that lies above all but 5% of the standard normal distribution, which can be found as 1.645 on a table ($\Phi(1.645) = 0.95$). Therefore, the cutoff is where

$$\frac{\overline{X} - 39.0}{0.64} = 1.645$$

which has solution

$$\overline{X} = 39.0 + 1.645 \cdot 0.64 = 40.05$$

A result is significant with a one-tailed test if it lies 1.645 standard errors above the mean.

To find the power of our test, we try different possible true means greater than 39.0 and compute how likely our test is to detect them. Suppose the true mean is 40.0. The probability making a type II error and failing to reject the null hypothesis is the probability of picking 25 plants from a distribution $N(40.0, 10.24)$ and getting an average height of less than 40.05. Let the random variable Y represent the mean of these 25 plants. Then

$$Y \sim N\left(40.0, \frac{10.24}{25}\right) = N(40.0, 0.4096)$$

from the Central Limit Theorem for Means, and

$$\Pr(Y < 40.05) = \Pr\left(\frac{Y - 40.0}{\sqrt{0.4096}} < \frac{40.05 - 40.0}{\sqrt{0.4096}}\right)$$

$$= \Pr(Z < 0.078)$$

$$= \Phi(0.078) = 0.531$$

The power of the test is $1 - 0.531 = 0.469$. Our test has a more than 50% chance of missing an effect this small. The area associated with this calculation is shown in Figure 8.30a.

Figure 8.30

The power of a normal test with different true means

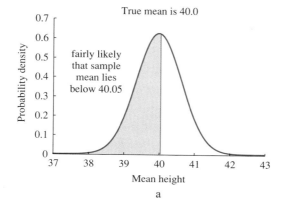
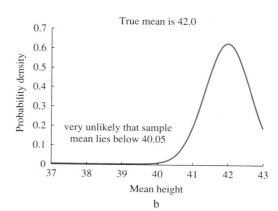

If the true mean of fertilized plants were 42.0,

$$Y \sim N\left(42.0, \frac{10.24}{25}\right) = N(42.0, 0.4096)$$

The probability of missing a result this large is

$$\Pr(Y < 40.05) = \Pr\left(\frac{Y - 42.0}{\sqrt{0.4096}} < \frac{40.05 - 42.0}{\sqrt{0.4096}}\right)$$

$$= \Pr(Z < -3.047)$$

$$= \Phi(-3.047) = 0.0016$$

(Figure 8.30b). The power is 0.9984, and we are unlikely to miss an effect this large.

These calculations of power depend on the true mean, which we do not know. What good are they? They have shown that our test is unlikely to detect a difference of 1.0 cm (true mean of 40.0) but quite likely to detect a difference of 3.0 cm (true mean of 42.0). If we are not concerned with such small effects as a 1.0-cm change in height, this test is appropriate.

If we are concerned with smaller effects, we must come up with a more precise test. Generally, this requires increasing the sample size to decrease the standard error and increase our resolution. For example, suppose we wish our test to have 90% power to detect a difference of 1.0 cm. How large a sample size do we need? Let the random variable Y represent the mean of these n plants. Then

$$Y \sim N\left(40.0, \frac{10.24}{n}\right)$$

We have found that results are significant with a one-tailed test if they lie 1.645 standard errors above the null hypothesis, or

$$\hat{\mu} \geq 39.0 + 1.645\sqrt{\frac{10.24}{n}} = 39.0 + \frac{5.264}{\sqrt{n}}$$

The probability of missing the signal is the probability that the sample mean is less than this value, or

$$\Pr\left(Y < 39.0 + \frac{5.264}{\sqrt{n}}\right) = \Pr\left(\frac{Y - 40.0}{3.2/\sqrt{n}} < \frac{39.0 + (5.264/\sqrt{n}) - 40.0}{3.2/\sqrt{n}}\right)$$

$$= \Pr\left(Z < 1.645 - \frac{\sqrt{n}}{3.2}\right) = 0.1$$

This has a solution where $\Phi(z) = 0.1$, or at -1.28. Solving for n, we get

$$1.645 - \frac{\sqrt{n}}{3.2} = -1.28$$

$$\sqrt{n} = 3.2 \cdot (1.645 + 1.28) = 9.36$$

$$n = 88$$

It would take 88 plants to detect a 1.0-cm change in height with 90% probability. Furthermore, the threshold for triggering the test is

$$39.0 + \frac{3.2}{\sqrt{88}} = 39.56$$

Decisions regarding the level of power desired, the degree of significance, and the smallest difference that can be detected are part of the discipline of **statistical design**. The underlying principle, however, is that of statistics itself; understanding the connection between measurements and the underlying probabilistic model.

Likelihood and the Normal Distribution

The criticisms of p-values made in Section 8.4 also apply here. We can use the method of support as an alternative. To do so, we use a fundamental property of the support function. If we draw independent random variables with values $X_1, \ldots, X_n$ from a p.d.f. $f(x)$, the likelihood L of this result can be found by multiplying the likelihood of each individual measurement (Theorem 6.3), so we have

$$L = f(X_1) \cdots f(X_n)$$

Letting $S = \ln(L)$, and using the fact that the logarithm converts multiplication into addition, we get

$$S = \sum_{i=1}^{n} \ln[f(X_i)]$$

We give the following theorem:

■ **THEOREM 8.4** The support of a model from a set of independent observations is the sum of the support from each observation. ■

The support of the parameter μ from an observation X_i from a normal distribution is

$$\ln[f(X_i)] = \ln\left(\frac{1}{\sqrt{2\pi}\sigma} e^{-(X_i-\mu)^2/2\sigma^2}\right)$$

$$= \frac{(X_i - \mu)^2}{2\sigma^2} - \ln(\sqrt{2\pi}\sigma)$$

The likelihood of μ is the product of these terms. Suppose that the variance σ^2 is known. The support of the data set for a hypothesized mean μ is

$$S(\mu) = \sum_{i=1}^{n} \left(\frac{-(X_i - \mu)^2}{2\sigma^2} - \ln(\sqrt{2\pi}\sigma)\right)$$

After much algebra (Exercise 9), we find

$$S(\mu) = -\frac{n(\overline{X} - \mu)^2}{2\sigma^2} + c$$

where $\overline{X}$ is the sample mean and c is a constant (depending on σ).
Because

$$S'(\mu) = \frac{n(\overline{X} - \mu)}{\sigma^2}$$

the maximum support occurs at $\mu = \overline{X}$. The maximum likelihood estimate of μ is the sample mean $\overline{X}$. We mistrust any hypothesis (a particular value of μ) for which

$$S(\mu) < S(\overline{X}) - 2$$

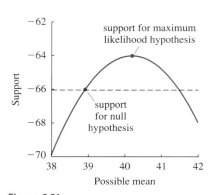

Figure 8.31

The support as a function of the mean

The crossing points are where

$$\frac{n(\overline{X} - \mu)^2}{2\sigma^2} = 2.0$$

which has solutions at

$$\mu = \overline{X} \pm \frac{2\sigma}{\sqrt{n}}$$

virtually identical to the test for a significance level of 0.05 using p-values with a large sample. A likelihood difference of 2 corresponds to means differing by 2 standard errors.

The support function for a sample of size 25 with known variance of 10.24 and sample mean 40.2 is plotted in Figure 8.31. The value $\mu = 39.0$ lies within these bounds and is not convincingly inconsistent with the data, matching the results of the two-tailed test.

SUMMARY

We have seen how to compute p-values when the underlying distribution is normal with known variance or large enough for us to estimate the sample variance accurately. In each case, we transform the measured sample mean into a random variable with a standard normal distribution. The same approach can be used to find the power of a test and to estimate the sample size needed to identify a particular difference. The method of support can be used as an alternative way to compare hypotheses.

8.5 EXERCISES

1. The weights, heights, and yields of 10 plants grown in an experimental plot are given in the table. Each measurement is known to be normally distributed with known variances of 9.0 for weight, 16.0 for height, and 6.25 for yield. Plants outside the plot have mean weight 10.0, mean height 38.0, and mean yield 9.0.

Number	Weight	Height	Yield
1	11.83	46.2	10.11
2	7.65	39.4	9.29
3	14.57	41.4	12.26
4	14.97	40.8	6.61
5	14.26	38.5	9.51
6	11.26	40.7	8.34
7	11.08	39.8	13.35
8	7.04	37.9	12.90
9	5.44	34.8	10.35
10	11.82	43.1	9.84

   a. State one and two-tailed hypotheses for weight and find their significance.
   b. Do part **a** for height.
   c. Do part **a** for yield.

2. For the circumstances in Exercise 1, find the smallest values of the sample mean for which the given hypotheses are rejected.
   a. The mean weight is 10.0 with a one-tailed test at the 0.01 level.
   b. The mean height is 38.0 with a two-tailed test at the 0.05 level.
   c. The mean yield is 9.0 with a two-tailed test at the 0.001 level.

3. Find the power of the tests from Exercise 2 assuming the following true means.
   a. The true mean weight is 13.0 for Exercise 1a.
   b. The true mean height is 43.0 for Exercise 1b.
   c. The true mean yield is 11.0 for Exercise 1c.

4. How many standard errors from the mean are the following? What are the corresponding p-values for a two-tailed test?
   a. The support for the null hypothesis is less than the maximum support by 2.
   b. The support for the null hypothesis is less than the maximum support by 3.
   c. The support for the null hypothesis is less than the maximum support by 4.
   d. The support for the null hypothesis is less than the maximum support by 5.

5. Consider again plants with the null hypothesis that mean height is 39.0. Assume that the standard deviation is known to be 3.2 cm.
   a. Show that a measured sample mean of 40.0 is highly significant if the sample size is $n = 88$.
   b. Why is the power with this sample size only 90% (as found in the text)?
   c. Without doing any new calculations, find the significance level with a one-tailed test if the sample mean is 39.44.

6. Of 100 patients tested with a new drug, 60 improve, but it is known that 45% improve without it.
   a. Find the significance of this result.
   b. How much better would the result be if 120 out of 200 patients improved?
   c. Why might a company not start out by testing 200 people?

7. A new method to reduce mutation rates is being tested. The number of mutations in a well-studied piece of DNA is known to have a Poisson distribution with mean 35.0. The new piece has only 27 mutations.
   a. Find the normal approximation to the null hypothesis.
   b. Find the significance level.
   c. What is the largest number of mutations which would reject the null hypothesis at the 0.01 level?

8. Recall the two populations in Exercise 6 in Section 7.8. Think of the assumptions there as the null hypothesis, with the alternative that the populations are doing better.
   a. Suppose population 1 was found to be 15.0 at time 20. Find the significance level.
   b. Suppose population 2 was found to be 15.0 at time 20. Find the significance level.

9. Use the method of support to check the hypotheses in Exercise 1.

10. Follow the steps to show that the support has the simple quadratic form given in the text.
    a. Show that
    $$S(\mu) = -\frac{1}{2\sigma^2}\left[(n-1)\sigma^2 + n(\bar{X} - \mu)^2\right] - n\ln\left(\sqrt{2\pi}\sigma\right)$$
    Expand the quadratic and substitute definitions of $\bar{X}$ and σ^2.
    b. Remove the terms that do not depend on μ.
    c. Show that the maximum occurs at $\mu = \bar{X}$.

8.6 Comparing Experiments: Unpaired Data

In Sections 8.4 and 8.5, we tested whether the results of an experiment are consistent with a previously known baseline. Often the baseline is unknown and must be estimated simultaneously with a controlled experiment. When a different sample is measured for each experiment, the data are said to be **unpaired**. We must test for a difference between the results of two experiments, each with some level of uncertainty. The general method is the same: set up null and alternative hypotheses and check whether the data are consistent with the null hypothesis. The method of support provides a flexible alternative to the normal distribution when sample sizes are small.

Unpaired Normal Distributions

A widely used and important hypothesis test uses p-values to compare the means of samples from two normally distributed populations. We consider only cases with known variance or large sample size (when the sample variance is a dependable estimate) and leave the case of unknown variance to a more advanced text.

Let $\bar{X}_1$ and $\bar{X}_2$ be random variables representing the means of samples of size n_1 and n_2 with known variances σ_1^2 and σ_2^2 and unknown means μ_1 and μ_2. Think of the first population as the control and the second as the experimental population. By the Central Limit Theorem for Means,

$$\overline{X}_1 \sim N\left(\mu_1, \frac{\sigma_1^2}{n_1}\right)$$

$$\overline{X}_2 \sim N\left(\mu_2, \frac{\sigma_2^2}{n_2}\right)$$

Let the new random variable D,

$$D = \overline{X}_1 - \overline{X}_2$$

represent the difference of the two means. Then

$$E(D) = E(\overline{X}_1) - E(\overline{X}_2) = \mu_1 - \mu_2$$

$$\text{Var}(D) = \text{Var}(\overline{X}_1) + \text{Var}(\overline{X}_2) = \frac{\sigma_1^2}{n_1} + \frac{\sigma_2^2}{n_2}$$

because the expectation of the difference is the difference of the expectations (Theorem 7.6) and the variance of the difference is the sum of the variances (Theorem 7.11). Furthermore, D has a normal distribution because of the additive property of the normal distribution (Section 7.8). Therefore,

$$D \sim N\left(\mu_1 - \mu_2, \frac{\sigma_1^2}{n_1} + \frac{\sigma_2^2}{n_2}\right)$$

If we took samples from each population, computed the sample means, and found their difference many times, these differences would follow this distribution.

We can use the new random variable D to express the null hypothesis in a convenient way. Our null hypothesis is that the treatment has no effect, or that $\mu_1 = \mu_2$. In this case $\mu_1 - \mu_2 = 0$ and the distribution of D is

$$D \sim N\left(0, \frac{\sigma_1^2}{n_1} + \frac{\sigma_2^2}{n_2}\right)$$

This is convenient because it reduces a null hypothesis about *two numbers* (μ_1 and μ_2) into a null hypothesis about *one number* (the difference $\mu_1 - \mu_2$). We use this mathematical version of the null hypothesis to test whether the difference between the sample means differ more than we would expect by chance.

Suppose we treat 25 plants with a new fertilizer and keep 50 untreated plants as controls. Assume that the variance with or without fertilizer has the known value of 10.24 cm². The known parameters describing the control population are

$$n_1 = 50 \quad \text{and} \quad \sigma_1^2 = 10.24$$

and the parameters describing the treatment population are

$$n_2 = 25 \quad \text{and} \quad \sigma_2^2 = 10.24$$

If we measure sample means of $\hat{\mu}_1 = 39.0$ in the control and $\hat{\mu}_1 = 40.2$ in the treated population, do we have reason to suspect that the underlying populations are really different?

Under the null hypothesis, the difference of the sample means has distribution

$$D \sim N\left(0, \frac{10.24}{50} + \frac{10.24}{25}\right) = N(0, 0.6144)$$

(Figure 8.32). What is the probability that the difference exceeds the measured difference of 1.2, the difference between the measured sample means $\hat{\mu}_2$ and $\hat{\mu}_1$? If we had no prior reason to expect the fertilizer to work and produce taller plants, we use a two-tailed test and compute

$$\Pr(D \geq 1.2) + \Pr(D \leq -1.2) = 1 - \Phi\left(\frac{1.2}{\sqrt{0.6144}}\right) + \Phi\left(\frac{-1.2}{\sqrt{0.6144}}\right)$$
$$= 1 - \Phi(1.53) + \Phi(-1.53) = 0.126$$

The result is not significant.

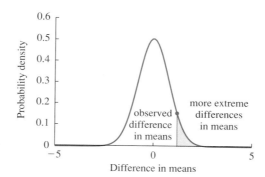

Figure 8.32
The null hypothesis: the difference is centered at 0

If we had predicted that the treatment would have a larger mean than the control, we could use a one-tailed test. In this case,

$$\Pr(D \geq 1.2) = 1 - \Phi(1.53) = 0.063$$

There is a trend toward significance, but not a convincing one. This loss of certainty (compared to the significant finding in Section 8.5) results from the uncertainly in the control population.

When both samples are sufficiently large (greater than 30), the sample variances s_1^2 and s_2^2 can be used in place of σ_1^2 and σ_2^2 in each of the above calculations. We summarize these methods in the following algorithm.

■ **Algorithm 8.5** **(Comparing the means of two experiments)**

Suppose n_1 individuals from a control population have sample mean $\hat{\mu}_1$ and n_2 individuals from a treatment population have sample mean $\hat{\mu}_2$. Suppose that either

1. the variances are known to be σ_1^2 in the control and σ_2^2 in the treatment, or
2. $n_1 \geq 30$ and $n_2 \geq 30$ with sample variances s_1^2 and s_2^2.

The null hypothesis is that the true means are equal, $\mu_1 = \mu_2$, or that the difference of the sample means has distribution

$$D \sim N\left(0, \frac{\sigma_1^2}{n_1} + \frac{\sigma_2^2}{n_2}\right)$$

if the variances are known or

$$D \sim N\left(0, \frac{s_1^2}{n_1} + \frac{s_2^2}{n_2}\right)$$

if they are unknown. We then test whether the measured difference is consistent with the null hypothesis by computing the probability of a result more extreme than $\hat{\mu}_2 - \hat{\mu}_1$ (using a one-tailed or two-tailed test as appropriate). ■

Comparing Population Proportions

The normal approximation provides a convenient way to compare population proportions. We measure two populations of cells for expression of a gene, and find 96 in a control group of $n_1 = 200$ cells, and 54 in a treatment group of $n_2 = 100$ cells. If the cells act independently, the number of cells expressing the gene will follow a binomial distribution in each treatment. We estimate the proportions p_1 and p_2 as

$$\hat{p}_1 = \frac{96}{200} = 0.48$$

$$\hat{p}_2 = \frac{54}{100} = 0.54$$

We can apply the normal approximation only if there are more than five successes and five failures in each sample, as there are in this case. The two proportions differ by 0.06. Is the difference significant?

Let p_1 and p_2 be the unknown true proportion for cells of each type and P_1 and P_2 be random variables describing the distribution of sample proportions. The normal approximations are

$$P_1 \sim N\left(p_1, \frac{p_1(1-p_1)}{200}\right)$$

$$P_2 \sim N\left(p_2, \frac{p_2(1-p_2)}{100}\right)$$

Let D be a random variable representing the difference in the two measured proportions. Using the additive relations for the mean and variance of independent normals, we get

$$D \sim N\left(p_1 - p_2, \frac{p_1(1-p_1)}{200} + \frac{p_2(1-p_2)}{100}\right)$$

Our null hypothesis is that $p_1 = p_2$, so that D has mean 0.

The variance of this distribution depends on the true unknown proportions p_1 and p_2. As in our computation of confidence limits for the binomial distribution (Section 8.3), we use the estimated proportions $\hat{p}_1$ and $\hat{p}_2$. Our null hypothesis is then that the distribution of the difference D between the sample proportions is

$$D \sim N\left(0.0, \frac{\hat{p}_1(1-\hat{p}_1)}{n_1} + \frac{\hat{p}_2(1-\hat{p}_2)}{n_2}\right)$$

For our example, we have

$$N\left(0.0, \frac{0.48 \cdot 0.52}{200} + \frac{0.54 \cdot 0.46}{100}\right) = N(0.0, 0.00373)$$

(Figure 8.33).

How does our measured difference of 0.06 compare? If we had prior reason to believe that the second group of cells would have a higher proportion, we use a one-tailed test and check $\Pr(D \geq 0.06)$:

$$\Pr(D \geq 0.06) = \Pr\left(\frac{D}{\sqrt{0.00373}} \geq \frac{0.06}{\sqrt{0.00373}}\right)$$

$$= \Pr(Z \geq 0.982) = 1 - \Phi(0.982) = 0.163$$

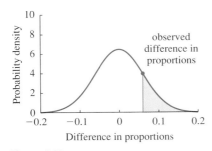

Figure 8.33

The null hypothesis describing the difference in proportions

8.6 Comparing Experiments: Unpaired Data

The difference in proportions of 0.06 differs from 0 by less than 1 standard error, and the p-value of 0.163 is nowhere close to significant with the one-tailed test. There is no evidence to reject the null hypothesis.

■ **Algorithm 8.6** (Comparing proportions from two experiments)

A control population and treatment population are tested for a trait. A proportion $\hat{p}_1$ out of n_1 individuals from the control population and a proportion $\hat{p}_2$ out of n_2 individuals from the treatment population are identified with the trait. The null hypothesis is that $p_1 = p_2$, or that the difference of the sample means has distribution

$$D \sim N\left(0.0, \frac{\hat{p}_1(1-\hat{p}_1)}{n_1} + \frac{\hat{p}_2(1-\hat{p}_2)}{n_2}\right)$$

We then test whether the measured difference is consistent with the null hypothesis by computing the probability of a result more extreme than $\hat{p}_2 - \hat{p}_1$ using a one-tailed or two-tailed test as appropriate. ■

Likelihood

We test two populations of ferocious bears for a parasite. At great risk, we capture ten bears from each population and find two infested bears in the first population and eight in the second. Do we have reason to think that the second population is in worse shape than the first? Our samples are too small for us to use the normal approximation. More advanced texts describe tests based on p-values. The method of support, as usual, provides a convenient way to compare hypotheses. The method follows the same steps as for a one-sample test:

1. Decide on two hypotheses to compare.
2. Write the support function for each hypothesis.
3. Find the parameter values for each hypothesis that are best supported by the data.
4. Compare the results.

The null hypothesis is that the two populations have the same proportion. If so, there is no difference between them, and we treat our data in one unit, as **pooled** (Figure 8.34). The maximum likelihood estimator of the **pooled proportion** p is

$$\hat{p} = \frac{10}{20} = 0.5$$

The support for the whole data set is the sum of the supports for each component (Theorem 8.4), so the support for p with the data as we observed it is

$S_n(p) = $ support for p with 2 out of 10 + support for p with 8 out of 10

$$= \ln\left[\binom{10}{2} p^2 (1-p)^8\right] + \ln\left[\binom{10}{8} p^8 (1-p)^2\right]$$

$$= \ln\binom{10}{2} + \ln\binom{10}{8} + 10\ln(p) + 10\ln(1-p)$$

The maximum occurs at $p = \hat{p} = 0.5$.

668 Chapter 8 Introduction to Statistical Reasoning

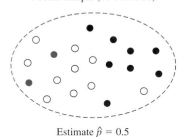

Figure 8.34

Unpooled and pooled comparison of bear data

The alternative is that the two populations are different. Let the parameters p_1 describing the first population and p_2 describing the second population take on different values. The maximum likelihood estimator of p_1 is the proportion in the first sample, or

$$\hat{p}_1 = \frac{2}{10} = 0.2$$

The maximum likelihood estimator of p_2 is the proportion in the second sample, or

$$\hat{p}_2 = \frac{8}{10} = 0.8$$

The support for the hypothesis that each sample behaves separately is again the sum of the supports for each hypothesis, or

$S_a(p_1, p_2) =$ support for p_1 with 2 out of 10 + support for p_2 with 8 out of 10

$$= \ln\left[\binom{10}{2} p_1^2 (1 - p_1)^8\right] + \ln\left[\binom{10}{8} p_2^8 (1 - p_2)^2\right]$$

$$= \ln\binom{10}{2} + \ln\binom{10}{8} + 2\ln(p_1) + 8\ln(1 - p_1)$$

$$+ 8\ln(p_2) + 2\ln(1 - p_2)$$

The maximum occurs at $p_1 = \hat{p}_1 = 0.2$ and $p_2 = \hat{p}_2 = 0.8$.

We compare the support of the two hypotheses by computing the difference. The constants involving binomial coefficients cancel, so we have

$$S_a(0.2, 0.8) - S_n(0.5) = 2\ln(0.2) + 8\ln(1 - 0.2) + 8\ln(0.8) + 2\ln(1 - 0.8)$$
$$- 10\ln(0.5) + 10\ln(1 - 0.5)$$
$$= 3.85$$

We again use the rule of thumb that a difference of more than 2 indicates that the alternative hypothesis is better supported by the data. There is good reason to believe that the two samples come from populations that are different.

■ **Algorithm 8.7** **(Comparing experiments with the method of support)**

Suppose two samples are measured. Let $\hat{\alpha}_1$ be the maximum likelihood estimator of some parameter in the first sample and $\hat{\alpha}_2$ be the maximum likelihood estimator of that parameter in the second sample. Let $\hat{\alpha}$ be the maximum likelihood estimator of that parameter in the pooled sample. The null hypothesis is that the true parameters in the two populations are equal to $\hat{\alpha}$. If the support functions in the two samples are $S_1(\alpha)$ and $S_2(\alpha)$, then the support for the null hypothesis is

$$S_n(\hat{\alpha}) = S_1(\hat{\alpha}) + S_2(\hat{\alpha})$$

and the support for the best alternative hypothesis is

$$S_a(\hat{\alpha}_1, \hat{\alpha}_2) = S_1(\hat{\alpha}_1) + S_2(\hat{\alpha}_2)$$

If

$$S_a(\hat{\alpha}_1, \hat{\alpha}_2) - S_n(\hat{\alpha}) > 2$$

then there is reason to place the null hypothesis in doubt. ■

Similarly, suppose one organism has 8 mutations and another has 18 in 1 million base pairs. We could use the normal approximation of the Poisson distribution to study whether these values are statistically different (Exercise 5). To use the method of support, follow Algorithm 8.7. The maximum likelihood estimator of the parameter Λ is $\Lambda_1 = 8$ in the first organism and $\Lambda_2 = 18$ in the second organism. The pooled sample consists of 26 mutations in 2 million base pairs, with maximum likelihood estimator of $\Lambda = 13$ per million.

The support for the null hypothesis $\Lambda = 13$ is

$$\begin{aligned}
S_n(\Lambda) &= S_1(\Lambda) + S_2(\Lambda) & &\text{support is sum of supports} \\
&= \ln[p(8; 13)] + \ln[p(18; 13)] & &\text{Poisson distribution} \\
&= \ln\left(\frac{e^{-13}13^8}{8!}\right) + \ln\left(\frac{e^{-13}13^{18}}{18!}\right) & &\text{formula for Poisson distribution} \\
&= -13 + 8\ln(13) - \ln(8!) - 13 & &\text{Laws 1 and 2 of logs} \\
&\quad + 18\ln(13) - \ln(18!)
\end{aligned}$$

The support for the alternative hypothesis $\Lambda_1 = 8$ and $\Lambda_2 = 18$ is

$$\begin{aligned}
S_a(\Lambda_1, \Lambda_2) &= S_1(\Lambda_1) + S_2(\Lambda_2) & &\text{support is sum of supports} \\
&= \ln[p(8; 8)] + \ln[p(18; 18)] & &\text{Poisson distribution} \\
&= \ln\left(\frac{e^{-8}8^8}{8!}\right) + \ln\left(\frac{e^{-18}18^{18}}{18!}\right) & &\text{formula for Poisson distribution} \\
&= -8 + 8\ln(8) - \ln(8!) - 18 & &\text{Laws 1 and 2 of logs} \\
&\quad + 18\ln(18) - \ln(18!)
\end{aligned}$$

The difference is

$$S_a(8, 18) - S_n(13) = 8\ln(8) + 18\ln(18) - 26\ln(13) = 1.97$$

The result is near the threshold. It might be worth doing more experiments to demonstrate whether these two organisms are really different.

This example only sketches the use of this method. When confronted with more complex data, consult a statistics text or a statistician before attempting to interpret differences in support for different hypotheses.

SUMMARY We have extended the methods for testing one sample against a known baseline to compare two samples with each other. When the underlying distribution is known to be normal, the null hypothesis of no difference can be phrased as a statement about the distribution of the difference between sample means. If the difference is large, we reject the null hypothesis. For sufficiently large samples, the normal approximation to the binomial distribution can be used to compare population proportions. With smaller samples, we can use the method of support to compare the maximum support for two hypotheses: the null hypothesis that the two samples have the same underlying parameter and the alternative hypothesis that two samples have different parameters.

8.6 EXERCISES

1. Test the null hypothesis that the means from two populations are equal for the following cases. Sample means $\hat{\mu}_1$ and $\hat{\mu}_2$ are found from samples with size n_1 and n_2 drawn from normal distributions with known variances σ_1^2 and σ_2^2. State the significance level of the test. Use a two-tailed test.
   a. $\hat{\mu}_1 = 20.0$, $\hat{\mu}_2 = 25.0$, $n_1 = 25$, $n_2 = 25$, $\sigma_1^2 = 25.0$, $\sigma_2^2 = 25.0$
   b. $\hat{\mu}_1 = 20.0$, $\hat{\mu}_2 = 21.0$, $n_1 = 25$, $n_2 = 50$, $\sigma_1^2 = 25.0$, $\sigma_2^2 = 25.0$
   c. $\hat{\mu}_1 = 20.0$, $\hat{\mu}_2 = 21.0$, $n_1 = 50$, $n_2 = 100$, $\sigma_1^2 = 16.0$, $\sigma_2^2 = 16.0$

2. Use p-values to test the null hypothesis of equal means against an alternative that $\mu_2 > \mu_1$ when sample means $\hat{\mu}_1$ and $\hat{\mu}_2$ are found from samples of size n_1 and n_2 with sample variances s_1^2 and s_2^2. Use a one-tailed test. State the significance level of the test.
   a. $\hat{\mu}_1 = 70.0$, $\hat{\mu}_2 = 74.0$, $n_1 = 50$, $n_2 = 50$, $s_1^2 = 25.0$, $s_2^2 = 25.0$
   b. $\hat{\mu}_1 = 70.0$, $\hat{\mu}_2 = 70.4$, $n_1 = 500$, $n_2 = 500$, $s_1^2 = 25.0$, $s_2^2 = 25.0$. Why is the significance level worse?
   c. $\hat{\mu}_1 = 70.0$, $\hat{\mu}_2 = 74.0$, $n_1 = 500$, $n_2 = 500$, $s_1^2 = 250.0$, $s_2^2 = 250.0$. Why does this match the answer to part **a**?

3. Use the normal approximation to test the null hypothesis of equal proportions for the following cases. State the significance level of a two-tailed test.
   a. Of 50 men, 35 believe that if dolphins were so smart they could find their way out of those nets; 40 out of 50 women believe this.
   b. Three hundred fifty out of 500 men and 400 out of 500 women believe as in part **a**.
   c. Thirty-five out of 50 men and 400 out of 500 women believe as in part **a**.
   d. Thirty-five out of 500 men and 40 out of 500 women believe as in part **a**.

4. A cell is placed in a medium with volume equal to that of the cell. One hundred marked molecules are placed inside. After a long time, 40 are found inside the cell and 60 are found outside. A scientist is attempting to determine whether the cell membrane acts as a filter.
   a. What is the null hypothesis?
   b. Should this be treated as a problem with a treatment and control (the inside and outside of the cell) or as a problem with known baseline?
   c. Conduct the appropriate test.

5. One organism has 8 mutations in 1 million base pairs and another has 18 in 1 million. Use the normal approximation to test whether the difference is significant.

6. Algorithm 8.6 uses $\hat{p}_1$ and $\hat{p}_2$ to estimate the variance under the null hypothesis.
   a. Why might it make more sense to use $\hat{p}$, the proportion in the pooled sample?
   b. What is the pooled proportion if 96 out of 200 are found in the control population and 54 out of 100 in the treatment population?
   c. Redo the test using $\hat{p}$. How different are the results?
   d. Under what circumstances might it make a larger difference?

7. Of eight mice treated with a chemical, five get cancer. Of six mice kept as controls, 1 gets cancer. Use the method of

support to see whether these data give us reason to think that the chemical is carcinogenic.
   a. State two hypotheses to compare.
   b. Find the maximum support for the null hypothesis.
   c. Find the maximum support for the alternative hypothesis.
   d. Compute the difference. Is there reason to suspect this chemical?

8. Two identical pieces of DNA are treated with a known mutagen at different temperatures. The first piece of DNA has 6 mutations, and the second has 15. Use the method of support to analyze whether this experiment provides reason to think that temperature affects the action of this mutagen. Remember that the number of mutations follows a Poisson distribution.
   a. State two hypotheses to compare.
   b. Find the maximum support for the null hypothesis.
   c. Find the maximum support for the alternative hypothesis.
   d. Compute the difference. Is there reason to think that temperature has an effect?

9. A cell is dipped in a salubrious bath; another remains in a standard culture. The first survives 30 min, the second only 5 min. Suppose that cell mortality follows an exponential model.
   a. State two hypotheses to compare.
   b. Find the maximum support for the null hypothesis.
   c. Find the maximum support for the alternative hypothesis.
   d. Compute the difference. Is there reason to think that the salubrious bath lengthens cell life?

10. Show that the two-sample test turns into the one-sample test as n_1 approaches infinity.

11. **COMPUTER:** Use the Monte Carlo method to test the three hypotheses in Exercise 1. How many simulations did you have to do to estimate the significance level in each case?

12. **COMPUTER:** Develop a Monte Carlo method to test the hypotheses in Exercise 8. What did you use as your null hypothesis? What defines a "more extreme" result? Can you think of more than one way to answer these questions? Could different answers provide different levels of significance? Are any of them correct?

8.7 Regression

In Section 8.6, we compared whole samples, treating each individual in the same way. More information is available when we have tracked individuals in a population, perhaps by measuring each before and after a treatment or by comparing two measurements of each. Here we introduce the method of **regression** to test specific relations between random variables. We examine **linear regression** and indicate its connection with correlation and maximum likelihood.

Linear Regression

Suppose we measure both mass (in micrograms) and toxin tolerance of ten cells, finding

Mass	Toxin tolerance
$x_1 = 0.1$	$y_1 = 2.015$
$x_2 = 0.2$	$y_2 = 1.589$
$x_3 = 0.3$	$y_3 = 1.784$
$x_4 = 0.4$	$y_4 = 1.782$
$x_5 = 0.5$	$y_5 = 2.732$
$x_6 = 0.6$	$y_6 = 2.079$
$x_7 = 0.7$	$y_7 = 2.019$
$x_8 = 0.8$	$y_8 = 2.317$
$x_9 = 0.9$	$y_9 = 3.028$
$x_{10} = 1.0$	$y_{10} = 3.786$

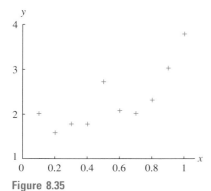

Figure 8.35
Data that might indicate a relationship between two measurements

(Figure 8.35). It looks as though larger cells are more able to tolerate toxin. Our goal is to quantify this hypothesis.

Linear regression is the process of finding a line through the data that lies "closest" to it. Before indicating how the best line is found, we show how "close" is measured. We have used the symbol x_i to denote the mass of cell i and y_i to denote its toxin tolerance. Suppose we predict that

$$\hat{y}_i = 1.3x_i + 1.7$$

as shown in the following table.

Mass	Toxin tolerance	Prediction $\hat{y}_i = 1.3x_i + 1.7$	Difference, or residual	Square of residual
0.1	2.015	1.83	0.185	0.034
0.2	1.589	1.96	−0.371	0.138
0.3	1.784	2.09	−0.306	0.094
0.4	1.782	2.22	−0.438	0.192
0.5	2.732	2.35	0.382	0.146
0.6	2.079	2.48	−0.401	0.161
0.7	2.019	2.61	−0.591	0.349
0.8	2.317	2.74	−0.423	0.179
0.9	3.028	2.87	0.158	0.025
1.0	3.786	3.00	0.786	0.618

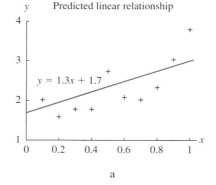
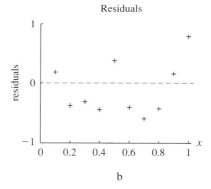

Figure 8.36
Testing a proposed linear relationship between two random variables

a b

This line passes pretty close to the data points (Figure 8.36a). If the prediction were exactly true, random variables X and Y describing the two measurements would obey

$$Y = 1.3X + 1.7$$

Our data, however, has some noise around it. In mathematical terms, we have

$$y_i = 1.3x_i + 1.7 + \epsilon_i$$

where ϵ_i represents the "error," or the degree to which tolerance is not predicted by mass for cell i. As a relation between random variables, we have

$$Y = 1.3X + 1.7 + \epsilon$$

If the random variable ϵ takes on small values, X is a good predictor of Y. If ϵ takes on large values, the error obscures the signal and X is a poor predictor of Y.

How do we evaluate our model? How well does it describe the data? We begin by subtracting the predicted values from the actual data (see the difference column in the table and Figure 8.36b). These values, called the **residuals**, indicate the portion of the result that is not explained by the model. A measure of the total error, in the same spirit as the variance, is the sum of the squares of these differences. Let $\hat{y}_i$ represent the predicted tolerance of cell i. The total error, called the **sum of squares of errors (SSE)**, is found by adding the last column of the table. In this case, we have

$$\text{SSE} = \sum_{i=1}^{10}(y_i - \hat{y}_i)^2 = 1.936$$

As with the variance, we can evaluate this number only by an appropriate comparison.

The fundamental idea of statistics is to compare a measure indicating an effect with a null hypothesis. Because the hypothesis is that two measurements have some relationship with each other, the null hypothesis is that they have *no relationship*. Graphically, no relationship is indicated by a horizontal prediction, where all the variation in the vertical variable y_i is caused by experimental noise (Figure 8.37a). The horizontal line that passes closest to the center of the data is the sample mean of the y values, $\overline{Y}$ (Exercise 4). For our example,

$$\overline{Y} = 2.313$$

The errors incurred by assuming that y_i has nothing to do with x_i are given in the following table.

Mass	Toxin tolerance	Prediction $\hat{y}_i = \bar{y}$	Residual	Square of residual
0.1	2.015	2.313	−0.298	0.089
0.2	1.589	2.313	−0.724	0.524
0.3	1.784	2.313	−0.529	0.279
0.4	1.782	2.313	−0.531	0.282
0.5	2.732	2.313	0.419	0.176
0.6	2.079	2.313	−0.234	0.055
0.7	2.019	2.313	−0.294	0.086
0.8	2.317	2.313	0.004	0.000
0.9	3.028	2.313	0.715	0.511
1.0	3.786	2.313	1.473	2.170

(Figure 8.37b). The sum of the squares of the residuals for the null hypothesis model is called the **total sum of squares (SST)**. Adding the last column in the table, we get

$$\text{SST} = \sum_{i=1}^{10}(y_i - 2.313)^2 = 4.172$$

We see that SST is much larger than SSE, indicating that the first model is much closer to the data than the null model. The most common way to compare

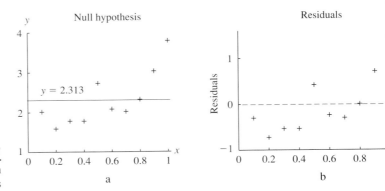

Figure 8.37
Testing the null relationship between two random variables

these values is called the **coefficient of determination**, r^2, defined by

$$r^2 = 1 - \frac{\text{SSE}}{\text{SST}}$$

The coefficient of determination r^2 can take on values between 0 (the horizontal line is as good as the proposed model) and 1 (a perfect linear relation). For our example, we have

$$r^2 = 1 - \frac{1.936}{4.172} = 0.535$$

We sometimes think of r^2 as the fraction of the variability in Y "explained" by X. About 53% of the variation in a cell's tolerance of toxin can be attributed to the cell's mass. The remainder could be due to unmeasured factors (age or other cell characteristics) or experimental noise. If the linear model describes the data well, SSE will be much smaller than SST and r^2 will be near 1. If not, SSE will be nearly as large as SST and r^2 will be near 0.

When a given model is known, the calculation of r^2 proceeds as follows.

■ **Algorithm 8.8** **(Calculation of the coefficient of determination)**

Suppose measurements $x_1, \ldots, x_n$ and $y_1, \ldots, y_n$ are thought to be related by the model

$$y_i = f(x_i)$$

1. To find SSE, compute

$$\text{SSE} = \sum_{i=1}^{n} [y_i - f(x_i)]^2$$

2. To find SST, compute

$$\text{SST} = \sum_{i=1}^{n} (y_i - \overline{Y})^2$$

where $\overline{Y}$ is the sample mean of the y_i.

3. The coefficient of determination r^2 is

$$r^2 = 1 - \frac{\text{SSE}}{\text{SST}}$$

■

Using Linear Regression

In the next section, we will see how to find the line through the data that minimizes SSE (and thus maximizes the value of r^2). If we have such results, how do we analyze the data? First, what do different values of r^2 mean in practice? When r^2 is near 1 (as in Figure 8.38a and b), the data fall nearly along the predicted line. As r^2 becomes smaller, the data appear more and more like a "cloud" (Figure 8.38c and d). However, r^2 alone cannot be used to tell whether the relationship between random variables X and Y is **significant**. Determining significance requires computing the slope of the best line through the data and checking whether that slope is significantly different from 0.

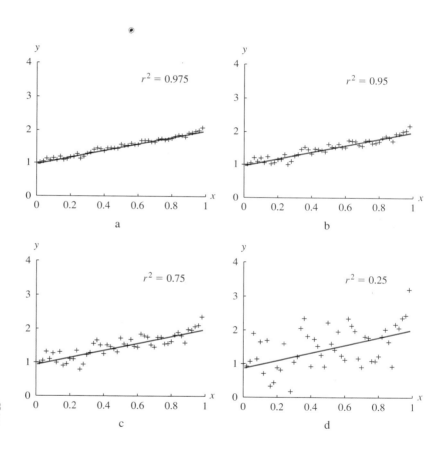

Figure 8.38
Regressions with different values of r^2

Two graphs should be examined in the process of making a regression: a plot of Y against X and a plot of the residuals. The data in Figure 8.39 fit the model well; the noise is distributed evenly around the prediction. In Figure 8.40, the noise is distributed evenly but increases in magnitude for larger values of X. The r^2 statistic weights all errors equally, which can be inappropriate in cases like this. If data follow such a pattern, consult a statistics text.

Figure 8.41 shows a curved pattern. The residuals are positive for large and small values of X and negative for medium values, rather than being evenly spread. A linear model is not appropriate. In such a case, we might want to use a different

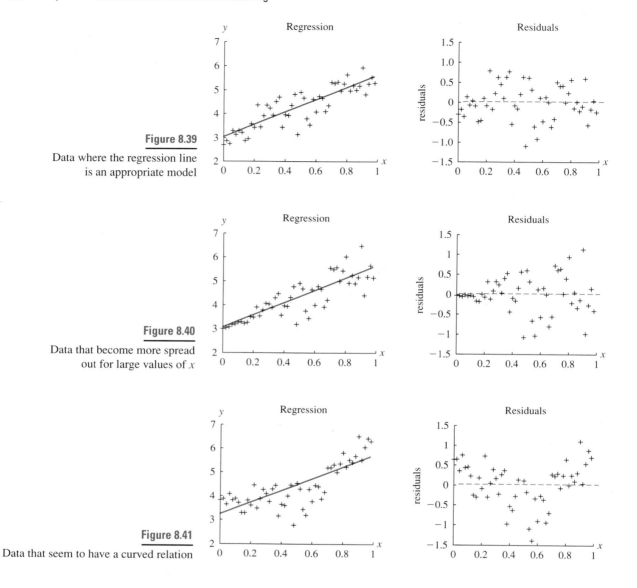

Figure 8.39 Data where the regression line is an appropriate model

Figure 8.40 Data that become more spread out for large values of x

Figure 8.41 Data that seem to have a curved relation

type of model, perhaps

$$Y = aX^2 + b + \epsilon$$

A statistics text should be consulted for more details about **nonlinear regression** techniques. However, the values of SSE, SST, and r^2 computed have the same meaning as for linear regression (Exercise 8).

The Theory of Linear Regression

Suppose we guess that

$$\hat{y}_i = ax_i + b$$

for n data points. The sum of the squared deviations of the measured y_i from the prediction $\hat{y}_i$ is a function of the slope a and the intercept b with formula

$$S(a,b) = \sum_{i=1}^{n}(y_i - \hat{y}_i)^2$$

A line right through the data has a small value of S because $\hat{y}_i$ is close to y_i. The best line minimizes the value of S over all possible slopes a and intercepts b. Because we are minimizing the squared deviation, the method is said to be based on the **principle of least squares**.

Finding the minimum value of S requires use of **partial derivatives**. Consult a more advanced text for the details of this computation. The resulting formulas can be written in terms of the sample mean, sample variance, and sample covariance. An estimate of the variance of X is the **uncorrected variance**,

$$\widehat{\text{Var}}(X) = E(X^2) - E(X)^2 = \frac{\sum_{i=1}^{n} x_i^2}{n} - \left(\frac{\sum_{i=1}^{n} x_i}{n}\right)^2$$

The uncorrected variance is like the sample variance (Definition 8.9) with n rather than $n-1$ in the denominator (it is also the maximum likelihood estimator of the variance). The **sample covariance** is the average of the product minus the product of the average, or

$$\widehat{\text{Cov}}(X,Y) = E(XY) - E(X)E(Y) = \frac{\sum_{i=1}^{n} x_i y_i}{n} - \frac{\sum_{i=1}^{n} x_i}{n} \frac{\sum_{i=1}^{n} y_i}{n}$$

Recall that the covariance measures the strength of the relation between two sets of measurements.

THEOREM 8.5 Suppose two measurements X and Y have sample means $\overline{X}$ and $\overline{Y}$, covariance $\widehat{\text{Cov}}(X,Y)$ and uncorrected variance $\widehat{\text{Var}}(X)$. The slope a and intercept b of the line that minimizes SSE are

$$\hat{a} = \frac{\widehat{\text{Cov}}(X,Y)}{\widehat{\text{Var}}(X)}$$

$$\hat{b} = \overline{Y} - \hat{a}\overline{X}$$

■

We place hats over a and b to indicate that these are **estimators** of the true relation. The estimated slope of the regression is positive if the covariance is positive, negative if the covariance is negative, and 0 if the covariance is 0. Although it is similar to the correlation coefficient, the slope of the regression can take on any value.

Using our data on size and toxin tolerance of cells, we get

$$\widehat{\text{Cov}}(X,Y) = 0.144$$
$$\widehat{\text{Var}}(X) = 0.0825$$
$$\overline{X} = 0.55$$
$$\overline{Y} = 2.313$$

From Theorem 8.5, we have

$$\hat{a} = \frac{0.144}{0.0825} = 1.745$$

$$\hat{b} = 2.313 - 1.745 \cdot 0.55 = 1.353$$

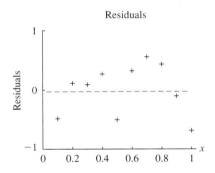

Figure 8.42
The best linear fit to a data set

(Figure 8.42). The line

$$y = \hat{a}x + \hat{b} = 1.745 \times 1.353$$

minimizes the SSE.

Mass	Tolerance	Best prediction	Residuals
0.1	2.015	1.528	−0.487
0.2	1.589	1.702	0.113
0.3	1.784	1.877	0.092
0.4	1.782	2.051	0.269
0.5	2.732	2.226	−0.506
0.6	2.079	2.400	0.321
0.7	2.019	2.574	0.556
0.8	2.317	2.749	0.432
0.9	3.028	2.923	−0.104
1.0	3.786	3.098	−0.688

By adding the squares of the residuals, we find SSE = 1.671, smaller than the 1.935 found with our initial guess (Figure 8.36). This line snakes right through the middle of the points, rather than being a shade high. Furthermore, we have

$$r^2 = 1 - \frac{\text{SSE}}{\text{SST}} = 1 - \frac{1.671}{4.172} = 0.599$$

The coefficient of determination indicates that this model "explains" about 60% of the variation in the data.

Computing SSE and r^2 is greatly simplified by the following computational formula.

■ **THEOREM 8.6** If the slope $\hat{a}$ and intercept $\hat{b}$ are chosen to minimize SSE as in Theorem 8.5, then

$$\text{SSE} = \sum_{i=1}^{n} y_i^2 - \hat{b} \sum_{i=1}^{n} y_i - \hat{a} \sum_{i=1}^{n} x_i y_i \qquad ■$$

Strictly speaking, minimizing S is justified by the method of maximum likelihood. Assume that the error ϵ in the equation

$$Y = aX + b + \epsilon$$

follows a normal distribution with mean 0 and known variance σ^2. We assume that the variance of the error is independent of X and that there is no error in the measurement of X. If these assumptions are not true, more complicated forms of regression must be used. With this model, we can compute maximum likelihood estimators of a and b by finding the values most likely to have produced the measured values. For a given guess of the slope a and the intercept b, the residual is

$$\epsilon_i = y_i - (ax_i + b)$$

which must be explained by the error term ϵ. If a and b produce large values of ϵ_i, the model is unlikely to have produced the data. The support for a particular pair a and b from data point i is

$$\ln\left(\frac{1}{\sqrt{2\pi}\sigma}e^{-(y_i-(ax_i+b))^2/(2\sigma^2)}\right) = \frac{-[y_i - (ax_i + b)]^2}{2\sigma^2} - \ln\left(\sqrt{2\pi}\sigma\right)$$

The support for the model by the whole data set is equal to the sum of the supports from each piece of independent data (Theorem 8.4) and is therefore

$$S(a,b) = \sum_{i=1}^{n} \frac{-[y_i - (ax_i + b)]^2}{2\sigma^2} - n\ln\left(\sqrt{2\pi}\sigma\right)$$

The log likelihood is maximized precisely where the numerator

$$\sum_{i=1}^{n}[y_i - (ax_i + b)]^2$$

is minimized. The slope and intercept found in Theorem 8.5 are thus the maximum likelihood estimates of these parameters under a specific set of assumptions. When these assumptions do not hold, this approach is not justified.

SUMMARY

When we believe that two sets of measurements are related by some function, we can quantify the fit by computing the **residuals** and **SSE**, the sum of the squared deviations. By comparing the SSE with the total sum of squares from the null model described by a horizontal line, we compute the **coefficient of determination** r^2, which can be thought of as indicating the fraction of variability in the data explained by the model. Examination of a graph of the residuals can reveal problems with the method, such as changes in the variance and nonlinearity of the data. By applying the **principle of least squares**, we found the line that minimizes SSE and passes most closely through a set of data points. This line is called the **linear regression** line. This linear regression line can be justified with the method of maximum likelihood when errors are normally distributed.

8.7 EXERCISES

1. Consider the data in the following table.

Weight	Yield	Height	Weight	Yield	Height
0.20	0.599	10.17	1.2	1.995	15.25
0.40	0.909	12.83	1.4	2.518	14.47
0.60	1.220	13.47	1.6	2.330	17.74
0.80	1.363	14.60	1.8	2.963	20.46
1.0	1.523	13.55	2.0	3.628	21.12

a. Plot yield (Y) against weight (W).
b. Suppose we tried the line $Y = W + 0.6$. Plot the line on your graph of yield against weight.
c. Find and plot the residuals.
d. Plot height (H) against weight.
e. Suppose we tried the line $H = 8W + 5$. Plot the line on your graph of height against weight.
f. Find and plot the residuals.

2. Find SSE for the two models of Exercise 1.

3. Use the data in Exercises 1 and 2 to compute r^2.
   a. Find the null model that best fits Y as a function of W.
   b. Find SST.
   c. Find r^2 for the model in Exercise 1b.
   d. Find the null model that best fits H as a function of W.
   e. Find SST.
   f. Find r^2 for the model in Exercise 1d.

4. Consider models of the form $Y = b$. Show that the sum of the squares of the residuals is minimized when $b = \overline{Y}$, the sample mean of the y_i.

5. Check the calculations of $\widehat{\text{Cov}}(X,Y)$, $\widehat{\text{Var}}(Y)$, $\overline{X}$, and $\overline{Y}$ in the text. Compute the correlation of X and Y. How does the correlation compare with r^2? Which is a more dependable measure of the strength of the relation?

6. Consider again the data in Exercise 1.
   a. Use Theorem 8.5 to compute the best fit line for yield on weight.
   b. Graph the line.
   c. Find the sum of the squared errors.
   d. Find r^2, and compare with Exercise 1b.

7. Follow the steps in Exercise 6 for height.

8. Consider the measurements

x	y
1.0	1.1
2.0	3.9
3.0	8.8
4.0	16.5

   a. Find the best linear fit. Plot the line and find r^2. How good is the model?
   b. Use the principle of least squares to write the expression you would use to fit a curve of the form $Y = aX^2 + b$. One easy way to solve is to think of a new measurement $Z = X^2$, and find the linear regression of Y on Z. Plot the linear regression of Y against Z and the curved regression of Y against X^2. Compute r^2.
   c. Which model does better?

9. Check the dimensions for each term of the regression equation for toxin tolerance as a function of mass, and show that the equation is consistent.

10. Both the best linear fit from Theorem 8.5 and the horizontal line pass through the center of the data in the sense that the sum of the residuals around each is 0. Prove it.

11. **COMPUTER:** The following table gives the winning Olympic times for men and women in the 400-m race.

Year	Men's record	Women's record
1896	54.2	
1900	49.4	
1904	49.2	
1906	53.2	
1908	50.0	
1912	48.2	
1920	50.0	
1924	47.6	
1928	47.8	
1932	46.2	
1936	46.5	
1948	46.2	
1952	45.9	
1956	46.7	
1960	44.9	
1964	45.1	52.0
1968	43.86	52.0
1972	44.66	51.08
1976	44.26	49.29
1980	44.60	48.88
1984	44.27	48.83
1988	43.87	48.65

   a. Use your computer to find the best linear regression for men and for women.
   b. Plot the residuals for each linear regression. Does the linear model fit well?
   c. Predict the times in the 2000 Olympics for women and men.
   d. Predict when women will outrun men. Do you believe this?

Supplementary Problems for Chapter 8

EXERCISE 1
Bacteria can be in one of two states: chemical producing (with probability 0.2) or chemical absorbing (with probability 0.8). Bacteria produce chemical at a rate of 2.0 fmol/s (or 10^{-15} mol) or absorb at a rate of 0.4 fmol/s. Suppose there are 100 bacteria in a culture, acting independently.

   a. Find the mean and variance of the number of bacteria in the producing and absorbing states.
   b. Find the mean and variance of the rate of change of total chemical.
   c. Write an exact expression for the probability that the amount of chemical is decreasing at any particular time. How would you evaluate it?

d. Suppose that instead of acting independently, all bacteria respond to the same external cue (but with the same probabilities). Find the answers to parts **b** and **c**.

EXERCISE 2
It is observed that a population grows by a factor R determined by the temperature T as follows:

$$R = \begin{cases} 1.5 \text{ with probability } 0.2 \text{ if } T = 10 \\ 0.5 \text{ with probability } 0.8 \text{ if } T = 10 \\ 1.5 \text{ with probability } 0.7 \text{ if } T = 20 \\ 0.5 \text{ with probability } 0.3 \text{ if } T = 20 \end{cases}$$

Furthermore $T = 10$ with probability 0.4 and $T = 20$ with probability 0.6.
  a. Will this population grow?
  b. Find the correlation of temperature and growth rate.

EXERCISE 3
One team hits 30 out of 100 shots and another hits 40 out of 100. Use the normal approximation to decide whether the second team really shoots better.

EXERCISE 4
Cosmic rays are thought to hit an object according to a Poisson process. During the first minute, 19 hit and during the second, 25. Use the normal approximation to find 98% confidence limits around the maximum likelihood estimate of the true rate.

EXERCISE 5
Molecules bind to a certain type of receptor at a rate of λ per second and never unbind.
  a. Suppose the receptor begins in an unbound state. Write and solve a differential equation for the probability the receptor is unbound at time t.
  b. A cell has two of these receptors. One binds at time $t = 1.5$ and another at $t = 2.5$. Find the maximum likelihood estimate of λ.

EXERCISE 6
Suppose T measures temperature above 37°C in degrees Centigrade and A measures an activity level. For three cells, $t_1 = 1$, $t_2 = 2$, $t_3 = 3$, $a_1 = 2$, $a_2 = 3$, and $a_3 = 3$.
  a. Plot A against T.
  b. Find the linear regression of A on T.
  c. Find the residuals.

EXERCISE 7
A model indicates that the probability a cell is healthy after a treatment is $1 - q$, that it is damaged is $2q/3$, and that it is moribund is $q/3$ for some unknown parameter q. Of 20 cells tested, 10 are found to be healthy, 6 damaged, and 4 moribund.
  a. Find the likelihood function and the maximum likelihood estimate of q.
  b. Without the model, the probabilities are $1 - q_1 - q_2$ (healthy), q_1 (damaged), and q_2 (moribund), with maximum likelihood estimates $q_1 = 0.3$ and $q_2 = 0.2$. Find the likelihood in this case. Do you think this model is better supported by the data than the one in part **a**?

EXERCISE 8
Although students struggled heroically to solve correctly the eight problems on their final exam, their professor assigned random grades independently for each problem and student, giving 10 points with probability 0.2, 15 points with probability 0.3, 20 points with probability 0.4, and 25 points with probability 0.1.
  a. Find the mean and variance of the total score.
  b. Find the probability that a student gets a perfect score.
  c. Use the normal approximation to find the probability that a student scores above 140 points. What is the lowest score that appears in the top 25%?
  d. Sketch the distribution of scores on this test.

EXERCISE 9
A new drug produces measurable reduction of a symptom in 75 of 100 patients tested. An older drug produces measurable reduction of the symptom 65% of the time.
  a. Use the normal approximation to find 99% confidence interval for the fraction of patients aided by the new drug.
  b. Test the hypothesis that the new drug is no better than the old drug. What is the significance level? Make sure to say whether you used a one- or two-tailed test and why.

EXERCISE 10
The following data for temperature T and height H are measured:

T	H
10.0	12.0
12.0	13.0
14.0	15.0
16.0	17.0
18.0	20.0

A proposed regression line is $H = T + 1.0$.
  a. Graph the data and proposed regression line.
  b. Find the residuals.
  c. Find SSE (with this line), SST, and r^2.

EXERCISE 11
The average density of trout along a stream is 0.2 per meter.
  a. Find the probability of three or more trout in 5 m.
  b. Find the expected number and coefficient of variation of the number of trout in 1 km.
  c. If the stream flows at 2 m/s (and the trout don't swim), find the rate at which trout pass a given point.

EXERCISE 12
In a standard calculus class, 50 students have scores on a standardized test normally distributed with mean 60; 30 students in an innovative calculus class score a mean of 63. The standard deviation of each and every student's score is known to be 5.0.

a. In mathematical language, give the null hypothesis that students from the innovative class did no better than those from the the standard class.
b. What is the significance level of the test?

EXERCISE 13
The following data describe ten measurements taken of height from control and treatment populations known to have normal distributions with standard deviation of 4.0. Test whether the treatment mean is different from the control mean. What is the significance level? What would you do if you did not know the standard deviation?

Control	Treatment
11.0	8.86
10.8	16.7
13.3	12.8
3.03	12.9
14.5	15.2
9.36	21.1
3.77	7.84
6.88	9.36
7.92	7.89
8.97	11.2

EXERCISE 14
A lazy scientist wants to estimate the rate at which wolves leave Yellowstone Park by waiting for the first one to leave. This happens after 3.0 months.
a. Find the maximum likelihood estimate of the rate.
b. Find 95% confidence limits around this estimate.
c. How would you use the method of support to approximate these confidence limits? Write the equation you would solve.
d. When is the most likely time for the second wolf to leave? The mean time?

EXERCISE 15
A company is testing a new insecticide. They run three experiments, measuring the number out of 100 bugs that survive at three dosages.

Experiment number	Dosage (x)	Number surviving (y)
1	18	3
2	15	4
3	9	5

A highly paid statistician finds the linear regression $y = 7 - 0.2x$.
a. Graph the data and the proposed regression line.
b. Find SSE, SST, and r^2.
c. Write the equation (using the data in the table) that the statistician solved to find this line.

d. Do you think that the pesticide works better at the higher dose? How would you check this statistically?

EXERCISE 16
After application of a proposed mutagen, five 100,000 base sequences of mitochondrial DNA have 24, 44, 29, 33, and 30 mutations, respectively. It is known that the mutation rate without the mutagen is 0.0002 per base.
a. Use the method of support to check whether the data are consistent with the null hypothesis.
b. Use the normal approximation to test the hypothesis that the mutagen increases the mutation rate. What does your significance level mean?
c. How would you use the Monte Carlo method to test the null hypothesis?

EXERCISE 17
The National Park Service has begun an intensive study of the elk in Yellowstone Park. As a first step, they wish to estimate elk density. A preliminary survey locates 36 elk droppings in 1 km².
a. What is the sample and what is the population?
b. What assumptions must you make to estimate the true density of elk droppings in the park?
c. What equations would you solve to find exact 99% confidence limits around an estimate of the density?
d. Use the normal distribution to find approximate 99% confidence limits.
e. How might you use this data to estimate the number of elk?

EXERCISE 18
To corroborate the intensive survey of Exercise 17, elk-dropping counters walk in straight lines at exactly 2 km/h, recording the first ten times they spot a dropping. The data from one surveyor are given in the table.

Dropping number i	Time since last dropping (t_i)	Dropping number i	Time since last dropping (t_i)
1	0.124	6	0.118
2	0.056	7	0.220
3	0.244	8	0.014
4	0.227	9	0.015
5	0.001	10	0.016

An untrained but trustworthy assistant computes that

$$\sum_{i=1}^{10} t_i = 1.035 \qquad \sum_{i=1}^{10} t_i^2 = 0.193$$

a. Sketch a graph of where the droppings are.
b. What is your best guess of the rate at which droppings are encountered?
c. Write and sketch the support function.

d. Sketch how you would use this graph to find approximate confidence limits, and write the equation you would solve to find them.
e. How might you relate this data to that in Exercise 17?

EXERCISE 19
In a second stage of the project of Exercise 17, managers decide to count both elk and their droppings. In five different 1 = km² regions, they find the data in the table.

Sample number i	Number of droppings d_i	Number of elk e_i
1	39	85
2	49	101
3	49	104
4	36	78
5	51	111

a. Graph these data.
b. A mathematician proposes two lines, $e_i = 2d_i + 5$ and $e_i = 5d_i + 2$. Sketch them. Which line gives a better fit to the data?
c. Find r^2 for the better of the two lines.
d. Interpret the relation described by this line. Does it make sense?

EXERCISE 20
In an even more advanced study, the managers of Yellowstone Park use portable scales to weight elk inside and outside the park to check whether they are different. They weigh 30 adult males inside the park (weights denoted by $x_1, \ldots, x_{30}$) and 40 adult males outside (weights denoted by $y_1, \ldots, y_{40}$), all in kilograms. They compute

$$\sum_{i=1}^{30} x_i = 1.829 \times 10^4 \qquad \sum_{i=1}^{30} x_i^2 = 1.146 \times 10^7$$

$$\sum_{i=1}^{40} y_i = 2.316 \times 10^4 \qquad \sum_{i=1}^{40} y_i^2 = 1.368 \times 10^7$$

a. Find the sample mean and sample variance for each group of elk.
b. Find the standard error for each.
c. State the null hypothesis in mathematical terms.
d. Test the null hypothesis statistically.

EXERCISE 21
Having estimated numbers and weights, the managers of Yellowstone Park begin working on health. They capture 40 elk and check them for parasites. They had reported to Congress that no more than 20% of the elk in the park were infested, but they find that 13 out of their 40 are infested.

a. What is the probability that their report to Congress was true but they got unlucky? What does this have to do with significance levels?
b. What would a Congressional representative compute if she were using the popular method of support?
c. Give one clever excuse the managers could give for their result. How might they argue for an expensive follow-up study?

Projects for Chapter 8

PROJECT 1
The statistical tests studied in the text are called **parametric** because they assume a particular form for the underlying distribution. For example, when we compared whether the means of two populations differed in Section 8.5, we assumed that the measurements came from a normal distribution. When there is no reason to believe that the distribution is normal, it would be nice to have **distribution-free** tests that work no matter where the measurements came from. This project studies some of the properties of one such test, the **Wilcoxon rank-sum test**.

Suppose that a set of 20 measurements in the control is

0.46894, 0.35273, 0.92801, 0.50284, 0.94539, 0.29496,
0.15446, 0.99030, 0.10896, 0.99643, 0.00097, 0.92011,
0.04307, 0.00001, 0.00024, 0.39957, 0.00077, 0.35379,
0.00004, 0.05706

and that the measurements in the treatment are

0.48770, 1.82495, 0.90703, 1.37176, 0.90001, 0.93359,
1.38502, 1.56536, 1.71434, 1.28598, 0.89246, 1.74330,
1.47014, 0.90556, 0.90001, 1.82305, 0.90394, 1.80227,
0.90044, 1.83284

Do these data look normally distributed? What result do you get if you assume that they are normally distributed and test for a difference in the means?

An alternative test for weird distributions instead uses the **ranks** of the two sets of data. Pool the data into a single data set, and sort it in order of increasing size. Give each value a rank, ranging from a rank of 1 for the smallest measurement 0.00001 from the control to a rank of 40 for the measurement 1.89246 from the treatment. If you add up the ranks of all measurements that come from the control, the sum is much smaller than the sum of the ranks of all measurements from the treatment.

To determine whether this result is significant, try the following experiment. Suppose that the control and treatment values come from the same distribution. Use a computer to sample 20 values for the control and 20 values for the treatment, and compute the rank sums. Do this many times. How often are the results more extreme than the values found?

Estimate the mean and variance of your simulations. The values should be roughly normally distributed with a mean of 410 and a variance of 1366.7. Use this result to estimate the significance. How well does it compare with the results from your Monte Carlo simulations?

PROJECT 2

Suppose the data in the table is supposed to fit the model

$$Y = \frac{1}{1 + aX}$$

for an unknown value of the parameter a.

X	Y
1	0.365
2	0.192
3	0.124
4	0.129
5	0.0829
6	0.0755
7	0.0738
8	0.0490
9	0.0548
10	0.0486

We want to use the method of least squares to estimate the parameter a. This problem requires the method of **nonlinear** least squares because a does not simply multiply the value of X. The idea is to set

$$\hat{y}_i = \frac{1}{1 + ax_i}$$

for some value of a. Our goal is to minimize the least squares measure

$$S(a) = \sum_{i=1}^{n} (y_i - \hat{y}_i)^2$$

Find a way to do this and plot a graph comparing the actual data with the value predicted by the model.

An alternative method transforms the data to make this into a linear least squares problem. Create a new measurement

$$Z = \frac{1}{Y}$$

Find Z as a function of X according to the model. Use the method of linear least squares to estimate a. Graph the transformed data along with the value predicted by the model.

Which estimate do you think is more accurate? What would happen to the two estimates if the last value in the data set became $y_{10} = 0.02$?

Bibliography for Chapter 8

Devore, J. L. *Statistics: The Exploration and Analysis of Data.* Duxbury Press, Belmont, Calif., 1997.

Edwards, A. W. F. *Likelihood.* Cambridge University Press, Cambridge, England, 1972.

Zar, J. H. *Biostatistical Analysis.* Prentice-Hall, Englewood Cliffs, N.J., 1984.

Answers

Chapter 1
Section 1.2

1. **a.** $f(0) = 0$, $f(1) = 1$, $f(4) = 16$
   b. $g(0.29) = 0.93$, $g(-1.18) = -3.80$, $g(7.32) = 23.57$
   c. $h(0.29) = 3.51$, $g(-1.18) = 2.04$, $g(7.32) = 10.54$

3. **a.** Only the point $(-2, 4)$ lies on the graph because $f(-2) = 4$.
   b. Only the point $(1.23, 3.96)$ lies on the graph because $g(1.23) = 3.96$.
   c. Only the point $(1.23, 4.55)$ lies on the graph because $h(1.23) = 4.55$.

5. **a.**

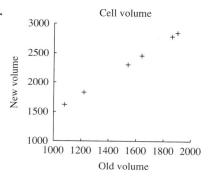

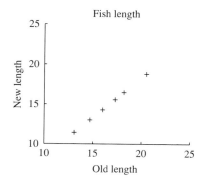

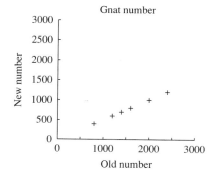

b. The updating function for cell volumes is

$$v_{new} = 1.5 v_{old}$$

The updating function is for fish length is

$$l_{new} = l_{old} - 1.7$$

The updating function is for gnat number is

$$n_{new} = 0.5 n_{old}$$

c. Because old cell volume is multiplied by 1.5, the new volume would be $1.5 \cdot 1420\,\mu\text{m} = 2130\,\mu\text{m}$.

d. Because old fish length is decreased by 1.7, the new length would be $1.5\,\text{cm} - 1.7\,\text{cm} = -0.2\,\text{cm}$. This does not make sense, because fish length must be positive. Therefore, I might modify the updating function to say that it is defined only on the domain $l_{old} \geq 1.7$. It could be that fish with length less than 1.7 cm would die.

7.

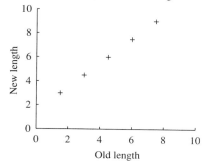

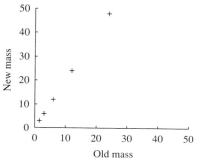

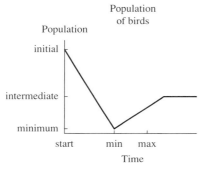

d. The length increases by 1.5 cm each half day, so $l_{new} = l_{old} + 1.5$ cm. The mass doubles each half day, so $m_{new} = 2m_{old}$. Age, not surprisingly, increases by one half day each half day, so $a_{new} = a_{old} + 0.5$ days.

9.

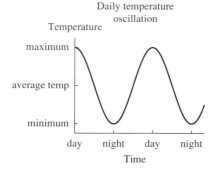

Section 1.3

1. **a.** Pressure has dimensions of $M/(LT^2)$.
   b. Rate of spread of bacteria on a plate has dimensions of L^2/T.
   c. Rate of spread of bacteria in a beaker has dimensions of L^3/T.
   d. In dimensions,
   $$G\frac{\text{mass}^2}{\text{length}^2} = \text{force} = \frac{\text{mass length}}{\text{time}^2}$$
   Solving for G, we find that G has dimensions
   $$\frac{\text{length}^3}{\text{mass time}^2}$$

3. **a.** 1 lb is about 454 g, so 3.4 lb is 1543.6 g.
   b. Using the conversions 1 yd = 36 in and 1 in ≈ 25.4 mm, we find that 1 yd ≈ 914.4 mm.
   c. 60 yr is about 525,600 h.
   d. About 2905 cm/s.
   e. About 143.5 lb/ft^3.

5. **a.** The tree with height 23.1 m will have volume of $\pi \cdot 23.1 \cdot 0.5^2 \approx 18.14$ m^3. When the height is 24.1 m, the volume is about 18.92 m^3.
   b. A tree with height 23.1 m has a trunk with radius 2.31 m, giving a volume of $\pi \cdot 23.1 \cdot 2.31^2 \approx 387$ m^3. The tree with height 24.1 m has a volume of 440 m^3.
   c. The bottom portion of the 23.1-m-tall tree will have half the volume in **a**, or 9.1 m^3. The top part is $23.1/2 = 11.55$ m high and is thus a sphere with radius of $11.55/2 = 5.78$ m. Plugging into the formula for the volume of a sphere, we find 807 m^3. The total volume is 816 m^3. The 24.1-m tree has a total volume of $9.46 + 916 \approx 926$ m^3.

7. **a.** $V_{new} = V_{old} + \pi \cdot 0.5^2$.
   b. Let V_{new} and V_{old} be the old and new volumes, respectively. We know that
   $$V_{old} = \frac{4\pi}{3}r_{old}^3$$
   $$V_{new} = \frac{4\pi}{3}r_{new}^3$$
   Therefore,
   $$V_{new} = \frac{4\pi}{3}(r_{old} + 1.0)^3$$
   We would like to express the right-hand side in terms of V_{old}. We can solve for r_{old} in terms of V_{old} as
   $$r_{old} = \sqrt[3]{\frac{3V_{old}}{4\pi}}$$
   Substituting, we get
   $$V_{new} = \frac{4\pi}{3}\left(\sqrt[3]{\frac{3V_{old}}{4\pi}} + 1.0\right)^3$$

9. My hair seems to grow about an inch a month. We have that

$$1 \text{ in./month} \approx 1 \text{ mi}/60{,}000 \text{ in.} \times 1 \text{ month}/30 \text{ d} \times 1 \text{ d}/24 \text{ h}$$

$$\approx \frac{1}{60{,}000 \times 700} \text{ mi/h}$$

$$\approx \frac{1}{4 \times 10^7} \text{ mi/h}$$

$$\approx 2 \times 10^{-8} \text{ mi/h}$$

Section 1.4

1. a. 1 cm = $\frac{1}{2.54}$ in. = 0.3937 in. The slope is 0.3937.
b. 1 in. = 2.54 cm. The slope is 2.54.

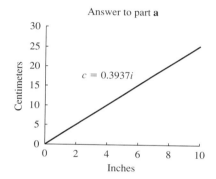

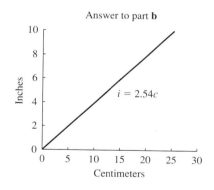

c. 1 lb = 454 g. The slope is 454.
d. 1 g = $\frac{1}{454}$ lb = 0.0022 lb. The slope is 0.0022.

3. a. 9600 ft.
b. 10,000 ft.
c. Let d represent distance moved and a represent altitude. Then, $a = -0.2d + 10{,}000$.

5. a. This line has slope -5 and passes through the point $(-7, 8)$.

b.

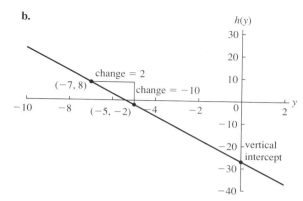

c. The point $(-5, -2)$ lies on the line.
d. The slope is

$$\text{slope} = \frac{\text{change in output}}{\text{change in input}} = \frac{-2 - 8}{-5 - (-7)} = -5$$

This checks.
e. The line is $h(y) = -5y - 27$, so the vertical intercept is -27.

7. If we used the last two rows, we find the slope to be $\frac{130-110}{50-40} = 2.0$. Then, using $(40, 110)$ as (x_0, y_0), we get

$$x_{new} = m(x_{old} - x_0) + y_0$$
$$= 2.0(x_{old} - 40) + 110 = 2.0 x_{old} + 30$$

9. a.

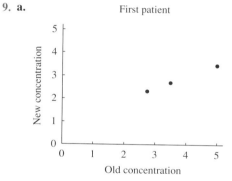

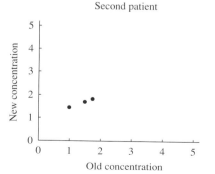

b. Let the level be M_{old} at the beginning of the day and M_{new} at the end of the day. Using the first two lines in the table, we find the two points to be $(5.0, 3.5)$ and $(3.5, 2.75)$. The slope is

$$\text{slope} = \frac{\text{change in output}}{\text{change in input}} = \frac{2.75 - 3.5}{3.5 - 5.0} = 0.5$$

In point-slope form, the equation of the updating function is

$$M_{new} = 0.5(M_{old} - 5.0) + 3.5$$

c. Again using the first two lines in the table, we find the two points to be $(1.0, 1.5)$ and $(1.5, 1.75)$. The slope is

$$\text{slope} = \frac{\text{change in output}}{\text{change in input}} = \frac{1.75 - 1.5}{1.5 - 1.0} = 0.5$$

In point-slope form, the equation of the updating function is

$$M_{new} = 0.5(M_{old} - 1.0) + 1.5$$

d. Both have the equation

$$M_{new} = 0.5 M_{old} + 1.0$$

in slope-intercept form. It is rather odd that the same updating function can produce such different effects.

Section 1.5

1. **a.** 1.05×10^{12}.
   b. 1.10×10^{18}.
   c. The volume of one bacterium is about 1.0×10^{-8} cm^3, so the population takes up roughly 10^4 cm^3 and weighs about 10 kg after 20 h. After 40 h, it would take up 10^{10} cm^3 and weigh 10^7 kg.

3. **a.** The difference is $2.0^t \cdot 0.7 \times 10^6$.
   b. The ratio is always 3.33.
   c. Both populations are growing at the same rate, but the first has a head start. It is always 3.33 times bigger, which is a larger difference as the populations become larger.

5. **a.** $v_t = 1.5^t \cdot 1220 \; \mu m^3$.
   b. $l_t = 13.1 - 1.7t$ cm.
   c. $n_t = 0.5^t \cdot 1200$.

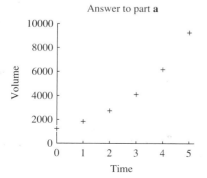

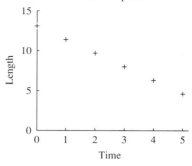

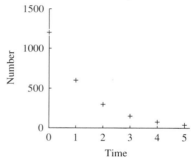

7. **a.** $b_1 = 3.0 \times 10^6$, $b_2 = 5.0 \times 10^6$, $b_3 = 9.0 \times 10^6$.
   b. 3.0×10^6.
   c. $b_{t+1} = 2.0 b_t - 1.0 \times 10^6$.
   d. The population would have doubled three times, so there would be 16.0×10^6 bacteria. There are 7.0×10^6 ready to harvest, 4.0×10^6 more than in part **b**. This is because the bacteria removed early (during the first hour) never had a chance to reproduce.

9. **a.** The ratio is 1.01.   **b.** 4.0804.
   c. $T_{t+1} = 1.01 T_t$. The tones heard are $T_0 = 400$, $T_1 = 404$, $T_2 = 408.04$, $T_3 = 412.12$, $T_4 = 416.24$, and $T_5 = 420.40$.
   d. The ratio is 1.00125. The tones are $T_0 = 400$, $T_1 = 400.5$, $T_2 = 401.00$, $T_3 = 401.5$, $T_4 = 402.0$, and $T_5 = 402.51$.

11. The increase is 40 the first week, 80 the second week, and 160 the third week. It looks like we add $40 \cdot 2^t$ each week. This gives the formula $x_t = -30 + 40 \cdot 2^t$. Alternatively, adding 30, we see that $x_0 + 30 = 40$, $x_1 + 30 = 80 = 40 \cdot 2$, $x_2 + 30 = 160 = 40 \cdot 2^2$, and $x_3 + 30 = 320 = 40 \cdot 2^3$. It looks like $x_t + 30 = 40 \cdot 2^t$, so $x_t = 40 \cdot 2^t - 30$.

Section 1.6

1. **a.** $(f + g)(x) = f(x) + g(x) = (3 + 2x) + (4 + 5x) = 7 + 7x$.
   b. $(f + g)(x) = x^2 + x + 1$.
   c. $(F + H)(y) = y^2 - y + \dfrac{y}{3}$.

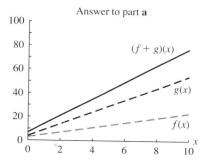

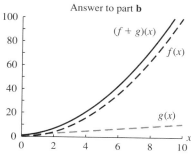

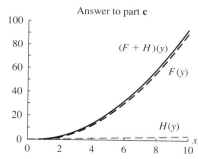

3. a. Denote the updating function by $f(v_t) = 1.5v_t$. Then,
$$(f \circ f)(v_t) = f(1.5v_t) = 1.5(1.5v_t) = 2.25v_t$$

b. Denote the updating function by $g(l_t) = l_t - 1.7$. Then,
$$(g \circ g)(l_t) = g(l_t - 1.7) = (l_t - 1.7) - 1.7 = l_t - 3.4$$

c. Denote the updating function by $h(n_t) = 0.5n_t$. Then,
$$(h \circ h)(n_t) = h(0.5n_t) = 0.5(0.5n_t) = 0.25n_t$$

5. a. Total of 10,000 squares

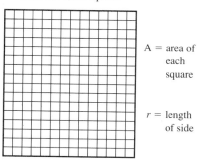

b. $A(r) = r^2$.
c. $M(A) = 0.4A$.
d. $T(M) = 10000M$.
e. $T = 10000M = 4000A = 4000r^2$.
f. $C(r) = 2.5 \times 10^2 \dfrac{4000r^2}{1000 + 4000r^2}$

g. Substituting very large values of r, the energy consumption seems to level out at 2.5×10^2 calories. If a bigger colony consumes no more energy, each individual must be doing worse. Perhaps the ones in the center are starving.

7. *Do not commute:* Beating the egg whites and mixing them into the cake; learning the material and taking a test. *Do commute:* Crossing your arms and turning around; starting your car and turning on the radio.

9. a. $h(x) = 8$ at $x = 2$.
b. Substituting $x = 1$, we have $h(1) = 5$. Using the point $(1, 5)$ on the line, we can write the equation $y = 3(x - 1) + 5$.
c. Two points on the line are $(1, 5)$ and $(2, 8)$. The ratio of output to input is 5 at the first point and 4 at the second point. These ratios would match if this were a proportional relation.
d. We can find the inverse by solving for x in terms of y:

$$3x + 2 = y$$
$$3x = y - 2$$
$$x = \frac{y - 2}{3}$$

Therefore,

$$h^{-1}(y) = \frac{y - 2}{3}$$

Substituting $y = 8$, we find $h^{-1}(8) = 2$.

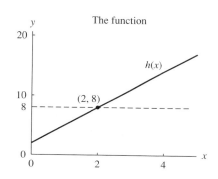

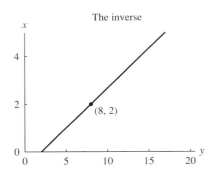

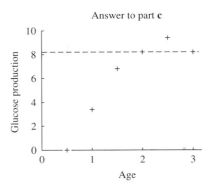

11. **a.** Because the mass is the same at ages 2.5 days and 3.0 days, the function mapping a to M has no inverse. Knowing the mass does not give enough information to estimate the age.
   b. Volume never has the same value twice and thus contains sufficient information to find the age.
   c. Glucose production is 8.2 mg at ages 2.0 and 3.0 days. We cannot go backward.

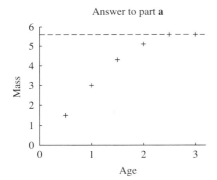

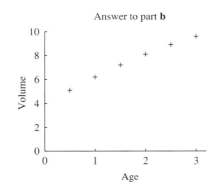

Section 1.7

1. **a.** Law 6: $e^0 = 1$.
   b. Law 2: $(e^{4.5})^{5.5} = e^{4.5 \cdot 5.5} = e^{24.75} = 5.61 \times 10^{10}$.
   c. No law applies, but value is 698.2
   d. Law 3: $e^{-3.5} = 1/e^{3.5} = 0.0302$.
   e. Law 1: $e^{4.5} \cdot e^{5.5} = e^{4.5+5.5}$.
   f. No law applies, but value is 334.71.
   g. Law 2: $(e^{3.5})^{6.5} = e^{3.5 \cdot 6.5} = 7.59 \times 10^9$.
   h. Law 3: $e^{-4.5} = 1/e^{4.5} = 0.011$.
   i. Law 1: $e^{6.5} \cdot e^{3.5} = e^{10.0} = 2.2 \times 10^4$.
   j. Law 5: $e^1 = e$.

3. **a.** Law 1: $\ln(1) = 0$.
   b. You can't take the log of a negative number, not defined.
   c. Law 1: $\ln(4.5) + \ln(5.5) = \ln(4.5 \cdot 5.5) = \ln(24.75) = 3.21$.
   d. You can't take the log of a negative number, not defined.
   e. No law applies, but $\ln(3.5 + 6.5) = \ln(10) = 2.302$.
   f. Law 3: $\ln(1/4.5) = -\ln(4.5) = -1.50$.
   g. Law 2: $\ln(4.5^{5.5}) = 5.5 \ln(4.5) = 8.27$.
   h. Law 1: $\ln(3.5) + \ln(6.5) = \ln(3.5 \cdot 6.5) = \ln(22.75) = 3.12$.
   i. No law applies, but $\ln(4.5 + 5.5) = \ln(10) = 2.302$.
   j. Law 2: $\ln(3.5^{6.5}) = 6.5 \ln(3.5) = 8.14$.

5. **a.** If $\sigma = 0.75$, $r = 1.5$. The population at time t is then

$$b_t = 1.0 \times 10^6 \cdot 1.5^t = 1.0 \times 10^6 \cdot e^{\ln(1.5)t}$$

Setting $b_t = 8.0 \times 10^6$, we get

$$1.0 \times 10^6 e^{\ln(1.5)t} = 8.0 \times 10^6$$
$$e^{\ln(1.5)t} = 8.0$$
$$\ln(1.5)t = \ln(8.0)$$
$$t = \frac{\ln(8.0)}{\ln(1.5)} = 5.12$$

The population will not have reached 8.0×10^6 in 5 h, but will have passed it in 6 h.
   b. 7.08 h.
   c. -9.3 h. This population is shrinking, and might have had the larger value between 9 and 10 h ago.

7. a. When is the size 2.0 cm? When

$$1.0e^{1.0t} = 2.0$$
$$e^{1.0t} = 2.0$$
$$1.0t = \ln(2.0)$$
$$t = \ln(2.0) = 0.693 \text{ d}$$

b. $t_d = 0.6931/1.0 = 0.6931$ d
c. $t_d = 0.6931/0.1 = 6.931$ h

9. a. $P(t) = 5.0 \times 10^9 e^{0.0289t}$.
b. It will have doubled twice and will be 2.0×10^{10}.

11. a. 1
b. $10^{4.5 \cdot 5.5} = 5.6 \times 10^{24}$
c. No law applies, but value is 3.165×10^6.
d. $1/10^{3.5} = 3.16 \times 10^{-4}$
e. 10^{10}
f. 10
g. 0
h. You cannot take the log of a negative number, not defined.
i. $\log_{10}(4.5 \cdot 5.5) = 1.393$
j. No law applies, but answer is 1.
k. $-\log_{10}(4.5) = -0.653$

Section 1.8

1. a. The proportions are 0.60, 0.64, 0.69, 0.74, 0.79 and 0.85.
b. This is not a proportional relation because the proportions change over time.
c. 87.7 cm. This would be an adult elk with little antlers.
d. 75.5 cm. This would be a baby elk with huge antlers.

3. b.

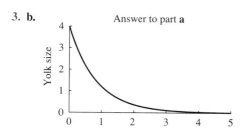

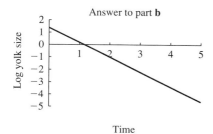

c. $S_1 = 0.022Y^{-0.833}$
d. $\ln(S_1) = -0.833\ln(Y) - 3.82$

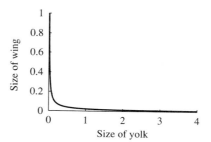

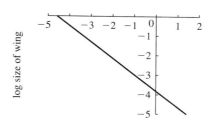

e. Half-life is 0.577 d.
f. $\ln(Y) = -1.2\ln(S_1) - 4.58$

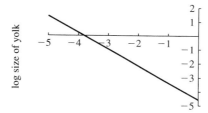

5. a. Density of a 1-kg animal would be 36 per km². Density of a 10-kg animal would be 2.6 per km². Density of a 100-kg animal would be 0.19 per km².

b.

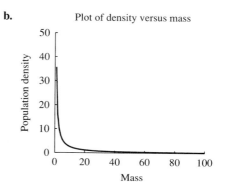

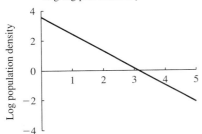

Log-log plot of density versus mass

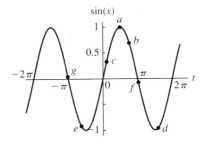

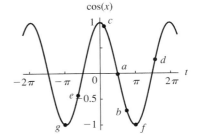

c. $\ln(Y) = 3.58 - 1.14 \ln(X)$
d. It should be 0.19 per km^2, giving a total of 1.75×10^5, or 175,000. This is a factor of about 1500 higher than other animals.
e. An animal with density $1500 \cdot 0.19 = 285$ has $\ln(Y) = 5.65$. Solving $5.65 = 3.58 - 1.14 \ln(X)$ for X gives $\ln(X) = -1.81$ and $X = 0.16$ kg. Humans have the population density of animals that weigh 160 g.

7. **a.** g/cm
   b. Four times as much
   c. 56.6 km/h

9. **a.** $1/(cm \cdot s)$
   b. 81 times as much
   c. The smaller vessels carry only 1/16 as much, so it would have to branch into 16 of the smaller vessels. The total cross-sectional area would increase by a factor of 4.

Section 1.9

1. **a.** $\pi/6$ rad
   b. $11\pi/6$ rad
   c. 0.017 rad
   d. $-\pi/6$ rad (the same angle as in part **b**).
   e. 114.6°
   f. 36°
   g. $-36° = 324°$
   h. 1718.9°. We subtract 360 degrees until the answer looks reasonable; this is the same as 268.9°.

3. **a.** $\sin(\pi/2) = 1$, $\cos(\pi/2) = 0$
   b. $\sin(3\pi/4) = \sqrt{2}/2 = 0.707$, $\cos(3\pi/4) = -\sqrt{2}/2 = 0.707$
   c. $\sin(\pi/9) = 0.342$, $\cos(\pi/9) = 0.940$
   d. $\sin(5.0) = -0.95$, $\cos(5.0) = 0.284$
   e. $\sin(-2.0) = -0.909$, $\cos(-2.0) = -0.416$
   f. $\sin(3.2) = -0.058$, $\cos(3.2) = -0.998$
   g. $\sin(-3.2) = 0.058$, $\cos(-3.2) = -0.998$

5.

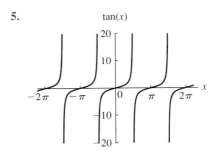

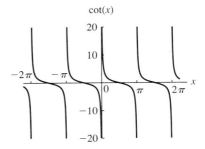

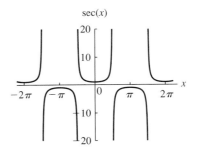

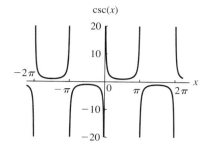

7.

Answer to part **a**

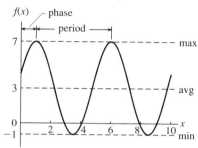

Answer to part **b**

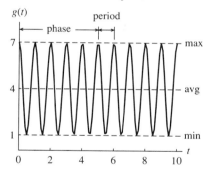

Answer to part **c**

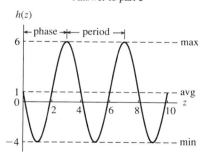

9. a. Let $S_c(t)$ represent the circadian rhythm. Then, $S_c(t) = \cos(2\pi t/24)$.
  b. Let $S_u(t)$ represent the ultradian rhythm. Then, $S_u(t) = 0.4\cos(2\pi t/4)$.
  c. To plot, compute the value of the combined function $S_c(t) + S_u(t)$ every hour and smoothly connect the dots.

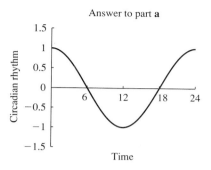

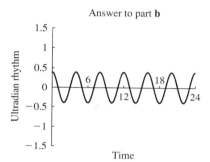

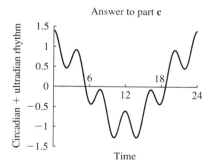

d. Right at midnight, when both cycles peak together.
e. At 10:00 A.M. and 2:00 P.M. There is not a minimum at 12:00 because the ultradian rhythm and the circadian rhythm are opposed.

11. a.

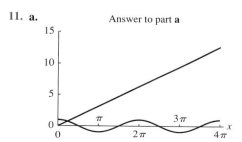

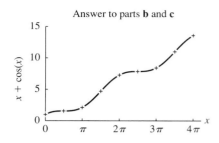

Answer to parts **b** and **c**

c. $f(0) = 1.0, f(\pi/2) = 1.571, f(\pi) = 2.142,$
$f(3\pi/2) = 4.712, f(2\pi) = 7.283, f(5\pi/2) = 7.854,$
$f(3\pi) = 8.424, f(7\pi/2) = 10.996, f(4\pi) = 13.566$
d. The size of an organism that grows on average, but grows most quickly during the day and slowly at night.

Section 1.10

1. **a.** amount = volume times concentration = 2.0 mmol
   b. 0.5 L at 1.0 mmol/L = 5.0 mmol
   c. 1.5 L at 1.0 mmol/L = 1.5 mmol
   d. 0.5 L at 5.0 mmol/L = 2.5 mmol
   e. 1.5 + 2.5 = 4.0 mmol
   f. 4.0 mmol/2.0 L = 2.0 mmol/L
   g. $q = 0.5/2.0 = 0.25$. Then,
   $$c_{t+1} = (1-q)c_t + q\gamma = 0.75 \cdot 1.0 + 0.25 \cdot 5.0$$
   $$= 2.0 \text{ mmol/L}$$

3. **a.** Total score is $50 \cdot 60 + 50 \cdot 90 = 7500$ points, an average of 75 per student. Alternatively, half the students got 60 and half got 90, so the average is
   $$0.5 \cdot 60 + 0.5 \cdot 90 = 30 + 45 = 75$$
   b. Total score is $70 \cdot 60 + 30 \cdot 90 = 6900$ points, an average of 69 per student. Alternatively, 70% of the students got 60 and 30% got 90, so the average is
   $$0.7 \cdot 60 + 0.3 \cdot 90 = 42 + 27 = 69$$
   c. Average is $60 \cdot 0.2 + 90 \cdot 0.8 = 84$.
   d. Average is $60 \cdot 0.2 + 70 \cdot 0.6 + 80 \cdot 0.2 = 70$.

5. **a.** Starts with 2.0 mmol, then 1.0 mmol is absorbed leaving 1.0 mmol and a concentration of 0.5 mmol/L. 0.5 L of air with this concentration are breathed out, taking with them 0.25 mmol, leaving 0.75 mmol. 0.5 L are breathed in with concentration 5.0, adding 2.5 to the remaining 0.75 for a total of 3.25 mmol, and a concentration of 1.625 mmol/L.
   b. Starts with $2.0c_t$, has $2.0c_t - 1.0$ after absorption, and a concentration of $c_t - 0.5$. Exhaled air then has $0.5c_t - 0.25$ mmol, leaving $1.5c_t - 0.75$ in the lung.

Inhaled air always carries 2.5 mmol in, for a total of $1.5c_t + 1.75$ and a concentration of $0.75c_t + 0.875$.
   c. $c_{t+1} = 0.75c_t + 0.875$
   d. The value $2.0c_t - 1.0$ is negative if $c_t < 0.5$. The lung cannot absorb 1.0 mmol if there aren't 1.0 mmol to absorb.

7.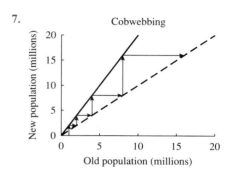

9. **a.** 1.8×10^6
   b. 2.8×10^6
   c. $b_{t+1} = 0.6b_t + 1.0 \times 10^6$
   d. We would need $q = 0.4$ and $\gamma = 2.5 \times 10^6$

Section 1.11

1.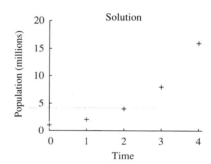

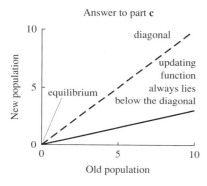

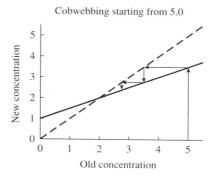

b. The equilibrium is $M^* = 2.0$.
c. The process is similar, except that half the old medication is used by the body rather than breathed out.

7. a.

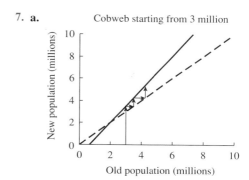

3. a. Because $q = 0.25$, the updating function is $c_{t+1} = 0.75c_t + 0.25 \cdot 5.0 = 0.75c_t + 1.25$. The equilibrium equation $c^* = 0.75c^* + 1.25$ has solution $c^* = 5.0$, as it should.
b. The updating function is $c_{t+1} = 0.9c_t + 0.8$, which has equilibrium 8.0. This checks.
c. The updating function is $c_{t+1} = 0.1c_t + 4.5$, which has equilibrium 5.0. This checks.

b. $b^* = 2.0 \times 10^6$
c. The cobwebs move away from the equilibrium, not toward it.

9. a. Starts with $c_t V$ mmol total. After A mmol are absorbed, this leaves $c_t V - A$ mmol with a concentration of $c_t - A/V$ mmol/L. This is multiplied by the volume exhaled to find the amount exhaled as $c_t W - AW/V$. Subtracting from the total leaves $c_t(V-W) - A(1-W/V)$. An amount γW is added, for a total of $c_t(V-W) - A(1-W/V) + \gamma W$. Dividing by V gives the new concentration.

5. a.

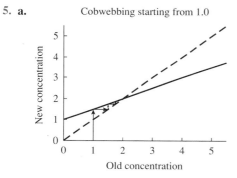

696 Answers

b. The updating function is

$$c_{t+1} = \frac{c_t(V - W) - A(1 - W/V) + \gamma W}{V}$$

$$= c_t(1 - q) - \frac{A}{V}(1 - q) + \gamma q$$

where $q = W/V$.

c. Setting $A = 0$ does give $c_{t+1} = (1 - q)c_t + q\gamma$.

d. $c^*(1 - q) - \frac{A}{V}(1 - q) + \gamma q = c^*$

the equation for the equilibrium

$c^* - c^*(1 - q) + \frac{A}{V}(1 - q) - \gamma q = 0$

move everything to one side

$qc^* + \frac{A}{V}(1 - q) - \gamma q = 0$

simplify a bit, but nothing to factor

$$c^* = \gamma - \frac{A}{V}\frac{1 - q}{q}$$

solve for c^*

e. The subtraction indicates that the equilibrium is smaller than γ if $A > 0$. This makes sense because absorption should reduce the chemical concentration.

Section 1.12

1. a. $p_t = \frac{m_t}{m_t + b_t} = \frac{1.2 \times 10^5}{1.2 \times 10^5 + 3.5 \times 10^6} = 0.033$

$m_{t+1} = 1.2 m_t = 1.44 \times 10^5$

$b_{t+1} = 2.0 b_t = 7.0 \times 10^6$

$p_{t+1} = \frac{m_{t+1}}{m_{t+1} + b_{t+1}} = \frac{1.44 \times 10^5}{1.44 \times 10^5 + 7.0 \times 10^6}$

$= 0.020$

b. $p_t = 0.074$, $m_{t+1} = 1.44 \times 10^5$, $b_{t+1} = 3.0 \times 10^6$, $p_{t+1} = 0.046$

c. $p_t = 0.033$, $m_{t+1} = 2.16 \times 10^5$, $b_{t+1} = 2.8 \times 10^6$, $p_{t+1} = 0.072$

d. $p_t = 0.033$, $m_{t+1} = 0.36 \times 10^5$, $b_{t+1} = 1.75 \times 10^6$, $p_{t+1} = 0.020$

e. $p_t = 0.033$, $m_{t+1} = 2.16 \times 10^5$, $b_{t+1} = 6.3 \times 10^6$, $p_{t+1} = 0.033$

3.

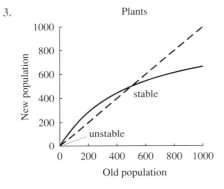

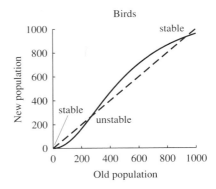

5. a. $0.2b_t$ mutate and $0.1m_t$ revert.

b. $b_{t+1} = b_t - 0.2b_t + 0.1m_t = 0.8b_t + 0.1m_t$
$m_{t+1} = m_t - 0.1m_t + 0.2b_t = 0.9m_t + 0.2b_t$

c. The total number before is $b_t + m_t$. The total number after is

$b_{t+1} + m_{t+1} = (0.8b_t + 0.1m_t) + (0.9m_t + 0.2b_t)$
$= 0.8b_t + 0.2b_t + 0.1m_t + 0.9m_t$
$= b_t + m_t$

d. Divide the updating function for m_{t+1} by $b_{t+1} + m_{t+1}$, we get

$p_{t+1} = \frac{m_{t+1}}{b_{t+1} + m_{t+1}} = \frac{0.9m_t + 0.2b_t}{b_{t+1} + m_{t+1}}$

$= \frac{0.9m_t + 0.2b_t}{b_t + m_t} = \frac{0.9m_t}{b_t + m_t} + \frac{0.2b_t}{b_t + m_t}$

$= 0.9p_t + 0.2(1 - p_t) = 0.2 + 0.7p_t$

e. The equilibrium solves $p^* = 0.2 + 0.7p^*$, or $p^* = 0.667$.

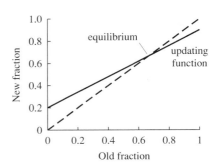

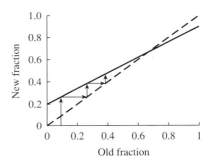

f. The equilibrium seems to be stable. The fraction of mutants will increase until it reaches about 66.7%.

7. a. $M_{t+1} = M_t - 0.5M_t + 1.0$ where 0.5 is the fraction used.
b. This fraction is always smaller than 0.5 and becomes extremely small when M_t is large. This means that the body uses a smaller fraction of the medication when the concentration is high, perhaps because there is an upper limit to how much it can use.
c. $M_{t+1} = M_t - \dfrac{0.5}{1.0 + 0.1M_t} M_t + 1.0$
d. The equilibrium is 2.5. It is larger because the body absorbs less efficiently.

9. a. $x_1 = 100 - 20 + 30 = 110$
$y_1 = 100 - 30 + 20 = 90$
$x_2 = 115$ and $y_2 = 85$
b. $x_{t+1} = x_t - 0.2x_t + 0.3y_t = 0.8x_t + 0.3y_t$
c. $y_{t+1} = y_t - 0.3y_t + 0.2x_t = 0.7y_t + 0.2x_t$
d. $x_{t+1} + y_{t+1} = x_t + y_t$. Dividing, then, we find that $p_{t+1} = 0.8p_t + 0.3(1 - p_t)$.
e. $p_1 = 0.8 \cdot 0.5 + 0.3 \cdot 0.5 = 0.55$. This matches the result of 110 out of 200.
$p_2 = 0.8 \cdot 0.55 + 0.3 \cdot 0.45 = 0.575$. This matches the result of 115 out of 200.

f.

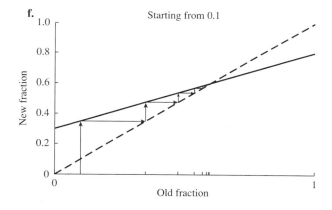

Starting from 0.1

g. The equilibrium is at $p_t = 0.6$. This is greater than 0.5 because a higher proportion of butterflies move from the second island to the first.

11. a. Consider the first island. After migration, there are 80 butterflies that started on the first island and 30 that started on the second. The 80 reproduce, making a total of $x_1 = 190$. On the second island, there are 20 from the first and 70 from the second after migration. The 70 reproduce, making a total of $y_1 = 160$. Following the same reasoning, $x_2 = 352$ and $y_2 = 262$.
b. $x_{t+1} = 2(x_t - 0.2x_t) + 0.3y_t = 1.6x_t + 0.3y_t$
c. $y_{t+1} = 2(y_t - 0.3y_t) + 0.2x_t = 1.4y_t + 0.2x_t$
d. $x_{t+1} + y_{t+1} = 1.8x_t + 1.7y_t$. Then,

$$\frac{x_{t+1}}{x_{t+1} + y_{t+1}} = \frac{1.6x_t + 0.3y_t}{x_{t+1} + y_{t+1}} = \frac{1.6x_t + 0.3y_t}{1.8x_t + 1.7y_t}$$

$$= \frac{\dfrac{1.6x_t}{x_t + y_t} + \dfrac{0.3y_t}{x_t + y_t}}{\dfrac{1.8x_t}{x_t + y_t} + \dfrac{1.7y_t}{x_t + y_t}}$$

$$= \frac{1.6p_t + 0.3(1 - p_t)}{1.8p_t + 1.7(1 - p_t)}$$

e. $p_1 = 0.543$, or 190 out of a total of 350.
$p_2 = 0.573$, or 352 out of a total of 614.

f.

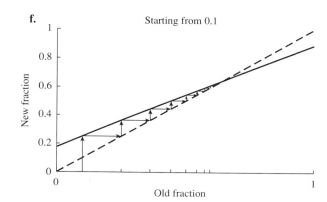

Starting from 0.1

g. We must solve the equation

$$p^* = \frac{1.6p^* + 0.3(1 - p^*)}{1.8p^* + 1.7(1 - p^*)} = \frac{0.3 + 1.3p^*}{1.7 + 0.1p^*}$$

Multiplying and using the quadratic formula, we find $p^* = 0.646$. This is between the previous two results.

Section 1.13

1. a. $\hat{V}_t = 0.5 \cdot 30.0 = 15.0 < V_c$. The heart will beat, and $V_{t+1} = 25.0$ mV.
b. $\hat{V}_t = 0.6 \cdot 30.0 = 18.0 < V_c$. The heart will beat, and $V_{t+1} = 28.0$ mV.
c. $\hat{V}_t = 0.7 \cdot 30.0 = 21.0 > V_c$. The heart will not beat, and $V_{t+1} = 21.0$ mV.
d. $\hat{V}_t = 0.8 \cdot 30.0 = 24.0 > V_c$. The heart will not beat, and $V_{t+1} = 24.0$ mV.

3. a. $c = e^{-\alpha\tau} = 0.3678$. The heart beats every time.
b. $c = e^{-\alpha\tau} = 0.6068$. The heart beats every time.
c. $c = e^{-\alpha\tau} = 0.3678$. The heart beats every time, twice as fast as in part **a**, but with recovery also twice as fast.
d. $c = e^{-\alpha\tau} = 0.7788$. The heart will not beat every time.

Answers to Supplementary Problems for Chapter 1

1. **a.** 3.0×10^{10} bacteria
   b. $2.0 \times 10^{10} \leq$ number $\leq 5.0 \times 10^{10}$

3. **a.** $f^{-1}(y) = -\ln(y)/2$, $g^{-1}(y) = \sqrt[3]{y-1}$
   $f(x) = 2$, when $x = f^{-1}(2) = -\ln(2)/2 = -0.345$.
   Similarly, we find $g(1) = 2$.
   b. $f \circ g(x) = e^{-2(x^3+1)}$, $g \circ f(x) = e^{-6x} + 1$, $f \circ g(2) = e^{-18} = 1.5 \times 10^{-8}$, $g \circ f(2) = 1 + e^{-12} = 1.00006$
   c. $(g \circ f)^{-1}(y) = -\ln(y-1)/6$, domain is $y > 1$.

5.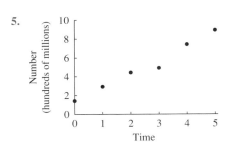

   a. Line has slope of 1.5 million bacteria/h, and equation $b(t) = 1.5 + 1.5t$. At $t^* = 3$, the point 5.0 lies below the line.
   b. Substituting $t = 3$ into the equation, we get 6.0 million.
   c. Substituting $t = 7$ into the equation, we get 12.0 million.

7. **a.** $\pi/3$ rad, $\sin(\theta) = \sqrt{3}/2$, $\cos(\theta) = 1/2$
   b. $-\pi/3$ rad, $\sin(\theta) = -\sqrt{3}/2$, $\cos(\theta) = 1/2$
   c. 1.919 rad, $\sin(\theta) = 0.9397$, $\cos(\theta) = -0.3420$
   d. -3.316 rad, $\sin(\theta) = 0.1736$, $\cos(\theta) = -0.9848$
   e. 20.25 rad, $\sin(\theta) = 0.9848$, $\cos(\theta) = 0.1736$

9. **a.** Let B_t be the number of butterflies in late summer. There are then $1.2B_t$ eggs, leading to $0.6B_t$ new butterflies from reproduction, plus 1000 from immigration. The updating function is $B_{t+1} = 0.6B_t + 1000$.

   b.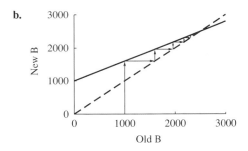

   c. Equilibrium has 2500 butterflies.

11. **a.**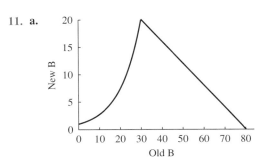

    b. At $t = 30$, the size is $1.0 \times e^{0.1 \cdot 30} = 20.08$. The size then decreases by 0.4 g/d, so it will take $20.08/0.4 = 50.2$ days to disappear. The tumor is then 80.2 days old.

13. **a.** 2.66 cm^2
    b. 1.82 cm^2
    c. 0.55
    d. When $2.0 \times 1.1^t = 10$ or in 16.9 h.
    e. The updating function is $A_{t+1} = 1.1A_t$.

    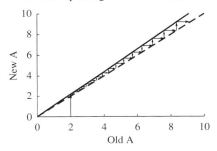

15. **b.**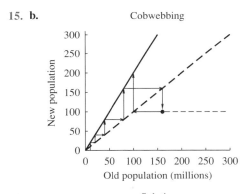

    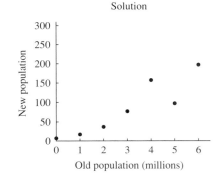

c. The only equilibrium is at 0.

17. a. $x_{t+1} = 0.5 + x_t/(1 + x_t)$

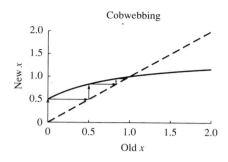

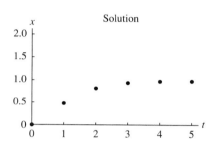

b. The equilibrium is at 1.0.

19. a. She has $1120, and the casino has $10,880.
b. Let g represent the money the gambler has and c the amount the casino has. Then,
$$g_{t+1} = g_t - 0.1g_t + 0.02c_t = 0.9g_t + 0.02c_t$$
$$c_{t+1} = c_t + 0.1g_t - 0.02c_t = 0.1g_t + 0.98c_t$$

c. $p_{t+1} = \dfrac{g_{t+1}}{g_{t+1} + c_{t+1}} = \dfrac{0.9g_t + 0.02c_t}{g_t + c_t}$
$= \dfrac{0.9p_t + 0.02(1 - p_t)}{p_t + (1 - p_t)} = 0.88p_t + 0.02$

d. The equilibrium is at $p = 0.167$.
e. They start out with $12,000, and the gambler ends up with 1/6, or $2000.

21. a. $D(0) = 10.0$, $H(0) = 50.0$, $D(7.5) = 12.5$, $H(7.5) = 98.2$, $D(15) = 15.7$, $H(15) = 193.9$

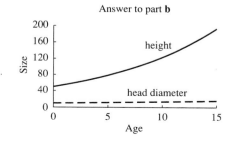

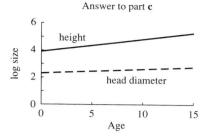

d. $H(t) = \dfrac{D(t)^3}{20}$.

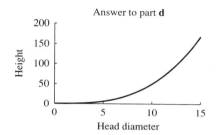

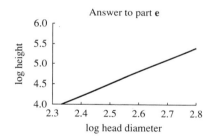

f. Doubling time is 23.1 for head diameter and 7.7 for height.

23. a. $148,643
b. In 16.88 yr, or in about the year 2012.
c. $M_{t+1} = 1.1M_t$
d. $M_t = 1.000001 \times 10^6 \cdot 1.1^t$ where t is measured in years before or after 1995.

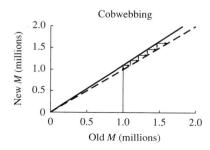

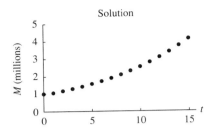

25. a. It will have $460 million.

b.

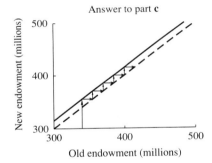

c. $M_{t+1} = M_t + 15$
d. I'd hire the second Texan—the money keeps piling up.

27. a. It decreases to 30, and it will beat.

b.

c. The equilibrium would be when $v^* = 0.75v^* + 30$ or $v^* = 120$, if it beats every time. But if it starts at 120, it decreases only to 90, too high to beat again. This heart will show sort of AV block.

Chapter 2

Section 2.1

1. a.

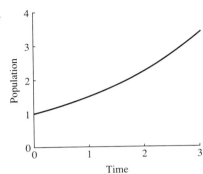

b. $b(0) = 1.0, b(1.0) = 1.5, b(2.0) = 2.25, b(3.0) = 3.375$
c. $\Delta b = 1.5 - 1.0 = 0.5$, so $\Delta b/\Delta t = 0.5$
d. $\Delta b = 2.25 - 1.5 = 0.75$, so $\Delta b/\Delta t = 0.75$
e. $b(0) = 1.0, b(0.5) = 1.225, b(1.0) = 1.5,$
  $b(1.5) = 1.837, b(2.0) = 2.25$
f. $\Delta b = 0.225, \Delta b/\Delta t = 0.45$
g. $\Delta b = 0.275, \Delta b/\Delta t = 0.55$

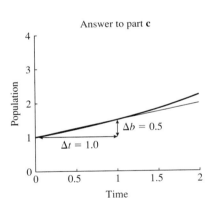

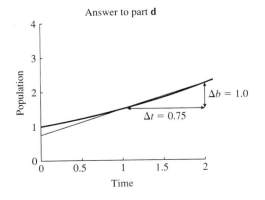

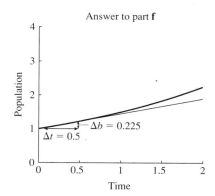

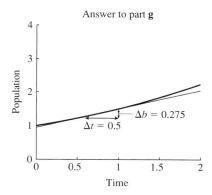

3. **a.** The slope is $(2.0^{1.0} - 1.0)/1.0 = 1.0$.
   b. The slope is $(2.0^{0.1} - 1.0)/0.1 = 0.718$.
   c. The slope is $(2.0^{0.01} - 1.0)/0.01 = 0.696$.
   d. After trying smaller values of Δt, it looks like the slope is 0.693.
   e. $\hat{b}(t) = 0.693t + 1.0$.

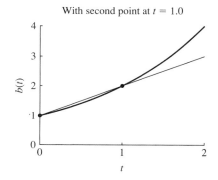

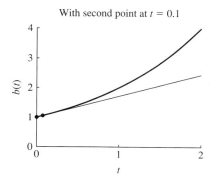

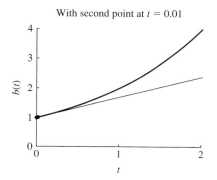

5. **a.**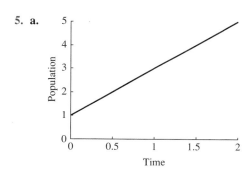
   b. 2.0 million bacteria/h
   c. 2.0 million bacteria/h
   d. 2.0 million bacteria/h
   e. 2.0 million bacteria/h
   f. 2.0 million bacteria/h
   g. 2.0 million bacteria/h

7. **a.**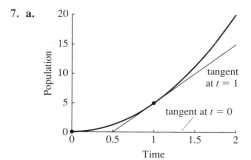

b. 5.0 m/s
c. 15.0 m/s
d. 2.5 m/s
e. 7.5 m/s
f. Average rate is $5\Delta t$ m/s. The limit must be 0. The tangent line is then $\hat{p}(t) = 0$.
g. Average rate is $10 + 5\Delta t$ m/s. The limit must be 10. The tangent line is $\hat{p}(t) = 10(t-1) + 5 = 10t - 5$.

9. a. Using the previous measurement as the point,

Age	Rate of change (m/yr)	Rate divided by height
1	1.07	0.0957
2	1.22	0.0984
3	1.34	0.0975
4	1.27	0.0846
5	1.60	0.0963
6	1.66	0.0909
7	1.90	0.0942
8	1.84	0.0836
9	2.44	0.0998
10	2.40	0.0894

b.
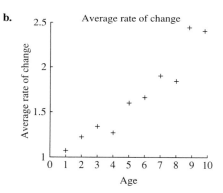
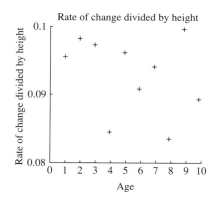

c. Both seem to jump around a bit, but the tree seems to grow by about 9% per year.

11. a. $1.05 \cdot \$1000 = \$1050.00$
b. $(1.025)^2 \cdot \$1000 = \$1050.62$
c. $(1.0041667)^{12} \cdot \$1000 = \$1051.16$
d. $(1.000137)^{365} \cdot \$1000 = \$1051.27$
e. Breaking into intervals of length Δt, we get
$$\lim_{\Delta t \to \infty} (1 + 0.05\Delta t)^{1/\Delta t} \$1000$$
It looks like the limit is $1051.27.

Section 2.2

1. a. At $x = 0.1$, value is 2.594; at $x = 0.01$, it is 2.705; at $x = 0.001$, it is 2.717; and at $x = 0.0001$, it is 2.718. The limit seems to be e.
b. At $x = 0.1$, value is 0.998; at $x = 0.01$, it is 0.999. The limit seems to be 1.

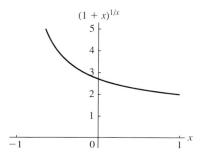

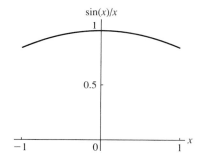

c. At $x = 0.1$, value is 0.499; at $x = 0.01$, it is 0.005; at $x = 0.001$, it is 0.0005. The limit seems to be 0.
d. $0.1^{0.1} = 0.794$, $0.01^{0.01} = 0.955$, $0.001^{0.001} = 0.993$. Also, $0.11^{0.11} = 0.784$, $0.011^{0.011} = 0.952$. The limit is 1.0.

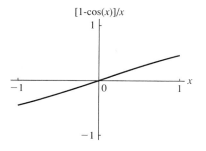

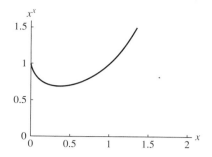

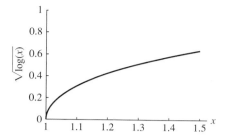

e. $0.1 \ln(0.1) = -0.230$, $0.01 \ln(0.01) = -0.046$, $0.001 \ln(0.001) = -0.006$. It seems to be going to 0.
f. This seems to go to negative infinity.

3. **a.** The limit is 3 times the limit of $g(x)$, or 3.
   b. The limit is the limit of $g(x)$ plus 3, or 4.
   c. The limit is $\lim_{x\to 0} x$ multiplied by the limit of $g(x)$. $\lim_{x\to 0} x = 0$, so the limit is $0 \cdot 3 = 0$.
   d. The limit is 0.
   e. The limit is ∞.

5. **a.** $\Delta f = 5\Delta x$, so the average rate of change is 5 (except for $\Delta x = 0$).

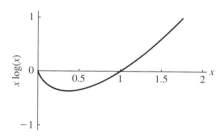

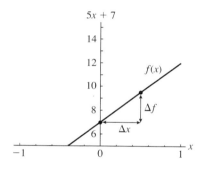

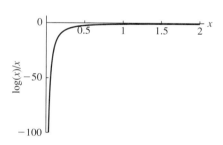

g. This seems to go to negative infinity.
h. $\sqrt{\ln(1.1)} = 0.309$, $\sqrt{\ln(1.01)} = 0.099$, $\sqrt{\ln(1.001)} = 0.032$. It seems to be going to zero, but rather slowly.

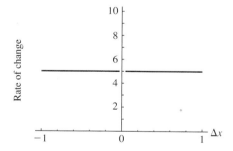

b. The average rate of change is 5 (except for $\Delta x = 0$).
c. $\Delta f = 5\Delta x^2$, so the average rate of change is $5\Delta x$ (except for $\Delta x = 0$).
d. $\Delta f = 5(1 + \Delta x)^2 - 5$, so the average rate of change is $10 + 5\Delta x$ (except for $\Delta x = 0$).
e. $\Delta f = 5(1 + \Delta x)^2 + 7(1 + \Delta x) - 12$, so the average rate of change is $17 + 5\Delta x$.

7. **a.** $\lim_{T\to 0} V(T) = 1.0$ cm^3
   $\lim_{T\to 0} H(T) = 8.1$
   b. $V(2.0) = 1.0$ and $H(2.0) = 6.94$
   c. $V(1.0) = 0.67$ and $H(1.0) = 7.23$

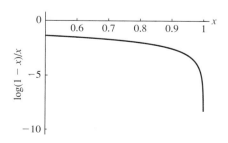

$V(0.1) = 0.84$ and $H(0.1) = 7.94$
$V(0.01) = 0.98$ and $H(0.01) = 8.08$

9. **a.** For g_1, x must be less than 0.0001. For g_2, x must be less than 0.01. For g_3, x must be less than 0.1.
   b. For g_1, x must be less than 10^{-8}. For g_2, x must be less than 10^{-4}. For g_3, x must be less than 10^{-2}.
   c. It looks like g_3 becomes huge the fastest.

Section 2.3

1. **a.** This is a linear function and is continuous everywhere.
   b. This is a polynomial and is continuous everywhere.
   c. This is constructed as the quotient of the continuous exponential function and a continuous linear function. It is guaranteed to be continuous everywhere that the denominator is not equal to 0. The only potential trouble point is $x = -1$.
   d. This is the product of a continuous polynomial (y^2) and a natural logarithm. The natural logarithm is the composition of a linear function ($y - 1$) with the natural log. The only potential trouble point is where $y - 1 = 0$, or at $y = 1$. The function is not defined for $y < 1$.
   e. This is a composition of the natural log with a linear function divided by a polynomial. The theorems guarantee continuity except at $z = 1$, where we are taking the natural log of 0, and $z = 0$, where the denominator is 0.
   f. This is a triple composition, of cosine with the exponential with a polynomial. Because each piece is continuous, the combination is.
   g. This is composition of the continuous cosine function with a continuous linear function ($x - (\pi/2)$), and is continuous everywhere.
   h. The two pieces of the function are a polynomial (t^2) and a linear function (0), both of which are continuous. The only trouble spot is where the pieces of the definition must match up. In this case, however, both pieces are equal to 0, so the function is continuous.
   i. This is the quotient of the constant 1 by the function $(1 + w)^4$, which is a polynomial. This is guaranteed to be continuous except where the denominator is 0, or when $w = -1$.

3. **a.** $4.8 < M < 5.2$
   $4.8 < 2.0V < 5.2$
   $2.4 < V < 2.6$

   V must be within 0.1 cm³ of 2.5 cm³ for M to be within 0.2 g of 5.0 g.

   b. $4\pi - 0.5 < A < 4\pi + 0.5$
   $12.07 < \pi r^2 < 13.07$
   $3.84 < r^2 < 4.16$
   $1.96 < r < 2.04$

   The radius must be within 0.04 of 2.0 to get within 0.5 of the desired area.

   c. $0.95 < F < 1.05$
   $0.95 < r^4 < 1.05$
   $0.987 < r < 1.012$
   The radius must be within about 1% of 1.0 to guarantee a flow within 5%.

   d. $2.618 < S < 2.818$
   $2.618 < e^{0.001t} < 2.818$
   $0.963 < 0.001t < 1.036$
   $963 < t < 1036$
   The time must be within about 37 s of 1000.

5. **a.** Between 0.96 and 1.04 g/L. Tolerance is 0.04 g/L.
   b. Between 15.36 and 16.64 g/L. Tolerance is 0.64 g/L.
   c. Between 491.52 and 532.48 g/L. Tolerance is 20.48 g/L.
   d. The tolerances become larger because a difference in initial conditions is decreased as the concentration moves toward from its stable equilibrium.

7. **a.** Denote this function by f. Then
$$f(V) = \begin{cases} 80 & \text{if } V < 50 \\ 2V & \text{if } V > 50 \end{cases}$$

b.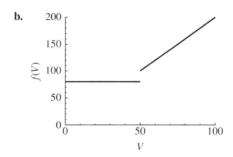

c.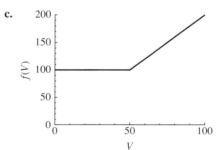

d.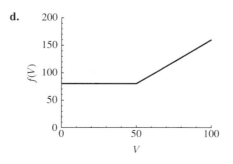

c. We need that $2V = V^*$ at $V = V_0 = 50$. In this case, $V^* = 100$.
d. We need that $kV = 80$ at $V = V_0 = 50$, or $k = 1.6$.

9. a.
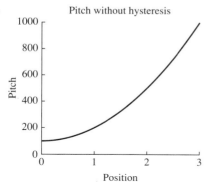

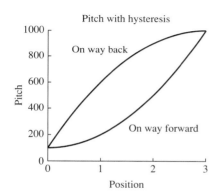

b. The second seems more likely. The pitch is usually different in different directions.
c. Both sound awful.

Section 2.4

1.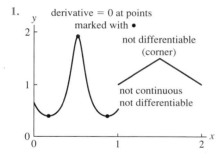

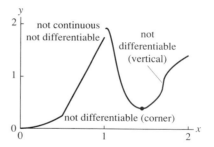

3. a. At $a = 1.0$ d, $M = 4.0$ g. Then $M(1.5) = 5.5$, so $\Delta M = M(1.5) - M(1) = 5.5 - 4 = 1.5$, and $\Delta M/\Delta a = 3.0$. Similarly, $M(1.0 - 0.5) = M(0.5) = 2.5$, so $\Delta M = M(0.5) - M(1) = 2.5 - 4 = -1.5$, and $\Delta M/\Delta a = 3.0$. The units are g per day.

b. At $a = 2.0$ d, $V = 8.4$ cm^3. Then $V(2.5) = 9.5$, so $\Delta V = V(2.5) - V(2) = 9.5 - 8.4 = 1.1$, and $\Delta V/\Delta a = 2.2$. Similarly, $V(2.0 - 0.5) = V(1.5) = 7.3$, so $\Delta V = V(1.5) - V(2) = 7.3 - 8.4 = -1.1$ and $\Delta V/\Delta a = 2.2$. The units are cm^3 per day.

c. At $a = 3.0$ d, $G = 10$ g. Then $G(3.5) = 6$, so $\Delta G = G(3.5) - G(3) = 6 - 10 = -4$, and $\Delta G/\Delta a = -8$. Similarly, $G(3.0 - 0.5) = G(2.5) = 14$, so $\Delta G = G(2.5) - G(3) = 14 - 10 = 4$ and $\Delta G/\Delta a = -8$. The units are g per day.

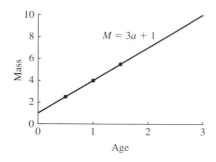

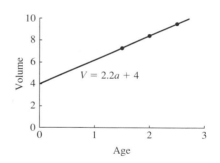

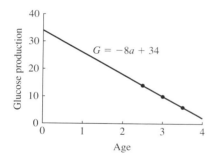

5.

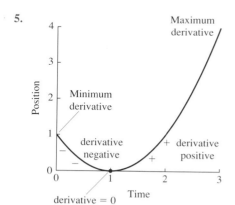

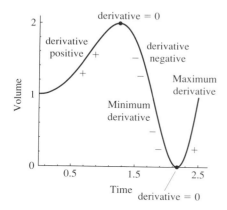

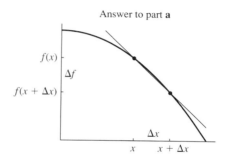

Answer to part **a**

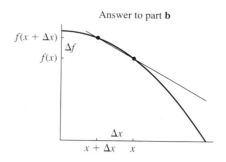

Answer to part **b**

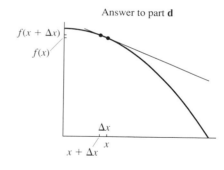

Answer to part **d**

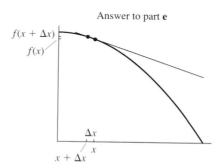

Answer to part **e**

7. a. $f(x) = f(1.0) = 3.0, f(x + \Delta x) = f(1.5) = 1.75,$
   $\Delta f = -1.25, \Delta f / \Delta x = -2.5$
b. $f(x) = f(1.0) = 3.0, f(x + \Delta x) = f(0.5) = 3.75,$
   $\Delta f = -0.75, \Delta f / \Delta x = -1.5$
c. $f'(1) = -2.0$.
d. $f(x) = f(0.5) = 3.75, f(x + \Delta x) = f(0.6) = 3.64,$
   $\Delta f = -0.11, \Delta f / \Delta x = -1.1$
e. $f(x) = f(0.5) = 3.75, f(x + \Delta x) = f(0.4) = 3.84,$
   $\Delta f = 0.09, \Delta f / \Delta x = -0.9$
f. $f'(0.5) = -1.0$
g. $\hat{f}(x) = -1.0(x - 0.5) + f(0.5) = -1.0(x - 0.5) + 3.75$
h. $f'(x) = -2x$

9. b. The slopes of both secants at $x = -1$ are -1. The slopes of both secants at $x = 1$ are 1. The slope of the secant at $x = 0$ with $\Delta x = 0.5$ is 1. The slope of the secant at $x = 0$ with $\Delta x = -0.5$ is -1.
c. The slope of the graph is -1 at $x = -1$ and 1 at $x = 1$.

d.

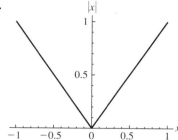

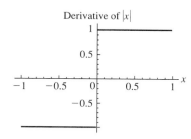

Section 2.5

1. **a.** $f'(x) = 10x$
   b. $g'(z) = 10z + 7$
   c. $h'(y) = 10y + 10y = 20y$

3. **a.** 90 mi/h
   b. Relative to the train, -25 mi/h. Relative to the ground, 65 mi/h.
   c. 70 mi/h

5. **a.** $\dfrac{db}{dt} = 2b(t)$.
   b. We can write
   $$\dfrac{db}{dt} = 2b_1(t) + 3b_2(t),$$
   but cannot get rid of b_1 and b_2. If the population consists of 10 from b_1 and none from the other, the growth rate is 20. If the total population is the same but consists entirely of type b_2, the growth rate is 30. Knowing the total population is thus not sufficient information to deduce its rate of change.

7. **a.** $A'(t) = 525t^2$, with units of people per year. The number of people getting infected is increasing.
   b. $F'(r) = 6.0r^3$
   c. $A'(r) = 2\pi r$. The units are centimeters. The derivative is equal to the perimeter of the circle, corresponding to the area of the little ring around the circle.
   d. $V'(r) = 4\pi r^2$. The derivative has units of area and is equal to the surface area of the sphere.

9. **a.** $5x^4$ **b.** $-5x^{-6}$ **c.** $0.2x^{-0.8}$
   d. $-0.2x^{-1.2}$ **e.** ex^{e-1} **f.** $-ex^{-e-1}$

g. $\dfrac{1}{e}x^{(1/e)-1}$ **h.** $\dfrac{-1}{e}x^{-(1/e)-1}$

11. **a.** $\ln(A) = 1.7\ln(L) - 3.65$ **b.** $\dfrac{d\ln(A)}{d\ln(L)} = 1.7$
    c. $\ln(L) = 0.589\ln(A) + 2.5$ **d.** $\dfrac{d\ln(L)}{d\ln(A)} = 0.589$
    e. Both of these derivatives are constants. We can tell that A is growing faster because that derivative is greater than 1.

13. **a.** The bigger it is, the faster it grows. But the older it is, the slower it grows. The other case does not slow down with age.
    b. $S(t) = t^{12}$ works.
    c. $S(t) = t^{24}$ works.
    d. At $t = 1$, both are the same. By $t = 2$, the second organism is more than 4000 times the size.

Section 2.6

1. **a.** $f'(x) = 2 \cdot (-3x + 2) + (-3) \cdot (2x + 3) = -12x - 5$
   b. $10z + 7$
   c. $\dfrac{dh}{dx} = \dfrac{d(x+2)}{dx}(2x+3)(-3x+2)$
   $\qquad + (x+2)\dfrac{d(2x+3)(-3x+2)}{dx}$
   $\qquad = 1 \cdot (2x+3)(-3x+2) + (x+2) \cdot (-12x - 5)$
   $\qquad = (2x+3)(-3x+2) - (x+2)(12x+5)$
   d. $5(y^2 - 1) + 2y(5y - 3)$
   e. $2t(3t^2 - 1) + 6t(t^2 + 2)$

3. **a.** $T(t) = P(t)W(t) = (2.0 \times 10^6 + 2.0 \times 10^4 t)(80 - 0.5t)$
   b.

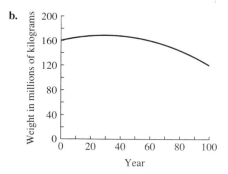

   c. $T'(t) = 6.0 \times 10^5 - 2.0 \times 10^4 t$
   d. $T'(t) = 0$ when $6.0 \times 10^5 - 2.0 \times 10^4 t = 0$ or when $t = 30$ yr.
   e. At time 60, the population is 2.6×10^6, the weight per person is 65 kg, and the total weight is 1.69×10^8 kg.

5. **a.** Total mass below ground is $(-1.0t + 40)(1.8 + 0.02t)$. The derivative is $-0.04t - 1.0$.
   b. Total mass above ground is $(3.0t + 20)(1.2 - 0.01t)$. The derivative is $-0.06t + 1.6$.

708 Answers

c. Total mass is $(-1.0t+40)(1.8+0.02t)+(3.0t+20)(1.2-0.01t)$. The derivative is $-0.04t - 1.0 - 0.06t + 1.6 = -0.1t + 0.6$.

7. a. $f'(x) = \dfrac{1}{(1+x)^2}$.

b. Now $u(x) = 1$ and $v(x) = 1 + 1/x$, so that $u'(x) = 0$ and $v'(x) = -1/x^2$. Therefore,

$$f'(x) = \dfrac{\left(1 + \dfrac{1}{x}\right) \cdot 0 - 1 \cdot \dfrac{-1}{x^2}}{\left(1 + \dfrac{1}{x}\right)^2} = \dfrac{1}{x^2 \left(1 + \dfrac{1}{x}\right)^2}$$

$$= \dfrac{1}{(x+1)^2}$$

c. $\dfrac{x}{1+x} = \dfrac{1+x-1}{1+x} = 1 - \dfrac{1}{1+x}$. The derivative of the constant is 0, and the derivative of the other piece is $1/(1+x)^2$, again matching the original result.

9. a. $S(1) = 0.8$, $S(5) = 3.0$, $S(10) = 2.0$. $S'(N) = 1 - 0.16N$. It looks like the bird does best by laying about six eggs.

b. $S(1) = 0.67$, $S(5) = 1.43$, $S(10) = 1.67$. $S'(N) = \dfrac{1}{(1+0.5N)^2}$, which is always positive. It looks like the bird does best by laying as many eggs as it can.

c. $S(1) = 0.91$, $S(5) = 1.43$, $S(10) = 0.91$.

$$S'(N) = \dfrac{1 - 0.1N^2}{(1 + 0.1N^2)^2}$$

It looks like the bird does best by laying $\sqrt{10}$ eggs.

Answer to part **a**

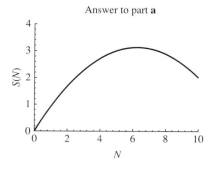

Answer to part **b**

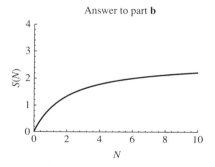

Answer to part **c**

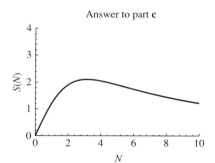

11. a. $h_n(0) = 0$ for each n and $h_n(1) = 0.5$ for each n. $h_2(2) = 0.8$, $h_5(2) = 0.97$, and $h_{10}(2) = 0.999$.

b. $h'_n(x) = \dfrac{nx^{n-1}}{(1 + x^n)^2}$

$h'_n(0) = 0$ for each n.
$h'_2(1) = 0.5$, $h'_5(1) = 1.25$, and $h'_{10}(1) = 2.5$.
$h'_2(2) = 0.16$, $h'_5(2) = 0.07$, and $h'_{10}(2) = 0.0049$.

c.

Answer to part **a**

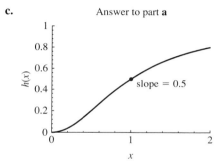

Answer to part **b**

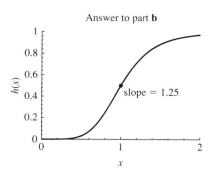

Answer to part **c**

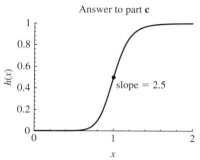

d. With small n, the response gets larger slowly for larger stimuli. With large n, the response gets larger very quickly right near $x = 1$. This describes something like a threshold, no response for $x < 1$ and a complete response for $x > 1$.

Section 2.7

1.

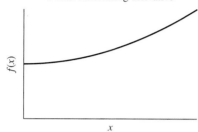
Positive increasing derivative

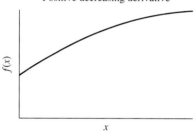
Positive decreasing derivative

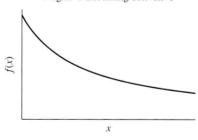
Negative increasing derivative

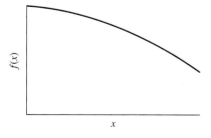
Negative decreasing derivative

3. a. The point of inflection has negative third derivative because the second derivative decreases from positive to negative values.
  b. The point of inflection at about $x = -1$ has positive third derivative because the curvature changes from negative to positive and is therefore increasing.

5. a. $f'(x) = -3x^{-4} < 0$ so the function is decreasing, $f''(x) = 12x^{-5} > 0$ so the function is always concave up.
  b. $g'(z) = 1 - z^{-2}$, so there is a critical point at $z = 1$. $g''(z) = 2z^{-3} > 0$ so the function is always concave up.

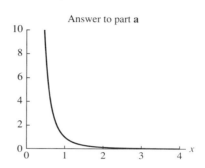
Answer to part **a**

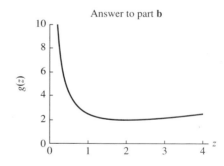
Answer to part **b**

  c. $h'(x) = -3x^2 + 12x - 11$, which has solutions at 2.577 and 1.422. $h''(x) = -6x + 12$ so the function is concave up for $x < 0.5$ and concave down for $x > 0.5$.
  d. $M'(t) = \dfrac{1}{(1+t)^2}$, which is positive, meaning that the function is increasing. By expanding the denominator, we can compute $M''(t) = -\dfrac{2}{(1+t)^3} < 0$, so the graph is concave down.

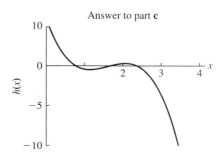
Answer to part **c**

Answer to part d

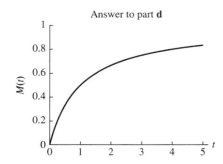

7. a. It is 0, because the degree will have been reduced to 0.
b. $5x^4, 20x^3, 60x^2, 120x, 120$ and 0.
c. Positive.

9. A function that never changes curvature can only intersect a straight line once. Suppose it crosses first from above to below, then from below to above, and then from above to below. Between the first two equilibria, the derivative must be increasing (at least some of the time), so the function has to be concave up. Between the second two equilibria, the derivative must be decreasing (at least some of the time), so the function has to be concave down. There must be a point of inflection in between.

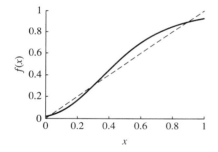

11. a. The tangent line is $\hat{M}(t) = t$.
b. The second derivative is $M''(t) = -\dfrac{2}{(1+t)^3}$, so $M''(0) = -2$.
c. By subtracting t^2, the quadratic $\hat{M}(t) = t - t^2$ matches both the first and second derivatives at $t = 0$.
d.

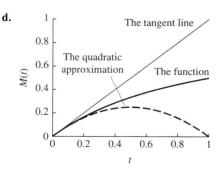

Section 2.8

1. a. $\dfrac{df}{dx} = \lim\limits_{h \to 0} \dfrac{5^{x+h} - 5^x}{h}$

b. $\dfrac{df}{dx} = \lim_{h \to 0} \dfrac{5^x 5^h - 5^x}{h}$
$= \lim_{h \to 0} \dfrac{5^x(5^h - 1)}{h}$
$= 5^x \left(\lim_{h \to 0} \dfrac{5^h - 1}{h} \right)$

c. For $h = 1$, $(5^1 - 5^0)/1 = 4.0$. For $h = 0.1$, $(5^{0.1} - 5^0)/0.1 = 1.746$. For $h = 0.01$, $(5^{0.01} - 5^0)/0.01 = 1.622$. For $h = 0.001$, $(5^{0.001} - 5^0)/0.001 = 1.611$. For $h = 0.0001$, $(5^{0.0001} - 5^0)/0.0001 = 1.609$.

d.

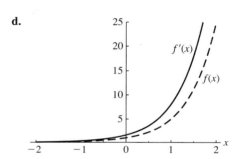

e. $e^{1.609} = 4.998$, which is suspiciously close to 5.

3. a. $f'(x) = (x^2 + 2x)e^x$
$f(0) = f'(0) = 0, f(1) = e, f'(1) = 3e$

b. $g'(x) = \dfrac{xe^x}{(1+x)^2}$
$g(0) = 1, g'(0) = 0, g(1) = e/2, g'(1) = e/4$

c. $F'(x) = 2x + e^x$
$F(0) = 1, F'(0) = 1, F(1) = 1 + e, F'(1) = 2 + e$

d. $H'(x) = \dfrac{10e^x}{(10 + e^x)^2}$
$H(0) = 0.09, H'(0) = 0.08, H(5) = 0.94, H'(5) = 0.06$

Answer to part a

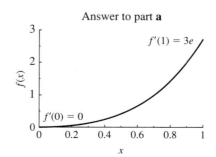

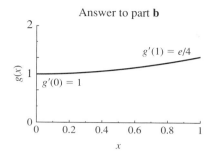

Answer to part b

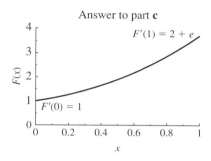

Answer to part c

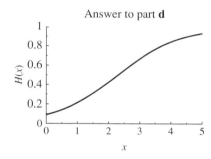

Answer to part d

5. **a.** The definition is
$$\frac{d\ln(x)}{dx} = \lim_{h\to 0}\frac{\ln(1+h)-\ln(1)}{h} = \lim_{h\to 0}\frac{\ln(1+h)}{h}$$
   b. The limit seems to be 1.0, meaning that the derivative is 1.0.
   c. The formula is that the derivative of $\ln(x)$ is $1/x$, which is 1.0 at $x = 1$.

7. **a.** $x + 2x\ln(x)$
   b. $e^x \ln(x) + \dfrac{e^x}{x}$
   c. $\ln(x^2) = 2\ln(x)$, so the derivative is $2/x$
   d. $\dfrac{1}{x} + \dfrac{1}{x^2}$

9. **a.** $\hat{f}(x) = x + 1$. The y-intercept is -1.
   b. $\hat{f}(x) = e^2(x-2) + e^2$. The y-intercept is e^2.
   c. $\hat{f}(x) = e^{-2}(x+2) + e^{-2}$. The y-intercept is $-3e^{-2}$.
   d. $\hat{f}(x) = e^{x_0}(x - x_0) + e^{x_0}$. The y-intercept is $e^{x_0}(x_0 - 1)$, which is zero when $x_0 = 1$. The line connecting the point $(0,0)$ with (x_0, e^{x_0}) has slope $\dfrac{e^{x_0}}{x_0}$, which is equal to the derivative e^{x_0} only if $x_0 = 1$.

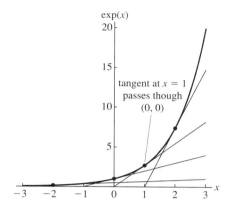

Section 2.9

1. **a.** $2x\sin(x) + x^2\cos(x)$
   b. $2x\cos(x) - x^2\sin(x)$
   c. $\cos(\theta) - \sin(\theta)$
   d. $-\sin^2(\theta) - \cos^2(\theta)$
   e. $\dfrac{2}{[1+\cos(\theta)]^2}$

3. It is the height of a triangle similar to (having the same shape as) the triangle with base $\cos(h)$ and height $\sin(h)$. Because the base is length 1,
$$\frac{\text{Length}}{1} = \frac{\sin(h)}{\cos(h)} = \tan(h)$$

5. **a.** With the product rule
$$\frac{d}{d\theta}\tan(\theta)\cos(\theta) = \sec^2(\theta)\cos(\theta) - \tan(\theta)\sin(\theta)$$
$$= \sec(\theta) - \frac{\sin^2(\theta)}{\cos(\theta)}$$
$$= \frac{1-\sin^2(\theta)}{\cos(\theta)} = \frac{\cos^2(\theta)}{\cos(\theta)} = \cos(\theta)$$
Directly, we see that $\tan(\theta)\cos(\theta) = \sin(\theta)$. The derivative is $\cos(\theta)$.
   b. Get $-\sin(\theta)$ both ways.
   c. Get $\sec(\theta)\tan(\theta)$ both ways.

7. **a.** This is an expanding oscillation. The average value is always around 0, but the amplitude gets bigger and bigger.
   b. This is like part **a**, but expands exponentially.
   c. This looks a population growing exponentially, but with wiggles.

712 Answers

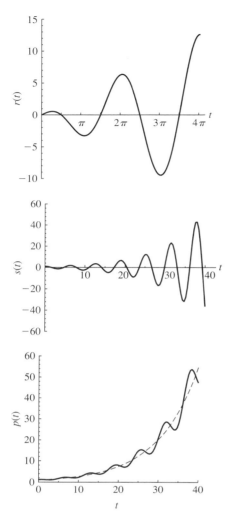

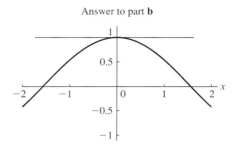

Answer to part **b**

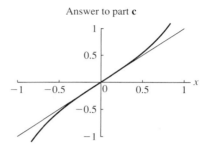

Answer to part **c**

d. Tangent line is $\widehat{\sin(x)} = \dfrac{\sqrt{2}}{2} + \dfrac{\sqrt{2}}{2}x$. Second derivative is $-\dfrac{\sqrt{2}}{2}$.

e. Tangent line is $\widehat{\cos(x)} = \dfrac{\sqrt{2}}{2} - \dfrac{\sqrt{2}}{2}x$. Second derivative is $-\dfrac{\sqrt{2}}{2}$.

f. Tangent line is $\widehat{\tan(x)} = 1 + \dfrac{1}{2}x$. Second derivative is 1.

9. **a.** Tangent line is $\widehat{\sin(x)} = x$. Second derivative is 0. The tangent line is very close.
   b. Tangent line is $\widehat{\cos(x)} = 1$. Second derivative is -1. The tangent line is not as good an approximation.
   c. Tangent line is $\widehat{\tan(x)} = x$. Second derivative is 0. The tangent line is very close. The tangent is closest when the second derivative is 0.

Answer to part **d**

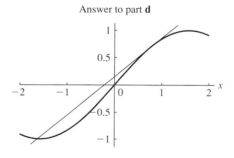

Answer to part **a**

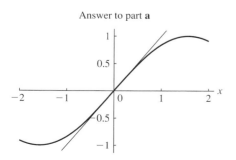

Answer to part **e**

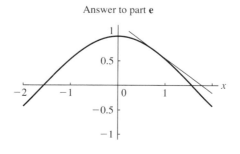

Answers 713

Answer to part **f**

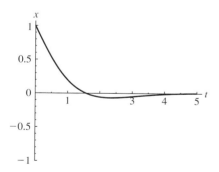

11. a. This says that acceleration has two pieces: one proportional to the negative of displacement (the $-2x$ term), and one in the proportional to the negative of velocity (the $2(dx/dt)$ term).

b. $\dfrac{dx}{dt} = -e^{-t}\cos(t) - e^{-t}\sin(t), \quad \dfrac{d^2x}{dt^2} = 2e^{-t}\sin(t)$

But

$$-2x - 2\dfrac{dx}{dt} = -2e{-t}\cos(t)$$
$$- 2[-e^{-t}\cos(t) - e^{-t}\sin(t)]$$
$$= 2e^{-t}\sin(t) = \dfrac{d^2x}{dt^2}$$

This is indeed a solution.

c. The result is an oscillation that gets smaller and smaller. Friction is bringing the object to a stop. The friction must be quite strong because the spring makes only one reasonably large bounce.

Section 2.10

1. a. $g(x) = f[h(x)]$, where $h(x) = 1 + 3x$ and $f(h) = h^2$. Then,

$$g'(x) = f'[h(x)]h'(x) = 2h \cdot 3 = 6(1 + 3x)$$

b. $F(z) = g[h(z)]$, where $h(z) = 1 + 2/(1 + z)$ and $g(h) = h^3$. To find $h'(z)$ we note that $h(z) = 1+2(1+z)^{-1}$ which has derivative $-2(1 + z)^{-2}$ using the chain rule as in **a**. Then,

$$F'(z) = g'[h(z)]h'(z) = 3h^2 \cdot [-2(1 + z)^{-2}]$$
$$= -6\dfrac{\{1 + [2/(1 + z)]\}^2}{(1 + z)^2}$$

c. $f_1'(t) = 30(1 + t)^{29}$
d. $f_2'(t) = 30t(1 + t^2)^{14}$
e. $h'(x) = 6(1 + 2x)^2$
f. $r'(x) = \dfrac{-6x(1 + 3x)}{(1 + 2x)^4}$

3. a. With the chain rule, set $f(x) = x^3$ and $l(f) = \ln(f)$. Then,

$$(l \circ f)'(x) = l'[f(x)]f'(x) = \dfrac{1}{f(x)}3x^{3-1} = \dfrac{1}{x^3}3x^2 = \dfrac{3}{x}$$

With law 2 of logs, $\ln(x^3) = 3\ln(x)$. Then,

$$\dfrac{d\ln(x^3)}{dx} = 3\dfrac{d\ln(x)}{dx} = \dfrac{3}{x}$$

The two answers check.

b. With the chain rule, set $f(x) = 3x$ and $l(f) = \ln(f)$. Then,

$$(l \circ f)'(x) = l'[f(x)]f'(x) = \dfrac{1}{f} \cdot 3 = \dfrac{1}{3x} \cdot 3 = \dfrac{1}{x}$$

With law 1 of logs, $\ln(3x) = \ln(3) + \ln(x)$. Then,

$$\dfrac{d\ln(3x)}{dx} = \dfrac{d\ln(3)}{dx} + \dfrac{d\ln(x)}{dx} = 0 + \dfrac{1}{x} = \dfrac{1}{x}$$

The two answers check.

5. a. If $F(x) = I[f(x)]$, then $F(x) = f(x)$ because the identity function does nothing. We expect that $F'(x) = f'(x)$. By the chain rule, $F'(x) = I'(f)f'(x)$. Because $I(f) = f$, $I'(f) = 1$. Therefore, $F'(x) = 1 \cdot f'(x) = f'(x)$.

b. If $G(x) = g[I(x)]$, then $G(x) = g(x)$ because the identity function does nothing. We expect that $G'(x) = g'(x)$. By the chain rule, $G'(x) = g'(I)I'(x)$. Because $I(x) = x$, $I'(x) = 1$. Therefore, $G'(x) = g'[I(x)] \cdot 1 = g'(x)$ as expected.

7. a. Set $H(y) = r[p(y)]$ where $p(y) = 1 + y^3$ and $r(p) = 1/p$. Then,

$$p'(y) = 3y^2, \quad r'(p) = \dfrac{-1}{p^2}$$

By the chain rule,

$$H'(y) = r'(p)p'(y) = \dfrac{-1}{p^2} \cdot 3y^2 = \dfrac{-3y^2}{(1 + y^3)^2}$$

b. With the quotient rule, set $u(y) = 1$ and $v(y) = 1 + y^3$. Then,

$$H'(y) = \left(\dfrac{u}{v}\right)'(y) = \dfrac{v(y)u'(y) - u(y)v'(y)}{v^2(y)}$$
$$= \dfrac{(1 + y^3) \cdot 0 - 1 \cdot 3y^2}{(1 + y^3)^2} = -\dfrac{3y^2}{(1 + y^3)^2}$$

714 Answers

c. I would have to go with the quotient rule.

9. a. $f'(x) = -1.6\pi \sin[2\pi(x - 1.0)/5.0]$.
   b. $g'(t) = -6.0\pi \sin[2\pi(x - 5.0)]$.
   c. $h'(z) = -2.5\pi \sin[2\pi(x - 3.0)/4.0]$.

Answer to part a

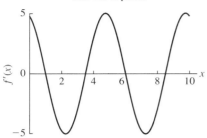

Answer to part b

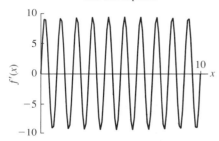

Answer to part c

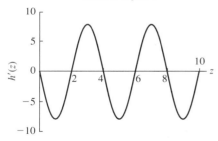

11. a. This population is growing, but per capita reproduction oscillates.
    b. $b'(t) = [0.1 + 0.1 \cos(0.8t)]b(t)$
       $= 0.1b(t)[1 + \cos(0.8t)]$.
    c.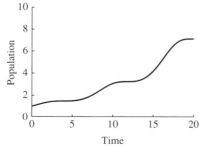

d. The oscillation in the per capita reproduction is larger for larger A. When A is larger than 1, there are times when the population decreases.
e. $b'(t) = [0.1 + 0.1A \cos(0.8t)]b(t) = 0.1b(t) \times [1 + A \cos(0.8t)]$.

f.

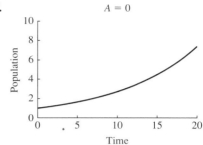

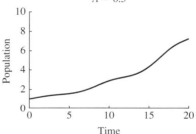

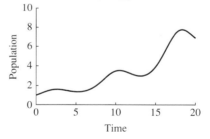

Answers to Supplementary Problems for Chapter 2

1. a. $F'(y) = 4y^3 + 10y$
   b. $H'(c) = 2c(1 + c)/(1 + 2c)^2$ for $c \neq -1/2$
   c. $G'(r) = 24r^2 + 28r + 4$
   d. $a'(x) = 28x^6 + 28x^3$
   e. $f'(x) = \dfrac{1 + \sqrt{x}/2}{(1 + \sqrt{x})^2}$ for $x \geq 0$
   f. $g'(y) = (9y^2 + 4y + 1)e^{3y^3 + 2y^2 + y}$
   g. $h'(z) = \dfrac{2 \ln(z) - 1}{[1 + \ln(z^2)]^2}$ for $z \geq 0$
   h. $b'(y) = -0.75y^{-1.75}$ for $y \geq 0$
   i. $c'(z) = \dfrac{2 - z^2}{(1 + z)^2(2 + z)^2}$

Answers   715

3. **a.** $f'(x) = (1-x)e^{-x}$
   $\hat{f}(x) = x$
   b. $g'(t) = (1-2t^2)/(1+2t^2)^2$. Because $g(1) = 1/3$ and $g'(1) = -1/9$, we have $\hat{g}(t) = 1/3 - 1/9(t-1)$.
   c. $h'(x) = 4x/(1+2x^2)$. Because $h(0) = h'(0) = 0$, the tangent line is we have $\hat{h}(x) = 0$.

5. **a.** The 1000 has units of individuals, and the 10 has units of individuals/yr^2.

   b.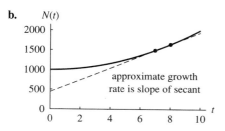

   c. The population is 1490 after 7 yr and 1640 after 8 yr. The approximate growth rate is 150 during this year.
   d. The derivative is $20t$.

   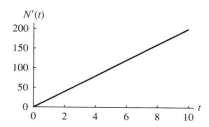

   e. The per capita rate of growth is $20t/(1000 + 10t^2)$. It starts at 0 and increases for a while. Eventually, however, it will decrease.

7. **a.**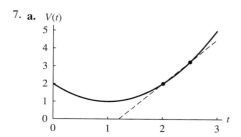

   b. The secant line has slope $[V(2.5) - V(2)]/0.5 = 2.5$. Because $V(2) = 2$, the equation is $V_s(t) = 2.5(t-2) + 2$.
   c. $V'(t) = 2t - 2$, so $V'(2) = 2$ in units of μm/min.
   d. At $t = 2$, cell volume is increasing by 2μm/min.

e.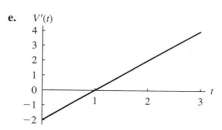

9. **a.** $\hat{f}(x) = 0.5 - 0.25x$. $\hat{f}(-0.03) = 0.5075$. In this case, $f(-0.03) = 0.5068$.
   b. $\hat{g}(y) = 81(y-1)$
   $\hat{g}(1.02) = 1.62$
   c. $\hat{h}(z) = 6.22(z-3) + 10.67$
   $\hat{h}(3.04) = 10.92$

11. **a.** $p_0 = 0.333$.
    b. p_0 would have to be between 0.243 and 0.428, or within about 0.1.
    c. It would get smaller and smaller also.

13. **a.**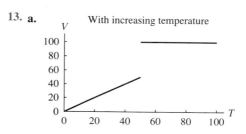

    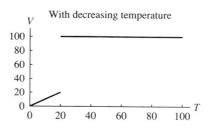

    b. $\lim_{T \to 50^+} V_i(T) = 100$
    $\lim_{T \to 50^-} V_i(T) = 50$
    $\lim_{T \to 50^+} V_d(T) = \lim_{T \to 50^-} V_d(T) = 100$
    c. $\lim_{T \to 20^+} V_i(T) = \lim_{T \to 20^-} V_i(T) = 20$
    $\lim_{T \to 20^+} V_d(T) = 100$
    $\lim_{T \to 20^-} V_d(T) = 20$

Chapter 3
Section 3.1

1. In the first picture, at the lower equilibrium, the updating function crosses the diagonal from below to above and should therefore be unstable. At the upper equilibrium, the

updating function crosses from above to below and should be unstable. In the second picture, the updating function crosses the diagonal from above to below at the first and last equilibria that should therefore be stable. At the middle equilibrium, the updating function crosses from below to above and should be unstable. In the third picture, the updating function crosses the diagonal from above to below at the lower equilibrium that should therefore be stable. At the upper equilibrium, the updating function crosses from below to above and should be unstable.

3. a. The equilibrium is 3.5 and is stable.
    b. The equilibrium is 3.125 and is stable.
    c. The equilibrium is 0.99 and is stable.
    d. The equilibrium is 2.0×10^6 and is unstable.

5. a. $f'(p) = \dfrac{3}{[1.5p + 2.0(1-p)]^2}$

    $f'(0) = 0.75$, $f'(1) = 1.333$
    The equilibrium at $p = 0$ is stable, the other is unstable.

    b. $f'(p) = \dfrac{2.25}{[1.5p + 1.5(1-p)]^2} = 1$

    $f'(0) = f'(1) = 1$
    Both derivatives are exactly 1, so we cannot tell. This updating function exactly matches to "do nothing" function, meaning that solutions move neither toward nor away from equilibria.

    c. $f'(p) = \dfrac{rs}{[sp + r(1-p)]^2}$

    $f'(0) = s/r$, $f'(1) = r/s$
    The equilibrium at $p = 0$ is stable when $r > s$ and unstable when $r < s$. The opposite conditions hold at $p = 1$. Both equilibria cannot be stable.

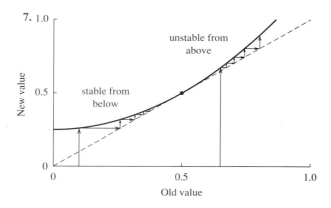

7.

9. a. There is no equilibrium and the cobwebbing creeps slowly past the point where the equilibrium used to be.
    b. There are two equilibria, the lower one unstable and the upper one stable.

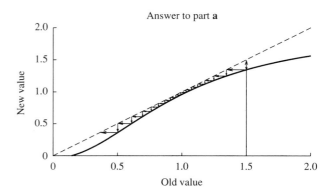

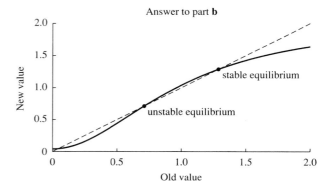

c. We go from 0 equilibria, to 1 that is "half" stable, to 2, one of which is stable.

11. a. $x_{t+1} = r \dfrac{x_t^2}{1.0 + x_t^2}$

    b. The equilibria are $x = 0$, and

    $$x^* = \dfrac{-r \pm \sqrt{r^2 - 4}}{2}$$

    The second two only make sense when $r^2 > 4$ or $r > 2$.

c.

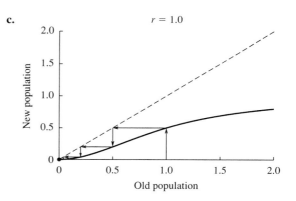

Answers **717**

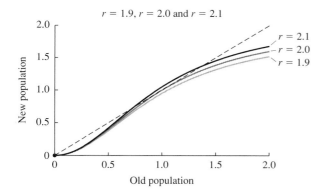

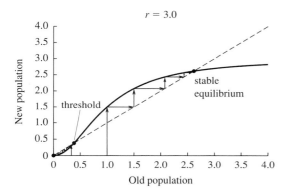

d. With $r < 2$, the population just goes extinct. With $r > 2$, it will only survive if the initial condition is larger than the middle equilibrium. There is a sort of threshold. Perhaps these organisms need a certain population density to find mates or protect each other from predators.

Section 3.2

1. a.

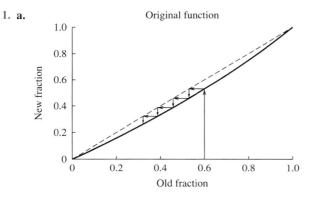

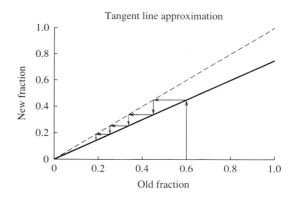

b.

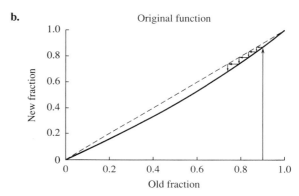

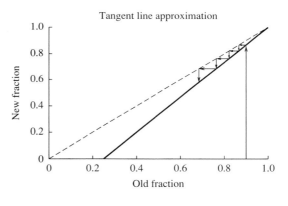

c.

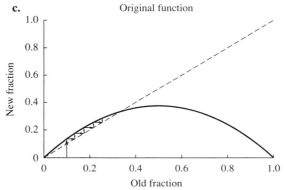

718 Answers

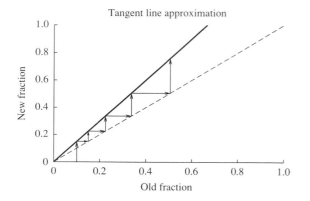

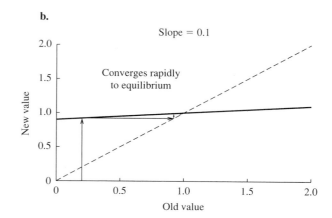

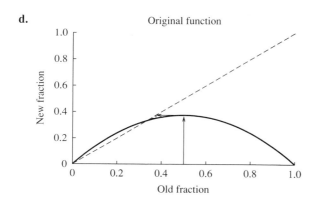

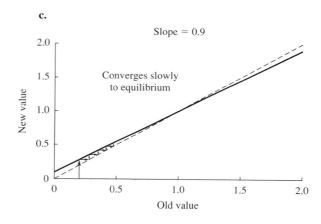

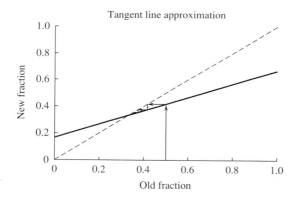

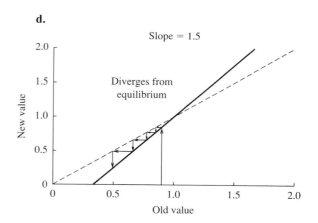

e. The tangent line is identical to the original function. The two pictures are the same.

3. a. The equilibrium is $y^* = 1.0$.

e.

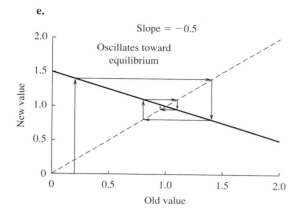

f.

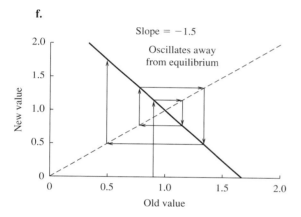

g. The results match perfectly!

5. a. This line passes through the points $(20, 21)$ and $(19, 18)$. Let z be the setting on the thermostat and T be the temperature. The equation is $T = 3(z - 20) + 21$.
  b. When it is $18°$, you set the thermometer to $22°$. This results in a temperature of $27°$. Setting the thermometer to $13°$ then results in a temperature of $0°$. Things are getting pretty chilly.
  c. $z_t = 40 - T_t$
  d. $T_{t+1} = 3(z_t - 20) + 21 = 3(40 - T_t - 20) + 21 = 81 - 3T_t$
  e. That slope of -3 means that the temperatures will oscillate more and more widely.
  f. The system would have to have a better correction mechanism than this one. I would have it estimate T as a function of z and correct based on that basis.

7. a. The updating function is $n_{t+1} = 100 - 0.5n_t$ with a stable equilibrium at 66.7.
  b. The updating function is $n_{t+1} = 100 - 2.0n_t$ with an unstable equilibrium at 33.3.

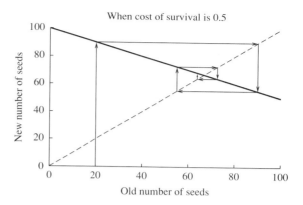

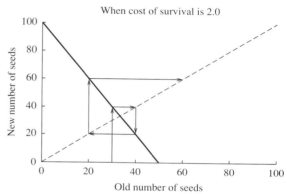

9. a. The updating function is $b_{t+1} = b_t(0.5 + 0.5b_t)$.
  b. The equilibria are at $b^* = 0$ and $b^* = 1$.
  c. The equilibrium at 0 looks stable and the one at 1 looks unstable.

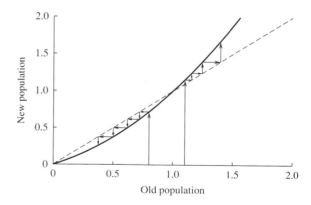

  d. Populations starting below 1 die out, while those starting above 1 blast off to infinity. This species does well with a little help from their friends.

11. a. $f'(x) = r - 3rx^2$
   $f''(x) = -6rx < 0$

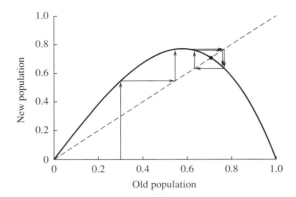

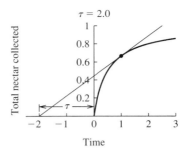

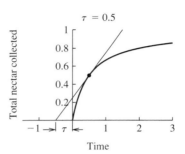

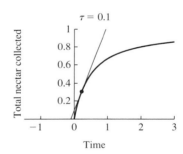

b. The equilibria are at $x = 0$ and $x^* = \sqrt{1 - \dfrac{1}{r}}$ (when $r \geq 1$).

c. $f'(0) = r$ and $f'(x^*) = 3 - 2r$. The first equilibrium is stable when $r < 1$ as usual, and the positive equilibria is stable when $r < 2$.

Section 3.3

1. a. $a'(x) = \dfrac{1}{(1 + x)^2}$, which is always positive. There are no critical points.

b. $f'(x) = 2 - 4x$. Critical point at $x = 1/2$.

c. $c'(w) = 3w^2 - 3$. Therefore, $c'(w) = 0$ if $3w^2 = 3$, which has solutions at $w = -1$ and $w = 1$.

d. $g'(y) = \dfrac{1 - y^2}{(1 + y^2)^2}$, which is zero at $y = -1$ and $y = 1$.

e. $h'(z) = 2ze^{z^2}$, which is zero only at $z = 0$.

f. $c'(z) = 2\pi \sin(2\pi\theta)$, which is zero when θ is an integer.

3. a. $a''(x) = \dfrac{-2}{(1 + x)^3}$, negative for $0 \leq x \leq 1$. The graph is increasing, concave down.

b. $f''(x) = -4 < 0$. The graph has a maximum at $x = 0.5$, consistent with the fact that it is concave down.

c. $c''(w) = 6w$. Therefore, $c''(1) = 6$ and $c''(-1) = -6$. This is consistent with the fact that this function has a minimum at 1 and a maximum at -1.

d. $g''(y) = \dfrac{2y(y^2 - 3)}{(1 + y^2)^3}$. Therefore, $g''(1) = -1/2$ and $g''(-1) = 1/2$, consistent with the local maximum at $y = 1$ and the local minimum at $y = -1$.

e. $h''(z) = (4z^2 + 2)e^{z^2}$, which is always positive. The critical point at $z = 0$ must be a minimum.

5. a. $t = 1.0$. The tangent line is $\hat{F}(t) = 2/3 + 2/9(t - 1)$. It is true that $\hat{F}(-2.0) = 0$.

b. $t = 0.5$. The tangent line is $\hat{F}(t) = 1/2 + 1/2(t - 0.5)$. It is true that $\hat{F}(-0.5) = 0$.

c. $t = \sqrt{0.05} = 0.223$. The tangent line is $\hat{F}(t) = 0.309 + 0.955(t - 0.223)$. It is true that $\hat{F}(-0.1) = 0$.

d. $t = \sqrt{0.5\tau}$

e. t approaches 0. The shorter the travel time, the less time the bee should spend on each flower.

7. a. $\tau = 0.5$
b. $\tau = 0.005$
c. $\tau = 8.0$
d. $\tau = t^2 c$

9. a. $R(n) = \dfrac{n}{P(n)}$.

b. $R'(n) = \dfrac{P(n) - nP'(n)}{P(n)^2}$. The critical point is where $P(n) = nP'(n)$ or $P'(n) = \dfrac{P(n)}{n}$.

c. When the slope of the tangent to $P(n)$ is equal to the slope of the line connecting the point to the origin $(0, 0)$.

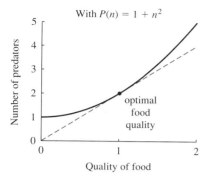

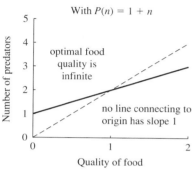

d. The maximum is where $n = 1$.
e. There is no maximum, the bees do better with the largest possible n. The result is different because the number of predators does not increase faster than the quality of the food.

11. The derivative of the updating function is $2.5(1 - 2N_t) - h$. At $h = 0.75$, the equilibrium is 0.25. The value of the derivative is then 0.5, indicating stability.

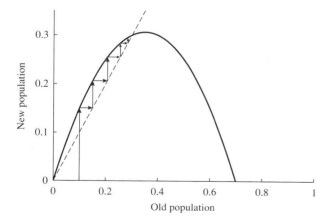

13. a. $h = 0.625$
    b. $h = 0.5$ when $c = 0.2$, $h = 0.125$ when $c = 0.5$, and $h = -0.5$ when $c = 1.0$
    c. That negative harvest looks suspicious. When c becomes too large, it is not worth fishing.

Section 3.4

1. a. $c_{t+1} = 0.5(1 - q)c_t + q\gamma$
   b. The equilibrium is at
   $$c^* = \frac{q\gamma}{1 - 0.5(1 - q)} = \frac{q\gamma}{0.5 + 0.5q}$$
   c. With $q = 0.5$ and $\gamma = 5$, the updating function is
   $$c_{t+1} = 0.25c_t + 2.5$$

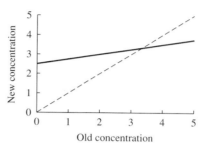

   d. With the values chosen, the equilibrium is 3.33, indeed less than γ.

3. a. The price of gasoline does not change continuously and therefore need not take on all intermediate values.
   b. The Intermediate Value Theorem guarantees this crossing. It is possible that that it crosses the larger value, but it need not.
   c. We cannot apply the theorem directly because the value of the function at the endpoints is 0. At $\theta = \pi/2$, however, $\sin(\theta) = 1$. Therefore, there must be one crossing of 0.34 between 0 and $\pi/2$ and another between $\pi/2$ and π.
   d. $x_{t+1} > x_t$ when $x_t = 0$, and $x_{t+1} < x_t$ when $x_t = \pi/2$. Because this updating function is continuous, there must be a crossing in between.

5. a. The Intermediate Value Theorem.
   b. The average rate of increase is the slope of the secant connecting age 0 and age 14, which has a slope of 4.0 kg/yr. Therefore, the Mean Value Theorem guarantees such a point.

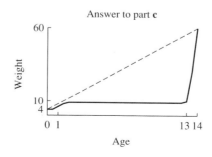

Answer to part c

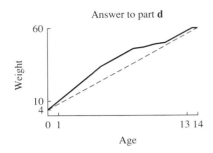

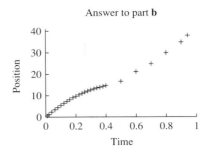

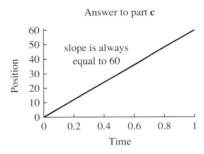

7.

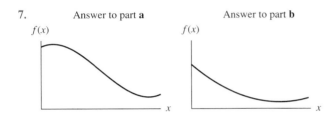

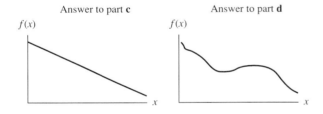

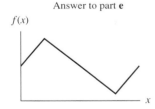

11. a. Per capita reproduction of type a is a decreasing function of p_t. This could be a consequence of the fact that type a individuals have to compete more when they are more frequent.

b. Per capita reproduction of type b is a decreasing function of p_t. This could be a consequence of the fact that type b individuals have to compete less when type a individuals are more frequent.

c. Substituting in the new definitions of s and r,

$$p_{t+1} = \frac{s(1-\alpha p_t)p_t}{s(1-\alpha p_t)p_t + r[1-\beta(1-p_t)](1-p_t)}$$

d. At $p_t = 0$, $p_{t+1} = 0$ so this is an equilibrium. Similarly, at $p_t = 1$, $p_{t+1} = 1$ so this is also an equilibrium.

e. Let $f(p)$ denote the updating function. It is of the form

$$f(p) = \frac{u(p)}{u(p) + v(p)}$$

where $u(p) = s(1-\alpha p)p$ and

$$v(p) = r[1-\beta(1-p)](1-p)$$

Then,

$$f'(p) = \frac{[u(p)+v(p)]u'(p) - u(p)[u'(p)+v'(p)]}{[u(p)+v(p)]^2}$$

$$= \frac{v(p)u'(p) - u(p)v'(p)}{[u(p)+v(p)]^2}$$

Note that $u(0) = v(1) = 0$. Then,

$$f'(0) = \frac{v(0)u'(0)}{v(0)^2} = \frac{u'(0)}{v(0)}$$

$$f'(1) = \frac{-u(1)v'(1)}{u(1)^2} = -\frac{v'(1)}{u(1)}$$

9.

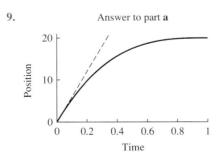

Substituting values, we find
$$f'(0) = \frac{s}{r(1-\beta)}$$
$$f'(1) = \frac{r}{s(1-\alpha)}$$

f. If $s > r$, require that $s(1-\alpha) < r$. If $s < r$, require that $r(1-\beta) < s$.

g. The curve must cross the diagonal from above somewhere in between. The new equilibrium corresponds to coexistence. The conditions in part **f** imply that this is possible when type a is a superior competitor ($s > r$) but competes strongly with itself (α is pretty big) or when type b is a superior competitor ($r > s$) but competes strongly with itself (β is pretty big).

h. The new equilibrium seems to be stable. The two types coexist because each competes more with its own type than with the other type.

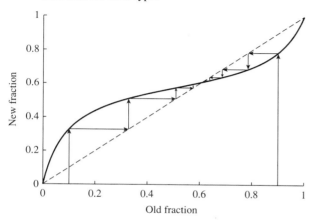

13. a. $c_{t+1} = 0.5 \left(\dfrac{1}{1+c_t^n}\right) c_t + 0.75$

b. At $c_t = 1$, $c_{t+1} = 1$, an equilibrium.

c. The derivative is
$$f'(c) = 0.5 \frac{(1+c^n) - nc^n}{(1+c^n)^2}$$
$$f'(1) = \frac{2-n}{8}$$

d. The equilibrium is unstable if $f'(1) < -1$, which occurs for $n > 10$.

e.

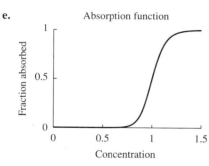

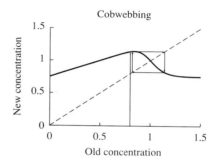

f. This lung has very high absorption when the concentration is high, and almost none when low. It therefore fills up with chemical on one breath, and is then totally depleted on the next, oscillating between high and low concentrations.

Section 3.5

1. a. 0 **b.** ∞ **c.** ∞
d. 1 **e.** ∞ **f.** ∞
g. 0
h. ∞

3. The order is b, e, f, d, a, c, h, g.

5. a. The Extreme Value Theorem guarantees a maximum for a function defined for $0 \leq c \leq X$. Pick X so large that $A(X)$ is tiny [smaller than the positive value $A(1)$, for example]. The maximum must then lie in between.

b. The maximum is at $c^* = 1/\beta$ for part **d** and at $c^* = \sqrt{k}$ for part **e**.

c. The Intermediate Value Theorem guarantees that it takes on a value equal to half the maximum between 0 and c^*, and again between c^* and X, where $A(X) < A(c^*)/2$. We know that such an X exists because the limit is 0.

7. a. To make sure $10^8 1.1^t > 10^{10}$, we need
$$1.1^t > 10^2$$
$$e^{\ln(1.1)t} > 10^2$$
$$\ln(1.1)t > \ln(10^2)$$
$$t > \ln(10^2)/\ln(1.1) = 48.3$$

The population exceeds the threshold after about 49 generations.

b. 96.6 **c.** 11.35 **d.** 6.6

9. a. Equilibrium is 2.0.
b. The 0.5^t term approaches 0, so the limit is also 2.0.
c. Need to find when it reaches 2.02, which happens after 4.3 days.
d. We could show that the slope of the updating function at the equilibrium is less than 1, or find the solution and prove it approaches the equilibrium as a limit.

11. a. The lung has a limited capacity to take up chemical.
b. When chemical concentration is too high, the lung is poisoned.

724 Answers

c. When chemical concentration becomes high, new absorption pathways are turned on.

Section 3.6

1. a. $f_0(x) = 1$, $f_\infty(x) = x$
   b. $g_0(y) = y$, $g_\infty(y) = y^3$
   c. $h_0(z) = e^z$, $h_\infty(z) = e^z$
   d. $m_0(z) = \dfrac{1}{z}$, $m_\infty(z) = 30z^2$
   e. $F_0(a) = -39$ because both exponential terms approach 1, $F_\infty(a) = -40e^{6a}$
   f. $G_0(c) = \dfrac{5}{c^5}$, $G_\infty(c) = \dfrac{5}{c^2}$

3. a. For A_1, the numerator has only one term, so the leading behavior is $2c^2$ at both 0 and ∞. The denominator has leading behavior 1 for c near 0 and c for c large. Therefore,

$$A_1(c)_0 = \frac{2c^2}{1} = 2c^2$$

$$A_1(c)_\infty = \frac{2c^2}{c} = 2c$$

$$\lim_{c \to 0} A_1(c) = 0$$

$$\lim_{c \to \infty} A_1(c) = \infty$$

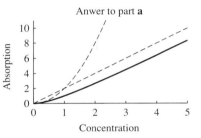

Anwer to part a

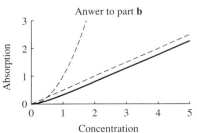

Anwer to part b

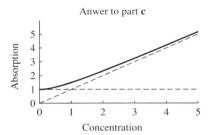

Anwer to part c

b. For A_2, the numerator has only one term, so the leading behavior is c^2 at both 0 and ∞. The denominator has leading behavior 1 for c near 0 and $2c$ for c large. Therefore,

$$A_2(c)_0 = \frac{c^2}{1} = c^2$$

$$A_2(c)_\infty = \frac{c^2}{2c} = \frac{c}{2}$$

$$\lim_{c \to 0} A_2(c) = 0$$

$$\lim_{c \to \infty} A_2(c) = \infty$$

c. For A_3, the numerator has leading behavior 1 near 0 and c^2 for c large. The denominator has leading behavior 1 for c near 0 and c for c large. Therefore,

$$A_3(c)_0 = \frac{1}{1} = 1$$

$$A_3(c)_\infty = \frac{c^2}{c} = c$$

$$\lim_{c \to 0} A_3(c) = 1$$

$$\lim_{c \to \infty} A_3(c) = \infty$$

d. For A_4, the numerator has leading behavior 1 near 0 and c for c large. The denominator has leading behavior 1 for c near 0 and c^2 for c large. Therefore,

$$A_4(c)_0 = \frac{1}{1} = 1$$

$$A_4(c)_\infty = \frac{c}{c^2} = \frac{1}{c}$$

$$\lim_{c \to 0} A_4(c) = 1$$

$$\lim_{c \to \infty} A_4(c) = 0$$

e. For A_5, the numerator has only one term, so the leading behavior is $3c$ at both 0 and ∞. The denominator has leading behavior 1 for c near 0 and $\ln(1+c)$ for c large. Therefore,

$$A_5(c)_0 = \frac{3c}{1} = 3c$$

$$A_5(c)_\infty = \frac{3c}{\ln(1+c)}$$

$$\lim_{c \to 0} A_5(c) = 0$$

$$\lim_{c \to \infty} A_5(c) = \infty$$

f. For A_6, neither the numerator nor the denominator can be simplified for c near 0. The leading behavior of the numerator is e^c and the leading behavior of the denominator is e^{2c} for c large. Therefore,

$$A_6(c)_0 = A_6(c)$$

$$A_6(c)_\infty = \frac{e^c}{e^{2c}} = e^{-c}$$

$$\lim_{c \to 0} A_6(c) = 1$$

$$\lim_{c \to \infty} A_6(c) = 0$$

Anwer to part d

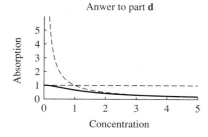

Anwer to part e

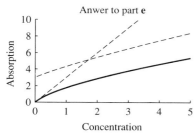

Anwer to part f

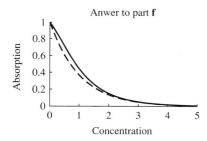

5. **a.** L'Hopital's rule not appropriate at $c = 0$. As $c \to \infty$,

$$\lim_{c \to \infty} \frac{2c^2}{1+c} = \lim_{c \to \infty} \frac{4c}{1} = \infty$$

b. $\lim_{c \to 0} B_2(c) = 2$, $\lim_{c \to \infty} B_2(c) = 0$

c. $\lim_{c \to 0} B_3(c) = 1$, $\lim_{c \to \infty} B_3(c) = \infty$

d. L'Hopital's rule not appropriate at $c = 0$. As $c \to \infty$,

$$\lim_{c \to \infty} B_4(c) = 0$$

e. $\lim_{c \to 0} B_5(c) = 0.5$, $\lim_{c \to \infty} B_5(c) = 1$

f. $\lim_{c \to 0} B_6(c) = 2$, $\lim_{c \to \infty} B_6(c) = 0$

7. **a.** The definition of the derivative at 0 is

$$\lim_{\Delta x \to 0} \frac{h(0 + \Delta x) - h(0)}{\Delta x} = \lim_{\Delta x \to 0} \frac{e^{\alpha \Delta x} - 1}{\Delta x}$$

b. $h'(x) = \alpha e^{\alpha x}$, and $h'(0) = \alpha$.
c. The definition of the derivative is

$$\lim_{\Delta x \to 0} \frac{f(x + \Delta x) - f(x)}{\Delta x}$$

As long as f is continuous, this is indeterminate. Taking derivative of top and bottom with respect to Δx (using the chain rule on the first term), we get

$$\lim_{\Delta x \to 0} \frac{f(x + \Delta x) - f(x)}{\Delta x} = \lim_{\Delta x \to 0} \frac{f'(x + \Delta x)}{1} = f'(x)$$

9. **a.** We found that the optimal t is $t = \sqrt{0.5\tau}$.

$$R\left(\sqrt{0.5\tau}\right) = \frac{F\left(\sqrt{0.5\tau}\right)}{\sqrt{0.5\tau} + \tau}$$

$$= \frac{\sqrt{0.5\tau}}{\left(0.5 + \sqrt{0.5\tau}\right)\left(\sqrt{0.5\tau} + \tau\right)}$$

b. This is an indeterminate form. The limit is 2.0.
c. This limit is the slope of F at $t = 0$. Because the travel time is nearly 0, the bee can take the nectar at the maximum rate for an instant and then move to the next flower.

Section 3.7

1. **a.** Let $f(x) = x^3$ with base point $a = 2.0$. $f(2) = 8.0$ and $f'(2) = 12.0$. $\hat{f}(x) = 8.0 + 12.0(x - 2.0)$ and $\hat{f}(2.02) = 8.0 + 12.0(2.02 - 2.0) = 8.0 + 12.0 \cdot 0.02 = 8.24$.
b. Let $f(x) = x^2$ with base point $a = 3.0$. $f(3) = 9.0$ and $f'(3) = 6.0$. $\hat{f}(x) = 9.0 + 6.0(x - 3.0)$ and $\hat{f}(3.03) = 9.0 + 6.0(3.03 - 3.0) = 9.18$.
c. Let $f(x) = \ln(x)$. $f(1) = \ln(1) = 0$. $f'(x) = 1/x$ and $f'(1) = 1$. So $\hat{f}(x) = 0 + 1(x - 1)$ and $\hat{f}(1.02) = 0.02$.
d. Let $f(x) = \sqrt{x}$. $f(4) = 2$. $f'(x) = \frac{1}{2}x^{-1/2}$ and $f'(4) = 0.25$. So $\hat{f}(x) = 2 + 0.25(x - 4)$ and $\hat{f}(4.01) = 2.0025$.
e. $e^{\ln(2) + 0.01} = 2e^{0.01}$. Let $f(x) = 2e^x$, with base point $a = 0$. $f(0) = 2$. $f'(x) = 2e^x$, so $f'(0) = 2$. $\hat{f}(x) = 2 + 2(x - 0)$ and $\hat{f}(0.01) = 2.02$.
f. Let $f(x) = \sin(x)$. $f(0) = \sin(0) = 0$. $f'(x) = \cos(x)$ and $f'(0) = 1$. So $\hat{f}(x) = 0 + 1(x - 0)$ and $\hat{f}(0.02) = 0.02$.
g. Let $f(x) = \cos(x)$. $f(0) = \cos(0) = 1$. $f'(x) = -\sin(x)$ and $f'(0) = 0$. So $\hat{f}(x) = 1 + 0(x - 0)$ and $\hat{f}(-0.02) = 1.0$.

3. **a.** At $a = 1.0$, $M_1(a) = 1.0$. $M'(a) = 3a^2$ so $M'(1.0) = 3.0$. Then $\hat{M}_1(a) = 1.0 + 3.0(a - 1.0)$ and $\hat{M}_1(1.25) = 1.0 + 3.0(0.25) = 1.75$.

b. At $a = 1.5$, $M(1.5) = 3.375$ and $M'(1.5) = 6.75$. Then $\hat{M}_{1.5}(a) = 3.375 + 6.75(a - 1.5)$ and $\hat{M}_{1.5}(1.25) = 3.375 + 6.75(-0.25) = 1.6875$.

c. The secant line is $M_s(a) = 3.375 + 4.75(a - 1.5)$ and $M_s(1.25) = 3.375 + 4.75(-0.25) = 2.1875$. The exact value is 1.953. None of the approximations are very close.

d. At 1.45, the first tangent gives 2.35, the second tangent gives 3.0375, and the secant gives 3.1375. The exact answer is 3.0486, closest to the second tangent. This is because we are close to the point of tangency.

5. a. The tangent line is $\hat{b}(t) = 1 + t$, so $\hat{b}(0.1) = 1.1$.

b. If $b(0.1)$ is about 1.1, then $b'(0.1)$ is also, because our differential equation says that the value and derivative are equal. The tangent line is then $\hat{b}(t) = 1.1 + 1.1(t - 0.1)$, so $\hat{b}(0.2) = 1.21$.

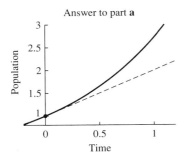

Answer to part **a**

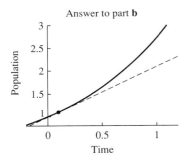

Answer to part **b**

c. It turns out that the population is increased by a factor of 1.1 each step, giving a final answer of $1.1^{10} = 2.594$. This is fairly close to the exact answer of $e = 2.718$.

7. a. Let $f(x) = x^3$ with base point $a = 2.0$. $f(2) = 8.0$ and $f'(2) = 12.0$ and $f''(2) = 12.0$. $P_2(x) = 8.0 + 12.0(x - 2.0) + 6.0(x - 2.0)^2$ and $P_2(2.02) = 8.2424$. This is very close to $f(2.02) = 8.242408$.

b. Let $f(x) = x^2$ near $a = 3.0$. $f(3) = 9.0$ and $f'(3) = 6.0$ and $f''(3) = 2$. $P_2(x) = 9.0 + 6.0(x - 3.0) + (x - 3.0)^2$ and $P_2(3.03) = 9.1809$. This is exactly right because the original function is quadratic already.

c. Let $f(x) = \ln(x)$. $f(1) = \ln(1) = 0$. $f'(1) = 1/1 = 1$ and $f''(1) = -1/1^2 = -1$. So $P_2(x) = 0 + 1(x - 1) - (x - 1)^2/2$ and $P_2(1.02) = 0.019800$. Exact answer is about 0.019803.

d. Let $f(x) = \sqrt{x}$. $f(4) = 2$. $f'(x) = \frac{1}{2}x^{-1/2}$ and $f'(4) = 0.25$. Also, $f''(4) = -1/32$, so $P_2(x) = 2 + 0.25(x - 4) + (x - 4)^2/64$ and $P_2(4.01) = 2.002498438$. Exact answer is 2.002498439.

e. $e^{\ln(2) + 0.01} = 2e^{0.01}$. Let $f(x) = 2e^x$, with base point $a = 0$. $f(0) = 2$. $f'(x) = 2e^x$, so $f'(0) = f''(0) = 2$. $P_2(x) = 2 + 2(x - 0) + (x - 0)^2$ and $P_2(0.01) = 2.0201$. Exact answer is 2.0201003.

f. Let $f(x) = \sin(x)$. $f(0) = \sin(0) = 0$. $f'(x) = \cos(x)$ and $f''(x) = -\sin(x)$. $f'(0) = 1$ and $f''(0) = 0$. So $P_2(x) = 0 + 1(x - 0)$, identical to the tangent line. $P_2(0.02) = 0.02$. Exact answer is 0.19998.

g. Let $f(x) = \cos(x)$. $f(0) = \cos(0) = 1$. $f'(x) = -\sin(x)$ and $f''(x) = -\cos(x)$. $f'(0) = 0$ and $f''(0) = -1$. So $P_2(x) = 1 + 0(x - 0) - (x - 0)^2/2$ and $P_2(-0.02) = 0.9998$. Exact answer is 0.999800007. This one is really close.

9. a. $\hat{A}_1(c) = 0$, $P_2(c) = 2c^2$. $P_2(c)$ matches the leading behavior.

b. $\hat{A}_2(c) = 0$, $P_2(c) = c^2$. $P_2(c)$ matches the leading behavior.

c. $\hat{A}_3(c) = 1$, $P_2(c) = 1 + c^2$. The tangent line matches.

d. $\hat{A}_4(c) = 1$, $P_2(c) = 1 - c^2$. The tangent line matches.

e. $\hat{A}_5(c) = 3c$, $P_2(c) = 3c - 3c^2$. The tangent line matches.

f. $\hat{A}_6(c) = 1 - c/2$, $P_2(c) = 1 - c/2 - c^2/4$. The method of leading behavior failed.

Section 3.8

1.

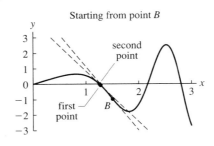

Answers 727

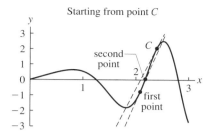

Starting from point C

3. **a.** Suppose initial guess is 3. Because $f'(x) = 3x^2$, $f(3) = 7$ and $f'(3) = 27$, the tangent line is $\hat{f}(x) = 7 + 27(x-3)$ which intersects the horizontal axis at $x = 74/27 = 2.74$. The Newton's method updating function is

$$x_{t+1} = x_t - \frac{f(x_t)}{f'(x_t)} = x_t - \frac{(x_t)^3 - 20}{3x_t^2}$$

If $x_0 = 3$, then $x_1 = 2.740740741$, $x_2 = 2.714669625$, and $x_3 = 2.714417640$. The exact answer is 2.714417616.

b. Suppose initial guess is $x_0 = 1$. Because $h'(x) = 0.5e^{x/2} - 1$, $h(1) = -0.351$ and $h'(3) = -0.1756$, the tangent line is $\hat{h}(x) = -0.351 - 0.1756(x-1)$, which intersects the horizontal axis at $x = -1$. The Newton's method updating function is

$$x_{t+1} = x_t - \frac{h(x_t)}{h'(x_t)} = x_t - \frac{e^{x/2} - x - 1}{0.5e^{x/2} - 1}$$

If $x_0 = 1$, then $x_1 = -1.00$, $x_2 = -0.1294668027$, and $x_3 = -0.0037771286$. This seems to be approaching 0. If we start from $x_0 = 2$, we get $x_1 = 2.7844$, $x_2 = 2.5479$, and $x_3 = 2.51355$. Running for more steps, the answer converges to 2.512862414.

c. Let $f(x) = \cos(x) - x$. Suppose initial guess is 1. Because $f'(x) = \sin(x) - 1$, $f(1) = -0.45969$ and $f'(1) = -1.84147$, the tangent line is $\hat{f}(x) = -0.45969 - 1.84147(x-1)$, which intersects the horizontal axis at $x = 0.75036$. The Newton's method updating function is

$$x_{t+1} = x_t - \frac{f(x_t)}{f'(x_t)} = x_t - \frac{\cos(x) - x}{\sin(x) - 1}$$

If $x_0 = 1$, then $x_1 = 0.7503638679$, $x_2 = 0.7391128909$, and $x_3 = 0.7390851334$. The exact answer is 0.7390851332.

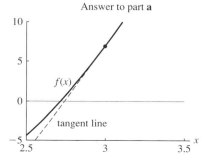

Answer to part a

Answer to part b

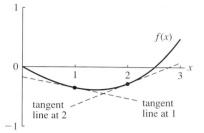

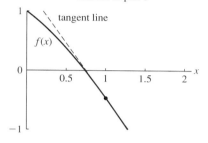

Answer to part c

5. The method fails if we start at a critical point of the function, which occur where $x = \pm 0.577$. All values below the lower critical point converge to the negative solution. None of the starting points seem to go straight to the exact answer.

7. **a.** $N^* = \ln\left(\frac{r}{1+h}\right)$.
   b. $P(h) = hN^* = h\ln\left(\frac{r}{1+h}\right)$. Then,

$$P'(h) = \ln\left(\frac{r}{1+h}\right) - \frac{h}{1+h} = 0$$

   c. We can use this function as f in Newton's method. The result turns out to be 0.6724.

9. **a.** The Newton's method updating function is

$$x_{t+1} = x_t - \frac{x_t^2}{2x_t} = \frac{x_t}{2}$$

This converges to 0 rather slowly, because the derivative of the function x^2 is 0 at the solution.

   b. The Newton's method updating function is

$$x_{t+1} = x_t - \frac{x_t^{1/2}}{0.5x_t^{-1/2}} = -x_t$$

This jumps back and forth and never approaches 0, presumably because the derivative of the function at $x = 0$ is infinity.

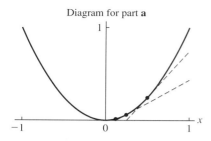

Diagram for part a

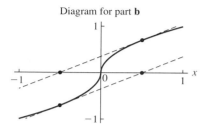

Diagram for part b

11. **a.** This is the secant line approximation.
    b. $x_{t+1} = x_t - \dfrac{e^{x_t} - x_t - 2}{e^{x_t+1} - 1 - e^{x_t}}$
    c. It took me 14 steps to get near the answer starting from $x_0 = 2$.

13. **a.** The distance from the equilibrium is b_t, and b_{t+1} is exactly a factor r closer.
    b. It still moves in by a factor of r. The figure shows an example with $r = 0.5$.

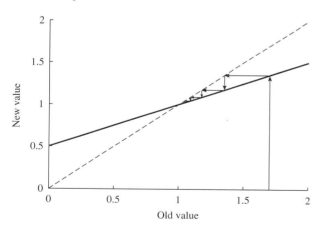

c. A nonlinear updating function acts much like its tangent line near the equilibrium. Therefore, it moves toward its equilibrium by a factor approximately equal to the slope of the tangent.

Section 3.9

3. Equation 3.19 says that

$$c^* = \dfrac{\gamma r T}{1 - (1-rT)[1 - \alpha(1 - e^{-kT})]}$$

when the updating function is

$$c_{t+1} = (1 - rT)[c_t - \alpha c_t(1 - e^{-kT})] + rT\gamma$$

The equilibrium is

$$c^* = (1 - rT)[c^* - \alpha c^*(1 - e^{-kT})] + rT\gamma$$

equation for equilibrium

$$r\gamma T = c^* - (1 - rT)[c^* - \alpha c^*(1 - e^{-kT})]$$

move all the c^*s to one side

$$r\gamma T = c^* \left\{1 - (1 - rT)[1 - \alpha(1 - e^{-kT})]\right\}$$

factor out a c^*

$$c^* = \dfrac{\gamma r T}{1 - (1 - rT)[1 - \alpha(1 - e^{-kT})]}$$

solve for c^*

Equation 3.21 says that

$$c^* = \dfrac{\gamma r T}{1 - (1 - rT)\left(1 - \alpha \dfrac{T^2}{k + T^2}\right)}$$

when the updating function is

$$c_{t+1} = \left[1 - rT\left(c_t - \alpha c_t \dfrac{T^2}{k + T^2}\right)\right] + rT\gamma$$

The equilibrium is

$$c^* = (1 - rT)\left(c^* - \alpha c^* \dfrac{T^2}{k + T^2}\right) + rT\gamma$$

equation for equilibrium

$$r\gamma T = c^* - (1 - rT)\left(c^* - \alpha c^* \dfrac{T^2}{k + T^2}\right)$$

move all the c^*s to one side

$$r\gamma T = c^*\left[1 - (1 - rT)\left(1 - \alpha \dfrac{T^2}{k + T^2}\right)\right]$$

factor out a c^*

$$c^* = \dfrac{\gamma r T}{1 - (1 - rT)\left(1 - \alpha \dfrac{T^2}{k + T^2}\right)}$$

solve for c^*

5. After some major algebra, we find that

$$\frac{d}{dT}\frac{c^*A(T)}{T} = \frac{d}{dT}\frac{\alpha\gamma r\frac{T^2}{k+T^2}}{1-(1-rT)\left(1-\alpha\frac{T^2}{k+T^2}\right)}$$

$$= \frac{[k-T^2(1-\alpha)]\gamma r^2\alpha}{(rT^2\alpha - rT^2 - rk - \alpha T)^2}$$

This is positive as $T=0$ and zero at $T=\sqrt{k/(1-\alpha)}$. Therefore, absorption is maximized. The optimal T becomes larger for larger values of α. This is consistent with our findings on the previous problem: Highly efficient lungs can wait longer between breaths.

7. **a.** With $A(T) = \alpha T$,

$$c^* = \frac{\gamma rT}{1-(1-rT)[1-\alpha T]}$$

$$\frac{c^*A(T)}{T} = \frac{\alpha\gamma rT}{1-(1-rT)[1-\alpha T]}$$

b. With $A(T) = \alpha(1-e^{-kT})$,

$$c^* = \frac{\gamma rT}{1-(1-rT)[1-\alpha(1-e^{-kT})]}$$

$$\frac{c^*A(T)}{T} = \frac{\alpha\gamma r(1-e^{-kT})}{1-(1-rT)[1-\alpha(1-e^{-kT})]}$$

c. With $A(T) = \dfrac{\alpha T^2}{k+T^2}$,

$$c^* = \frac{\gamma rT}{1-(1-rT)\left(1-\alpha\frac{T^2}{k+T^2}\right)}$$

$$\frac{c^*A(T)}{T} = \frac{\alpha\gamma r\frac{T^2}{k+T^2}}{1-(1-rT)\left(1-\alpha\frac{T^2}{k+T^2}\right)}$$

9. The equilibrium value is

$$c^* = \frac{r\gamma(k+T)}{\alpha + rk + rT - rTa}$$

The equilibrium absorption rate is then

$$R(T) = \frac{\alpha r\gamma}{\alpha + rk + rT - rT\alpha}$$

As long as $\alpha < 1$, this is a decreasing function of T, meaning that the organism would do best to pant no matter what the parameter values.

Answers to Supplementary Problems for Chapter 3

1. **a.** $S'(t) = -t^3 + 180t^2 - 8000t + 96000$, which is zero at $t = 20$, $t = 40$, and $t = 120$.
   b. Substituting the endpoints ($t = 0$ and $t = 150$) and the critical points into the function $S(t)$, we find a maximum of 576,000 at $t = 120$.
   c. $S''(t) = -3t^3 + 360t - 8000$. Then $S''(20) = -2000$, $S''(40) = 1600$, and $S''(120) = -8000$. The first and last are maxima and the middle one is a minimum.

   d.

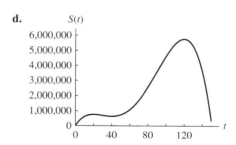

3. **a.** Let $f(x) = x^3$. Then, $f'(x) = 3x^2$. Substituting $x = -1$, we find $f(-1) = -1$ and $f'(-1) = 3$, so $\hat{f}(x) = -1 + 3(x+1)$, and $\hat{f}(-0.97) = -1 + 3(-0.97 + 1) = -0.91$.
   b. Let $f(x) = 1/(3+x^2)$. Then, $f'(x) = -2x/(3+x^2)^2$. Substituting $x = 1$, we find $f(1) = 1/4$ and $f'(1) = -1/8$, so $\hat{f}(x) = 1/4 - 1/8(x-1)$, and $\hat{f}(1.01) = 1/4 - 1/8(1.01 - 1) = 0.24875$.
   c. Let $f(x) = e^{3x^2+2x}$. Then, $f'(x) = (6x+2)e^{3x^2+2x}$. Substituting $x = 1$, we find $f(1) = e^5$ and $f'(1) = 8e^5$, so $\hat{f}(x) = e^5 + 8e^5(x-1)$ and $\hat{f}(1.02) = e^5 + 8e^5(1.02 - 1) = 1.16e^5$.

5. **a.** $c_{t+1} = 0.75(1-\alpha)c_t + 0.25\gamma$
   b. $c^* = 0.25\gamma/(0.25 + 0.75\alpha)$
   c. $\alpha c^* = 0.25\alpha\gamma/(0.25 + 0.75\alpha)$
   d. This has its maximum at $\alpha = 1$.
   e. Sure, absorb as much as you can as fast as you can.

7. **a.**

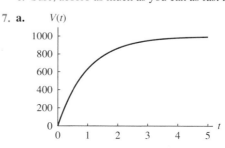

b. The fraction outside is $1 - H(t) = 1/(1 + e^t)$, so the total volume outside is

$$1000(1-e^{-t})/(1+e^t)\,\mu m^3$$

Call this function $V_o(t)$.

c. $V_o'(t) = \dfrac{1000(2 + e^{-t} - e^t)}{(1+e^t)^2}\,\dfrac{\mu m^3}{d}$

d. This is a bit tricky. The maximum is the critical point where $V_o'(t) = 0$, or where $(2 + e^{-t} - e^t) = 0$. Letting $x = e^t$, this is $2 + 1/x - x = 0$. Multiplying both sides by x, we get the quadratic $2x + 1 - x^2 = 0$, which can be solved with the quadratic formula to give $x = 1 + \sqrt{2}$, so that $t = \ln(1 + \sqrt{2}) = 0.88$ days.

9. a. It looks like $x_0 = 1$ is a good guess.

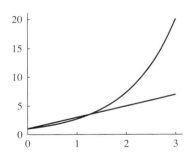

b. We need to solve $g(x) = e^x - 2x - 1 = 0$, so
$$x_{t+1} = x_t - \frac{g(x_t)}{g'(x_t)} = x_t - \frac{e^x - 2}{e^x - 2x - 1}$$

c. $x_1 = 1.3922$.

d. The derivative of the updating function is
$$\frac{e^x(e^x - 2x - 1)}{(e^x - 2)^2}$$
which is 0 when $e^x - 2x - 1 = 0$.

11. a. At $x = 1$, the left-hand side is smaller. At $x = e$, the left-hand side is larger. At $x = e^2$, the left-hand side is smaller again. There must be at least two solutions, one between 1 and e and another between e and e^2. To find the lower one, I bet that 2 is a good guess.

b. Our function is $f(x) = \ln(x) - x/3$. The Newton's method updating function is
$$x_{t+1} = x_t - \frac{f(x_t)}{f'(x_t)} = x_t - \frac{\ln(x_t) - x_t/3}{1/x_t - 1/3}$$
Substituting $x_0 = 2$, we find $x_1 = 1.8411$, and then that $x_2 = 1.8570$. The exact answer is 1.8571.

c. $f'(3) = 0$. This would be a rather bad guess.

13. a. Let $N(t)$ be net energy gain. Then $N(t) = F(t) - 2t$.

b. At $t = \sqrt{(3/2)} - 1 = 0.225$.

c. We need to maximize $\frac{N(t)}{1+t}$, which occurs when $t = 0.2$.

d. The answer to part **c** is smaller because the bee has other options besides sucking as much nectar as possible out of the flower.

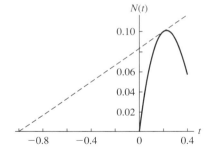

15. a. $\hat{f}(x) = 1 + x - x^2$

b. $\hat{f}(x) = 1 - \frac{1}{2}(x - 1)$.

c. $\hat{g}(x) = \frac{1}{2} + \frac{1}{4}x - \frac{1}{4}x^2$

d. $\hat{g}(x) = x + x^2$

17. a. $\lim_{t \to 0} F(t) = 1$

b. $\lim_{t \to 0} F'(t) = \frac{-3}{2}$

c. $\lim_{t \to 0} F''(t) = \frac{11}{3}$

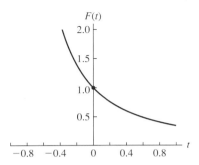

19. a. The updating function f is $f(x) = 4x^2/(1 + 3x^2)$, multiplying the per capita reproduction by x, the number of individuals.

b. First, find that 0 is a solution. The rest is a quadratic, which has solutions at $x = 1/3$ and $x = 1$.

c. We find that $f'(x) = 8x/(1 + 3x^2)^2$. Then $f'(0) = 0$, $f'(1/3) = 3/2$ and $f'(1) = 1/2$. Therefore, the equilibria at 0 and 1 are stable and the one at $1/3$ is unstable.

d. The tangent line at 0 is $\hat{f}(x) = 0$. The tangent line at $x = 1/3$ is $\hat{f}(x) = 1/3 + 3/2(x - 1/3)$. The tangent line at $x = 1$ is $\hat{f}(x) = 1 + 1/2(x - 1)$.

e. This dynamical system shoots off to infinity for starting points greater than $1/3$, and off to negative infinity for starting points less than $1/3$. This is different from the behavior of the original system, in which such solutions approach $x = 1$ and $x = 0$, respectively.

21. a. It is $\frac{r}{1 + N_t^2}$.

b. $N^* = \sqrt{r - 1}$.

c. The derivative is $\frac{2 - r}{r}$.

d. No, this is always greater than -1.

Chapter 4

Section 4.1

1. **a.** You could measure position with a map even if your speedometer was broken. You could measure mass with a balance, total sodium or total chemical with a destructive device that separated out the chemical.
    b. You could measure speed but not position if your speedometer worked but you were lost. You could measure growth rate but not total mass if you check how much carbon a growing plant absorbed from the atmosphere. You could measure sodium entering a cell but not total sodium if you could track the change in charge, and rate of chemical production but not the total if the reaction produced an easily measured byproduct such as heat or carbon dioxide.

3. **a.** This is a pure-time differential equation.
    b.

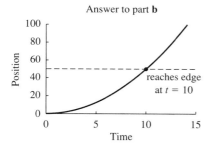

   c. $\dfrac{dp}{dt} = t$

   d. The function $p(t) = t^2$ has derivative $2t$, so we divide by 2 to guess $p(t) = t^2/2$. By good luck, this matches the initial condition $p(0) = 0$.
   f. We want to find when $p(t) = t^2/2 = 50$. This has a solution of 10 min. The speed is 10 cm/min at this time.

5. **a.** $-e^{-t}$
   b. Need to multiply by -1, so the derivative of $-e^{-t}$ is e^{-t}.
   c. If $P(t) = -e^{-t}$, then $P(0) = -1$, which is 1 too low. We need to add 1.
   d. $P(t) = 1 - e^{-t}$. The derivative is $\dfrac{dP}{dt} = e^{-t}$, and $P(0) = 1 - 1 = 0$. This checks.

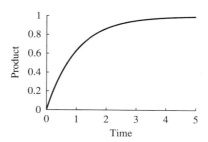

   e. The limit is $\lim_{t \to \infty} P(t) = 1$. Even if we waited forever, we would not get more than 1 mol out of this reaction.

7. **a.** This is an autonomous differential equation.
   b. Taking the derivative, we get
   $$\frac{d}{dt} 100 e^{-3t} = -300 e^{-3t} = -3 \cdot b(t)$$
   Also, $b(0) = 100 e^{-3 \cdot 0} = 100$, matching the initial condition.
   c. The solution gets smaller, consistent with the negative derivative. Furthermore, it gets smaller more slowly the smaller it gets.

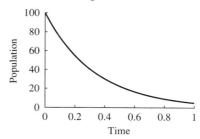

9. We have that
   $$\hat{b}(t + \Delta t) = \hat{b}(t) + 2\hat{b}(t)\Delta t$$
   so
   $$\hat{b}(t + 0.1) = \hat{b}(t) + 2\hat{b}(t)0.1 = 1.2\hat{b}(t)$$
   To update for one time step, we multiply by 1.2. After 10 time steps, the approximate population is
   $$\hat{b}(1.0) = 1.2^{10} \cdot 1.0 \times 10^6 = 6.19 \times 10^6$$

Section 4.2

1. **a.** $H'(0) = 0.2[10 - (-20)] = 6.0$. Then $\hat{H}(2.0) = H(0) + 2.0 \cdot 6.0 = -8.0°C$.
   b. $H'(0) = 0.2(10 - 40) = -6.0$. Then $\hat{H}(2.0) = H(0) + 2.0 \cdot (-6.0) = 28.0°C$.
   c. $\hat{H}(1.0) = H(0) + 1.0 \cdot (-6.0) = 34.0°C$.
   d. $H'(1.0) = 0.2(10 - 34) = -4.8$. Then, $\hat{H}(2.0) = \hat{H}(1.0) + 1.0 \cdot (-4.8) = 29.2°C$.
   e. The object 30°C cooler than the ambient temperature (at $-20°C$) warmed up by approximately 12°C (part **a**) and the object 30°C warmer than the ambient temperature

(at 40°C) cooled off by approximately 12°C (part **b**). We imagine that the estimate in part **d** is more accurate than in part **b**, and the actual temperature after 2 min is somewhat warmer than 29.2°C.

5. **a.**

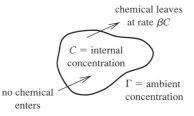

b. The equation is
$$\frac{dC}{dt} = -\beta C$$

c. This is the same as the differential equation for a shrinking bacterial population, and has solution
$$C(t) = C_0 e^{-\beta t}$$

d. The concentration inside decays exponentially to zero because there is no source of chemical.

7. **a.** $\hat{p}(2.0) = 0.1 + 0.1 \cdot 0.9 \cdot 2 = 0.28$.
b. We need to take four steps.
$$\hat{p}(0.5) = 0.1 + 0.1 \cdot 0.9 \cdot 0.5 = 0.145$$
$$\hat{p}(1.0) = 0.145 + 0.145 \cdot 0.855 \cdot 0.5 = 0.207$$
$$\hat{p}(1.0) = 0.207 + 0.207 \cdot 0.793 \cdot 0.5 = 0.289$$
$$\hat{p}(1.0) = 0.289 + 0.289 \cdot 0.711 \cdot 0.5 = 0.392$$

c. The exact solution is
$$p(t) = \frac{0.1e^{2.0t}}{0.1e^{2.0t} + 0.9e^{1.0t}}$$
and $p(2) = 0.45$.

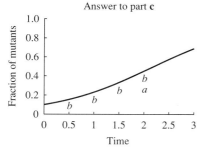

9. The change in p is
$$\Delta p = p_{t+1} - p_t$$
$$= \frac{sp_t}{sp_t + r(1 - p_t)} - p_t$$
$$= \frac{sp_t - sp_t^2 - rp_t(1 - p_t)}{sp_t + r(1 - p_t)}$$
$$= \frac{(s - r)p_t(1 - p_t)}{sp_t + r(1 - p_t)}$$

The numerator looks a lot like the rate of change in Equation 4.7, but the denominator is new.

Section 4.3

1. **a.** This is actually both a pure-time and an autonomous differential equation.
   b. Neither   **c.** Autonomous
   d. Pure-time   **e.** Pure-time
   f. Neither

3. **a.** $\dfrac{7x^3}{3} + c$ **b.** $36t^2 + 5t + c$
   c. $\dfrac{y^5}{5} + \dfrac{5y^4}{4} + c$ **d.** $t^{10} + t^6 + c$
   e. $\dfrac{-5x^{-2}}{2} + c$ **f.** $\dfrac{21}{10}z^{10/7} + c$
   g. $\dfrac{4}{3}t^{2/3} + 3t + c$

5. **a.** $V_1(t) = t^2 + 5.0t + 10.0$, $V_2(t) = 2.5t^2 + 2.0t + 10.0$
   b. $\dfrac{dV}{dt} = 7.0t + 7.0$
   c. With initial condition $V(0) = V_1(0) + V_2(0) = 20$, $V(t) = 3.5t^2 + 7.0t + 20.0$. This is the sum of $V_1(1)$ and $V_2(t)$.
   d. $\int 7.0t + 7.0\, dt = \int 2.0t + 5.0\, dt + \int 5.0t + 2.0\, dt$

7. **a.** The units of each side are grams per day.
   b. $M(t) = 4.0\sqrt{t} + c$. Substituting the initial conditions, we have $M(0) = c = 5.0$ so $M(t) = 4.0\sqrt{t} + 5.0$.
   c.

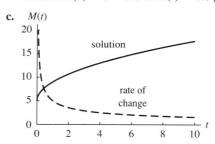

   d. The mass approaches infinity, but does so at a slower and slower rate.

9. **a.**

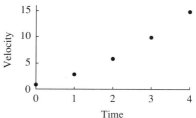

   b. The tangent line is $\hat{p}(t) = 10.0 + 1.0t$, and $\hat{p}(1) = 11.0$.
   c. At $t = 5$, we estimate the position is 45.0.

Section 4.4

1. **a.** $e^x + \ln(x) - \cos(x) + \sin(x) + c$
   b. $3e^x + \dfrac{x^4}{2} + c$
   c. $2\ln(t) + \dfrac{t^2}{4} + c$
   d. $\dfrac{-3}{z} + \dfrac{z^3}{9} + c$
   e. $-2\cos(x) + 3\sin(x) + c$

3. **a.** Substitute $y = e^x$, finding $dy = e^x\, dx$ so
   $$\int \frac{e^x}{1+e^x}\,dx = \int \frac{1}{1+y}\,dy$$
   $$= \ln(1+y) + c = \ln(1+e^x) + c$$
   b. The integral is $\frac{2}{3}(1+y^2)^{3/2} + c$.
   c. The integral is $-e^{\sin(x)} + c$.
   d. The integral is $-\ln(|\cos(x)|)$.
   e. The integral is $\frac{1}{5}(1+e^t)^5 + c$.
   f. Following the hint, we get
   $$\int \frac{t}{1+t}\,dt = \int 1 - \frac{1}{1+t}\,dt$$
   The last piece can be integrated with the substitution $y = 1 + t$,
   $$\int \frac{t}{1+t}\,dt = t - \ln(1+t) + c$$

5. **a.** $P'(0) = 5.0$, so $\hat{P}(0.1) = P(0) + 5.0 \cdot 0.1 = 0.5$.
   b. Integrating, $P(t) = -2.5e^{-2.0t} + c$. Substituting the initial condition, we find $c = 2.5$.
   c. $P(0.1) = 2.5(1 - e^{-0.2}) = 0.453$.
   d. The limit is 2.5 as t approaches infinity.
   e.

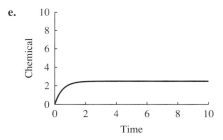

   f. In this problem, the rate of chemical production decreases to 0 much more quickly (exponentially) than in Exercise 4.

7. **a.** $L(t) = 54.0(1 - e^{1.19t})$
   b. The limit is 54.0 cm.
   c. These walleye reach maturity when $L(t) = 45$, or at $t = 1.5$ yr.
   d. These walleye reach maturity much more quickly, but then more or less stop growing.

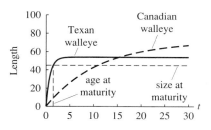

9. **a.**

   (Temperature vs. Day graph)

   b. The solution is
   $$L(t) = \int 2.0 + 1.0\cos\left[\frac{2\pi(t - 190.0)}{365}\right]dt$$
   $$= 2.0t + \sin\left[\frac{2\pi(t - 190.0)}{365}\right]\frac{365}{2\pi} + c$$

   Using the initial condition $T(0) = 0.1$, we find $c = -7.38$. At $t = 30$, $L(30.0) = 20.2$.
   c. Using the initial condition $T(150) = 0.1$, we find $c = -263.0$. After 30 days, at $t = 180$, $L(180.0) = 87.1$.
   d. The maximum growth rate is on day 190, so the best day to hatch is about 15 days before (so development covers the whole warmest period).

Section 4.5

1. **a.** $\Delta t = 0.4$, $t_0 = 0$, $t_1 = 0.4$, $t_2 = 0.8$, $t_3 = 1.2$, $t_4 = 1.6$, $t_5 = 2.0$
   b. $\Delta t = 0.2$, $t_0 = 0$, $t_1 = 0.2$, $t_2 = 0.4$, $t_3 = 0.6$, $t_4 = 0.8$, $t_5 = 1.0$, $t_6 = 1.2$, $t_7 = 1.4$, $t_8 = 1.6$, $t_9 = 1.8$, $t_{10} = 2.0$
   c. $\Delta t = 0.2$, $t_0 = 2.0$, $t_1 = 2.2$, $t_2 = 2.4$, $t_3 = 2.6$, $t_4 = 2.8$, $t_5 = 3.0$
   d. $\Delta t = 0.01$, $t_0 = 2.0$, $t_1 = 2.01, \ldots, t_{99} = 2.99$, $t_{100} = 3.0$
   e. $\Delta t = 0.4$, $t_0 = -2.0$, $t_1 = -1.6$, $t_2 = -1.2$, $t_3 = -0.8$, $t_4 = -0.4$, $t_5 = 0.0$, $t_6 = 0.4$, $t_7 = 0.8$, $t_8 = 1.2$, $t_9 = 1.6$, $t_{10} = 2.0$

3. **a.** $I_l = \sum_{i=0}^{4} 2t_i \cdot 0.2$, $I_r = \sum_{i=1}^{5} 2t_i \cdot 0.2$
   b. $I_l = \sum_{i=0}^{4} 2t_i \cdot 0.4$, $I_r = \sum_{i=1}^{5} 2t_i \cdot 0.4$

c. $I_l = \sum_{i=0}^{4} t_i^2 \cdot 0.4$, $I_r = \sum_{i=1}^{5} t_i^2 \cdot 0.4$
d. $I_l = \sum_{i=0}^{4}(1 + t_i^3) \cdot 0.2$, $I_r = \sum_{i=1}^{5}(1 + t_i^3) \cdot 0.2$

5. In each case, Euler's method gives exactly the same answer as the left-hand estimate. This is because both methods make the same approximation: Assume that the rate during an interval (or between two steps) is equal to the rate at the beginning of the interval.

7. a.

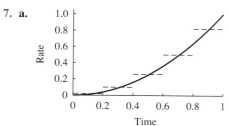

b. $I_a = \sum_{i=0}^{4} \dfrac{t_i^2 + t_{i+1}^2}{2} \cdot \Delta t$

c. $\dfrac{0.0^2 + 0.2^2}{2} \cdot 0.2 + \dfrac{0.2^2 + 0.4^2}{2} \cdot 0.2$
$+ \dfrac{0.4^2 + 0.6^2}{2} \cdot 0.2 + \dfrac{0.6^2 + 0.8^2}{2} \cdot 0.2$
$+ \dfrac{0.8^2 + 1.0^2}{2} \cdot 0.2 = 0.004$
$+ 0.02 + 0.052 + 0.1 + 0.164 = 0.34$

d. Approximating the value by the average is much more accurate than approximating it by the value at the beginning or the end of the interval.

9. a.

(graph: Rate vs Time, 0 to 1)

b. $\sum_{i=0}^{n-1} f\left(\dfrac{t_i + t_{i+1}}{2}\right) \Delta t$

c. $I_m = (0.1^2 + 0.3^2 + 0.5^2 + 0.7^2 + 0.9^2)0.2 = 0.33$.
d. Same reason as before. We are making a much better approximation of the value during the interval.

Section 4.6

1. a. $\int_0^1 2t\, dt = t^2\big|_0^1 = 1.0$. This is close to the answer to Exercise 2 in Section 4.5 and matches the answer to Exercise 4 in Section 4.5.

b. $\int_0^2 2t\, dt = t^2\big|_0^2 = 4.0$

c. $\int_0^2 t^2\, dt = \dfrac{t^3}{3}\bigg|_0^2 = 2.667$

d. $\int_0^1 t^3\, dt = \dfrac{t^4}{4}\bigg|_0^1 = 0.25$

3. a. $\int_1^2 f(t)\, dt = 3$, $\int_2^3 f(t)\, dt = 5$, and $\int_1^3 f(t)\, dt = 8$, which is equal to $3 + 5$.

b. $\int_1^2 g(t)\, dt = \dfrac{7}{3}$, $\int_2^3 g(t)\, dt = \dfrac{19}{3}$, and $\int_1^3 g(t)\, dt = \dfrac{26}{3}$, which is equal to $\dfrac{7}{3} + \dfrac{19}{3}$.

c. $\int_1^2 h(t)\, dt = 3.75$, $\int_2^3 h(t)\, dt = 16.25$, and $\int_1^3 h(t)\, dt = 20$, which is equal to $3.75 + 16.25$.

5. a. $p(3) - p(1) = -49.2$, $p(5) - p(2) = -88.4$. These add to -137.6.

b. $L(3) - L(1) = 10.84$, $L(5) - L(3) = 9.05$. These add to 19.89.

c. $L(1.0) - L(0.5) = 13.36$, $L(1.5) - L(1.0) = 7.37$. These add to 20.72.

d. $A(1986) - A(1985) = 10{,}650$, $A(1987) - A(1986) = 15{,}888$. These add to $26{,}539$.

e. $P(7.5) - P(5.0) = 0.937$, $P(10.0) - P(7.5) = 0.680$. These add to 1.616.

f. $P(7.5) - P(5.0) = 0.0001$, $P(10.0) - P(7.5) = 7.5 \times 10^{-7}$. These add to 0.0001; nothing much happens during the second half of the interval.

7. a. $p(t) = 100 + \int_0^t -9.8s - 5.0\, ds = 100 - 4.9t^2 - 5.0t$

b. $L(t) = 5.0 + \int_0^t 6.48 e^{-0.09s}\, ds = 77.0 - 72.0 e^{-0.09t}$

c. $L(t) = 5.0 + \int_0^t 64.3 e^{-1.19s}\, ds = 59.0 - 54.0 e^{-1.19t}$

d. $A(t) = 34{,}000 + \int_{1981}^{t} 523.8(s - 1981)^2\, ds = 34{,}000 + 174.6(t - 1981)^3$

e. $P(t) = 2.0 + \int_0^t \dfrac{5}{1 + 2.0s}\, ds = 2.0 + 2.5\ln(1 + 2.0t)$

f. $P(t) = 2.0 + \int_0^t 5.0 e^{-2.0s}\, ds = 2.0 + 2.5(1 - e^{-2.0t})$

9. The integral is always zero. The change "between" times a and a is zero because there was no time to change. In terms of the Fundamental Theorem, if $F(x)$ is an indefinite integral of $f(x)$, then
$$\int_a^a f(x)\, dx = F(a) - F(a) = 0$$

Section 4.7

1. a. $\int_0^3 3x^3\, dx = 3x^4/4\big|_0^3 = 60.75$.

b. $\int_0^{\ln 2} e^x\, dx = e^x\big|_0^{\ln 2} = 2 - 1 = 1$.

c. Use the substitution $y = x/2$. Then, $dx = 2\, dy$ and the limits of integration are $y = 0$ to $y = \dfrac{\ln 2}{2}$.

$$\int_0^{\ln 2} e^{x/2}\,dx = \int_0^{(\ln 2)/2} 2e^y\,dy$$
$$= 2(\sqrt{2}-1) = 0.828$$

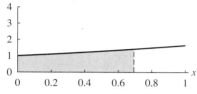

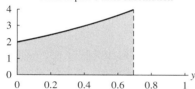

d. Set $u = (1+3t)$. Then $dt = du/3$, and the limits of integration are from $u=1$ to $u=7$:

$$\int_0^2 (1+3t)^3\,dt = \int_1^7 \frac{u^3}{3}\,du = \left.\frac{u^4}{12}\right|_1^7 = 200$$

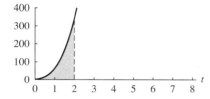

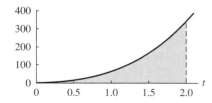

e. Set $z=(3+4y)$. Then, $dy = dz/4$, and the limits of integration are from $z=3$ to $z=11$:

$$\int_0^2 (3+4y)^{-2}\,dy = \int_3^{11} \frac{z^{-2}}{4}\,dz = \left.\frac{-z^{-1}}{4}\right|_3^{11} = 0.061$$

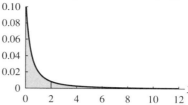

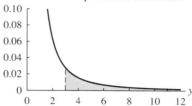

f. $\int_0^\pi \sin(z)\,dz = -\cos(z)|_0^\pi = 2$

3.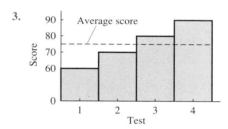

The average is $(60+70+80+90)/4 = 75$. The function is

$$f(x) = \begin{cases} 60 & \text{for } 0 \le x < 1 \\ 70 & \text{for } 1 \le x < 2 \\ 80 & \text{for } 2 \le x < 3 \\ 90 & \text{for } 3 \le x < 4 \end{cases}$$

$$\text{total score} = \int_0^4 f(x)\,dx = 60+70+80+90 = 300$$

$$\text{average score} = \frac{\text{total score}}{\text{width of interval}} = \frac{300}{4} = 75$$

5. **a.** The critical points are at $x=0$ and $x=200$ (both endpoints). Because $\rho(0) = 1.0$ and $\rho(200) = 1.08$, the first is the minimum and the second is the maximum.
b. $\int_0^{200} \rho(x)\,dx = 208$ gm.
c. The average is 1.04 g/cm, which lies right between the minimum and the maximum.
d.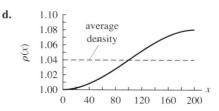

7. **a.** $\int_0^1 t^3\, dt = 0.25$ cm^3. The average rate is 0.25 cm^3/s.
   b. $\int_0^1 \sqrt{t}\, dt = 0.667$ cm^3. The average rate is 0.667 cm^3/s.
   c. $\int_0^1 t\, dt = 0.5$ cm^3. The average rate is 0.5 cm^3/s.
   d. In the first vessel, the rate at time 0.5 is 0.125, less than the average rate during the first second. In the second vessel, the rate at time 0.5 is $\sqrt{0.5} = 0.707$, greater than the average rate during the first second. In the third, the average matches the rate at the average time.
   e. In the first vessel, the rate has a positive second derivative.

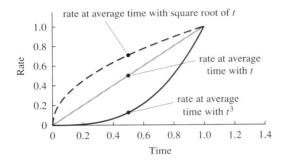

9. **a.**

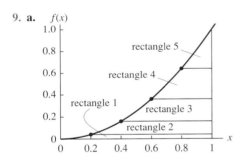

   b. Rectangle 1: lower size estimate is $0.8 \cdot f(0.2) = 0.0032$; upper size estimate is $1.0 \cdot f(0.2) = 0.004$, Rectangle 2: lower size estimate is $0.6 \cdot [f(0.4) - f(0.2)] = 0.072$; upper size estimate is $0.8 \cdot [f(0.4) - f(0.2)] = 0.096$. Rectangle 3: lower size estimate is $0.4 \cdot [f(0.6) - f(0.4)] = 0.08$; upper size estimate is $0.6 \cdot [f(0.6) - f(0.4)] = 0.08$. Rectangle 4: lower size estimate is $0.2 \cdot [f(0.8) - f(0.6)] = 0.056$; upper size estimate is $0.4 \cdot [f(0.8) - f(0.6)] = 0.112$. Rectangle 5: lower size estimate is 0.0; upper size estimate is $0.2 \cdot [f(1.0) - f(0.8)] = 0.072$.
   c. Lower estimate is 0.211; upper estimate is 0.364.
   d. A rectangle at height y goes from the point where $x^2 = y$, or $x = \sqrt{y}$, to $x = 1$. Its length is $1 - \sqrt{y}$.
   e. Area $= \int_0^1 (1 - \sqrt{y})\, dy$.
   f. $\int_0^1 (1 - \sqrt{y})\, dy = (y - 2y^{3/2}/3)|_0^1 = 1/3$. It checks.

11. **a.** The limits of integration are both equal to 1, so the definite integral is 0.

b. $l(6) - l(3) = \int_1^6 \frac{1}{x}\, dx - \int_1^3 \frac{1}{x}\, dx = \int_3^6 \frac{1}{x}\, dx$

$= \int_1^2 \frac{1}{y}\, dy = l(2)$

where we set $y = x/3$, so $dy = dx/3$ and the limits of integration go from 1 to 2.
c. Set $y = x/a$, so that $dx = a\, dy$. The integrand becomes dy/y and the limits of integration go from 1 to 2. The area from a to $2a$ is equal to the area from 1 to 2, and thus is equal to $l(2)$.
d. We have that $\int_1^{a^b} 1/x\, dx = l(a^b)$. Substituting $y = \sqrt[b]{x}$, we find that

$$\frac{dy}{dx} = \frac{1}{b} x^{1/b - 1} = \frac{1}{b} \frac{y}{x}$$

Then the integrand becomes

$$\frac{1}{x}\, dx = \frac{b}{y}\, dy$$

and the limits of integration go from 1 to a. So

$$\int_1^{a^b} \frac{1}{x}\, dx = \int_1^a \frac{b}{y}\, dy = b\,l(a)$$

This matches our law of logs.

Section 4.8

1. **a.** $\int_0^\infty e^{-3t}\, dt = 1/3$.
   b. $\int_0^\infty e^t\, dt$ does not converge because the integrand does not decrease to 0.
   c. Does not converge because the integrand approaches 0 too slowly.
   d. $\int_5^\infty \frac{1}{x^2}\, dx = -\frac{1}{x}\Big|_5^\infty = \frac{1}{5}$.
   e. Using the substitution $u = 1 + 3x$, we find that this integral is $2/3$.

3. **a.** The quotient is

$$\frac{e^{-x}}{1/x} = \frac{x}{e^x}$$

Using L'Hôpital's rule, the limit of this quotient (which is an indeterminate form because both the numerator and denominator approach infinity) is

$$\lim_{x \to \infty} \frac{x}{e^x} = \lim_{x \to \infty} \frac{1}{e^x} = 0$$

Therefore, e^{-x} is much smaller than $1/x$ as x approaches infinity.
b. Algebraically,

$$\frac{1/x^2}{1/x} = \frac{1}{x}$$

This approaches 0, so $1/x^2$ is much smaller than $1/x$ as x approaches infinity.

c. Using the same algebra, we see that $1/x$ approaches infinity as x approaches zero, so $1/x^2$ is much larger than $1/x$ as x approaches zero.

d. The quotient is

$$\frac{1/\ln(x)}{1/x} = \frac{x}{\ln(x)}$$

Using L'Hôpital's rule,

$$\lim_{x \to \infty} \frac{x}{\ln(x)} = \lim_{x \to \infty} \frac{1}{1/x} = \lim_{x \to \infty} x = \infty$$

Therefore, $1/x$ is much larger than $1/\ln(x)$ as x approaches infinity.

e. The integral of $1/x$ diverges, and $1/\ln(x)$ is larger. It seems likely that it will diverge.

5. **a.** Diverges because the power is not less than 1.
   b. Diverges because the power is greater than 1.
   c. $\int_0^{.001} \frac{1}{\sqrt[3]{x}} dx = 0.015$.
   d. $\int_0^{\infty} \frac{1}{\sqrt[3]{x}} dx$ does not converge as x approaches infinity because the power is less than 1. It would converge for x near 0, but the whole integral diverges.

7. **a.** With $y = 1/x$, $dy = -y^2 dx$, and

$$\int_0^1 \frac{1}{\sqrt{x}} dx = \int_1^{\infty} \frac{\sqrt{y}}{y^2} dy = \int_1^{\infty} y^{-3/2} dy = 2$$

b. It turns out, after some algebra, that

$$\int_0^1 \frac{1}{x^p} dx = \int_1^{\infty} \frac{1}{y^{2-p}} dy$$

The original integral converges if $p < 1$. The new integral converges if $2 - p > 1$ or $p < 1$. They match!

c. The original power was p, the new power is $2 - p$. They are equal when $p = 1$.

Answers to Supplementary Problems for Chapter 4

1. **a.**

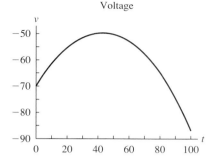

b. Solving with the indefinite integral, we get

$$v(t) = t + 50\ln(1 + 0.02t) - 100e^{0.01t} + c$$

Substituting the initial condition, we have that $c = 30$. Therefore, $v(100) = 100 + 50\ln(3.0) - 100e^1 + 30 = -86.9$.

3. **a.** At a speed of 10 m/s = 1000 cm/s, it takes $t = 80$ cm/(1000 cm/s) = 0.08 s to reach your hand. Similarly, your elbow seems to be 50 cm from the brain, so it takes the signal 0.05 s to get there.

b.

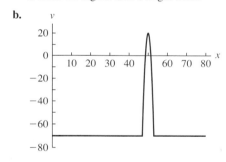

c. Average $= \frac{1}{6}\int_{47}^{53} -70.0 + 10.0[9.0 - (x - 50.0)^2] dx$. Substituting $y = x - 50$, we have that $dy = dx$ and limits of integration from $y = -3$ to $y = 3$, or

$$\text{average} = \frac{1}{6}\int_{-3}^{3} -70.0 + 10.0(9.0 - y^2) dy$$

$$= \frac{1}{6}\int_{-3}^{3} 20.0 - 10.0y^2 dy$$

$$= \frac{1}{6} 20.0y - \frac{10.0}{3} y^3 \Big|_{-3}^{3}$$

$$= \frac{1}{6}\left(60.0 - \frac{10.0}{3} 3^3\right)$$

$$- \left(-60.0 - \frac{10.0}{3} - 3^3\right)$$

$$= -10$$

d. The total voltage along the 6-cm piece is -60. Along the rest (74 cm) the total is $74 \cdot (-70) = -5180$. The total along the whole thing is $-5180 - 60 = -5240$, so the average is $-5240/80 = -65.6$.

5. a. The second is a pure-time differential equation. The first might describe a population of bacteria that are autonomously reproducing, and the second might describe a population being supplemented from outside at an ever-increasing rate.
  b. $\hat{b}(0.1) = b(0) + b'(0) \cdot 0.1 = 1 + 2 \cdot 0.1 = 1.2$. Similarly, $\hat{B}(0.1) = B(0) + B'(0) \cdot 0.1 = 1 + 2 \cdot 0.1 = 1.2$.
  c. $\hat{b}(0.2) = \hat{b}(0.1) + 2\hat{b}(0.1) \cdot 0.1 = 1.2 + 2 \cdot 1.2 \cdot 0.1 = 1.44$. Similarly, $\hat{B}(0.2) = \hat{B}(0.1) + B'(0.1) \cdot 0.1 = 1.2 + 1.2 \cdot 0.1 = 1.32$.

7. a. The volume is increasing until time $t = 2$ and decreases thereafter.
  b. The volume is a maximum when the rate of change is 0, or at $t = 2$.
  c. LHE $= V(0) \cdot 1 + V(1) \cdot 1 + V(2) \cdot 1 = 7$ and RHE $= V(1) \cdot 1 + V(2) \cdot 1 + V(3) \cdot 1 = -2$
  d. $V(3) = \int_0^3 4 - t^2 \, dt = 4t - t^3/3 |_0^3 = 3$

9. a. The function hits 0 when $x = 10$. The derivative is zero at $x = 2.5$, where the density is 56.25. At the endpoints, we have densities of 50 (at $x = 0$) and 250 (at $x = 20$). The maximum is thus at $x = 20$, with the minimum at $x = 10$.

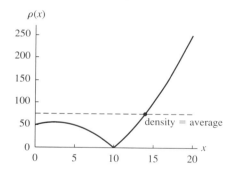

  b. Taking into account the absolute values, the total number is
$$\text{total} = \int_0^{20} |-x^2 + 5x + 50| \, dx$$
$$= \int_0^{10} -x^2 + 5x + 50 \, dx + \int_{10}^{20} x^2 - 5x - 50 \, dx$$
$$= -\frac{x^3}{3} + \frac{5x^2}{2} + 50x \Big|_0^{10}$$
$$+ \frac{x^3}{3} - \frac{5x^2}{2} - 50x \Big|_{10}^{20} = 1500$$
  c. The average density is 75.0.

11. a. $A'(t) = -15/(2 + .3t)^2 + 0.125e^{0.125t}$, so $A'(0) = -3.625$. Also, $A(0) = 35.0$ and $A(120) = 46.13$. The graph thus begins decreasing and then increases. The maximum is at $t = 120$.

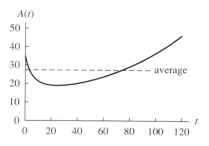

  b. $\int_0^{120} A(t) \, dt = \frac{50}{0.3} \ln(2 + 0.3t) + 800e^{0.0125t} \Big|_0^{120}$
  $= 3276$
  c. The average is $3276/120 = 27.3$.
  d. The minimum value must be less than 27.3.
  e. RHE $= \sum_{i=1}^{6} A(20i) 20$. I bet the estimate is high.

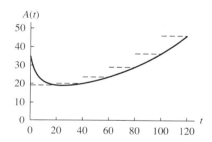

Chapter 5
Section 5.1

1. a. Nonautonomous
  b. Autonomous
  c. Pure-time
  d. Nonautonomous

3. a. $b(t) = e^t$
  b. $b(1) = e$
  c. $b(t) = e^{t-1}$, so $b(2) = e^{2-1} = e$
  d.

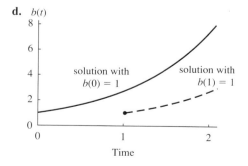

  e. They match because what happens does not depend on *when* it happens, only on the population size itself.

5. a.

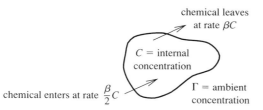

b. $\dfrac{dC}{dt} = \dfrac{\beta}{2}\Gamma - \beta C$

c. The only equilibrium is at $C^* = \Gamma/2$ (or at $\beta = 0$). If $\Gamma = 2.0$ mol/L, then $C^* = 1.0$ mol/L. The equilibrium is less than the external concentration because the cell is excluding half of the chemical from entering.

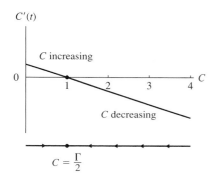

7. a. The equilibrium are at $N = 0$ and $N = K$.

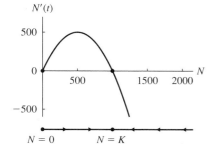

b.

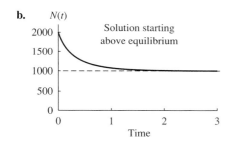

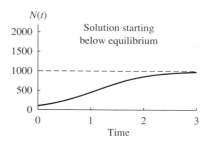

c. If it starts at a positive value, it always ends up at K.

9. a. Everything has units of meters per second squared.
b. The equilibrium is $v^* = 55.3$ m/s. This is the terminal velocity of a sky-diver in free fall.
c. This is about 120 mph.

11. a. The equilibrium is at $S^* = 0$. Eventually, all substrate will be used.
b. In both cases the rate is always negative. However, the graph of the rate in Torricelli's law of draining is much steeper near a value of 0 for the state variable.

c. $\dfrac{dS}{dt} = R - \dfrac{S}{1+S}$

d. $S^* = \dfrac{R}{1-R}$. This only makes sense if $R < 1$. Otherwise, there seems to be no equilibrium.

e. The lack of an equilibrium when $R > 1$ occurs because the maximum rate at which substrate is used is less than R.

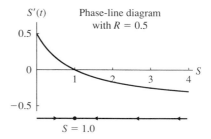

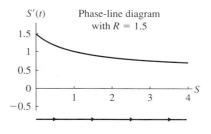

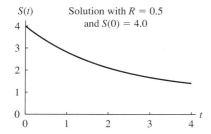

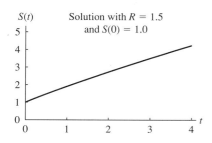

Section 5.2

1. **a.**

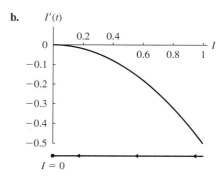

   b. Denoting the rate of change by $f(p)$, we get

   $$f'(p) = \frac{d}{dp}(\mu - \lambda)p(1-p) = (\mu - \lambda)(1 - 2p)$$

   Then $f'(0) = \mu - \lambda < 0$, implying that the equilibrium $p = 0$ is stable. $f'(1) = \lambda - \mu > 0$, implying that the equilibrium $p = 1$ is unstable.

3. **a.** With $\alpha = \mu$, the equation is

   $$\frac{dI}{dt} = \alpha I(1 - I) - \alpha I = -\alpha I^2$$

   The only equilibrium is at $I = 0$.

   b. $I'(t)$

   c. The derivative of the rate of change is $-2\alpha I$, which is equal to 0 at $I = 0$. From our phase-line diagram, however, we see that the rate of change is always negative, implying that the equilibrium is stable.

5. Suppose there are two stable equilibria in a row. Then the rate of change must cross from negative to positive at each. In particular, just above the lower one, the value of the rate of change is positive. Just below the upper one, the value of the rate of change is negative. By the Intermediate Value Theorem, there must be another crossing in between.

Section 5.3

1. **a.** $\dfrac{db}{b} = 0.01\,dt$, so $\ln(b) = 0.01t + c$ and $b(t) = Ke^{0.01t}$ with $K = 1000$.

   b. $\dfrac{db}{b} = -3\,dt$, so $\ln(b) = -3t + c$ and $b(t) = Ke^{-3t}$ with $K = 1.0 \times 10^6$.

   c. $\dfrac{dN}{1+N} = dt$, so $\ln(1 + N) = t + c$, $1 + N(t) = Ke^t$ and $N(t) = Ke^t - 1$. Substituting $t = 0$, we find $K = 2$, so $N(t) = 2e^t - 1$.

   d. $\dfrac{db}{1000 - b} = dt$, so $-\ln(1000 - b) = t + c$, $1000 - b(t) = Ke^{-t}$ and $b(t) = 1000 - Ke^{-t}$. Substituting $t = 0$, we find $K = 500$, so $b(t) = 1000 - 500e^{-t}$.

   e. As before, $b(t) = 1000 - Ke^{-t}$. Plugging in $t = 0$, we find $K = 0$, so $b(t) = 1000$. This is constant, consistent with the fact that $b = 1000$ is an equilibrium of this equation.

3. **a.** $C(t) = \Gamma + e^{-\beta t}[C(0) - \Gamma]$
   $= 2.0 + e^{-0.01t}(5.0 - 2.0)$
   $= 2.0 + 3.0e^{-0.01t}$
   $C(10) = 2.0 + 3.0e^{-0.1}$
   $= 4.715$

   The equilibrium is 2.0, and it is halfway to the equilibrium when $C(t) = 3.5$. Solving for t, we get

   $3.5 = 2.0 + 3.0e^{-0.01t}$
   $1.5 = 3.0e^{-0.01t}$

$$0.5 = e^{-0.01t}$$
$$\ln(0.5) = -0.01t$$
$$100 \cdot \ln(0.5) = t = 69.3$$

  b. $C(t) = 2.0 - 1.0e^{-0.01t}$, $C(10) = 1.095$. Again, the time to reach 1.5, halfway to the equilibrium, is 69.3 s.
  c. $C(t) = 2.0 + 3.0e^{-0.1t}$, $C(10) = 3.104$. The time to reach 3.5, halfway to the equilibrium, is 6.93 s.
  d. $C(t) = 2.0 - 1.0e^{-0.1t}$, $C(10) = 1.632$. The time to reach 1.5, halfway to the equilibrium, is again 6.93 s.
  e. The solutions approach the equilibrium exponentially with parameter β, regardless of the initial conditions. This time is equal to the half-life of something following the equation $x(t) = e^{-\beta t}$.

5. When $\alpha = \mu = 2.0$, the differential equation is

$$\frac{dI}{dt} = -2.0I^2$$

Separating variables, we get

$$\frac{dI}{I^2} = -2\,dt \qquad \text{separating variables}$$

$$-\frac{dI}{I} + c_1 = -2t + c_2 \qquad \text{integrating}$$

$$-\frac{dI}{I} = -2t + c \qquad \text{combining constants}$$

$$I = \frac{1}{2t - c} \qquad \text{solving for } I$$

Substituting $t = 0$, we find $c = -2$, so

$$I(t) = \frac{1}{2t + 2}$$

This solution declines to 0, consistent with the finding of stability of the equilibrium in Exercise 3 in Section 5.2. However, this function decreases to 0 more slowly than an exponential.

7. a. The per capita reproduction is an increasing function of time, meaning that things are getting better and better for this population. $\frac{db}{b} = t\,dt$, so $\ln(b) = t^2/2 + c$ and $b = Ke^{t^2/2}$ with $K = 10^6$. This population grows faster than an exponential.
  b. The per capita reproduction is a decreasing function of time, meaning that things are getting worse and worse for this population. $\frac{db}{b} = \frac{1}{1+t}\,dt$, so $\ln(b) = \ln(1+t) + c$ and $b = K(1+t)$ with $K = 10^6$. This population grows linearly, which is much slower than an exponential.

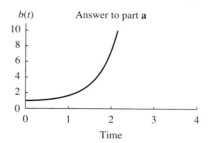

Answer to part a

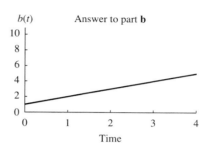

Answer to part b

  c. The per capita reproduction is an oscillating function of time. $\frac{db}{b} = \cos(t)\,dt$, so $\ln(b) = \sin(t) + c$ and $b = Ke^{\sin(t)}$ with $K = 10^6$. This population oscillates around its initial value.
  d. The per capita reproduction is a rapidly decreasing function of time, meaning that things are getting worse and worse for this population. $\frac{db}{b} = e^{-t}dt$, so $\ln(b) = -e^{-t} + c$ and $b = Ke^{-e^{-t}}$ with $K = e \times 10^6$. This population actually stops growing at $b = 2.718 \times 10^6$.

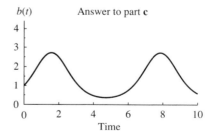

Answer to part c

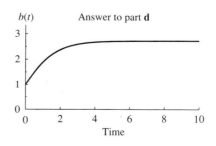

Answer to part d

9. a. $\dfrac{1}{p(1-p)}\,dp = dt$

b. Put them over a common denominator.

c. $\displaystyle\int \dfrac{1}{p} + \dfrac{1}{1-p}\,dp = \ln(p) - \ln(1-p) + c_1$

d. $\ln\left(\dfrac{p}{1-p}\right) + c_1$.

e. $\ln\left(\dfrac{p}{1-p}\right) = t + c$.

f. $\dfrac{p}{1-p} = e^{t+c} = Ke^t$. Solving for p, we get $p = \dfrac{Ke^t}{1 + Ke^t}$.

g. $0.01 = \dfrac{K}{1+K}$, so $K = 0.0101$.

h. The limit as t approaches infinity is $p = 1$.

Section 5.4

1. a. $\dfrac{db}{dt} = (1.0 - 0.05p)b$

$\dfrac{dp}{dt} = (-1.0 + 0.02b)p$

b. $\dfrac{db}{dt} = (2.0 - 0.01p)b$

$\dfrac{dp}{dt} = (1.0 + 0.01b)p$

These are different because the predators can reproduce even without the prey.

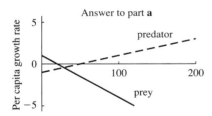

Answer to part a

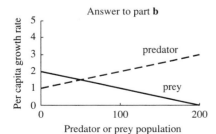

Answer to part b

c. $\dfrac{db}{dt} = (2.0 - 0.0001p^2)b$

$\dfrac{dp}{dt} = (1.0 + 0.01b)p$

d. $\dfrac{db}{dt} = (2.0 - 0.01p)b$

$\dfrac{dp}{dt} = (1.0 + 0.0001b^2)p$

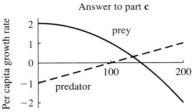

Answer to part c

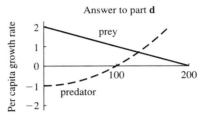

Answer to part d

3. The general form is

$$\dfrac{dH}{dt} = \alpha(A - H)$$

$$\dfrac{dA}{dt} = \alpha_2(H - A)$$

α_2 will be 10 times smaller due to the size of the room, but 5 times larger due to the specific heat. Therefore, we expect that

$$\dfrac{dH}{dt} = \alpha(A - H)$$

$$\dfrac{dA}{dt} = \dfrac{\alpha}{2}(H - A)$$

5. The equation for a is

$$\dfrac{da}{dt} = \mu\left(1 - \dfrac{a}{K_a}\right) a$$

with equilibria at $a = 0$ and $a = K_a$. If $\mu > 0$, the equilibrium at 0 is unstable and that at K_a is stable. The equation for b is

$$\dfrac{db}{dt} = \lambda\left(1 - \dfrac{b}{K_b}\right) b$$

with equilibria at $b = 0$ and $b = K_b$. If $\lambda > 0$, the equilibrium at 0 is unstable and that at K_b is stable.

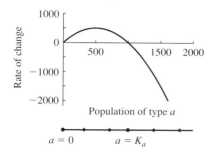

7. a. $v = \dfrac{dx}{dt}$

b. $a = \dfrac{dv}{dt}$. But we know that $a = -x$, so $\dfrac{dv}{dt} = -x$.

c. $\dfrac{dx}{dt} = v$

$\dfrac{dv}{dt} = -x$

d. The only equilibrium is at $x = 0$ and $v = 0$. The object is sitting still, and the spring is neither stretched nor compressed.

e. $v(t) = -\sin(t)$ and $\dfrac{dv}{dt} = -\cos(t) = -x(t)$.

9. a.

t	$\hat{H}$	$\hat{A}$
0.0	60.0	20.0
0.1	58.8	20.4
0.2	57.6	20.8

b.

t	$\hat{H}$	$\hat{A}$
0.0	0.0	20.0
0.1	0.6	19.8
0.2	1.18	19.6

c.

t	$\hat{H}$	$\hat{A}$
0.0	20.0	60.0
0.1	21.2	59.6
0.2	22.3	59.2

Section 5.5

1. a. Place p on the vertical axis. The p-nullcline consists of the two pieces $p = 0$ and $b = 600$. The b-nullcline consists of the two pieces $b = 0$ and $p = 500$. The equilibria are $(0, 0)$ and $(600, 500)$.

b. The p-nullcline consists of the two pieces $p = 0$ and $b = 900$. The b-nullcline consists of the two pieces $b = 0$ and $p = 200$. The equilibria are $(0, 0)$ and $(900, 200)$.

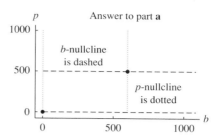

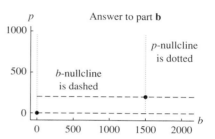

c. Both nullclines are the line $H = A$, which consists entirely of equilibria.

d. Place b on the vertical axis. The a-nullcline is the two pieces $a = 0$ and $b = 10^6 - a$. The b-nullcline is the two pieces $b = 0$ and $b = 10^7 - a$. the equilibria are $(0, 0)$, $(10^6, 0)$ and $(0, 10^7)$.

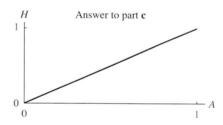

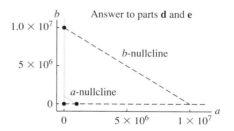

e. The equilibria and nullclines are the same. The growth rates have no effect on where these populations end up because the results are governed by the ability to compete rather than by high growth rates.

3. a. With p on the vertical axis, the b-nullcline is $b = 0$ and $p = 200$. The p-nullcline is $p = 0$ or $b = -100$ (which is absurd). The only equilibrium is at $(0, 0)$.

b. With p on the vertical axis, the b-nullcline is $b = 0$ and $p = \sqrt{20{,}000} = 141$. The p-nullcline is $p = 0$ or

$b = -100$ (which is absurd). The only equilibrium is at $(0, 0)$.

c. With p on the vertical axis, the b-nullcline is $b = 0$ and $p = 200$. The p-nullcline is $p = 0$. Again, the only equilibrium is at $(0, 0)$.

5. a. The p-nullcline is the same, $p = 0$ or $b = 1000$. The b-nullcline is $b = 0$ or $p = 1000 - 0.2b$.

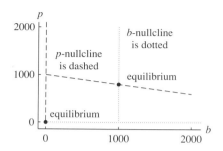

b. There are two intersections, one at $(0, 0)$ and one where the line $p = 1000 - 0.2b$ crosses the vertical line at $b = 1000$. This occurs at the point $(1000, 800)$.

7. a. The per capita growth of a decreases half as quickly as a function of b as it does as a function of a. Therefore, individuals of type b decreases reproduction of individuals of type a only half as much.

b. The a-nullcline is $a = 0$ and $b = 2(K_a - a)$. The b-nullcline is $b = 0$ and $b = K_b - \dfrac{a}{2}$.

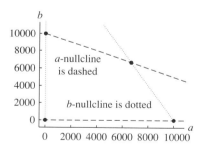

c. Besides the old equilibria at $(a, b) = (0, 0)$, $(a, b) = (K_a, 0)$ and $(a, b) = (0, K_b)$, there is a new one where

$$\mu\left(1 - \frac{a + b/2}{K_a}\right) = 0$$

$$\lambda\left(1 - \frac{a/2 + b}{K_b}\right) = 0$$

The two lines cross when

$$2(K_a - a) = K_b - \frac{a}{2}$$

or

$$2(10^4 - a) = 10^4 - \frac{a}{2}$$

$$2 \times 10^4 - 2a = 10^4 - \frac{a}{2}$$

$$\frac{3a}{2} = 10^4$$

$$a = 6.667 \times 10^3$$

We can find the value of b as $10^4 - \dfrac{a}{2} = 6.667 \times 10^3$.

d. Because the two types interfere weakly with each other, they can coexist.

9. a. The only individuals who can reproduce are the susceptibles, and all of their offspring are also susceptible.

b. $\dfrac{dS}{dt} = bS - \alpha IS - kS$

$\dfrac{dI}{dt} = \alpha IS - \mu I - kI$

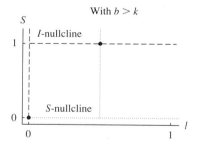

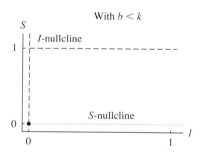

Section 5.6

1. a.

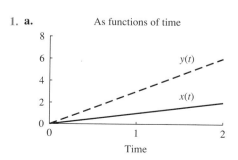

Answers **745**

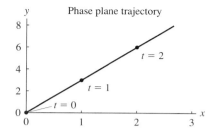

b.
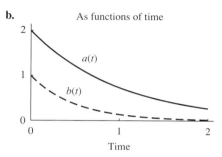

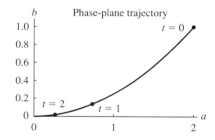

c.
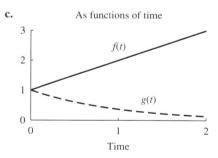

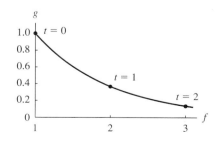

d.
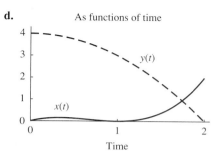

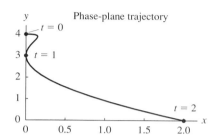

3. a.
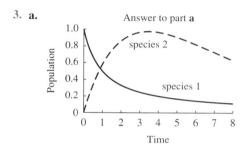

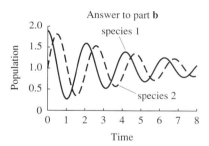

5. a.
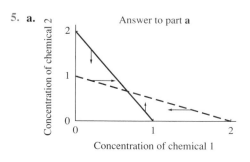

746 Answers

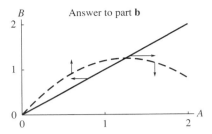
Answer to part **b**

7. a.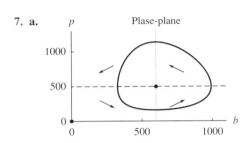

b.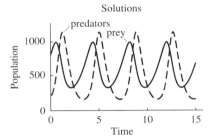

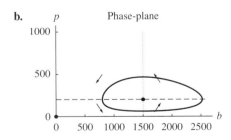

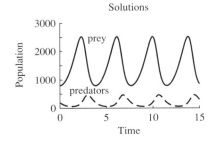

c.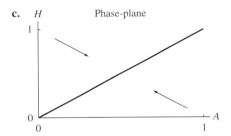

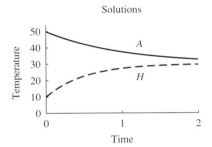

d.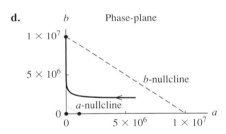

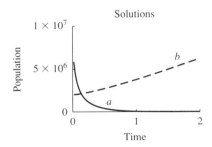

e.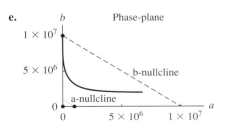

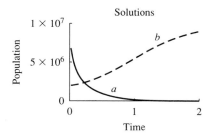

Section 5.7

1. One such equation is

$$\frac{dv}{dt} = -v(v - a_1)(v - e_1)(v - a_2)(v - e_2)$$

where a_1 and a_2 are the two thresholds and 0, e_1 and e_2 are the three equilibria. The figure uses $a_1 = 0.3$, $a_2 = 0.7$, $e_1 = 0.5$ and $e_2 = 1.0$.

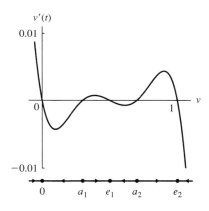

3. $$\frac{dw}{v - \gamma w} = \epsilon \, dt$$

$$\frac{-1}{\gamma} \ln |v - \gamma w| = \epsilon t + c$$

If $w < v/\gamma$,

$$w = \frac{v}{\gamma} - K e^{-\epsilon \gamma t}$$

If $w > v/\gamma$,

$$w = \frac{v}{\gamma} + K e^{-\epsilon \gamma t}$$

The smaller the value of ϵ, the slower the decay toward the equilibrium at $\dfrac{v}{\gamma}$.

5.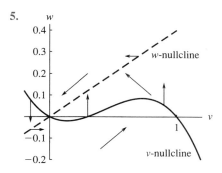

7. **a.** With ϵ small, the phase-plane trajectory moves very slowly in the vertical direction, and follows the v-nullcline. The response is slow but sharp.

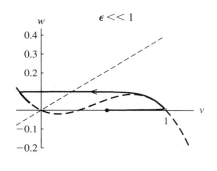

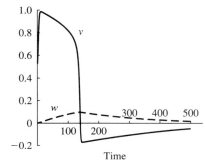

b. With ϵ large, the phase-plane trajectory sort of loops around, producing a poor excuse for an action potential.

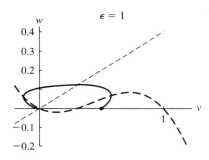

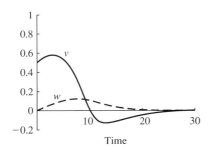

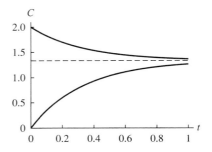

9. If the cell gets pushed far enough *below* the equilibrium, it creates a downward "action potential" before returning to equilibrium.

3. a. The first terms look like diffusion. The factor of 3 means that C changes more quickly than Γ, so it must have a smaller volume. The $\Gamma^2/3$ term means that Γ is being depleted the larger its concentration. The $+1$ is an outside supplement to C.

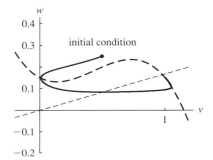

b.

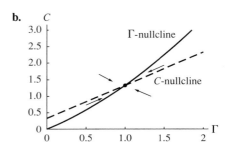

11. Translating level of terror into speed of moving legs might be a useful thing.

5. a. Looks like a pure-time differential equation, because the rate of change depends only on the time.

Answers to Supplementary Problems for Chapter 5

1. a. If Γ is constant, this is an autonomous differential equation. The first term represents diffusion of chemical, and the second term describes a constant increase from an outside source.

b.

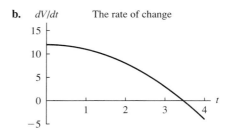

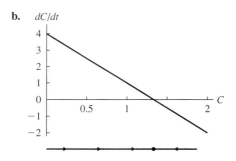

c. The equilibrium is the solution of $3(\Gamma - C) + 1 = 0$, or $c = \Gamma + 1/3$. If $f(C) = 3(\Gamma - C) + 1$, then $f'(C) = -3$, which is clearly negative. Therefore, the equilibrium is stable.

d. With $\Gamma = 1.0$,

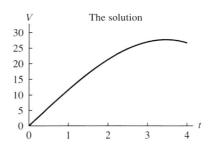

c. V takes on a maximum when $dV/dt = 0$, or at $t = \sqrt{12}$.

d. $\hat{V}(0.1) = V(0) + V'(0) \cdot (0.1) = 0 + 12 \cdot 0.1 = 1.2$

e. Integrating, we find $V(t) = 12t - t^3/3 + c$, and the constant c must be 0. Solving $V(t) = 0$ gives $t = 6$.

f. The average volume is

$$\frac{1}{6}\int_0^6 12t - \frac{t^3}{3}\,dt = \frac{1}{6}\left(6t^2 - \frac{t^4}{12}\Big|_0^6\right) = 18$$

7. a. $\dfrac{db}{dt} = \dfrac{1}{\sqrt{b}} \cdot b = \sqrt{b}$.

b. Using separation of variables,

$$\frac{db}{\sqrt{b}} = dt$$

$$2\sqrt{b} = t + c$$

$$b = \left(\frac{t+c}{2}\right)^2$$

Checking, $b'(t) = (t + c)/2$, which is indeed $\sqrt{b}$.

c. If $b(0) = 10{,}000$, $(c/2)^2 = 10{,}000$, so $c = 200$. Substituting $t = 2$, we find $b = 101^2 = 10{,}201$.

d. A quadratic increases more slowly than an exponential, so this population grows more slowly. This occurs because per capita reproduction decreases with larger populations.

9. a. This is a pure-time differential equation because the rate of change depends only on time, not on N.

b.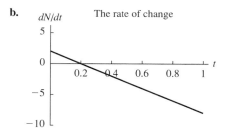

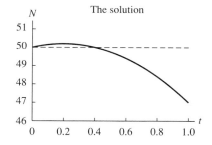

d. $\hat{N}(0.1) = N(0) + N'(0) \cdot 0.1 = 50 + 2 \cdot 0.1 = 50.2$

e. Integrating, $N(t) = 2t - 5t^2 + c$. The constant is 50 from the initial condition. Substituting, we find that $N(1) = 47$.

f. $N(t) = 50$ at $t = 0$ and at $t = 0.4$.

11. a. A system of coupled autonomous differential equations. This looks like a system where type a has a lower per capita reproduction, and the per capita reproduction of type a is reduced by its own presence and the presence of the other, and rather more by b, which probably eats a. The per capita reproduction of type b is increased by type a and decreased by itself. Perhaps bs eat as but fight with each other.

b. The a-nullcline is $a = 0$ and $b = 200 - 0.4a$. The b-nullcline is $b = 0$ and $b = 100 + 0.1a$. The equilibria are at $(0, 0)$, $(0, 100)$, $(500, 0)$, $(200, 120)$.

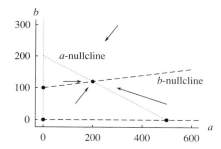

13. a. Because distances are given in millimeters, we must convert densities into moles per millimeter, so $s(x) = 0.0012/(1.0 + 0.2x)$. Integrating to find the total, we get

$$\text{total} = \int_0^{20} \frac{0.0012}{1.0 + 0.2x}\,dx$$
$$= 0.006\ln(1.0 + 0.2x)\big|_0^{20} = 0.006[\ln(5) - \ln(1)]$$
$$= 0.00966 \text{ mol}$$

b. The average density is $0.00966/20 = 0.00048$ mol/mm.

c. The density decreases along the tongue from a maximum of 0.0012 at $x = 0$ to a minimum of 0.00024 at $x = 20$. The average lies between these values, as it must.

15. a. The first term in the L equation indicates that microtubules break down faster when they are larger, but build up faster in the presence of the component E. The component E is constantly supplemented from outside and seems to be converted into microtubule.

b. Putting E on the vertical axis, the L-nullcline is at $L = 0$ or $E = 2.0/5.4 = 0.37$. The E nullcline is $E = 3.5/L$. L increases when E is above the nullcline, and E increases when E is below the nullcline.

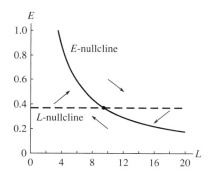

Chapter 6

Section 6.1

1. **a.** It takes $t = \ln(100)/\ln(1.1) = 48.3$ yr.
   b. Simulation 1 takes about 44 yr and simulation 2 never makes it.
   c. Simulation 2 takes about 42 yr and simulation 1 never even gets close.

3. Suppose the population starts at 1000. It will be 600 after one generation, 900 after that, 540 after that, 810 after that. It looks like this population is shrinking. This might be surprising because the average per capita reproduction is $(0.6 + 1.5)/2 = 1.05 > 1$.

5. **a.**

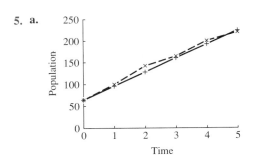

b.

Generation	Number of immigrants 1	Number of immigrants 2
1	32	36
2	32	42
3	32	22
4	32	36
5	32	20

   c. The first acts deterministically and is described by the updating function $N_{t+1} = N_t + 32$.
   d. The other one follows the stochastic updating function $N_{t+1} = N_t + I(t)$, where $I(t)$ averages about 32.

7. The number of immigrants in a given year might be more than 2 or less than 0. The number of immigrants does not depend on the population size. There is no reproduction.

9. **a.**

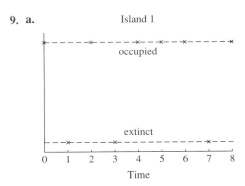

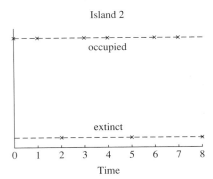

   b. The second acts deterministically, occupied for 2 yr and unoccupied for 1 yr.
   c. It jumps back and forth, but seems to be occupied more often than unoccupied.
   d. The second is always occupied on years that are multiples of 3. It seems that the first is more likely to be occupied than unoccupied, but we cannot guess which.

Section 6.2

1. **a.** $0.8^3 = 0.512$
   b. $0.512 \cdot 5000 = 2560$
   c. Need to solve $0.8^t = 0.1$, finding $t = \ln(0.1)/\ln(0.8) = 10.3$ hours.

3. **a.** We multiply the two probabilities together, or

$$\text{(probability both up)} = \text{(probability first up)}$$
$$\times \text{(probability second up)}$$
$$= 0.8^t \cdot 0.8^t$$
$$= 0.8^{2t}$$

   b. The probability that each has jumped is $1 - p_t = 1 - 0.8^t$. The probability that both have jumped is $(1 - 0.8^t)^2$.
   c. This is the only other possibility, so

$$\text{probability exactly one up}$$
$$= 1 - 0.8^{2t} - (1 - 0.8^t)^2$$
$$= 1 - 0.8^{2t} - (1 - 2 \cdot 0.8^t + 0.8^{2t})$$
$$= 2 \cdot 0.8^t - 2 \cdot 0.8^{2t}$$
$$= 2 \cdot 0.8^t(1 - 0.8^t)$$

5. **a.** $\dfrac{dP}{dt} = -0.001P$.
   b. $P(t) = e^{-0.001t}$, so $P(500) = e^{-0.5} = 0.607$.
   c. The average time would be around when $P(t) = 0.5$, which happens when

$$t = \frac{\ln(0.5)}{-0.001} = 693$$

7. **a.** $p_t + 1 = 0.9p_t + 0.2(1 - p_t)$.
   b. $p^* = 0.9p^* + 0.2(1 - p^*)$, so $p^* = 0.67$.
   c. About 67 of the 100 islands would be occupied.

a.

b. $c_{t+1} = 0.8c_t + 0.1d_t$, $d_{t+1} = 0.2c_t + 0.9d_t$.

c. $p_{t+1} = \dfrac{c_{t+1}}{c_{t+1} + d_{t+1}}$

$= \dfrac{0.8c_t + 0.1d_t}{0.8c_t + 0.1d_t + 0.2c_t + 0.9d_t}$

$= \dfrac{0.8c_t + 0.1d_t}{c_t + d_t} = 0.8p_t + 0.1(1 - p_t)$

d. This describes the same process but is about fluids, which are effectively infinite numbers of molecules. There is nothing random about the results of this discrete-time dynamical system.

Section 6.3

1. a. Half will be heterozygous, and hence half will be tall.
b. Again, half will be tall.
c. It depends. If the two have the same allele, then none will be tall. If the two have different alleles, then all will be tall.

3. a. All the **AA** offspring (0.25) + one-fourth of the heterozygous offspring (0.25 · 0.5) gives 0.25 + 0.125 = 0.375.
b. All the **AA** grand-offspring (0.375) + one-fourth of the heterozygous offspring (0.25 · 0.25) gives 0.375 + 0.0625 = 0.4375.
c. Let A_t be the fraction with genotype **AA** in generation t. The updating function is

$$A_{t+1} = A_t + 0.25 h_t = A_t + 0.25 \cdot 0.5^t$$

This equation adds up factors of $0.25 \cdot 0.5^t$, so

$$A_t = \sum_{j=0}^{t} 0.25 \cdot 0.5^j = 0.25 \cdot (2 - 0.5^{t-1})$$

We used the fact that $\sum_{j=0}^{t} 0.5^j = 2 - 0.5^{t-1}$. Alternatively, we could realize that the homozygotes are evenly split between **aa** and **AA**. After taking off the 0.5^t heterozygotes, we get

$$A_t = 0.5 \cdot (1 - 0.5^t)$$

matching the other formula.

5. a. All the offspring have height 54 cm.
b. Of the offspring, 25% have height 52 cm, 50% have height 54 cm, and 25% have height 55 cm.
c. Half will have height 52 cm, and the other half have height 54 cm.

7. a. All offspring have height 42 cm.

b. 1/16 have height 40 cm, 4/16 have height 41 cm, 6/16 have height 42 cm, 4/16 have height 43 cm, and 1/16 have height 44 cm.

9. a. 45 cm, 50 cm, and 55 cm.
b. 47.5 cm, 50 cm, and 52.5 cm.
c. $50 - \dfrac{10}{2^n}$ cm, 50 cm, and $50 + \dfrac{10}{2^n}$ cm.

11. a. Could be **AA**, **AB**, **BB**, **AN**, **MB**, or **NM** where **N** and **M** are new alleles created by mutation.
b. Multiply 0.5 (probability it started out as **A**) times the probability it did not mutate (0.99) to get 0.495.
c. 0.495   **d.** 0.245025
e. 0.245025   **f.** 0.49005
g. 0.50995

Section 6.4

1. a. Three possible genotypes, **bb**, **Bb** and **BB**.
b. Six possibilities: both **bb**, one **bb** and the other **Bb**, both **Bb**, one **bb** and the other **BB**, one **bb** and the other **BB**, and both **BB**.
c. There are eight simple events: {In, In, In}, {In, In, Out}, {In, Out, In}, {In, Out, Out}, {Out, In, In}, {Out, In, Out}, {Out, Out, In}, {Out, Out, Out}.
d. There are 17 simple events, the numbers ranging from 0 to 16.
e. There are an infinite number of simple events.

3. a. $\{t_1 = 1, t_2 = 1, t_3 = 1, t_4 = 1\}$, $\{t_1 = 1, t_2 = 2, t_3 = 3, t_4 = 4\}$, $\{t_1 = 4, t_2 = 3, t_3 = 2, t_4 = 1\}$.
b. If $\{t_1 = 1, t_2 = 2, t_3 = 3, t_4 = 4\}$, then molecules 3 and 4 left between times 3 and 7.
If $\{t_1 = 4, t_2 = 3, t_3 = 2, t_4 = 1\}$, then molecules 1 and 2 left between times 3 and 7.
If $\{t_1 = 5, t_2 = 5, t_3 = 2, t_4 = 2\}$, then molecules 1 and 2 left between times 3 and 7.
c. The prime numbered molecules are 2 and 3. Possibilities are $\{t_1 = 1, t_2 = 2, t_3 = 3, t_4 = 4\}$, $\{t_1 = 1, t_2 = 2, t_3 = 2, t_4 = 1\}$, $\{t_1 = 4, t_2 = 5, t_3 = 7, t_4 = 6\}$.

5. a. $\mathbf{a} \cap \mathbf{b}$ = $\{N = 91\}$, $\{N = 93\}$, $\{N = 94\}$, $\{N = 97\}$, $\{N = 99\}$. $\mathbf{a} \cup \mathbf{b}$ = all odd numbers plus $\{N = 92\}$, $\{N = 94\}$, $\{N = 96\}$, $\{N = 98\}$, $\{N = 100\}$.
b. $\mathbf{a} \cap \mathbf{c}$ is empty. $\mathbf{a} \cup \mathbf{c}$ is $N = 30, 31, 32, 68, 69, 70$ and $N > 90$.
c. $\mathbf{b} \cap \mathbf{c}$ = $\{N = 31\}$, $\{N = 69\}$. $\mathbf{b} \cup \mathbf{c}$ = all odd numbers plus $N = 30, 32, 68$ or 70.

7. a. The biologically reasonable assignment is Pr(**bb**) = 0.25, Pr(**Bb**) = 0.5, Pr(**BB**) = 0.25. Another possibility is Pr(**bb**) = 0.5, Pr(**Bb**) = 0.25, Pr(**BB**) = 0.25.
b. The biologically reasonable assignment is

Pr(both are **bb**) = 1/16

Pr(one is **bb** and the other is **Bb**) = 1/4

Pr(both are **Bb**) = 1/4

Pr(one is **BB** and the other is **Bb**) = 1/4

752 Answers

$$\Pr(\text{both are } \mathbf{BB}) = 1/16$$
$$\Pr(\text{one is } \mathbf{bb} \text{ and the other is } \mathbf{BB}) = 1/8$$

These add to 1. Another possibility is that

$$\Pr(\text{both are } \mathbf{bb}) = \Pr(\text{both are } \mathbf{BB}) = 0.5$$

and all the rest are 0.

c. *First assignment:* Each simple event has probability 0.125. These add up to 1, and correspond to the case where each result is equally likely. *Second assignment:* The simple event {In, In, In} has probability 1, the rest have probability 0. These also add to 1, and correspond to the case where the molecule never leaves. Both seem reasonable.

9.

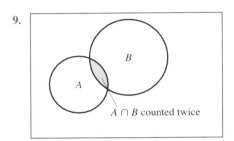

By adding together the area in **A** and **B**, we count the area in the intersection $A \cap B$ twice. Subtracting it off corrects for this. In Exercise 8, $A \cap B = \{N = 3\}$.
a. $\Pr(A \cap B) = 0.1$ and $0.6 = 0.4 + 0.3 - 0.1$.
b. $\Pr(A \cap B) = 0.0$, which is why the formula worked.
c. $\Pr(A \cap B) = 0.25$, and $0.75 = 0.5 + 0.5 - 0.25$.

Section 6.5

1. **a.** There are only three events, **bb**, **Bb**, **BB**. This is the only possible choice.
   b. At least one **bb**, both **BB** and the rest (both **Bb** or one **BB** and the other **Bb**).
   c. We could break the eight elements in the sample space into those with an "Out" at time 2 (four simple events), those with an "In" at time 2 and an "Out" at time 5 (two simple events), and the rest (two events).
   d. Let N denote the number of plants taller than 50 cm. We could use $N = 0$, $0 < N < 16$, and $N = 16$.
   e. All taller than 40 cm, all shorter than 20 cm and all the rest (at least one shorter than 40 cm and at least one taller than 20 cm).

3. **a.**

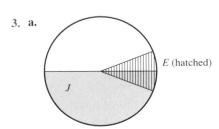

b. Let J be the event of seeing a jackrabbit, E the event of seeing an eagle. Then

$$\Pr(J \mid E) = \frac{\Pr(J \cap E)}{\Pr(E)} = \frac{0.15}{0.2} = 0.75$$

It is more likely that she will see a rabbit when she sees an eagle. Perhaps rabbits enjoy watching the eagles.

c. $\Pr(E \mid J) = \dfrac{\Pr(E \cap J)}{\Pr(J)} = \dfrac{0.15}{0.5} = 0.3$

It is more likely that she will see an eagle when she sees a rabbit. Perhaps eagles like to eat rabbits.

5. Let S be the event "stains," A_1 the event "less than 1 day old," A_2 the event "between 1 and 2 days old," A_3 the event "between 2 and 3 days old," A_4 the event "more than 3 days old."
   a. From the law of total probability,

$$\Pr(S) = \Pr(S \mid A_1)\Pr(A_1) + \Pr(S \mid A_2)\Pr(A_2)$$
$$+ \Pr(S \mid A_3)\Pr(A_3) + \Pr(S \mid A_4)\Pr(A_4)$$
$$= 0.95 \cdot 0.4 + 0.9 \cdot 0.3 + 0.8 \cdot 0.2 + 0.5 \cdot 0.1$$
$$= 0.76$$

   b. After 3-day-old cells are eliminated, the probabilities become

$$\Pr(\text{cell is less than 1 day old}) = 0.4444$$
$$\Pr(\text{cell is between 1 and 2 days old}) = 0.3333$$
$$\Pr(\text{cell is between 2 and 3 days old}) = 0.2222$$

Then,

$$\Pr(S) = \Pr(S \mid A_1)\Pr(A_1) + \Pr(S \mid A_2)\Pr(A_2)$$
$$+ \Pr(S \mid A_3)\Pr(A_3)$$
$$= 0.95 \cdot 0.4444 + 0.9 \cdot 0.3333 + 0.8 \cdot 0.2222$$
$$= 0.9$$

   c. After 2- and 3-day-old cells are eliminated, the probabilities become

$$\Pr(\text{cell is less than 1 day old}) = 0.57$$
$$\Pr(\text{cell is between 1 and 2 days old}) = 0.43$$

Then,

$$\Pr(S) = \Pr(S \mid A_1)\Pr(A_1) + \Pr(S \mid A_2)\Pr(A_2)$$
$$= 0.95 \cdot 0.57 + 0.9 \cdot 0.43 = 0.93$$

   d. Only the young cells are left. The probability is 0.95.

7. **a.** This is 0. The test always catches sick people.
   b. First, find $\Pr(P)$ with the law of total probability,

$$\Pr(P) = \Pr(P \mid D)\Pr(D) + \Pr(P \mid N)\Pr(N)$$
$$= (1.0)(0.2) + (0.05)(0.8) = 0.24$$

Then, find the conditional probability with Bayes' Theorem,

$$\Pr(D \mid P) = \frac{\Pr(P \mid D) \Pr(D)}{\Pr(P)}$$

$$= \frac{1.0 \cdot 0.2}{0.24} = 0.833$$

This is much better, the test is now identifying mainly people with the disease. This is because the fraction with the disease is significantly greater than the fraction of false positives.

  c. First, find $\Pr(P)$ with the law of total probability,

$$\Pr(P) = \Pr(P \mid D) \Pr(D) + \Pr(P \mid N) \Pr(N)$$
$$= (0.95)(0.2) + (0.05)(0.8) = 0.23$$

Then, find the conditional probability with Bayes' theorem,

$$\Pr(D \mid P) = \frac{\Pr(P \mid D) \Pr(D)}{\Pr(P)}$$

$$= \frac{0.95 \cdot 0.2}{0.23} = 0.826$$

Also,

$$\Pr(D \mid P^c) = \frac{\Pr(P^c \mid D) \Pr(D)}{\Pr(P^c)}$$

$$= \frac{0.05 \cdot 0.2}{0.77} = 0.013$$

The false negatives do not significantly reduce the fraction of positive tests with the disease, but 1.3% of those who test negative are indeed sick.

9. Let F be the event "first red," S the event "second red," O the event "one red" and B the event "both red."
  a. $\Pr(S \mid F) = \dfrac{\Pr(S \cap F)}{\Pr(F)}$

  The probability that both are red is $1/6$. The probability that the first is red is $1/2$. So the probability is $1/3$.
  b. $\Pr(B \mid O) = \dfrac{\Pr(B \cap O)}{\Pr(O)}$

  Because B is a subset of O, $\Pr(B \cap O) = \Pr(B) = 1/6$. $\Pr(O) = 5/6$, so the probability is $1/5$.
  c. I have no idea.
  d. The first is $1/2$ and the second is $1/3$.

Section 6.6

1. The probability of both is the product of the probabilities, or 0.1.

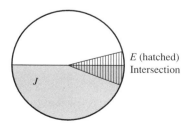

3. a. $0 = \Pr(A \cap B) \neq \Pr(A) \Pr(B)$, as long as $\Pr(A) > 0$ and $\Pr(B) > 0$.
   b. $\Pr(A \cap C) = \Pr(A) \geq \Pr(A) \Pr(C)$, as long as $\Pr(A) > 0$ and $0 < \Pr(C) < 1$.

5. Label the three cells **a**, **b** and **c**, and let the events a_t, b_t and c_t mean that it was in cell **a**, **b**, or **c**, respectively, at time t.
   a. Suppose it always goes to **a** with probability 0.8, and to **b** or **c** with probability 0.1. Then,

   $\Pr(a_{t+1} \mid a_t) = 0.8, \Pr(a_{t+1} \mid b_t) = 0.8, \Pr(a_{t+1} \mid c_t) = 0.8$
   $\Pr(b_{t+1} \mid a_t) = 0.1, \Pr(b_{t+1} \mid b_t) = 0.1, \Pr(b_{t+1} \mid c_t) = 0.1$
   $\Pr(c_{t+1} \mid a_t) = 0.1, \Pr(c_{t+1} \mid b_t) = 0.1, \Pr(c_{t+1} \mid c_t) = 0.1$

   b. Suppose that the probability of the molecule staying put is 0.8.

   $\Pr(a_{t+1} \mid a_t) = 0.8, \Pr(a_{t+1} \mid b_t) = 0.1, \Pr(a_{t+1} \mid c_t) = 0.1$
   $\Pr(b_{t+1} \mid a_t) = 0.1, \Pr(b_{t+1} \mid b_t) = 0.8, \Pr(b_{t+1} \mid c_t) = 0.1$
   $\Pr(c_{t+1} \mid a_t) = 0.1, \Pr(c_{t+1} \mid b_t) = 0.1, \Pr(c_{t+1} \mid c_t) = 0.8$

   c. Suppose **b** is to the right of **a**, **c** to the right of **b**, and **a** to the right of **c**.

   $\Pr(a_{t+1} \mid a_t) = 0.8, \Pr(a_{t+1} \mid b_t) = 0.0, \Pr(a_{t+1} \mid c_t) = 0.2$
   $\Pr(b_{t+1} \mid a_t) = 0.2, \Pr(b_{t+1} \mid b_t) = 0.8, \Pr(b_{t+1} \mid c_t) = 0.0$
   $\Pr(c_{t+1} \mid a_t) = 0.0, \Pr(c_{t+1} \mid b_t) = 0.2, \Pr(c_{t+1} \mid c_t) = 0.8$

   d. Suppose **b** is in the middle.

   $\Pr(a_{t+1} \mid a_t) = 0.8, \Pr(a_{t+1} \mid b_t) = 0.1, \Pr(a_{t+1} \mid c_t) = 0.0$
   $\Pr(b_{t+1} \mid a_t) = 0.2, \Pr(b_{t+1} \mid b_t) = 0.8, \Pr(b_{t+1} \mid c_t) = 0.2$
   $\Pr(c_{t+1} \mid a_t) = 0.0, \Pr(c_{t+1} \mid b_t) = 0.1, \Pr(c_{t+1} \mid c_t) = 0.8$

7. a. Pr(mates with red) = 0.2, Pr(mates with blue) = 0.3, Pr(mates with green) = 0.5.
   b. Assume she runs into males independently. She will only mate with a green bird if both are green, with probability $0.5 \cdot 0.5 = 0.25$. She will mate with a blue bird if she meets two blue males (probability $0.3 \cdot 0.3 = 0.09$), a blue then a green (probability $0.3 \cdot 0.5 = 0.15$), or a green then a blue (probability $0.5 \cdot 0.3 = 0.15$). Adding these up gives 0.39. This leaves a 0.36 chance of her mating with a red bird.

9. The long-term probability is 0.529. The probability for two minutes straight is

$$\Pr(I_{t+1} \mid I_t) \Pr I_t = 0.2 \cdot 0.529 = 0.106$$

11. Let M denote "rain in the morning," A "rain in the afternoon," and U "brought umbrella." The assumptions are $\Pr(M) = \Pr(A) = 0.4$, $\Pr(A \mid M) = 0.75$ and $\Pr(U \mid M) = 0.75$.
   a. Set M^c to be the event "no rain in the morning." By the law of total probability

   $$\Pr(A) = \Pr(A \mid M) \Pr(M) + \Pr(A \mid M^c) \Pr(M^c)$$

so

$$0.4 = 0.75 \cdot 0.4 + \Pr(A \mid M^c) \cdot 0.6$$

We can solve for $\Pr(A \mid M^c) = 0.167$.

b. If independent, the probability of getting rained on with an umbrella is exactly the probability of rain in the afternoon conditional on it raining in the morning, or 0.75.

c. By the law of total probability,

$$\Pr(U) = \Pr(U \mid M)\Pr(M) + \Pr(U \mid M^c)\Pr(M^c)$$
$$= 0.75 \cdot 0.4 + 0.167 \cdot 0.6 = 0.4$$

d. Yes, he could bring his umbrella on the 30% of rainy mornings when it will rain in the afternoon, and on the 10% of dry mornings when it will rain in the afternoon.

e. Half the time.

Section 6.7

1.

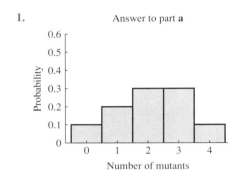

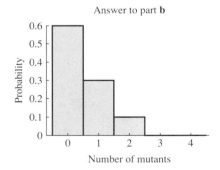

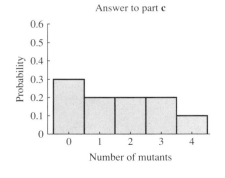

3. For the first histogram, **a.** 0.38, **b.** 0.12, **c.** 0.34. For the second histogram, **a.** 0.37, **b.** 0.19, **c.** 0.08. For the third histogram, **a.** 0.38, **b.** 0.12, **c.** 0.64.

5. a. Most likely is a measurement of 1, least a measurement of 0. Probability less than 0.2 is about 0.04, greater than 0.8 about 0.35.

b. Most likely is a measurement of 0.5, least a measurement of 0 or 1. This is symmetric. Probability less than 0.2 is about 0.1, greater than 0.8 is also about 0.1.

c. Most likely is a measurement of 1, least a measurement of 0. Probability less than 0.2 is about 0.1, greater than 0.8 about 0.3.

7. a.

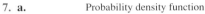

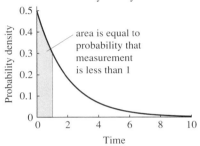

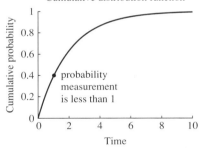

b. $G'(x) = 0 - (-0.5e^{-0.5x}) = 0.5e^{-0.5x}$
c. This is $G(1.0) = 0.393$.
d. In terms of the p.d.f., this is $\int_1^\infty g(x)\,dx$. In terms of the c.d.f., it is $1 - G(1.0) = 0.607$.
e. This is $G(3) - G(1) = 0.383$.
f. This is $G(1.01) - G(1) = 0.0030$. Dividing by 0.01 gives 0.30, very close to $g(1.0) = 0.30$.

9. a. The c.d.f. is $F(x) = x$. $F(1) = 1$, as it should. The probability that the measurement is less than 0.3 is $F(0.3) = 0.3$.
b. The c.d.f. is $F(x) = 2x - x^2$. $F(1) = 2 - 1 = 1$. The probability that the measurement is less than 0.3 is $F(0.3) = 0.51$.
c. The c.d.f. is $F(x) = 3x^2 - 2x^3$. $F(1) = 3 - 2 = 1$. The probability that the measurement is less than 0.3 is $F(0.3) = 0.216$.

Section 6.8

1. **a.** First, count whether one molecule is in (probability 0.9) or out (probability 0.1) at time $t = 1$. Second, count whether the same molecule is in (probability 0.81) or out (probability 0.19) at time $t = 2$. The probabilities correspond to what would happen if it leaves with probability 0.1 each minute.
   b. The measurement could be the number out of three inside at time 1 or the number out of five inside at time 1. Not very many would have left, but it is probably more likely that at least one out of five would have left.

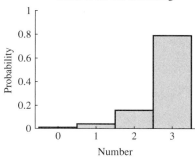

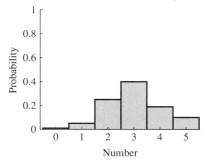

   c. The time the first out of three left, or the time the first out of five left. It seems likely that this time would be smaller with more molecules.

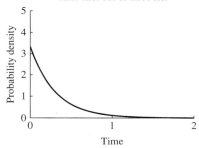

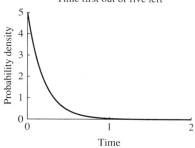

3. **a.** Expectation is $9 \cdot 0.3 + 7 \cdot 0.7 = 7.6$.
   b. Expectation is $9 \cdot 0.7 + 7 \cdot 0.3 = 8.4$.
   c. Expectation is $9 \cdot 1.0 + 7 \cdot 0.0 = 9.0$.

5. **a.** The expectation is 0.1, so about 1 would arrive in 10 years. The expectation is pretty close to the middle of the distribution.
   b. The expectation is 9.9, the average number arriving per year. About 99 individuals would arrive in 10 yr, meaning that this population will grow. The expectation seems quite far from the middle of the distribution, having been pulled way up by the rare years with many immigrants.
   c. The expectation is -3.5, or -35 in 10 yr. This population will decline, pulled down in the rare negative years.

7. Breaking into eight pieces, starting at 0, 0.5, 1, 1.5 2, 2.5, 3, and 3.5, we get

   a.

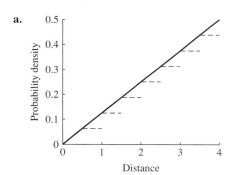

   b. The probability that the distance is between 0 and 0.5 is approximately $h(0) \cdot 0.5 = 0.0$, between 0.5 and 1 is approximately $h(0.5) \cdot 0.5 = 0.031$, between 1 and 1.5 is approximately $h(1) \cdot 0.5 = 0.062$, between 1.5 and 2 is approximately $h(1.5) \cdot 0.5 = 0.093$, between 2 and 2.5 is approximately $h(2) \cdot 0.5 = 0.124$, between 2.5 and 3 is approximately $h(2.5) \cdot 0.5 = 0.156$, between 3 and 3.5 is approximately $h(3) \cdot 0.5 = 0.187$, between 3.5 and 4 is approximately $h(3.5) \cdot 0.5 = 0.219$.
   c. The approximate expectation 2.18, which seems a bit low.
   d. The probabilities add to 0.875, which is low also, again because the approximate probability distribution lies below the p.d.f.

9. a. This approaches infinity at $x = 0$. However,

$$\int_0^1 \frac{1}{2\sqrt{x}} \, dx = \sqrt{x}\Big|_0^1 = 1$$

so this is a good p.d.f. The expectation is

$$\int_0^1 x \frac{1}{2\sqrt{x}} \, dx = \frac{x^{3/2}}{3}\Big|_0^1 = 1/3$$

b. $\int_1^\infty \frac{1}{t^2} \, dt = -\frac{1}{t}\Big|_1^\infty = 1$

so this is a good p.d.f. The expectation is

$$\int_1^\infty t \frac{1}{t^2} \, dt = -\ln(t)\Big|_1^\infty = \infty$$

This measurement has an average value of infinity, which does seem rather odd.

c. This approaches infinity at $y = 1$. However,

$$\int_0^1 \frac{1}{2\sqrt{1-y}} \, dy = -\sqrt{1-y}\Big|_0^1 = 1$$

so this is a good p.d.f. The expectation is

$$\int_0^1 \frac{y}{2\sqrt{1-y}} \, dy$$

To integrate this, change variables to $z = 1 - y$. Then $dz = -dy$, the limits are integration change to 1 to 0, so

$$\int_1^0 -\frac{1-z}{2\sqrt{z}} \, dz = \int_0^1 \frac{1-z}{2\sqrt{z}} \, dz$$
$$= \int_0^1 \frac{1}{2\sqrt{z}} - \frac{z}{2\sqrt{z}} \, dz$$
$$= \sqrt{z} - \frac{z^{3/2}}{3}\Big|_0^1 = 2/3$$

Section 6.9

1. a. The mean is $57,000, the median is $30,000, and the mode is $20,000.

b. The mode tells what you will probably get, and the median gives a good idea of what happens in the middle. The mean is heavily affected by the top salary.

c. The mean increases to $219,000 and the rest stay the same.

3. a. The mean and median are 1. All values are equally likely, so there is no mode.

b. The mean is $2/3$, the median is $\sqrt{2}/2$, the mode is 1.

c. The mean, median and mode of this symmetric distribution are all equal to 0.5.

d. The cumulative distribution function $F(t)$ is

$$F(t) = \int_0^t f(s) \, ds = -e^{-0.1s}\Big|_0^t = 1 - e^{-0.1t}$$

The median $\tilde{T}$ satisfies

$$F(\tilde{T}) = 1 - e^{-0.1\tilde{T}} = 0.5$$
$$e^{-0.1\tilde{T}} = 0.5$$
$$\tilde{T} = -10\ln(0.5) = 6.93$$

Half the measurements governed by this p.d.f. lie below 6.93. The mean of a random variable with this p.d.f. is 10.0 (see Exercise 8 in Section 6.8). The mean is larger than the median because the expectation weights large values of T.

5. a. The geometric mean is found by computing

$$\int_{1.0}^{1.2} 5.0 \ln(x) \, dx = 5.0x \ln(x) - x\Big|_{1.0}^{1.2} = 0.0939$$

and exponentiating as $e^{0.0939} = 1.0984$. The arithmetic mean is 1.1, slightly greater.

b. Geometric mean is 1.075, less than the arithmetic mean of 1.1.

c. Geometric mean is found by computing

$$\int_{0.5}^{1.5} \ln(x) x \, dx = \frac{x^2}{2}\ln(x) - \frac{x^2}{4}\Big|_{0.5}^{1.5} = 0.0428$$

and exponentiating as $e^{0.0428} = 1.0437$. The arithmetic mean is

$$\int_{0.5}^{1.5} x \cdot x \, dx = \frac{x^3}{3}\Big|_{0.5}^{1.5} = 1.0833$$

The geometric mean is indeed less.

7. a. $50, $75, $37.50, $56.25

b. After an even number of weeks, the price p_t is

$$p_t = 0.75^{t/2} \cdot 100$$

After an odd number of weeks, the price is

$$p_t = 0.5 \cdot 0.75^{(t-1)/2} \cdot 100$$

c. The low price manager wins out.

d. The geometric mean of the population is 0.866. In two years, on average, the population goes down by a factor of 0.75, like the price. However, this is only an average.

Section 6.10

1. a. The range is from -10 to 15. The quartiles are roughly at -3 and 10. MAD $= 7.245$, $\sigma^2 = 78.32$, $\sigma = 8.85$, CV $= 2.64$.

b. The range is from -10 to 15. The quartiles are roughly at -3 and 5.

$$\text{MAD} = |-10 - 3.75| \cdot 0.1 + |-3 - 3.75| \cdot 0.2$$
$$+ |4 - 3.75| \cdot 0.15 + |5 - 3.75| \cdot 0.25$$
$$+ |10 - 3.75| \cdot 0.2 + |15 - 3.75| \cdot 0.1$$
$$= 2.9$$

The variance is 39.89, the standard deviation is 6.316 and CV = 1.684.

c. The range is from −10 to 15. The quartiles do not mean much. MAD = 0.813, σ^2 = 7.44, σ = 2.73, CV = 0.67.

3. With the top salary of \$450,000, MAD = \$45,888, σ^2 = 7,441,338 square dollars. σ = \$86,000 dollars and CV = 1.51. With the top salary of \$4,500,000, MAD = \$342,506, σ^2 = 764,733,018 square dollars. σ = 874,490 dollars and CV = 3.99. The variance is hugely sensitive to the one large value.

5. a. The mean is $10p$, so MAD = $20p(1-p)$. The variance is $100p(1-p)$, the standard deviation $10\sqrt{p(1-p)}$, and the CV is $\sqrt{(1-p)/p}$. The coefficient of variation remains the same.

b. The mean is 0, MAD = $20p(1-p)$. The variance is $100p(1-p)$, the standard deviation $10\sqrt{p(1-p)}$, and the CV is undefined (negative values in the range). MAD, the variance and the standard deviation remained the same.

7.

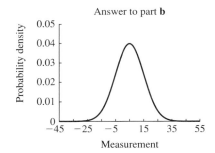

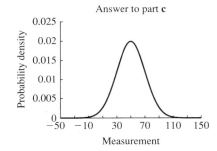

9. a. N_1 is exactly the same as R_0. The mean is 1.1 and the variance is

$$\text{Var } N_1 = 1.5^2 \cdot 0.6 + 0.5^2 \cdot 0.4 - 1.1^2 = 0.24$$

b. $\ln(N_1) = \ln(1.5) = 0.405$ with probability 0.6 and $\ln(N_1) = \ln(0.5) = -0.693$ with probability 0.4. Therefore,

$$E[\ln(N_1)] = 0.405 \cdot 0.6 - 0.693 \cdot 0.4 = -0.034$$

The variance is

$$\text{Var}[\ln(N_1)] = 0.405^2 \cdot 0.6 + 0.693^2 \cdot 0.4 - 0.034^2 = 0.290$$

Answers to Supplementary Problems for Chapter 6

1. a.

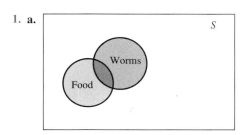

b. To apply Bayes' theorem, we must compute Pr(W) using the law of total probability, finding

$$\Pr(W) = \Pr(W \mid F)\Pr(F) + \Pr(W \mid F^c)\Pr(F^c)$$
$$= 0.1 \cdot 0.02 + 0.04 \cdot 0.98 = 0.0396$$

Then,

$$\Pr(F \mid W) = \frac{\Pr(W \mid F)\Pr(F)}{\Pr(W)} = \frac{0.1 \cdot 0.02}{0.0396} = 0.051$$

c. Not independent. You are more likely to find worms on the food.

d. The plate is 10^7 mm^2 in area. There will be about $0.002 \cdot 10^7 = 2.0 \times 10^4$ worms on the food and $0.0392 \cdot 10^7 = 3.92 \times 10^5$ off the food.

3. a.

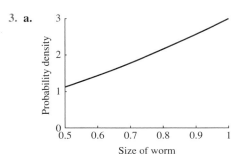

b. $\Pr(L < 0.75) = \int_0^{0.75} 1.5(l + l^2)\, dl = 0.38$.

c. About $0.01 \cdot f(0.75) = 0.0197$.

d. If 38% are less than 0.75 mm long, 62% will be more, or about 620 out of 1000.

5. a. There are eight: MMM, MMN, MNM, MNN, NMM, NMN, NNM, NNN.
   b. MMM, MMN, NMM, NMN.
   c. The simple events in "do not match at first" is NMM, NMN, NNM, NNN. Intersection: NMM, NMN. Union: MMM, MMN, NMM, NMN, NNM, NNN.
   d. There would have to have been no changes at any of the sites. The probability of no change is 0.9 at the first, 0.8 at the second, and 0.6 at the third. Multiplying (with independence), we get 0.432.

7. a.

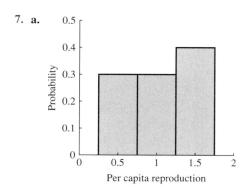

   b. $0.5 \cdot 0.3 + 1.0 \cdot 0.3 + 1.5 \cdot 0.4 = 1.05$

9. a. Let P represent the event positive, M represent mutant, and W represent non-mutant ("wild type"). Then
$$\Pr(P) = \Pr(P \mid M) \Pr(M) + \Pr(P \mid W) \Pr(W)$$
$$= 0.9 \cdot 0.01 + 0.02 \cdot 0.99$$
$$= 0.0288$$

   b. By Bayes theorem,
$$\Pr(M \mid P) = \Pr(P \mid M) \Pr(M) / \Pr(P) = 0.31$$

11. a. The p.d.f. is positive, and
$$\int_0^e f(z) \, dz = \ln(z) \Big|_0^e = 1$$

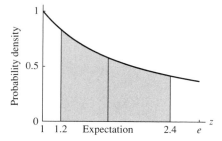

   b. $\Pr(1.2 \le Z \le 2.4) = \int_{1.2}^{2.4} \ln(z) \, dz = 0.693$

   c. $E(Z) = \int_1^e z f(z) \, dz = \int_1^e 1 \, dz = e - 1 = 1.718$

13. a. Let I_t represent infested and N_t not infested at time t. Then $\Pr(I_{t+1} \mid I_t) = 0.90$, $\Pr(I_{t+1} \mid N_t) = 0.05$, $\Pr(N_{t+1} \mid I_t) = 0.10$, $\Pr(N_{t+1} \mid N_t) = 0.95$.
   b. Let p_t be the probability infested at t. Then $p_{t+1} = 0.9 p_t + 0.05(1 - p_t)$.
   c. 0.333
   d. $p_{t+1} = 0.8[0.9 p_t + 0.05(1 - p_t)] = 0.72 p_t + 0.04(1 - p_t)$. The equilibrium is at 0.125. The parasite levels are reduced by a lot more than 80%.

15. a. Let p_t be the probability of making a shot. Then $p_{t+1} = 0.3 p_t + 0.6(1 - p_t)$. This has an equilibrium of 0.46.
   b. Let S represent made shot, M represent missed, and C represent complained. Then $\Pr(C) = \Pr(C \mid S) \Pr(S) + \Pr(C \mid M) \Pr(M) = 0.47$.
   c. By Bayes' theorem,
$$\Pr(S \mid C) = \Pr(C \mid S) \Pr(S) / \Pr(C) = 0.196$$

17. a. Calling the cumulative distribution $F(t)$, $F(1) = 1$, which means that the last frost certainly occurs before April 30.
   b. The median solves $F(t) = 0.5$, or $t = 1 - \sqrt{2/2} = 0.293$.
   c. $F(2/3) = 1/9$
   d. $f(t) = 2 - 2t$
   e. $\int_0^1 t f(t) \, dt = 1/3$
   f. I would plant on May 1. Why take chances?

Chapter 7
Section 7.1

1. a. $\Pr(S_1 \text{ and } M_2) = 0.3 - 0.03 = 0.27$
   b. $\Pr(M_1 \text{ and } S_2) = 0.3 - 0.03 = 0.27$
   c. $\Pr(M_1 \text{ and } M_2) = 1.0 - 0.03 - 0.27 - 0.27 = 0.43$

3.

	D	N
P	0.01	0.0495
P^c	0.00	0.9405

5. a. The random variables are age A, taking on values 0, 1, 2, and 3, and stain S, taking on value 1 for proper staining and 0 otherwise.

   b.

	$S = 1$	$S = 0$
$A = 0$	0.38	0.02
$A = 1$	0.27	0.03
$A = 2$	0.16	0.04
$A = 3$	0.05	0.05

c. The marginal distribution for cell age is given, and for staining is $\Pr(S = 0) = 0.86$ and $\Pr(S = 1) = 0.14$.

d.

	$S = 1$	$S = 0$
$A = 0$	0.44	0.14
$A = 1$	0.31	0.21
$A = 2$	0.19	0.29
$A = 3$	0.06	0.36

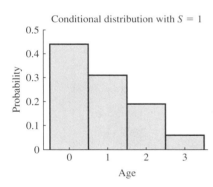

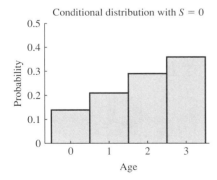

The conditional distribution with $S = 1$ looks a lot like the marginal, but with $S = 0$ it is very different. Cells that do not stain properly tend to be much older.

7. **a.** Let A_t represent the event "A made a good shot at time t" and B_t represent the event "B made a good shot at time t". Then

$$\Pr(A_{t+1} \mid B_t) = 0.8$$
$$\Pr(A_{t+1} \mid B_t^c) = 0.2$$
$$\Pr(B_{t+1} \mid A_t) = 0.8$$
$$\Pr(B_{t+1} \mid A_t^c) = 0.2$$

b. This is different because the behavior of player A depends on what B did on the last step, not on what A did.

c. By the law of total probablity,

$$\Pr(A_{t+1} \mid A_{t-1}) = \Pr(A_{t+1} \mid B_t)\Pr(B_t \mid A_{t-1})$$
$$+ \Pr(A_{t+1} \mid B_t^c)\Pr(B_t^c \mid A_{t-1})$$
$$= 0.8 \cdot 0.8 + 0.2 \cdot 0.2 = 0.68$$

d. By the law of total probablity,

$$\Pr(A_{t+1} \mid A_{t-1}^c) = \Pr(A_{t+1}^c \mid B_t)\Pr(B_t \mid A_{t-1})$$
$$+ \Pr(A_{t+1}^c \mid B_t^c)\Pr(B_t^c \mid A_{t-1})$$
$$= 0.2 \cdot 0.8 + 0.8 \cdot 0.2 = 0.32$$

e. Let p_{t+1} be the probability that A makes a good shot at time $t + 1$. From the previous two parts,

$$p_{t+1} = 0.68 p_{t-1} + 0.32(1 - p_{t-1})$$

The equilibrium of this occurs at $p^* = 0.5$.

f. Using the fact that As shot was good with probability 0.5, we find

	A_t	A_t^c
B_{t+1}	0.4	0.1
B_{t+1}^c	0.1	0.4

9.

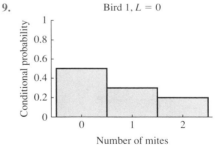

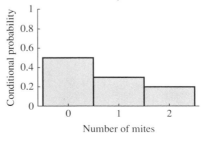

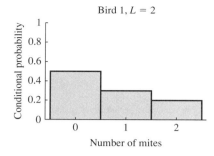
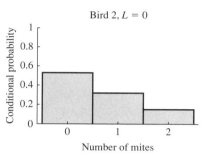
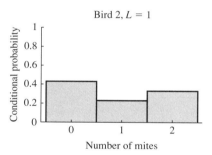
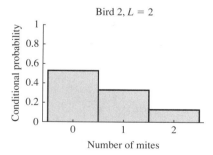
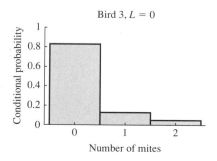
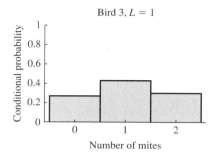
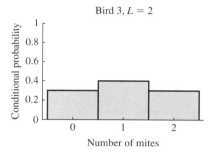
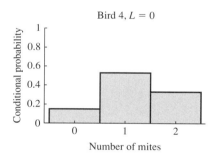
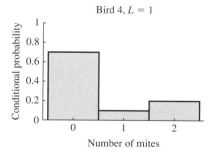
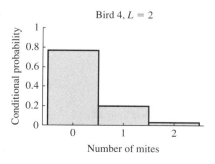

11. We need to show that

$$\sum_{i=1}^{n} \Pr(X = x_i \mid Y = y_j) = 1$$

From Definition 7.4, we have that

$$\sum_{i=1}^{n} \Pr(X = x_i \mid Y = y_j) = \sum_{i=1}^{n} \frac{\Pr(X = x_i \text{ and } Y = y_j)}{\Pr(Y = y_j)}$$

$$= \frac{\Pr(Y = y_j)}{\Pr(Y = y_j)} = 1$$

where we used the fact that the events $X = x_i$ and $Y = y_j$ for all i partition the event $Y = y_j$.

Section 7.2

1. For bird 1, covariance and correlation are 0. For bird 2, the covariance and correlation are also 0. For bird 4, the covariance is -0.12 and the correlation is -0.185.

3. **a.** We must find the mean and standard deviation of each marginal distribution. Using the solution of Exercise 2 in Section 7.1, and the fact that both J and E are Bernoulli random variables, we have that

$$\bar{J} = 0.5, \bar{E} = 0.2$$

$$\sigma_J = \sqrt{0.25} = 0.5, \sigma_E = \sqrt{0.16} = 0.4$$

The covariance is

$$\text{Cov}(J, E) = 0 \cdot 0.35 + 0 \cdot 0.15 + 0 \cdot 0.45$$
$$+ 1 \cdot 0.05 - 0.5 \cdot 0.2$$
$$= -0.05$$

The correlation is -0.25.

b. Let A represent cell age, then $E(A) = 1.1$, $\text{Var}(A) = 2.0$, $\sigma_A = 1.41$. Also, if S represents staining (1 for proper, 0 otherwise), then $E(S) = 0.86$, $\text{Var}(S) = 0.12$, $\sigma_S = 0.347$. Also, $\text{Cov}(A, S) = -0.21$, $\rho_{A,S} = -0.43$. The correlation is negative because older cells (larger value of A) tend to get smaller values of S.

5. **a.** $E(S) = 33.5$, $\text{Var}(S) = 965.25$, $E(T) = 20.0$, $\text{Var}(T) = 60.0$, $\text{Cov}(S, T) = 150$, $\rho = 0.623$
   b. $E[\ln(S)] = 2.99$, $\text{Var}[\ln(S)] = 1.15$, $\text{Cov}[\ln(S), T] = 5.54$, $\rho = 0.667$

7. **a.**

	$X = 0$	$X = 1$	$X = 2$
$Y = 1$	0.2	0.0	0.0
$Y = 4$	0.0	0.3	0.0
$Y = 7$	0.0	0.0	0.5

b. $\bar{X} = 1.3$, $\bar{Y} = 4.9$,

$$\text{Cov}(X, Y) = 0 \cdot 1 \cdot 0.2 + 1 \cdot 4 \cdot 0.3 + 2$$
$$\cdot 7 \cdot 0.5 - 1.3 \cdot 4.9$$
$$= 1.83$$

c. $\sigma_X = \sqrt{0.61}$, $\sigma_Y = \sqrt{5.49}$, so

$$\rho_{X,Y} = \frac{1.83}{\sqrt{0.61}\sqrt{5.49}} = 1$$

In accord with Theorem 7.3, these linearly related random variables have a correlation of 1.

9. **a.** Covariance is 0.17.
   b. Covariance is -0.27.

Section 7.3

1. **a.**

Number	Probability for bird 2	Probability for bird 4
0	0.21	0.06
1	0.26	0.42
2	0.29	0.39
3	0.20	0.12
4	0.04	0.01

b. $E(P_2) = 0 \cdot 0.21 + 1 \cdot 0.26 + 2 \cdot 0.29$
$+ 3 \cdot 0.20 + 4 \cdot 0.04 = 1.6$
$E(P_4) = 0 \cdot 0.06 + 1 \cdot 0.42 + 2 \cdot 0.39$
$+ 3 \cdot 0.12 + 4 \cdot 0.01 = 1.6$

c. The expectations of the sum must be the sum of the expectations, or $0.9 + 0.7$, as for birds 1 and 3.

$\text{Var}(P_2) = 0^2 \cdot 0.21 + 1^2 \cdot 0.26 + 2^2 \cdot 0.29$
$+ 3^2 \cdot 0.20 + 4^2 \cdot 0.04 - 1.6^2 = 1.3$
$\text{Var}(P_4) = 0^2 \cdot 0.06 + 1^2 \cdot 0.42 + 2^2 \cdot 0.39$
$+ 3^2 \cdot 0.12 + 4^2 \cdot 0.01 - 1.6^2 = 1.06$

d. Even though L and M are not independent for bird 2, the covariance is 0, so $\text{Var}(P_2) = \text{Var}(L) + \text{Var}(M) + 0 = 1.3$. For bird 4, the covariance is -0.12, so $\text{Var}(P_4) = \text{Var}(L) + \text{Var}(M) - 2 \cdot 0.12 = 1.06$.

3. Let I_1, I_2 and I_3 represent the number of immigrant, of each species. Then $E(I_1) = 0.8$, $E(I_2) = 0.6$, $E(I_1) = 1.5$. The total number of immigrants is $I = I_1 + I_2 + I_3$. Therefore,

$E(I) = E(I_1) + E(I_2) + E(I_3) = 0.8 + 0.6 + 1.5 = 2.9$

5. **a.**

	Species 1		
Species 2	0	1	2
0	0.18	0.36	0.06
1	0.06	0.12	0.02
2	0.06	0.12	0.02

b. $\Pr(0 \text{ immigrants}) = 0.18$
$\Pr(1 \text{ immigrants}) = 0.06 + 0.36 = 0.42$
$\Pr(2 \text{ immigrants}) = 0.06 + 0.12 + 0.06 = 0.24$
$\Pr(3 \text{ immigrants}) = 0.12 + 0.02 = 0.14$
$\Pr(4 \text{ immigrants}) = 0.02$

c. The expected total is 1.4.
d. The variances of the individual species are

$$\text{Var}(S_1) = 0^2 \cdot 0.2 + 1^2 \cdot 0.6 + 2^2 \cdot 0.1 - 0.8^2 = 0.36$$
$$\text{Var}(S_2) = 0^2 \cdot 0.6 + 1^2 \cdot 0.2 + 2^2 \cdot 0.2 - 0.6^2 = 0.64$$

The variance of the total number is

$$\text{Var}(S_1 + S_2) = 0^2 \cdot 0.18 + 1^2 \cdot 0.42 + 2^2 \cdot 0.24$$
$$+ 3^2 \cdot 0.14 + 4^2 \cdot 0.14 - 1.4^2 = 1.0$$

matching the sum of the variances.

7. a. Let $R_1 = 1000 S_1$, $R_2 = 1000 S_2$ and $R_3 = -500 S_3$ be the revenues from each species. Then

$$E(R_1) = 1000 E(S_1) = 800$$
$$E(R_2) = 1000 E(S_2) = 600$$
$$E(R_3) = -500 E(S_3) = -750$$

The total revenue $R = R_1 + R_2 + R_3$ has expectation

$$E(R) = E(R_1) + E(R_1) + E(R_1) = 650$$

b. We found (in the answer to Exercise 5) that $\text{Var}(S_1) = 0.36$ and $\text{Var}(S_1) = 0.64$. Similarly, $\text{Var}(S_3) = 0.45$. Therefore,

$$\text{Var}(R_1) = 1000^2 \, \text{Var}(S_1) = 360{,}000$$
$$\text{Var}(R_2) = 1000^2 \, \text{Var}(S_2) = 640{,}000$$
$$\text{Var}(R_3) = -500^2 \, \text{Var}(S_3) = 112{,}500$$

c. If they immigrate independently the variance is the sum, or 1,112,500.
d. Taking the square root, the standard deviation is $1054 and the coefficient of variation is 1.62. Tourism is a chancy business.

9. a.

D	Probability for bird 1	Probability for bird 3
-2	0.08	0.02
-1	0.18	0.14
0	0.35	0.55
1	0.24	0.20
2	0.15	0.09

b. $E(D) = 0.2$ for each. For bird 1, $\text{Var}(D) = 1.6$. For bird 3, $\text{Var}(D) = 1.04$.
c. $E(D) = E(L) - E(M) = 0.9 - 0.7 = 0.2$.

d. $\text{Var}(L - M)$
$$= E[(L - M)^2] - [E(L - M)]^2$$
$$= E(L^2 - 2LM + M^2) - E(L)^2$$
$$\quad + 2E(L)E(M) - E(M)^2$$
$$= E(L^2) - 2E(LM) + E(M^2) - E(L)^2$$
$$\quad + 2E(L)E(M) - E(M)^2$$
$$= E(L^2) - E(L)^2 - 2E(LM) + 2E(L)E(M)$$
$$\quad + E(M^2) - E(M)^2$$
$$= \text{Var}(L) + \text{Var}(M) - 2\,\text{Cov}(L, M)$$

Therefore, because $\text{Cov}(L, M) = 0.28$ for the second bird, $\text{Var}(D) = 0.9 + 0.7 - 2 \cdot 0.28 = 1.04$. The variance is smaller when the two measurements are positively correlated because the two random variables tend to have similar values.

11. Let X and Y be our random variables. The geometric mean of the product $\widetilde{XY}$ then satisfies

$$\widetilde{XY} = e^{E[\ln(XY)]} \quad \text{definition}$$
$$= e^{E[\ln(X) + \ln(Y)]} \quad \text{Law 1 of logs}$$
$$= e^{E[\ln(X)] + E[\ln(Y)]} \quad \text{Theorem 7.4}$$
$$= e^{E[\ln(X)]} e^{E[\ln(Y)]} \quad \text{Law 1 of exponents}$$
$$= \tilde{X}\tilde{Y} \quad \text{definition}$$

Section 7.4

1. Number of molecules out of 20 that remain in a cell after 10 min. Need to assume that each molecule has the same probability of having left, and that they are independent. 2. Number of cells out of 50 that stain properly. Need to assume that each cell has the same probability of staining, and that they are independent. 3. Number of birds in a population of 200 that have lice. Need to assume that each bird has the same probability of being infested, and that they are independent.

3. See table.

5. a. There are $5! = 120$ ways to order the prescriptions, only one of which is right. The probability is $1/120$.
b. Each patient has a $1/5$ chance.
c. The events are not independent. If the first patient gets the right medicine, that improves the odds for the second.
d. The conditional probability is $1/4$.

7. Let H be the number. Then,

$$\Pr(H = 0) = b(0; 5, 0.5) = 0.031$$
$$\Pr(H = 1) = b(1; 5, 0.5) = 0.156$$
$$\Pr(H = 2) = b(2; 5, 0.5) = 0.313$$
$$\Pr(H = 3) = b(3; 5, 0.5) = 0.313$$
$$\Pr(H = 4) = b(4; 5, 0.5) = 0.156$$
$$\Pr(H = 5) = b(5; 5, 0.5) = 0.031$$

The expectation is 2.5, and the variance is 1.25.

Answers **763**

Table for Section 7.4, Exercise 3

Year 1	Year 2	Year 3	Year 4	Total immigrants k	Probability	$b(k; n, p)$
0	0	0	0	0	$(1-p)^4$	$b(0; 4, p) = (1-p)^4$
1	0	0	0	1	$p(1-p)^3$	
0	1	0	0	1	$p(1-p)^3$	
0	0	1	0	1	$p(1-p)^3$	
0	0	0	1	1	$p(1-p)^3$	$b(1; 4, p) = 4p(1-p)^3$
1	1	0	0	2	$p^2(1-p)^2$	
1	0	1	0	2	$p^2(1-p)^2$	
1	0	0	1	2	$p^2(1-p)^2$	
0	1	1	0	2	$p^2(1-p)^2$	
0	1	0	1	2	$p^2(1-p)^2$	
0	0	1	1	2	$p^2(1-p)^2$	$b(2; 4, p) = 6p^2(1-p)^2$
1	1	1	0	3	$p^3(1-p)$	
1	1	0	1	3	$p^3(1-p)$	
1	0	1	1	3	$p^3(1-p)$	
0	1	1	1	3	$p^3(1-p)$	$b(3; 4, p) = 4p^3(1-p)$
1	1	1	1	4	p^4	$b(4; 4, p) = p^4$

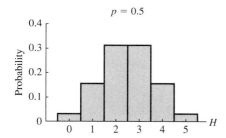

9. **a.** Mean = 50, variance = 25, CV = 0.1.
   b. Mean = 12.5, variance = 6.25, CV = 0.2. The coefficient of variation measures uncertainly about the result. With fewer tries, we are more uncertain of being close to the mean number.
   c. Mean = 22.5, variance = 2.25, CV = 0.067.
   d. Mean = 2.5, variance = 2.25, CV = 0.6. The coefficient is larger because the mean is smaller and the relative uncertainty is higher.

Section 7.5

1. **a.** $b(3; 12, 0.25) = \binom{12}{3}0.25^3 0.75^9 = \dfrac{12 \cdot 11 \cdot 10}{3 \cdot 2 \cdot 1}0.25^3 0.75^9 = 0.258$
   b. $b(6; 12, 0.5) = \binom{12}{6}0.5^6 0.5^6 = \dfrac{12 \cdot 11 \cdot 10 \cdot 9 \cdot 8 \cdot 7}{6 \cdot 5 \cdot 4 \cdot 3 \cdot 2 \cdot 1}0.5^6 0.5^6 = 0.226$
   c. 0.684
   d. 0.191

3. **a.** 10 with probability 2/3 and 0 with probability 1/3.
   b. The mean number occupied is $10 \cdot 2/3 + 0 \cdot 1/3 = 6.67$.
   c. The mode is 10.
   d. The islands are not independent—all respond to the same weather.

5. **a.** At time 1, this is a binomial distribution with $n = 5$ and $p = 0.8$. The probabilities are 0.0003, 0.0064, 0.051, 0.205, 0.410, 0.328.
   b. At time 2, this is a binomial distribution with $n = 5$ and $p = 0.8^2 = 0.64$.

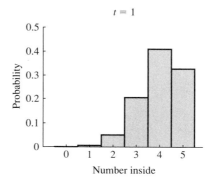

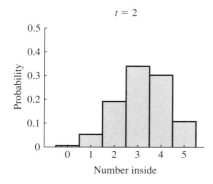

c. At time 5, this is a binomial distribution with $n = 5$ and $p = 0.8^5 = 0.32$.
d. At time 10, this is a binomial distribution with $n = 5$ and $p = 0.8^5 = 0.107$.

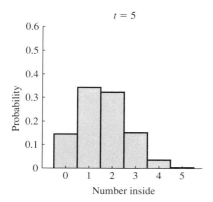

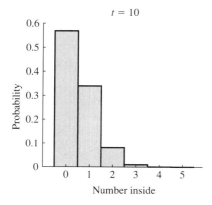

7. We use the fact that
$$\frac{d}{dt}0.9^t = \ln(0.9)0.9^t$$

Then, with the product rule
$$\frac{dv}{dt} = 100 \ln(0.9)0.9^t(1 - 0.9^t) - 100 \cdot 0.9^{2t} \ln(0.9)$$
$$= 100 \cdot \ln(0.9) \cdot 0.9^t(1 - 0.9^t - 0.9^t)$$
$$= 100 \cdot \ln(0.9) \cdot 0.9^t(1 - 2 \cdot 0.9^t)$$

Setting equal to 0 and solving, we get
$$1 - 2 \cdot 0.9^t = 0$$
$$0.9^t = 0.5$$
$$t = \frac{\ln(0.5)}{\ln(0.9)} = 6.58$$

This is when exactly half are inside.

9. Let P be the random variable describing the number of immigrants. Because they arrive in pairs, $P = 2I$, where I is a binomial random variable with $p = 0.5$ and $n = 100$.

11. a. $E(P) = E(2I) = 2E(I) = 2 \cdot 100 \cdot 0.5 = 100$
    b. $\text{Var}(P) = \text{Var}(2I) = 2^2 \text{Var}(I) = 4 \cdot 100 \cdot 0.5 \cdot 0.5 = 100$
    c. $\Pr(P = 106) = \Pr(I = 53) = b(53; 100, 0.5)$

Section 7.6

1. a. The probability of leaving during the first second is 0.3, during the second second is $0.3 \cdot 0.7 = 0.21$, and during the third second is $0.3 \cdot 0.7^2 = 0.147$. In general, the probability of leaving during second t is $0.3 \cdot 0.7^{t-1}$. The expectation is $1/0.3 = 3.33$ min, and the variance is $(1 - 0.3)/0.3^2 = 7.78$ min^2.
   b. Probability is 0.006, expectation is 100, variance is 9900 d^2.

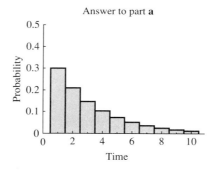

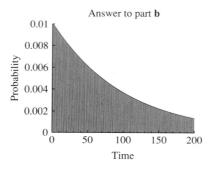

c. 0.0156. Expectation is 2, variance is 2.
d. Probability is 0.066. The expectation is 10 and the variance is 90.

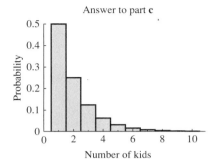

Answer to part c

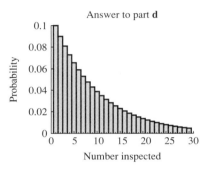

Answer to part d

e. Probability is 1.28×10^{-5}. Expected lifespan is 1.25 d, variance is 0.3125 d^2.

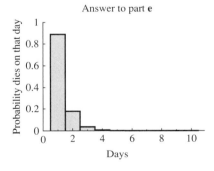

Answer to part e

3. a. The first one had to be okay (probability 0.9) and the second one bad (probability 0.2), for a total of 0.18.
   b. The first one had to be okay (probability 0.9) and the second one okay (probability 0.8), and the third one bad (probability 0.3), for a total of 0.216.
   c. $\Pr(T = 4) = 0.9 \cdot 0.8 \cdot 0.7 \cdot 0.4 = 0.201$
   $\Pr(T = 5) = 0.9 \cdot 0.8 \cdot 0.7 \cdot 0.6 \cdot 0.5 = 0.151$
   $\Pr(T = 6) = 0.9 \cdot 0.8 \cdot 0.7 \cdot 0.6 \cdot 0.5 \cdot 0.6 = 0.091$
   $\Pr(T = 7) = 0.9 \cdot 0.8 \cdot 0.7 \cdot 0.6 \cdot 0.5 \cdot 0.4 \cdot 0.7$
   $= 0.043$

$\Pr(T = 8) = 0.9 \cdot 0.8 \cdot 0.7 \cdot 0.6 \cdot 0.5 \cdot 0.4 \cdot 0.3 \cdot 0.8$
$= 0.014$
$\Pr(T = 9) = 0.9 \cdot 0.8 \cdot 0.7 \cdot 0.6 \cdot 0.5 \cdot 0.4 \cdot 0.3 \cdot 0.2$
$\cdot 0.9 = 0.005$
$\Pr(T = 10) = 0.9 \cdot 0.8 \cdot 0.7 \cdot 0.6 \cdot 0.5 \cdot 0.4 \cdot 0.3 \cdot 0.2$
$\cdot 0.1 \cdot 1.0 = 0.0005$

5. a. $P(t) = e^{-2.0t}$
   b. A binomial distribution with $n = 100$ and $P(t) = e^{-2.0t}$
   c. $E(N) = 100e^{-2.0t}$
   $\text{Var}(N) = 100(1 - e^{-2.0t})e^{-2.0t}$
   d. Variance is at a maximum when the population is 50, or at $t = 0.347$.

7. a. $\dfrac{dS}{dt} = -tS$

   b.
   $\dfrac{dS}{S} = -t\,dt$ separating variables

   $\ln(S) + c_1 = -\dfrac{t^2}{2} + c_2$ integrating

   $\ln(S) = -\dfrac{t^2}{2} + c$ combining constants

   $S = e^{-t^2/2} e^c$ solve for S

   $S = e^{-t^2/2}$ use initial condition $S(0) = 1$ to show $e^c = 1$

   c.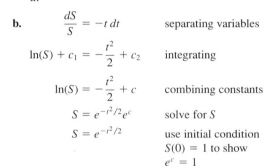

   d. The c.d.f. is $F(t) = 1 - S(t) = 1 - e^{-t^2/2}$. The p.d.f. is the derivative of the c.d.f., or
   $$f(t) = te^{-t^2/2}$$

   e. The probability it survives to age 1 is $S(1) = 0.606$. The probability it survives to age 2 conditional on reaching age 1 is $\dfrac{S(2)}{S(1)} = 0.223$. This animal is much less likely to make it through its second year.

9. a. If you multiply $\sum_{i=0}^{n} x^i$ by $1 - x$, all terms cancel except the 1 and the $-x^{n+1}$.

b. Plugging in $1 - q$ for x,

$$G_t = \sum_{i=1}^{t} q(1-q)^{i-1} = q\sum_{i=1}^{t}(1-q)^{i-1}$$
$$= q\sum_{j=0}^{t-1}(1-q)^j = q\frac{1-(1-q)^t}{1-(1-q)} = 1 - (1-q)^t$$

11. a. Probability it leaves immediately is q, the time is 1.
   b. Probability it does not leave is $1 - q$. The expected wait after not leaving is still $E(T)$. However, it already waited one time unit, giving a total expected wait of $1 + E(T)$.
   c. $E(T) = 1 \cdot q + [1 + E(T)](1 - q)$
   d. Solving for $E(T)$ gives $1/q$.

Section 7.7

1. There were 10 hits in 12 s in Figure 7.27, or an average of 0.833 per second. This is pretty close to the expected number of 1 per second. There were 46 hits in 12 s in Figure 7.28, or an average of 3.833 per second. This is pretty close to the expected number of 4.5 per second.

3. a. $p(0; 0.9) + p(1; 0.9) + p(2; 0.9) + p(3; 0.9) + p(4; 0.9) = 0.998$
   b. This is 1 minus the probability of 4 or less, or $1 - p(0; 1.8) + p(1; 1.8) + p(2; 1.8) + p(3; 1.8) + p(4; 1.8) = 0.036$.
   c. $p(5; 8.4) + p(6; 8.4) + p(7; 8.4) + p(8; 8.4) + p(9; 8.4) + p(10; 1.8) = 0.695$

5. a. $\Lambda = 1.2 \cdot 60 = 72.0$, equal to the mean and the variance. The standard deviation is $\sqrt{72.0} = 8.49$.
   b. $p(70, 72.0) = 0.046$
   c. $\Lambda = 1.2 \cdot 60/7 = 10.3$, equal to the mean and the variance. The standard deviation is $\sqrt{10.3} = 3.21$.
   d. $p(10, 10.3) = 0.125$

7. a. The random variable has $n = 100{,}000$ and $p = 0.0067$.
   b. It has $\Lambda = np = 100{,}000 \cdot 0.0067 = 670$.
   c. 670
   d. The variance of the binomial is $np(1-p) = 100{,}000 \cdot 0.0067 \cdot 0.9933 = 665.5$. The variance of the Poisson is $\Lambda = 670.0$.

9. a. Poisson with $\Lambda = 6.11$.
   b. Poisson with $\Lambda = 10.34$.
   c. Poisson with $\Lambda = 6.11 + 10.34 = 16.45$.
   d. By breaking up the equation, we see that

$$\Lambda = \left(\int_0^2 1.3 + 0.9t\, dt\right) \cdot 4.7 = 20.68$$

11. a. $P_1(t + \Delta t) \approx (1 - \lambda \Delta t)P_1(t) + \lambda \Delta t P_0(t)$.
   b. $\dfrac{P_1(t + \Delta t) - P_1(t)}{\Delta t} = -\lambda P_1(t) + \lambda P_0(t)$

$$\frac{dP_1}{dt} = -\lambda P_1(t) + \lambda P_0(t)$$

c. $\dfrac{d\lambda t e^{-\lambda t}}{dt} = \lambda e^{-\lambda t} dt - \lambda^2 t e^{-\lambda t} dt$
$= \lambda P_0(t) - \lambda P_1(t)$

Section 7.8

1. a. Five 1s and five 0s, or two 2.5s and eight 0s.
   b. There are $\binom{10}{5} = 252$ ways to get five 1s and five 0s and $\binom{10}{8} = 45$ ways to get two 2.5s and eight 0s.
   c. The probability of the five 1s and five 0s is $252 \cdot 0.5^5 0.25^5 = 0.0077$. The probability of two 2.5s and eight 0s is $45 \cdot 0.25^2 0.25^8 = 0.0004$.
   d. The total is 0.0081.

3. a. $3X \sim N(15.0, 144.0)$. The variance is multiplied by 9.
   b. $X + Y \sim N(15.0, 25.0)$

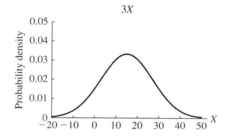

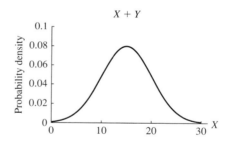

c. Let S_9 represent the sum of 9 independent samples from X. Then $S_9 \sim N(45.0, 144.0)$.
d. Let A_9 represent the average of 9 independent samples from Y. Then $A_9 \sim N(10.0, 1.0)$.

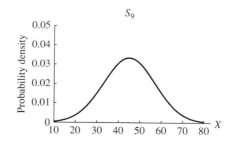

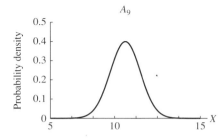

5. **a.** Each locus gives a mean of 1.875 points, with variance of 1.172. The normal approximation is $IQ \sim N(98.75, 11.72)$.
   b. Each locus gives a mean of 0.45 points, with variance of 0.067. The normal approximation is $IQ \sim N(89.0, 1.34)$.
   c. Each factor gives a mean of 0.45 points, with variance of 0.202. The normal approximation is $IQ \sim N(93.5, 6.06)$.
   d. The total is $IQ \sim N(121.24, 19.12)$.
   e. The maximum possible is 25 from genes of large effect plus 12 from genes of small effect plus 27 from the environment, or $80 + 25 + 12 + 27 = 144$.

7. **a.** If the second event occurs at time t, the first had to occur at some time s before that, and the second wait had to make up the difference exactly.
   b. $\Pr(S_2 = t) = \int_{s=0}^{t} e^{-s} e^{-t+s} ds = \int_{s=0}^{t} e^{-t} ds$
   $= se^{-t}\big|_0^t = te^{-t}$
   c. $\Pr(S_3 = t) = \int_{s=0}^{t} \Pr(S_2 = s) \Pr(T_3 = t - s) ds$
   $= \int_{s=0}^{t} se^{-s} e^{-t+s} ds = \int_{s=0}^{t} se^{-t} ds$
   $= \frac{s^2}{2} e^{-t} \big|_0^t = \frac{t^2}{2} e^{-t}$
   d. It sure looks like the p.d.f. for S_n is $\frac{t^n}{n!} e^{-t}$. Therefore, the probability that the nth event occurs at time t is equal to the probability that exactly n events have occurred by time t.

9. **a.** Let M represent the amount of money won. $M = 1$ with probability 0.5, $M = 2$ with probability 0.25, $M = 4$ with probability 0.125, and so forth. In general, $M = 2^{n-1}$ with probability 2^{-n}. Multiplying and adding, we get
   $$E(M) = \sum_{n=1}^{\infty} 2^{n-1} \cdot 2^{-n} = \sum_{n=1}^{\infty} 0.5 = \infty$$
   b. Even if I trusted my opponent, I would not chip in more than \$20.
   c. When you only play once, the mean does not mean very much. The mode is to get \$1 (and lose the entrance fee).

Section 7.9

1. **a.** Let M represent the mass. Then,
   $$\Pr(M \leq 0.4) = \Pr\left(\frac{M - 0.38}{0.09} \leq \frac{0.4 - 0.38}{0.09}\right)$$
   $$= \Pr\left(Z \leq \frac{0.4 - 0.38}{0.09}\right)$$
   $$= \Pr(Z \leq 0.222) = \Phi(0.222) = 0.588$$
   b. 0.933 **c.** 0.866 **d.** 0.516

3. **a.** Let I represent the number of insects caught in an hour. $\Lambda = 0.21 \cdot 60 = 12.6$. The mean is then 12.6 and the standard deviation is $\sqrt{12.6} = 3.55$. Let X be a normally distributed random variable with matching mean and standard deviation. Then,
   $$\Pr(10 \leq I \leq 15) \approx \Pr(9.5 \leq X \leq 15.5)$$
   $$= \Phi\left(\frac{9.5 - 12.6}{3.55}\right)$$
   $$- \Phi\left(\frac{15.5 - 12.6}{3.55}\right)$$
   $$= \Phi(0.816) - \Phi(-0.873) = 0.6020$$
   b. 0.8357 **c.** 0.1572 **d.** 0.0877

5. **a.** The scores have mean 12.55 and variance 154.7. Therefore, the variance of the average of 5 scores is $154.7/5 = 30.94$ and the standard deviation is $\sqrt{30.94} = 5.56$. Let X be a normally distributed random variable with mean 12.55 and standard deviation 5.56. Then,
   $$\Pr(10 \leq S \leq 15) \approx \Pr(10.0 \leq X \leq 15.0)$$
   $$= \Phi\left(\frac{10.0 - 12.55}{5.56}\right)$$
   $$- \Phi\left(\frac{15.0 - 12.55}{5.56}\right)$$
   $$= \Phi(0.441) - \Phi(-0.459) = 0.347$$
   b. The average of 10 scores has mean 12.55 and variance 15.47. The probability is 0.554.
   c. The average of 100 scores has mean 12.55 and variance 1.547. The probability is 0.984.
   d. The numbers get larger because the average of many scores is more and more likely to be near the true mean.
   e. The probabilities would be smaller if the scores were positively correlated, because a long string of good or bad scores would be very different from the average.
   f. The probabilities might be larger because good and bad scores would average out faster.

7. **a.** $\Pr(IQ \geq 100) = 1 - \Phi\left(\frac{100.0 - 98.75}{\sqrt{11.72}}\right)$
   $= 1 - \Phi(0.365) = 1 - 0.6425$
   $= 0.3575$

b. $\Pr(IQ \geq 100) = 1 - \Phi\left(\dfrac{100.0 - 89.0}{\sqrt{1.34}}\right)$
$= 1 - \Phi(9.502) \approx 0.0$

c. $\Pr(IQ \geq 100) = 1 - \Phi\left(\dfrac{100.0 - 93.5}{\sqrt{6.06}}\right)$
$= 1 - \Phi(2.644) = 1 - 0.9959$
$= 0.0041$

d. $\Pr(IQ \geq 100) = 1 - \Phi\left(\dfrac{100.0 - 121.24}{\sqrt{19.12}}\right)$
$= 1 - \Phi(-4.857) = 1 - 5.9 \times 10^{-6}$
$= 0.9999994$

9. For the expectation,
$$E(Z) = E\left(\dfrac{X-\mu}{\sigma}\right) = E\left(\dfrac{X}{\sigma}\right) - E\left(\dfrac{\mu}{\sigma}\right)$$
$$= \dfrac{E(X)}{\sigma} - \dfrac{\mu}{\sigma}$$
$$= \dfrac{\mu}{\sigma} - \dfrac{\mu}{\sigma} = 0$$

We used Theorem 7.6 in Section 7.3 and the fact that the expectation of a constant is the constant itself. For the variance,
$$\mathrm{Var}(Z) = \mathrm{Var}\left(\dfrac{X-\mu}{\sigma}\right) = \mathrm{Var}\left(\dfrac{X}{\sigma}\right) - \mathrm{Var}\left(\dfrac{\mu}{\sigma}\right)$$
$$= \dfrac{\mathrm{Var}(X)}{\sigma^2} - 0 = \dfrac{\sigma^2}{\sigma^2} = 1$$

We used Theorem 7.11 in Section 7.3 and the fact that the variance of a constant is 0.

Answers to Supplementary Problems for Chapter 7

1. **a.** If she asks about them independently.
   b. This is the geometric distribution, with value $\Pr(T = 4) = 0.4^3 \cdot 0.6 = 0.0384$.
   c. This is the binomial distribution, with value
   $$\Pr(N = 4) = b(4; 6, 0.6) = \binom{6}{4} 0.6^4 \cdot 0.4^2 = 0.138.$$
   d. The expectation is $np = 4.8$ and the variance is $np(1-p) = 1.92$.

3. **a.** Using the Poisson distribution, $E(N) = \lambda t = 12.5 \cdot 0.5 = 6.25$.
   b. $p(5; 6.25) = 6.25^5 e^{-6.25}/5! = 0.153$.
   c. In this case $\Lambda = 12.5 \cdot 0.167 = 2.08$. The probability of zero drops is $e^{-2.08} = 0.124$.
   d. The expected number in an hour is $12.5 \cdot 60 = 750$. The variance is then 750 and the standard deviation is $\sqrt{750} = 27.4$. The value 700 is within two standard deviations of 750 and would not be too surprising.

5. **a.** They must play independently.
   b. Expected number of As is 3, the variance is 0.75.
   c. $b(2; 4, 0.75) = \binom{4}{2} 0.75^2 0.25^2 = 0.21$
   d. They could either all play A or all play D, so the probability is
   $$b(0; 4, 0.75) + b(4; 4, 0.75) = 0.75^4 + 0.25^4 = 0.32$$
   e. This is $b(2; 3, 0.32) = \binom{3}{2} 0.32^2 0.68 = 0.21$

7. **a.** The volume is reduced to 80 with probability 0.3, to 90 with probability 0.2 and increased to 120 with probability 0.5. The expectation is $80 \cdot 0.3 + 90 \cdot 0.2 + 120 \cdot 0.5 = 102$.
   b. Because this is a multiplicative process, we must use the geometric mean. Let V be the random variable representing the change in volume (0.8 with probability 0.2, 0.9 with probability 0.3 and 1.2 with probability 0.5). Then the geometric mean is
   $$e^{E(\ln V)} = e^{\ln(0.8) \cdot 0.2 + \ln(0.9) \cdot 0.3 + \ln(1.2) \cdot 0.5}$$
   $$= e^{-0.039} = 0.96$$
   Because this is less than 1, the band will, on average, become quieter by about 4% per song.

9. **a.** Let D be the event "delivered," B be the event "ordered from Baxter," F be the event "ordered from Fisher," Σ be the event "ordered from Sigma." By the law of total probability,
   $$\Pr(D) = \Pr(D \mid B)\Pr(B) + \Pr(D \mid F)\Pr(F)$$
   $$+ \Pr(D \mid \Sigma)\Pr(\Sigma)$$
   $$= 0.9 \cdot 0.2 + 0.95 \cdot 0.7 + 0.8 \cdot 0.1 = 0.939$$
   The average time $\bar{T}$ (conditional on delivery) is
   $$\bar{T} = \Pr(B \mid D) \cdot 4 + \Pr(F \mid D) \cdot 6 + \Pr(\Sigma \mid D) \cdot 3$$
   $$= \dfrac{\Pr(D \mid B)\Pr(B) \cdot 4 + \Pr(D \mid F)\Pr(F) \cdot 6 + \Pr(D \mid \Sigma)\Pr(\Sigma) \cdot 3}{\Pr(D)}$$
   $$= 5.27$$
   b. By Bayes' theorem,
   $$\Pr(F \mid D) = \dfrac{\Pr(D \mid F)\Pr(F)}{\Pr(D)} = \dfrac{0.95 \cdot 0.7}{0.939} = 0.708$$
   c. Let N be the event "not delivered."

	B	F	Σ
D	0.18	0.665	0.08
N	0.02	0.035	0.02

   d. Ordering everything from Fisher. However, this would result in slower delivery.

11. **a.** The attempts must be independent. $p = 0.2$ and $n = n$.
    b. $E(S_{10}) = np = 2$. $\Pr(S_{10} = 2) = b(2 : 10, 0.2) = 45 \cdot 0.2^2 \cdot 0.8^8 = 0.302$.

13. a. About 0.0012.
  b. The area is 28.27 cm^2, so B has a Poisson distribution with $\Lambda = 0.12 \cdot 28.27 = 3.39$. Both expectation and variance are 3.39.
  c. $e^{-3.39} = 0.034$.

15. a. $\Pr(X \leq 11) = \Pr\left(\dfrac{X - 10}{3} \leq \dfrac{11 - 10}{3}\right)$
$= \Pr\left(Z \leq \dfrac{11 - 10}{3}\right)$
$= \Pr(Z \leq 0.33) = \Phi(0.33) = 0.629$
  b. $E(X + Y) = E(X) + E(Y) = 30$ and $\text{Var}(X + Y) = \text{Var}(X) + \text{Var}(Y) = 13$.
  c. The standard deviation is $\sqrt{13} = 3.605$.
$\Pr(X + Y \leq 13) = \Pr\left(\dfrac{X + Y - 30}{3.605} \leq \dfrac{13 - 30}{3.605}\right)$
$= \Pr\left(Z \leq \dfrac{13 - 30}{3.605}\right)$
$= \Pr(Z \leq -4.715) = \Phi(-4.715)$
This is not on the chart, but the computer says 1.2×10^{-6}.

17. a. Only 0.00069.
  b. Expect only 0.021 students, or one every 50 times the experiment is run. The variance is also 0.021, meaning that this is pretty uncertain.
  c. Using the binomial distribution, this is 0.192.
  d. The normal approximation to the binomial has mean 24.9 and variance 4.23. We need to compute the probability that this exceeds 24.5, or that $Z > -0.1944$, which is 0.5771.

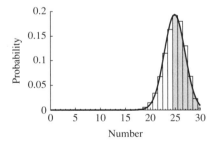

19. a.

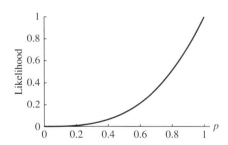

  b. The Poisson model would have 150 elk, with variance 50. The second model would have 148 elk with no variance.
  c. In the long run, the second model would be faster because it grows exponentially rather than linerly.
  d. The Poisson model deals better with randomness, but the actual data seems to increase faster as time goes along, more consistent with the second model.
  e. I would try to include both stochastic immigration and stochastic reproduction in each year.

Chapter 8

Section 8.1

1. a. This is terrible, because it samples mainly the winter (or the summer in Australia).
  b. This is terrible, because it samples mainly the summer (or the winter in Australia).
  c. This is terrible, because March 15 is not representative of the whole year.
  d. For convenience, sample every Tuesday and Friday for 1 yr (this would require 104 samples).

3. a. $L(p) = b(2; 4, p) = 6p^2(1 - p)^2$. For a fair coin, $p = 0.5$ and $L(0.5) = 0.375$.
  b. $L(p) = b(2; 4, p) = 6p^2(1 - p)^2$. For a fair die, $p = 0.167$ and $L(0.167) = 0.116$. It is less likely that a die will give the expected result for a coin.
  c. $L(p) = b(2; 12, p) = 66p^2(1 - p)^{10}$. For a fair coin, $p = 0.5$ and $L(0.5) = 0.016$.
  d. $L(p) = b(2; 12, p) = 66p^2(1 - p)^{10}$. For a fair die, $p = 0.167$ and $L(0.167) = 0.296$. It is more likely that a coin will give the expected result for a die.
  e. If the coin and the die are independent, the likelihood is the product, or $0.375 \cdot 0.296 = 0.111$.

5. a. $L(p) = p^3$

  b. The maximum is at $p = 1$.
  c. You are certain to get three out of three if everybody has the gene. However, it is perfectly possible to get three out of three if 95% of people have the gene.

7. a. $L(\lambda) = \lambda^{10} e^{-8.12\lambda}$
  b. The maximum is at $\lambda = 0.812$. The likelihood is 5.69×10^{-6}, which looks pretty small.
  c. The likelihood of $\lambda = 1.0$ is 4.51×10^{-6}, which is pretty close. This seems perfectly plausible.

9. **a.** $L(\Lambda) = \dfrac{\Lambda^{14} e^{-\Lambda}}{14!}$.

   b. Taking the derivative, we get
   $$L'(\Lambda) = \dfrac{14\Lambda^{13} e^{-\Lambda} - \Lambda^{14} e^{-\Lambda}}{14!}$$
   This is 0 when $\Lambda = 0$ or $\Lambda = 14$. The maximum is at $\Lambda = 14$, which is reassuring.

Section 8.2

1. **a.** The probability of 1 or more out of two is equal to $2p(1 - p) + p^2$. The lower confidence limit satisfies
   $$2p(1 - p) + p^2 = 0.025$$
   which has solution $p_l = 0.0126$ (using the quadratic formula). Similarly,

   The probability of 1 or fewer out of two
   $$= 2p(1 - p) + (1 - p)^2$$

   The upper confidence limit satisfies
   $$2p(1 - p) + (1 - p)^2 = 0.025$$
   which has solution $p_h = 0.987$.

   b. To find the 99% confidence limits, solve
   $$2p(1 - p) + p^2 = 0.005$$
   for $p_l = 0.0025$ and
   $$2p(1 - p) + (1 - p)^2 = 0.005$$
   for $p_h = 0.9975$.

   c. To find p_l, I would try the experiment 1000 times with small values of p, seeking one so small that a result of one or more out of two occurred only ten times. To find p_h, I would try the experiment 10 times with large values of p, seeking one so large that a result of 1 or fewer out of 2 occurred only ten times.

3. **a.** The probability of three or more out of three equals p^3. The lower confidence limit solves $p^3 = 0.025$, which has solution $p_l = 0.292$. Similarly,

   The probability of three or fewer out of three $= 1$

   This is always true, so the upper confidence limit is $p_h = 1$.

   b. To find the 99% confidence limits, solve
   $$p^3 = 0.005$$
   for $p_l = 0.171$. Again, the upper confidence limit is 1.

   c. The maximum likelihood estimator is already the largest possible value.

5. **a.** Not until the 11th try.

   b. The probability of five or six out of six is
   $$b(5; 6, 0.5) + b(6; 6, 0.5) = 6 \cdot 0.5^6 + 1 \cdot 0.5^6 = 0.109$$
   According to the geometric distribution with $q = 0.109$, the average wait is $1/q = 9.14$, or about nine sets of six flips.

7. **a.** The confidence limits Λ_l and Λ_h satisfy
   $$\Pr(14 \text{ or more}) = \sum_{k=14}^{\infty} p(k; \Lambda_l) = 0.025$$
   $$\Pr(14 \text{ or fewer}) = \sum_{k=0}^{14} p(k; \Lambda_h) = 0.025$$
   By substituting various values into the computer, we find $\Lambda_l = 7.65$ and $\Lambda_h = 23.5$.

   b. Remember the link between the Poisson and the binomial distributions. We could pick different tiny values of p (in the ball park of 14 out of a million or 0.000014), and simulate DNA until we found a value $p_l < 0.000014$ such that 14 or more mutations occurred with probability 0.025 and a value $p_h > 0.000014$ such that 14 or fewer mutations occurred with probability 0.025.

   c. The likelihood and support are
   $$L(\Lambda) = \dfrac{e^{-\Lambda} \Lambda^{14}}{14!}$$
   $$S(\Lambda) = \ln\left(\dfrac{e^{-\Lambda} \Lambda^{14}}{14!}\right)$$
   $$= -\Lambda + 14\ln(\Lambda) - \ln(14!)$$
   We need to solve for where $S(14) - S(\Lambda) = 2$ or
   $$2 = -14 + 14\ln(14) + \Lambda - 14\ln(\Lambda)$$
   By substituting in values for Λ, using Newton's method, or asking a friendly computer, we find $\Lambda = 7.79$ and $\Lambda = 22.9$.

9. With experimentation, we plug various values into the equation
   $$f(p) = 20\ln(0.2) + 80\ln(0.8) - 20\ln(p) + 80\ln(1 - p)$$
   looking for a value p where $f(p) = 2$. We find that $f(0.10) = 4.44$, $f(0.12) = 2.59$, $f(0.13) = 1.90$. The right answer is pretty close to 0.13. Also $f(0.30) = 2.57$, $f(0.28) = 1.70$ and $f(0.29) = 2.11$. The right answer is pretty close to 0.29. We can also use Newton's method to solve $f(p) = 2$. Define the function $g(p) = f(p) - 2$. We want to find where $g(p) = 0$. Following the method in Section 3.8 on Newton's method, our updating function is
   $$p_{t+1} = p_t - \dfrac{g(p_t)}{g'(p_t)}$$
   Starting from a guess of $p_t = 0.15$, we get subsequent estimates of 0.122, 0.128, 0.128. It has converged. Starting from a guess of 0.25, we get estimates 0.299, 0.288, 0.287, 0.287.

Section 8.3

1. a. For W, $\bar{W} = 67.43$, $\tilde{W} = 43.8$, $\bar{W}_{tr(5)} = 57.92$,
$\bar{W}_{tr(10)} = 51.17$, $\bar{W}_{tr(20)} = 42.57$, $s^2 = 5726.6$, $s = 75.7$.
b. For H, $\bar{H} = 7.39$, $\tilde{H} = 7.05$, $\bar{H}_{tr(5)} = 7.14$,
$\bar{H}_{tr(10)} = 6.98$, $\bar{H}_{tr(20)} = 6.93$, $s^2 = 12.88$, $s = 3.59$.
c. For Y, $\bar{Y} = 0.442$, $\tilde{Y} = 0.304$, $\bar{Y}_{tr(5)} = 0.392$,
$\bar{Y}_{tr(10)} = 0.369$, $\bar{Y}_{tr(20)} = 0.360$, $s^2 = 0.152$, $s = 0.390$.

3. a. Lower limit is $0.4 - 1.96 \cdot \sqrt{0.0083} = 0.221$. Upper limit is $0.4 + 1.96 \cdot \sqrt{0.0083} = 0.579$. Yes, these do include the true mean of 0.5.
b. Lower limit is $0.6 - 1.96 \cdot \sqrt{0.0083} = 0.421$. Upper limit is $0.6 + 1.96 \cdot \sqrt{0.0083} = 0.779$. These include the true mean of 0.5.
c. $\mu_l = 23.4 - 1.96 \cdot \sqrt{0.4} = 22.2$. $\mu_h = 23.4 + 1.96 \cdot \sqrt{0.4} = 24.6$. These do not include the true mean of 25.0.
d. $\mu_l = 0.7 - 1.96 \cdot \sqrt{0.0083} = 0.521$. $\mu_h = 0.7 + 1.96 \cdot \sqrt{0.0083} = 0.879$. These do not include the true mean of 0.5.

5. a. $\hat{p} = 0.44$ and $s^2 = 0.44 \cdot 0.56 = 0.246$. Then,

$$p_l = 0.44 - 1.96 \frac{\sqrt{0.246}}{\sqrt{100}} = 0.44 - 0.097 = 0.537$$

$$p_h = 0.44 + 1.96 \frac{\sqrt{0.246}}{\sqrt{100}} = 0.44 + 0.097 = 0.343$$

b. $\hat{p} = 0.32$ and $s^2 = 0.32 \cdot 0.68 = 0.218$. Then,

$$p_l = 0.32 - 1.96 \frac{\sqrt{0.218}}{\sqrt{1000}} = 0.32 - 0.029 = 0.291$$

$$p_h = 0.32 + 1.96 \frac{\sqrt{0.218}}{\sqrt{1000}} = 0.32 + 0.029 = 0.349$$

This is where those $\pm 3\%$ come from when poll results are reported.
c. $\hat{p} = 0.2$ and $s^2 = 0.2 \cdot 0.8 = 0.16$. Then,

$$p_l = 0.2 - 1.96 \frac{\sqrt{0.16}}{\sqrt{30}} = 0.2 - 0.14 = 0.06$$

$$p_h = 0.2 + 1.96 \frac{\sqrt{0.16}}{\sqrt{30}} = 0.2 + 0.14 = 0.34$$

7. a. The number per million has a Poisson distribution, so the standard deviation should be $\sqrt{13.5} = 3.67$.
b. The standard error is $\frac{3.67}{\sqrt{30}} = 0.67$.
c. The 99% confidence limits are $13.5 - 2.576 \cdot 0.67 = 10.77$ and $13.5 + 2.576 \cdot 0.67 = 16.23$.
d. More measurements give more confidence.

9. a. The 95% confidence limits are 1.96 standard errors above and below the sample mean, for a total width of 3.92 standard errors. Therefore, we require that standard error =

$\frac{2.0}{3.92} = 0.51$. But

$$\text{standard error} = \frac{3.2}{\sqrt{n}}$$

so $n = (3.2/0.51)^2 = 39$, and 39 plants are required.
b. Improving by a factor of 4 takes 16 times more plants, or about 630.
c. 68 plants
d. 1087 plants

11. a. The variance of each measurement is 0.25.
b. There are four possibilities: $(0, 0), (0, 1), (1, 0), (1, 1)$. Each has probability 0.25.
c. The variance of the first and last is 0. The mean squared deviation from the mean (0.5) of the others is

$$\frac{(0 - 0.5)^2 + (1 - 0.5)^2}{2} = 0.25$$

The average is $0 \cdot 0.25 + 0.25 \cdot 0.25 + 0.25 \cdot 0.25 + 0 \cdot 0.25 = 0.125$.
d. This is half what it should be. If we had used a denominator of 1 rather than 2 in part **c**, our answer would have been right.
e. The first and last possible samples miss the variance in the population. Samples tend not to capture all the variability in the full distribution.

Section 8.4

1. If any abnormal cell triggers a positive response, the test will catch all cancers but make many type I errors when cells look a bit odd for other reasons. If only grossly abnormal cells triggered a positive response, the test would give very few false positives, but would miss many less obvious cancers. Deciding where to set the threshold would depend on the risk from that particular cancer, the cost and difficulty of follow-up testing, and the abundance of the cancer.

3. a. Null: $\lambda = 3.5$, alternative: $\lambda > 3.5$
b. $p(7; 3.5) = 0.0385$
c. $\sum_{k=7}^{\infty} p(k; 3.5) = 0.065$
d. The result is not quite significant.

5. a. The null hypothesis is that you bulb was ordinary but you got lucky ($\lambda = 0.001$). The alternative is that the bulb was improved ($\lambda < 0.001$).
b. Pr(bulb lasts more than 4000 hours) = $e^{-0.001 \cdot 4000} = 0.0183$. This is the significance level, and the result looks significant.
c. Where

$$\Pr(\text{bulb lasts more than } t \text{ hours}) = e^{-0.001t} = 0.05$$

which has solution $t = 2996$.

7. a. Null is $\lambda = 0.001$, and the alternative is $\lambda > 0.001$.
b. Pr(bulb lasts less than 40 h) = $1 - e^{-0.001 \cdot 40} = 0.039$.

c. This seems odd because the probability that the bulb blows out after 40 h is greater than the probability of the bulb blowing out after 1000 h even if the null hypothesis is true. This occurs because the mode of the exponential distribution is nowhere near the mean.

d. Not much.

9. **a.** $S(p) = \ln[b(9; 10, p)]$. The alternative has maximum at $p = 0.9$, so the difference is $S(0.9) - S(0.5) = 3.68$.

b. Support for a particular λ is

$$-\lambda + 7\ln(\lambda) - \ln(7!)$$

Ignoring the $\ln(k!)$ term, which will cancel when we subtract, the support for the null is $-3.5 + 7\ln(3.5) = 5.269$. The maximum is at $\lambda = 7$, with support $-7 + 7\ln(7.0) = 6.62$. The difference is 1.35. The null looks okay.

c. $S(\lambda) = \ln(\lambda e^{-4000\lambda}) = \ln(\lambda) - 4000\lambda$

The maximum likelihood hypothesis is $\lambda = 0.00025$, with a null of $\lambda = 0.001$. Then $S(0.00025) - S(0.001) = 1.61$.

d. $S(\lambda) = \ln(\lambda e^{-40\lambda}) = \ln(\lambda) - 40\lambda$
The maximum likelihood hypothesis is $\lambda = 0.025$, with a null of $\lambda = 0.001$. Then $S(0.025) - S(0.001) = 2.26$.

e. Support for a particular q is

$$\ln(q) + (25 - 1)\ln(1 - q)$$

At $q = 0.1$, we get $\ln(0.1) + (25 - 1)\ln(1 - 0.1) = -4.831$. The maximum occurs where

$$\frac{1}{q} - \frac{24}{1-q} = 0$$

which has solution $q = 0.025$. The support is $\ln(0.025) + (25 - 1)\ln(1 - 0.025) = -4.297$. The difference is only 0.534.

Section 8.5

1. **a.** Let ω represent the mean weight of experimental plants. The one-tailed alternative hypothesis is $\omega > 10.0$. The two-tailed alternative hypothesis is $\omega \neq 10.0$. Let W be a random variable representing the mean of 10 plants each with mean size 10.0 and variance 9.0 (the null hypothesis). The standard error of W is $\sqrt{(9.0)}/\sqrt{10} = 0.949$. The sample mean of the numbers in the table is 10.99. For the one-tailed test,

$$\Pr(W > 10.99) = \Pr\left(\frac{W - 10.00}{0.949} > \frac{10.99 - 10.0}{0.949}\right)$$
$$= \Pr(Z > 1.044)$$
$$= 1 - \Phi(1.044) = 0.197$$

This is far from significant. For the two-tailed test,

$$\Pr(W > 10.99 \text{ or } W < 9.01)$$

$$= \Pr\left(\frac{W - 10.00}{0.949} > \frac{10.99 - 10.0}{0.949}\right)$$

$$+ \Pr\left(\frac{W - 10.00}{0.949} < \frac{9.01 - 10.0}{0.949}\right)$$
$$= \Pr(Z < 1.044) + \Pr(Z < -1.044)$$
$$= 1 - \Phi(1.044) + \Phi(-1.044) = 0.394$$

This is even worse. We have no reason to think our treatment does anything.

b. The sample mean is 40.26. The standard error is $4.0/\sqrt{10} = 1.265$. The difference is $2.26/1.265 = 1.787$ standard errors above the mean. With a one-tailed test, the significance is 0.037. With a two-tailed test, it is 0.074.

c. The sample mean is 10.26. The standard error is $2.5/\sqrt{10} = 0.791$. The difference is $1.26/0.791 = 1.594$ standard errors above the mean. With a one-tailed test, the significance is 0.055 which is not quite significant. With a two-tailed test, it is 0.110.

3. **a.** We need the probability that the mean weight of 10 plants exceeds the critical value of 12.21 if the true mean is 13.0. Let X represent the mean weight of ten plants with true mean 10.0. The standard error is still 0.949. So

$$\Pr(X > 12.21) = \Pr\left(\frac{X - 13.0}{0.949} > \frac{12.21 - 13.0}{0.949}\right)$$
$$= \Pr(Z > -0.833)$$
$$= 1 - \Phi(-0.833) = 0.798$$

We have nearly an 80% chance of finding a difference of 3.0 in the mean height with a one-tailed test at the 0.01 significance level.

b. We need the probability that the mean height is between 35.52 and 40.48. This is 0.023, for a power of 0.977.

c. We need the probability that the mean yield is between 11.60 and 6.4. This is 0.038, for a power of 0.962.

5. **a.** The significance is 0.0017, which is highly significant.

b. The probability of getting a sample mean of 40 or more with the null hypothesis is tiny. Suppose that the true mean is 40.0. The test fails not only when the sample mean is less than 39, but when it is less than the threshold 39.56, which is much more probable.

c. This will match the probability of a type II error if the true mean is 40.0, or a significance level of 0.1.

7. **a.** $X \sim N(35.0, 35.0)$.

b. $\Pr(X \leq 27.5) = 0.102$. Not significant.

c. It would have to be at least 2.326 standard deviations below 35, or $35 - 2.326\sqrt{35} = 21.2$. There would have to be no more than 21 mutations.

9. **a.** The difference in likelihood is 1.044 (the number of standard errors separating the two hypotheses). The data provides no reason to prefer one over the other.

b. The difference in likelihood is 1.787.

c. The difference in likelihood is 1.594.

Section 8.6

1. a. The distribution of the sample means from the first sample is $N(\mu_1, 1.0)$ and for the second is $N(\mu_2, 1.0)$, because the sample error is $\sigma^2/n = 1.0$ in both cases. The distribution of the difference D is then $D \sim N(\mu_1 - \mu_2, 2.0)$. We want to find the probability that $D \geq 5.0$ or $D \leq -5.0$. Transforming into a standard normal distribution, we get

$\Pr(D \geq 5.0 \text{ or } D \leq -5.0)$

$= \Pr\left(\dfrac{D}{\sqrt{2.0}} \geq \dfrac{5.0}{\sqrt{2.0}} \text{ or } \dfrac{D}{\sqrt{2.0}} \leq \dfrac{-5.0}{\sqrt{2.0}}\right)$

$= \Pr(Z \geq 3.53) + \Pr(Z \leq -3.53)$

$= 1 - \Phi(3.53) + \Phi(-3.53) = 0.0046$

The difference is very highly significant.
b. 0.414   **c.** 0.149

3. a. The distribution of the sample proportion from the first sample is $N[p_1, p_1(1 - p_1)/n_1]$ and for the second is $N[p_2, p_2(1 - p_2)/n_2]$. Using $\hat{p}_1 = 0.7$, $\hat{p}_2 = 0.8$ and $n_1 = n_2 = 50$, the distributions of the sample proportions are $N(p_1, 0.0042)$ and $N(p_2, 0.0032)$. The distribution of the difference D is then $D \sim N(p_1 - p_2, 0.0074)$. We want to find the probability that $D \geq 0.1$ or $D \leq -0.1$. Transforming into a standard normal distribution, we get

$\Pr(D \geq 0.1 \text{ or } D \leq -0.1)$

$= \Pr\left(\dfrac{D}{\sqrt{0.0074}} \geq \dfrac{0.1}{\sqrt{0.0074}}\right.$

$\left. \text{ or } \dfrac{D}{\sqrt{0.0074}} \leq \dfrac{-0.1}{\sqrt{0.0074}}\right)$

$= \Pr(Z \geq 1.16) + \Pr(Z \leq -1.16)$

$= 1 - \Phi(1.16) + \Phi(-1.16) = 0.245$

The difference is not even close to significant.
b. 0.00024   **c.** 0.137   **d.** 0.548

5. If the mutation rates are Λ_1 and Λ_2, the numbers M_1 and M_2 are approximately

$M_1 \sim N(\Lambda_1, \Lambda_1), \quad M_2 \sim N(\Lambda_2, \Lambda_2)$

The difference D is then

$D \sim N(\Lambda_1 - \Lambda_2, \Lambda_1 + \Lambda_2)$

The null hypothesis has $\Lambda_1 = \Lambda_2$. For the sum, we have no better guess than the sum of the estimators $\hat{\Lambda}_1 + \hat{\Lambda}_2 = 8 + 18 = 26$, for a null hypothesis of

$D \sim N(0, 26)$

The measured difference is 10. With a two-tailed test, the significance level is 0.0498. As found with the method of support, this result is right on the borderline of being significant.

7. a. The null is that the chemical has no effect, with the alternative that the chemical causes cancer.
b. The pooled proportion is $\hat{p} = \frac{6}{14} = 0.428$. The support for the null is

$S_n(0.428) = \ln[b(5; 8, 0.428)] + \ln[b(1; 6, 0.428)]$
$= -3.74$

c. The proportion in the treated group is 0.625, and in the control is 0.167. Therefore,

$S_a(0.625, 0.167) = \ln[b(5; 8, 0.625)] + \ln[b(1; 6, 0.167)]$
$= -2.18$

d. The difference is only 1.56, not enough to provide much evidence that the chemical causes cancer.

9. a. The null is that the two leave at the same rate λ, with an alternative that the rates of different.
b. The average leaving rate is $1/17.5 = 0.057$ (see Section 8.1). The support is

$S_n(\lambda) = \ln(\lambda e^{-30\lambda}) + \ln(\lambda e^{-5\lambda})$

With $\lambda = 0.057$, this is -7.72.
c. The best guess for λ_1 is $1/30 = 0.0333$, and and the best guess for λ_2 is $1/5 = 0.2$. The support is

$S_a(\lambda_1, \lambda_2) = \ln(\lambda_1 e^{-30\lambda_1}) + \ln(\lambda_2 e^{-5\lambda_2})$

With the maximum likelihood estimators of λ_1 and λ_2, this is -7.01.
d. The difference is far too small to be convincing.

Section 8.7

1. a.

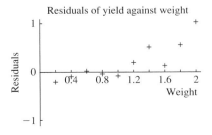

b. For weight and yield, the residuals are

w	y	$\hat{y}$	$y - \hat{y}$
0.2	0.599	0.8	−0.201
0.4	0.909	1.0	−0.091
0.6	1.220	1.2	0.020
0.8	1.363	1.4	−0.037
1.0	1.523	1.6	−0.077
1.2	1.995	1.8	0.195
1.4	2.518	2.0	0.518
1.6	2.330	2.2	0.130
1.8	2.963	2.4	0.563
2.0	3.628	2.6	1.028

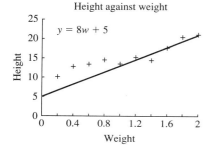

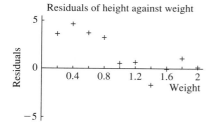

c. For weight and height, the residuals are

w	h	$\hat{h}$	$h - \hat{h}$
0.20	10.17	6.6	3.57
0.40	12.83	8.2	4.63
0.60	13.47	9.8	3.67
0.80	14.60	11.4	3.20
1.0	13.55	13.0	0.55
1.2	15.25	14.6	0.65
1.4	14.47	16.2	−1.73
1.6	17.74	17.8	−0.06
1.8	20.46	19.4	1.06
2.0	21.12	21.0	0.12

3. a. The null model uses the mean yield, or $Y = \bar{Y} = 1.905$.

b. The sum of the squares of the residuals is SST = 8.259.

c. $r^2 = 1 - \dfrac{\text{SSE}}{\text{SST}} = 0.788$.

d. The null model uses the mean height or $H = \bar{H} = 15.37$.

e. The sum of the squares of the residuals is 106.4.

f. $r^2 = 0.41$.

5. To compute the correlation, we need to find the variance of Y. Using the formula for the uncorrected variance, we find $\widehat{\text{Var}}(Y) = 0.417$. The correlation is then

$$\rho_{X,Y} = \dfrac{\widehat{\text{Cov}}(X,Y)}{\sqrt{\widehat{\text{Var}}(X)}\sqrt{\widehat{\text{Var}}(Y)}}$$

$$= \dfrac{0.144}{\sqrt{0.0825} \cdot \sqrt{0.417}} = 0.776$$

The correlation coefficient seems a bit larger. I would trust r^2 because it is more clear which two models are being compared.

7. a. The best fit line is found by computing

$$\widehat{\text{Cov}}(W, H) = 1.746$$
$$\widehat{\text{Var}}(W) = 0.330$$
$$\bar{W} = 1.10$$
$$\bar{H} = 15.34$$

From Theorem 8.5, we have

$$\hat{a} = \dfrac{1.746}{0.330} = 5.29$$
$$\hat{b} = 15.34 - 5.29 \cdot 1.10 = 9.55$$

The line is then

$$H = 5.29W + 9.55$$

b.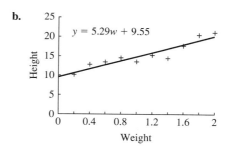

c. Using Theorem 8.6, we get SSE = 14.01.

d. $r^2 = 1 - \dfrac{14.01}{106.4} = 0.87$, again larger than before.

9. The units of x and y are mass and toxin tolerance. $\hat{a}$ and $\hat{b}$ have dimensions

$$\hat{a} = \frac{\widehat{\text{Cov}}(X,Y)}{\widehat{\text{Var}}(X)}$$

$$= \frac{\text{mass} \cdot \text{tolerance}}{\text{mass}^2}$$

$$= \frac{\text{tolerance}}{\text{mass}}$$

$$\hat{b} = \bar{Y} - \hat{a}\bar{X}$$

$$= \text{tolerance} - \frac{\text{tolerance}}{\text{mass}} \text{mass}$$

$$= \text{tolerance}$$

The regression equation has dimensions

$$\hat{y} = \hat{a}x + b$$

$$\text{tolerance} = \frac{\text{tolerance}}{\text{mass}} \text{mass} + \text{tolerance} = \text{tolerance}$$

These check.

Answers to Supplementary Problems for Chapter 8

1. **a.** The number that are producing has a binomial distribution with $n = 100$ and $p = 0.2$, so the mean is 20 and the variance is 16. The number that are absorbing has a binomial distribution with $n = 100$ and $p = 0.8$, so the mean is 80 and the variance is 16.

   b. Let P be a random variable giving the number that are producing. The number absorbing is $100 - P$. The rate R is

   $$R = 2.0P - 4.0(100 - P) = 6.0P - 400$$

   Then, $E(R) = 6.0 E(P) - 400 = -280$ and $\text{Var}(R) = 6.0^2 \text{Var}(P) = 576$.

   c. It is decreasing if $R < 0$ or if $6.0P - 400 < 0$ or if $P < 67$. I would evaluate it by finding the cumulative probability $B(66; 100, 0.2)$.

   d. Either all are producing (with probability 0.2) or all are absorbing (with probability 0.8). In the first case, $R = 200$, and in the second $R = -400$. Then

   $$E(R) = 0.2 \cdot 200 + 0.8 \cdot (-400) = 280$$

   as before. But $\text{Var}(R) = 57{,}600$, one hundred times larger. The probability that the chemical is decreasing is 0.8.

3. Let P_1 and P_2 be random variables representing the fraction of shots hit by each team. The variance is approximately $0.3 \cdot 0.7/100 = 0.0021$ for the first and $0.4 \cdot 0.6/100 = 0.0024$ for the second. Then P_1 has approximately an $N(p_1, 0.0021)$ distribution and P_2 has distribution approximately $N(p_2, 0.0024)$. Therefore $D = P_1 - P_2$ has distribution $N(p_1 - p_2, 0.0045)$. Under the null hypothesis, $p_1 = p_2$, giving D distribution $N(0, 0.0045)$. We use a one-tailed test because we suspected that the second team was better.

   $$\Pr(D \geq 0.1) = \Pr\left(\frac{D}{\sqrt{0.0045}} \geq \frac{0.1}{\sqrt{0.0045}}\right)$$

   $$= \Pr(Z \geq 1.49) = 0.068$$

   This is not a significant result, meaning that this could well have happened by chance. Nonetheless, it would be heavily interpreted by sports pundits.

5. **a.** Let $P(t)$ be the probability unbound at time t. Then, $P(0) = 1$ and

   $$\frac{dP}{dt} = -\lambda P$$

   which has solution $P(t) = e^{-\lambda t}$.

   b. The p.d.f. is $\lambda e^{-\lambda t}$. Then

   $$L(\lambda) = \Pr(\text{data} \mid \lambda)$$
   $$= \lambda e^{-1.5\lambda} \lambda e^{-2.5\lambda}$$
   $$= \lambda^2 e^{-4.0\lambda}$$

   The maximum is at $\lambda = 0.5$.

7. **a.** $L(q) = (1-q)^{10} \left(\frac{2q}{3}\right)^6 \left(\frac{q}{3}\right)^4$

   $$= \frac{2^6}{3^{10}}(1-q)^{10} q^6$$

   This is maximized at $q = 0.5$.

   b. The likelihood function is

   $$L_2(q_1, q_2) = (1 - q_1 - q_2)^{10} q_1^6 q_2^4$$

   Taking logs, we find that the support $S_2(0.3, 0.2) = -20.59$. With the first model $S(0.5) = -20.69$. This is too small a difference to enable us to choose between the models.

9. **a.** The fraction has distribution

   $$N\left(\hat{p}, \frac{\hat{p}(1-\hat{p})}{100}\right) = N(0.75, 0.001875)$$

   where $\hat{p} = 0.75$. The 99% confidence limits are

   $$0.75 - 2.576\sqrt{0.001875} = 0.64$$
   $$0.75 + 2.576\sqrt{0.001875} = 0.86$$

   b. Under the null hypothesis, the fraction P has distribution

   $$N\left(0.65, \frac{0.65(1-0.65)}{100}\right) = N(0.65, 0.00227)$$

   Then,

   $$\Pr(P \geq 0.75) = \Pr\left(\frac{P - 0.65}{\sqrt{0.00227}} \geq \frac{0.75 - 0.65}{\sqrt{0.00227}}\right)$$

   $$= \Pr(Z \geq 2.096) = 0.018$$

   I used a one-tailed test because I have faith that new drugs are better, and found a significant result.

11. a. The number of trout in 5 m has a Poisson distribution with $\Lambda = 0.2 \cdot 5 = 1$. Let T represent the number of trout.

$$\Pr(T \geq 3) = 1 - \Pr(T \leq 2)$$
$$= 1 - p(0; 1) - p(1; 1) - p(2; 1) = 0.08$$

b. The expected number in 1000 m is 200, the coefficient of variation is then $\sqrt{200}/200 = 0.071$.

c. Multiply speed of stream by density of trout to get 0.4 trout per second.

13. The mean of the control is 8.95 and of the treatment is 12.38. The null hypothesis is that the treatment came from distribution $N(8.95, 16/\sqrt{10}) = N(8.95, 5.06)$. Using a two-tailed test,

$$\Pr(H \geq 12.38) + \Pr(H \leq 5.52)$$
$$= \Pr\left(\frac{H - 8.95}{\sqrt{5.06}} \geq \frac{3.43}{\sqrt{5.06}}\right)$$
$$+ \Pr\left(\frac{H - 8.95}{\sqrt{5.06}} \leq \frac{-3.43}{\sqrt{5.06}}\right)$$
$$= \Pr(Z \geq 1.52) + \Pr(Z \leq -1.52) = 0.128$$

Not significant. If I did not know the standard deviation, I would estimate it from the data, and then use a more advanced statistics book to figure out how to test the null hypothesis.

15. a.

[Plot of Number of survivors vs Dosage]

b.

D	S	$\hat{S}$	$\hat{S} - S$	$\bar{S} - S$
18.0	3.0	3.4	0.4	−1.0
15.0	4.0	4.0	0.0	0.0
9.0	5.0	5.2	0.2	1.0

Therefore, SST = 2, SSE = 0.2, $r^2 = 0.9$.

c. I have no idea what this statistician did other than incorrectly analyze an inadequate data set.

d. I'm not convinced it works. I would do more experiments.

17. a. The sample is the elk droppings in 1 km². The population is all droppings in the park.

b. We must assume that the density of droppings is the same throughout the park.

c. The lower confidence limit is the largest value of k, for which 36 or more elk droppings are found with probability less than 0.005. The upper confidence limit is the smallest value of k, for which 36 or fewer elk droppings are found with probability less than 0.005.

d. With a mean and variance of 36, the confidence limits are 20.54 to 51.45.

e. Too disgusting to think about. Figure the number of droppings per elk and how long they last.

19. a.

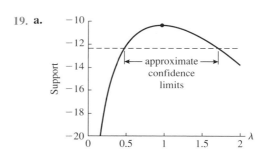

b. The first line gives a better fit.

c. SST = 758.8, and SSE = 26.0, so $r^2 = 0.966$.

d. The slope means that there is one elk for every two droppings. But I do not know what the 5 is for.

21. a. The probability of 13 or more is 0.0432. The value of $p = 0.2$ would lie just inside the 95% confidence limit (which allows for an error of 0.025).

b. The support for $p = 13/40$ is −25.22, and the support for $p = 0.2$ is −26.95. This is within 2, so $p = 0.2$ is reasonably well supported by the data.

c. It might be easier to capture infected elk. They need to try different trapping methods and compare the results.

Index

2:1 AV block, **109**

Absorption function(s), 258
  for breathing rate, 301
  leading behavior, 271
  limits at infinity, 264
Acceleration, **181**, 334–336
Action potential, **436**
Adaptation, 445
Addition of functions, **40**
  preserves continuity, 141
Additive genes, **464**
AIDS cases, 167, 337
Allele(s), **462**
Allometry, **61**–**67**, 115
  of surface area and volume, 66, 163, 394
Alternative hypothesis, **647**
Amplitude of an oscillation, **71**–**76**
Antiderivative, **330**
Applied mathematics, **2**
  variable names in, 45
Approximating functions, 280–287
  with leading behavior, 268
  with normal distribution, 594
  with quadratics, 284–286
  with secant line, 282
  with tangent line, 124, 281–284
  with Taylor polynomials, 287
Arbitrary constant of integration, **331**
Archimedes, 366, 385
Area under a curve, 366–369
  as a probability, 490
Argument (of function), **9**
Arithmetic mean, **501**, 513
Arithmetic-Geometric Inequality, 513
Associated function, **265**
Autocatalytic reaction, 425

Autonomous differential equation(s), **312**, 387
  equilibria, 388, 389
  Euler's method, 319
  separation of variables, 404
  systems of, 409
AV node, 106
Average
  of a function, **370**
  of an oscillation, **71**–**76**
Average rate of change, **120**
Avogradro's number, 454

Bacterial population growth, 6
  average rate of change, 119
  derivative of, 205
  differential equation, 314, 315
  equilibria of differential equation, 389
  equilibria of updating function, 90, 93
  input and output tolerances, 143
  measured continuously, 119
  phase-line diagram, 397
  solution of differential equation, 402–404
  solution of updating function, 32, 33, 50
  stability of differential equation, 397
  stability of updating function, 213, 218
  stochastic, 449
  updating function, 7, 49
  with negative per capita reproduction, 228, 229
Base point
  of line, **26**
  of secant line, **121**
Bayes' Theorem, **477**
Bell curve, 492
  and the normal distribution, 594
  percentiles, 523

  points of inflection, 522
  rules of thumb, 522, 594
Bernoulli random variable, **498**
  and binomial distribution, 558
  expectation and variance, 522
Bertalanffy growth equation, 346
Biased estimator, **620**
Bifurcation
  of autonomous differential equation, 402
  diagram, 402
  of discrete-time dynamical system, 223
Binomial coefficient, **566**
Binomial distribution, 465, 557–573
  application to genetics, 567
  coefficient of variation, 566, 568
  comparing population proportions, 666, 667
  conditions for, 558
  confidence limits, 645
  cumulative distribution, 570
  expectation, 558
  formula, 564
  mode, 569
  normal approximation, 607–609
  parameter estimation, 619
  relation to Poisson, 590
  variance, 559
Binomial theorem, **161**
Birth process, 616
Bisection, 299
Breathing rate, 299–305

Carbon-14, 58
Carrying capacity, 105, 393, 527
Cartesian coordinates, **10**
Central Limit Theorem
  for averages, 598–600
  for sums, 596, 597

Central Limit Theorem (*cont.*)
  used for computing confidence limits, 638, 644
Chain rule, **199**–202
  and integration by substitution, 341
  with more than two functions, 201
Chaos, 105, 231
Chebyshev's inequality, 525
Cherry, Josh, 446
Circadian rhythm, 77
Circle(s)
  area, 17, 21, 167, 385
  degrees and radians, 70
  perimeter, 17
Cobwebbing, 83–89
  of competition model, 100
  from an equilibrium, 87, 90
  of heart model, 108–110
  of lung model, 83
  and stability, 213–217
Coefficient of determination, **674**
Coefficient of variation, **523**
  of binomial distribution, 568
  of geometric distribution, 577
  of Poisson distribution, 587
Coin tossing, 450
  and binomial distribution, 558
  as a Markov chain, 484
Commuting functions, **42**
Competition differential equation, 326
  compared with updating function, 326
  equilibria, 390
  phase-line diagram, 391, 392
  separation of variables, 408
  solution, 327, 328, 390
  stability, 398
Competition model, 96–101
  derivative of updating function, 172
  equilibria, 100, 102
  general formulation, 99
  with mutation, 115
  stability, 214, 218, 219
  updating function, 98, 99
Competition system of differential equations, 411
  direction arrows, 431
  equilibria and nullclines, 420–422
  phase plane trajectory, 432
  solution, 432
Complement, **469**
Composition of functions, **42**
  derivative of, 198
  preserves continuity, 141
Concave down, **178**, 241
Concave up, **178**, 241
Conditional distribution, **535**

Conditional probability, **474**–478
  and Markov chains, 483
  rare disease, 477, 478
Confidence interval, 627, **630**
Confidence limits, 628–635
  estimated with method of support, 634
  exact, 629
  links with hypothesis testing, 650
  of sample mean, 638–642
Constant function, **133**
  continuity of, 140
  limit of, 134
Constant of proportionality, **23**
Constant product rule
  for derivatives, 164, 169
  for integrals, 332
Constant sum rule for derivatives, **159**, 160
Continuity correction, 608–611
Continuous function(s), **139**–144
  and the Intermediate Value Theorem, 249–253
  basic, 140, 141
  combining, 140, 141
Continuous random variable, **497**
  expectation, 502
  geometric mean, 513
  joint probability density function, 532
  median, 507
  probability density function, 499
  variance, 520
Convergence of improper integral, **376**
Conversion factor, **15**
  degrees to radians, 69
  Fahrenheit to Celsius, 27
Correlation, **542**–546
  contrasted with linear regression, 677
  perfect, 544–546
Cosine function, 69–76
  derivative of, 191–194, 206
  integral, 341
Coupled autonomous differential equations, 409–413
Covariance, 539–542
  computational formula, 539
Critical point(s), **150**, 238–241
Cumulative distribution, **487**
  of binomial distribution, 570
  of discrete random variable, 498
  of geometric distribution, 576
Cumulative distribution function, **493**, 494
  of continuous random variable, 500
  of exponential distribution, 579
  of standard normal, 603

Darts
  covariance of scores, 540
  distance from center as random variable, 499
  as metaphor for probability, 468
  score as random variable, 497
Darwin, Charles, 5, 465
Decreasing function, 30, 150
Definite integral, **353**
  as area under curve, 366–368
  and average value of function, 369, 370
  link with indefinite integral, 356–359
  and mass of a bar, 371
  summation property, 361, **362**
Delta (Δ) notation, **25**, 119
Density
  of a bar, 370, 371
  effect on flight, 66, 67
  of an object, 16–18
  of otters, 371, 372
  of probability, 490
  as slope of proportional relation, 25
Dependent variable, **9**
Derivative, **124**
  of basic power function, 161
  of composition, 199
  of cosine function, 206
  of exponential function, 184, 205
  first, 177
  of general power function, 162
  of inverse function, 203
  of linear function, 150
  of a product, 169
  of quadratic function, 152
  of a quotient, 172
  second, 177
  of a sum, 158
  of updating function, 216
Descriptive statistics, 448
  coefficient of variation, 523
  correlation, 542
  covariance, 539
  expectation, 501
  mean absolute deviation, 517
  median, 507
  mode, 510
  percentiles, 516
  range, 515
  standard deviation, 522
  variance, 519
Deterministic dynamical system, 5, 9, 447
Differentiable function, **148**
Differential equation(s), 6, 125–127, 311
  autonomous, 312
  competition, 325
  for exponential distribution, 578

falling rock, 153
Newton's law of cooling, 322
nonlinear, 327
polynomial, 334
population growth, 314
solution, 312, 327
stochastic diffusion, 456–458, 578
systems, 409
Differential notation, **124**
Diffusion, 4
across a membrane, 324, 325
assigning probabilities, 472
binomial distribution, 571, 572
Markov chain, 483
stochastic, 454–456
Dimensional analysis, **20**
Dimensions, **16**–22
Diploid, **462**
Direction arrows, 427–433
Discontinuous function(s), **139**
recognizing, 141
Discrete logistic model, 105, 224–226, 230–232, 308
equilibria, 224
stability, 225, 226, 231, 232
Discrete random variable, **497**
binomial, 557
correlation, 543
expectation, 501
geometric, 575
geometric mean, 513
independence, 534, 536, 537
joint probability distribution, 532
median, 508
variance, 520
Discrete-time dynamical system, **5**
bacterial population growth, 6, 49
cobwebbing, 83
compared with differential equation, 127
competition model, 96–101
equilibrium, 86
logistic, 105
lung model, 78–80
medication concentration, 31
mites on lizards, 28, 29
solution, 32
Disease model, 398
autonomous differential equation, 399
threshold theorem, 400
with recovery, 416
Disjoint sets, **469**
Distribution-free tests, 683
Divergence of improper integral, **376**
Do nothing updating function, 87
Domain, **9**

Dominant allele, **464**, 567
Doubling time, **55**
Dummy variable, **364**
Dynamical system, 2
chaotic, 105
continuous-time, 6
deterministic, 5, 9, 447
discrete-time, 5
nonlinear, 95
probabilistic, 6, 9, 447

e (the number), 52, **186**
Egg-laying strategy, 165
Endemic disease, 400
Equilibria
of autonomous differential equation, 389
of bacterial population growth differential equation, 389
of bacterial population growth model, 93
of competition differential equation, 390
of competition model, 100
of competition system of differential equations, 420
of discrete-time dynamical system, 86–95
and Intermediate Value Theorem, 250
and limits of sequences, 265
of lung model, 88, 90, 93
with negative slope, 226
of Newton's law of cooling, 390
with parameters, 92
stable, 102, 392
of system of autonomous differential equations, 417–420
unstable, 102, 392
Estimation, 20–22
Estimator, **620**
maximum likelihood, 622
of slope and intercept of linear regression, 677
Euler's method, 315–321
implicit, 384
in the phase plane, 425, 426
refinements, 384
systems of autonomous differential equations, 413
Event(s), **468**
assigning probabilities, 470
independent, 480
mutually exclusive, 469
mutually exclusive and collectively exhaustive, 476
simple, 468

Expectation, **501**–504
of binomial distribution, 558
constants come outside, 550
of continuous random variable, 503
of discrete random variable, 501
of exponential distribution, 579
of geometric distribution, 576
of Poisson distribution, 588
of a product, 551, 552
of a sum, 547
Exponential distribution, 578–581
memoryless property, 581
parameter estimation, 624
relation to Poisson, 585
Exponential function, 49, 52
to the base a, 51
to the base e, 52
derivative of, 184–186, 206
improper integrals, 378
integral of, 340
laws of, 52
limit at infinity, 261, 263
relation to allometry, 59, 60
tangent line approximation, 282
Taylor approximation, 287
Extreme Value Theorem, **252**

Factoids, 210
Factorial, **287**, 564
Stirling approximation, 566
Falling rock, 130, 151, 334–336
acceleration of, 182
differential equation, 153
drag, 394
and Fundamental Theorem of Calculus, 359, 360
and summation property of definite integral, 363
Fish length, 13
and Fundamental Theorem of Calculus, 360, 361
described by differential equation, 345–347
general solution, 364
Fisher, Sir Ronald, 5
FitzHugh-Nagumo equations, **439**–442
Flight, 65–67
Flow through a pipe, 68, 147
Fourier series, 75, 78
Fractal, 156
Frequency-dependent selection, 256
Function(s), **9**
absorption, 257
addition, 40
approximation, 280
commuting, 42

Function(s) (*cont.*)
    comparing at infinity, 260–263
    composition, 42
    continuous, 139
    decreasing, 30, 150
    differentiable, 148
    dimensions and units, 19, 43
    exponential, 49–52, 184
    graphing, 10
    increasing, 30, 150
    inverse, 45
    likelihood, 621
    linear, 22–27, 148
    logarithmic, 53
    maxima and minima, 237
    natural logarithm, 53
    polynomial, 134
    power, 61
    random variable, 497
    step, 349
    trigonometric, 69
    updating, 7
Functional response, **257**, 267
Fundamental relation, **17**
    amount, concentration and volume, 79
    mass, density and volume, 17
    number and mass, 17
Fundamental Theorem of Calculus, 356–364
    applied to cumulative distribution function, 494, 603

Gamma distribution, **188**
    as sum of exponential distributions, 597
    derivative of, 188
Generalized polynomials, 190
Genetics, 4
    and probability, 462
    binomial distribution, 567
    conditional probability, 474
    genetic drift, 5
    mutation, 616
Geometric distribution, 574–578
    cumulative distribution, 576
    formula, 575
    memoryless property, 581
Geometric mean, 511–514
    of product, 557
Geometric series, 576
Global maximum, **240**
Gompertz model, 69
Graph, **10**
    of exponential function, 52
    of linear function, 29, 30
    log-log, 63

of natural logarithm, 53
second derivative, 179
semilog, 62
of sum of functions, 40

Haldane, J.B.S., 5
Half-life, **57**
Harvesting, 39, 95
    maximization, 244–246, 251, 252
Heart model, 106–111
Heaviside function, 137
Heterozygous, **462**
Hill function, **173**, 211, 256
Histogram, **451**, 486
Hodgkin, Sir Alan, 4, 442
Homozygous, **462**
Horizontal line test, **46**
Huxley, Sir Andrew, 4, 442
Hypothesis testing, 647–653
    comparing means, 665
    with method of support, 652, 653
    normal theory, 654–658
    unpaired data, 663–665
Hysteresis, **144**–146

Identity function, **133**
    continuity of, 140
    limit of, 134
Imaginal disks, 68
Immigration
    assigning probabilities, 471
    binomial distribution, 557
    stochastic, 450, 451
Improper integral, 374–381
    with infinite integrand, 380, 381
    with infinite limits of integration, 375–379
    and leading behavior, 379
Inbreeding, **462**
Increasing function, 30, 150
Indefinite integral, **330**
    link with definite integral, 356–359
Independence, **480**–484
    of discrete random variables, 534
    and joint probability distribution, 530
    multiplication rule, 481
    mutual, 482
Independent and identically distributed, **596**
Independent random variables, **534**
    expectation of a product, 553
    variance of a sum, 553
Independent variable, **9**
Indeterminate form, **276**
Inferential statistics, 448, **618**

Infinitely large measurement, 135
Information
    and independence, 537
    and inverse functions, 46
Initial condition(s), **33**
    and the arbitrary constant, 331
    and cobwebbing, 83
    for differential equation, 312
    for discrete-time dynamical system, 33
    effects on solution, 34–37
    sensitive dependence on, 105
Inner function, 42
Input and output tolerances, **143**
Instantaneous rate of change, **122**
Integral(s), 330
    approximation with sums, 348–350
    connection between definite and indefinite, 356–359
    definite, 353
    improper, 374
    indefinite, 330
    of probability density function, 491
    Riemann, 353, 354
Integrand, **330**
    infinite, 380, 381
Integration, **330**
    constant product rule, 332
    of cosine, 341
    of exponential function, 340
    limits of, 354
    of natural logarithm, 340
    by partial fractions, 408
    power rule, 331
    of sine, 341
    by substitution, 341–346
    sum rule, 333
Intermediate Value Theorem, **250**
Interpolation, **29**, 283
Intersection (of sets), **469**
Inverse function, **45**
    derivative of, 203

Joint probability distribution, 530–539

Kotar, Mirjam, 617
Kurtosis, 602

L'Hôpital's rule, 276–278
Law of total probability, 476, 531
Laws of exponents, **52**
Laws of logarithms, **53**
Leading behavior, **268**–276
    and improper integrals, 379–381
Left-hand estimate, 349
Left-hand limit, **132**

Light bulb, 461, 583, 654
Likelihood function, **621**
Limit(s), **124**, 129–136
 of basic functions, 133
 combining, 133, 134
 and continuous functions, 139
 of functions, 129
 infinite, 135
 at infinity, 257–264
 of integration, 354
 left-hand, 132
 properties of, 133
 right-hand, 132
 scientific definition, 130
 of sequences, 265, 266
Limits of integration, **354**
 infinite, 374
Line(s)
 equation of, 25
 finding equation from data, 28
 graphing, 29, 30
 log-log plot of power relation, 63
 point-slope form, 26
 semilog plot of exponential growth, 62
 slope-intercept form, 27
Linear function(s), 22–27
 derivative, 148, 150
 and perfect correlation, 545
 regression, 672
 secant and tangent line, 149
Linear interpolation, 29, **283**
Linear regression, 671–679
 estimating slope and intercept, 677
Local maximum, 239
Log-log plot, **63**
Logarithm(s), **53**
Logistic differential equation, 393
Logistic model (see also discrete logistic model), 105
Lognormal distribution, 613
Lung model, 78–85
 with absorption, 85, 95, 289, 295
 cobwebbing, 83, 84
 equilibria, 88, 90, 93
 general formulation, 82
 solution, 84
 stability, 214, 218
 updating function, 81
Luria, Salvador, 616

Malaria, 3
Marginal probability distribution, **533**–537
Marginal Value Theorem, **244**, 291, 617

Markov chain, **451**, 483
 and binomial distribution, 569
 coin tossing, 484
 presence and absence, 452
 stochastic diffusion, 458
Mathematical crime
 felony, 122, 276
 misdemeanor, 122
Mathematical induction, 160, 175
Maximization, 237–246, 251–253
Maximum
 of a function, 237
 of an oscillation, 71–75
Maximum likelihood, **622**–625
 connection with linear regression, 678
Mean, **501**
 arithmetic, 501
 geometric, 511
Mean absolute deviation, 517–519
Mean Value Theorem, **254**–256
Median, **508**
 of exponential distribution, 579
 of geometric distribution, 576
Medication concentration model, 31
 equilibria, 94
 input and output tolerances, 143
 nonlinear, 104, 222
 run backwards, 148, 219
 solution, 36–38
 stability, 219
 as a sum of terms, 40
 updating function, 36
Meiotic drive, 466
Memoryless property, 581, 582
Mendel, Gregor, 4, 464, 466
Meteorology, 486, 526
Method of leading behavior, **268**
Method of matched leading behaviors, **274**, 275
Michaelis-Menten reaction kinetics, 257, 264, 394
Midpoint method, 384
Minimum
 of a function, 237
 of an oscillation, 71–75
Mites on lizards model, 28, 29
 equilibria, 92
 inverse, 45
 run backwards, 148
 solution, 35
Mode, 509, 510
 of binomial distribution, 569
 of geometric distribution, 576
 of Poisson distribution, 588

Model(s), **2**
 do the impossible, 98
 role in statistics, 618
Monte Carlo method, 632, 633
 confidence limits, 633
 hypothesis testing, 650
Multiplication rule for independent events, 481, 533
Mutation, 39, 104, 115, 588, 616
Mutual independence, **482**
Mutually exclusive and collectively exhaustive events, **476**
Mutually exclusive events, **469**
 conditional probability, 475

n choose k, **563**, 564
Natural logarithm, **53**
 approaches infinity slowly, 261
 defined with definite integral, 374
 derivative of, 187, 203
 integral of, 340
 limit of, 136
Natural selection, 5, 96, 465
Negative binomial distribution, 583
Negative feedback, 437
Neuron, 4, 435
 action potential, 436
 periodic bursting, 442
 resting potential, 436
Newton's law of cooling, **322**–325
 direction arrows, 433
 equilibria, 390, 395
 equilibria and nullclines, 422, 423
 Euler's method, 323, 324
 with oscillating temperature, 329
 perverse version, 396
 phase-line diagram, 391
 separation of variables, 405, 406
 system of autonomous differential equations, 412, 413
Newton's Law of Motion, 182, 195
Newton's method, 289–297
 updating function, 294
Newton, Sir Isaac, 2, 127
Nonautonomous differential equation, **387**
 separation of variables, 408
Nonlinear regression, 676, 684
Normal approximation, 594–598
 of binomial, 607–609, 658
 comparing population proportions, 666, 667
 of gamma distribution, 597
 of Poisson, 610, 611
 power of tests, 658–660

Normal approximation (*cont.*)
    used for computing confidence limits, 638, 645
    used for hypothesis testing, 654
Normal distribution, 198, 492, 592–600
    additive property, 597
    probability density function, 594
    standard, 602–606
Null hypothesis, **647**
    with linear regression, 673
Nullclines, 417–420
    of competition system of differential equations, 422
    of Newton's law of cooling, 423
    of predator-prey equations, 420
    predator-prey equations, 417–419
Numerical analysis, 321

One-tailed test, **648**
Ordered pair, **11**
Oscillation(s), 69–75
    of discrete-time dynamical system, 228
    induced by thermostat, 235
    of neuron, 442
    of predator-prey equations, 426
    sums of, 76
Othmer, Hans, 445
Otters, 371, 372
Outer function, **42**
Overcompensation, **264**, 271

p-values, **648**
Parameter estimation, **619**
Parameter(s), **49**
    in solving for equilibria, 92
Parametric graph, 427
Partial fractions, 408
Per capita growth rate, 326
Per capita rate of change, **126**
Per capita reproduction, **49**
    stochastic, 449, 512
Percentiles, 516
    of bell-shaped curve, 523
Period of an oscillation, **71**–76
Periodic hematopoiesis, 211
Petersburg paradox, 601
Phase
    of an autonomous differential equation, 391
    of an oscillation, 71–75
Phase plane, **418**–420
Phase plane trajectory, **427**
Phase-line diagram, **390**
    and stable and unstable equilibria, 396
    for bacterial population growth, 397
    for competition differential equation, 392
    for disease model, 400
    for Newton's law of cooling, 391
    for perverse version of Newton's law of cooling, 396
    for sodium channels, 437
Plant height, 464, 593
Point of inflection, **178**
    of bell-shaped curve, 522
Point-slope form of a line, **26**
Poisson distribution, 584–591
    additive property, 589
    comparing parameters, 669
    conditions, 586
    formula, 586
    normal approximation, 610, 611
    relation to binomial, 590
    in space, 588
Poisson process, **584**–586
Polymerase chain reaction, 267
Polynomial function, **134**
    degree of, 164
    derivative of, 164
    limit of, 134
Pooled data, 667
Population, **619**
Population growth, 6
    geometric mean, 511
Positive feedback, 437
Power function(s)
    behavior near infinity, 273
    behavior near zero, 273
    improper integrals, 377, 378, 381
    leading behavior, 272
    limit at infinity, 261, 263
Power of statistical test, **648**, 651, 652
    with normal approximation, 658–660
Power relation, 59–66
    derivative of, **161**, 162
    graphed with second derivative, 179, 180
Power rule for derivatives, 161, 162
    general powers, 162
    integer powers, **161**
Power rule for integrals, **331**
Predator-prey equations, 409–411
    direction arrows, 427–430
    Euler's method, 414, 425
    nullclines, 417–420
    phase plane trajectory, 427
    solutions, 426
Prime notation, **124**
Principle of least squares, 677
Principle of mass action, **409**
Probability, 447, 468–473
    area interpretation, 471
    assigning to events, 470
Probability density function, 488–492
    definition, 490
    of exponential distribution, 579
    of normal distribution, 492
    of random variable, 499
Probability distribution, **486**
    binomial, 557
    geometric, 574
    Poisson, 584
    symmetric, 487
Product rule for derivatives, **169**
Proper integral, **375**
Proportional relation, **23**
Pure-time differential equation, **311**
    Euler's method, 315
    general solution, 364
    separation of variables, 404
    solving with antiderivative, 330
Pythagorean theorem, 192

Quadratic function
    approximation with, 284–286
    derivative of, 152
    graphed with second derivative, 179
Qualitative reasoning
    about discrete-time dynamical systems, 226
    about graphs, 12
Quartiles, **516**
Quotient rule for derivatives, 171–173

Radians, **69**
Ragsdale, Janice, 115
Random number generator, 453
Random variable, **497**–504
    Bernoulli, 498
    binomial, 557
    continuous, 497
    discrete, 497
    exponential, 579
    Poisson, 585
Range
    of a function, **10**
    of a random variable, **515**
Rare disease, 477, 478
Rate of change, 126
    and differential equations, 311
    probabilistic, 578, 584
Rational function, **134**
Regression, 671–679
Replication, 4
Residuals, **673**

Ricker model, 231, 233–235
   equilibria, 233
   stability, 233–235
   stochastic, 527
   with two delays, 307
Riemann integral, **353**, 354
Riemann sum, **351**, 352
   for area under curve, 366
   for expectation of a continuous random variable, 502
   for mass of a bar, 371
Right-hand estimate, **349**
Right-hand limit, **132**
Rolle's Theorem, **254**
Ross, Sir Ronald, 3
Runge-Kutta method, 384, 426

SA node, 106
Salinity, 85
Sample, **619**
Sample covariance, **677**
Sample mean(s), **636**
   comparing, 665
   confidence limits, 638–642
Sample median, **637**
Sample space, **468**
Sample standard deviation, **644**
Sample variance, **643**
Saturation, 258, 302
Schnakenberg reaction, 425
Secant line, **121**
Secant line approximation, 282
Second degree block, **109**
Second derivative, 176–178
   used to find local maxima and minima, 241, 242
Selfing, **462**
Semilog plot, **62**, 63
Separation of variables, 402–407
Sequences, 265
Set theory, 469
Significance level, **647**–649
   high, 649
   of linear regression, 675
   very high, 649
Signum function, 133, 140
Simple event, **468**
Simple harmonic oscillator, **195**
   as system of autonomous differential equations, 416
Simulation, 2
   of diffusion, 456–460
   of immigration, 451
   of Markov chain, 452, 484
   Monte Carlo method, 632, 650

of plant height, 465
of Poisson process, 585
of population size, 449, 512, 513
Simultaneous equations, 417
Sine function, 69–71
   derivative of, 191–194
   integral, 341
Sinusoidal oscillation, **71**–76, 195
Skewness, 602
Sleep, 77
Slope, **23**–30
   of line, 26
   of linear regression, 675
   of power relation, 65
   of proportional relation, 23–25
Slope-intercept form of a line, **27**
Sodium channels, 436
Solution
   of differential equation, 312
   of discrete-time dynamical system, 32
   in phase plane, 427
   of pure-time differential equation, 364
   of system of autonomous differential equations, 426
Specific heat, **323**
Speed, 122
   as limit of average rate of change, 131
   computed with Mean Value Theorem, 254
Sphere
   surface area, 17, 66
   volume, 17, 66, 167
Spring, 195, 196
   as system of autonomous differential equations, 416
   with friction, 197, 209
Square wave, 78
Stability Theorem
   for Autonomous Differential Equations, **397**
   for Discrete-Time Dynamical Systems, **230**
Stable equilibrium
   of autonomous differential equation, 392
   of discrete-time dynamical system, 102, 215–217
   and hysteresis, 145
Standard deviation, **522**
Standard error, **644**
Standard normal distribution, 602–606
State variable, **312**
Statistic, **448**
Statistical design, 620, 661
Statistical test, 647

Statistics, 447, **618**
Step function, 349
Stirling approximation, 566
Stochastic, **448**
Stochastic dynamical system, 447, 451
Sum of squares of errors, **673**
Sum of squares total, **673**
Sum rule for derivatives, **158**–160
Sum rule for integrals, **333**
Summation notation, **351**
Summation property of definite integral, **362**
   area interpretation, 367
Superstable equilibrium, **296**
Support, **634**, 635
   used for hypothesis testing, 652, 653, 661, 662
   used to compare Poisson distributions, 669
   used to compare population proportions, 667, 668
Surface area, 65, 66, 163
Survivorship function, 579
Symmetric probability distribution, **487**, 510
Systems of autonomous differential equations, 409–413
   competition, 411
   direction arrows, 427
   equilibria, 417–420
   Euler's method, 413
   FitzHugh-Nagumo, 439
   Newton's law of cooling, 412
   nullclines, 418
   predator-prey, 409
   simple harmonic oscillator, 416

t distribution, 644, 656
Tangent line, **123**
Tangent line approximation, **124**, 281–284
   and Euler's method, 316
   and L'Hôpital's rule, 278
   and Newton's method, 292
   to updating function, 226
Taylor polynomial, **287**
Temperature
   daily and monthly cycles, 74–76, 207
   expanding oscillations, 235
   Fahrenheit to Celsius conversion, 27
Time step, **317**
Torricelli's law of draining, 394, 408
Total sum of squares, **673**
Trapezoid, 355
Tree growth model, 8
   lack of equilibria, 91

Tree growth model (*cont.*)
    solution, 34
    updating function, 8
Trigonometric function(s), 69, 76, 77
    derivatives of, 195
Trimmed mean, **638**
Two-tailed test, **650**
Type I error, **647**
Type II error, **647**

Ultradian rhythm, 77
Unbiased estimator, **620**
    sample variance, 643
Union (of sets), **469**
Units, 14–16
    of coefficient of variation, 523
    converting between, 16
    of correlation, 542
    of standard deviation, 522
Unpaired data, 663–665
Unstable equilibrium
    of autonomous differential equation, 392
    of discrete-time dynamical system, 102, 215–217

Updating function, **7**
    bacterial population growth, 7
    changing dimensions, 18–20
    cobwebbing, 83
    compared with differential equation, 127
    competition model, 98, 99
    composed with itself, 41, 44
    discontinuous, 144
    do nothing, 87
    lung model, 81
    medication concentration, 31
    mites on lizards, 28, 29
    Newton's method, 294
    nonlinear, 98
    selfing, 462
    slope and stability, 216, 217
    stochastic diffusion, 459
    tree growth, 8
Updating system, **416**

Value (of function), **9**
Variable, **49**
Variance, **520**
    of binomial distribution, 559

computational formula, 520
constants come outside, 554
of a difference, 555
of exponential distribution, 579
of geometric distribution, 577
of Poisson distribution, 588
of a sum, 553
uncorrected, 677
Velocity, **152**
    as limit, 131
Venn diagram, **469**
Voltage-gated channels, 436
Volume, 65, 163

Waiting times, 574
Weber-Fechner law, 39
Weighted average, **82**
Wenckebach phenomenon, **110**, 111
Wilcoxon rank-sum test, 683
Wright, Sewall, 5

y-intercept, **27**–29

z-scores, 605

Rules of Differentiation

Suppose $f(x)$ and $g(x)$ are differentiable functions and c is a constant.

Rule	Function	Derivative
Sum rule	$f(x) + g(x)$	$f'(x) + g'(x)$
Constant sum rule	$f(x) + c$	$f'(x)$
Power rule	x^n	nx^{n-1}
Product rule	$f(x)g(x)$	$f(x)g'(x) + g(x)f'(x)$
Constant product rule	$cf(x)$	$cf'(x)$
Quotient rule	$\dfrac{f(x)}{g(x)}$	$\dfrac{f'(x)g(x) - g'(x)f(x)}{(g(x))^2}$
Chain rule	$f(g(x))$	$f'(g(x))g'(x)$
Exponential	e^x	e^x
Natural logarithm	$\ln(x)$	$\dfrac{1}{x}$
Sine	$\sin(x)$	$\cos(x)$
Cosine	$\cos(x)$	$-\sin(x)$

Basic Indefinite Integrals

$$\int x^n dx = \frac{x^{n+1}}{n+1} + c \quad \text{if } n \neq -1$$

$$\int \frac{1}{x} dx = \ln(x) + c$$

$$\int e^{\alpha x} dx = \frac{e^{\alpha x}}{\alpha} + c \quad \text{if } \alpha \neq 0$$

$$\int \sin(\alpha x) dx = -\frac{\cos(\alpha x)}{\alpha} + c \quad \text{if } \alpha \neq 0$$

$$\int \cos(\alpha x) dx = \frac{\sin(\alpha x)}{\alpha} + c \quad \text{if } \alpha \neq 0$$

Definitions in Probability Theory

Conditional Probability: The probability of event A conditional on event B is $\Pr(A \mid B) = (\Pr(A \cap B))/(\Pr(B))$ if $\Pr(B) \neq 0$.

Independence: Event A is **independent** of event B if $\Pr(A \mid B) = \Pr(A)$.

Expectation: The mean of a random variable found as the sum of the values weighted by their probabilities.

Median: The value of a random variable is exceeded by exactly 50% of the measurements.

Mode: The most probable value of a random variable.

Variance: A measure of the spread of a random variable equal to the mean squared deviation from the mean.

Standard Deviation: The square root of the variance.

Theorems of Probability

Law of Total Probability: Suppose that $E_1, E_2, \ldots, E_n$ form a set of mutually exclusive and collectively exhaustive events. Then for any event A, $\Pr(A) = \sum_{i=1}^{n} \Pr(A \mid E_i) \Pr(E_i)$.

Baye's Theorem: For any events A and B where $\Pr(A) \neq 0$, $\Pr(B \mid A) = \dfrac{\Pr(A \mid B) \Pr(B)}{\Pr(A)}$.

Multiplication Rule for Independent Events: Suppose A and B are any two independent events. Then $\Pr(A \cap B) = \Pr(A) \Pr(B)$.

Expectation of the Sum of Random Variables: The expectation of the sum of random variables is equal to the sum of the expectations.

Variance of the Sum of Independent Random Variables: The variance of the sum of independent random variables is equal to the sum of the variances.

Central Limit Theorem: The sum of independent, identically distributed random variables has approximately a normal distribution.

Probability Distributions

Distribution	Type of random variable	When to use	Formula	Expectation	Variance
Binomial	Discrete	Number of successes N in n trials each with probability p	$\Pr(N = k) = b(k : n, p) = \binom{n}{k} p^k (1-p)^{n-k}$	np	$np(1-p)$
Poisson	Discrete	Number of events N in time t at rate λ	$\Pr(N = k) = p(k; \lambda t) = \dfrac{e^{-\lambda t}(\lambda t)^k}{k!}$	λt	λt
Geometric	Discrete	Time T before a success in a series of trials each with probability q	$\Pr(T = t) = g_t = q(1-q)^{t-1}$	$\dfrac{1}{q}$	$\dfrac{1-q}{q^2}$
Exponential	Continuous	Waiting time T for an event that occurs at rate λ	p.d.f. $= \lambda e^{-\lambda t}$	$\dfrac{1}{\lambda}$	$\dfrac{1}{\lambda^2}$
Normal	Continuous	Value X of many continuous measurements	p.d.f. $= \dfrac{1}{\sqrt{2\pi}\sigma} e^{-\frac{(x-\mu)^2}{2\sigma^2}}$	μ	σ^2